Hans H. Maurer, Karl Pfleger, Armin A. Weber

Mass Spectral and GC Data

of Drugs, Poisons, Pesticides, Pollutants
and Their Metabolites

1807–2007 Knowledge for Generations

Each generation has its unique needs and aspirations. When Charles Wiley first opened his small printing shop in lower Manhattan in 1807, it was a generation of boundless potential searching for an identity. And we were there, helping to define a new American literary tradition. Over half a century later, in the midst of the Second Industrial Revolution, it was a generation focused on building the future. Once again, we were there, supplying the critical scientific, technical, and engineering knowledge that helped frame the world. Throughout the 20th Century, and into the new millennium, nations began to reach out beyond their own borders and a new international community was born. Wiley was there, expanding its operations around the world to enable a global exchange of ideas, opinions, and know-how.

For 200 years, Wiley has been an integral part of each generation's journey, enabling the flow of information and understanding necessary to meet their needs and fulfill their aspirations. Today, bold new technologies are changing the way we live and learn. Wiley will be there, providing you the must-have knowledge you need to imagine new worlds, new possibilities, and new opportunities.

Generations come and go, but you can always count on Wiley to provide you the knowledge you need, when and where you need it!

William J. Pesce
President and Chief Executive Officer

Peter Booth Wiley
Chairman of the Board

Hans H. Maurer, Karl Pfleger, Armin A. Weber

Mass Spectral and GC Data

of Drugs, Poisons, Pesticides, Pollutants and Their Metabolites

Volume 1: Methods and Tables

Third, Revised and Enlarged Edition

WILEY-VCH Verlag GmbH & Co. KGaA

The Authors

Prof. Dr. Hans H. Maurer
Prof. Dr. Karl Pfleger
Armin A. Weber
Department of Experimental and Clinical
Toxicology
Saarland University
66421 Homburg (Saar)
Germany

All books published by Wiley-VCH are carefully produced. Nevertheless, authors, editors, and publisher do not warrant the information contained in these books, including this book, to be free of errors. Readers are advised to keep in mind that statements, data, illustrations, procedural details or other items may inadvertently be inaccurate.

Library of Congress Card No.:
applied for

British Library Cataloguing-in-Publication Data
A catalogue record for this book is available from the British Library.

Bibliographic information published by the Deutsche Nationalbibliothek
Die Deutsche Nationalbibliothek lists this publication in the Deutsche Nationalbibliografie; detailed bibliographic data are available in the Internet at <http://dnb.d-nb.de>.

© 2007 WILEY-VCH Verlag GmbH & Co. KGaA, Weinheim

All rights reserved (including those of translation into other languages). No part of this book may be reproduced in any form – by photoprinting, microfilm, or any other means – nor transmitted or translated into a machine language without written permission from the publishers. Registered names, trademarks, etc. used in this book, even when not specifically marked as such, are not to be considered unprotected by law.

Printing: Strauss GmbH, Mörlenbach
Bookbinding: Litges & Dopf Buchbinderei GmbH, Heppenheim
Wiley Bicentennial Logo: Richard J. Pacifico

Printed in the Federal Republic of Germany
Printed on acid-free paper

ISBN: 978-3-527-31538-3

Dedicated to

Claudia, Christine and Johannes

Ursula, Maria, Hildegard and Franziska

Karin, Christian, Thorsten and Daniel

"Unicum signum certi dati veneni

est analysis chimica…"

Josef Jakob von Plencks:
Toxicologica seu doctrina de venenis est antidoti
Vienna, 1785

2 Experimental Section

All mass spectral and gas chromatographic data have been recorded in our laboratory, according to our standard operation procedures outlined below.

2.1 Origin and choice of samples

Most of the metabolite data have been recorded from samples (plasma, gastric contents, or urine) of patients suspected of poisoning or intoxication and admitted to various anti-poison centers, from in-patients of several clinics, particularly the University Hospital at Homburg, or from volunteers treated with therapeutic dosages of drugs. If suitable samples from humans were not available, samples from rats were used. Rats were administered the corresponding drugs in aqueous suspension, by gastric intubation [1].

The choice of biosample for toxicological analysis depends on the toxicological problem. Concentrations of drugs are relatively high in urine, so that urine is the sample of choice for a comprehensive screening and identification of unknown drugs or poisons [2-4]. However, the metabolites of these drugs must be identified in addition, or even exclusively. Blood (plasma, serum) is the sample of choice for quantification. However, if the blood concentration is high enough, screening can also be performed therein. This may be advantageous, as sometimes only blood samples are available and some procedures allow simultaneous screening and quantification [5-9]. GC-MS analysis of drugs in alternative matrices such as hair [10], sweat and oral fluid [11,12], meconium [13], or nails [14] was also described, but the toxicological interpretation of the analytical results may be difficult.

2.2 Sample preparation

Independent of the biosample, a suitable sample preparation is necessary before GC-MS analysis. This may involve cleavage of conjugates, isolation, clean-up steps and/or derivatization of the drugs and their metabolites. In the following sections, sample preparation and derivatization procedures are described for systematic toxicological analysis, as well as for sensitive procedures for particular drugs of toxicological relevance.

Compounds were isolated by liquid-liquid extraction (LLE) or by solid phase extraction (SPE) preceded or followed by clean-up steps. Universal LLE procedures have still been used for emergency analyses and for systematic toxicological analysis (STA), because substances with very different physico-chemical properties had to be isolated from heterogeneous matrices. SPE is preferred for screening and quantification of lower-dosed drugs in plasma, as this results in considerably cleaner extracts. For example, GC-MS procedures were published for screening and quantification of classical and newer designer drugs in plasma [8,9]. The information of which compound can be detected in which sample after which type of sample preparation (column *Detected* in Table 1-8-1 and in the legends of the mass spectra) is based on the sample preparation procedures described in Section 1-2.2. Special procedures are not described here, as series of different (new) extraction procedures are in use around the world, and these have been recently reviewed [2,4,15-17].

Many drugs and poisons are excreted in urine in a completely metabolized and conjugated form. Cleavage of conjugates by enzymatic or acid hydrolysis is necessary before extraction, as conjugates can only be detected by relatively elaborate procedures [2,18]. The gentle but time-consuming method of enzymatic hydrolysis is preferable in metabolism studies or in forensic and doping analysis, if the drugs are destroyed during acid hydrolysis and the time of analysis is not limited (Section 1-2.2.1.3). For systematic toxicological analysis - especially in emergency cases - it is preferable to cleave the conjugates by rapid acid hydrolysis (Section 1-2.2.1.2). However, the formation of artifacts during this procedure must be considered (Section 1-4).

Typical derivatization procedures are given in Section 1-2.2.2.

2.2.1 Standard extraction procedures

2.2.1.1 Standard liquid-liquid extraction (LLE) for plasma, urine or gastric contents (P, U, G)

Plasma (1 mL) mixed with 0.1 mL internal standard (IS, 0.01 mg/mL trimipramine-D3 in methanol) or 2.5 mL of urine or gastric contents were extracted with 5 mL of a mixture of diethyl ether:ethyl acetate (1:1; v/v) after addition of 2 mL saturated sodium sulfate solution. After phase separation by centrifugation, the organic extract was transferred into a flask and evaporated to dryness. The aqueous residue was then basified with 0.5 mL of 1 M sodium hydroxide and extracted a second time with 5 mL of

"Unicum signum certi dati veneni

est analysis chimica…"

Josef Jakob von Plencks:
Toxicologica seu doctrina de venenis est antidoti
Vienna, 1785

Preface to the Third Edition

After publication of the first edition in 1985 with 1500 data sets, the second in 1992 with 4700, and with the Supplement Volume in 2000 with 6700 data sets, the publisher and the authors decided to publish a completely revised and enlarged new edition. We decided to reduce the number of volumes to one small volume with all the Tables and one with all the spectra. This was possible by enlarging the number of spectra per page from five to seven and by depicting them only once under the (pseudo) molecular mass. This seemed advisable in order to save costs for the combined electronic and printed version, the most sold form. The former concept of depicting the spectra several times under molecular ion, base peak and group-specific ions in order to simplify searching of the reference spectrum seemed no longer up-to-date, because everybody uses the electronic version for searching. Finally, the measured compounds were listed only in alphabetical order and the list of potential poisons was removed, because such data can be found today in (on-line) data bases.

Since publication of the Supplement Volume in 2000, we have recorded and collected in our laboratory mass spectral and gas chromatographic data of new drugs, drugs of abuse, poisons and their metabolites bringing the total number of mass spectra to **7840** in this new edition, among which 1900 data sets have been revised or added. Over 3300 data sets are from metabolites, over 2300 from acetylated, over 1000 from methylated, over 700 from trimethylsilylated, over 400 from trifluoroacetylated, over 200 each from pentafluoropropionylated or heptafluorobutyrylated compounds. All formulas were redrawn in the molefile format allowing their use in electronic databases.

The sections on sample preparation and GC-MS method have been updated, but only our standard procedures have been described in detail, as series of papers on sample preparation have been published during the past years. Current review articles have been cited in order to rapidly find a suitable procedure for a particular case.

The Tables cover all the data with directions on which page in Volume 2 the data may be found, as well as the entry numbers of the electronic versions (Table 1-8-1). The numbering of the Tables, Figures and Sections indicates whether they can be found in Volume 1 or 2 by the corresponding prefix '1-' or '2-'. The middle number indicates the corresponding Section.

The GC and MS data are also available in electronic form. The combination of the handbook with the full data and the electronic version has proven to be very efficient in analytical toxicology. Using macros, the data evaluation can be automated, but the final decision on the identity of a compound will always rest in the hands of the experienced toxicologist or analyst who undertakes a visual comparison between the full mass spectrum of the measured compound and the reference spectra, by considering further chemical, analytical and toxicological aspects for plausibility.

Again, while many colleagues throughout the world have acknowledged the value of our collection, some have also drawn our attention to errors and have provided suggestions, and for this we would like to express our sincere gratitude.

It is our hope that the new edition will be accepted as well as the previous ones by our colleagues in the fields of clinical and forensic toxicology, doping control, drug metabolism and in all other areas of analytical toxicology.

Homburg (Saar), February 2007   **Hans H. Maurer**

Acknowledgements

The authors are indebted to many pharmaceutical companies and to many colleagues for supplying reference substances or biosamples containing rare compounds and/or metabolites.

Professor Maurer thanks the following former or present co-workers for the recording and interpretation of new mass spectra during their scientific studies and further support: J. Arlt, J. Beyer, J. Bickeboeller-Friedrich, A. Bierl, K. Bock, A.H. Ewald, C. Fritz, P. Goepfert, C. Haas, J. Jung, C. Gerber, I. Kleff, N. Kneller, T. Kraemer, C. Kratzsch, O. Ledvinka, N. Makkinejad, M.R. Meyer, L.D. Paul, F.T. Peters, A.A. Philipp, D. Remane, S. Roditis, C. Sauer, S. Schaefer, C.J. Schmitt, C. Schroeder, F. Sick, D. Springer, R.F. Staack, E. Strack, G. Theis, S.W. Toennes, F.X. Tauvel, D.S. Theobald, G. Ulrich, I. Vernaleken.

Furthermore, the authors wish to thank their families for forbearance and sacrifice of personal interests. Finally, they wish to thank the Wiley-VCH editors Dr. Frank Weinreich and Dr. Reinhard Neudert for their understanding and the agreeable collaboration.

Contents of Volume 1 (Methods, Tables)

Methods

1	**Introduction** 3	
2	**Experimental Section** 4	
2.1	**Origin and choice of samples** 4	
2.2	**Sample preparation** 4	
2.2.1	Standard extraction procedures 4	
2.2.1.1	Standard liquid-liquid extraction (LLE) for plasma, urine or gastric contents (P, U, G) 4	
2.2.1.2	STA procedure (hydrolysis, extraction and microwave-assisted acetylation) for urine (U+UHYAC) 5	
2.2.1.3	Extraction of urine after cleavage of conjugates by glucuronidase and arylsulfatase (UGLUC) 5	
2.2.1.4	Extractive methylation procedure for urine or plasma (UME, PME) 5	
2.2.1.5	Solid-phase extraction for plasma or urine (PSPE, USPE) 5	
2.2.1.6	LLE of plasma for determination of drugs for brain death diagnosis 6	
2.2.1.7	Extraction of ethylene glycol and other glycols from plasma or urine followed by microwave-assisted pivalylation (PEGPIV or UEGPIV) 6	
2.2.2	Derivatization procedures 6	
2.2.2.1	Acetylation (AC) 7	
2.2.2.2	Methylation (ME) 7	
2.2.2.3	Ethylation (ET) 7	
2.2.2.4	tert.-Butyldimethylsilylation (TBDMS) 7	
2.2.2.5	Trimethylsilylation (TMS) 7	
2.2.2.6	Trimethylsilylation followed by trifluoroacetylation (TMSTFA) 7	
2.2.2.7	Trifluoroacetylation (TFA) 7	
2.2.2.8	Pentafluoropropionylation (PFP) 7	
2.2.2.9	Pentafluoropropylation (PFPOL) 7	
2.2.2.10	Heptafluorobutyrylation (HFB) 7	
2.2.2.11	Pivalylation (PIV) 8	
2.2.2.12	Heptafluorobutyrylprolylation (HFBP) 8	
2.3	**GC-MS Apparatus** 8	
2.3.1	Apparatus and operation conditions 8	
2.3.2	Quality assurance of the apparatus performance 8	
2.4	**Determination of retention indices** 10	
2.5	**Systematic toxicological analysis (STA) of several classes of drugs and their metabolites by GC-MS** 10	
2.5.1	Screening for 200 drugs in blood plasma after LLE 10	
2.5.2	Screening for most of the basic and neutral drugs in urine after acid hydrolysis, LLE and acetylation 10	
2.5.3	Systematic toxicological analysis procedures for the detection of acidic drugs and/or their metabolites 14	
2.5.4	General screening procedure for zwitterionic compounds after SPE and silylation 14	
2.6	**Application of the electronic version of this handbook** 15	
2.7	**Quantitative determination** 15	
3	**Correlation between Structure and Fragmentation** 16	
3.1	**Principle of electron-ionization mass spectrometry (EI-MS)** 16	
3.2	**Correlation between fundamental structures or side chains and fragment ions** 16	

4	**Formation of Artifacts 17**	
4.1	**Artifacts formed by oxidation during extraction with diethyl ether 17**	
4.1.1	N-Oxidation of tertiary amines 17	
4.1.2	S-Oxidation of phenothiazines 17	
4.2	**Artifacts formed by thermolysis during GC (GC artifact) 17**	
4.2.1	Decarboxylation of carboxylic acids 17	
4.2.2	Cope elimination of N-oxides ($-(CH_3)_2NOH$, $-(C_2H_5)_2NOH$, $-C_6H_{14}N_2O_2$) 17	
4.2.3	Rearrangement of bis-deethyl flurazepam ($-H_2O$) 17	
4.2.4	Elimination of various residues 17	
4.2.5	Methylation of carboxylic acids in methanol ((ME), ME in methanol) 18	
4.2.6	Formation of formaldehyde adducts using methanol as solvent (GC artifact in methanol) 18	
4.3	**Artifacts formed by thermolysis during GC and during acid hydrolysis (GC artifact, HY artifact) 18**	
4.3.1	Dehydration of alcohols ($-H_2O$) 18	
4.3.2	Decarbamoylation of carbamates 18	
4.3.3	Cleavage of morazone to phenmetrazine 18	
4.4	**Artifacts formed during acid hydrolysis 18**	
4.4.1	Cleavage of the ether bridge in beta-blockers and alkanolamine antihistamines (HY) 18	
4.4.2	Cleavage of 1,4-benzodiazepines to aminobenzoyl derivatives (HY) 18	
4.4.3	Cleavage and rearrangement of N-demethyl metabolites of clobazam to benzimidazole derivatives (HY) 19	
4.4.4	Cleavage and rearrangement of bis-deethyl flurazepam (HY $-H_2O$) 19	
4.4.5	Cleavage and rearrangement of tetrazepam and its metabolites 19	
4.4.6	Dealkylation of ethylenediamine antihistamines (HY) 19	
4.4.7	Hydration of a double bond ($+H_2O$) 19	
5	**Table of Atomic Masses 20**	
6	**Abbreviations 21**	
7	**References 24**	

Tables

8	**Table of Compounds in Order of Names 35**	
8.1	**Explanatory notes 35**	
8.2	**Table of compounds in order of names 37**	
9	**Table of Compounds in Order of Categories 189**	
9.1	**Explanatory notes 189**	
9.2	**Table of compounds in order of categories 189**	

Contents of Volume 2 (Mass Spectra)

1 **Explanatory Notes** 1
1.1 Arrangement of spectra 1
1.2 Lay-out of spectra 1

2 **Abbreviations** 3

3 **Compound Index** 7

Mass Spectra 89

Methods

1 Introduction

After publication of the first edition in 1985 with 1500 data sets, the second in 1992 with 4700, and with the Supplement Volume in 2000 with 6700 data sets, the new edition contains **7840 data sets**, among which 1900 data sets have been revised or added. Over 3300 data sets are from metabolites, over 2300 from acetylated, over 1000 from methylated, over 700 from trimethylsilylated, over 400 from trifluoroacetylated, over 200 each from pentafluoropropionylated or heptafluorobutyrylated compounds. All formulas were redrawn in the molefile format allowing their use in electronic databases.

Section 1-2 on sample preparation and GC-MS methods has been updated, but only our standard operation procedures have been described in detail, as they were used for recording the data.

The structures of the compounds are also reproduced to facilitate the identification of unknown metabolites by correlating the fragmentation patterns with the probable structure, as described in Section 1-3. Further data are also included, as outlined in the Explanatory notes.

Several artifacts have been detected which are formed during sample preparation and/or the GC procedure. Their formation is described in detail in Section 1-4.

Section 1-5 contains the table of atomic masses used for molecular mass calculations and Section 1-6 the actualized list of abbreviations used.

All the compounds are listed alphabetically in Table 1-8-1 in order to assist the search for data concerning specific compounds with directions on which page in Volume 2 the data can be found, as well as the entry numbers of the electronic versions. In Table 1-9-1, the parent compounds are listed in order of category to assist the search of data of all the compounds measured within a particular category. The numbering of the Tables, Figures and Sections indicates whether they can be found in Volume 1 or 2 by the corresponding prefix '1-' or '2-'. The middle number indicates the corresponding Section.

The extremely large amount of data presented makes it likely that some errors will be present in this handbook. The authors cannot be held responsible for such errors, nor for any consequences arising from the employment of the published data. Users are requested to report any errors that are found and to suggest other data and other groups of compounds, which are of interest (e-mail: hans.maurer@uniklinikum-saarland.de or hans.maurer@uks.eu). If possible, these will be included in future editions, along with corrections of any errors.

2 Experimental Section

All mass spectral and gas chromatographic data have been recorded in our laboratory, according to our standard operation procedures outlined below.

2.1 Origin and choice of samples

Most of the metabolite data have been recorded from samples (plasma, gastric contents, or urine) of patients suspected of poisoning or intoxication and admitted to various anti-poison centers, from in-patients of several clinics, particularly the University Hospital at Homburg, or from volunteers treated with therapeutic dosages of drugs. If suitable samples from humans were not available, samples from rats were used. Rats were administered the corresponding drugs in aqueous suspension, by gastric intubation [1].

The choice of biosample for toxicological analysis depends on the toxicological problem. Concentrations of drugs are relatively high in urine, so that urine is the sample of choice for a comprehensive screening and identification of unknown drugs or poisons [2-4]. However, the metabolites of these drugs must be identified in addition, or even exclusively. Blood (plasma, serum) is the sample of choice for quantification. However, if the blood concentration is high enough, screening can also be performed therein. This may be advantageous, as sometimes only blood samples are available and some procedures allow simultaneous screening and quantification [5-9]. GC-MS analysis of drugs in alternative matrices such as hair [10], sweat and oral fluid [11,12], meconium [13], or nails [14] was also described, but the toxicological interpretation of the analytical results may be difficult.

2.2 Sample preparation

Independent of the biosample, a suitable sample preparation is necessary before GC-MS analysis. This may involve cleavage of conjugates, isolation, clean-up steps and/or derivatization of the drugs and their metabolites. In the following sections, sample preparation and derivatization procedures are described for systematic toxicological analysis, as well as for sensitive procedures for particular drugs of toxicological relevance.

Compounds were isolated by liquid-liquid extraction (LLE) or by solid phase extraction (SPE) preceded or followed by clean-up steps. Universal LLE procedures have still been used for emergency analyses and for systematic toxicological analysis (STA), because substances with very different physico-chemical properties had to be isolated from heterogeneous matrices. SPE is preferred for screening and quantification of lower-dosed drugs in plasma, as this results in considerably cleaner extracts. For example, GC-MS procedures were published for screening and quantification of classical and newer designer drugs in plasma [8,9]. The information of which compound can be detected in which sample after which type of sample preparation (column *Detected* in Table 1-8-1 and in the legends of the mass spectra) is based on the sample preparation procedures described in Section 1-2.2. Special procedures are not described here, as series of different (new) extraction procedures are in use around the world, and these have been recently reviewed [2,4,15-17].

Many drugs and poisons are excreted in urine in a completely metabolized and conjugated form. Cleavage of conjugates by enzymatic or acid hydrolysis is necessary before extraction, as conjugates can only be detected by relatively elaborate procedures [2,18]. The gentle but time-consuming method of enzymatic hydrolysis is preferable in metabolism studies or in forensic and doping analysis, if the drugs are destroyed during acid hydrolysis and the time of analysis is not limited (Section 1-2.2.1.3). For systematic toxicological analysis - especially in emergency cases - it is preferable to cleave the conjugates by rapid acid hydrolysis (Section 1-2.2.1.2). However, the formation of artifacts during this procedure must be considered (Section 1-4).

Typical derivatization procedures are given in Section 1-2.2.2.

2.2.1 Standard extraction procedures

2.2.1.1 Standard liquid-liquid extraction (LLE) for plasma, urine or gastric contents (P, U, G)

Plasma (1 mL) mixed with 0.1 mL internal standard (IS, 0.01 mg/mL trimipramine-D3 in methanol) or 2.5 mL of urine or gastric contents were extracted with 5 mL of a mixture of diethyl ether:ethyl acetate (1:1; v/v) after addition of 2 mL saturated sodium sulfate solution. After phase separation by centrifugation, the organic extract was transferred into a flask and evaporated to dryness. The aqueous residue was then basified with 0.5 mL of 1 *M* sodium hydroxide and extracted a second time with 5 mL of the solvent mixture. This organic extract was transferred to the same flask and evaporated at 60 °C under reduced pressure. The combined residues were dissolved in 100 μL methanol [19].

2.2.1.2 STA procedure (hydrolysis, extraction and microwave-assisted acetylation) for urine (U+UHYAC)

Some analytes are completely destroyed and therefore not detectable after acid hydrolysis. Therefore, the following modification was developed [20]. The same volume of untreated urine was added to the hydrolyzed sample before extraction and derivatization. In addition, the derivatization time could be reduced from 30 to 5 min using microwave irradiation. The reduction of the extract concentrations of compounds excreted in conjugated form could be compensated by the higher sensitivity of newer GC-MS apparatus [1,21-29].

A 5-mL portion of urine was divided in two equal parts. One 2.5-mL part was refluxed with 1 mL of 37% hydrochloric acid for 15 min. Following hydrolysis, the sample was basified with 2 mL of 2.3 M aqueous ammonium sulfate and 1.5 mL of 10 M aqueous sodium hydroxide to obtain a pH between 8 and 9. Before extraction, the second 2.5 mL part of untreated urine was added. This solution was extracted with 5 mL of a dichloromethane:isopropanol:ethyl acetate mixture (1:1:3; v/v/v). After phase separation by centrifugation, the organic layer was transferred to a flask and evaporated to dryness at 60 °C under reduced pressure. The residue was derivatized by acetylation with 100 μL of an acetic anhydride:pyridine mixture (3:2; v/v) for 5 min under microwave irradiation at about 400 W (for details see ref. [30]). After evaporation of the derivatization mixture at 60 °C under reduced pressure, the residue was dissolved in 100 μL of methanol and 1–2 μL were injected into the gas chromatograph [20].

2.2.1.3 Extraction of urine after cleavage of conjugates by glucuronidase and arylsulfatase (UGLUC)

Several enzymatic hydrolysis procedures have been described using solutions of beta-glucuronidase and arylsulfatase from Helix pomatia or Patella vulgata and incubation times from 2 to 24 h [1,31-34]. In our experience the following procedure is an acceptable compromise between time consumption and cleavage efficiency.

A 5-mL portion of urine was adjusted to pH 5.2 with acetic acid (1 M) and incubated at 50 °C for 1.5 h with 100 μL of a mixture (100 000 Fishman units per mL) of glucuronidase (EC no. 3.2.1.31) and arylsulfatase (EC no. 3.1.6.1) from *Helix Pomatia*. The mixture was then adjusted to pH 8-9 with a mixture of 1 mL of 37% hydrochloric acid, 2 mL of 2.3 M aqueous ammonium sulfate and 1.5 mL of 10 M aqueous sodium hydroxide and extracted with 5 mL of a dichloromethane:isopropanol:ethyl acetate mixture (1:1:3; v/v/v). After phase separation by centrifugation, the organic layer was carefully evaporated to dryness at 60 °C under a stream of nitrogen. The residue was derivatized if necessary. The residue was dissolved in 50 μL of the corresponding solvent and 2 μL were injected into the gas chromatograph [22].

2.2.1.4 Extractive methylation procedure for urine or plasma (UME, PME)

A 2-mL portion of urine or plasma was mixed in a centrifuge tube with 2 mL of the phase-transfer reagent consisting of 0.02 M tetrahexylammonium hydrogen sulfate in 1 M phosphate buffer pH 12. After addition of 6 mL of freshly prepared 1 M methyl iodide in toluene, the closed tube was shaken in a 50 °C water bath for 30 min. After phase separation by centrifugation at 3000 × g for 3 min, the organic phase, containing the analytes and tetrahexylammonium salts, was transferred to a diol SPE column, which was conditioned as follows: 5 mL of methanol at a slow flow rate, drying the column under vacuum for 15 seconds, 5 mL of toluene at a slow flow rate. The organic phase was rinsed through the sorbent at a flow rate of 3 mL/min to adsorb the tetrahexylammonium salts. The part of the analytes also adsorbed on the sorbent was selectively eluted with 5 mL of diethyl ether:ethyl acetate (95:5, v/v) at a flow rate of 3 mL/min. The combined eluates were carefully evaporated to dryness at 60 °C under reduced pressure. The residue was dissolved in 50 μL of ethyl acetate and a 1–2 μL aliquot of this extract was injected into the GC-MS [35].

2.2.1.5 Solid-phase extraction for plasma or urine (PSPE, USPE)

SPE is preferred for analysis of lower-dosed drug.

Plasma or urine (0.5-1 mL) was diluted with 2 mL of purified water. After addition of 50-100 μL of a methanolic solution of suitable IS, the samples were mixed for 15 s on a rotary shaker, centrifuged for 3 min at 1000 × g and loaded on mixed-mode (Bond Elute Certify, 130 mg, 3 mL: C8 and strong cation exchange sorbents) SPE cartridges previously conditioned with 1 mL of methanol and 1 mL of purified water. After extraction, the cartridges were washed with 1 mL of purified water, 1 mL of 0.01 M aqueous hydrochloric acid and 2 mL of methanol. Reduced pressure was applied until the cartridges were dry, and the analytes were eluted with 1 mL of a freshly prepared mixture of methanol-aqueous ammonia (98:2, v/v) into 1.5 mL polypropylene reaction vials. The eluates were evaporated to dryness under a stream of nitrogen at 60 °C and the residue was dissolved in 50-100 µL of methanol [8,9,36-38].

2.2.1.6 LLE of plasma for determination of drugs for brain death diagnosis

After addition of 50 µL of IS solution (4.0 mg/L of pentobarbital-D5, 2.0 mg/L of methohexital-D5, 40.0 mg/L of phenobarbital-D5, 0.8 mg/L of each diazepam-D5 and nordazepam-D5) and 50 µL of butyl acetate, plasma samples (200 µL) were extracted for 2 min on a rotary shaker. After phase separation by centrifugation (1 min, 10 000 × g, ambient temperature), the organic layer (upper) was transferred to autosampler vials and 2 µL were injected into the GC-MS [39].

2.2.1.7 Extraction of ethylene glycol and other glycols from plasma or urine followed by microwave-assisted pivalylation (PEGPIV or UEGPIV)

Among the different methods for the determination of ethylene glycol in plasma or blood, gas chromatographic procedures require isolation and derivatization, e.g., by esterification with boronic acid derivatives. However, derivatization with substituted boronic acids cannot be used for diethylene glycol because the distance between the hydroxy groups is too great to form stable cyclic esters. Furthermore, in our experience capillary columns drastically lose their resolving power after injection of the boronic acid reagent. Our GC-MS procedure for identification and quantification of ethylene glycol and diethylene glycol using pivalylation [40] was markedly improved by microwave-assisted derivatization and further slight modifications [41].

Plasma or urine (200 µL) was mixed in a reaction vessel with 50 µL of an aqueous solution of 1,3-propylene glycol (1.2 g/L) as a suitable IS. After addition of 1 mL acetone, the sample was shaken for 5 min and centrifuged for 1 min at 10 000×g. A 1-mL portion of the supernatant was transferred into a flask and evaporated at 70 °C under reduced pressure. The residue was dissolved in 50 µL of a freshly prepared (!) mixture of pivalic acid anhydride:triethylamine:methanol (20:1:1; v/v/v) and incubated for 10 min under microwave irradiation at about 400 W. This solution was diluted with 200 µL methanol and a 1–2 µL aliquot of this was injected into the GC-MS in splitless mode (column temperature: 80 °C, 6 min; then raised to 310 °C at a rate of 50 °C/min, solvent delay 6 min).

The glycols were first identified by a comparison of the full mass spectrum with reference spectra, and then quantified. The peak area ratio in the total ion chromatogram (ethylene glycol or diethylene glycol/1,3-propane diol) of the sample was compared with the calibration curve in which the peak area ratios of the standards (0.1, 0.25, 0.5, 0.75 and 1 g/L), prepared in the same way, were plotted versus their concentrations.

Typical fragment ions for the detection and quantification of ethylene glycol, diethylene glycol and 1,3-propane diol: m/z 85, 129 and 143. The mass spectra are depicted in Volume 2 (cf. Table 1-8-1). The method was linear, at least from 0.1 to 1 g/L, with a detection limit of 0.01 g/L [41].

2.2.2 Derivatization procedures

Derivatization steps are necessary to improve the gas chromatographic characteristics of polar compounds. Furthermore, mass spectra of several compounds are altered on derivatization so that they contain more typical ions, e.g. the molecular ion (cf. the mass spectra of amfetamine and its acetyl derivative). The sensitivity of the method can be improved by introduction of halogen atoms into the molecule if negative-ion chemical ionization (NICI) is used. Finally, enantiomers can be separated on achiral columns after derivatization to the diastereomers with chiral reagents, e.g. S(-)-trifluoroacetylprolylchloride [42] or S(-)-heptafluorobutyrylprolyl chloride [37,43,44].

The use of **microwave irradiation** drastically reduces the incubation time (e.g., of acetylation from 30 min to 5 min [20,30]). However, microwave irradiation should not be used for chiral derivatization, because the enantiomers may racemize. Therefore, derivatization should no longer be renounced due to time consumption. An overview on derivatization procedures for GC-MS can be found in the review of Segura et al. [45]. The following procedures have been used for recording the mass spectra presented in Volume 2.

2.2.2.1 Acetylation (AC)

The evaporated sample was acetylated with 50 µL of an acetic anhydride:pyridine mixture (3:2; v/v) for 5 min under microwave irradiation at about 400 W. After evaporation of the derivatization mixture, the residue was dissolved in 50 µL of methanol and 1–2 µL were injected into the GC-MS.

2.2.2.2 Methylation (ME)

A 50-µL aliquot of extract in methanol was methylated for 30 min (carboxylic acids) or for about 12 h (phenols) at room temperature with 100 µL of an ethanol-free solution of diazomethane in diethyl ether synthesized according to [46]. This solution was stable for several months when stored in sealed 10 mL flasks in the freezer at -20 °C. After evaporation of the methylation mixture the residue was dissolved in 50 µL methanol and 1-2 µL were injected into the GC-MS.

(Extractive methylation was described in Section 1-2.2.1.4.)

2.2.2.3 Ethylation (ET)

The procedure was as for methylation, but using diazoethane synthesized according to the procedure of McKay et al. using 1-methyl-3-nitro-1-nitroso-3-nitroguanidine, KOH and diethyl ether [46].

2.2.2.4 tert.-Butyldimethylsilylation (TBDMS)

A 50-µL aliquot of extract was evaporated and then derivatized with 50 µL of N-methyl-N-(tert.-butyldimethylsilyl)-trifluoroacetamide (MTBSTFA) for 10 min at 60 °C. Aliquots of this mixture (1–2 µL) were injected into the GC-MS with an alcohol- and water-free syringe [47].

2.2.2.5 Trimethylsilylation (TMS)

A 50-µL aliquot of extract was evaporated and then silylated with 50 µL N-methyl-N-trimethylsilyl-trifluoroacetamide (MSTFA) for 30 min at 60 °C. Aliquots (1-2 µL) of this mixture were injected into the GC-MS with an alcohol- and water-free syringe.

2.2.2.6 Trimethylsilylation followed by trifluoroacetylation (TMSTFA)

A 50-µL aliquot of extract was evaporated and then silylated with 100 µL MSTFA for 30 min at 60 °C, and then derivatized with 20 µL N-methyl-bis-trifluoroacetamide (MBTFA) for 10 min at 60 °C. After evaporation of the derivatization mixture the residue was dissolved in 50 µL alcohol- and water-free ethyl acetate and 1–2 µL of this mixture were injected into the GC-MS with an alcohol- and water-free syringe [33].

2.2.2.7 Trifluoroacetylation (TFA)

A 50-µL aliquot of extract was evaporated and then derivatized with 50 µL trifluoroacetic anhydride and 50 µL ethyl acetate for 3 min under microwave irradiation at about 400 W. After evaporation of the derivatization mixture the residue was dissolved in 50 µL alcohol- and water-free ethyl acetate and 1–2 µL were injected into the GC-MS.

2.2.2.8 Pentafluoropropionylation (PFP)

A 50-µL aliquot of extract was evaporated and then derivatized with 50 µL pentafluoropropionic anhydride for 5 min under microwave irradiation at about 400 W. After evaporation of the reagent the residue was dissolved in 50 µL alcohol- and water-free ethyl acetate and 1–2 µL were injected into the GC-MS.

2.2.2.9 Pentafluoropropylation (PFPOL)

A 50-µL aliquot of extract was evaporated and then derivatized with 50 µL pentafluoropropanol (PFPOL) for 5 min under microwave irradiation at about 400 W. After evaporation of the reagent the residue was dissolved in 50 µL alcohol- and water-free ethyl acetate and 1–2 µL were injected into the GC-MS.

2.2.2.10 Heptafluorobutyrylation (HFB)

A 50-µL aliquot of extract was evaporated and then derivatized with 50 µL heptafluorobutyric anhydride for 5 min under microwave irradiation at about 400 W. After evaporation of the reagent the residue was dissolved in 50 µL alcohol- and water-free ethyl acetate and 1–2 µL were injected into the GC-MS [8,9,18,37].

2.2.2.11 Pivalylation (PIV)

The sample was dissolved in 50 µL of a freshly prepared (!) mixture of pivalic acid anhydride:triethylamine:methanol (20:1:1; v/v/v) and incubated for 10 min under microwave irradiation at about 400 W. This solution was diluted by 200 µL methanol and 1–2 µL were injected into the GC-MS [41].

2.2.2.12 Heptafluorobutyrylprolylation (HFBP)

The evaporated sample was derivatized by addition of 0.2 mL of aqueous carbonate buffer (sodium bicarbonate/sodium carbonate, 7:3; 5%, w/v; pH 9) and 6 µL of derivatization reagent (0.1 M S-HFBPCl in dichloromethane). The reaction vials were sealed and left on a rotary shaker at room temperature for 30 min. Thereafter, 0.1 mL of cyclohexane were added, the reaction vials were sealed again and left on a rotary shaker for 1 min. The phases were separated by centrifugation (14 000 × g, 1 min) and the cyclohexane phase (upper) was transferred to autosampler vials. Aliquots (3 µL) were injected into the GC-MS [37,43,44,48].

2.3 GC-MS Apparatus

2.3.1 Apparatus and operation conditions

The following GC-MS apparatus were used for recording the GC and MS data in this handbook:

Agilent Technologies (Waldbronn, Germany) 5890 Series II gas chromatograph combined with an HP 5989B MS Engine mass spectrometer or and HP 5970 MSD, or Agilent Technologies HP 6890 gas chromatograph combined with an HP 5973 MSD or an HP 5972 MSD.

The operating conditions were as follows:

Column:	cross-linked methylsilicone capillary Varian Factor Four VF-1ms, 12 m × 0.2 I.D., film thickness 0.33 µm
Column temperature:	solvent delay: 3 min at 100 °C programmed from 100 to 310 °C in 30 °C/min, 5 min at maximum temperature
Injector port temperature:	280 °C
Carrier gas:	helium, flow rate 1 mL/min
Ionization mode:	electron ionization (EI)
Ionization energy:	70 eV
Ion source temperature:	200 °C
Scan rate:	1 scan/sec

2.3.2 Quality assurance of the apparatus performance

For daily checking of apparatus performance, a methanolic solution of typical drugs covering a wide range of physicochemical properties and relevant retention time was injected into the GC-MS. Fig. 1-2-1 shows a typical total ion chromatogram of this standard solution containing 50 ng/µL of metamfepramone, acetylated amfetamine, pentobarbital, diphenhydramine, phenobarbital, methaqualone, codeine, morphine, nalorphine, quinine, haloperidol, strychnine and the hydrocarbon C40 (injection volume 1 µL).

The quality criteria are as follows: all compounds should clearly be separated and the peaks should be sharp and of sufficient abundance. The relation of the peak areas of underivatized (!) morphine to codeine should be at least 1:10. If these demands are not fulfilled, the liner should be removed, and the GC column should be shortened (10–20 cm) or even replaced. Last, not least, the ion source should be cleaned.

Of course, the actual performance should be tested in an analysis series using calibrators and negative control samples. The use of negative control samples is indispensable, as analyte carry-over is a major problem in trace analysis. Fig. 1-2-2 (upper part) shows a total ion chromatogram of a urine sample (U+UHYAC, cf. Section 1-2.2.1.2) containing, besides other drugs, dihydrocodeine and its metabolites. After 20 (!) rinsing steps of the autosampler needle, dihydrocodeine could still be detected (lower part). False-positive GC-MS results reported by proficiency test organizers were typically caused by analyte carry-over.

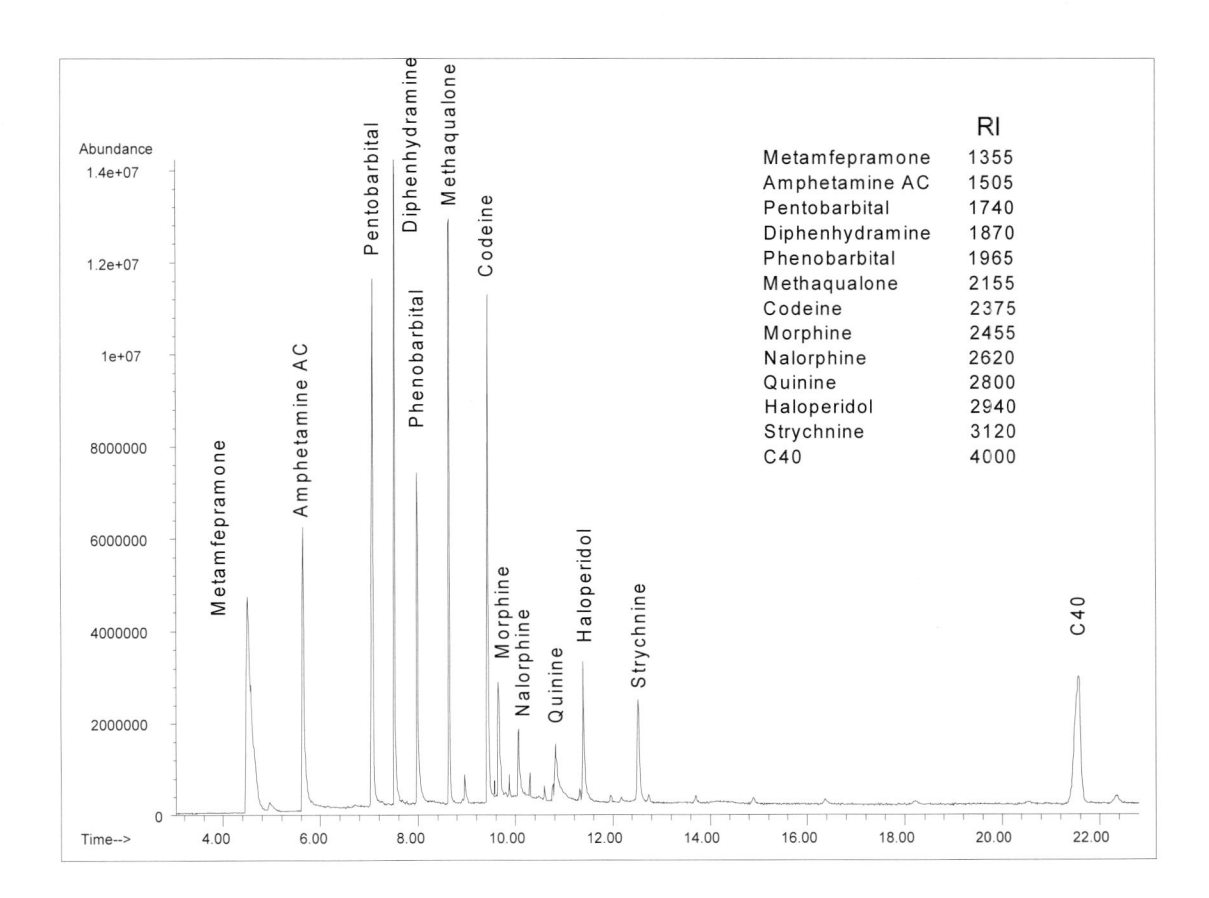

Fig. 1-2-1: Typical total ion chromatogram of the standard test solution containing 50 ng/µL of the given compounds (injection volume 1 µL).

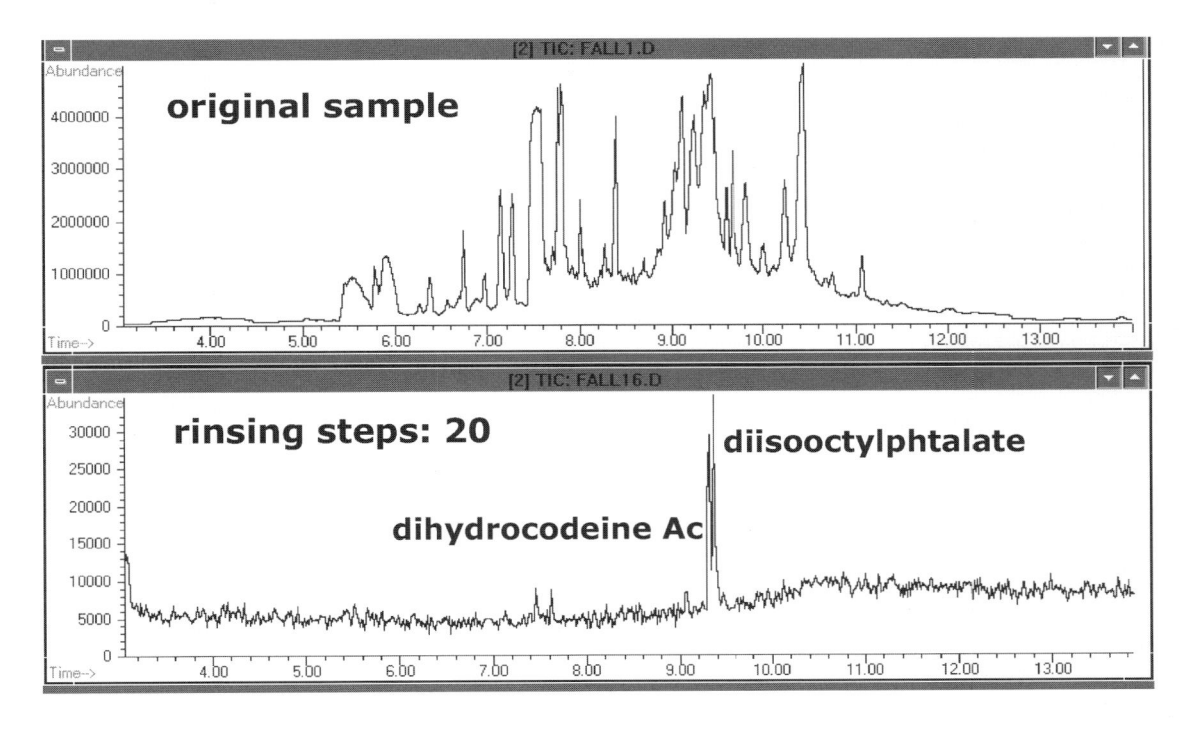

	RI
Metamfepramone	1355
Amphetamine AC	1505
Pentobarbital	1740
Diphenhydramine	1870
Phenobarbital	1965
Methaqualone	2155
Codeine	2375
Morphine	2455
Nalorphine	2620
Quinine	2800
Haloperidol	2940
Strychnine	3120
C40	4000

Fig. 1-2-2: (Upper part) Total ion chromatogram of a urine sample (U+UHYAC) containing, besides other drugs, dihydrocodeine and its metabolites. Total ion chromatogram of a methanol injection after 20 (!) rinsing steps of the autosampler needle after the injection of the urine sample (lower part).

2.4 Determination of retention indices

The retention indices (RI) were measured by GC-MS on methyl silicone capillary columns using a temperature program. They were calculated in correlation with the Kovats' indices [49] of the components of a standard solution of typical drugs (Section 1-2.3.2, Fig. 1-2-1) which is measured daily for testing the GC-MS performance [50]. The RIs of compounds with an asterisk (*) are not detectable by nitrogen-selective flame-ionization detection (N-FID).

2.5 Systematic toxicological analysis (STA) of several classes of drugs and their metabolites by GC-MS

The screening strategy of the systematic toxicological analysis (STA) must be very extensive, because several thousands of drugs or pesticides are to be considered. GC-MS, especially in the full scan electron ionization (EI) mode, is still the best method for such STA procedures [1-4,21-27]. Most of the STA procedures cover basic (and neutral) drugs, which are toxicants that are more important. For example, most of the psychotropic drugs have, like neurotransmitters, basic properties. Nevertheless, some classes of acidic drugs or drugs producing acidic metabolites, e.g., angiotensin converting enzyme (ACE) inhibitors, angiotensin receptor II (AT-II) blockers, dihydropyridine calcium channel blockers (metabolites), diuretics, laxatives, coumarin anticoagulants, antidiabetic sulfonylureas, barbiturates, or non-steroidal anti-inflammatory drugs (NSAIDs), are relevant to clinical and forensic toxicology or doping control.

Concentrations of drugs and/or their metabolites are relatively high in urine, so that urine is the sample of choice for a comprehensive screening and identification of unknown drugs or poisons. However, the metabolites of these drugs must be identified in addition, or even exclusively. In horse doping control, urine is also the common sample for screening. If the blood concentration is high enough, screening can also be performed therein. This may be advantageous, as sometimes only blood samples are available and some procedures allow simultaneous screening and quantification [5,6].

Our four screening procedures will be presented here, one for screening for 200 drugs in blood plasma after LLE (Section 1-2.5.1), one for most of the basic and neutral drugs in urine after acid hydrolysis, LLE and acetylation (Section 1-2.5.2), one for most of the acidic drugs and poisons in urine after extractive methylation (Section 1-2.5.3), and one for zwitterionic compounds in urine after SPE and silylation.

2.5.1 Screening for 200 drugs in blood plasma after LLE

GC-MS full-scan screening procedures in blood principally allow to detect a wide range of analytes although their concentrations are generally lower than those in urine [2]. Maurer has described a rather comprehensive plasma screening procedure based on LLE with diethyl ether-ethyl acetate at pH 7 and 12 after addition of the universal IS trimipramine-D3 (Section 1-2.2.1.1) [19,51,52]. This universal extract can be used for GC-MS as well as for LC-MS screening, identification and quantification [19,52,53]. The full scan GC-MS screening is based on reconstructed mass chromatography using macros for selection of suspected drugs followed by identification of the unknown spectra by library search. The selected ions for screening in plasma (and gastric content) have recently been updated using experiences from the daily routine work with this procedure and are summarized in Table 1-2-1 [2]. Generation of the mass chromatograms can be started by clicking the corresponding pull down menu which executes the user defined macros [54]. The about 200 compounds detected so far by this procedure can be found with a 'P' in the column *Detected* in Table 1-8-1 and in the legends to the mass spectra.

A rather universal SPE procedure (Section 1-2.2.1.5) has proved to be a good alternative for the LLE procedure leading to cleaner extracts [8,9,37,44,55,56]. With exception of neutral and acidic drugs, this SPE can also be used for the described plasma screening.

Of course, further compounds can be detected, if they are extractable under the conditions applied, volatile in GC and if their mass spectra are in this mass spectral database and reference library. In order to widen the screening window, comprehensive urine screening by full-scan GC-MS allowing detection of several thousand compounds is strongly recommended.

Table 1-2-1: Classes of basic and neutral drugs simultaneously screened for and identified by the GC-MS STA with LLE of plasma (Section 1-2.2.1.1) and the ions selected for reconstructed mass chromatography after full-scan GC-MS analysis [2].

Drug classes	Ions (*m/z*) selected for monitoring
Psychotropics I	58, 61 (trimipramine-D3), 70, 72, 86, 98, 112, 192
Psychotropics II	100, 123, 210, 242, 243, 276, 297, 303, 329
Benzodiazepines	239, 253, 269, 283, 300, 305, 308, 310, 312, 315
Other sedative-hypnotics	83, 105, 156, 163, 167, 172, 180, 235, 248, 261
Anticonvulsants	102, 113, 180, 185, 190, 193, 204, 280, 288, 394
Analgesics	120, 139, 151, 161, 188, 214, 217, 230, 231, 299

2.5.2 Screening for most of the basic and neutral drugs in urine after acid hydrolysis, LLE and acetylation

A screening method for the detection of most of basic, neutral and some acidic drugs in urine after acid hydrolysis, LLE and acetylation was developed and has been improved during the past years. Acid hydrolysis has proved to be very rapid and efficient for the cleavage of conjugates, particularly in emergency toxicology. However, some compounds were destroyed or altered during acid hydrolysis (Section 1-4.4). Therefore, the standard procedure had to be modified [20]. Before extraction, half of the untreated urine volume is added to the previously hydrolyzed part. The extraction solvent used has proved to be very efficient in extracting compounds with very different chemical properties from biomatrices, so that it has been used for a STA procedure for basic and neutral analytes [1-3,21-27]. Acetylation has proved to be very suitable for robust derivatization in order to improve the GC properties and thereby the detection limits of thousands of drugs and their metabolites. The use of microwave irradiation reduced the incubation time from 30 min to 5 min [20,30], so that derivatization should no longer be renounced due to expense of time.

As listed in Table 1-2-2, this comprehensive full scan GC-MS screening procedure allows within one run the simultaneous screening and library-assisted identification of the following categories of drugs: tricyclic antidepressants [57,58], selective serotonin reuptake inhibitors (SSRIs) [20], butyrophenone neuroleptics [59], phenothiazine neuroleptics [60], benzodiazepines [2,61], barbiturates [62] and other sedative-hypnotics [2,62], anticonvulsants [63], antiparkinsonian drugs [64], phenothiazine antihistamines [65], alkanolamine antihistamines [66], ethylenediamine antihistamines [67], alkylamine antihistamines [68], opiates and opioids [69], non-opioid analgesics [70], stimulants and hallucinogens [2,42,71-76], designer drugs of the amphetamine-type [21-23,71,77], the piperazine-type [77-83,83,84,84], the phenethylamine-type [1,25-29], the phencyclidine-type [24], *Eschscholtzia californica* ingredients [85], nutmeg ingredients [86], beta-blockers [87], antiarrhythmics [87], and diphenol laxatives [88]. In addition, series of further compounds can be detected [89], if they are present in the extract and their mass spectra are contained in the used reference libraries [90-92]. As shown in Table 1-2-2, several ions per category were individually selected from the mass spectra of the corresponding drugs and their metabolites identified in authentic urine samples. Generation of the mass chromatograms can be started by clicking the corresponding pull down menu which executes the user defined macros [54].

The more than 2000 compounds detected so far by this procedure can be found with a 'U+UHYAC' in the column *Detected* in Table 1-8-1 and in the legends to the mass spectra.

The procedure is illustrated in Figs. 1-2-3 and 1-2-4. Typical reconstructed mass chromatograms with the given ions of an acetylated extract of a rat urine sample collected over 24 hours after application of 0.8 mg/kg BM of 2C-D which corresponds to a common users' dose of about 50 mg. The identity of peaks in the mass chromatograms was confirmed by computerized comparison of the underlying mass spectrum with reference spectra published in this handbook. The selected ions were used for indication of the 2C-D and its metabolites. Fig. 1-2-4 illustrates the mass spectrum underlying the marked peak in Fig. 1-2-3, reference spectrum, structure, and the hit list found by computer library search [90].

Table 1-2-2: Classes of basic and neutral drugs simultaneously screened for and identified by the GC-MS STA with acid hydrolysis, LLE and acetylation of urine, the monitoring ions, and the corresponding references [2,93].

Drug classes	Ions (*m/z*) selected for monitoring	References
Antidepressants, SSRIs	58, 72, 86, 173, 176, 234, 238, 290	[20]
Antidepressants, tricyclic	58, 84, 86, 100, 191, 193, 194, 205, and	[57,58]
	120, 182, 195, 235, 261, 276, 284, 293	
Neuroleptics, butyrophenones	112, 123, 134, 148, 169, 257, 321 and	[59]
	189, 191, 223, 233, 235, 245, 287, 297	
Neuroleptics, phenothiazine	58, 72, 86, 98, 100, 113, 114, 141 and	[60]
	132, 148, 154, 191, 198, 199, 243, 267	
Benzodiazepines	111, 205, 211, 230, 241, 245, 249, 257, 308, 312, 333, 340, 357	[2,61]
Barbiturates	83, 117, 141, 157, 167, 207, 221, 235	[62]
Other sedative-hypnotics	83, 105, 156, 163, 167, 172, 216, 235, 248, 261	[62]
Anticonvulsants	102, 113, 146, 185, 193, 204, 208, 241	[63]
Antiparkinsonian drugs	86.98.136.150.165.196.197.208	[64]
Phenothiazine antihistamines	58.72.100.114.124.128.141.199	[65]
Alkanolamine antihistamines	58.139.165.167.179.182.218.260	[66]
Ethylenediamine antihistamines	58.72.85.125.165.183.198.201	[67]
Alkylamine antihistamines	58.169.203.205.230.233.262.337	[68]
Opiates and opioids	111, 138, 187, 245, 259, 327, 341, 343, 359, 420	[69]
Non-opioid analgesics	120, 139, 151, 161, 188, 217, 230, 231, 258, 308	[70]
Stimulants/Hallucinogens	58, 72, 86, 82, 94, 124, 140, 192, 250	[2,42,71-76]
Designer drugs, amphetamine-type	58, 72, 86, 150, 162, 164, 176, 178	[21-23,71,77]
Designer drugs, piperazine-type	BZP: 91, 107, 137, 146, 191, 204	[78]
	MDBP: 135, 137, 170, 262, 306	[82]
	mCPP: 143, 145, 166, 182, 238, 254	[80]
	TFMPP: 157, 161, 174, 200, 216, 330	[79]
	MeOPP: 109, 148, 151, 162, 234, 262	[81]
	Drugs with metabolites common with such designer drugs:	
	Dropropizine 132, 148, 175, 233, 320, 378	[84]
	Fipexide: 135, 137, 141, 170, 262, 306	[83]
Designer drugs, phenethylamine-type	2C-B: 228, 287, 288	[29]
	2C-D: 178, 238, 164, 223, 236, 295	[28]
	2C-E: 192, 251, 178, 237	[27]
	2C-I: 290, 349, 276, 335	[26]
	2C-T-2: 224, 283, 256, 315	[25]
	2C-T-7: 238, 297, 296, 355	[1]
Designer drugs, phencyclidine-type	232, 274, 273, 290	[24]
Eschscholtzia californica ingredients	136, 148, 165, 174, 188, 190	[85]
Nutmeg ingredients	150, 164, 165, 180, 194, 252, 266	[86]
Beta-blockers	72, 86, 98, 140, 151, 159, 200, 335	[87]
Antiarrhythmics	72, 86, 98, 140, 151, 159, 200, 335	[87]
Laxatives	349, 360, 361, 379, 390, 391, 402, 432	[88]

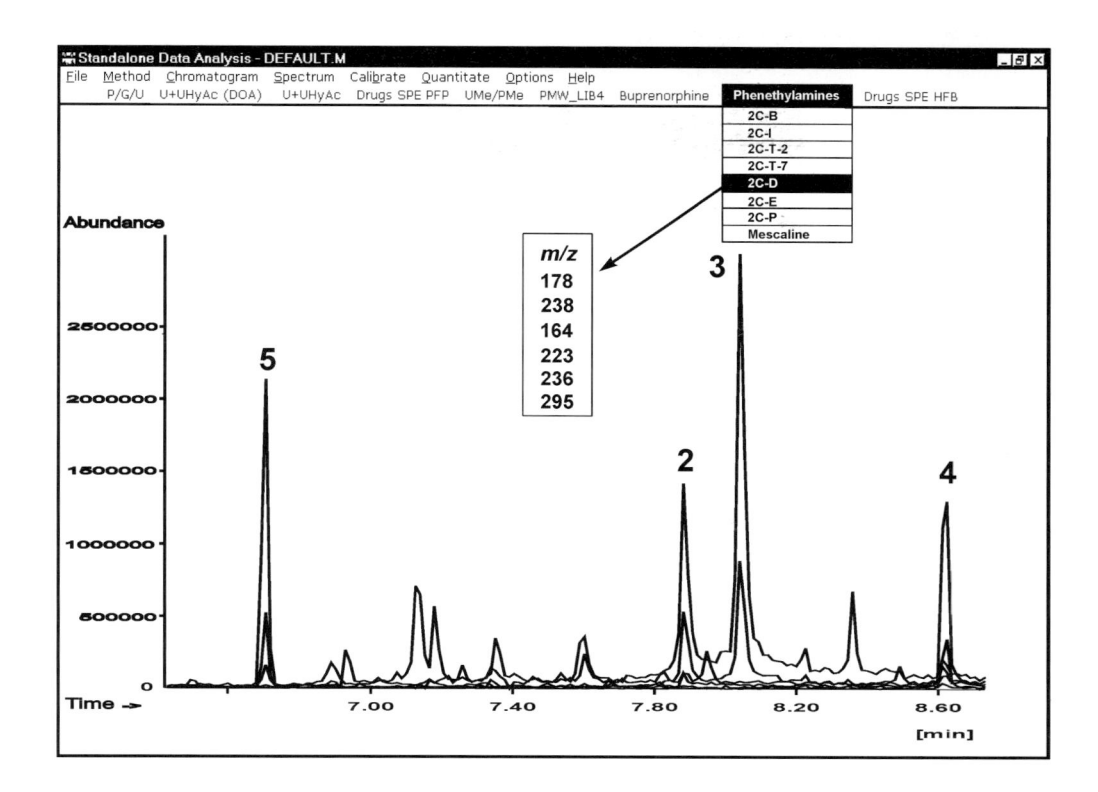

Fig. 1-2-3: Selective mass chromatograms indicating the new designer drugs 2C-D in a urine extract after acid hydrolysis and acetylation (U+UHYAC). Peaks 2-5 indicate 2C-D metabolites (taken from reference [28]). The merged ion chromatograms can be differentiated by their colors on a color screen.

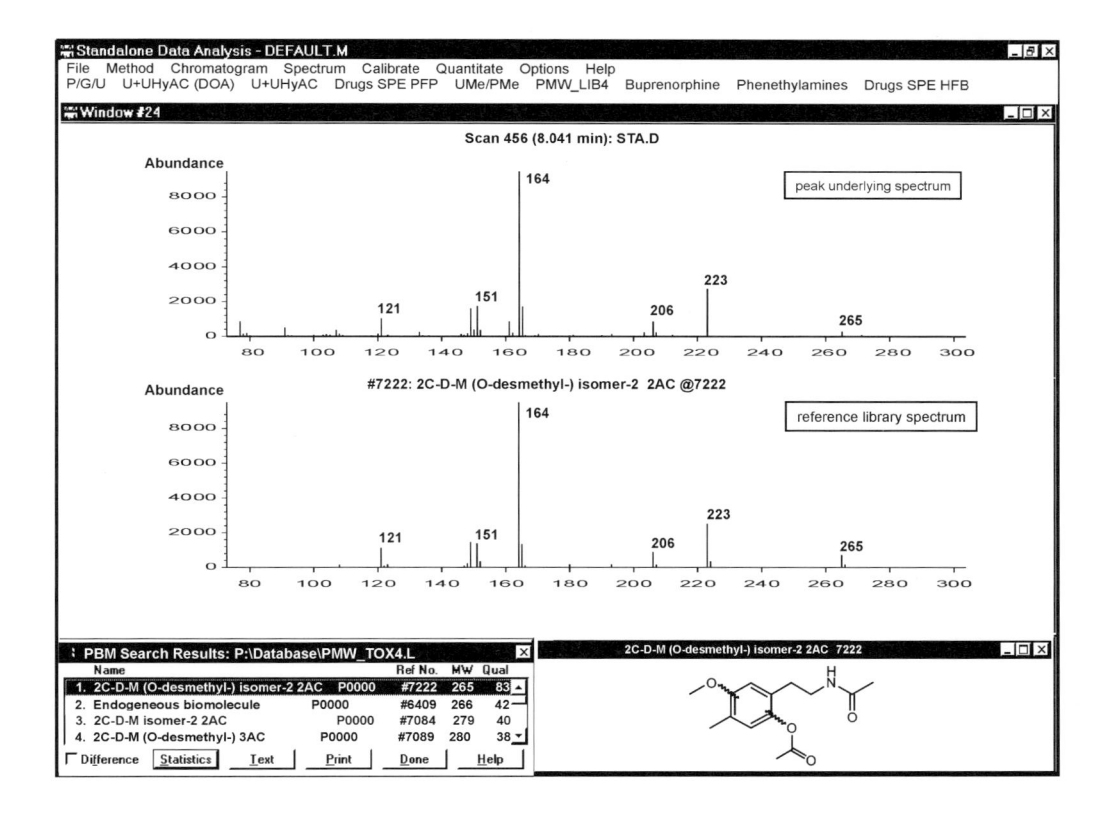

Fig. 1-2-4: Mass spectrum underlying peak 3 in Fig. 1-2-3, the reference spectrum, the structure and the hit list found by library search [90] (taken from reference [28]).

2.5.3 Systematic toxicological analysis procedures for the detection of acidic drugs and/or their metabolites

Extractive alkylation has proved to be a powerful procedure for simultaneous extraction and derivatization of acidic compounds. The extracted and derivatized analytes were separated by GC and identified by GC-MS in the full-scan mode. As already described in Section 1-2.5.2, the possible presence of acidic drugs and/or their metabolites could be indicated using mass chromatography with selective ions followed by peak identification using library search [90].

As listed in Table 1-2-3, this full scan GC-MS screening procedure allows within one run the simultaneous screening and library-assisted identification of the following categories of drugs: ACE inhibitors and AT_1 blockers [94], coumarin anticoagulants of the first generation [95], dihydropyridine calcium channel blockers [96], NSAIDs [35], barbiturates [97], diuretics [98], antidiabetics of the sulfonylurea type (sulfonamide part), and finally after enzymatic cleavage of the acetalic glucuronides, anthraquinone and diphenol laxatives [99] or buprenorphine [100]. In addition, various other acidic compounds could also be detected (Table 1-8-1).

Table 1-2-3: Classes of acidic drugs simultaneously screened for and identified by GC-MS after extractive methylation of urine, the monitoring ions, and the corresponding references.

Drug classes	Ions (*m/z*) selected for monitoring	References
ACE inhibitors and AT-II blocker	157, 160, 172, 192, 204, 220, 234, 248, 249, 262	[94]
Anticoagulants	291, 294, 295, 309, 313, 322, 324, 336, 343, 354	[95]
Calcium channel blockers (dihydropyridines)	139, 284, 297, 298, 310, 312, 313, 318, 324, 332	[96]
NSAIDs	119, 135, 139, 152, 165, 229, 244, 266, 272, 326	[35]
Barbiturates	117, 169, 183, 185, 195, 221, 223, 232, 235, 249	[97]
Diuretics (thiazides)	267, 352, 353, 355, 386, 392	[98]
Diuretics (loop diuretics)	77, 81, 181, 261, 270, 295, 406, 438	[98]
Diuretics (others)	84, 85, 111, 112, 135, 161, 249, 253, 289, 363	[98]
Laxatives (after enzymatic cleavage of the acetalic glucuronides)	305, 290, 335, 320, 365, 350, 311, 326, 271, 346	[99]
Buprenorphine (after enzymatic cleavage of the acetalic glucuronides)	352, 384, 392, 424, 441, 481	[100]

2.5.4 General screening procedure for zwitterionic compounds after SPE and silylation

Pyrrolidinophenone derivatives such as *R,S*-α-pyrrolidinopropiophenone (PPP), *R,S*-4'-methyl-α-pyrrolidinopropiophenone (MPPP), 4'-methyl-α-pyrrolidinobutyrophenone (MPBP), *R,S*-4'-methyl-α-pyrrolidinohexanophenone (MPHP), *R,S*-3',4'-methylenedioxy-α-pyrrolidinopropiophenone (MDPPP), *R,S*-4'-methoxy-α-pyrrolidinopropiophenone (MOPPP), and α-Pyrrolidinovalerophenone (PVP) are new designer drugs which have appeared on the illicit drug market [101-107]. Unfortunately, these drugs cannot be detected by common screening procedures due to the zwitterionic structure of their main metabolites. Mixed-mode SPE (Section 1-2.2.1.5) has proven to be suitable even for the extraction of their zwitterionic metabolites. Trimethylsilylation (Section 1-2.2.2.5) led to good GC properties. This comprehensive full scan GC-MS screening procedure allows within one run the simultaneous screening and library-assisted identification of the following drugs and/or their metabolites: PPP [102], MPPP [102], MPBP [106], MPHP [103], MDPPP [105], MOPPP [104], and PVP [107]. This procedure has proven to suitable also for detection of many other basic drugs.

2.6 Application of the electronic version of this handbook

The mass spectral and GC data are also available in electronic form with reduced data in the library formats of nearly all GC-MS manufacturers. The combination of the handbook presenting the full data with the electronic version allows their efficient use in analytical toxicology. Using macros, the data evaluation can be automated [2,4,54,108].

An efficient alternative is the use of AMDIS (Automated Mass Spectral Deconvolution and Identification System), a new, easy to use, sophisticated software for GC-MS data interpretation from the U.S. National Institute of Standards and Technology (NIST). It helps to analyze GC-MS data of complex mixtures, even with strong background and coeluting peaks. Further information is available at: (http://www.amdis.net/What_is_AMDIS/what_is_amdis_.htmL).
The library can also be used for the fully automated GC-MS procedure of Polettini [5,7] and for the comprehensive screening procedure used in doping control [33], as all compounds were recorded also after trimethylsilylation and after combined trimethylsilylation plus trifluoroacetylation.

2.7 Quantitative determination

GC-MS, especially in the selected ion monitoring (SIM) mode, can be employed for precise and sensitive quantification in biosamples. In contrast to therapeutic drug, monitoring or testing for drugs of abuse, in clinical toxicology several hundred drugs or poisons must be quantified, ideally using one or a few standard procedures (e.g., after standard extraction; Section 1-2.2.1.1). Even if such procedures cannot be fully validated for all compounds, they can be used in emergency toxicology with sufficient precision. However, when ever possible, quantification procedures should be validated before use in analytical toxicology. In emergency cases, one-point (linear through zero) calibration is often used as a compromise between necessary calibration, workload, and time. After comparing bias and precision data obtained with full and one-point calibration, Peters and Maurer could show that one-point calibration with a calibrator close to the center of the full calibration range can be a feasible alternative to full calibration, especially at medium or high concentrations [109].

Several guidelines for quality assurance and/or method validation have been developed and recently reviewed [110,111].

Besides such general procedures, selective GC-MS procedures have been described for the quantification of particular analytes in blood using isotopically labeled analogs as ISs [8,9,112-114].

3 Correlation between Structure and Fragmentation

This section deals with the correlation between chemical structure and mass fragmentation pattern. Only descriptive explanations are included for illustrating the identification of metabolites using electron-ionization mass spectrometry. Detailed mechanistic explanations for the interpretation of mass spectra are included in specialist texts [115,116].

3.1 Principle of electron-ionization mass spectrometry (EI-MS)

The substance, dispersed in the helium carrier gas, is introduced into the ion source of the mass spectrometer through a GC-MS interface (on-line). Compounds of low volatility can be introduced through a direct insert system (DIS) (off-line). A small fraction of the evaporated substance is ionized by electron bombardment in the high vacuum of the ion source. The resulting radical cations ($M^{+\cdot}$) decompose to defined fragment ions. All the positive ions are accelerated by the positive pusher of the ion source, focused and then separated according to their mass to charge ratio (m/z) in an electrodynamic quadrupole field (quadrupole mass spectrometer). The current of each separated ion is measured. The mass spectrum represents the relation between the mass to charge ratio of the several fragment ions and the relative intensity of their ion currents. The mass spectrum is usually represented as a bar graph, in which the abscissa represents the mass to charge ratio (m/z) in atomic mass units (u) and the ordinate represents the relative intensities of the ion currents of the several fragment ions in %. The fragmentation pattern is reproducible and characteristic for each organic compound.

For this reason, mass spectrometry is the most specific method for the identification of organic compounds.

3.2 Correlation between fundamental structures or side chains and fragment ions

Fundamental structures or side chains of organic compounds can be correlated to fragment ions by observing the fragmentation pattern of analogous compounds or drugs. The elemental composition of fragment ions can be calculated from accurate mass measurements (Section 1-5). This allows confirmation of suspected correlations. These empirical observations are useful for the identification of metabolites of known drugs or poisons. Metabolites usually have the same fundamental structure and/or the same side chains as their parent compounds and therefore their mass spectra contain similar or identical fragment ions. Chemical changes to fundamental structures and side chains by metabolism and/or derivatization procedures lead to typical shifts in the mass spectrum. The most important shifts are summarized in Table 1-3-1. Consideration of these correlations and of the fundamental principles of the metabolism of xenobiotics allow the identification of metabolites [1,21-28,71-73,77-80,82-86,102-106,117-125]. The precise chemical identification can then be confirmed by synthesis followed by further spectroscopic studies [126].

Table 1-3-1: Shifts of the fragment ions of derivatized metabolites.

Metabolite	Underivatized	AC	ME	TMS	TFA	PFP	HFB
		+42	+14	+72	+96	+146	+196
HO-	+16	+58	+30	+88	+112	+162	+212
Di-HO-	+32	+116	+60	+176	+224	+324	+424
HO-Methoxy-	+46	+88	+60	+118	+142	+192	+242
Demethyl-	-14	+28	±0	+58	+82	+132	+182
Deethyl-	-28	+14	-14	+44	+68	+118	+168
Bis-demethyl-	-28	+14	±0	+44	+68	+118	+168
Bis-deethyl-	-56	-14	-28	+16	+30	+90	+130
Demethyl-HO-	+2	+86	+30	+146	+194	+294	+394
Demethyl-HO-Methoxy-	+32	+116	+60	+176	+224	+324	+424
Bis-demethyl-HO-Methoxy-	+18	+102	+60	+162	+210	+310	+410
Deethyl-HO-Methoxy-	+18	+102	+46	+162	+210	+310	+410
Bis-deethyl-HO-Methoxy-	-10	+74	+32	+134	+172	+282	+382

4 Formation of Artifacts

Several of the compounds studied were modified during the analytical procedure employed. Sometimes this occurred thermally in the GC and sometimes during acid hydrolysis. Since the artifacts were formed reproducibly and they have been identified by mass spectrometry, they can be used for detection of the parent compounds. The artifacts are indicated in Table 1-8-1) and in the legends of the mass spectra in Volume 2 (-CO2, -H2O, HY etc.; cf. abbreviation in Section 1-6). It cannot be excluded that further compounds were modified.

It should be noted that compounds can be acetylated by acetylsalicylic acid in the gastric contents if acetylsalicylic acid was simultaneously ingested.

4.1 Artifacts formed by oxidation during extraction with diethyl ether

N-Oxidation and S-oxidation are metabolic pathways catalyzed by cytochrome p-450. Nevertheless, N- and S-oxidation were also observed during extraction with diethyl ether, which contained traces of peroxides.

4.1.1 N-Oxidation of tertiary amines

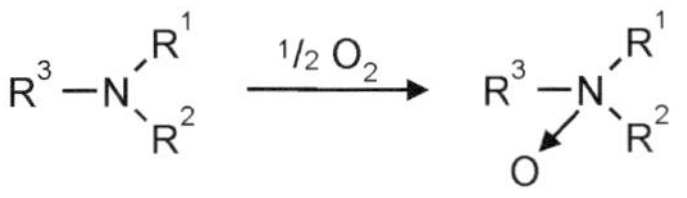

(e.g. amitriptyline)

N-Oxides undergo Cope elimination during GC (Section 1-4.2.2).

4.1.2 S-Oxidation of phenothiazines

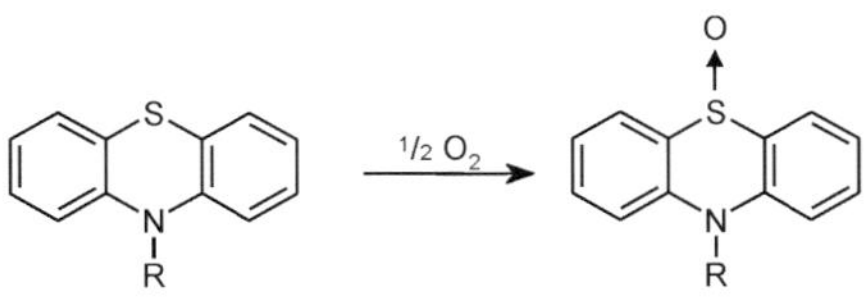

(e.g. promazine)

4.2 Artifacts formed by thermolysis during GC (GC artifact)

4.2.1 Decarboxylation of carboxylic acids

$-CO_2$ decarboxylation

$$R-COOH \xrightarrow[-CO_2]{\Delta T} R-H$$

$-C_2H_2O_2$ decarboxylation after hydrolysis of methyl carboxylate

$-C_3H_4O_2$ decarboxylation after hydrolysis of ethyl carboxylate

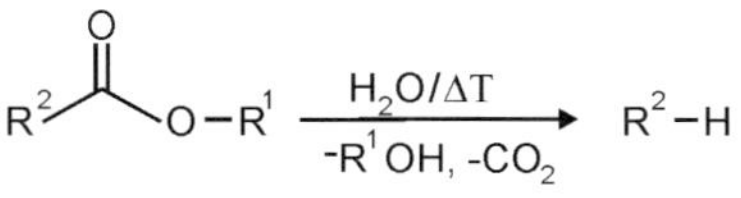

$$R^2-CO-O-R^1 \xrightarrow[-R^1OH, -CO_2]{H_2O/\Delta T} R^2-H$$

(e.g. Carbendazim)

4.2.2 Cope elimination of N-oxides
$(-(CH_3)_2NOH, -(C_2H_5)_2NOH, -C_6H_{14}N_2O_2)$

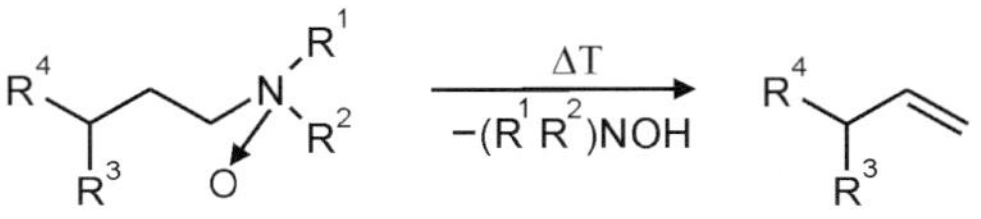

4.2.3 Rearrangement of bis-deethyl flurazepam ($-H_2O$)

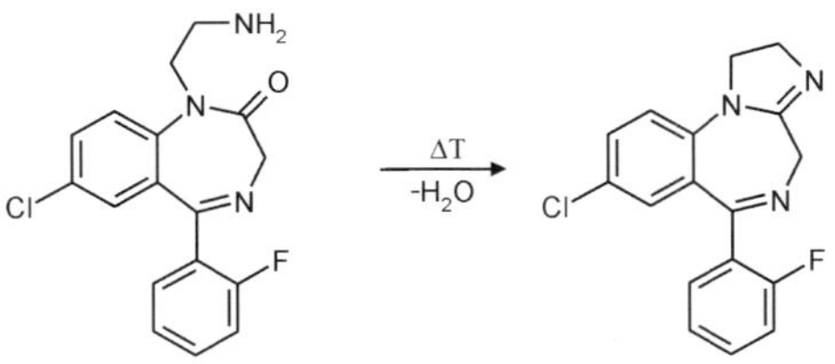

Reference: [127]

4.2.4 Elimination of various residues

Several artifacts were observed by elimination of various residues.

$-CH_3Br$	elimination of methyl bromide
$-CH_2O$	elimination of formaldehyde
$-HCl$	elimination of hydrochloric acid
$-HCN$	elimination of prussic acid
$-NH_3$	elimination of ammonia
$-SO_2NH$	elimination of a sulfonamide group

4.2.5 Methylation of carboxylic acids in methanol ((ME), ME in methanol)

Some carboxylic acids are methylated when their methanolic solutions are injected into the GC-MS system.

$$R\text{-COOH} \xrightarrow[-H_2O]{+CH_3OH/\Delta T} R\text{-COOCH}_3$$

4.2.6 Formation of formaldehyde adducts using methanol as solvent (GC artifact in methanol)

Primary amines (e.g. amfetamine), beta-blockers and some local anesthetics (e.g. flecainide, prilocaine) are altered during GC by reaction with formaldehyde, which is probably formed by thermal dehydrogenation of methanol in the injection port of the GC. This was confirmed using deuteromethanol [1,25,128].

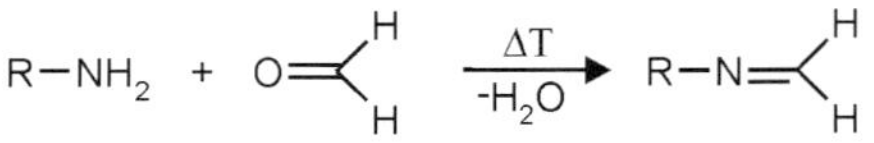

(e.g. amfetamine)

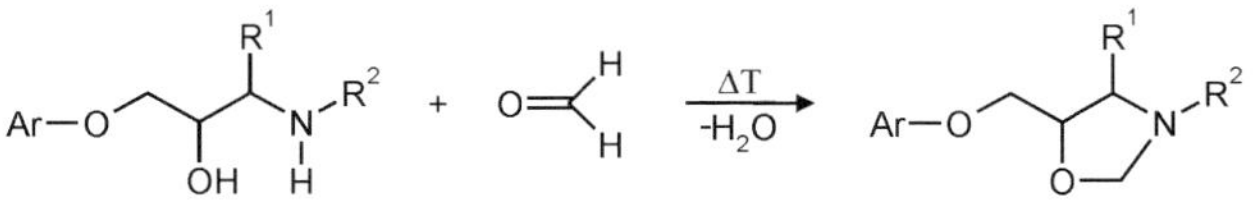

(e.g. acebutolol)

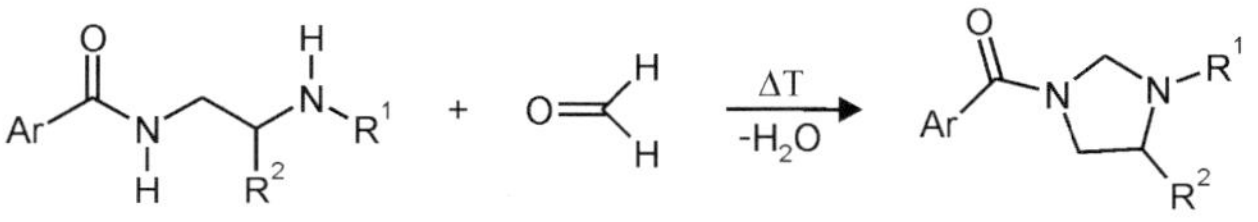

(e.g. flecainide)

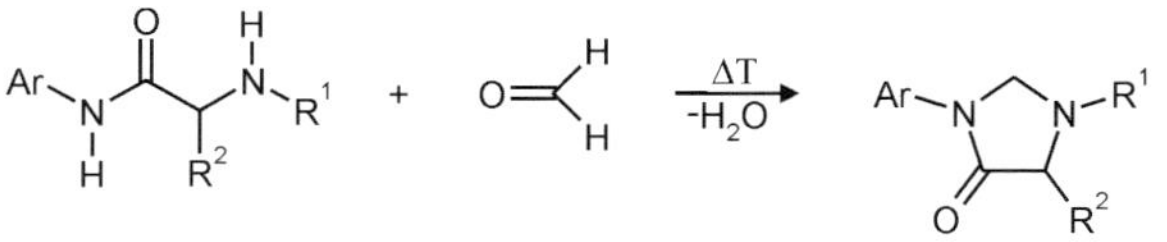

(e.g. prilocaine)

References: [1,25,87,128]

4.3 Artifacts formed by thermolysis during GC and during acid hydrolysis (GC artifact, HY artifact)

4.3.1 Dehydration of alcohols (-H$_2$O)

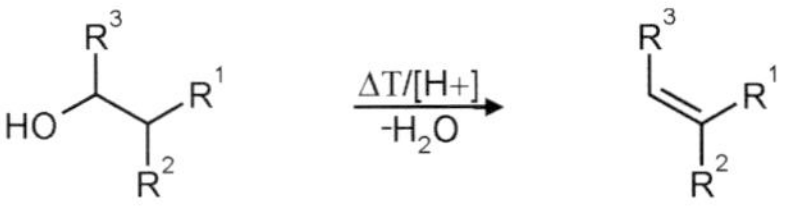

4.3.2 Decarbamoylation of carbamates

Carbamates (e.g. insecticides) are easily cleaved during gas chromatography and/or acid hydrolysis.

-CHNO	decarbamoylation
-C$_2$H$_3$NO	N-methyl decarbamoylation
-C$_3$H$_5$NO	N,N-dimethyl decarbamoylation
-C$_5$H$_9$NO	N-isobutyl decarbamoylation

4.3.3 Cleavage of morazone to phenmetrazine

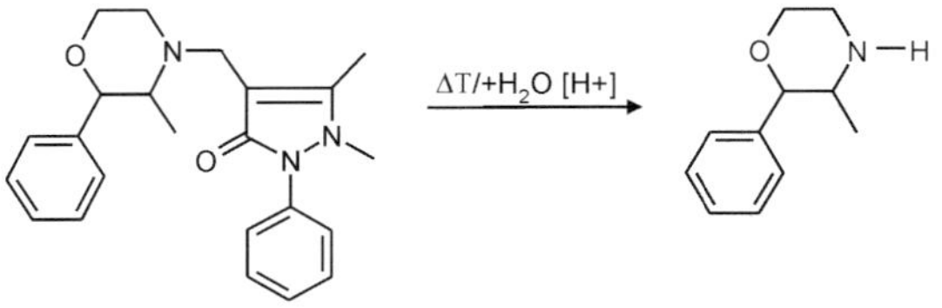

References: [129] (GC)
[130] (HY)

4.4 Artifacts formed during acid hydrolysis

4.4.1 Cleavage of the ether bridge in beta-blockers and alkanolamine antihistamines (HY)

$$R^1\text{-O-}R^2 \xrightarrow{+H_2O\ [H+]} R^1\text{-OH} + \text{HO-}R^2$$

(e.g. acebutolol or diphenhydramine)

References: [66,87]

4.4.2 Cleavage of 1,4-benzodiazepines to aminobenzoyl derivatives (HY)

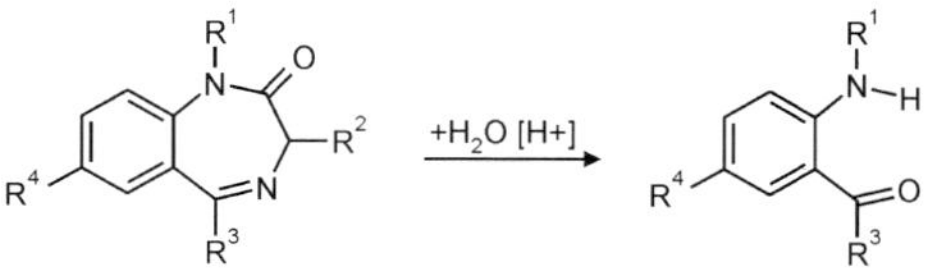

(e.g. diazepam)

Reference: [61]

4.4.3 Cleavage and rearrangement of N-demethyl metabolites of clobazam to benzimidazole derivatives (HY)

Reference: [131]

4.4.4 Cleavage and rearrangement of bis-deethyl flurazepam (HY -H$_2$O)

Reference: [131]

4.4.5 Cleavage and rearrangement of tetrazepam and its metabolites

Tetrazepam and its metabolites are transformed into two *cis/trans* isomeric hexahydroacridone derivatives each.

References: [61,132]

4.4.6 Dealkylation of ethylenediamine antihistamines (HY)

(e.g. antazoline)

Reference: [67]

4.4.7 Hydration of a double bond (+H$_2$O)

(e.g. pentazocine, alprenolol)

Reference: [87]

5 Table of Atomic Masses

The accurate atomic masses of the most abundant isotopes of the elements, which were employed for the calculations in this handbook, are listed in Table 1-5-1.

Table 1-5-1: Atomic masses of elements.

Symbol	Element	Atomic mass
As	Arsenic	74.921595
B	Boron	11.009305
Br	Bromine	78.918336
C	Carbon	12.000000
Cl	Chlorine	34.968853
D	Deuterium	2.014102
F	Fluorine	18.998403
Fe	Iron	55.934939
H	Hydrogen	1.007825
Hg	Mercury	201.970632
I	Iodine	126.904477
N	Nitrogen	14.003074
O	Oxygen	15.994915
P	Phosphorus	30.973763
Pb	Lead	207.976641
S	Sulfur	31.972072
Si	Silicon	27.976928

Reference: [133]

6 Abbreviations

The abbreviations used in this handbook and library are listed in Table 1-6-1. The given Sections correspond to those of Volume 1.

Table 1-6-1: Abbreviations.

Abbreviation	Meaning	see Section (Volume 1)
A	Artifact	4
AC	Acetylated	2.2.2.1
(AC)	Possibly acetylated	
ACE	Angiotensin converting enzyme	
ALHY	Extract after alkaline hydrolysis	
Altered during HY	Altered compound detectable in UHY	4
Artifact ( )	( ) artifact	4
BPH	Benzophenone	
BZP	N-Benzylpiperazine	2.5.2
CI	Chemical ionization	
$-CH_3Br$	Artifact formed by elimination of methyl bromide	4.2.4
-CHNO	Artifact formed by decarbamoylation	4.3.2
$-C_2H_3NO$	Artifact formed by N-methyl decarbamoylation	4.3.2
$-C_3H_5NO$	Artifact formed by N,N-dimethyl decarbamoylation	4.3.2
$-C_5H_9NO$	Artifact formed by N-isobutyl decarbamoylation	4.3.2
$-(CH_3)_2NOH$	Artifact formed by Cope elimination of the N-oxide	4.2.2
$-(C_2H_5)_2NOH$	Artifact formed by Cope elimination of the N-oxide	4.2.2
$-C_6H_{14}N_2O_2$	Artifact formed by Cope elimination of the N-oxide	4.2.2
$-CH_2O$	Artifact formed by elimination of formaldehyde	4.2.4
$-C_2H_2O_2$	Artifact formed by decarboxylation after hydrolysis of methyl carboxylate	4.2.1
$-C_3H_4O_2$	Artifact formed by decarboxylation after hydrolysis of ethyl carboxylate	4.2.1
$-CO_2$	Artifact formed by decarboxylation	4.2.1
mCPP	1-(3-Chlorophenyl)piperazine	2.5.2
DIS	Direct insert system used for recording the spectrum	
EG	Ethylene glycol	
EI	Electron ionization	3.1
ET	Ethylated	2.2.2.3
FID	Flame-ionization detector	
G	Standard extract of gastric contents	2.2.1.1
GC	Gas chromatographic, -graph, -graphy	2.3
GC artifact	Artifact formed during GC	4.2-3
GC artifact in methanol	Artifact (of beta-adrenergic blocking agents) formed by reaction with methanol during GC	4.2.6
-HCl	Artifact formed by elimination of hydrogen chloride	4.2.4
-HCN	Artifact formed by elimination of hydrogen cyanide	4.2.4
HFB	Heptafluorobutyrylated	2.2.2.10
HFBP	Heptafluorobutyrylprolylated	2.2.2.12
HO-	Hydroxy	

Abbreviation	Meaning	see Section (Volume 1)
$+H_2O$	Artifact formed by hydration (of an alkene)	4.4.7
$-H_2O$	Artifact formed by dehydration (of an alcohol or	4.3.1
	with rearrangement of an amino oxo compound)	4.4.4
HOOC-	Carboxy	
HY	Acid-hydrolyzed or acid hydrolysis	2.2.1.2
HY artifact	Artifact formed during acid hydrolysis	4.4
-I	Intoxication; this compound is only detectable after a toxic dosage	
I.D.	Internal diameter	
INN	International non-proprietary name (WHO)	
IS	Internal standard	
iso	Isomer	
LLE	Liquid-liquid extraction	2.2
LM	Low-resolution mass spectrum	
LM/Q	Low-resolution mass spectrum recorded on a quadrupole MS	
LOD	Limit of detection	
LS	Background subtracted low-resolution mass spectrum	
LS/Q	Background subtracted low-resolution mass spectrum recorded on a quadrupole MS	
M	1 mol/L	
M^+	Molecular ion	
-M	Metabolite	
-M ( )	( ) metabolite	
-M (HO-)	Hydroxy metabolite	
-M (HOOC-)	Carboxylated metabolite	
-M (nor-)	N-Demethyl metabolite	
-M (ring)	Ring compound as metabolite (e.g., of phenothiazines)	
-M artifact	Artifact of a metabolite	
-M/artifact	Metabolite or artifact	
m/z	Mass to charge ratio	3.1
MBTFA	N-methyl-bis-trifluoroacetamide	2.2.2.6-7
MDBP	1-(3,4-Methylenedioxybenzyl)piperazine	2.5.2
MDPPP	R,S-3',4'-Methylenedioxy-α-pyrrolidinopropiophenone	2.5.4
ME	Methylated	2.2.2.2
(ME)	Methylated by methanol during GC	4.2.5
ME in methanol	Methylated by methanol during GC	4.2.5
MeOPP	1-(4-Methoxyphenyl)piperazine	2.5.2
MOPPP	R,S-4'-Methoxy-α-pyrrolidinopropiophenone	2.5.4
MPBP	4'-Methyl-α-pyrrolidinobutyrophenone	2.5.4
MPHP	R,S-4'-Methyl-α-pyrrolidinohexanophenone	2.5.4
MPPP	R,S-4'-Methyl-α-pyrrolidinopropiophenone	2.5.4
MS	Mass spectrometric, -meter, -metry, mass spectrum	
MSTFA	N-Methyl-N-trimethylsilyl-trifluoroacetamide	2.2.2.5
MTBSTFA	N-Methyl-N-(tert.-butyldimethylsilyl)-trifluoroacetamide	2.2.2.4
N-FID	Nitrogen-sensitive flame-ionization detector	
$-NH_3$	Artifact formed by elimination of ammonia	4.2.4
NICI	Negative-ion chemical ionization	
NIST	National Institute of Standards and Technology	2.6
Not detectable after HY	Compound destroyed during acid hydrolysis	2.2.1.2

Abbreviation	Meaning	see Section (Volume 1)
P	Standard extract of plasma	2.2.1.1
PEGPIV	Pivalylated extract of plasma for determination of glycols	2.2.1.7
PFP	Pentafluoropropionylated	2.2.2.8
PFPA	Pentafluoropropionic anhydride	2.2.2.8
PFPOH	Pentafluoropropanol	2.2.2.9
PIV	Pivalylated	2.2.2.11
PPP	R,S-α-Pyrrolidinopropiophenone	2.5.4
PS	Pure substance	
PSPE	Solid-phase extract of plasma	2.2.1.5
PTHCME	Extract of plasma for detection of tetrahydrocannabinol metabolites [134]	
PVP	α-Pyrrolidinovalerophenone	2.5.4
R	Any unknown substituent	
Rat	Compound found in the urine of rats	2.1
RI	Retention index	2.4
-SO$_2$NH	Artifact formed by elimination of a sulfonamide group	4.2.4
SIM	Selected ion mode	
SPE	Solid phase extraction	2.2.1.5
STA	Systematic toxicological analysis	2.5
TBDMS	Tertiary butyl dimethyl silylated	2.2.2.4
TFA	Trifluoroacetylated	2.2.2.7
TFMPP	1-(3-Trifluoromethylphenyl)piperazine	2.5.2
THC	Tetrahydrocannabinol	
THC-COOH	11-Nor-delta-9-tetrahydrocannabinol-9-carboxylic acid	
TM	Trade mark	
TMS	Trimethylsilylated	2.2.2.5
TMSTFA	Trimethylsilylated followed by trifluoroacetylation	2.2.2.6
u	(Atomic mass) Unit, 1/12 of the mass of the nuclide ^{12}C (*SI* unit)	5
U	Standard extract of urine	2.2.1.1
UA	Extract of urine for detection of amphetamines [42]	
UCO	Extract of urine for detection of cocaine [135]	
UGLUC	Extract of urine after cleavage of conjugates by glucuronidase and arylsulfatase	2.2.1.3
UHY	Extract of urine after acid hydrolysis	2.2.1.2
ULSD	Extract of urine for detection of lysergide (LSD) [136]	
UMAM	Extract of urine for detection of 6-monoacteyl morphine [137]	
USPE	Solid-phase extract of urine	2.2.1.5
UTHCME	Extract of urine for detection of tetrahydrocannabinol metabolites after methylation [134]	
U+UHYAC	Extract of urine with and without acid hydrolysis and acetylation	2.2.1.2
*	Compound contains no nitrogen and cannot be detected by N-FID	
----	RI not determined	
9999	RI > 4000, compound not detectable by GC (-MS).	

7 References

1. D.S. Theobald, S. Fehn and H.H. Maurer. New designer drug 2,5-dimethoxy-4-propylthiophenethylamine (2C-T-7): studies on its metabolism and toxicological detection in rat urine using gas chromatography/mass spectrometry. *J. Mass Spectrom.* 40, 105-116 (2005).

2. H.H. Maurer. Position of chromatographic techniques in screening for detection of drugs or poisons in clinical and forensic toxicology and/or doping control [review]. *Clin. Chem. Lab. Med.* 42, 1310-1324 (2004).

3. H.H. Maurer and F.T. Peters. Towards High-throughput Drug Screening Using Mass Spectrometry. *Ther. Drug Monit.* 27, 686-688 (2005).

4. H.H. Maurer. Hyphenated mass spectrometric techniques - indispensable tools in clinical and forensic toxicology and in doping control [review]. *J. Mass Spectrom.* 41, 1399-1413 (2006).

5. A. Polettini, A. Groppi, C. Vignali and M. Montagna. Fully-automated systematic toxicological analysis of drugs, poisons, and metabolites in whole blood, urine, and plasma by gas chromatography-full scan mass spectrometry. *J. Chromatogr. B* 713, 265-279 (1998).

6. A. Polettini. A simple automated procedure for the detection and identification of peaks in gas chromatography--continuous scan mass spectrometry. Application to systematic toxicological analysis of drugs in whole human blood. *J. Anal. Toxicol.* 20, 579-586 (1996).

7. A. Polettini. Systematic toxicological analysis of drugs and poisons in biosamples by hyphenated chromatographic and spectroscopic techniques [review]. *J. Chromatogr. B* 733, 47-63 (1999).

8. F.T. Peters, S. Schaefer, R.F. Staack, T. Kraemer and H.H. Maurer. Screening for and validated quantification of amphetamines and of amphetamine- and piperazine-derived designer drugs in human blood plasma by gas chromatography/mass spectrometry. *J. Mass Spectrom.* 38, 659-676 (2003).

9. V. Habrdova, F.T. Peters, D.S. Theobald and H.H. Maurer. Screening for and validated quantification of phenethylamine-type designer drugs and mescaline in human blood plasma by gas chromatography/mass spectrometry. *J. Mass Spectrom.* 40, 785-795 (2005).

10. F. Pragst and M.A. Balikova. State of the art in hair analysis for detection of drug and alcohol abuse. *Clin. Chim. Acta* 370, 17-49 (2006).

11. O.H. Drummer. Review: Pharmacokinetics of illicit drugs in oral fluid. *Forensic Sci. Int.* 150, 133-142 (2005).

12. H.H. Maurer. Advances in analytical toxicology: Current role of liquid chromatography-mass spectrometry for drug quantification in blood and oral fluid [review]. *Anal. Bioanal. Chem.* 381, 110-118 (2005).

13. M.A. Huestis and R.E. Choo. Drug abuse's smallest victims: in utero drug exposure. *Forensic Sci. Int.* 128, 20-30 (2002).

14. A. Palmeri, S. Pichini, R. Pacifici, P. Zuccaro and A. Lopez. Drugs in nails: physiology, pharmacokinetics and forensic toxicology. *Clin. Pharmacokinet.* 38, 95-110 (2000).

15. H.H. Maurer. Systematic toxicological analysis procedures for acidic drugs and/or metabolites relevant to clinical and forensic toxicology or doping control [review]. *J. Chromatogr. B Biomed. Sci. Appl.* 733, 3-25 (1999).

16. F. Pragst. Application of solid-phase microextraction (SPME) in analytical toxicology [review]. *Anal. Bioanal. Chem.* (2007), in press.

17. S.M. Wille and W.E. Lambert. Recent developments in extraction procedures relevant to analytical toxicology [review]. *Anal. Bioanal. Chem,.* (2007), in press.

18. H.H. Maurer, J. Bickeboeller-Friedrich, T. Kraemer and F.T. Peters. Toxicokinetics and analytical toxicology of amphetamine-derived designer drugs ("Ecstasy"). *Toxicol. Lett.* 112, 133-142 (2000).

19. H.H. Maurer, C. Kratzsch, T. Kraemer, F.T. Peters and A.A. Weber. Screening, library-assisted identification and validated quantification of oral antidiabetics of the sulfonylurea-type in plasma by atmospheric pressure chemical ionization liquid chromatography-mass spectrometry (APCI-LC-MS). *J. Chromatogr. B Analyt. Technol. Biomed. Life Sci.* 773, 63-73 (2002).

20. H.H. Maurer and J. Bickeboeller-Friedrich. Screening procedure for detection of antidepressants of the selective serotonin reuptake inhibitor type and their metabolites in urine as part of a modified systematic toxicological analysis procedure using gas chromatography-mass spectrometry. *J. Anal. Toxicol.* 24, 340-347 (2000).

21. A.H. Ewald, F.T. Peters, M. Weise and H.H. Maurer. Studies on the metabolism and toxicological detection of the designer drug 4-methylthioamphetamine (4-MTA) in human urine using gas chromatography-mass spectrometry. *J. Chromatogr. B Analyt. Technol. Biomed. Life Sci.* 824, 123-131 (2005).

22. A.H. Ewald, G. Fritschi, W.R. Bork and H.H. Maurer. Designer drugs 2,5-dimethoxy-4-bromoamphetamine (DOB) and 2,5-dimethoxy-4-bromomethamphetamine (MDOB): Studies on their metabolism and toxicological detection in rat urine using gas chromatographic/mass spectrometric techniques. *J. Mass Spectrom.* 41, 487-498 (2006).

23. A.H. Ewald, G. Fritschi and H.H. Maurer. Designer drug 2,4,5-trimethoxyamphetamine (TMA-2): Studies on its metabolism and toxicological detection in rat urine using gas chromatographic/mass spectrometric techniques. *J. Mass Spectrom.* 41, 1140-1148 (2006).

24. C. Sauer, F.T. Peters, R.F. Staack, G. Fritschi and H.H. Maurer. New designer drug (1-(1-Phenylcyclohexyl)-3-ethoxypropylamine (PCEPA): Studies on its metabolism and toxicological detection in rat urine using gas chromatography/mass spectrometry. *J. Mass Spectrom.* 41, 1014-1029 (2006).

25. D.S. Theobald, R.F. Staack, M. Puetz and H.H. Maurer. New designer drug 2,5-dimethoxy-4-ethylthio-beta-phenethylamine (2C-T-2): studies on its metabolism and toxicological detection in rat urine using gas chromatography/mass spectrometry. *J. Mass Spectrom.* 40, 1157-1172 (2005).

26. D.S. Theobald, M. Putz, E. Schneider and H.H. Maurer. New designer drug 4-iodo-2,5-dimethoxy-beta-phenethylamine (2C-I): studies on its metabolism and toxicological detection in rat urine using gas chromatographic/mass spectrometric and capillary electrophoretic/mass spectrometric techniques. *J. Mass Spectrom.* 41, 872-886 (2006).

27. D.S. Theobald and H.H. Maurer. Studies on the metabolism and toxicological detection of the designer drug 4-ethyl-2,5-dimethoxy-beta-phenethylamine (2C-E) in rat urine using gas chromatographic-mass spectrometric techniques. *J. Chromatogr. B Analyt. Technol. Biomed. Life Sci.* 842, 76-90 (2006).

28. D.S. Theobald and H.H. Maurer. Studies on the metabolism and toxicological detection of the designer drug 2,5-dimethoxy-4-methyl-beta-phenethylamine (2C-D) in rat urine using gas chromatographic-mass spectrometric techniques. *J. Mass Spectrom.* 41, 1509-1519 (2006).

29. D.S. Theobald, G. Fritschi and H.H. Maurer. Studies on the toxicological detection of the designer drug 4-bromo-2,5-dimethoxy-beta-phenethylamine (2C-B) in rat urine using gas chromatography-mass spectrometry. *J. Chromatogr. B Analyt. Technol. Biomed. Life Sci.* 846, 374-377 (2007).

30. T. Kraemer, A.A. Weber and H.H. Maurer. Improvement of sample preparation for the STA - Acceleration of acid hydrolysis and derivatization procedures by microwave irradiation. In *Proceedings of the Xth GTFCh Symposium in Mosbach*, F. Pragst, Ed. Helm-Verlag, Heppenheim, 1997, pp. 200-204.

31. S.W. Toennes and H.H. Maurer. Efficient cleavage of urinary conjugates of drugs or poisons in analytical toxicology using purified and immobilized β-glucuronidase and arylsulfatase packed in columns. *Clin. Chem.* 45, 2173-2182 (1999).

32. R. Meatherall. Optimal enzymatic hydrolysis of urinary benzodiazepine conjugates. *J. Anal. Toxicol.* 18, 382-384 (1994).

33. A. Solans, M. Carnicero, R. de-la-Torre and J. Segura. Comprehensive screening procedure for detection of stimulants, narcotics, adrenergic drugs, and their metabolites in human urine. *J. Anal. Toxicol.* 19, 104-114 (1995).

34. R.W. Romberg and L. Lee. Comparison of the hydrolysis rates of morphine-3-glucuronide and morphine-6-glucuronide with acid and beta-glucuronidase. *J. Anal. Toxicol.* 19, 157-162 (1995).

35. H.H. Maurer, F.X. Tauvel and T. Kraemer. Screening procedure for detection of non-steroidal antiinflammatory drugs (NSAIDs) and their metabolites in urine as part of a systematic toxicological analysis (STA) procedure for acidic drugs and poisons by gas chromatography-mass spectrometry (GC-MS) after extractive methylation. *J. Anal. Toxicol.* 25, 237-244 (2001).

36. J. Beyer, F.T. Peters, T. Kraemer and H.H. Maurer. Detection and validated quantification of herbal phenalkylamines and methcathinone in human blood plasma by LC/MS/MS. *J. Mass Spectrom.* 42, 150-160 (2007).

37. F.T. Peters, T. Kraemer and H.H. Maurer. Drug testing in blood: validated negative-ion chemical ionization gas chromatographic-mass spectrometric assay for determination of amphetamine and methamphetamine enantiomers and its application to toxicology cases. *Clin. Chem.* 48, 1472-1485 (2002).

38. J. Beyer, F.T. Peters, T. Kraemer and H.H. Maurer. Detection and validated quantification of toxic alkaloids in human plasma - Comparison of LC-APCI-MS with LC-ESI-MS/MS. *J. Mass Spectrom.* 42 (2007), Feb 26; [Epub ahead of print].

39. F.T. Peters, J. Jung, T. Kraemer and H.H. Maurer. Fast, simple, and validated gas chromatographic-mass spectrometric assay for quantification of drugs relevant to diagnosis of brain death in human blood plasma samples. *Ther. Drug Monit.* 27, 334-344 (2005).

40. H. Maurer and C. Kessler. Identification and quantification of ethylene glycol and diethylene glycol in plasma using gas chromatography-mass spectrometry. *Arch. Toxicol.* 62, 66-69 (1988).

41. H.H. Maurer, F.T. Peters, L.D. Paul and T. Kraemer. Validated GC-MS assay for determination of the antifreezes ethylene glycol and diethylene glycol in human plasma after microwave-assisted pivalylation. *J. Chromatogr. B Biomed. Sci. Appl.* 754, 401-409 (2001).

42. H.H. Maurer and T. Kraemer. Toxicological detection of selegiline and its metabolites in urine using fluorescence polarization immunoassay (FPIA) and gas chromatography-mass spectrometry (GC-MS) and differentiation by enantioselective GC-MS of the intake of selegiline from abuse of methamphetamine or amphetamine. *Arch. Toxicol.* 66, 675-678 (1992).

43. F.T. Peters, N. Samyn, T. Kraemer, G. de Boeck and H.H. Maurer. Concentrations and Ratios of Amphetamine, Methamphetamine, MDA, MDMA, and MDEA Enantiomers determined in Plasma Samples From Clinical Toxicology and Driving Under the Influence of Drugs Cases by GC-NICI-MS. *J. Anal. Toxicol.* 27, 552-559 (2003).

44. F.T. Peters, N. Samyn, C. Lamers, W. Riedel, T. Kraemer, G. de Boeck and H.H. Maurer. Drug Testing in Blood: Validated Negative-Ion Chemical Ionization Gas Chromatographic-Mass Spectrometric Assay for Enantioselective Determination of the Designer Drugs MDA, MDMA (Ecstasy) and MDEA and Its Application to Samples from a Controlled Study with MDMA. *Clin. Chem.* 51, 1811-1822 (2005).

45. J. Segura, R. Ventura and C. Jurado. Derivatization procedures for gas chromatographic-mass spectrometric determination of xenobiotics in biological samples, with special attention to drugs of abuse and doping agents [review]. *J. Chromatogr. B* 713, 61-90 (1998).

46. A.F. McKay, W.L. Ott, G.W. Taylor, M.N. Buchanan and J.F. Crooker. Diazohydrocarbons. *Can. J. Res.* 28, 683-688 (1950).

47. S.W. Toennes, A.S. Fandino and G. Kauert. Gas chromatographic-mass spectrometric detection of anhydroecgonine methyl ester (methylecgonidine) in human serum as evidence of recent smoking of crack. *J. Chromatogr. B* 127-132 (1999).

48. F.T. Peters, N. Samyn, T. Kraemer, W. Riedel and H.H. Maurer. Negative-ion chemical ionization gas chromatographic-mass spectrometric assay for enantioselective determination of amphetamines in oral fluid: Application to a controlled study with MDMA and driving under the influence of drugs cases. *Clin. Chem.* (2007), Mar 1; [Epub ahead of print].

49. E. Kovats. Gaschromatographische Charakterisierung organischer Verbindungen. Teil 1. Retentionsindices aliphatischer Halogenide, Alkohole, Aldehyde und Ketone. *Helv. Chim. Acta* 41, 1915-1932 (1958).

50. R.A.de-Zeeuw, J.P.Franke, H.H.Maurer and K.Pfleger. *Gas Chromatographic Retention Indices of Toxicologically Relevant Substances and their Metabolites (Report of the DFG commission for clinical toxicological analysis, special issue of the TIAFT bulletin),* 3rd ed. VCH publishers, Weinheim, New York, Basle, (1992).

51. H.H. Maurer, C. Kratzsch, A.A. Weber, F.T. Peters and T. Kraemer. Validated assay for quantification of oxcarbazepine and its active dihydro metabolite 10-hydroxy carbazepine in plasma by atmospheric pressure chemical ionization liquid chromatography/mass spectrometry. *J. Mass Spectrom.* 37, 687-692 (2002).

52. C. Kratzsch, O. Tenberken, F.T. Peters, A.A. Weber, T. Kraemer and H.H. Maurer. Screening, library-assisted identification and validated quantification of 23 benzodiazepines, flumazenil, zaleplone, zolpidem and zopiclone in plasma by liquid chromatography/mass spectrometry with atmospheric pressure chemical ionization. *J. Mass Spectrom.* 39, 856-872 (2004).

53. H.H. Maurer, T. Kraemer, C. Kratzsch, F.T. Peters and A.A. Weber. Negative ion chemical ionization gas chromatography-mass spectrometry (NICI-GC-MS) and atmospheric pressure chemical ionization liquid chromatography-mass spectrometry (APCI-LC-MS) of low-dosed and/or polar drugs in plasma. *Ther. Drug Monit.* 24, 117-124 (2002).

54. H.H. Maurer. Toxicological analysis of drugs and poisons by GC-MS [review]. *Spectroscopy Europe* 6, 21-23 (1994).

55. H.H. Maurer, O. Tenberken, C. Kratzsch, A.A. Weber and F.T. Peters. Screening for, library-assisted identification and fully validated quantification of twenty-two beta-blockers in blood plasma by liquid chromatography-mass spectrometry with atmospheric pressure chemical ionization. *J. Chromatogr. A* 1058, 169-181 (2004).

56. C. Kratzsch, A.A. Weber, F.T. Peters, T. Kraemer and H.H. Maurer. Screening, library-assisted identification and validated quantification of fifteen neuroleptics and three of their metabolites in plasma by liquid chromatography/mass spectrometry with atmospheric pressure chemical ionization. *J. Mass Spectrom.* 38, 283-295 (2003).

57. J. Bickeboeller-Friedrich and H.H. Maurer. Screening for detection of new antidepressants, neuroleptics, hypnotics, and their metabolites in urine by GC-MS developed using rat liver microsomes. *Ther. Drug Monit.* 23, 61-70 (2001).

58. H. Maurer and K. Pfleger. Screening procedure for detection of antidepressants and their metabolites in urine using a computerized gas chromatographic-mass spectrometric technique. *J. Chromatogr.* 305, 309-323 (1984).

59. H. Maurer and K. Pfleger. Screening procedure for detecting butyrophenone and bisfluorophenyl neuroleptics in urine using a computerized gas chromatographic-mass spectrometric technique. *J. Chromatogr.* 272, 75-85 (1983).

60. H. Maurer and K. Pfleger. Screening procedure for detection of phenothiazine and analogous neuroleptics and their metabolites in urine using a computerized gas chromatographic-mass spectrometric technique. *J. Chromatogr.* 306, 125-145 (1984).

61. H. Maurer and K. Pfleger. Identification and differentiation of benzodiazepines and their metabolites in urine by computerized gas chromatography-mass spectrometry. *J. Chromatogr.* 422, 85-101 (1987).

62. H.H. Maurer. Identification and differentiation of barbiturates, other sedative-hypnotics and their metabolites in urine integrated in a general screening procedure using computerized gas chromatography- mass spectrometry. *J. Chromatogr.* 530, 307-326 (1990).

63. H.H. Maurer. Detection of anticonvulsants and their metabolites in urine within a "general unknown" analysis procedure using computerized gas chromatography-mass spectrometry. *Arch. Toxicol.* 64, 554-561 (1990).

64. H. Maurer and K. Pfleger. Screening procedure for the detection of antiparkinsonian drugs and their metabolites in urine using a computerized gas chromatographic-mass spectrometric technique. *Fresenius' Z. Anal. Chem.* 321, 363-370 (1985).

65. H. Maurer and K. Pfleger. Identification of phenothiazine antihistamines and their metabolites in urine. *Arch. Toxicol.* 62, 185-191 (1988).

66. H. Maurer and K. Pfleger. Screening procedure for the detection of alkanolamine antihistamines and their metabolites in urine using computerized gas chromatography-mass spectrometry. *J. Chromatogr.* 428, 43-60 (1988).

67. H. Maurer and K. Pfleger. Toxicological detection of ethylenediamine and piperazine antihistamines and their metabolites in urine by computerized gas chromatography-mass spectrometry. *Fresenius' Z. Anal. Chem.* 331, 744-756 (1988).

68. H. Maurer and K. Pfleger. Identification and differentiation of alkylamine antihistamines and their metabolites in urine by computerized gas chromatography-mass spectrometry. *J. Chromatogr.* 430, 31-41 (1988).

69. H. Maurer and K. Pfleger. Screening procedure for the detection of opioids, other potent analgesics and their metabolites in urine using a computerized gas chromatographic-mass spectrometric technique. *Fresenius' Z. Anal. Chem.* 317, 42-52 (1984).

70. H. Maurer and K. Pfleger. Screening procedure for detecting anti-inflammatory analgesics and their metabolites in urine. *Fresenius' Z. Anal. Chem.* 314, 586-594 (1983).

71. H.H. Maurer. On the metabolism and the toxicological analysis of methylenedioxyphenylalkylamine designer drugs by gas chromatography-mass spectrometry. *Ther. Drug Monit.* 18, 465-470 (1996).

72. T. Kraemer, I. Vernaleken and H.H. Maurer. Studies on the metabolism and toxicological detection of the amphetamine-like anorectic mefenorex in human urine by gas chromatography-mass spectrometry and fluorescence polarization immunoassay. *J. Chromatogr. B Biomed. Sci. Appl.* 702, 93-102 (1997).

73. T. Kraemer, G.A. Theis, A.A. Weber and H.H. Maurer. Studies on the metabolism and toxicological detection of the amphetamine-like anorectic fenproporex in human urine by gas chromatography-mass spectrometry and fluorescence polarization immunoassay (FPIA). *J. Chromatogr. B Biomed. Sci. Appl.* 738, 107-118 (2000).

74. T. Kraemer, R. Wennig and H.H. Maurer. The antispasmodic mebeverine leads to positive amphetamine results with the fluorescence polarization immuno assay (FPIA) - Studies on the toxicological detection in urine by GC-MS and FPIA. *J. Anal. Toxicol.* 25, 1-6 (2001).

75. T. Kraemer, S.K. Roditis, F.T. Peters and H.H. Maurer. Amphetamine concentrations in human urine following single-dose administration of the calcium antagonist prenylamine – Studies using FPIA and GC-MS. *J. Anal. Toxicol.* 27, 68-73 (2003).

76. H.H. Maurer, T. Kraemer, O. Ledvinka, C.J. Schmitt and A.A. Weber. Gas chromatography-mass spectrometry (GC-MS) and liquid chromatography-mass spectrometry (LC-MS) in toxicological analysis. Studies on the detection of clobenzorex and its metabolites within a systematic toxicological analysis procedure by GC-MS and by immunoassay and studies on the detection of alpha- and beta-amanitin in urine by atmospheric pressure ionization electrospray LC-MS. *J. Chromatogr. B Biomed. Sci. Appl.* 689, 81-89 (1997).

77. R.F. Staack, J. Fehn and H.H. Maurer. New designer drug para-methoxymethamphetamine: Studies on its metabolism and toxicological detection in urine using gas chromatography-mass spectrometry. *J. Chromatogr. B Analyt. Technol. Biomed. Life Sci.* 789, 27-41 (2003).

78. R.F. Staack, G. Fritschi and H.H. Maurer. Studies on the metabolism and the toxicological analysis of the new piperazine-like designer drug N-benzylpiperazine in urine using gas chromatography-mass spectrometry. *J. Chromatogr. B Analyt. Technol. Biomed. Life Sci.* 773, 35-46 (2002).

79. R.F. Staack, G. Fritschi and H.H. Maurer. New designer drug 1-(3-trifluoromethylphenyl)piperazine (TFMPP): gas chromatography/mass spectrometry and liquid chromatography/mass spectrometry studies on its phase I and II metabolism and on its toxicological detection in rat urine. *J. Mass Spectrom.* 38, 971-981 (2003).

80. R.F. Staack and H.H. Maurer. Piperazine-derived designer drug 1-(3-chlorophenyl)piperazine (mCPP): GC-MS studies on its metabolism and its toxicological detection in urine including analytical differentiation from its precursor drugs trazodone and nefazodone. *J. Anal. Toxicol.* 27, 560-568 (2003).

81. R.F. Staack and H.H. Maurer. Toxicological detection of the new designer drug 1-(4-methoxyphenyl)piperazine and its metabolites in urine and differentiation from an intake of structurally related medicaments using gas chromatography-mass spectrometry. *J. Chromatogr. B Analyt. Technol. Biomed. Life Sci.* 798, 333-342 (2003).

82. R.F. Staack and H.H. Maurer. New designer drug 1-(3,4-methylenedioxybenzyl) piperazine (MDBP): studies on its metabolism and toxicological detection in rat urine using gas chromatography/mass spectrometry. *J. Mass Spectrom.* 39, 255-261 (2004).

83. R.F. Staack and H.H. Maurer. Studies on the metabolism and the toxicological analysis of the nootropic drug fipexide in rat urine using gas chromatography-mass spectrometry. *J. Chromatogr. B Analyt. Technol. Biomed. Life Sci.* 804, 337-343 (2004).

84. R.F. Staack, D.S. Theobald and H.H. Maurer. Studies on the Human Metabolism and the Toxicologic Detection of the Cough Suppressant Dropropizine in Urine Using Gas Chromatography-Mass Spectrometry. *Ther. Drug Monit.* 26, 441-449 (2004).

85. L.D. Paul and H.H. Maurer. Studies on the metabolism and toxicological detection of the *Eschscholtzia californica* alkaloids californine and protopine in urine using gas chromatography-mass spectrometry. *J. Chromatogr. B Analyt. Technol. Biomed. Life Sci.* 789, 43-57 (2003).

86. J. Beyer, D. Ehlers and H.H. Maurer. Abuse of Nutmeg (*Myristica fragrans* Houtt.): Studies on the Metabolism and the Toxicological Detection of its Ingredients Elemicin, Myristicin and Safrole in Rat and Human Urine Using Gas Chromatography/Mass Spectrometry. *Ther. Drug Monit.* 28, 568-575 (2006).

87. H. Maurer and K. Pfleger. Identification and differentiation of beta-blockers and their metabolites in urine by computerized gas chromatography-mass spectrometry. *J. Chromatogr.* 382, 147-165 (1986).

88. H.H. Maurer. Toxicological detection of the laxatives bisacodyl, picosulfate, phenolphthalein and their metabolites in urine integrated in a "general-unknown" analysis procedure using gas chromatography-mass spectrometry. *Fresenius' J. Anal. Chem.* 337, 144(1990).

89. H.H.Maurer, T.Kraemer, C.Kratzsch, L.D.Paul, F.T.Peters, D.Springer, R.F.Staack and A.A.Weber. What is the appropriate analytical strategy for effective management of intoxicated patients? In *Proceedings of the 39th International TIAFT Meeting in Prague, 2001*, M.Balikova and E.Navakova, Eds. Charles University, Prague, 2002, pp. 61-75.

90. H.H.Maurer, K.Pfleger and A.A.Weber. *Mass Spectral Library of Drugs, Poisons, Pesticides, Pollutants and their Metabolites,* 4th ed. Agilent Technologies, Palo Alto (CA), (2007).

91. U.S.Department of Commerce. *NIST/EPA/NIH Mass Spectral Library 2005,* John Wiley & Sons, New York NY, (2005).

92. F.W.McLafferty. *Registry of Mass Spectral Data, 7th Ed.,* John Wiley & Sons, New York NY, (2001).

93. H.H.Maurer. Forensic Screening with GC-MS. In *Handbook of analytical separation sciences: Forensic Sciences, 2nd Ed.*, M.Bogusz, Ed. Elsevier Science, Amsterdam, (2007), in press.

94. H.H. Maurer, T. Kraemer and J.W. Arlt. Screening for the detection of angiotensin-converting enzyme inhibitors, their metabolites, and AT II receptor antagonists. *Ther. Drug Monit.* 20, 706-713 (1998).

95. H.H. Maurer and J.W. Arlt. Detection of 4-hydroxycoumarin anticoagulants and their metabolites in urine as part of a systematic toxicological analysis procedure for acidic drugs and poisons by gas chromatography-mass spectrometry after extractive methylation. *J. Chromatogr. B Biomed. Sci. Appl.* 714, 181-195 (1998).

96. H.H. Maurer and J.W. Arlt. Screening procedure for detection of dihydropyridine calcium channel blocker metabolites in urine as part of a systematic toxicological analysis procedure for acidics by gas chromatography-mass spectrometry (GC-MS) after extractive methylation. *J. Anal. Toxicol.* 23, 73-80 (1999).

97. H.H.Maurer, F.X.Tauvel and T.Kraemer. Detection of non-steroidal anti-inflammatory drugs (NSAIDs), barbiturates and their metabolites in urine as part of a systematic toxicological analysis (STA) procedure for acidic drugs and poisons by GC-MS. In I.Rasanen, Ed. TIAFT, Helsinki, 2001, pp. 316-323.

98. J. Beyer, A. Bierl, F.T. Peters and H.H. Maurer. Screening procedure for detection of diuretics and uricosurics and/or their metabolites in human urine using gas chromatography-mass spectrometry after extractive methylation. *Ther. Drug Monit.* 27, 509-520 (2005).

99. J. Beyer, F.T. Peters and H.H. Maurer. Screening procedure for detection of stimulant laxatives and/or their metabolites in human urine using gas chromatography-mass spectrometry after enzymatic cleavage of conjugates and extractive methylation. *Ther. Drug Monit.* 27, 151-157 (2005).

100. A.M. Lisi, R. Kazlauskas and G.J. Trout. Gas chromatographic-mass spectrometric quantitation of urinary buprenorphine and norbuprenorphine after derivatization by direct extractive alkylation. *J. Chromatogr. B* 692, 67-77 (1997).

101. P. Roesner, T. Junge, G. Fritschi, B. Klein, K. Thielert and M. Kozlowski. Neue synthetische Drogen: Piperazin-, Procyclidin- und alpha-Aminopropiophenonderivate. *Toxichem. Krimtech.* 66, 81-90 (1999).

102. D. Springer, G. Fritschi and H.H. Maurer. Metabolism of the new designer drug alpha-pyrrolidinopropiophenone (PPP) and the toxicological detection of PPP and 4'methyl-alpha-pyrrolidinopropiophenone (MPPP) studied in urine using gaschromatography-mass spectrometry. *J. Chromatogr. B Analyt. Technol. Biomed. Life Sci.* 796, 253-266 (2003).

103. D. Springer, F.T. Peters, G. Fritschi and H.H. Maurer. New designer drug 4'-methyl-alpha-pyrrolidinohexanophenone: Studies on its metabolism and toxicological detection in urine using gaschromatography-mass spectrometry. *J. Chromatogr. B Analyt. Technol. Biomed. Life Sci.* 789, 79-91 (2003).

104. D. Springer, G. Fritschi and H.H. Maurer. Metabolism and toxicological detection of the new designer drug 4'-methoxy-α-pyrrolidinopropiophenone studied in rat urine using gas chromatography-mass spectrometry. *J. Chromatogr. B Analyt. Technol. Biomed. Life Sci.* 793, 331-342 (2003).

105. D. Springer, G. Fritschi and H.H. Maurer. Metabolism and toxicological detection of the new designer drug 3',4'-methylenedioxy-alpha-pyrrolidinopropiophenone studied in urine using gas chromatography-mass spectrometry. *J. Chromatogr. B Analyt. Technol. Biomed. Life Sci.* 793, 377-388 (2003).

106. F.T. Peters, M.R. Meyer, G. Fritschi and H.H. Maurer. Studies on the metabolism and toxicological detection of the new designer drug 4'-methyl-alpha-pyrrolidinobutyrophenone (MPBP) in urine using gas chromatography-mass spectrometry. *J. Chromatogr. B Analyt. Technol. Biomed. Life Sci.* 824, 81-91 (2005).

107. C. Sauer, F.T. Peters, C. Haas, M.R. Meyer, G. Fritschi and H.H. Maurer. New designer drug alpha-pyrrolidinovalerophenone: Studies on its metabolism and toxicological detection in rat urine using gas chromatographic/mass spectrometric techniques. *J. Mass Spectrom.* (2007), in preparation.

108. H.H.Maurer and F.T.Peters. Analyte Identification Using Library Searching in GC-MS and LC-MS. In *Encyclopedia of Mass Spectrometry*, M.Gross and R.M.Caprioli, Eds. Elsevier Science, Oxford, (2006), pp. 115-121.

109. F.T. Peters and H.H. Maurer. Systematic comparison of bias and precision data obtained with multiple-point and one-point calibration in six validated assays for quantification of drugs in human plasma. *Anal. Chem.* (2007), in press.

110. F.T. Peters and H.H. Maurer. Bioanalytical method validation and its implications for forensic and clinical toxicology - A review [review]. *Accred. Qual. Assur.* 7, 441-449 (2002).

111. F.T. Peters, O.H. Drummer and F. Musshoff. Validation of new methods. *Forensic Sci. Int.* 165, 216-224 (2007).

112. T. Gunnar, K. Ariniemi and P. Lillsunde. Validated toxicological determination of 30 drugs of abuse as optimized derivatives in oral fluid by long column fast gas chromatography/electron impact mass spectrometry. *J. Mass Spectrom.* 40, 739-753 (2005).

113. T. Gunnar, K. Ariniemi and P. Lillsunde. Fast gas chromatography-negative-ion chemical ionization mass spectrometry with microscale volume sample preparation for the determination of benzodiazepines and alpha-hydroxy metabolites, zaleplon and zopiclone in whole blood. *J. Mass Spectrom.* 41, 741-754 (2006).

114. A. Kankaanpaa, T. Gunnar, K. Ariniemi, P. Lillsunde, S. Mykkanen and T. Seppala. Single-step procedure for gas chromatography-mass spectrometry screening and quantitative determination of amphetamine-type stimulants and related drugs in blood, serum, oral fluid and urine samples. *J. Chromatogr. B Analyt. Technol. Biomed. Life Sci.* 810, 57-68 (2004).

115. F.W. McLafferty and F. Turecek. *Interpretation of Mass Spectra,* 4th ed. University Science Books, Mill Valley, CA, (1993).

116. R.M. Smith and K.L. Busch. *Understanding mass spectra - A basic approach,* Wiley, New York (NY), (1999).

117. R.F. Staack and H.H. Maurer. Metabolism of Designer Drugs of Abuse [review]. *Curr. Drug Metab.* 6, 259-274 (2005).

118. H.H. Maurer, T. Kraemer, D. Springer and R.F. Staack. Chemistry, Pharmacology, Toxicology, and Hepatic Metabolism of Designer Drugs of the Amphetamine (Ecstasy), Piperazine, and Pyrrolidinophenone Types, a Synopsis [review]. *Ther. Drug Monit.* 26, 127-131 (2004).

119. D. Springer, F.T. Peters, G. Fritschi and H.H. Maurer. Studies on the metabolism and toxicological detection of the new designer drug 4'-methyl-alpha-pyrrolidinopropiophenone in urine using gas chromatography-mass spectrometry. *J. Chromatogr. B Analyt. Technol. Biomed. Life Sci.* 773, 25-33 (2002).

120. H.K. Ensslin, H.H. Maurer, E. Gouzoulis, L. Hermle and K.A. Kovar. Metabolism of racemic 3,4-methylenedioxy-ethylamphetamine in humans. Isolation, identification, quantification, and synthesis of urinary metabolites. *Drug Metab. Dispos.* 24, 813-820 (1996).

121. H. Maurer and P. Wollenberg. Urinary metabolites of benzbromarone in man. *Arzneim. -Forsch.* 40, 460-462 (1990).

122. H.H. Maurer and C.F. Fritz. Metabolism of pholcodine in man. *Arzneim. -Forsch.* 40, 564-566 (1990).

123. H. Maurer. Metabolism of trimipramine in man. *Arzneim. -Forsch.* 39, 101-103 (1989).

124. J. Knabe, H.P. Buch, H.H. Maurer and P. Christensen. Acylanilide, 3. Mitteilung: Untersuchungen zur Methämoglobinbildung und zum Metabolismus bei der Ratte. *Arch. Pharm. (Weinheim)* 321, 739-741 (1988).

125. H. Maurer and I. Kleff. On the metabolism of ditazole in man. *Arzneim. -Forsch.* 38, 1843-1845 (1988).

126. F.T. Peters, C.A. Dragan, D.R. Wilde, M.R. Meyer, M. Bureik and H.H. Maurer. Biotechnological synthesis of drug metabolites using human cytochrome P450 2D6 heterologously expressed in fission yeast exemplified for the designer drug metabolite 4'-hydroxymethyl-alpha-pyrrolidinobutyrophenone. *Biochem. Pharmacol.* (2007), submitted.

127. A.J. Clatworthy, L.V. Jones and M.J. Whitehouse. The gas chromatography mass spectrometry of the major metabolites of flurazepam. *Biomed. Mass Spectrom.* 4, 248-254 (1977).

128. C. Koppel, J. Tenczer and K.M. Peixoto Menezes. Formation of formaldehyde adducts from various drugs by use of methanol in a toxicological screening procedure with gas chromatography-mass spectrometry. *J. Chromatogr.* 563, 73-81 (1991).

129. G.P. Cartoni, A. Cavalli, A. Giarusso and F. Rosati. A gas chromatographic study of the metabolism of tarugan. *J. Chromatogr.* 84, 419-422 (1973).

130. G. Bohn, G. Rucker and H. Kroger. [Investigations of the decomposition and detection of morazone by thin-layer- and gas-liquid-chromatography]. *Arch. Toxicol.* 35, 213-220 (1976).

131. H. Maurer and K. Pfleger. Determination of 1,4- and 1,5-benzodiazepines in urine using a computerized gas chromatographic-mass spectrometric technique. *J. Chromatogr.* 222, 409-419 (1981).

132. H. Schutz, S. Ebel and H. Fitz. [Screening and detection of tetrazepam and its major metabolites]. *Arzneimittelforschung.* 35, 1015-1024 (1985).

133. A.H. Wapstra and K. Bos. *Atomic Data and Nuclear Data Tables* 19, 177-214 (1977).

134. S. Steinmeyer, D. Bregel, S. Warth, T. Kraemer and M.R. Moeller. Improved and validated method for the determination of tetrahydrocannabinol (THC), 11-hydroxy-THC and 11-nor-9-carboxy-THC in serum, and in human liver microsomal preparations using gas chromatography-mass spectrometry. *J. Chromatogr. B Analyt. Technol. Biomed. Life Sci.* 722, 239-248 (2002).

135. M.R. Moeller, P. Fey and R. Wennig. Simultaneous determination of drugs of abuse (opiates, cocaine and amphetamine) in human hair by GC/MS and its application to a methadone treatment program. *Forensic Sci Int* 63, 185-206 (1993).

136. E.D. Clarkson, D. Lesser and B.D. Paul. Effective GC-MS procedure for detecting iso-LSD in urine after base-catalyzed conversion to LSD. *Clin. Chem.* 44, 287-292 (1998).

137. M.R. Moeller and C. Mueller. The detection of 6-monoacetylmorphine in urine, serum and hair by GC/MS and RIA. *Forensic Sci Int* 70, 125-133 (1995).

138. ABDATA. *List of pharmaceutical substances,* 12th Ed. ed. Werbe- und Vertriebsgesellschaft Deutscher Apotheker, Eschborn (Germany), (2000).

139. M.J.O'Neil, P.E.Heckelman, C.B.Koch, K.J.Roman, C.M.Kenny and M.A.D'Arecca. *The Merck Index: An Encyclopedia of Chemicals, Drugs, and Biologicals,* 13th Ed. ed. Merck & Co., Whitehouse Station NY, (2006).

140. M.Negwer and H.G.Scharnow. *Organic-chemical drugs and their synonyms,* 5th Ed. ed. Wiley-VCH, Weinheim, (2001).

141. P.Perkow and H.Ploss. *Wirksubstanzen der Pflanzenschutz- und Schädlingsbekämpfungsmittel,* 3rd Ed. ed. Verlag Paul Parey, Berlin, (2005).

Tables

8 Table of Compounds in Order of Names

8.1 Explanatory notes

This Table is arranged in order to facilitate the search for the data of particular compounds. Derivatives, metabolites and derivatized metabolites are listed under their parent compounds. Metabolites or derivatives common to several substances are listed under all their parent compounds. In electronic versions not using the NIST search algorithm, the name of only one parent compound can be given, but the symbol '@' indicates, that further compounds can form this particular metabolite, derivative or artifact. This information can be found at the corresponding mass spectrum on the page in Volume 2 given by the library search result.

The first column contains the compound names (INN for drugs, common names for pesticides, chemical names for chemicals, abbreviations in Table 1-6). If necessary, a synonym index, e.g. [138-141], should be used in conjunction with this list.

The second column contains the information in which biosample and after which sample preparation (Section 1-2.2, abbreviations in Table 1-6) compounds could be detected. These data have been evaluated from about 80 000 clinical and forensic cases. It should be recognized that the plasma samples were analyzed by our most sensitive GC-MS (HP 5973, details cf. Section 1-2.3.1).

The third column lists the retention indices (RI, Section 1-2.4) and the fourth column fragment ions typical for the particular compound and their relative intensities.

The fifth column indicates the page in Volume 2 on which the mass spectrum is reproduced under the molecular or pseudomolecular mass. The sixth column indicates the library entry number of the electronic versions.

8.2 Table of compounds in order of names

Table 1-8-1: Compounds in order of names   Abacavir

Name	Detected	RI	Typical ions and intensities					Page	Entry
Abacavir		2745	286 $_{31}$	271 $_{12}$	189 $_{60}$	175 $_{100}$	162 $_{16}$	544	5867
Abacavir AC		2780	328 $_{46}$	313 $_{18}$	189 $_{81}$	175 $_{100}$	162 $_{15}$	750	5868
Abacavir 2AC		3210	370 $_{100}$	231 $_{77}$	189 $_{100}$	173 $_{62}$	79 $_{90}$	920	6558
Abacavir 2HFB		2565	678 $_{23}$	385 $_{100}$	371 $_{74}$	200 $_{21}$	79 $_{22}$	1201	6148
Abacavir 2PFP		2605	578 $_{14}$	335 $_{100}$	321 $_{82}$	200 $_{32}$	79 $_{66}$	1191	6133
Abacavir 2TMS		3090	430 $_{64}$	415 $_{35}$	261 $_{46}$	247 $_{100}$	73 $_{31}$	1081	5869
Abacavir 3HFB		2460	706 $_{15}$	413 $_{45}$	385 $_{100}$	331 $_{11}$	169 $_{17}$	1205	6149
Acebutolol	G U	2955	336 $_{1}$	321 $_{4}$	221 $_{60}$	151 $_{22}$	72 $_{100}$	789	1562
Acebutolol 2TMSTFA		2780	504 $_{1}$	284 $_{100}$	218 $_{19}$	129 $_{61}$	73 $_{61}$	1167	6159
Acebutolol 3TMS		2800	552 $_{5}$	537 $_{10}$	365 $_{41}$	350 $_{69}$	72 $_{100}$	1185	5465
Acebutolol 4TMS		2870	624 $_{1}$	609 $_{2}$	437 $_{13}$	144 $_{100}$	73 $_{79}$	1197	5466
Acebutolol formyl artifact	U	3055	348 $_{18}$	333 $_{73}$	221 $_{93}$	151 $_{100}$	86 $_{70}$	839	1563
Acebutolol -H2O	G U P	2850	318 $_{19}$	303 $_{56}$	151 $_{26}$	140 $_{67}$	98 $_{100}$	704	4
Acebutolol -H2O AC		3100	360 $_{21}$	259 $_{80}$	230 $_{78}$	151 $_{100}$	98 $_{73}$	885	1345
Acebutolol -H2O HY	UHY	2010	248 $_{23}$	233 $_{24}$	140 $_{39}$	98 $_{100}$	56 $_{73}$	371	1565
Acebutolol -H2O HY2AC	U+UHYAC	3055	332 $_{8}$	289 $_{30}$	231 $_{100}$	202 $_{60}$	98 $_{30}$	769	1570
Acebutolol HY	UHY	2240	266 $_{6}$	151 $_{100}$	72 $_{33}$			452	1567
Acebutolol-M/artifact (phenol)	G U P	2450	221 $_{38}$	151 $_{100}$	136 $_{24}$			276	1
Acebutolol-M/artifact (phenol) HY	UHY	1530	151 $_{100}$	136 $_{63}$	108 $_{37}$	80 $_{34}$		128	1564
Acebutolol-M/artifact (phenol) HYAC	U+UHYAC	1850	193 $_{98}$	151 $_{100}$	136 $_{89}$	133 $_{44}$		199	1568
Acecarbromal	P G U	1720	250 $_{18}$	208 $_{6}$	165 $_{18}$	129 $_{62}$	69 $_{100}$	505	2
Acecarbromal artifact	P	1115	157 $_{2}$	129 $_{97}$	114 $_{14}$	87 $_{19}$	57 $_{100}$	137	1026
Acecarbromal artifact-1		1210	180 $_{5}$	129 $_{61}$	69 $_{99}$			173	1328
Acecarbromal artifact-2	G P U	1480	223 $_{29}$	191 $_{5}$	149 $_{11}$	102 $_{15}$	69 $_{100}$	282	1880
Acecarbromal artifact-3		1510	165 $_{11}$	113 $_{99}$	98 $_{81}$	69 $_{74}$		146	1329
Acecarbromal-M (carbromal)	P G U	1515	208 $_{41}$	191 $_{4}$	165 $_{16}$	114 $_{14}$	69 $_{100}$	327	652
Acecarbromal-M (desbromo-carbromal)		1380	143 $_{2}$	130 $_{59}$	113 $_{95}$	87 $_{99}$	71 $_{62}$	138	655
Acecarbromal-M/artifact (carbromide)	P G U	1215	165 $_{67}$	150 $_{18}$	114 $_{52}$	69 $_{100}$	55 $_{38}$	199	653
Aceclidine		1460	169 $_{18}$	141 $_{6}$	126 $_{100}$	110 $_{16}$	98 $_{37}$	156	2785
Aceclofenac ME	P(ME) G(ME)	2540	367 $_{26}$	277 $_{8}$	242 $_{40}$	214 $_{100}$	179 $_{12}$	907	6489
Aceclofenac-M (diclofenac)	G P	2205	295 $_{17}$	242 $_{54}$	214 $_{100}$	179 $_{19}$	108 $_{30}$	588	4469
Aceclofenac-M (diclofenac) ME	P(ME) G(ME)	2195	309 $_{48}$	277 $_{10}$	242 $_{48}$	214 $_{100}$	179 $_{20}$	658	717
Aceclofenac-M (diclofenac) TMS		2170	367 $_{19}$	352 $_{11}$	242 $_{38}$	214 $_{100}$	73 $_{86}$	907	5467
Aceclofenac-M (diclofenac) 2ME		2220	323 $_{53}$	264 $_{10}$	228 $_{100}$	214 $_{21}$		725	2323
Aceclofenac-M (diclofenac) -H2O	P G U+UHYAC	2135	277 $_{72}$	242 $_{62}$	214 $_{99}$	179 $_{29}$	89 $_{25}$	498	716
Aceclofenac-M (diclofenac) -H2O ET		2130	305 $_{100}$	290 $_{69}$	270 $_{97}$	242 $_{65}$	227 $_{42}$	637	6390
Aceclofenac-M (diclofenac) -H2O ME	G P	2300	291 $_{19}$	263 $_{31}$	228 $_{100}$	200 $_{33}$	109 $_{8}$	567	2324
Aceclofenac-M (diclofenac) -H2O TMS		2180	349 $_{100}$	314 $_{27}$	241 $_{11}$	190 $_{48}$	73 $_{81}$	841	4538
Aceclofenac-M (HO-diclofenac) -H2O	P U+UHYAC	2400	293 $_{39}$	258 $_{100}$	230 $_{46}$	195 $_{34}$	166 $_{40}$	578	6467
Aceclofenac-M (HO-diclofenac) -H2O ME	G P	2365	307 $_{65}$	209 $_{15}$	201 $_{47}$	272 $_{97}$	244 $_{100}$	647	6490
Aceclofenac-M (HO-diclofenac) -H2O iso-1 AC	U+UHYAC	2520	335 $_{88}$	293 $_{100}$	258 $_{93}$	230 $_{93}$	166 $_{23}$	782	2321
Aceclofenac-M (HO-diclofenac) -H2O iso-2 AC	U+UHYAC	2540	335 $_{22}$	293 $_{100}$	258 $_{29}$	230 $_{48}$	195 $_{17}$	782	1212
Aceclofenac-M/artifact	U+UHYAC	2980	355 $_{88}$	320 $_{100}$	292 $_{10}$	228 $_{16}$	75 $_{6}$	863	2322
Acemetacin ET		3220	443 $_{39}$	312 $_{12}$	158 $_{5}$	139 $_{100}$	111 $_{16}$	1105	3167
Acemetacin ME	PME UME	3150	429 $_{44}$	312 $_{14}$	158 $_{6}$	139 $_{100}$	111 $_{13}$	1077	1374
Acemetacin artifact-1 ME	PME UME	2130	233 $_{38}$	174 $_{100}$				315	1230
Acemetacin artifact-1 2ME	PME UME	2090	247 $_{38}$	188 $_{100}$	173 $_{11}$	145 $_{8}$		367	6294
Acemetacin artifact-2 ME		2390	291 $_{17}$	233 $_{16}$	174 $_{100}$	159 $_{13}$	131 $_{14}$	568	1384
Acemetacin-M (chlorobenzoic acid)	G UHY U+UHYAC	1400*	156 $_{61}$	139 $_{100}$	111 $_{54}$	85 $_{4}$	75 $_{39}$	136	2726
Acemetacin-M/artifact (HO-indometacin) 2ME	UME	2880	401 $_{27}$	262 $_{4}$	139 $_{100}$	111 $_{23}$		1020	6293
Acemetacin-M/artifact (indometacin)	G P-I	2550	313 $_{30}$	139 $_{101}$	111 $_{24}$			871	1038
Acemetacin-M/artifact (indometacin) ET		2820	385 $_{40}$	312 $_{19}$	158 $_{6}$	139 $_{100}$	111 $_{20}$	969	3168
Acemetacin-M/artifact (indometacin) ME	P(ME) G(ME) U(ME)	2770	371 $_{9}$	312 $_{6}$	139 $_{100}$	111 $_{17}$		922	1039
Acemetacin-M/artifact (indometacin) TMS		2650	429 $_{23}$	370 $_{4}$	312 $_{17}$	139 $_{100}$	73 $_{22}$	1078	5462
Acenaphthene		1440*	154 $_{100}$	153 $_{93}$	126 $_{3}$	87 $_{5}$	76 $_{31}$	133	3700
Acenaphthylene		1380*	152 $_{100}$	126 $_{3}$	98 $_{1}$	76 $_{9}$	63 $_{5}$	131	2558
Acenocoumarol AC	U+UHYAC	3105	395 $_{4}$	353 $_{26}$	335 $_{12}$	310 $_{100}$	121 $_{13}$	1001	4788
Acenocoumarol ET	UET	3040	381 $_{18}$	338 $_{100}$	310 $_{93}$	189 $_{9}$	121 $_{37}$	955	4781
Acenocoumarol ME	UME UGLUCME	3035	367 $_{15}$	324 $_{100}$	278 $_{7}$	189 $_{8}$	121 $_{10}$	908	1372
Acenocoumarol TMS		3110	425 $_{23}$	382 $_{82}$	261 $_{16}$	219 $_{24}$	73 $_{100}$	1069	4885
Acenocoumarol-M (acetamido-) ME	UME UGLUCME	3520	379 $_{33}$	336 $_{100}$	322 $_{13}$	280 $_{11}$	201 $_{20}$	947	4433
Acenocoumarol-M (acetamido-) 2ET	UET	3200	421 $_{29}$	378 $_{100}$	350 $_{84}$	292 $_{8}$	121 $_{19}$	1063	4787
Acenocoumarol-M (acetamido-) 2ME	UME UGLUCME	3265	393 $_{21}$	350 $_{100}$	336 $_{9}$	278 $_{17}$	56 $_{30}$	994	4434
Acenocoumarol-M (amino-) 2ET	UET	3040	379 $_{74}$	322 $_{100}$	308 $_{30}$	148 $_{29}$	121 $_{23}$	949	4784
Acenocoumarol-M (amino-) 2ME	UME UHYME	2980	351 $_{34}$	294 $_{100}$	278 $_{24}$	120 $_{23}$		848	4430
Acenocoumarol-M (amino-) 3ET	UET	3070	407 $_{57}$	392 $_{26}$	350 $_{100}$	306 $_{25}$	121 $_{10}$	1034	4785
Acenocoumarol-M (amino-) 3ME	UME UHYME	2985	365 $_{19}$	308 $_{100}$	292 $_{26}$	249 $_{5}$	121 $_{7}$	900	4431
Acenocoumarol-M (amino-dihydro-) 3ET	UET	3065	409 $_{64}$	394 $_{40}$	362 $_{28}$	350 $_{100}$	176 $_{6}$	1039	4786
Acenocoumarol-M (amino-dihydro-) 3ME	UME	3060	367 $_{51}$	334 $_{25}$	308 $_{100}$	292 $_{35}$		910	4432
Acenocoumarol-M (HO-) iso-1 2ET	UET	3435	425 $_{16}$	382 $_{100}$	354 $_{65}$	233 $_{6}$	165 $_{7}$	1069	4782
Acenocoumarol-M (HO-) iso-1 2ME	UME UGLUCME	3350	397 $_{11}$	354 $_{100}$	308 $_{4}$	219 $_{2}$	151 $_{8}$	1009	4428
Acenocoumarol-M (HO-) iso-2 2ET	UET	3630	425 $_{15}$	382 $_{100}$	354 $_{61}$	233 $_{6}$	165 $_{9}$	1069	4783
Acenocoumarol-M (HO-) iso-2 2ME	UME UGLUCME	3500	397 $_{14}$	354 $_{100}$	308 $_{3}$	219 $_{3}$	151 $_{5}$	1009	4429
Acephate		1470	183 $_{5}$	142 $_{13}$	136 $_{100}$	94 $_{57}$	79 $_{17}$	181	3504

Acephate -C2H2O TFA Table 1-8-1: Compounds in order of names

Name	Detected	RI	Typical ions and intensities	Page	Entry
Acephate -C2H2O TFA		1110	237_{15} 168_{22} 125_{55} 96_{100} 69_{58}	331	4031
Acepromazine	G U UHY U+UHYAC	2755	326_4 241_3 198_2 86_8 58_{100}	740	3
Acepromazine-M (dihydro-) AC	U+UHYAC	2765	370_3 310_3 225_7 86_{28} 58_{100}	920	1307
Acepromazine-M (dihydro-) -H2O	U+UHYAC	2720	310_6 225_6 86_{10} 58_{100}	667	1306
Acepromazine-M (HO-) AC	U+UHYAC	3040	384_7 256_2 86_{23} 58_{100}	967	1309
Acepromazine-M (HO-dihydro-) 2AC	U+UHYAC	3000	428_3 343_{16} 154_{14} 86_{25} 58_{100}	1076	1308
Acepromazine-M (nor-) AC	U+UHYAC	3145	354_{20} 241_{17} 114_{99} 100_{34}	861	1235
Acepromazine-M (nor-dihydro-) -H2O AC	U+UHYAC	3150	338_5 114_{100} 100_{13}	797	1310
Acepromazine-M (ring)	UHY U+UHYAC	2525	241_{100} 226_{18} 198_{45} 166_4 154_5	346	6804
Aceprometazine		2625	326_5 255_{10} 222_7 197_7 72_{100}	740	5
Aceprometazine-M (dihydro-) AC	U+UHYAC	2690	370_3 299_6 224_8 72_{100}	920	1236
Aceprometazine-M (HO-) AC	U+UHYAC	3025	384_3 313_3 256_3 72_{100}	967	1238
Aceprometazine-M (methoxy-dihydro-) AC	U+UHYAC	3165	400_1 329_1 270_1 225_1 72_{100}	1018	1239
Aceprometazine-M (methoxy-dihydro-) -H2O	U+UHYAC	2920	340_3 271_1 238_6 72_{100}	808	1237
Aceprometazine-M (nor-) AC	U+UHYAC	2940	354_6 254_{23} 114_{24} 72_{23} 58_{100}	861	1311
Aceprometazine-M (nor-HO-) 2AC	U+UHYAC	3205	412_3 254_{16} 114_{45} 100_{32} 58_{100}	1046	1312
Aceprometazine-M (ring)	UHY U+UHYAC	2525	241_{100} 226_{18} 198_{45} 166_4 154_5	346	6804
Acetaldehyde		<1000*	44_{83} 29_{100}	90	4193
Acetaminophen	G P U	1780	151_{34} 109_{100} 81_{16} 80_{23}	129	825
Acetaminophen AC	U+UHYAC PAC	1765	193_{10} 151_{53} 109_{100} 80_{24}	199	188
Acetaminophen HFB	UHYHFB PHFB	1735	347_{24} 305_{39} 169_{13} 108_{100} 69_{31}	833	5099
Acetaminophen ME	PME UME	1630	165_{59} 123_{74} 108_{100} 95_{10} 80_{20}	147	5046
Acetaminophen PFP		1675	297_{19} 255_{31} 119_{38} 108_{100} 80_{28}	601	5095
Acetaminophen TFA		1630	247_{11} 205_{30} 108_{100} 80_{19} 69_{34}	365	5092
Acetaminophen 2AC	U+UHYAC	2085	235_{10} 193_{11} 151_{30} 109_{100}	323	827
Acetaminophen 2TMS		1780	295_{50} 280_{68} 206_{83} 116_{15} 73_{100}	592	4578
Acetaminophen Cl-artifact AC	U+UHYAC	2030	227_6 185_{74} 143_{100} 114_4 79_{12}	296	2993
Acetaminophen HY	UHY	1240	109_{99} 80_{41} 52_{90}	102	826
Acetaminophen HYME	UHYME	1100	123_{27} 109_{100} 94_7 80_{96} 53_{47}	108	3766
Acetaminophen-M (HO-) 3AC	U+UHYAC	2150	251_6 209_{23} 167_{87} 125_{100}	383	2384
Acetaminophen-M (HO-methoxy-) AC	U+UHYAC	2170	239_{12} 197_{86} 155_{100} 140_{42} 110_9	339	2383
Acetaminophen-M (methoxy-) AC	U+UHYAC	1940	223_{12} 181_{79} 139_{100}	283	201
Acetaminophen-M (methoxy-) Cl-artifact AC	U+UHYAC	2060	257_6 215_{77} 173_{100} 158_{21} 130_5	409	2994
Acetaminophen-M 2AC	U+UHYAC	2270	262_{20} 220_{35} 188_{17} 160_{74} 146_{100}	428	2387
Acetaminophen-M 3AC	U+UHYAC	2340	304_{15} 261_{31} 219_{46} 160_{100} 146_{72}	632	2388
Acetaminophen-M conjugate 2AC	U+UHYAC	3050	396_{20} 354_7 246_{100} 204_{73} 162_{71}	1004	2389
Acetaminophen-M conjugate 3AC	U+UHYAC	3030	438_{35} 353_{40} 246_{72} 204_{97} 162_{100}	1094	1387
Acetaminophen-M iso-1 3AC	U+UHYAC	2200	305_{26} 263_{57} 221_{14} 160_{69} 146_{100}	637	2385
Acetaminophen-M iso-2 3AC	U+UHYAC	2220	305_{34} 263_{100} 221_{82} 162_{54} 146_{99}	637	2386
Acetanilide	G	1380	135_{35} 93_{100}	115	222
Acetazolamide ME	UEXME	1995	236_{11} 129_{21} 108_9 88_{16} 70_{100}	326	6843
Acetazolamide 3ME	UEXME	2040	264_{22} 249_{100} 108_{21} 92_2 83_{18}	438	6844
Acetic acid		<1000*	60_{47} 45_{78} 43_{100}	91	1548
Acetic acid ET		<1000*	88_5 70_{14} 61_{16} 43_{100} 29_{33}	98	60
Acetic acid ME		<1000*	74_{23} 59_{10} 43_{100} 29_{23}	94	3777
Acetic acid anhydride		<1000*	102_1 60_3 43_{100}	101	2756
Acetochlor		1845	269_8 223_{39} 174_{39} 146_{64} 59_{100}	465	3507
Acetone		<1000*	58_{44} 43_{100}	91	1547
Acetonitrile		<1000	41_{100}	89	2752
N-Acetyl-2-amino-octanoic acid ME	UME	1560	215_1 172_5 156_{41} 114_{100} 88_{21}	258	4941
Acetylmethadol	G P U+UHYAC	2230	353_1 338_1 225_4 91_6 72_{100}	859	5616
N-Acetyl-proline ME		1465	171_8 128_1 112_{32} 70_{100} 68_7	158	2708
Acetylsalicylic acid	G P-I U+UHYAC	1545*	180_1 138_{79} 120_{99} 92_{37}	173	1443
Acetylsalicylic acid ME	P(ME)	1400*	194_{60} 179_{40} 138_{15} 91_{10}	203	2637
Acetylsalicylic acid-M	U	1825	195_{32} 177_{17} 121_{98} 120_{100} 92_{43}	206	956
Acetylsalicylic acid-M (deacetyl-)	G P UHY	1295*	138_{40} 120_{90} 92_{100} 64_{52}	118	953
Acetylsalicylic acid-M (deacetyl-) ET		1350*	166_{34} 120_{99} 92_{40} 65_{19}	149	955
Acetylsalicylic acid-M (deacetyl-) ME	P U+UHYAC	1200*	152_{39} 120_{94} 92_{100} 65_{53}	130	954
Acetylsalicylic acid-M (deacetyl-) 2ME	PME UME	1210*	166_{28} 135_{100} 133_{47} 92_{30} 77_{52}	150	6391
Acetylsalicylic acid-M (deacetyl-) 2TMS		1195*	267_{37} 193_4 135_{17} 91_{11} 73_{60}	526	4523
Acetylsalicylic acid-M (deacetyl-) artifact (trimer)	G U+UHYAC	3190*	360_{39} 240_{58} 152_{36} 120_{100} 92_{75}	883	4496
Acetylsalicylic acid-M (deacetyl-3-HO-) 3ME	UME	1385*	196_{73} 165_{81} 163_{100} 122_{26} 107_{20}	210	6393
Acetylsalicylic acid-M (deacetyl-5-HO-) 3ME	UME	1530*	196_{100} 181_{34} 165_{72} 163_{66} 107_{29}	210	6394
Acetylsalicylic acid-M (deacetyl-HO-) 2ME	PME UME	1210*	182_{36} 150_{100} 122_{12} 107_{30} 79_{18}	179	6392
Acetylsalicylic acid-M ME	U	1810	209_{20} 149_{12} 121_{100} 92_{17} 65_{22}	240	957
Acetylsalicylic acid-M MEAC	U+UHYAC	1885	251_3 209_{69} 177_{23} 149_{29} 121_{100}	383	2976
Acetylsalicylic acid-M 2ME	UME	1845	223_{12} 135_{100} 90_{58} 77_{44}	283	958
Acetyltriethylcitrate		1880*	318_1 273_8 213_{20} 203_{55} 157_{100}	703	4478
Adeptolon	U UHY U+UHYAC	2375	347_1 263_3 169_{11} 86_{11} 72_{100}	834	7
Adeptolon-M (HO-)	UHY	2760	363_2 325_7 169_{29} 90_{20} 72_{100}	894	2164
Adeptolon-M (HO-) AC	U+UHYAC	2780	405_1 333_1 169_{13} 135_3 72_{100}	1030	2160
Adeptolon-M (N-dealkyl-)	UHY U+UHYAC	1920	262_{97} 184_{100} 169_{32} 90_{67} 78_{70}	429	2156
Adeptolon-M (N-dealkyl-) AC	U+UHYAC	2200	304_{12} 261_{100} 245_{24} 90_{23} 78_{27}	632	2157
Adeptolon-M (N-dealkyl-HO-)	UHY	2510	278_{99} 184_{83} 169_{100} 90_{94}	504	2163

Table 1-8-1: Compounds in order of names — Alprenolol-M (HO-) 3AC

Name	Detected	RI	Typical ions and intensities					Page	Entry
Adeptolon-M (N-dealkyl-HO-) AC	U+UHYAC	2500	320_{26}	278_{100}	184_{53}	169_{79}	90_{55}	714	2158
Adeptolon-M (N-deethyl-) AC	U+UHYAC	2470	283_{57}	198_{56}	169_{100}	100_{53}	90_{60}	886	2165
Adeptolon-M (N-deethyl-HO-) 2AC	U+UHYAC	3010	419_{10}	333_{40}	177_{64}	169_{100}	100_{15}	1060	2162
Adeptolon-M (nor-) AC	U+UHYAC	2530	297_{18}	253_{9}	198_{20}	169_{39}	58_{100}	935	2159
Adeptolon-M (nor-HO-) 2AC	U+UHYAC	3030	433_{7}	333_{35}	177_{61}	169_{100}	58_{36}	1086	2161
Adinazolam		2955	351_{1}	308_{100}	280_{5}	205_{5}	58_{8}	847	3068
Adiphenine		2215	311_{1}	239_{2}	167_{15}	99_{20}	86_{100}	673	6
Adiphenine-M/artifact (HOOC-) (ME)		1715*	226_{24}	167_{100}	152_{13}			294	120
Air		<1000	44_{2}	40_{2}	32_{26}	28_{100}		89	3773
Air with Helium and Water		<1000	44_{1}	40_{2}	32_{26}	28_{100}	18_{36}	89	4251
Ajmaline		2880	326_{65}	297_{8}	220_{8}	182_{9}	144_{100}	741	2719
Ajmaline 2AC	U+UHYAC	2890	410_{100}	368_{16}	353_{30}	307_{15}	182_{24}	1042	2720
Ajmaline 2TMS		2565	470_{120}	455_{100}	246_{38}	182_{52}	73_{92}	1139	6273
Ajmaline-M (dihydro-) 3AC	U+UHYAC	3065	454_{100}	412_{87}	397_{37}	184_{31}	146_{15}	1120	2858
Ajmaline-M (HO-) iso-1 3AC	U+UHYAC	3100	468_{100}	426_{93}	384_{14}	198_{16}	160_{15}	1136	2859
Ajmaline-M (HO-) iso-2 3AC	U+UHYAC	3130	468_{100}	426_{64}	197_{29}	160_{65}		1137	6786
Ajmaline-M (HO-methoxy-) 3AC	U+UHYAC	3160	498_{100}	456_{56}	441_{67}	399_{12}	228_{20}	1162	6785
Ajmaline-M (nor-) 3AC	U+UHYAC	2980	438_{40}	396_{100}	354_{63}	222_{53}	196_{44}	1095	2857
Alachlor		1850	269_{10}	237_{23}	188_{94}	160_{100}	77_{32}	465	3505
Albendazole ME		2485	279_{100}	236_{41}	204_{21}	178_{44}	150_{16}	510	6071
Albendazole artifact (decarbamoyl-)		2510	207_{100}	178_{18}	165_{100}	134_{23}	122_{21}	233	6073
Albendazole artifact (decarbamoyl-) AC		2410	249_{71}	207_{86}	165_{100}	164_{99}	134_{26}	374	6072
Aldicarb		1320	144_{55}	100_{47}	86_{130}	76_{42}	58_{96}	193	3316
Aldrin		1945*	362_{1}	329_{2}	293_{11}	263_{39}	66_{100}	889	1330
Alfentanil	P-I	2990	416_{1}	359_{5}	289_{100}	268_{36}	140_{20}	1055	1773
Alimemazine	P G U+UHYAC	2315	298_{7}	198_{12}	100_{7}	84_{5}	58_{100}	607	8
Alimemazine-M (bis-nor-) AC	U+UHYAC	2765	312_{60}	212_{100}	114_{57}			675	1240
Alimemazine-M (HO-)	UHY	2650	314_{19}	228_{4}	214_{5}	100_{10}	58_{100}	685	11
Alimemazine-M (HO-) AC	U+UHYAC	2600	356_{15}	228_{3}	214_{7}	100_{15}	58_{100}	869	13
Alimemazine-M (nor-)	UHY	2335	284_{91}	252_{21}	212_{55}	199_{100}	180_{39}	535	2243
Alimemazine-M (nor-) AC	U+UHYAC	2710	326_{73}	212_{63}	198_{38}	180_{31}	128_{100}	740	14
Alimemazine-M (nor-HO-) 2AC	U+UHYAC	2930	384_{74}	270_{17}	228_{29}	214_{22}	128_{100}	967	15
Alimemazine-M (ring)	P G U UHY U+UHYAC	2010	199_{100}	167_{45}				216	10
Alimemazine-M AC	U+UHYAC	2550	257_{23}	215_{100}	183_{7}			409	12
Alimemazine-M 2AC	U+UHYAC	2865	315_{54}	273_{34}	231_{100}	202_{11}		688	2618
Alimemazine-M/artifact (sulfoxide)	G P U	2665	314_{2}	298_{5}	212_{32}	199_{10}	58_{100}	686	9
Alizapride		2855	190_{7}	162_{5}	147_{11}	132_{9}	110_{100}	690	7816
Alizapride AC		2855	176_{9}	148_{4}	133_{15}	110_{100}	70_{9}	873	7817
Alizapride ME		2700	329_{4}	190_{44}	147_{14}	110_{200}	70_{28}	755	7818
Alizapride TMS		2785	387_{1}	372_{1}	248_{5}	162_{4}	110_{100}	977	7819
Allethrin		2105*	302_{2}	168_{6}	136_{23}	123_{100}	79_{34}	625	2786
Allidochlor		1140	173_{4}	138_{31}	132_{28}	70_{37}	56_{100}	161	4041
Allobarbital	G P U UHY U+UHYAC	1595	208_{2}	193_{23}	167_{100}	124_{94}	80_{68}	237	16
Allobarbital 2ME	UME	1505	236_{7}	195_{100}	138_{100}	80_{49}		329	643
Allopurinol	U+UHYAC	2700	136_{100}	120_{3}	109_{6}	80_{4}	67_{7}	116	5241
Allylestrenol		2370*	300_{33}	259_{53}	241_{99}	201_{37}	91_{80}	617	1376
Aloe-emodin		2660*	270_{100}	241_{92}	213_{12}	139_{17}	121_{19}	467	3552
Aloe-emodin AC		2735*	312_{25}	270_{100}	241_{38}	139_{7}	121_{4}	674	3559
Aloe-emodin ME		2900*	284_{100}	266_{32}	238_{28}	209_{23}	139_{19}	531	3561
Aloe-emodin TMS		2695*	342_{16}	311_{100}	225_{5}	139_{6}	75_{18}	816	3575
Aloe-emodin 2AC		3000*	354_{45}	312_{27}	270_{100}	241_{13}	139_{4}	860	3560
Aloe-emodin -2H		2530*	268_{100}	239_{31}	183_{8}	155_{13}	127_{9}	458	3553
Aloe-emodin 2ME		2705*	298_{100}	267_{60}	239_{29}	209_{6}	155_{9}	605	3562
Aloe-emodin 2TMS		2785*	399_{100}	310_{6}	184_{9}	95_{21}	73_{21}	1050	3577
Aloe-emodin 3TMS		2900*	471_{100}	399_{9}	367_{3}	220_{2}	73_{58}	1154	3576
Alphamethrin		2790	415_{1}	209_{23}	181_{84}	163_{100}	91_{43}	1052	3509
Alprazolam	G P-I U+UHYAC-I	3100	308_{61}	279_{100}	273_{45}	239_{22}	204_{96}	654	1730
Alprazolam-M (HO-)		3245	324_{19}	322_{58}	293_{12}	287_{100}		730	1704
Alprazolam-M (HO-) AC	U+UHYAC-I	3180	366_{33}	323_{100}	295_{12}	271_{18}	77_{13}	903	1765
Alprazolam-M (HO-) artifact HYAC	U+UHYAC-I	2580	399_{2}	356_{83}	312_{50}	284_{100}	77_{58}	1013	2046
Alprazolam-M (HO-) -CH2O		3070	294_{82}	259_{100}	239_{43}	205_{61}	101_{54}	584	2392
Alprazolam-M/artifact HY	U+UHYAC-I	2500	341_{3}	298_{100}	105_{4}	77_{9}		810	2045
Alprenolol	G U	1825	249_{6}	234_{2}	205_{3}	100_{7}	72_{100}	378	17
Alprenolol AC		2185	291_{1}	273_{1}	158_{34}	116_{13}	72_{90}	571	1348
Alprenolol TFATMS		2080	402_{1}	284_{60}	228_{3}	126_{26}	73_{50}	1057	6153
Alprenolol TMS		1940	321_{4}	306_{6}	205_{3}	101_{18}	72_{100}	720	5449
Alprenolol 2AC	U+UHYAC	2275	333_{1}	273_{3}	200_{100}	98_{21}	72_{35}	775	1575
Alprenolol 2TMS		2205	393_{1}	378_{1}	144_{100}	101_{7}	73_{60}	997	5450
Alprenolol-M (deamino-di-HO-) +H2O 4AC	U+UHYAC	2450*	410_{5}	350_{9}	159_{100}	99_{8}		1041	1576
Alprenolol-M (deamino-di-HO-) 3AC	U+UHYAC	2220*	350_{4}	308_{1}	159_{100}	99_{16}		845	1574
Alprenolol-M (deamino-HO-) +H2O 3AC	U+UHYAC	2100*	352_{1}	292_{15}	159_{100}	99_{12}	91_{7}	854	1573
Alprenolol-M (deamino-HO-) 2AC	U+UHYAC	1850*	292_{6}	159_{100}	131_{13}	99_{14}		575	1572
Alprenolol-M (HO-) 2AC	U+UHYAC	2510	349_{2}	331_{1}	200_{8}	158_{100}	98_{13}	843	1577
Alprenolol-M (HO-) 3AC	U+UHYAC	2575	391_{1}	332_{7}	200_{100}	98_{19}	72_{17}	989	1578

Alprenolol-M/artifact (phenol) AC   **Table 1-8-1:** Compounds in order of names

Name	Detected	RI	Typical ions and intensities					Page	Entry
Alprenolol-M/artifact (phenol) AC	U+UHYAC	1520*	176 $_{26}$	134 $_{100}$	119 $_{26}$	107 $_{34}$	77 $_{17}$	163	1571
Amantadine	G P U UHY	1240	151 $_{13}$	134 $_4$	94 $_{100}$	57 $_{24}$		130	18
Amantadine AC	PAC U+UHYAC	1640	193 $_{34}$	136 $_{98}$	94 $_{55}$			202	22
Amantadine TMS		1525	223 $_{20}$	208 $_{19}$	166 $_{100}$	150 $_{19}$	73 $_{43}$	286	4524
Ambroxol	P G U UHY	2665	376 $_8$	279 $_{77}$	264 $_{100}$	262 $_{53}$	114 $_{69}$	938	19
Ambroxol AC		2850	418 $_7$	279 $_{100}$	262 $_{41}$	156 $_{48}$	97 $_{30}$	1058	2226
Ambroxol TMS		2665	448 $_4$	319 $_{23}$	279 $_{100}$	264 $_{92}$	186 $_{64}$	1110	4527
Ambroxol 2AC	U+UHYAC	3015	460 $_{22}$	419 $_{100}$	417 $_{51}$	279 $_{74}$	264 $_{56}$	1127	20
Ambroxol 2TMS		2800	520 $_1$	391 $_5$	351 $_{100}$	186 $_{50}$	73 $_{66}$	1175	4528
Ambroxol 3AC	U+UHYAC	3100	502 $_4$	461 $_{100}$	459 $_{47}$	401 $_{22}$	279 $_{41}$	1165	2228
Ambroxol formyl artifact	P G U UHY	2780	387 $_{29}$	331 $_{100}$	329 $_{53}$	289 $_{42}$	195 $_{82}$	978	6315
Ambroxol -H2O	P G U UHY	2395	358 $_{13}$	289 $_{42}$	264 $_{94}$	262 $_{55}$	68 $_{100}$	875	6314
Ambroxol -H2O 2AC	U+UHYAC	3030	444 $_{100}$	442 $_{50}$	303 $_{70}$	301 $_{34}$	81 $_{63}$	1102	2227
Ambroxol-M (HO-) 4AC	U+UHYAC	3375	560 $_3$	518 $_8$	303 $_{91}$	279 $_{73}$	264 $_{100}$	1187	4446
Ambroxol-M (HOOC-) ME	P U	1770	309 $_{68}$	307 $_{36}$	277 $_{100}$	275 $_{53}$	249 $_{27}$	646	5131
Ambroxol-M/artifact AC	U+UHYAC	1890	319 $_{97}$	317 $_{48}$	304 $_{11}$	277 $_{55}$		695	21
Ambucetamide		2330	292 $_1$	248 $_{100}$	192 $_6$	164 $_6$	136 $_{11}$	577	2287
Ametryne		1890	227 $_{88}$	212 $_{61}$	170 $_{41}$	68 $_{91}$	58 $_{100}$	298	3308
Amfebutamone		1695	239 $_1$	224 $_{11}$	139 $_{16}$	111 $_{25}$	100 $_{100}$	340	4699
Amfebutamone AC		2210	264 $_{40}$	225 $_{19}$	208 $_{52}$	183 $_{100}$	57 $_{43}$	521	5700
Amfebutamone formyl artifact		1755	237 $_5$	222 $_4$	139 $_{19}$	98 $_{38}$	57 $_{100}$	384	4700
Amfebutamone-M (3-chlorobenzoic acid)		1430*	156 $_{75}$	139 $_{100}$	111 $_{56}$	75 $_{44}$	65 $_6$	136	6024
Amfebutamone-M (3-chlorobenzyl alcohol)		1560*	77 $_{100}$	142 $_{53}$	113 $_{28}$	107 $_{33}$		121	6025
Amfebutamone-M (HO-)		2040	240 $_3$	224 $_{30}$	166 $_7$	139 $_{29}$	116 $_{100}$	401	7660
Amfebutamone-M (HO-) AC		2130	224 $_{12}$	166 $_4$	158 $_{100}$	115 $_{44}$	98 $_8$	602	7661
Amfebutamone-M (HO-) TMS		2075	240 $_{17}$	224 $_{79}$	188 $_{97}$	145 $_{53}$	73 $_{100}$	744	7662
Amfepramone	G U+UHYAC	1505	205 $_1$	160 $_1$	100 $_{100}$	77 $_9$	72 $_{12}$	229	25
Amfepramone-M (deethyl-)	SPE	1355	105 $_{11}$	77 $_{37}$	72 $_{100}$			165	6685
Amfepramone-M (deethyl-) AC	SPEAC	1705	219 $_2$	114 $_{50}$	105 $_{19}$	77 $_{62}$	72 $_{100}$	270	6691
Amfepramone-M (deethyl-) HFB	SPEHFB	1565	373 $_1$	268 $_{100}$	240 $_{34}$	105 $_{64}$	77 $_{57}$	930	6689
Amfepramone-M (deethyl-dihydro-) TMS	SPETMS	1435	236 $_2$	179 $_2$	163 $_8$	149 $_8$	72 $_{100}$	387	6684
Amfepramone-M (deethyl-dihydro-) 2AC	SPEAC	1845	263 $_1$	114 $_{62}$	105 $_{16}$	77 $_{20}$	72 $_{100}$	435	6690
Amfepramone-M (deethyl-dihydro-) 2HFB	SPEHFB	1540	358 $_5$	268 $_{100}$	240 $_{20}$	169 $_{15}$	105 $_9$	1189	6688
Amfepramone-M (deethyl-hydroxy-) HFB	SPEHFB	1910	476 $_5$	268 $_{55}$	240 $_{24}$	169 $_7$	121 $_{100}$	982	6679
Amfepramone-M (deethyl-hydroxy-) 2AC	SPEAC	2095	277 $_1$	192 $_2$	121 $_{21}$	114 $_{74}$	72 $_{100}$	501	6681
Amfepramone-M (deethyl-hydroxy-) 2HFB	SPEHFB	1725	516 $_2$	317 $_{17}$	268 $_{100}$	240 $_{31}$	169 $_{27}$	1192	6680
Amfepramone-M (deethyl-hydroxy-methoxy-) HFB	SPEHFB	1890	419 $_2$	268 $_{71}$	240 $_{37}$	151 $_{48}$	121 $_{100}$	1060	6677
Amfepramone-M (deethyl-hydroxy-methoxy-) 2AC	SPEAC	2190	307 $_1$	151 $_9$	123 $_5$	114 $_{65}$	72 $_{100}$	648	6682
Amfepramone-M (deethyl-hydroxy-methoxy-) 2HFB	SPEHFB	1830	476 $_4$	347 $_{20}$	268 $_{100}$	240 $_{17}$	169 $_{11}$	1196	6678
Amfepramone-M (dihydro-)	SPE	1565	206 $_1$	105 $_4$	100 $_{100}$	77 $_{14}$	72 $_{13}$	235	6683
Amfepramone-M (dihydro-) AC	SPEAC	1605	248 $_1$	117 $_6$	105 $_9$	100 $_{100}$	77 $_9$	378	6692
Amfepramone-M (dihydro-) HFB	SPEHFB	1525	403 $_1$	303 $_1$	190 $_{10}$	169 $_{24}$	100 $_{100}$	1026	6687
Amfepramone-M (dihydro-) TMS	SPETMS	1550	264 $_1$	179 $_1$	163 $_4$	149 $_5$	100 $_{100}$	515	6686
Amfetamine		1160	134 $_1$	120 $_1$	91 $_6$	77 $_1$	65 $_4$	115	54
Amfetamine	U	1160	134 $_1$	120 $_1$	91 $_6$	65 $_4$	44 $_{100}$	115	5514
Amfetamine AC	U+UHYAC	1505	177 $_1$	118 $_{19}$	91 $_{11}$	86 $_{31}$	65 $_5$	165	55
Amfetamine AC	UAAC U+UHYAC	1505	177 $_1$	118 $_{19}$	91 $_{11}$	86 $_{31}$	44 $_{100}$	165	5515
Amfetamine HFB		1355	331 $_1$	240 $_{79}$	169 $_{21}$	118 $_{100}$	91 $_{53}$	762	5047
Amfetamine PFP		1330	281 $_1$	190 $_{73}$	118 $_{100}$	91 $_{36}$	65 $_{12}$	521	4379
Amfetamine TFA		1095	231 $_1$	140 $_{100}$	118 $_{92}$	91 $_{45}$	69 $_{19}$	309	4000
Amfetamine TMS		1190	192 $_6$	116 $_{100}$	100 $_{10}$	91 $_{11}$	73 $_{87}$	235	5581
Amfetamine formyl artifact		1100	147 $_2$	146 $_6$	125 $_5$	91 $_{12}$	56 $_{100}$	125	3261
Amfetamine intermediate		1560	163 $_{16}$	146 $_{12}$	115 $_{100}$	105 $_{59}$	91 $_{81}$	142	2839
Amfetamine precursor (phenylacetone)		<1000*	134 $_{13}$	91 $_{54}$	65 $_{22}$			114	3240
Amfetamine precursor (phenylacetone)		<1000*	134 $_{13}$	91 $_{54}$	65 $_{22}$	43 $_{100}$		114	5516
Amfetamine R-(-)-enantiomer HFBP		1160	337 $_{15}$	294 $_{16}$	266 $_{100}$	118 $_{11}$	91 $_{12}$	1075	6514
Amfetamine S-(+)-enantiomer HFBP		1190	337 $_{13}$	294 $_{21}$	266 $_{100}$	118 $_{16}$	91 $_{19}$	1076	6515
Amfetamine-D5 AC	UAAC U+UHYAC	1480	182 $_3$	122 $_{46}$	92 $_{30}$	90 $_{100}$	66 $_{16}$	181	5690
Amfetamine-D5 HFB		1330	244 $_{100}$	169 $_{14}$	122 $_{46}$	92 $_{41}$	69 $_{40}$	787	6316
Amfetamine-D5 PFP		1320	194 $_{100}$	123 $_{42}$	119 $_{32}$	92 $_{46}$	69 $_{11}$	543	5566
Amfetamine-D5 TFA		1085	144 $_{100}$	123 $_{53}$	122 $_{56}$	92 $_{51}$	69 $_{28}$	330	5570
Amfetamine-D5 TMS		1180	212 $_1$	197 $_8$	120 $_{100}$	92 $_{11}$	73 $_{57}$	251	5582
Amfetamine-D11 PFP		1610	292 $_1$	194 $_{100}$	128 $_{82}$	98 $_{43}$	70 $_{14}$	575	7284
Amfetamine-D11 TFA		1615	242 $_1$	144 $_{100}$	128 $_{82}$	98 $_{43}$	70 $_{14}$	353	7283
Amfetamine-D11 R-(-)-enantiomer HFBP		1995	341 $_{66}$	294 $_{49}$	266 $_{100}$	128 $_{28}$	98 $_{31}$	1097	6518
Amfetamine-D11 S-(+)-enantiomer HFBP		2000	341 $_{51}$	294 $_{99}$	266 $_{100}$	128 $_{19}$	98 $_{26}$	1097	6519
Amfetamine-M (3-HO-) TMSTFA		1630	319 $_2$	206 $_{86}$	191 $_2$	140 $_{100}$	73 $_{58}$	708	6141
Amfetamine-M (3-HO-) 2AC	U+UHYAC	1930	235 $_2$	176 $_{48}$	134 $_{52}$	107 $_{21}$	86 $_{100}$	324	4387
Amfetamine-M (3-HO-) 2HFB		1620	330 $_{30}$	303 $_{11}$	240 $_{100}$	169 $_{15}$	69 $_{29}$	1182	5737
Amfetamine-M (3-HO-) 2PFP		1520	280 $_{36}$	253 $_9$	190 $_{100}$	119 $_{19}$	69 $_8$	1104	5738
Amfetamine-M (3-HO-) 2TFA		<1000	230 $_4$	203 $_1$	140 $_{14}$	115 $_1$		819	6224
Amfetamine-M (3-HO-) 2TMS	UHYTMS	1850	280 $_{11}$	179 $_3$	116 $_{100}$	100 $_{12}$	73 $_{72}$	594	5693
Amfetamine-M (3-HO-) formyl artifact ME		1290	177 $_2$	162 $_4$	121 $_6$	77 $_5$	56 $_{100}$	165	5129
Amfetamine-M (4-HO-)		1480	151 $_{10}$	107 $_{69}$	91 $_{10}$	77 $_{42}$	56 $_{100}$	129	1802

Table 1-8-1: Compounds in order of names   Aminocarb

Name	Detected	RI	Typical ions and intensities					Page	Entry
Amfetamine-M (4-HO-) AC	U+UHYAC	1890	193_1	134_{100}	107_{26}	86_{25}	77_{16}	201	1803
Amfetamine-M (4-HO-) ME		1225	165_1	122_{16}	107_2	91_3	77_7	148	3249
Amfetamine-M (4-HO-) ME		1225	165_1	122_{16}	91_3	77_7	44_{100}	148	5517
Amfetamine-M (4-HO-) TFA		1670	247_4	140_{15}	134_{54}	107_{100}	77_{15}	366	6335
Amfetamine-M (4-HO-) 2AC	U+UHYAC	1900	235_1	176_{72}	134_{100}	107_{47}	86_{71}	324	1804
Amfetamine-M (4-HO-) 2HFB		<1000	330_{48}	303_{15}	240_{100}	169_{44}	69_{42}	1182	6326
Amfetamine-M (4-HO-) 2PFP		<1000	280_{77}	253_{16}	190_{100}	119_{56}	69_{16}	1104	6325
Amfetamine-M (4-HO-) 2TFA		<1000	230_{94}	203_{14}	140_{130}	92_{15}	69_{76}	819	6324
Amfetamine-M (4-HO-) 2TMS		<1000	280_7	179_9	149_8	116_{100}	73_{78}	594	6327
Amfetamine-M (4-HO-) formyl art.		1220	163_3	148_4	107_{30}	77_{13}	56_{100}	142	6323
Amfetamine-M (4-HO-) formyl artifact ME		1255	177_6	162_4	121_{60}	77_{12}	56_{100}	166	3250
Amfetamine-M (deamino-oxo-di-HO-) 2AC	U+UHYAC	1735*	250_3	208_{15}	166_{45}	123_{100}		379	4210
Amfetamine-M (deamino-oxo-HO-methoxy-)	UHY	1510*	180_{19}	137_{100}	122_{19}	107_2	94_{16}	175	4247
Amfetamine-M (deamino-oxo-HO-methoxy-) ME	UHYME	1540*	194_{25}	151_{100}	135_4	107_{18}	65_4	204	4353
Amfetamine-M (di-HO-) 3AC	U+UHYAC	2150	293_2	234_{54}	192_{48}	150_{99}	86_{100}	579	3725
Amfetamine-M (HO-methoxy-)	UHY	1465	181_9	138_{100}	122_{18}	94_{24}	77_{16}	178	4351
Amfetamine-M (HO-methoxy-deamino-HO-) 2AC	U+UHYAC	1820*	266_3	206_9	164_{100}	150_{10}	137_{30}	450	6409
Amfetamine-M (norephedrine) TMSTFA		1890	240_8	198_3	179_{100}	117_5	73_{88}	708	6146
Amfetamine-M (norephedrine) 2AC	U+UHYAC	1805	235_1	107_{13}	86_{130}	176_5	134_7	325	2476
Amfetamine-M (norephedrine) 2HFB	UHYHFB	1455	543_1	330_{14}	240_{100}	169_{44}	69_{57}	1183	5098
Amfetamine-M (norephedrine) 2PFP	UHYPFP	1380	443_1	280_9	190_{100}	119_{59}	105_{26}	1104	5094
Amfetamine-M (norephedrine) 2TFA	UTFA	1355	343_1	230_6	203_5	140_{100}	69_{29}	819	5091
Amfetamine-M AC	U+UHYAC	1600*	222_1	180_{22}	137_{100}			280	4211
Amfetamine-M ME	UHYME	1550	195_1	152_{100}	137_{17}	107_{16}	77_{14}	208	4352
Amfetamine-M 2AC	U+UHYAC	2065	265_3	206_{27}	164_{100}	137_{23}	86_{33}	445	3498
Amfetamine-M 2HFB	UHFB	1690	360_{82}	333_{15}	240_{100}	169_{42}	69_{39}	1189	6512
Amfetamine-N-formyl		1490	163_1	118_{72}	91_{30}	72_{100}	65_{23}	143	6428
Amfetaminil		1755	132_{100}	105_{51}	91_{38}	77_{17}	65_{18}	381	56
Amfetaminil-M/artifact (AM)		1160	134_1	120_1	91_6	77_1	65_4	115	54
Amfetaminil-M/artifact (AM)	U	1160	134_1	120_1	91_6	65_4	44_{100}	115	5514
Amfetaminil-M/artifact (AM) AC	U+UHYAC	1505	177_1	118_{19}	91_{11}	86_{31}	65_5	165	55
Amfetaminil-M/artifact (AM) AC	UAAC U+UHYAC	1505	177_1	118_{19}	91_{11}	86_{31}	44_{100}	165	5515
Amfetaminil-M/artifact (AM) HFB		1355	331_1	240_{79}	169_{21}	118_{100}	91_{53}	762	5047
Amfetaminil-M/artifact (AM) PFP		1330	281_1	190_{73}	118_{100}	91_{36}	65_{12}	521	4379
Amfetaminil-M/artifact (AM) TFA		1095	231_1	140_{100}	118_{92}	91_{45}	69_{19}	309	4000
Amfetaminil-M/artifact (AM) TMS		1190	192_6	116_{100}	100_{10}	91_{11}	73_{87}	235	5581
Amfetaminil-M/artifact (AM) formyl artifact		1100	147_2	146_6	125_5	91_{12}	56_{100}	125	3261
Amfetaminil-M/artifact-D5 AC	UAAC U+UHYAC	1480	182_3	122_{46}	92_{30}	90_{100}	66_{16}	181	5690
Amfetaminil-M/artifact-D5 HFB		1330	244_{100}	169_{14}	122_{46}	92_{41}	69_{40}	787	6316
Amfetaminil-M/artifact-D5 PFP		1320	194_{100}	123_{42}	119_{32}	92_{46}	69_{11}	543	5566
Amfetaminil-M/artifact-D5 TFA		1085	144_{100}	123_{53}	122_{56}	92_{51}	69_{28}	330	5570
Amfetaminil-M/artifact-D5 TMS		1180	212_1	197_8	120_{100}	92_{11}	73_{57}	251	5582
Amfetaminil-M/artifact-D11 TFA		1615	242_1	144_{100}	128_{82}	98_{43}	70_{14}	353	7283
Amfetaminil-M/artifact-D11 PFP		1610	292_1	194_{100}	128_{82}	98_{43}	70_{14}	575	7284
Amidithion		1930	273_6	131_{74}	125_{100}	93_{94}	59_{77}	480	3317
Amidotrizoic acid 2ME	UME	3000	642_8	569_3	515_{100}	483_{21}	314_2	1198	3708
Amidotrizoic acid 3ME		2920	656_{10}	625_5	529_{100}	471_4	386_3	1199	3709
Amidotrizoic acid -CO2 ME		2725	584_6	516_{28}	457_{100}	389_{39}	288_3	1192	3710
Amidotrizoic acid -CO2 2ME		2680	598_5	471_{24}	403_{27}	328_3	287_1	1194	3711
Amiloride-M/artifact (HOOC-) ME		1840	202_{100}	171_{51}	144_{68}	116_{32}	101_{22}	221	2628
Amiloride-M/artifact (HOOC-) 2ME		1860	216_2	187_{100}	170_{33}	142_{20}	116_{17}	259	2629
Amiloride-M/artifact (HOOC-) 3ME		1930	230_{100}	201_{55}	169_{26}	129_{21}	114_7	306	6878
Amineptine (ME)AC		2885	393_{65}	250_{29}	208_{40}	192_{110}	178_{47}	997	6050
Amineptine ME		2610	351_1	192_{100}	178_{21}	165_{12}	115_{11}	851	6041
Amineptine TMS		2750	409_1	309_7	218_{12}	192_{100}	178_{19}	1040	6051
Amineptine TMSTFA		2770	505_1	304_{19}	300_{19}	193_{100}	178_{39}	1168	6052
Amineptine 2ME		2570	365_2	192_{100}	178_{27}	174_{21}	165_{15}	902	6042
Amineptine artifact (ring)		1775*	194_{100}	179_{74}	165_{23}	152_{12}	115_{32}	205	6036
Amineptine HY(ME)		1930	223_2	192_{100}	178_{35}	165_{36}	115_{14}	285	6046
Amineptine-M (dealkyl-) ME		1930	223_2	192_{100}	178_{35}	165_{36}	115_{14}	285	6046
Amineptine-M (N-pentanoic acid) ME		2550	323_1	192_{100}	178_{27}	165_{17}	115_{13}	729	6043
Amineptine-M (N-pentanoic acid) 2ME		2490	337_1	192_{100}	178_{34}	165_{20}	115_{22}	795	6049
Amineptine-M (N-pentanoic acid) -H2O		2585	291_{10}	206_6	192_{100}	178_{28}	165_{21}	570	6045
Amineptine-M (N-propionic acid) (ME)AC		2585	337_{12}	294_{100}	208_{34}	192_{68}	178_{39}	793	6044
Amineptine-M (N-propionic acid) ME		2400	295_1	192_{100}	178_{24}	165_{17}	115_{12}	593	6047
Amineptine-M (N-propionic acid) 2ME		2350	309_2	192_{100}	178_{36}	165_{26}	115_{19}	662	6048
4-Aminobenzoic acid AC	U+UHYAC	2145	179_{31}	137_{100}	120_{92}	92_{16}	65_{24}	169	3298
4-Aminobenzoic acid ET	G	1820	165_{37}	137_{11}	120_{100}	92_{19}	65_{22}	147	1457
4-Aminobenzoic acid ETAC		1990	207_{63}	165_{67}	137_{27}	120_{100}	92_{18}	233	1440
4-Aminobenzoic acid ME		1550	151_{55}	120_{100}	92_{28}	65_{26}		128	23
2-Aminobenzoic acid ME		1290	151_{60}	119_{100}	92_{68}	65_{21}		128	4939
4-Aminobenzoic acid MEAC		1985	193_{33}	151_{60}	120_{100}	92_{18}	65_{19}	199	24
4-Aminobenzoic acid 2TMS		1645	281_{50}	236_7	148_3	73_{55}		522	5487
Aminocarb		1720	208_{12}	151_{100}	136_{51}	120_{22}	77_{17}	238	3753

Table 1-8-1: Compounds in order of names

Name	Detected	RI	Typical ions and intensities					Page	Entry
Aminocarb TFA		1700	304_{35}	247_{34}	232_{13}	150_{100}	69_{80}	634	4032
Aminocarb -C2H3NO		1215	151_{100}	150_{83}	136_{56}	120_{15}	77_{20}	129	3911
Aminoethanol		<1000	61_{5}	42_{9}	30_{100}			92	4189
4-(1-Aminoethyl-)phenol		<1000	137_{100}	121_{29}	103_{4}	91_{8}	77_{7}	117	7597
4-(1-Aminoethyl-)phenol TFA		1430	233_{45}	218_{100}	148_{11}	120_{32}	95_{16}	314	7603
4-(1-Aminoethyl-)phenol TMS		1125	209_{100}	193_{19}	177_{8}	151_{5}	73_{20}	242	7598
4-(1-Aminoethyl-)phenol 2AC		1740	221_{14}	179_{46}	164_{59}	136_{17}	122_{100}	276	7600
4-(1-Aminoethyl-)phenol 2HFB		1370	529_{52}	514_{100}	319_{37}	316_{42}	169_{38}	1178	7605
4-(1-Aminoethyl-)phenol 2PFP		1225	429_{57}	414_{100}	269_{35}	266_{33}	119_{49}	1077	7604
4-(1-Aminoethyl-)phenol 2TFA		1200	329_{79}	314_{100}	219_{25}	216_{22}	103_{10}	752	7602
4-(1-Aminoethyl-)phenol 2TMS		1125	281_{1}	266_{100}	223_{7}	194_{3}	73_{30}	524	7599
Aminoglutethimide	P-I	2340	232_{49}	203_{100}	175_{56}	132_{56}	117_{20}	312	2741
Aminoglutethimide AC	U+UHYAC	2900	274_{78}	245_{56}	203_{100}	175_{26}	132_{21}	485	2249
Aminoglutethimide ME		2310	246_{55}	217_{72}	189_{100}	132_{44}	117_{20}	363	2742
Aminoglutethimide MEAC		2880	288_{100}	259_{49}	231_{84}	217_{41}	189_{54}	553	2250
Aminophenazone	P G U-I	1895	231_{36}	123_{7}	111_{17}	97_{56}	56_{100}	310	189
Aminophenazone-M (bis-nor-)	P U UHY	1955	203_{23}	93_{14}	84_{59}	56_{100}		224	219
Aminophenazone-M (bis-nor-) AC	P U U+UHYAC	2270	245_{30}	203_{13}	84_{50}	56_{100}		359	183
Aminophenazone-M (bis-nor-) 2AC	U+UHYAC	2280	287_{8}	245_{31}	203_{15}	84_{56}	56_{100}	547	3333
Aminophenazone-M (bis-nor-) artifact	U UHY	1945	180_{13}	119_{99}	91_{45}			173	424
Aminophenazone-M (deamino-HO-)	U UHY	1855	204_{35}	120_{18}	85_{100}	56_{50}		226	218
Aminophenazone-M (deamino-HO-) AC	U+UHYAC	2095	246_{2}	204_{19}	119_{1}	91_{3}	56_{100}	362	190
Aminophenazone-M (nor-)	P U UHY	1980	217_{17}	123_{14}	98_{7}	83_{17}	56_{100}	261	220
Aminophenazone-M (nor-) AC	P U+UHYAC	2395	259_{20}	217_{7}	123_{8}	56_{99}		416	184
3-Aminophenol	U UHY	1290	109_{86}	80_{100}				102	216
3-Aminophenol AC		1860	151_{33}	109_{100}	81_{52}	80_{54}		128	223
4-Aminophenol	UHY	1240	109_{99}	80_{41}	52_{90}			102	826
4-Aminophenol ME	UHYME	1100	123_{27}	109_{100}	94_{7}	80_{96}	53_{47}	108	3766
4-Aminophenol 2AC	U+UHYAC PAC	1765	193_{10}	151_{53}	109_{100}	80_{24}		199	188
4-Aminophenol 3AC	U+UHYAC	2085	235_{10}	193_{11}	151_{30}	109_{100}		323	827
Aminorex		2065	162_{14}	145_{6}	118_{22}	91_{11}	56_{100}	141	3197
Aminorex iso-1 2AC		1990	246_{4}	203_{100}	189_{14}	161_{94}	72_{33}	362	3203
Aminorex iso-2 2AC		2115	246_{6}	231_{25}	189_{43}	146_{100}	56_{86}	362	3204
4-Aminosalicylic acid ME		1600	167_{60}	135_{100}	107_{81}	79_{100}		152	214
4-Aminosalicylic acid 2ME		1735	181_{100}	149_{80}	121_{51}			177	215
4-Aminosalicylic acid acetyl conjugate ME		1995	209_{39}	167_{32}	135_{100}			240	213
4-Aminosalicylic acid-M (3-aminophenol)	U UHY	1290	109_{86}	80_{100}				102	216
4-Aminosalicylic acid-M acetyl conjugate		1860	151_{33}	109_{100}	81_{52}	80_{54}		128	223
4-Aminothiophenol		1025	125_{100}	98_{12}	93_{23}	80_{28}	65_{7}	109	6351
Amiodarone artifact		2800*	420_{10}	294_{91}	265_{83}	142_{99}	121_{93}	1183	1386
Amiodarone artifact AC		2965*	588_{83}	546_{100}	517_{57}	461_{50}		1193	7587
Amiodarone artifact HFB		3670*	517_{1}	268_{100}	240_{8}	201_{4}		1204	7591
Amiodarone artifact PFP		3650*	517_{1}	391_{1}	218_{100}	190_{9}	119_{4}	1202	7590
Amiodarone artifact TFA		3740*	642_{1}	529_{2}	251_{3}	168_{100}	140_{18}	1198	7589
Amiodarone artifact TMS		3055*	618_{71}	335_{28}	320_{45}	201_{53}	73_{80}	1196	7588
Amiodarone-M (N-deethyl-) artifact AC		2965*	588_{83}	546_{100}	517_{57}	461_{50}		1193	7587
Amiodarone-M (N-deethyl-) artifact HFB		3670*	517_{1}	268_{100}	240_{8}	201_{4}		1204	7591
Amiodarone-M (N-deethyl-) artifact PFP		3650*	517_{1}	391_{1}	218_{100}	190_{9}	119_{4}	1202	7590
Amiodarone-M (N-deethyl-) artifact TFA		3740*	642_{1}	529_{2}	251_{3}	168_{100}	140_{18}	1198	7589
Amiodarone-M (N-deethyl-) artifact TMS		3055*	618_{71}	335_{28}	320_{45}	201_{53}	73_{80}	1196	7588
Amiphenazole		2170	191_{100}	149_{24}	121_{58}	104_{30}	77_{38}	194	34
Amiphenazole 2AC		2575	275_{27}	233_{59}	191_{100}	121_{42}		488	35
Amiphenazole 2ME		1925	219_{2}	191_{100}	147_{50}	121_{71}	77_{62}	269	36
Amisulpride	U+UHYAC	3260	369_{1}	242_{6}	196_{3}	149_{3}	98_{100}	917	5409
Amisulpride PFP	U+UHYPFP	2880	515_{1}	388_{2}	266_{2}	98_{100}	70_{12}	1172	5838
Amisulpride TFA	U+UHYTFA	2905	338_{2}	216_{2}	187_{1}	98_{100}	70_{6}	1133	5837
Amisulpride TMS		3400	441_{1}	426_{1}	314_{5}	111_{5}	98_{100}	1101	5840
Amisulpride 2TMS	U+UHYTMS	3065	513_{1}	498_{2}	314_{10}	196_{4}	98_{100}	1172	5839
Amisulpride-M (O-demethyl-)	U+UHYAC	2960	355_{1}	228_{4}	182_{3}	135_{4}	98_{100}	865	5410
Amitraz artifact-1		1570	162_{29}	149_{63}	120_{100}	106_{63}	77_{31}	141	4043
Amitraz artifact-2		2570	252_{14}	132_{6}	121_{100}	106_{18}	77_{13}	387	4042
Amitriptyline	P G U UHY U+UHYAC	2205	277_{1}	215_{16}	202_{31}	189_{14}	58_{300}	503	37
Amitriptyline-M (bis-nor-HO-) -H2O AC	U+UHYAC	2710	289_{15}	230_{100}	215_{70}	202_{31}	189_{5}	558	1873
Amitriptyline-M (di-HO-N-oxide) -H2O -(CH3)2NOH	UHY	2280*	246_{100}	228_{35}	215_{50}	202_{36}	178_{33}	363	2698
Amitriptyline-M (di-HO-N-oxide) -H2O -(CH3)2NOH AC	U+UHYAC	2530*	288_{35}	246_{100}	229_{58}	215_{89}	202_{37}	552	2541
Amitriptyline-M (HO-)	P-I U UGLUC	2380	293_{1}	215_{3}	202_{2}	91_{1}	58_{100}	582	27
Amitriptyline-M (HO-) AC	UGLUCAC	2500	335_{1}	273_{1}	215_{5}	202_{5}	58_{100}	784	44
Amitriptyline-M (HO-) -H2O	UHY U+UHYAC	2235	275_{3}	215_{41}	202_{23}	189_{14}	58_{300}	492	40
Amitriptyline-M (HO-N-oxide) -(CH3)2NOH AC	U+UHYAC	2490*	290_{31}	248_{100}	230_{50}	215_{36}	202_{30}	564	1874
Amitriptyline-M (HO-N-oxide) -H2O -(CH3)2NOH	U UHY U+UHYAC	2000*	230_{100}	215_{40}				307	46
Amitriptyline-M (nor-)	P-I G U UHY	2255	263_{27}	220_{67}	202_{100}	189_{39}	91_{30}	436	38
Amitriptyline-M (nor-) AC	PAC U+UHYAC	2660	305_{10}	232_{100}	219_{11}	202_{8}	86_{19}	640	41
Amitriptyline-M (nor-) HFB		2420	459_{5}	240_{100}	232_{49}	217_{36}	202_{36}	1126	7685
Amitriptyline-M (nor-) PFP		2405	409_{3}	232_{100}	217_{71}	203_{69}	190_{69}	1038	7684

Table 1-8-1: Compounds in order of names Androsterone

Name	Detected	RI	Typical ions and intensities					Page	Entry
Amitriptyline-M (nor-) TFA		2410	359_3	232_{76}	217_{54}	202_{70}	140_{100}	880	7683
Amitriptyline-M (nor-) TMS		2340	335_1	320_1	203_5	116_{100}	73_{52}	785	5440
Amitriptyline-M (nor-)-D3		2250	266_6	220_{41}	215_{51}	202_{100}	189_{45}	453	7794
Amitriptyline-M (nor-)-D3 AC		2655	308_{11}	232_{100}	217_{46}	202_{47}	89_{23}	657	7795
Amitriptyline-M (nor-)-D3 HFB		2415	462_2	243_{58}	232_{100}	217_{40}	203_{33}	1130	7798
Amitriptyline-M (nor-)-D3 PFP		2400	412_2	232_{100}	217_{53}	203_{47}	193_{46}	1047	7797
Amitriptyline-M (nor-)-D3 TFA		2405	362_2	232_{100}	217_{53}	202_{48}	143_{47}	891	7796
Amitriptyline-M (nor-)-D3 TMS		2335	338_1	323_{10}	202_{33}	119_{100}	73_{73}	800	7799
Amitriptyline-M (nor-HO-)	U-I UGLUC	2390	279_8	261_6	218_{100}	203_{39}	91_{10}	513	39
Amitriptyline-M (nor-HO-) -H2O	UHY	2600	261_{14}	218_{99}	215_{100}	202_{66}	189_{23}	427	2270
Amitriptyline-M (nor-HO-) -H2O AC	U+UHYAC	2670	303_{20}	230_{100}	215_{74}	202_{35}	86_{18}	629	42
Amitriptyline-M (N-oxide) -(CH3)2NOH	P G U UHY U+UHYAC	1975*	232_{120}	217_{88}	202_{86}	189_{47}	165_{31}	313	45
Amitriptylinoxide -(CH3)2NOH	P G U UHY U+UHYAC	1975*	232_{120}	217_{88}	202_{86}	189_{47}	165_{31}	313	45
Amitriptylinoxide-M (deoxo-bis-nor-HO-) -H2O AC	U+UHYAC	2710	289_{15}	230_{100}	215_{70}	202_{31}	189_5	558	1873
Amitriptylinoxide-M (deoxo-HO-)	P-I U UGLUC	2380	293_1	215_3	202_2	91_1	58_{100}	582	27
Amitriptylinoxide-M (deoxo-HO-) AC	UGLUCAC	2500	335_1	273_1	215_5	202_5	58_{100}	784	44
Amitriptylinoxide-M (deoxo-HO-) -H2O	UHY U+UHYAC	2235	275_3	215_{41}	202_{23}	189_{14}	58_{300}	492	40
Amitriptylinoxide-M (deoxo-nor-HO-) -H2O	UHY	2600	261_{14}	218_{99}	215_{100}	202_{66}	189_{23}	427	2270
Amitriptylinoxide-M (deoxo-nor-HO-) -H2O AC	U+UHYAC	2670	303_{20}	230_{100}	215_{74}	202_{35}	86_{18}	629	42
Amitriptylinoxide-M (di-HO-) -H2O -(CH3)2NOH	UHY	2280*	246_{100}	228_{35}	215_{50}	202_{36}	178_{33}	363	2698
Amitriptylinoxide-M (di-HO-) -H2O -(CH3)2NOH AC	U+UHYAC	2530*	288_{35}	246_{100}	229_{58}	215_{89}	202_{37}	552	2541
Amitriptylinoxide-M (HO-) -(CH3)2NOH AC	U+UHYAC	2490*	290_{31}	248_{100}	230_{50}	215_{36}	202_{30}	564	1874
Amitriptylinoxide-M (HO-) -H2O -(CH3)2NOH	U UHY U+UHYAC	2000*	230_{100}	215_{40}				307	46
Amitrole		<1000	84_{100}	75_3	57_{14}			96	4509
Amitrole AC	U+UHYAC	1010	126_{18}	108_3	84_{100}	57_{35}		110	4233
Amitrole 2ME		1050	112_{25}	111_{28}	98_{100}	84_9	56_{75}	103	3121
Amlodipine AC	U+UHYAC	3170	450_1	347_{20}	339_{87}	208_{39}	86_{100}	1113	4844
Amlodipine ME		2820	422_3	311_{100}	254_{53}	208_{12}	88_{20}	1065	4843
Amlodipine TMS		2935	480_{20}	369_{100}	326_{53}	208_{17}	73_{17}	1148	5013
Amlodipine 2ME		2815	436_2	325_{52}	208_7	72_{100}	58_{62}	1092	4842
Amlodipine 2TMS		3130	552_{16}	441_{38}	174_{100}	116_{79}	73_{74}	1185	5014
Amlodipine-M (deamino-COOH) ME	PME UME	2830	437_3	326_{100}	312_{30}	280_{18}	208_{48}	1093	4846
Amlodipine-M (deethyl-deamino-COOH) 2ME	UME	2800	423_3	392_4	312_{40}	280_{15}	222_{52}	1065	4847
Amlodipine-M (dehydro-2-HOOC-) ME	UME	2430	391_1	356_{100}	296_{85}	268_{18}	224_{20}	987	4850
Amlodipine-M (dehydro-deamino-HOOC-) ME	UME	2635	435_1	400_{16}	347_{98}	318_{61}	260_{100}	1090	4848
Amlodipine-M (dehydro-deethyl-O-dealkyl-) -H2O	UME	2300	317_1	282_{100}	267_8	250_6	139_5	695	4849
Amlodipine-M/artifact (dehydro-) AC	U+UHYAC	2910	448_3	363_{34}	347_{100}	260_{83}	86_{41}	1110	4851
Amlodipine-M/artifact (dehydro-) 2ME	PME	2825	434_5	323_{100}	277_{12}	88_{20}	86_{21}	1088	4845
Amlodipine-M/artifact (dehydro-) 2TMS		2925	550_{11}	535_{58}	477_{100}	447_9	359_{10}	1184	5015
Amobarbital	P G U UHY U+UHYAC	1710	211_2	198_6	197_3	156_{100}	141_{73}	295	47
Amobarbital 2ME	UME	1595	239_2	225_7	184_{100}	169_{90}		400	51
Amobarbital 2TMS		1530	370_1	355_{95}	300_{38}	100_{51}	73_{90}	920	5498
Amobarbital-M (HO-)	U	1915	227_{26}	195_6	157_{73}	156_{100}	141_{61}	352	49
Amobarbital-M (HO-) 2ME	UME	1750	270_5	255_{22}	184_{95}	169_{64}	137_{100}	471	52
Amobarbital-M (HO-) -H2O	UHY U+UHYAC	1830	224_9	195_{13}	156_{73}	141_{37}	69_{100}	289	48
Amobarbital-M (HOOC-)	U	1960	212_{24}	183_{24}	156_{100}	141_{93}	55_{59}	406	50
Amobarbital-M (HOOC-) 3ME	UME	1850	240_8	184_{99}	169_{100}	137_{72}		607	53
Amodiaquine AC		2875	326_{17}	286_{29}	284_{100}	205_7	99_{11}	1010	6889
Amodiaquine ME		3030	369_{20}	354_{17}	297_{100}	269_{29}	252_{19}	916	7840
Amodiaquine TMS		3090	427_9	412_8	355_{76}	86_{25}	73_{100}	1163	7787
Amodiaquine TMS		3090	427_9	412_8	355_{76}	86_{25}	73_{100}	1074	7836
Amodiaquine 2AC		3000	356_{13}	314_{23}	282_{100}	253_{27}	218_{23}	1096	7838
Amodiaquine artifact		2850	284_{100}	268_{11}	248_{26}	234_{14}		531	7459
Amodiaquine artifact AC		2875	326_{13}	284_{100}	248_9	205_4	99_5	737	7839
Amodiaquine artifact ME		2905	296_{100}	260_{14}	232_7	99_{17}		595	7191
Amoxicilline-M/artifact ME2AC		1930	265_5	233_{37}	164_{43}	122_{100}	120_{59}	443	7653
Amoxicilline-M/artifact ME2AC		1930	314_{35}	230_{38}	198_{100}	156_{43}	97_{77}	684	7652
Amoxicilline-M/artifact ME2TFA		1755	422_{28}	326_{100}	267_{36}	196_{54}	165_{32}	1064	7656
Amoxicilline-M/artifact ME3AC		2025	307_5	265_{39}	233_{67}	180_{100}	120_{69}	648	7654
Amoxicilline-M/artifact MEAC		1980	272_{40}	230_{91}	215_{38}	100_{53}	97_{100}	478	7651
Amoxicilline-M/artifact MEPFP		1750	376_{100}	317_{65}	246_{70}	243_{62}	215_{53}	939	7657
Amoxicilline-M/artifact 4TMS		1215	455_8	440_7	216_{23}	172_{65}	73_{100}	1121	7655
Amperozide artifact (methylpiperazine)		2415	344_{14}	201_9	183_{12}	113_{100}	70_{35}	826	6097
Amperozide-M (deamino-carboxy-)	P-I UHY U+UHYAC	2230*	276_7	216_{17}	203_{100}	183_{22}		496	169
Amperozide-M (deamino-carboxy-) ME	P-I UHYME U+UHYAC	2125*	290_{12}	258_{13}	216_{30}	203_{100}	183_{22}	563	3372
Amperozide-M (deamino-HO-) AC	U+UHYAC	2150*	304_7	244_{22}	216_{41}	203_{100}	183_{22}	635	307
Amperozide-M (N-dealkyl-) AC	U+UHYAC	2970	141_{100}	109_9	372_{29}	300_{13}	201_{11}	928	3370
Amylnitrite		<1000	85_5	70_{18}	57_{54}	41_{100}		105	58
Ancymidol		2220	256_4	228_{76}	215_{17}	121_{61}	107_{100}	407	4144
Androst-4-ene-3,17-dione		2600*	286_{100}	244_{39}	148_{26}	124_{59}	79_{36}	544	3762
Androst-4-ene-3,17-dione enol 2TMS		2650*	430_{98}	415_{11}	234_{16}	209_{15}	73_{100}	1081	3803
Androstane-3,17-dione		2555*	288_{100}	255_{22}	244_{32}	217_{27}	124_{21}	555	3761
Androstane-3,17-dione enol 2TMS		2600*	432_{60}	417_{41}	290_{47}	275_{71}	73_{100}	1085	3801
Androsterone	UHY	2475*	290_{100}	246_{24}	147_{16}	107_{39}	67_{36}	566	59

Androsterone AC

Table 1-8-1: Compounds in order of names

Name	Detected	RI	Typical ions and intensities					Page	Entry
Androsterone AC	U+UHYAC	2580*	332 $_{27}$	272 $_{99}$	257 $_{36}$	201 $_{23}$	79 $_{53}$	771	61
Androsterone enol 2TMS		2500*	434 $_{46}$	419 $_{52}$	329 $_{35}$	169 $_{37}$	73 $_{100}$	1089	3208
Androsterone -H2O	UHY U+UHYAC	2240*	272 $_{100}$	218 $_{87}$	190 $_{39}$	161 $_{50}$	79 $_{49}$	480	2481
Anilazine		2050	274 $_{8}$	239 $_{100}$	178 $_{31}$	143 $_{26}$	75 $_{11}$	483	3426
Aniline		<1000	93 $_{100}$	66 $_{28}$				98	1550
Aniline AC	G	1380	135 $_{35}$	93 $_{100}$				115	222
4-Anisic acid ET		1415*	180 $_{21}$	77 $_{21}$	107 $_{12}$	152 $_{19}$	135 $_{100}$	174	6447
4-Anisic acid ME		1270*	166 $_{34}$	135 $_{100}$	107 $_{15}$	92 $_{16}$	77 $_{26}$	150	6446
p-Anisidine		<1000	123 $_{50}$	108 $_{100}$	95 $_{3}$	80 $_{44}$	65 $_{8}$	108	7638
p-Anisidine AC	PME UME	1630	165 $_{59}$	123 $_{74}$	108 $_{100}$	95 $_{10}$	80 $_{20}$	147	5046
p-Anisidine HFB	U+UHYHFB	1400	319 $_{100}$	304 $_{8}$	300 $_{6}$	150 $_{6}$	122 $_{78}$	706	6620
p-Anisidine TFA	U+UHYTFA	1335	219 $_{100}$	204 $_{16}$	149 $_{11}$	122 $_{76}$	109 $_{19}$	269	6615
p-Anisidine TMS		<1000	195 $_{57}$	180 $_{100}$	164 $_{9}$	147 $_{8}$	73 $_{46}$	207	7640
p-Anisidine formyl artifact		1080	135 $_{84}$	120 $_{100}$	92 $_{33}$	77 $_{8}$	65 $_{11}$	115	7639
Antazoline		2350	265 $_{1}$	182 $_{7}$	91 $_{22}$	84 $_{100}$		446	62
Antazoline +H2O AC	U+UHYAC	2650	325 $_{4}$	196 $_{30}$	182 $_{15}$	91 $_{100}$	77 $_{8}$	736	2068
Antazoline AC		2610	307 $_{4}$	274 $_{13}$	182 $_{45}$	91 $_{100}$	84 $_{82}$	650	2053
Antazoline TMS		2450	322 $_{1}$	180 $_{5}$	155 $_{100}$	91 $_{43}$	73 $_{53}$	795	5459
Antazoline artifact AC	U+UHYAC	2260	255 $_{3}$	196 $_{14}$	104 $_{7}$	91 $_{100}$	77 $_{14}$	402	2067
Antazoline HY	UHY	1930	183 $_{25}$	106 $_{16}$	91 $_{100}$	77 $_{25}$	65 $_{15}$	182	2065
Antazoline HYAC	U+UHYAC	2080	225 $_{22}$	183 $_{23}$	106 $_{10}$	91 $_{100}$	77 $_{25}$	291	2066
Antazoline-M (HO-) AC	U+UHYAC	2620	323 $_{1}$	254 $_{6}$	212 $_{7}$	91 $_{68}$	84 $_{100}$	728	2069
Antazoline-M (HO-) HY	UHY	1920	199 $_{53}$	163 $_{21}$	91 $_{100}$	76 $_{24}$	65 $_{34}$	217	2143
Antazoline-M (HO-) HY2AC	U+UHYAC	2340	283 $_{18}$	241 $_{16}$	199 $_{34}$	91 $_{100}$	65 $_{22}$	528	2072
Antazoline-M (HO-methoxy-) HY2AC	U+UHYAC	2460	313 $_{11}$	271 $_{8}$	254 $_{10}$	212 $_{18}$	91 $_{100}$	680	2074
Antazoline-M (HO-methoxy-) HYAC	U+UHYAC	2370	271 $_{12}$	212 $_{17}$	120 $_{9}$	91 $_{100}$	65 $_{18}$	474	2073
Antazoline-M (methoxy-) HYAC	U+UHYAC	2290	255 $_{22}$	213 $_{64}$	136 $_{22}$	122 $_{53}$	91 $_{100}$	402	2070
Anthracene		1760*	178 $_{100}$	176 $_{20}$	152 $_{7}$	89 $_{6}$	76 $_{6}$	167	2562
Anthranilic acid ME		1290	151 $_{60}$	119 $_{100}$	92 $_{68}$	65 $_{21}$		128	4939
Anthraquinone		2090*	208 $_{93}$	180 $_{100}$	152 $_{79}$	126 $_{10}$	76 $_{47}$	236	4048
ANTU 3ME		2090	244 $_{26}$	197 $_{100}$	182 $_{38}$	154 $_{50}$	127 $_{40}$	356	3972
Apomorphine		2715	267 $_{64}$	266 $_{100}$	224 $_{21}$	220 $_{16}$	152 $_{11}$	454	3988
Apomorphine 2AC	U+UHYAC	2830	351 $_{100}$	308 $_{67}$	266 $_{78}$	224 $_{12}$	165 $_{9}$	848	2286
Apomorphine 2TMS		2715	411 $_{67}$	410 $_{100}$	368 $_{15}$	322 $_{45}$	73 $_{47}$	1046	4525
Aprindine	G U UHY U+UHYAC	2460	322 $_{2}$	249 $_{8}$	206 $_{13}$	113 $_{74}$	86 $_{100}$	724	1378
Aprindine-M (4-aminophenol)	UHY	1240	109 $_{99}$	80 $_{41}$	52 $_{90}$			102	826
Aprindine-M (4-aminophenol) 2AC	U+UHYAC PAC	1765	193 $_{10}$	151 $_{53}$	109 $_{100}$	80 $_{24}$		199	188
Aprindine-M (aniline) AC	G	1380	135 $_{35}$	93 $_{100}$				115	222
Aprindine-M (deethyl-HO-) 2AC	U+UHYAC	3220	394 $_{25}$	280 $_{60}$	190 $_{53}$	117 $_{100}$	58 $_{66}$	1000	2889
Aprindine-M (deindane) AC	U+UHYAC	1880	248 $_{2}$	219 $_{7}$	176 $_{8}$	86 $_{100}$	72 $_{19}$	373	2881
Aprindine-M (deindane-HO-) 2AC	U+UHYAC	2205	306 $_{1}$	277 $_{5}$	219 $_{9}$	86 $_{100}$	58 $_{29}$	646	2883
Aprindine-M (dephenyl-) AC	U+UHYAC	2300	288 $_{4}$	216 $_{5}$	117 $_{9}$	86 $_{100}$	72 $_{19}$	555	2884
Aprindine-M (dephenyl-HO-) 2AC	U+UHYAC	2680	346 $_{2}$	187 $_{35}$	128 $_{15}$	116 $_{43}$	86 $_{100}$	833	2886
Aprindine-M (HO-) AC	U+UHYAC	2850	380 $_{4}$	307 $_{6}$	264 $_{5}$	113 $_{50}$	86 $_{100}$	954	2887
Aprindine-M (HO-methoxy-) AC	U+UHYAC	2995	410 $_{1}$	206 $_{26}$	162 $_{45}$	113 $_{47}$	86 $_{100}$	1042	2888
Aprindine-M (N-dealkyl-)	UHY U+UHYAC	1920	209 $_{48}$	166 $_{1}$	104 $_{100}$	94 $_{18}$	77 $_{33}$	242	2882
Aprindine-M (N-dealkyl-HO-) 2AC	U+UHYAC	2410	267 $_{36}$	225 $_{100}$	120 $_{74}$	115 $_{29}$	91 $_{21}$	454	2885
Aprobarbital	P G U UHY U+UHYAC	1610	210 $_{6}$	195 $_{18}$	167 $_{100}$	124 $_{43}$		245	63
Aprobarbital 2ME		1540	238 $_{2}$	195 $_{100}$	138 $_{67}$	111 $_{33}$		337	644
Aprobarbital 2TMS		1620	354 $_{3}$	339 $_{65}$	297 $_{32}$	100 $_{37}$	73 $_{80}$	862	5458
Aprobarbital-M (HO-)	U	1800	226 $_{4}$	183 $_{100}$	154 $_{62}$	97 $_{49}$	69 $_{76}$	293	2960
Arabinose 4AC		1760*	259 $_{8}$	170 $_{61}$	128 $_{100}$	115 $_{73}$	103 $_{50}$	702	1963
Arabinose 4HFB		1235*	478 $_{12}$	465 $_{8}$	293 $_{64}$	265 $_{11}$	169 $_{100}$	1206	5799
Arabinose 4PFP		1310*	411 $_{4}$	378 $_{13}$	243 $_{33}$	219 $_{43}$	119 $_{100}$	1203	5798
Arabinose 4TFA		1290*	311 $_{10}$	278 $_{9}$	265 $_{11}$	169 $_{63}$	69 $_{100}$	1180	5797
Arachidonic acid-M (15-HETE) METFA		2390*	430 $_{1}$	316 $_{6}$	131 $_{43}$	117 $_{66}$	91 $_{100}$	1080	4354
Arachidonic acid-M (15-HETE) -H2O ME		2360*	316 $_{4}$	189 $_{37}$	119 $_{54}$	105 $_{100}$	91 $_{92}$	694	4355
Aramite		2400*	334 $_{23}$	319 $_{37}$	185 $_{10}$	135 $_{30}$	63 $_{83}$	779	4049
Aramite -C2H3ClSO2		1650*	208 $_{14}$	193 $_{45}$	135 $_{100}$	107 $_{9}$	91 $_{10}$	239	4050
Arecaidine		1325	141 $_{39}$	96 $_{100}$	81 $_{12}$	68 $_{11}$	53 $_{17}$	120	5938
Arecaidine ME		<1000	155 $_{55}$	140 $_{100}$	124 $_{33}$	96 $_{90}$	81 $_{64}$	135	5870
Arecaidine TMS		1460	213 $_{32}$	198 $_{48}$	155 $_{78}$	96 $_{100}$		252	5939
Arecoline		<1000	155 $_{55}$	140 $_{100}$	124 $_{33}$	96 $_{90}$	81 $_{64}$	135	5870
Arecoline-M/artifact (HOOC-)		1325	141 $_{39}$	96 $_{100}$	81 $_{12}$	68 $_{11}$	53 $_{17}$	120	5938
Arecoline-M/artifact (HOOC-) TMS		1460	213 $_{32}$	198 $_{48}$	155 $_{78}$	96 $_{100}$		252	5939
Aripiprazole		3400	447 $_{1}$	285 $_{6}$	243 $_{100}$	84 $_{26}$		1110	7261
Aripiprazole-M (N-dealkyl-) AC	U+UHYAC	2255	272 $_{12}$	229 $_{9}$	200 $_{100}$	188 $_{35}$	56 $_{46}$	477	7123
Articaine		2170	284 $_{2}$	171 $_{12}$	139 $_{4}$	86 $_{100}$	56 $_{4}$	534	2342
Articaine AC	U+UHYAC	2455	295 $_{1}$	171 $_{2}$	156 $_{40}$	128 $_{33}$	86 $_{100}$	740	4442
Articaine artifact	U	2230	296 $_{28}$	281 $_{100}$	86 $_{86}$	84 $_{58}$	56 $_{79}$	598	4443
Articaine -CO2 AC	U+UHYAC	2250	268 $_{2}$	222 $_{5}$	156 $_{50}$	128 $_{42}$	86 $_{100}$	461	4444
Articaine-M (HO-) 2AC	U+UHYAC	2470	369 $_{1}$	229 $_{2}$	156 $_{27}$	128 $_{18}$	86 $_{100}$	967	4445
Artifact of roasted food (cyclo (Phe-Pro))	U+UHYAC P-I	2375	244 $_{36}$	153 $_{50}$	125 $_{100}$	91 $_{38}$	70 $_{50}$	356	4495
Artifact of roasted food (cyclo (Phe-Pro)) AC		2360	286 $_{29}$	153 $_{62}$	125 $_{100}$	91 $_{57}$	70 $_{73}$	544	5217

Table 1-8-1: Compounds in order of names **Azidocilline-M/artifact ME2AC**

Name	Detected	RI	Typical ions and intensities					Page	Entry
Ascorbic acid	U	2120*	176_7	116_{100}	85_{25}			162	64
Ascorbic acid 2AC		2065*	260_2	242_{18}	200_{100}	158_{42}	85_{44}	419	3307
Ascorbic acid 2ME		1700*	204_7	144_{100}	129_{29}	117_9	101_{14}	225	2634
Ascorbic acid iso-1 3ME		1600*	218_6	200_4	144_{100}	129_{28}	101_{12}	264	2635
Ascorbic acid iso-2 3ME		1720*	218_{11}	158_{100}	130_{77}	115_{21}	101_{16}	264	2636
Astemizole	G	3900	458_5	337_{88}	294_{13}	109_{19}	96_{100}	1125	1774
Astemizole-M (N-dealkyl-) AC	U+UHYAC	3150	366_{14}	268_{17}	242_{50}	109_{100}	82_{32}	905	4506
Astemizole-M (N-dealkyl-) 2AC	U+UHYAC	3170	408_1	366_{21}	268_{22}	242_{52}	109_{100}	1037	4505
Astemizole-M/artifact (N-dealkyl-)		2470	241_{61}	132_{14}	109_{100}	83_{11}		347	1775
Astemizole-M/artifact (N-dealkyl-) AC	U+UHYAC	2490	283_{30}	241_{13}	240_{17}	109_{100}	83_{11}	528	1776
Asulam -C2H2O2	G P UHY	2185	172_{57}	156_{55}	108_{50}	92_{74}	65_{100}	159	973
Asulam -COOCH3 4ME		2095	228_{44}	184_{30}	136_{100}	120_{70}	77_{29}	300	4098
Atenolol	G P-I U	2380	251_2	222_5	107_6	72_{100}		452	1721
Atenolol TMSTFA		2600	434_1	377_4	332_{15}	284_{100}	73_{65}	1088	6037
Atenolol 2TMS		2250	410_1	395_{23}	294_{22}	188_{19}	72_{100}	1042	5471
Atenolol 3TMS (amide/amide/HO-)		2220	467_2	295_{32}	188_{57}	73_{70}	72_{41}	1150	5474
Atenolol 3TMS (amide/amine/HO-)		2460	467_5	295_{27}	144_{100}	101_7	73_{90}	1150	5473
Atenolol 4TMS		2430	539_8	277_4	188_{19}	144_{100}	73_{99}	1185	5472
Atenolol artifact (formyl-HOOC-) ME	U	2175	293_{36}	278_{100}	127_{75}	112_{74}	56_{66}	581	2682
Atenolol artifact (HOOC-) ME		2140	281_1	267_1	237_5	107_{11}	72_{100}	523	2681
Atenolol formyl artifact	G P U	2400	278_9	263_{38}	127_{67}	86_{82}	56_{100}	508	65
Atenolol -H2O	U	2150	248_{17}	218_2	190_3	98_{100}	56_{76}	371	2680
Atenolol -H2O AC	U+UHYAC	2975	290_{27}	205_{87}	188_{38}	140_{100}	98_{53}	565	1349
Atomoxetine		2000	255_{12}	148_{33}	104_{48}	91_{39}	77_{100}	403	7192
Atomoxetine		2000	255_1	151_2	148_5	77_{13}	44_{100}	403	7247
Atomoxetine AC		2310	297_1	190_{65}	117_{34}	86_{100}	77_{16}	603	7236
Atomoxetine HFB		2190	451_{30}	209_{33}	197_{100}	169_{46}	72_{14}	1114	7239
Atomoxetine ME		1950	269_2	163_2	115_3	77_{13}	58_{100}	466	7193
Atomoxetine PFP		2250	401_{15}	223_{11}	209_{20}	197_{100}	119_{33}	1021	7238
Atomoxetine TFA		2000	351_1	244_{44}	140_{100}	117_{78}	77_{19}	848	7237
Atomoxetine TMS		2055	327_7	208_5	116_{100}	104_5	73_{44}	746	7245
Atomoxetine -H2O HYAC	U+UHYAC-I	1680	189_6	146_{30}	115_{56}	98_{100}	70_8	192	4339
Atomoxetine -H2O HYHFB		1470	343_9	252_{100}	174_{63}	146_5	115_{25}	820	7240
Atomoxetine -H2O HYPFP		1450	293_{17}	202_{87}	174_{100}	117_{51}	115_{90}	578	7242
Atomoxetine -H2O HYTMS		1580	219_{27}	204_{44}	161_{30}	103_{49}	75_{100}	271	7246
Atomoxetine HY2AC	U+UHYAC	1890	249_2	206_{100}	146_{36}	98_{78}	86_{34}	375	4340
Atomoxetine HY2HFB		1490	557_1	434_4	360_7	343_{31}	241_{100}	1186	7241
Atomoxetine HY2PFP		1430	457_2	334_4	310_{16}	239_{28}	190_{100}	1123	7243
Atomoxetine HY2TFA		1435	357_4	243_{20}	174_{32}	140_{100}	117_{42}	871	7244
Atomoxetine-D6 HY2PFP		1420	463_3	334_9	298_{41}	190_{100}	119_{87}	1131	7791
Atomoxetine-M (nor-) HY2PFP		1400	443_5	296_{40}	280_7	239_{100}	177_{14}	1104	7711
Atracurium-M (O-bisdemethyl-)/artifact 2AC	U+UHYAC	3370	427_{50}	385_9	354_{55}	312_{100}	137_{76}	1074	6789
Atracurium-M (O-bisdemethyl-)/artifact 2AC	U+UHYAC	3020	413_1	178_{29}	151_7	262_{34}	220_{43}	1048	6788
Atracurium-M (O-demethyl-)/artifact AC	U+UHYAC	3210	399_{59}	326_{47}	313_{19}	295_{25}	151_{70}	1015	6790
Atracurium-M (O-demethyl-)/artifact AC	U+UHYAC	2595	385_1	234_{88}	192_{100}	177_{16}	151_{17}	970	6787
Atracurium-M/artifact	P U+UHYAC	2575	357_1	206_{100}	190_{23}	162_8	151_7	874	6106
Atrazine	P G	1720	215_{44}	200_{75}	173_{27}	68_{68}	58_{100}	257	66
Atrazine-M (deethyl-)	U	1680	187_{23}	172_{73}	70_{77}	58_{100}		189	68
Atrazine-M (deethyl-deschloro-methoxy-)	U	1670	183_{49}	168_{95}	141_{54}	70_{79}	58_{100}	182	67
Atrazine-M (deisopropyl-)	U	1730	173_{100}	158_{97}	145_{77}	130_{18}	68_{78}	160	4236
Atropine	P G U	2215	289_9	272_1	140_5	124_{100}	94_6	559	69
Atropine AC	U+UHYAC	2275	331_5	140_8	124_{100}	94_{22}	82_{35}	765	71
Atropine TMS		2295	361_5	140_5	124_{100}	82_{20}	73_{50}	888	4526
Atropine -CH2O	U+UHYAC	1980	259_{25}	221_4	140_6	124_{100}	91_{31}	417	2343
Atropine -H2O	P G U UHY U+UHYAC	2085	271_{33}	140_8	124_{100}	96_{21}	82_{13}	475	70
Atropine-M/artifact (tropine) AC		1240	183_{25}	140_{11}	124_{100}	94_{42}	82_{63}	183	5125
Azamethiphos artifact		1655	184_{100}	143_{12}	129_{12}	101_{26}	64_{21}	183	4038
Azaperone		2650	327_6	309_{10}	233_{23}	165_{22}	107_{100}	745	6098
Azaperone enol TMS		2655	399_1	176_{100}	147_{22}	121_{36}	107_{22}	1016	6277
Azaperone-M (dihydro-)		2730	329_7	235_{10}	165_{16}	121_{29}	107_{100}	756	6115
Azaperone-M (dihydro-) AC		2775	371_3	311_3	222_1	121_{25}	107_{100}	925	6116
Azaperone-M (dihydro-) -H2O		2625	311_3	176_{100}	147_{62}	121_{69}	107_{38}	672	6117
Azapropazone		2610	300_{75}	189_{44}	160_{100}	145_{36}		616	1955
Azatadine	U UHY U+UHYAC	2375	290_{85}	246_{100}	232_{69}	96_{15}	70_{25}	565	1379
Azatadine-M (di-HO-aryl-) 2AC	U+UHYAC	2620	406_4	346_{13}	304_{15}	287_{100}	230_{13}	1032	2105
Azatadine-M (HO-alkyl-) AC	U+UHYAC	2520	348_{55}	305_{55}	288_{100}	244_{62}	230_{53}	839	2103
Azatadine-M (HO-alkyl-) -H2O	U+UHYAC	2410	288_{66}	244_{100}	230_{46}	216_{19}	70_{24}	554	2102
Azatadine-M (HO-alkyl-HO-aryl-) 2AC	U+UHYAC	2640	406_{100}	363_{75}	347_{27}	304_{70}	287_{40}	1033	2106
Azatadine-M (HO-aryl-) AC	U+UHYAC	2540	348_{100}	305_{76}	262_{72}	244_{74}	230_{54}	839	2104
Azatadine-M (nor-) AC	U+UHYAC	2720	318_{100}	258_{49}	246_{52}	232_{62}	217_{19}	704	2107
Azatadine-M (nor-HO-alkyl-) 2AC	U+UHYAC	2810	376_{55}	316_{90}	256_{52}	244_{100}	230_{67}	940	2109
Azatadine-M (nor-HO-alkyl-) -H2O AC	U+UHYAC	2750	316_{100}	256_{37}	244_{45}	230_{47}	217_{16}	694	2108
Azelastine		3180	381_6	271_{24}	256_{22}	130_{31}	110_{100}	956	4626
Azidocilline-M/artifact ME2AC		1930	314_{35}	230_{38}	198_{100}	156_{43}	97_{77}	684	7652

Azidocilline-M/artifact ME2TFA Table 1-8-1: Compounds in order of names

Name	Detected	RI	Typical ions and intensities					Page	Entry
Azidocilline-M/artifact ME2TFA		1755	422$_{28}$	326$_{100}$	267$_{36}$	196$_{54}$	165$_{32}$	1064	7656
Azidocilline-M/artifact MEAC		1980	272$_{40}$	230$_{91}$	215$_{38}$	100$_{53}$	97$_{100}$	478	7651
Azidocilline-M/artifact MEPFP		1750	376$_{100}$	317$_{65}$	246$_{70}$	243$_{62}$	215$_{53}$	939	7657
Azinphos-ethyl		2570	345$_{1}$	186$_{10}$	160$_{77}$	132$_{100}$	77$_{57}$	826	1380
Azinphos-methyl	G P-I U+UHYAC	2460	317$_{1}$	160$_{35}$	132$_{48}$	93$_{34}$	77$_{60}$	695	1412
Aziprotryne		1765	225$_{100}$	182$_{67}$	139$_{85}$	115$_{42}$	68$_{75}$	290	3506
Azosemide-M (N-dealkyl-) -SO2NH ME		1960	209$_{64}$	180$_{44}$	152$_{100}$	138$_{83}$	102$_{55}$	239	4279
Azosemide-M (thiophenecarboxylic acid)		<1000*	128$_{5}$	127$_{55}$	111$_{100}$	83$_{7}$		110	4282
Azosemide-M (thiophenecarboxylic acid) glycine conjugate ME		1720	199$_{10}$	167$_{5}$	140$_{17}$	111$_{100}$	83$_{7}$	216	4281
Azosemide-M (thiophenylmethanol)		<1000*	114$_{100}$	97$_{62}$	85$_{78}$	81$_{24}$		104	4280
Baclofen ME	PME UME	1715	196$_{30}$	138$_{100}$	103$_{19}$	77$_{10}$	63$_{8}$	297	4457
Baclofen -H2O	U	1990	195$_{48}$	138$_{100}$	103$_{26}$	77$_{14}$	63$_{7}$	206	4456
Baclofen -H2O AC	UMEAC	1975	237$_{5}$	195$_{5}$	138$_{100}$	103$_{23}$	77$_{23}$	332	4458
Bambuterol		2930	367$_{2}$	352$_{1}$	282$_{2}$	86$_{89}$	72$_{100}$	911	7546
Bambuterol AC		2900	409$_{4}$	394$_{2}$	334$_{6}$	86$_{100}$	72$_{120}$	1039	7548
Bambuterol TFA		2395	389$_{4}$	267$_{38}$	212$_{2}$	153$_{4}$	72$_{100}$	1131	7553
Bambuterol TMS		2600	439$_{3}$	354$_{9}$	282$_{5}$	86$_{94}$	72$_{130}$	1097	7554
Bambuterol -C2H8 AC		2760	335$_{8}$	293$_{1}$	221$_{1}$	72$_{100}$	55$_{16}$	783	7549
Bambuterol formyl artifact		2930	379$_{1}$	364$_{14}$	334$_{12}$	99$_{65}$	72$_{100}$	949	7547
Bambuterol HY2AC		2500	380$_{2}$	305$_{7}$	211$_{2}$	86$_{100}$	72$_{63}$	953	7550
Bambuterol HY3AC		2200	351$_{1}$	336$_{1}$	276$_{10}$	234$_{3}$	86$_{70}$	849	7551
Bamethan 3AC		2330	335$_{1}$	275$_{67}$	233$_{100}$	191$_{82}$	148$_{61}$	784	1402
Bamethan 3TMS		1865	410$_{3}$	267$_{4}$	158$_{100}$	116$_{17}$	73$_{90}$	1071	5483
Bamethan formyl artifact		2020	148$_{7}$	120$_{7}$	107$_{11}$	98$_{84}$	57$_{100}$	278	4654
Bamethan -H2O 2AC	U+UHYAC	2310	275$_{67}$	233$_{100}$	191$_{82}$	148$_{61}$	98$_{53}$	490	1385
Bamipine	G P U	2250	280$_{4}$	182$_{13}$	97$_{100}$	91$_{61}$	70$_{39}$	519	28
Bamipine-M (HO-)	UHY	2580	296$_{46}$	198$_{40}$	97$_{100}$	91$_{75}$	70$_{70}$	600	2139
Bamipine-M (HO-) AC	U+UHYAC	2620	338$_{20}$	240$_{16}$	97$_{100}$	91$_{44}$	70$_{55}$	799	2138
Bamipine-M (N-dealkyl-)	UHY	1930	183$_{25}$	106$_{16}$	91$_{100}$	77$_{25}$		182	2065
Bamipine-M (N-dealkyl-) AC	U+UHYAC	2080	225$_{22}$	183$_{23}$	106$_{10}$	91$_{100}$	77$_{25}$	291	2066
Bamipine-M (N-dealkyl-HO-)	UHY	1920	199$_{53}$	163$_{21}$	91$_{100}$	76$_{24}$	65$_{34}$	217	2143
Bamipine-M (N-dealkyl-HO-) 2AC	U+UHYAC	2340	283$_{18}$	241$_{16}$	199$_{34}$	91$_{100}$	65$_{22}$	528	2072
Bamipine-M (nor-) AC	U+UHYAC	2675	308$_{20}$	182$_{43}$	91$_{100}$	77$_{33}$		657	2141
Bamipine-M (nor-HO-) 2AC	U+UHYAC	3020	366$_{34}$	324$_{7}$	240$_{26}$	199$_{19}$	91$_{100}$	905	2142
Barban ME		2335	271$_{25}$	256$_{100}$	152$_{11}$	111$_{26}$	75$_{18}$	472	4091
Barban-M/artifact (chloroaniline) AC	U+UHYAC	1580	169$_{31}$	127$_{100}$	111$_{2}$	99$_{8}$		155	6593
Barban-M/artifact (chloroaniline) HFB		1310	323$_{60}$	304$_{8}$	154$_{100}$	126$_{68}$	111$_{59}$	724	6607
Barban-M/artifact (chloroaniline) ME		1100	141$_{74}$	140$_{100}$	111$_{9}$	105$_{11}$	77$_{31}$	120	4089
Barban-M/artifact (chloroaniline) TFA		1125	223$_{99}$	154$_{100}$	126$_{51}$	111$_{55}$	69$_{42}$	282	4124
Barban-M/artifact (chloroaniline) 2ME		1180	155$_{62}$	154$_{100}$	140$_{11}$	118$_{16}$	75$_{17}$	134	4090
Barban-M/artifact (HOOC-) ME		1500	185$_{100}$	153$_{64}$	140$_{87}$	99$_{46}$	59$_{69}$	185	4123
Barbital	G P U UHY U+UHYAC	1500	156$_{100}$	141$_{97}$	112$_{20}$	98$_{22}$	83$_{12}$	184	72
Barbital ME	P G U UHY U+UHYAC	1455	170$_{100}$	155$_{97}$	126$_{12}$	112$_{34}$		215	73
Barbital 2ME		1420	184$_{96}$	169$_{100}$	126$_{38}$	112$_{25}$		249	74
Barbituric acid 3ME		1645	170$_{100}$	113$_{12}$	98$_{23}$	82$_{47}$	55$_{75}$	157	75
Barnidipine	G	4140	491$_{1}$	315$_{2}$	210$_{17}$	159$_{100}$	91$_{50}$	1158	4507
2,3-BDB		1550	193$_{1}$	164$_{2}$	135$_{2}$	77$_{5}$	58$_{100}$	200	5414
2,3-BDB AC		1895	235$_{7}$	176$_{43}$	135$_{10}$	100$_{21}$	58$_{100}$	323	5504
2,3-BDB HFB		1660	389$_{23}$	345$_{8}$	254$_{59}$	176$_{100}$	135$_{57}$	982	5505
2,3-BDB PFP		1615	339$_{4}$	204$_{7}$	176$_{43}$	135$_{100}$	119$_{14}$	801	5544
2,3-BDB TFA		1705	289$_{30}$	176$_{100}$	154$_{74}$	135$_{62}$	77$_{24}$	556	5506
2,3-BDB TMS		1670	250$_{11}$	236$_{14}$	135$_{23}$	130$_{100}$	73$_{61}$	445	5603
2,3-BDB formyl artifact		1575	205$_{9}$	176$_{3}$	135$_{9}$	77$_{9}$	70$_{100}$	228	5415
BDB		1570	193$_{1}$	164$_{2}$	136$_{1}$	77$_{6}$	58$_{100}$	201	3253
BDB AC	U+UHYAC	1950	235$_{3}$	176$_{23}$	162$_{14}$	100$_{13}$	58$_{100}$	324	3262
BDB HFB		1690	389$_{4}$	254$_{6}$	176$_{42}$	135$_{100}$	77$_{11}$	982	5288
BDB PFP		1700	339$_{4}$	204$_{7}$	176$_{43}$	135$_{100}$	119$_{14}$	801	5287
BDB TFA		1705	289$_{4}$	176$_{33}$	154$_{11}$	135$_{100}$	77$_{12}$	557	5286
BDB formyl artifact		1585	205$_{8}$	176$_{3}$	135$_{25}$	77$_{16}$	70$_{100}$	228	3246
BDB intermediate-1 (1-(1,3-benzodioxol-5-yl)-butan-1-ol)		1560*	194$_{17}$	151$_{100}$	123$_{21}$	93$_{72}$	65$_{37}$	205	3290
BDB intermediate-1 AC		1670*	236$_{21}$	193$_{10}$	151$_{100}$	135$_{30}$	93$_{16}$	328	3294
BDB intermediate-2		1385*	176$_{59}$	131$_{100}$	103$_{80}$	77$_{41}$	63$_{26}$	163	3291
BDB intermediate-3 (1-(1,3-benzodioxol-5-yl)-butan-2-one)		1525*	192$_{20}$	135$_{100}$	105$_{6}$	77$_{22}$	57$_{21}$	197	3292
BDB precursor (piperonal)		1160*	150$_{80}$	149$_{100}$	121$_{34}$	91$_{12}$	63$_{49}$	127	3275
BDB-M (demethylenyl-) 3AC	U+UHYAC	2235	307$_{1}$	248$_{2}$	164$_{6}$	100$_{29}$	58$_{100}$	649	5551
BDB-M (demethylenyl-methyl-) 2AC	U+UHYAC	2140	279$_{4}$	220$_{32}$	178$_{74}$	100$_{24}$	58$_{100}$	512	5550
BDMPEA		1785	259$_{15}$	230$_{100}$	215$_{29}$	199$_{10}$	77$_{37}$	415	3254
BDMPEA AC		2180	301$_{15}$	242$_{100}$	229$_{31}$	199$_{12}$	148$_{39}$	617	3267
BDMPEA HFB		2030	455$_{32}$	242$_{98}$	229$_{28}$	199$_{20}$	148$_{100}$	1120	6941
BDMPEA PFP		1995	405$_{36}$	242$_{100}$	229$_{85}$	199$_{33}$	148$_{35}$	1029	6936
BDMPEA TFA		2000	355$_{42}$	242$_{99}$	229$_{83}$	199$_{37}$	148$_{36}$	864	6931
BDMPEA TMS		1935	331$_{1}$	272$_{3}$	229$_{2}$	102$_{100}$	73$_{50}$	761	6925
BDMPEA 2AC		2230	343$_{7}$	242$_{99}$	229$_{32}$	201$_{12}$	148$_{29}$	819	6924
BDMPEA 2TMS		2195	403$_{1}$	388$_{14}$	272$_{7}$	207$_{11}$	174$_{100}$	1025	6926

Table 1-8-1: Compounds in order of names — Benazepril-M/artifact (deethyl-HO-) iso-1 3ET

Name	Detected	RI	Typical ions and intensities					Page	Entry
BDMPEA formyl artifact		1840	271 $_{12}$	240 $_{65}$	229 $_{27}$	199 $_{11}$	77 $_{24}$	473	3245
BDMPEA formyl artifact		1840	271 $_{12}$	240 $_{65}$	229 $_{27}$	199 $_{11}$	42 $_{100}$	473	5522
BDMPEA intermediate-1 (2,5-dimethoxyphenyl-2-nitroethene)		1900	209 $_{100}$	162 $_{51}$	147 $_{52}$	133 $_{61}$	77 $_{62}$	240	3286
BDMPEA intermediate-2 (2,5-dimethoxyphenethylamine)		1630	181 $_{15}$	162 $_{33}$	152 $_{100}$	137 $_{47}$	121 $_{27}$	177	3287
BDMPEA intermediate-2 (2,5-dimethoxyphenethylamine)		1630	181 $_{15}$	162 $_{33}$	152 $_{100}$	137 $_{47}$	44 $_{95}$	178	5523
BDMPEA intermediate-2 (2,5-dimethoxyphenethylamine) AC		1935	223 $_{15}$	164 $_{100}$	149 $_{29}$	121 $_{31}$	91 $_{15}$	284	3288
BDMPEA intermediate-2 (2,5-dimethoxyphenethylamine) formyl artifact		1540	193 $_{12}$	162 $_{100}$	151 $_{30}$	121 $_{43}$	91 $_{29}$	200	3293
BDMPEA intermediate-2 (2,5-dimethoxyphenethylamine) formyl artifact		1540	193 $_{12}$	162 $_{100}$	151 $_{30}$	121 $_{43}$	42 $_{64}$	200	5524
BDMPEA precursor (2,5-dimethoxybenzaldehyde)		1345*	166 $_{100}$	151 $_{39}$	120 $_{36}$	95 $_{61}$	63 $_{62}$	150	3278
BDMPEA-M (deamino-COOH) ME		2030*	288 $_{100}$	273 $_{12}$	241 $_8$	229 $_{67}$	199 $_{19}$	550	7212
BDMPEA-M (deamino-di-HO-) 2AC	U+UHYAC	2230*	360 $_{13}$	300 $_{27}$	258 $_{43}$	245 $_{100}$	138 $_{17}$	882	7214
BDMPEA-M (deamino-di-HO-) 2TFA		1790*	468 $_{100}$	354 $_{14}$	341 $_{77}$	311 $_{14}$	276 $_{76}$	1136	7208
BDMPEA-M (deamino-HO-) AC	U+UHYAC	2300*	302 $_{20}$	242 $_{100}$	227 $_{25}$	183 $_{33}$	148 $_{42}$	622	7198
BDMPEA-M (deamino-HO-) TFA		1880*	356 $_{31}$	341 $_{10}$	242 $_{64}$	229 $_{46}$	148 $_{46}$	867	7209
BDMPEA-M (deamino-oxo-)		2020*	258 $_{35}$	229 $_{100}$	215 $_9$	199 $_{35}$	186 $_8$	411	7215
BDMPEA-M (O-demethyl- N-acetyl-) iso-1 TFA		2090	383 $_9$	324 $_{100}$	255 $_7$	148 $_{21}$	72 $_{23}$	961	7204
BDMPEA-M (O-demethyl- N-acetyl-) iso-2 TFA		2130	383 $_7$	311 $_{11}$	324 $_{101}$	227 $_{11}$	72 $_{17}$	961	7205
BDMPEA-M (O-demethyl-) iso-1 2AC	U+UHYAC	2410	329 $_4$	287 $_{33}$	228 $_{101}$	215 $_{16}$	165 $_5$	752	7196
BDMPEA-M (O-demethyl-) iso-1 2TFA		1900	437 $_{12}$	324 $_{100}$	311 $_{10}$	255 $_7$	148 $_{32}$	1092	7206
BDMPEA-M (O-demethyl-) iso-2 2AC	U+UHYAC	2440	329 $_8$	287 $_{21}$	228 $_{100}$	215 $_{17}$	72 $_{13}$	752	7197
BDMPEA-M (O-demethyl-) iso-2 2TFA		1950	437 $_6$	324 $_{100}$	311 $_{32}$	253 $_3$	227 $_{15}$	1092	7207
BDMPEA-M (O-demethyl-deamino-COOH) MEAC		2120*	316 $_{14}$	274 $_{101}$	242 $_{92}$	214 $_{88}$	186 $_{18}$	691	7213
BDMPEA-M (O-demethyl-deamino-COOH) METFA		1890*	370 $_{100}$	311 $_{46}$	257 $_{55}$	241 $_{60}$	148 $_{21}$	918	7211
BDMPEA-M (O-demethyl-deamino-COOH) -H2O	U+UHYAC	1980*	242 $_{56}$	214 $_{70}$	186 $_{100}$			349	7203
BDMPEA-M (O-demethyl-deamino-di-HO-) 3AC	U+UHYAC	2280*	388 $_6$	346 $_{43}$	286 $_{55}$	244 $_{67}$	231 $_{100}$	978	7201
BDMPEA-M (O-demethyl-deamino-HO-) 2TFA		1800*	438 $_{31}$	341 $_{16}$	324 $_{100}$	311 $_{35}$	227 $_{55}$	1094	7210
BDMPEA-M (O-demethyl-deamino-HO-) iso-1 2AC	U+UHYAC	2160*	330 $_5$	288 $_{17}$	270 $_6$	228 $_{100}$	213 $_{12}$	757	7199
BDMPEA-M (O-demethyl-deamino-HO-) iso-2 2AC	U+UHYAC	2180*	330 $_4$	288 $_{15}$	246 $_{10}$	228 $_{100}$	213 $_{14}$	757	7200
BDMPEA-M (O-demethyl-deamino-HO-oxo-) 2AC	U+UHYAC	2160*	344 $_{11}$	302 $_{74}$	260 $_{46}$	242 $_{100}$	214 $_{25}$	823	7202
BDMPEA-M (O-demethyl-N-acetyl-) iso-1 AC	U+UHYAC	2410	329 $_4$	287 $_{33}$	228 $_{101}$	215 $_{16}$	165 $_5$	752	7196
BDMPEA-M (O-demethyl-N-acetyl-) iso-2 AC	U+UHYAC	2440	329 $_8$	287 $_{21}$	228 $_{100}$	215 $_{17}$	72 $_{13}$	752	7197
Beclamide	U	1720	197 $_{20}$	162 $_{15}$	148 $_{10}$	106 $_{80}$	91 $_{100}$	212	76
Beclamide TMS		1690	269 $_3$	254 $_7$	234 $_{37}$	91 $_{100}$	73 $_{47}$	464	5468
Beclamide artifact (-HCl)		1680	161 $_{90}$	117 $_{39}$	106 $_{46}$	91 $_{53}$	55 $_{100}$	139	104
Beclamide artifact (-HCl) TMS		1160	233 $_{28}$	232 $_{30}$	218 $_{22}$	91 $_{100}$	73 $_{64}$	316	5469
Beclobrate		2430*	346 $_{22}$	273 $_{18}$	218 $_{83}$	183 $_{100}$	125 $_{35}$	832	2247
Befunolol		2610	291 $_9$	276 $_2$	247 $_5$	161 $_3$	72 $_{100}$	569	2400
Befunolol TMSTFA		2430	444 $_1$	402 $_3$	284 $_{95}$	129 $_{45}$	73 $_{80}$	1126	6181
Befunolol formyl artifact		2630	303 $_3$	288 $_4$	247 $_8$	161 $_2$	72 $_{100}$	628	2401
Befunolol -H2O AC		2730	315 $_{13}$	230 $_{22}$	140 $_{41}$	98 $_{100}$	56 $_{67}$	689	2427
Behenic acid ME		2460*	354 $_{13}$	311 $_6$	143 $_{21}$	87 $_{63}$	74 $_{100}$	863	2669
Bemegride		1350	155 $_{18}$	127 $_{24}$	113 $_{26}$	82 $_{37}$	55 $_{100}$	75	77
Bemetizide 2ME		3100	429 $_{70}$	333 $_{36}$	324 $_{100}$			1077	2854
Bemetizide 3ME		3070	445 $_{36}$	443 $_{100}$	348 $_{17}$	338 $_{75}$	240 $_{36}$	1104	2855
Bemetizide 4ME	UEXME	3700	457 $_1$	352 $_{100}$	244 $_{15}$	145 $_6$	105 $_7$	1123	6845
Bemetizide -SO2NH ME		2800	336 $_8$	240 $_8$	231 $_{100}$	105 $_9$	77 $_8$	787	2853
Benactyzine		2270	327 $_2$	239 $_9$	182 $_{19}$	105 $_{50}$	86 $_{100}$	745	1391
Benactyzine TMS		2230	399 $_1$	384 $_2$	255 $_{58}$	100 $_{64}$	86 $_{100}$	1016	6272
Benactyzine-M (HOOC-) ME		1840*	242 $_2$	183 $_{100}$	105 $_{72}$	77 $_{65}$		351	78
Benazepril ET	UET	3080	452 $_4$	406 $_9$	379 $_{100}$	218 $_{23}$	91 $_{16}$	1117	4722
Benazepril ME	G PME UME	3030	438 $_3$	392 $_8$	365 $_{100}$	204 $_{45}$	91 $_{53}$	1095	4714
Benazepril TMS		3070	496 $_5$	423 $_{100}$	262 $_{35}$	91 $_{42}$	73 $_{38}$	1162	4973
Benazepril 2ET	UET	3040	480 $_2$	407 $_{100}$	289 $_{10}$	218 $_{42}$	91 $_{30}$	1148	4723
Benazepril 2ME	UME	3015	452 $_1$	379 $_{100}$	204 $_{60}$	144 $_{37}$	91 $_{70}$	1117	4715
Benazepril isopropylester		3165	466 $_3$	420 $_6$	393 $_{100}$	232 $_{32}$	91 $_{44}$	1134	4724
Benazeprilate 2ET	UET	3080	452 $_4$	406 $_9$	379 $_{100}$	218 $_{23}$	91 $_{16}$	1117	4722
Benazeprilate 2ME	UME	2975	424 $_4$	392 $_{12}$	365 $_{100}$	204 $_{68}$	91 $_{60}$	1068	4716
Benazeprilate 2TMS	UTMS	3130	540 $_1$	525 $_3$	423 $_{100}$	262 $_{22}$	73 $_{42}$	1181	4974
Benazeprilate 3ET	UET	3040	480 $_2$	407 $_{100}$	289 $_{10}$	218 $_{42}$	91 $_{30}$	1148	4723
Benazeprilate 3ME	UME	2985	438 $_2$	379 $_{100}$	204 $_{39}$	144 $_{12}$	91 $_{16}$	1095	4717
Benazeprilate-M (HO-) iso-1 2ET	UET	3330	496 $_4$	423 $_{100}$	361 $_{18}$	218 $_{22}$	135 $_{33}$	1162	4725
Benazeprilate-M (HO-) iso-1 3ME	UME	3160	454 $_4$	395 $_{100}$	333 $_{39}$	204 $_{60}$	121 $_{65}$	1119	4718
Benazeprilate-M (HO-) iso-2 2ET	UET	3330	496 $_6$	423 $_{71}$	361 $_{20}$	218 $_{31}$	135 $_{80}$	1162	4726
Benazeprilate-M (HO-) iso-2 3ME	UME	3235	454 $_3$	395 $_{29}$	333 $_{17}$	204 $_{34}$	121 $_{50}$	1119	4719
Benazeprilate-M (HO-) iso-2 4ME	UME	3240	468 $_5$	409 $_{89}$	261 $_{17}$	347 $_{11}$	204 $_{60}$	1137	4721
Benazepril-M (HO-) iso-1 2ET	UET	3330	496 $_4$	423 $_{100}$	361 $_{18}$	218 $_{22}$	135 $_{33}$	1162	4725
Benazepril-M (HO-) iso-1 4ME	UME	3165	468 $_4$	409 $_{100}$	204 $_{65}$	121 $_{25}$	347 $_8$	1137	4720
Benazepril-M (HO-) iso-2 2ET	UET	3330	496 $_6$	423 $_{71}$	361 $_{20}$	218 $_{31}$	135 $_{80}$	1162	4726
Benazepril-M/artifact (deethyl-) 2ET	UET	3080	452 $_4$	406 $_9$	379 $_{100}$	218 $_{23}$	91 $_{16}$	1117	4722
Benazepril-M/artifact (deethyl-) 2ME	UME	2975	424 $_4$	392 $_{12}$	365 $_{100}$	204 $_{68}$	91 $_{60}$	1068	4716
Benazepril-M/artifact (deethyl-) 2TMS	UTMS	3130	540 $_1$	525 $_3$	423 $_{100}$	262 $_{22}$	73 $_{42}$	1181	4974
Benazepril-M/artifact (deethyl-) 3ET	UET	3040	480 $_2$	407 $_{100}$	289 $_{10}$	218 $_{42}$	91 $_{30}$	1148	4723
Benazepril-M/artifact (deethyl-) 3ME	UME	2985	438 $_2$	379 $_{100}$	204 $_{39}$	144 $_{12}$	91 $_{16}$	1095	4717
Benazepril-M/artifact (deethyl-HO-) iso-1 3ET	UET	3330	496 $_4$	423 $_{100}$	361 $_{18}$	218 $_{22}$	135 $_{33}$	1162	4725

Benazepril-M/artifact (deethyl-HO-) iso-1 3ME Table 1-8-1: Compounds in order of names

Name	Detected	RI	Typical ions and intensities					Page	Entry
Benazepril-M/artifact (deethyl-HO-) iso-1 3ME	UME	3160	454_4	395_{100}	333_{39}	204_{60}	121_{65}	1119	4718
Benazepril-M/artifact (deethyl-HO-) iso-1 4ME	UME	3165	468_4	409_{100}	204_{65}	121_{25}	347_8	1137	4720
Benazepril-M/artifact (deethyl-HO-) iso-2 3ET	UET	3330	496_6	423_{71}	361_{20}	218_{31}	135_{80}	1162	4726
Benazepril-M/artifact (deethyl-HO-) iso-2 3ME	UME	3235	454_3	395_{29}	333_{17}	204_{34}	121_{50}	1119	4719
Benazepril-M/artifact (deethyl-HO-) iso-2 4ME	UME	3240	468_5	409_{89}	261_{17}	347_{11}	204_{60}	1137	4721
Benazolin		2055	243_{61}	198_{59}	170_{100}	134_{48}	108_{16}	353	3623
Benazolin ME		2000	257_{61}	198_{59}	170_{100}	134_{48}	108_{16}	408	3624
Benazolin-ethyl		2045	271_{40}	198_{54}	170_{100}	134_{47}	108_{20}	472	3625
Bencyclane	G U	2120	289_1	198_6	102_{34}	86_{44}	58_{100}	561	79
Bencyclane-M (bis-nor-) AC	UAAC	2545	100_{100}	72_{24}				631	2306
Bencyclane-M (bis-nor-HO-) iso-1 2AC	UAAC	2670	114_{19}	100_{100}	91_{15}	72_{37}		889	2307
Bencyclane-M (bis-nor-HO-) iso-2 2AC	UAAC	2700	114_5	100_{100}	91_{12}	72_{26}		889	2308
Bencyclane-M (deamino-di-HO-) iso-1 2AC	UAAC	2640*	129_2	101_{100}	91_9	73_7		892	2310
Bencyclane-M (deamino-di-HO-) iso-2 2AC	UAAC	2660*	129_1	101_{100}	91_8	73_6		892	2311
Bencyclane-M (deamino-HO-) AC	UAAC	2345*	128_1	101_{100}	91_{13}	73_6		636	2309
Bencyclane-M (deamino-HO-oxo-) iso-1 2AC	UAAC	2440*	115_2	101_{100}	91_{19}			704	2312
Bencyclane-M (deamino-HO-oxo-) iso-2 2AC	UAAC	2560*	129_1	101_{100}	91_{10}			704	2313
Bencyclane-M (HO-) iso-1	U	2350	214_{18}	102_{69}	86_{56}	58_{100}		642	2297
Bencyclane-M (HO-) iso-1 AC	U+UHYAC	2420	256_{12}	129_1	102_{62}	86_{57}	58_{100}	837	2301
Bencyclane-M (HO-) iso-2	P U	2370	214_{12}	185_5	102_{52}	86_{39}	58_{100}	642	80
Bencyclane-M (HO-) iso-2 AC	U+UHYAC	2430	256_4	117_2	102_{51}	86_{48}	58_{100}	837	2302
Bencyclane-M (HO-oxo-) -H2O HYAC	U+UHYAC	1920*	258_{13}	227_{58}	190_{81}	129_{54}	91_{100}	414	2318
Bencyclane-M (HO-oxo-) HY	UHY	2280*	234_{71}	190_{29}	147_{23}	107_{100}	77_{24}	320	2320
Bencyclane-M (HO-oxo-) HY2AC	U+UHYAC	2240*	318_{17}	276_{100}	229_{15}	187_{27}	107_{15}	703	2317
Bencyclane-M (HO-oxo-) HYAC	U+UHYAC	2080*	276_{13}	234_{50}	206_{53}	127_{30}	107_{100}	496	2319
Bencyclane-M (nor-)	U	2130	198_3	184_{22}	91_{32}	88_{37}	72_{100}	493	2300
Bencyclane-M (nor-) AC	UAAC	2570	130_2	114_{100}	91_{13}	86_{14}		700	2303
Bencyclane-M (nor-HO-) iso-1 2AC	U+UHYAC	2690	130_2	114_{100}	91_{15}	86_{27}		937	2304
Bencyclane-M (nor-HO-) iso-2 2AC	U+UHYAC	2730	130_2	114_{100}	91_7	86_{11}		937	2305
Bencyclane-M (oxo-) iso-1	U	2340	212_6	102_{40}	86_{21}	58_{100}		631	2298
Bencyclane-M (oxo-) iso-1 HY	UHY	1380*	218_{34}	190_6	127_{42}	99_{30}	91_{100}	265	81
Bencyclane-M (oxo-) iso-1 HYAC	U+UHYAC	1750*	260_{12}	200_{33}	171_{100}	109_{56}	91_{71}	422	83
Bencyclane-M (oxo-) iso-2	U	2380	303_1	212_{20}	102_{63}	86_{38}	58_{100}	631	2299
Bencyclane-M (oxo-) iso-2 HY	UHY	1415*	218_{57}	189_{10}	107_{100}	77_{14}		265	82
Bencyclane-M (oxo-) iso-2 HYAC	U+UHYAC	1780*	260_{26}	218_{100}	189_{10}	107_{39}		422	2316
Bendiocarb		1640	223_6	166_{33}	151_{100}	126_{47}	58_{22}	284	3912
Bendiocarb TFA		1560	319_{52}	247_{100}	222_{19}	125_{40}	69_{88}	706	3607
Bendiocarb -C2H3NO		1110*	166_{42}	151_{92}	126_{100}	108_{18}	80_{15}	150	3913
Bendiocarb -C2H3NO TFA		<1000*	262_{47}	247_{100}	205_{34}	125_{51}	79_{25}	429	4131
Bendroflumethiazide 3ME		3360	463_{16}	372_{100}	264_9	260_6	91_{10}	1130	3106
Bendroflumethiazide 4ME		3360	477_1	386_{100}	278_{20}	145_3	91_7	1145	6890
Benfluorex		2175	350_1	216_6	192_{100}	159_{24}	105_6	848	4707
Benfluorex AC		2530	374_1	234_{57}	192_{100}	159_{22}	105_{51}	994	4709
Benfluorex ME		2220	364_1	230_8	206_{100}	159_{33}	149_{82}	900	4708
Benfluorex-M (-COOH) MEAC	U+UHYAC	1870	298_1	258_2	158_{38}	116_{100}	56_{37}	697	4711
Benfluorex-M (deamino-oxo-HO-) enol 2AC		2150*	303_1	216_{10}	159_{40}	143_{29}	101_{100}	623	4712
Benfluorex-M (hippuric acid)	U	1745	179_1	161_2	135_{22}	105_{100}	77_9	169	96
Benfluorex-M (hippuric acid) ME	UME	1660	193_5	161_7	134_{19}	105_{100}	77_{45}	199	97
Benfluorex-M (hippuric acid) TMS	UTMS	1925	251_2	236_8	206_{71}	105_{100}	73_{85}	383	5813
Benfluorex-M (hippuric acid) 2TMS	UTMS	2070	323_{10}	308_{16}	280_{11}	206_{50}	105_{100}	727	5812
Benfluorex-M (N-dealkyl-) AC	U+UHYAC	1510	245_2	226_5	186_6	159_{13}	86_{100}	358	782
Benfluorex-M/artifact (alcohol) 2AC	U+UHYAC	1890	312_1	250_5	172_{50}	130_{100}	87_{35}	764	4710
Benfluorex-M/artifact (benzoic acid)	P U UHY	1235*	122_{77}	105_{100}	77_{72}			107	95
Benfluorex-M/artifact (benzoic acid) ME	P(ME)	1180*	136_{30}	105_{99}	77_{73}			116	1211
Benfluorex-M/artifact (benzoic acid) TBDMS		1295*	221_1	179_{100}	135_{30}	105_{55}	77_{51}	330	6247
Benomyl artifact (desbutylcarbamoyl-) 2ME		1875	219_{37}	160_{100}	132_{23}	119_{16}	77_{13}	270	4078
Benomyl-M/artifact (aminobenzimidazole) 3ME		1715	175_{78}	160_{100}	146_{80}	131_{71}	119_{22}	162	4101
Benoxaprofen	P	2550	301_{52}	256_{100}	119_{31}	91_{47}	65_{25}	618	1458
Benoxaprofen ME		2485	315_{39}	256_{99}	119_{25}	91_{46}		688	1392
Benoxaprofen-M (HO-) ME	UME	2580	331_8	272_{93}	230_{100}	195_6	91_{12}	761	6286
Benperidol	G U+UHYAC	3440	381_2	363_{50}	230_{100}	109_{88}	82_{65}	957	84
Benperidol-M	U+UHYAC	1490*	180_{25}	125_{49}	123_{33}	95_{17}	56_{100}	174	85
Benperidol-M (N-dealkyl-)	UHY	2415	217_9	134_{99}	106_{42}	79_{87}		261	87
Benperidol-M (N-dealkyl-) AC	U+UHYAC	2770	259_{60}	216_{15}	134_{64}	125_{42}	82_{100}	417	89
Benperidol-M (N-dealkyl-) ME	UHY	2290	231_{12}	134_{99}	106_{37}	79_{81}		310	86
Benperidol-M (N-dealkyl-) 2AC	U+UHYAC	2750	301_{28}	259_{43}	134_{28}	125_{28}	82_{100}	620	88
Benproperine		2425	309_2	181_2	165_3	112_{100}	91_5	663	1749
Bentazone		2040	240_1	198_{61}	161_{31}	119_{100}	92_{43}	342	3626
Bentazone ME		1910	254_{23}	212_{100}	175_{22}	133_{48}	105_{83}	397	3628
Bentazone artifact		1675	178_{38}	120_{100}	92_{43}	65_{38}	58_{54}	168	3627
Benzaldehyde		<1000*	106_{76}	105_{75}	77_{100}	51_{48}		101	4215
Benzalkonium chloride compound-1 -C7H8Cl	G P U	1380	213_2	156_1	114_1	84_1	58_{100}	253	1057
Benzalkonium chloride compound-1 -CH3Cl	G P U	1965	289_3	160_1	134_{100}	91_{50}	58_5	561	1059
Benzalkonium chloride compound-2 -C7H8Cl	G P U	1595	241_1	170_1	128_1	84_1	58_{100}	349	1058

Table 1-8-1: Compounds in order of names   Benzoctamine-M (nor-HO-) iso-1 2AC

Name	Detected	RI	Typical ions and intensities					Page	Entry
Benzalkonium chloride compound-2 -CH3Cl	G P U	2150	317_1	253_3	206_3	134_{100}	91_{47}	700	1060
Benzamide		1400	121_{83}	105_{100}	77_{98}			107	90
Benzarone	UHY	2405*	266_{100}	251_{57}	223_{15}	121_{66}	93_{27}	448	1978
Benzarone AC	U+UHYAC	2405*	308_{100}	266_{81}	249_{80}	224_{27}	121_{58}	654	1986
Benzarone-M (di-HO-) 3AC	U+UHYAC	2550*	424_{28}	294_{26}	267_{100}	223_8	101_8	1067	2644
Benzarone-M (di-HO-) -H2O 2AC	U+UHYAC	2840*	364_{16}	322_{100}	280_{91}	173_{54}	121_{31}	896	2647
Benzarone-M (HO-) iso-1 2AC	U+UHYAC	2650*	366_5	324_{100}	282_{22}	187_{95}	121_{25}	903	2649
Benzarone-M (HO-) iso-2 2AC	U+UHYAC	2680*	366_{33}	324_{17}	282_{100}	265_{37}	137_{22}	904	2650
Benzarone-M (HO-) iso-3 2AC	U+UHYAC	2730*	366_{76}	324_{100}	282_{98}	265_{67}	121_{52}	904	2651
Benzarone-M (HO-) iso-4 2AC	U+UHYAC	2790*	366_{41}	324_{100}	282_{69}	265_{34}	121_{45}	904	2652
Benzarone-M (HO-ethyl-) -H2O AC	U+UHYAC	2440*	306_{49}	264_{100}	235_{30}	171_{14}	115_{21}	644	2643
Benzarone-M (HO-methoxy-) iso-1 2AC	U+UHYAC	2710*	396_9	354_{93}	312_{66}	187_{100}	145_{17}	1005	2654
Benzarone-M (HO-methoxy-) iso-2 2AC	U+UHYAC	2740*	396_{27}	354_{57}	312_{100}	197_4	120_{20}	1005	2655
Benzarone-M (HO-methoxy-) iso-3 2AC	U+UHYAC	2910*	396_{47}	354_{64}	312_{100}	269_{25}	151_{19}	1005	2656
Benzarone-M (HO-methoxy-) iso-4 2AC	U+UHYAC	2950*	396_{46}	354_{100}	312_{100}	187_{21}	151_{25}	1005	2657
Benzarone-M (methoxy-) AC	U+UHYAC	2570*	338_{84}	296_{100}	279_{55}	253_{54}	151_{57}	797	2645
Benzarone-M (oxo-) AC	U+UHYAC	2620*	322_{20}	280_{100}	237_{54}	187_{29}	121_{35}	721	2646
Benzatropine	G U	2315	307_4	167_{26}	140_{59}	124_{27}	83_{100}	651	91
Benzatropine HY	UHY	1645*	184_{45}	165_{14}	152_7	105_{100}	77_{63}	184	1333
Benzatropine HYAC	U+UHYAC	1700*	226_{20}	184_{20}	165_{100}	105_{14}	77_{35}	294	1241
Benzatropine HYME	UHY	1655*	198_{70}	167_{94}	121_{100}	105_{56}	77_{71}	215	6779
Benzbromarone	G U UHY	2750*	424_{99}	422_{51}	344_{31}	279_{32}	264_{58}	1064	1393
Benzbromarone AC	U+UHYAC	2820*	464_{17}	424_{100}	422_{51}	264_{28}	173_{15}	1132	2255
Benzbromarone ET		2760*	452_{100}	450_{54}	423_{19}	264_{74}	173_{46}	1113	2262
Benzbromarone ME		2730*	438_{100}	436_{52}	342_{14}	278_{57}	173_{21}	1091	2258
Benzbromarone-M (HO-aryl-) iso-1 2AC	U+UHYAC	2950*	522_1	482_{100}	480_{55}	440_{57}	187_{57}	1175	2659
Benzbromarone-M (HO-aryl-) iso-2 2AC	U+UHYAC	3080*	522_9	482_{66}	440_{100}	438_{53}	279_{32}	1175	2660
Benzbromarone-M (HO-ethyl-) -H2O AC	U+UHYAC	2850*	462_{17}	422_{100}	420_{51}	297_{65}	255_{45}	1130	2257
Benzbromarone-M (HO-methoxy-) 2AC	U+UHYAC-I	3120*	552_7	512_{56}	510_{28}	470_{90}	468_{49}	1184	2256
Benzbromarone-M (methoxy-) AC	U+UHYAC	3070*	494_5	454_{74}	452_{41}	372_{77}	284_{100}	1160	2661
Benzbromarone-M (oxo-) AC	U+UHYAC	2900*	478_3	438_{100}	436_{51}	395_{40}	187_{79}	1146	2261
Benzene		<1000*	78_{100}	51_{20}	39_{12}	63_3	50_{17}	95	1542
1,4-Benzenediamine	G	1280	108_{100}	91_3	80_{35}	53_{13}		102	5330
1,4-Benzenediamine ME		1000	122_{38}	108_{100}	93_6	80_{40}		108	5333
1,4-Benzenediamine 2AC	U+UHYAC	2690	192_{39}	150_{15}	108_{100}	80_{59}	52_{45}	198	5331
1,4-Benzenediamine 2HFB		1775	500_{57}	481_{16}	331_{12}	303_{100}	108_{51}	1164	5332
1,4-Benzenediamine 2ME		1060	136_{16}	122_{10}	108_{100}	93_6	80_{36}	116	5334
1,4-Benzenediamine 2PFP		1600	400_{65}	281_{17}	253_{82}	119_{28}	108_{100}	1017	5858
1,4-Benzenediamine 2TFA		1800	300_{100}	203_{59}	133_{16}	108_{54}	69_{30}	613	5397
Benzene-M (hydroquinone)	UHY	<1000*	110_{100}	81_{27}				103	814
Benzene-M (hydroquinone) 2AC	U+UHYAC	1395*	194_8	152_{26}	110_{100}			203	815
Benzene-M (hydroquinone) 2ME		<1000*	138_{56}	123_{100}	95_{54}	63_{22}		118	3282
Benzene-M (hydroxyhydroquinone)	UHY	1460*	126_{100}	109_{18}	79_{26}	53_9		110	3163
Benzene-M (hydroxyhydroquinone) 3AC	U+UHYAC	1710*	252_1	210_7	168_{46}	126_{100}	97_7	388	4336
Benzene-M (methoxyhydroquinone) 2AC	U+UHYAC	1450*	224_3	182_{23}	140_{100}	125_{71}	97_9	287	4337
Benzene-M (phenol)	UHY	<1000*	94_{100}	66_{10}				98	4219
Benzhydrol	UHY	1645*	184_{45}	165_{14}	152_7	105_{100}	77_{63}	184	1333
Benzhydrol AC	U+UHYAC	1700*	226_{20}	184_{20}	165_{100}	105_{14}	77_{35}	294	1241
Benzhydrol ME	UHY	1655*	198_{70}	167_{94}	121_{100}	105_{56}	77_{71}	215	6779
Benzil	U UHY U+UHYAC	1825*	210_3	105_{100}	77_{45}	51_{26}		244	1233
Benzilic acid ME		1840*	242_2	183_{100}	105_{72}	77_{65}		351	78
Benzil-M (HO-) AC	U+UHYAC	2160*	268_1	226_1	163_{31}	121_{100}	105_{30}	459	2546
Benzil-M (HO-) ME	UHYME	2290*	240_3	135_{100}	128_{10}	77_{10}		342	2545
Benzo[a]anthracene		2410*	228_{100}	164_{10}	131_{10}	114_{20}		300	3701
Benzo[a]pyrene		2775*	252_{100}	224_3	126_{17}	113_7		389	3703
Benzo[b]fluoranthene		2815*	252_{100}	224_3	126_{19}	113_{11}		389	3704
Benzo[g,h,i]perylene		3125*	276_{100}	138_{23}	124_4			495	3707
Benzo[k]fluoranthene		2750*	252_{100}	224_3	126_{19}	113_9		389	3702
Benzocaine	G	1820	165_{37}	137_{11}	120_{100}	92_{19}	65_{22}	147	1457
Benzocaine AC		1990	207_{63}	165_{67}	137_{27}	120_{100}	92_{18}	233	1440
Benzocaine TMS		1500	237_{72}	222_{100}	192_{50}	149_{61}	73_{83}	334	5486
Benzocaine-M (PABA) AC	U+UHYAC	2145	179_{31}	137_{100}	120_{92}	92_{16}	65_{24}	169	3298
Benzocaine-M (PABA) ME		1550	151_{55}	120_{100}	92_{28}	65_{26}		128	23
Benzocaine-M (PABA) MEAC		1985	193_{33}	151_{60}	120_{100}	92_{18}	65_{19}	199	24
Benzocaine-M (PABA) 2TMS		1645	281_{50}	236_7	148_{13}	73_{55}		522	5487
Benzoctamine	UHY	2070	249_2	218_{100}	203_{29}	191_{68}	178_{55}	377	94
Benzoctamine AC	U+UHYAC	2540	291_{39}	263_{100}	218_{83}	191_{77}		570	1245
Benzoctamine TMS		2240	306_1	218_4	191_9	116_{60}	73_{42}	720	5460
Benzoctamine-M (deamino-di-HO-) 2AC	U+UHYAC	2470*	336_7	266_{30}	249_{100}	191_{36}		787	1244
Benzoctamine-M (deamino-di-HO-methoxy-) 2AC	U+UHYAC	2685*	366_{27}	324_{51}	296_{100}	249_{56}	237_{52}	904	1246
Benzoctamine-M (deamino-HO-) AC	U+UHYAC	2145*	278_{29}	250_{100}	191_{86}			507	1242
Benzoctamine-M (HO-) 2AC	U+UHYAC	2890	349_{23}	321_{70}	279_{101}	207_{68}		843	1250
Benzoctamine-M (nor-) AC	U+UHYAC	2420	277_{12}	249_{100}	207_{17}	191_{13}	178_{13}	501	1243
Benzoctamine-M (nor-HO-) iso-1 2AC	U+UHYAC	2725	335_5	293_{33}	265_{100}			783	1247

Benzoctamine-M (nor-HO-) iso-2 2AC **Table 1-8-1:** Compounds in order of names

Name	Detected	RI	Typical ions and intensities					Page	Entry
Benzoctamine-M (nor-HO-) iso-2 2AC	U+UHYAC	2790	335_{11}	307_{29}	265_{100}	207_{16}		783	1248
Benzoctamine-M (nor-HO-methoxy-) 2AC	U+UHYAC	2875	365_{15}	323_{24}	295_{100}			901	1249
1-(1,3-Benzodioxol-6-yl)butane-2-yl-azane		1550	193_1	164_2	135_2	77_5	58_{100}	200	5414
1-(1,3-Benzodioxol-6-yl)butane-2-yl-azane AC		1895	235_7	176_{43}	135_{10}	100_{21}	58_{100}	323	5504
1-(1,3-Benzodioxol-6-yl)butane-2-yl-azane HFB		1660	389_{23}	345_8	254_{59}	176_{100}	135_{57}	982	5505
1-(1,3-Benzodioxol-6-yl)butane-2-yl-azane PFP		1615	339_4	204_7	176_{43}	135_{100}	119_{14}	801	5544
1-(1,3-Benzodioxol-6-yl)butane-2-yl-azane TFA		1705	289_{30}	176_{100}	154_{74}	135_{62}	77_{24}	556	5506
1-(1,3-Benzodioxol-6-yl)butane-2-yl-azane formyl artifact		1575	205_9	176_3	135_9	77_9	70_{100}	228	5415
1-(1,3-Benzodioxol-6-yl)butane-2-yl-dimethylazane		1660	192_3	135_2	96_4	86_{100}	71_8	278	5418
1-(1,3-Benzodioxol-6-yl)butane-2-yl-ethylazane		1670	192_5	135_5	86_{100}	77_6	58_{11}	278	5417
1-(1,3-Benzodioxol-6-yl)butane-2-yl-ethylazane AC		2000	263_3	192_4	176_{18}	128_{42}	86_{100}	435	5511
1-(1,3-Benzodioxol-6-yl)butane-2-yl-ethylazane HFB		1790	417_1	282_{100}	176_{25}	135_{24}	77_{17}	1056	5594
1-(1,3-Benzodioxol-6-yl)butane-2-yl-ethylazane PFP		1755	367_3	232_{100}	176_{16}	119_{30}	69_{10}	908	5595
1-(1,3-Benzodioxol-6-yl)butane-2-yl-ethylazane TFA		1780	317_5	182_{100}	176_{37}	154_{16}	135_{14}	697	5512
1-(1,3-Benzodioxol-6-yl)butane-2-yl-ethylazane TMS		1825	278_2	264_5	158_{100}	135_{16}	73_{48}	583	5596
1-(1,3-Benzodioxol-6-yl)butane-2-yl-methylazane		1610	178_3	135_3	89_4	72_{100}	57_7	234	5416
1-(1,3-Benzodioxol-6-yl)butane-2-yl-methylazane AC		1965	249_4	176_{17}	135_7	114_{35}	72_{100}	375	5507
1-(1,3-Benzodioxol-6-yl)butane-2-yl-methylazane HFB		1735	403_3	268_{100}	210_{17}	176_8	135_5	1025	5591
1-(1,3-Benzodioxol-6-yl)butane-2-yl-methylazane PFP		1710	353_8	218_{100}	176_{23}	160_{14}	135_{12}	857	5592
1-(1,3-Benzodioxol-6-yl)butane-2-yl-methylazane TFA		1725	303_4	176_{43}	168_{100}	135_{14}	110_{23}	627	5508
1-(1,3-Benzodioxol-6-yl)butane-2-yl-methylazane TMS		1730	264_2	250_5	144_{100}	135_{13}	73_{54}	514	5593
Benzoflavone		2810	272_{64}	244_8	170_{100}	122_{12}	114_{51}	478	6460
Benzofluorene		2220*	216_{100}	215_{70}	213_{19}	108_8	95_6	260	2568
Benzoic acid	P U UHY	1235*	122_{77}	105_{100}	77_{72}			107	95
Benzoic acid anhydride		1880*	226_7	198_{32}	105_{100}	77_{92}		293	1742
Benzoic acid butylester		1275*	178_2	123_{70}	105_{100}	77_{37}	56_{19}	167	98
Benzoic acid ethylester		1225*	150_{20}	122_{27}	105_{100}	77_{52}		127	99
Benzoic acid glycine conjugate	U	1745	179_1	161_2	135_{22}	105_{100}	77_9	169	96
Benzoic acid glycine conjugate TMS	UTMS	1925	251_1	236_8	206_{71}	105_{100}	73_{85}	383	5813
Benzoic acid glycine conjugate 2TMS	UTMS	2070	323_{10}	308_{16}	280_{11}	206_{50}	105_{100}	727	5812
Benzoic acid methylester	P(ME)	1180*	136_{30}	105_{99}	77_{73}			116	1211
Benzoic acid TBDMS		1295*	221_1	179_{100}	135_{30}	105_{55}	77_{51}	330	6247
Benzoic acid-M (glycine conjugate ME)	UME	1660	193_5	161_2	134_{19}	105_{100}	77_{45}	199	97
Benzophenone	U+UHYAC	1610*	182_{31}	152_3	105_{100}	77_{70}	51_{39}	180	1624
Benzoresorcinol	UHY	2280*	214_{61}	213_{83}	137_{100}	105_{21}	77_{33}	255	3660
Benzoresorcinol 2AC	U+UHYAC	2315*	298_3	256_{45}	213_{100}	137_{21}	77_{18}	605	3661
Benzquinamide	U U+UHYAC	2980	404_{17}	345_{30}	244_{44}	205_{100}	100_{10}	1029	1777
Benzquinamide artifact		2880	339_{100}	325_{10}	268_2	224_2		800	1778
Benzquinamide HY	UHY	3000	362_{12}	317_6	218_{27}	205_{100}	100_{10}	892	2135
Benzquinamide-M (N-deethyl-)	UHY U+UHYAC	2960	376_{17}	317_{70}	244_{100}	205_{75}	176_{16}	940	2136
Benzquinamide-M (O-demethyl-)	U+UHYAC	2990	390_7	303_{48}	272_{73}	230_{100}	191_{67}	985	2137
Benzthiazuron 2ME		1985	235_8	162_1	136_4	109_4	72_{100}	323	3941
Benzydamine	U UHY U+UHYAC	2400	309_2	225_5	91_{24}	85_{54}	58_{100}	662	1394
Benzydamine-M (deamino-HO-) AC	U+UHYAC	2450	324_1	273_1	162_1	101_{100}	91_{47}	731	4375
Benzydamine-M (HO-) AC	U+UHYAC	2670	367_3	283_4	265_7	85_{82}	58_{100}	911	4376
Benzydamine-M (nor-) AC	U+UHYAC	2780	337_1	114_{100}	91_{28}	86_{13}		794	1875
Benzydamine-M (nor-HO-) 2AC	U+UHYAC	3220	395_1	269_2	158_2	114_{100}	91_{25}	1003	4377
Benzydamine-M (O-dealkyl-) AC	U+UHYAC	2150	266_4	224_{47}	146_7	117_6	91_{100}	449	4378
Benzylacetamide	U+UHYAC	1410	149_{62}	106_{100}	91_{33}	79_{19}	77_{17}	126	5160
Benzylacetamide AC		1450	191_5	148_{37}	106_{100}	91_{24}	79_{17}	195	5161
Benzylalcohol		<1000*	108_{74}	107_{55}	91_{18}	79_{100}	77_{64}	102	4447
Benzylamine		<1000	107_{100}	91_{50}	79_{72}	77_{39}	65_{15}	102	100
Benzylamine AC	U+UHYAC	1410	149_{62}	106_{100}	91_{33}	79_{19}	77_{17}	126	5160
Benzylamine HFB		1220	303_{56}	184_6	169_6	134_{11}	91_{100}	626	6577
Benzylamine TFA		1155	203_{69}	134_{35}	104_{11}	91_{100}	69_{46}	223	6572
Benzylamine 2AC		1450	191_5	148_{37}	106_{100}	91_{24}	79_{17}	195	5161
Benzylamine artifact		1730	195_{22}	194_{26}	117_9	91_{100}	65_{20}	207	5159
Benzylbenzoate		1740*	212_{15}	194_6	105_{100}	91_{50}	77_{42}	248	4450
Benzylbutanoate		1065*	178_{17}	108_{63}	91_{100}	79_{14}	71_{36}	168	4448
Benzylbutylphthalate		2270*	312_1	206_{26}	149_{100}	91_{85}	65_{31}	676	3540
Benzylether		1600*	107_{13}	92_{100}	91_{79}	79_{18}	65_{25}	215	4449
N-Benzylethylenediamine 3TMS		2215	366_1	351_4	259_9	192_{100}	174_{29}	906	7635
N-Benzylidenebenzylamine		1730	195_{22}	194_{26}	117_9	91_{100}	65_{20}	207	5159
Benzylnicotinate		1800	213_{48}	168_3	106_{93}	91_{100}		252	1400
2-Benzylphenol		1680*	184_{99}	165_{33}	106_{53}	78_{40}		184	1395
4-Benzylphenol		1720*	184_{66}	165_{19}	106_{23}	91_{18}	77_1	184	1396
Benzylpiperazine		1530	176_9	146_5	134_{55}	91_{100}	56_{29}	164	5880
Benzylpiperazine AC		1915	218_4	146_{27}	132_{17}	91_{100}	85_{20}	266	5881
Benzylpiperazine HFB		1730	372_{13}	295_6	281_{15}	175_{13}	91_{100}	926	5884
Benzylpiperazine PFP		1690	322_{18}	245_9	231_{21}	175_{15}	91_{100}	722	5883
Benzylpiperazine TFA		1665	272_3	195_4	181_{10}	146_7	91_{100}	479	5882
Benzylpiperazine TMS		1860	248_{31}	157_{27}	102_{100}	91_{69}	73_{47}	372	5885
Benzylpiperazine-M (benzylamine)		<1000	107_{100}	91_{50}	79_{72}	77_{39}	65_{15}	102	100
Benzylpiperazine-M (benzylamine) AC	U+UHYAC	1410	149_{62}	106_{100}	91_{33}	79_{19}	77_{17}	126	5160

Table 1-8-1: Compounds in order of names Bisoprolol AC

Name	Detected	RI	Typical ions and intensities					Page	Entry
Benzylpiperazine-M (benzylamine) HFB		1220	303_{56}	184_6	169_6	134_{11}	91_{100}	626	6577
Benzylpiperazine-M (benzylamine) TFA		1155	203_{69}	134_{35}	104_{11}	91_{100}	69_{46}	223	6572
Benzylpiperazine-M (benzylamine) 2AC		1450	191_5	148_{37}	106_{100}	91_{24}	79_{17}	195	5161
Benzylpiperazine-M (deethylene-) HFB	U+UHYHFB	1870	302_5	295_{61}	190_8	119_{12}	91_{100}	831	7637
Benzylpiperazine-M (deethylene-) 2AC	U+UHYAC	2080	234_1	191_{49}	175_{17}	120_{83}	91_{100}	320	6507
Benzylpiperazine-M (deethylene-) 2HFB	U+UHYHFB	1705	542_1	524_1	345_{79}	226_6	91_{100}	1182	6576
Benzylpiperazine-M (deethylene-) 2PFP	U+UHYTFA	1875	311_9	295_{20}	190_3	119_2	91_{100}	1102	7636
Benzylpiperazine-M (deethylene-) 2TFA	U+UHYTFA	1670	342_1	324_{10}	245_{33}	126_8	91_{80}	816	6571
Benzylpiperazine-M (deethylene-) 3AC	U+UHYAC	2125	276_2	233_{57}	175_{11}	120_{130}	91_{99}	497	6513
Benzylpiperazine-M (HO-) iso-1 2AC	U+UHYAC	2245	276_9	204_{43}	149_{73}	107_{100}	85_{36}	497	6506
Benzylpiperazine-M (HO-) iso-1 2HFB	U+UHYHFB	1930	584_{10}	387_{31}	358_{20}	303_{100}	169_{95}	1192	6574
Benzylpiperazine-M (HO-) iso-1 2TFA	U+UHYTFA	1830	384_{30}	287_{23}	203_{100}	181_{37}	69_{95}	966	6569
Benzylpiperazine-M (HO-) iso-2 2AC	U+UHYAC	2290	276_3	204_{16}	149_{20}	107_{100}	85_{57}	497	6505
Benzylpiperazine-M (HO-) iso-2 2HFB	U+UHYHFB	1970	584_6	387_{42}	358_{25}	303_{100}	281_{36}	1192	6573
Benzylpiperazine-M (HO-) iso-2 2TFA	U+UHYTFA	1870	384_{100}	287_{47}	258_{29}	203_{92}	181_{28}	966	6568
Benzylpiperazine-M (HO-methoxy-) AC	U+UHYAC	2410	264_9	192_8	137_{100}	122_{18}	85_{42}	440	6509
Benzylpiperazine-M (HO-methoxy-) HFB	U+UHYHFB	2135	418_{22}	295_{15}	281_{27}	138_{87}	137_{100}	1058	6575
Benzylpiperazine-M (HO-methoxy-) TFA	U+UHYTFA	2120	318_8	181_{10}	137_{100}	122_{21}	69_{12}	703	6570
Benzylpiperazine-M (HO-methoxy-) 2AC	U+UHYAC	2380	306_2	234_9	179_{13}	137_{100}	85_{64}	645	6508
Benzylpiperazine-M (piperazine) 2AC		1750	170_{15}	85_{33}	69_{25}	56_{100}		157	879
Benzylpiperazine-M (piperazine) 2HFB		1290	478_3	459_9	309_{100}	281_{22}	252_{41}	1146	6634
Benzylpiperazine-M (piperazine) 2TFA		1005	278_{10}	209_{59}	152_{25}	69_{56}	56_{100}	505	4129
Betahistine AC		1575	178_{13}	135_{40}	106_{100}	93_{45}	86_{36}	168	5173
Betahistine impurity/artifact-1 AC		1700	192_{13}	149_{58}	120_{100}	107_{48}	86_{25}	197	5174
Betahistine impurity/artifact-2 AC		1755	206_{11}	163_{56}	134_{100}	121_{45}	86_{28}	230	5175
Betamethasone		2795*	312_{23}	268_7	160_{10}	122_{100}	91_{29}	992	5220
Betamethasone -2H2O		2910*	356_{21}	253_6	147_8	122_{100}	91_{19}	870	5221
Betaxolol	G	2355	307_2	292_5	263_{11}	100_4	72_{100}	652	1579
Betaxolol TMS		2220	364_4	263_5	188_3	101_{12}	72_{80}	951	5493
Betaxolol TMSTFA		2485	460_1	284_{100}	129_{46}	73_{69}	55_{88}	1144	6179
Betaxolol 2AC	U+UHYAC	2770	331_{15}	200_{100}	140_{52}	98_{71}	72_{49}	990	1582
Betaxolol 2TMS		2400	436_2	264_7	144_{80}	101_9	73_{61}	1115	5494
Betaxolol formyl artifact	P-I G	2410	319_{34}	304_{67}	127_{99}	112_{58}	55_{100}	712	1580
Betaxolol -H2O		2400	289_2	158_2	98_{30}	72_{100}	55_{16}	560	1583
Betaxolol -H2O AC		2720	331_3	288_5	140_{85}	98_{46}	55_{100}	766	1581
Betaxolol-M (O-dealkyl-) 3AC	U+UHYAC	2620	319_{12}	200_{100}	140_{55}	98_{50}	72_{60}	949	1585
Betaxolol-M (O-dealkyl-) -H2O 2AC	U+UHYAC	2570	319_{46}	234_{40}	217_{56}	140_{61}	98_{100}	709	1584
Bezafibrate	G U UHY U+UHYAC	3100	316_3	269_7	205_{53}	139_{44}	120_{100}	886	2494
Bezafibrate ME	PME UME	2910	375_3	316_{16}	220_{58}	139_{56}	120_{100}	936	1746
Bezafibrate -CO2	G P U+UHYAC	2800	317_1	275_2	139_{35}	120_{100}	107_{16}	697	1745
Bezafibrate-M (chlorobenzoic acid)	G UHY U+UHYAC	1400*	156_{61}	139_{10}	111_{54}	85_4	75_{39}	136	2726
Bifenox		2500	341_{52}	310_{17}	189_{28}	173_{37}	75_{70}	809	5685
Bifonazole		3070	310_3	243_{100}	228_8	165_{14}	91_6	667	2347
Binapacryl		2270	292_4	210_{46}	133_2	83_{100}	55_{19}	722	3510
Bioallethrin		2105*	302_2	168_6	136_{23}	123_{100}	79_{34}	625	2786
Bioresmethrin		2300*	338_7	171_{50}	143_{34}	128_{44}	123_{100}	799	4035
Biperiden	P-I G U+UHYAC	2280	311_1	218_{15}	98_{100}			673	101
Biperiden TMS		2420	383_1	294_2	205_3	98_{100}	73_{12}	965	4529
Biperiden-M (HO-)	U UHY	2645	327_3	218_5	114_6	98_{100}		746	102
Biperiden-M (HO-) AC	U+UHYAC	2620	369_4	257_8	112_5	98_{100}	84_5	918	103
Biphenyl		1320*	154_{100}	128_6	102_6	76_{25}	63_{16}	134	3318
Biphenylol	G P U+UHYAC	1550*	170_{100}	141_{31}	115_{26}	77_{16}		157	217
Biphenylol AC	U+UHYAC	1690*	212_7	170_{100}	141_{15}	115_{20}		248	2280
Biphenylol ME		1540*	184_1	170_{100}	141_{35}	115_{30}		185	2281
Biphenylol-M (HO-) 2AC	U+UHYAC	1900*	270_4	228_{21}	186_{100}	105_{36}		468	2349
2,2'-Bipyridine		1460	156_{100}	128_{34}	102_6	78_{31}	51_{54}	136	105
Bisacodyl	G U+UHYAC PAC-I	2835	361_{99}	319_{63}	277_{74}	199_{45}		887	106
Bisacodyl HY	UHY	2655	277_{99}	199_{51}				500	107
Bisacodyl HY2ME	UGLUCEXME	2595	305_{100}	290_{27}	227_{49}	182_6	169_6	639	6811
Bisacodyl-M (bis-deacetyl-)	UHY	2655	277_{99}	199_{51}				500	107
Bisacodyl-M (bis-deacetyl-) 2ME	UGLUCEXME	2595	305_{100}	290_{27}	227_{49}	182_6	169_6	639	6811
Bisacodyl-M (bis-methoxy-bis-deacetyl-)	UHY	2820	337_{100}	322_{69}	307_8	259_{14}		792	2458
Bisacodyl-M (bis-methoxy-bis-deacetyl-) 2AC	U+UHYAC	2950	421_{83}	379_{100}	364_{54}	337_{25}	322_{46}	1063	2456
Bisacodyl-M (bis-methoxy-bis-deacetyl-) 2ME	UGLUCEXME	2760	365_{100}	350_{61}	287_{41}	249_{13}	220_{11}	901	6813
Bisacodyl-M (bis-methoxy-deacetyl-)	U+UHYAC	2890	379_{100}	364_{34}	336_{25}	322_{41}	259_8	948	2457
Bisacodyl-M (deacetyl-)	U+UHYAC	2750	319_{100}	277_{65}	276_{97}	199_{31}	153_7	707	2459
Bisacodyl-M (methoxy-bis-deacetyl-)	UHY	2680	307_{100}	306_{49}	292_{19}	229_{35}	69_{22}	648	109
Bisacodyl-M (methoxy-bis-deacetyl-) 2AC	U+UHYAC	2870	391_{46}	349_{100}	307_{48}	292_{12}	229_{23}	988	1750
Bisacodyl-M (methoxy-bis-deacetyl-) 2ME	UGLUCEXME	2695	335_{100}	320_{40}	257_{57}	220_7	139_{13}	783	6812
Bisacodyl-M (methoxy-deacetyl-)	U+UHYAC	2810	349_{100}	307_{43}	306_{54}	292_{17}	229_{30}	842	210
Bisacodyl-M (trimethoxy-bis-deacetyl-) 2AC	U+UHYAC	3060	451_{72}	409_{100}	367_{81}	329_{77}	203_{62}	1114	3425
Bisoctylphenylamine		2910	393_7	378_5	322_{100}	250_{27}		998	4950
Bisoprolol	G P U	2570	325_1	310_1	281_2	116_{15}	72_{100}	737	2787
Bisoprolol AC	U+UHYAC	2880	367_1	352_1	158_8	98_{29}	72_{100}	911	2790

Bisoprolol TMSTFA Table 1-8-1: Compounds in order of names

Name	Detected	RI	Typical ions and intensities					Page	Entry
Bisoprolol TMSTFA		2570	493_1	332_{10}	284_{100}	221_{22}	73_{60}	1160	6134
Bisoprolol 2AC	U+UHYAC	2770	349_1	245_{15}	200_{100}	98_{23}	72_{55}	1040	2791
Bisoprolol formyl artifact	G P U	2595	337_{19}	322_{34}	234_{23}	127_{100}	112_{77}	796	2788
Bisoprolol -H2O	U	2400	307_9	220_1	204_2	98_{100}	56_{88}	652	2933
Bisoprolol -H2O AC	U+UHYAC	2900	349_1	306_4	262_{14}	140_{24}	98_{100}	844	2789
Bisoprolol N-AC		2730	349_1	245_6	158_{100}	139_9	72_{57}	912	6408
Bisoprolol-M (phenol)	U	1690*	210_4	167_9	123_{42}	107_{100}	77_{19}	245	2932
Bisphenol A	G U UHY	2155*	228_{33}	213_{100}				301	108
Bisphenol A 2AC	U+UHYAC	2380*	312_{11}	270_{28}	228_{31}	213_{100}	119_{11}	676	3360
Bitertanol		2650	337_2	170_{100}	141_{11}	112_{17}	57_{27}	794	4146
Bornaprine	G U+UHYAC	2260	329_{10}	314_9	257_2	171_4	86_{100}	756	110
Bornaprine-M (deethyl-HO-) iso-1 2AC	U+UHYAC	2790	401_6	358_4	142_{17}	112_{22}	58_{40}	1021	1252
Bornaprine-M (deethyl-HO-) iso-2 2AC	U+UHYAC	2875	401_5	358_5	169_{19}	128_{24}	58_{50}	1022	1253
Bornaprine-M (deethyl-HO-) iso-3 2AC	U+UHYAC	2890	401_5	358_5	169_{19}	128_{24}	58_{50}	1022	918
Bornaprine-M (HO-) iso-1 AC	U+UHYAC	2385	387_3	372_8	169_3	143_5	86_{100}	978	1251
Bornaprine-M (HO-) iso-2 AC	U+UHYAC	2465	387_3	372_9	169_2	91_3	86_{100}	978	632
Bornaprine-M (HO-) iso-3 AC	U+UHYAC	2565	387_3	372_9	233_4	169_5	86_{100}	978	683
Bornyl salicylate		1870*	274_4	137_{55}	121_{58}	81_{100}		486	1403
Bornyl salicylate ME		2110*	288_{10}	135_{100}	81_{12}			554	1405
Brallobarbital	P G U UHY U+UHYAC	1850	245_1	207_{100}	165_{18}	124_{15}	91_{14}	541	111
Brallobarbital (ME)	P	1780	259_1	221_{100}	176_1	136_{15}	91_{29}	613	3996
Brallobarbital 2ET		1830	263_{100}	221_{23}	121_4	91_7	77_6	815	2598
Brallobarbital 2ME		1725	235_{100}	193_{30}				683	645
Brallobarbital-M (desbromo-HO-)	U UHY U+UHYAC	1795	224_6	181_{13}	167_{100}	141_{13}	124_{19}	287	114
Brallobarbital-M (dihydro-)	U UHY U+UHYAC	1970	209_{99}	167_{49}	141_{45}	120_{38}	67_{38}	550	119
Brallobarbital-M (HO-)	U UHY U+UHYAC	2040	223_{99}					622	118
Brassidic acid ME		2610*	352_3	320_{20}	97_{27}	69_{55}	55_{100}	856	3795
Brofaromine AC	U+UHYAC	2780	351_{59}	308_{38}	266_{41}	125_{30}	56_{100}	846	2405
Brofaromine-M (HO-) 2AC	U+UHYAC	2980	409_{33}	369_{97}	367_{100}	324_{20}	284_{11}	1037	2710
Brofaromine-M (O-demethyl-) 2AC	U+UHYAC	2830	379_{49}	337_{73}	294_{52}	125_{28}	56_{100}	947	2404
Brofaromine-M/artifact (pyridyl-)AC	U+UHYAC	2650	331_{12}	291_{98}	289_{100}	182_7	153_{10}	761	2406
Brolamfetamine		1800	273_1	232_6	230_6	199_1	77_7	480	2548
Brolamfetamine		1800	273_1	230_6	105_3	77_7	44_{100}	480	5527
Brolamfetamine AC		2150	315_3	256_{20}	229_1	162_4	86_{15}	688	2549
Brolamfetamine AC		2150	315_3	256_{20}	162_4	86_{22}	44_{100}	688	5528
Brolamfetamine HFB		1945	469_{24}	256_{88}	240_{55}	229_{100}	199_{29}	1137	6008
Brolamfetamine PFP		1905	419_{21}	256_{69}	229_{94}	190_{55}	119_{87}	1059	6007
Brolamfetamine TFA		1935	369_{28}	256_{81}	229_{100}	199_{40}	69_{88}	914	6006
Brolamfetamine TMS		1920	345_1	272_2	229_1	116_{80}	73_{63}	827	6009
Brolamfetamine formyl artifact		1790	285_3	254_{15}	229_5	199_3	56_{100}	537	3242
Brolamfetamine precursor		1345*	166_{100}	151_{39}	120_{36}	95_{61}	63_{62}	150	3278
Brolamfetamine-M (bis-O-demethyl-) 3AC	U+UHYAC	2325	371_2	329_{35}	287_{54}	228_{100}	86_{44}	921	7075
Brolamfetamine-M (bis-O-demethyl-) artifact 2AC	U+UHYAC	2225	311_{23}	269_{98}	227_{59}	212_{39}	133_{18}	670	7184
Brolamfetamine-M (deamino-HO-) AC	U+UHYAC	1950*	316_7	274_{22}	214_{96}	186_{17}		692	7061
Brolamfetamine-M (deamino-oxo-)		1835*	272_7	229_{11}				477	7062
Brolamfetamine-M (HO-) 2AC	U+UHYAC	2270	373_3	86_{100}	313_{14}	271_{37}		929	7081
Brolamfetamine-M (HO-) -H2O	U+UHYAC	1960*	273_{31}	271_{29}	258_{43}	256_{42}		473	7073
Brolamfetamine-M (HO-) -H2O AC	U+UHYAC	2130*	313_{24}	271_{79}	256_{100}			678	7074
Brolamfetamine-M (O-demethyl-) iso-1 AC	U+UHYAC	2120	301_{18}	242_{100}	215_{11}	185_{13}	86_{20}	617	7070
Brolamfetamine-M (O-demethyl-) iso-1 2AC	U+UHYAC	2235	343_{12}	301_{23}	284_{57}	242_{100}	86_{81}	819	7065
Brolamfetamine-M (O-demethyl-) iso-2 AC	U+UHYAC	2180	301_{29}	242_{100}	215_{14}	86_{36}		618	7071
Brolamfetamine-M (O-demethyl-) iso-2 2AC	U+UHYAC	2275	343_2	284_{56}	242_{100}	215_{13}	86_{23}	819	7066
Brolamfetamine-M (O-demethyl-deamino-oxo-) AC	U+UHYAC	1930*	300_8	258_{94}	215_{100}			612	7063
Brolamfetamine-M (O-demethyl-deamino-oxo-) iso-1		1870*	260_{66}	258_{72}	217_{99}	215_{100}		411	7068
Brolamfetamine-M (O-demethyl-deamino-oxo-) iso-2		1885*	260_{99}	258_{100}	217_{97}	215_{93}		411	7069
Brolamfetamine-M (O-demethyl-HO-) 3AC	U+UHYAC	2385	401_1	359_5	317_7	258_{10}	86_7	1019	7067
Brolamfetamine-M (O-demethyl-HO-) -H2O 2AC	U+UHYAC	2280	341_{32}	299_{62}	257_{100}	242_{72}		809	7072
Brolamfetamine-M (O-demethyl-HO-deamino-HO-) 3AC	U+UHYAC	2145*	402_2	360_9	315_{13}	300_8	231_{24}	1022	7064
Bromacil	G U	1900	260_4	231_9	205_{100}	188_{19}	162_{16}	419	124
Bromadiolone artifact		1985*	260_{99}	258_{100}	178_{71}	152_{21}	76_{39}	412	3629
Bromantane		2420	305_{66}	184_{27}	171_{34}	135_{90}	130_{52}	638	6130
Bromantane AC		2515	347_{23}	288_{39}	213_{20}	135_{100}	67_{69}	834	6202
Bromantane HFB		2305	501_6	367_3	169_{12}	135_{100}	67_{59}	1164	6145
Bromantane ME		2310	319_{44}	198_{35}	135_{100}	93_{79}	67_{95}	706	6201
Bromantane PFP		2295	451_{14}	317_7	155_{16}	135_{100}	93_{74}	1114	6131
Bromantane TFA		2250	401_{10}	267_5	155_{11}	135_{100}	67_{57}	1019	6203
Bromazepam	P G U UGLUC	2670	315_{91}	286_{56}	236_{100}	208_{46}	179_{43}	687	125
Bromazepam TMS		2450	387_{68}	372_{12}	272_{10}	179_{19}	73_{100}	975	4530
Bromazepam artifact-3	U+UHYAC	2500	303_{99}	301_{100}	222_{82}			617	2117
Bromazepam HY	UHY	2250	276_{27}	247_{100}	198_{17}	168_{28}		494	127
Bromazepam HYAC	U+UHYAC	2490	318_9	289_9	247_{66}	121_{100}	78_{50}	701	129
Bromazepam iso-1 ME		2385	329_{84}	250_{100}	208_{33}	179_{48}		751	130
Bromazepam iso-2 ME		2540	329_{100}	300_{53}	250_{59}	78_{50}		751	131
Bromazepam-M (3-HO-)	UGLUC-I	2470	313_{100}	284_{22}	206_{44}	179_{13}		761	126

Table 1-8-1: Compounds in order of names

Name	Detected	RI	Typical ions and intensities					Page	Entry
Bromazepam-M (3-HO-) 2TMS		2475	475_7	460_2	386_{28}	360_{18}	73_{100}	1143	5441
Bromazepam-M (3-HO-) artifact-1	UHY-I U+UHYAC-I	2255	285_{100}	206_{97}	179_{18}			537	128
Bromazepam-M (3-HO-) artifact-2	U+UHYAC	2265	299_{100}	220_{90}	179_{19}	152_7		609	2116
Bromazepam-M (3-HO-) HY	UHY	2250	276_{27}	247_{100}	198_{17}	168_{28}		494	127
Bromazepam-M (3-HO-) HYAC	U+UHYAC	2490	318_9	289_9	247_{66}	121_{100}	78_{50}	701	129
Bromazepam-M (HO-) HYAC	U+UHYAC	2580	334_{54}	292_{66}	264_{17}	247_{100}	78_{36}	778	1876
Bromazepam-M (HO-) HYME	UEXME	2250	306_{21}	277_6	247_{101}	184_{23}	78_{34}	643	7703
Bromazepam-M/artifact	U+UHYAC	2310	316_{55}	288_{32}	248_{100}	238_{85}	210_{40}	691	2700
Bromazepam-M/artifact	U+UHYAC	2670	352_{100}	325_{31}	296_{16}	273_{34}	216_{17}	851	3059
Bromazepam-M/artifact AC	U+UHYAC	2260	319_{24}	277_{100}	249_{46}	198_{23}	170_{58}	705	1877
Bromhexine		2375	374_{12}	293_{50}	262_{38}	112_{66}	70_{100}	932	132
Bromhexine-M (HO-)	UHY	2660	390_9	293_{60}	262_{35}	128_{101}	86_{38}	984	133
Bromhexine-M (HO-) 2AC	U+UHYAC	2930	474_5	417_{26}	335_{98}	304_{34}	264_{29}	1142	134
Bromhexine-M (HOOC-) ME	P U	1770	309_{68}	307_{36}	277_{100}	275_{53}	249_{27}	646	5131
Bromhexine-M (nor-HO-)	P G U UHY	2665	376_8	279_{77}	264_{100}	262_{53}	114_{69}	938	19
Bromhexine-M (nor-HO-)	P G U UHY	2780	387_{29}	331_{100}	329_{53}	289_{42}	195_{82}	978	6315
Bromhexine-M (nor-HO-) TMS		2665	448_4	319_{23}	279_{100}	264_{92}	186_{64}	1110	4527
Bromhexine-M (nor-HO-) 2TMS		2800	520_1	391_5	351_{100}	186_{50}	73_{66}	1175	4528
Bromhexine-M (nor-HO-) -H2O	P G U UHY	2395	358_{13}	289_{42}	264_{94}	262_{55}	68_{100}	875	6314
Bromhexine-M (nor-HO-) iso-1 2AC	U+UHYAC	2935	460_{14}	417_{37}	279_{47}	262_{25}	81_{100}	1127	135
Bromhexine-M (nor-HO-) iso-2 2AC	U+UHYAC	3015	460_{22}	419_{100}	417_{51}	279_{74}	264_{56}	1127	20
Bromhexine-M (nor-HO-) iso-3 2AC	U+UHYAC	3165	460_{17}	417_{30}	279_{48}	262_{24}	81_{100}	1127	136
Bromisoval	P-I G U	1540	222_{16}	163_{73}	70_{81}	55_{100}		280	137
Bromisoval artifact	P G U	1510	137_{100}	120_{13}	100_{70}			116	138
Bromisoval-M (Br-isovalerianic acid)		1190*	165_1	140_{100}	138_{100}	120_9	101_{20}	172	2395
Bromisoval-M (HO-isovalerianic acid)		1140*	118_1	89_1	76_{100}	73_{66}	55_{35}	105	2394
Bromisoval-M (isovalerianic acid carbamide)		1850	129_4	112_5	102_{100}	85_{46}	59_{71}	123	139
Bromisoval-M/artifact (bromoisovalerianic acid)	G	1570*	180_{29}	163_{85}	137_{54}	70_{93}	55_{100}	173	2393
Bromobenzene		<1000*	158_{59}	156_{59}	77_{100}	51_{36}		135	3611
3-Bromo-d-camphor		1450*	230_{12}	151_{35}	123_{100}	83_{82}	55_{55}	306	2985
2-Bromo-4-cyclohexylphenol		1915*	254_{100}	198_{33}	185_{56}	132_{78}	107_{28}	397	5165
2-Bromo-4-cyclohexylphenol AC		1925*	296_6	254_{100}	198_{17}	185_{18}	132_{24}	596	5169
2-Bromo-4-cyclohexylphenol ME		1800*	268_{100}	199_{54}	146_{92}	118_{63}	90_{40}	459	5172
4-Bromo-2,5-dimethoxyphenylethylamine		1785	259_{15}	230_{100}	215_{29}	199_{10}	77_{37}	415	3254
4-Bromo-2,5-dimethoxyphenylethylamine AC		2180	301_{15}	242_{100}	229_{31}	199_{12}	148_{39}	617	3267
4-Bromo-2,5-dimethoxyphenylethylamine HFB		2030	455_{32}	242_{98}	229_{79}	199_{28}	148_{32}	1120	6941
4-Bromo-2,5-dimethoxyphenylethylamine PFP		1995	405_{36}	242_{100}	229_{85}	199_{33}	148_{35}	1029	6936
4-Bromo-2,5-dimethoxyphenylethylamine TFA		2000	355_{42}	242_{99}	229_{83}	199_{37}	148_{36}	864	6931
4-Bromo-2,5-dimethoxyphenylethylamine TMS		1935	331_1	272_3	229_2	102_{100}	73_{50}	761	6925
4-Bromo-2,5-dimethoxyphenylethylamine 2AC		2230	343_7	242_{99}	229_{32}	201_{12}	148_{29}	819	6924
4-Bromo-2,5-dimethoxyphenylethylamine 2TMS		2195	403_1	388_{14}	272_7	207_{11}	174_{100}	1025	6926
4-Bromo-2,5-dimethoxyphenylethylamine formyl artifact		1840	271_{12}	240_{65}	229_{27}	199_{11}	77_{24}	473	3245
4-Bromo-2,5-dimethoxyphenylethylamine formyl artifact		1840	271_{12}	240_{65}	229_{27}	199_{11}	42_{100}	473	5522
4-Bromo-2,5-dimethoxyphenylethylamine intermediate-1		1900	209_{100}	162_{51}	147_{52}	133_{61}	77_{62}	240	3286
4-Bromo-2,5-dimethoxyphenylethylamine intermediate-2		1630	181_{15}	162_{33}	152_{100}	137_{47}	121_{27}	177	3287
4-Bromo-2,5-dimethoxyphenylethylamine intermediate-2		1630	181_{15}	162_{33}	152_{100}	137_{47}	44_{95}	178	5523
4-Bromo-2,5-dimethoxyphenylethylamine intermediate-2 AC		1935	223_{15}	164_{100}	149_{29}	121_{31}	91_{15}	284	3288
4-Bromo-2,5-dimethoxyphenylethylamine intermediate-2 formyl artifact		1540	193_{12}	162_{100}	151_{30}	121_{43}	91_{29}	200	3293
4-Bromo-2,5-dimethoxyphenylethylamine intermediate-2 formyl artifact		1540	193_{12}	162_{100}	151_{30}	121_{43}	42_{64}	200	5524
4-Bromo-2,5-dimethoxyphenylethylamine-M (deamino-COOH) ME		2030*	288_{100}	273_{12}	241_3	229_{67}	199_{19}	550	7212
4-Bromo-2,5-dimethoxyphenylethylamine-M (deamino-di-HO-) 2AC	U+UHYAC	2230*	360_{13}	300_{27}	258_{43}	245_{100}	138_{17}	882	7214
4-Bromo-2,5-dimethoxyphenylethylamine-M (deamino-di-HO-) 2TFA		1790*	468_{100}	354_{14}	341_{77}	311_{14}	276_{76}	1136	7208
4-Bromo-2,5-dimethoxyphenylethylamine-M (deamino-HO-) AC	U+UHYAC	2300*	302_{20}	242_{100}	227_{25}	183_{33}	148_{42}	622	7198
4-Bromo-2,5-dimethoxyphenylethylamine-M (deamino-HO-) TFA		1880*	356_{31}	341_{10}	242_{64}	229_{46}	148_{46}	867	7209
4-Bromo-2,5-dimethoxyphenylethylamine-M (deamino-oxo-)		2020*	258_{35}	229_{100}	215_9	199_{35}	186_8	411	7215
4-Bromo-2,5-dimethoxyphenylethylamine-M (O-demethyl- N-acetyl-) iso-1 AC	U+UHYAC	2410	329_4	287_{33}	228_{101}	215_{16}	165_5	752	7196
4-Bromo-2,5-dimethoxyphenylethylamine-M (O-demethyl- N-acetyl-) iso-2 AC	U+UHYAC	2440	329_8	287_{21}	228_{100}	215_{17}	72_{13}	752	7197
4-Bromo-2,5-dimethoxyphenylethylamine-M (O-demethyl- N-acetyl-) iso-1 TFA		2090	383_9	324_{100}	255_7	148_{21}	72_{23}	961	7204
4-Bromo-2,5-dimethoxyphenylethylamine-M (O-demethyl- N-acetyl-) iso-2 TFA		2130	383_7	311_{11}	$324_{·01}$	227_{11}	72_{17}	961	7205
4-Bromo-2,5-dimethoxyphenylethylamine-M (O-demethyl-) iso-1 2AC	U+UHYAC	2410	329_4	287_{33}	$228_{·01}$	215_{16}	165_5	752	7196
4-Bromo-2,5-dimethoxyphenylethylamine-M (O-demethyl-) iso-2 2AC	U+UHYAC	2440	329_8	287_{21}	$228_{·00}$	215_{17}	72_{13}	752	7197
4-Bromo-2,5-dimethoxyphenylethylamine-M (O-demethyl-) iso-1 2TFA		1900	437_{12}	324_{100}	311_{10}	255_7	148_{32}	1092	7206
4-Bromo-2,5-dimethoxyphenylethylamine-M (O-demethyl-) iso-2 2TFA		1950	437_6	324_{100}	311_{32}	253_3	227_{15}	1092	7207
4-Bromo-2,5-dimethoxyphenylethylamine-M (O-demethyl-deamino-COOH) MEAC		2120*	316_{14}	274_{101}	242_{92}	214_{88}	186_{18}	691	7213
4-Bromo-2,5-dimethoxyphenylethylamine-M (O-demethyl-deamino-COOH) METFA		1890*	370_{100}	311_{46}	257_{55}	241_{60}	148_{21}	918	7211
4-Bromo-2,5-dimethoxyphenylethylamine-M (O-demethyl-deamino-COOH) -H2O	U+UHYAC	1980*	242_{56}	214_{70}	186_{26}			349	7203
4-Bromo-2,5-dimethoxyphenylethylamine-M (O-demethyl-deamino-di-HO-) 3AC	U+UHYAC	2280*	388_6	346_{43}	286_{65}	244_{67}	231_{100}	978	7201
4-Bromo-2,5-dimethoxyphenylethylamine-M (O-demethyl-deamino-HO-) 2TFA		1800*	438_{31}	341_{16}	324_{100}	311_{35}	227_{55}	1094	7210
4-Bromo-2,5-dimethoxyphenylethylamine-M (O-demethyl-deamino-HO-) iso-1 2AC	U+UHYAC	2160*	330_5	288_{17}	270_6	228_{100}	213_{12}	757	7199
4-Bromo-2,5-dimethoxyphenylethylamine-M (O-demethyl-deamino-HO-) iso-2 2AC	U+UHYAC	2180*	330_4	288_{15}	246_{10}	228_{100}	213_{14}	757	7200
4-Bromo-2,5-dimethoxyphenylethylamine-M (O-demethyl-deamino-HO-oxo-) 2AC	U+UHYAC	2160*	344_{11}	302_{74}	260_{46}	242_{100}	214_{25}	823	7202
Bromofenoxim artifact-1		1520	184_{100}	154_{28}	107_{41}	91_{45}	63_{88}	183	728
Bromofenoxim artifact-2		1690	277_{100}	275_{56}	168_{18}	117_{14}	88_{59}	487	3630
Bromofenoxim artifact-2 ME		1650	289_{15}	248_{12}	202_{26}	86_{100}	72_{59}	555	3631

5-Bromonicotinic acid   Table 1-8-1: Compounds in order of names

Name	Detected	RI	Typical ions and intensities					Page	Entry
5-Bromonicotinic acid		1020	201 $_{84}$	183 $_{49}$	156 $_{30}$	76 $_{68}$	51 $_{100}$	220	5252
5-Bromonicotinic acid ME	UME	1095	215 $_{42}$	184 $_{79}$	156 $_{70}$	136 $_{30}$	76 $_{100}$	256	5250
4-Bromophenol		1310*	174 $_{94}$	172 $_{100}$	93 $_{21}$	65 $_{56}$		159	1995
Bromophos	P-I	1995*	364 $_{1}$	331 $_{100}$	329 $_{74}$	213 $_{5}$	125 $_{69}$	896	1406
Bromophos-ethyl		2060*	359 $_{79}$	357 $_{60}$	301 $_{69}$	240 $_{20}$	97 $_{100}$	991	3508
Bromopride		2850	343 $_{1}$	245 $_{3}$	228 $_{11}$	99 $_{23}$	86 $_{100}$	820	1407
Bromopride AC		3080	385 $_{1}$	313 $_{1}$	270 $_{2}$	228 $_{4}$	86 $_{100}$	969	2607
Bromopropylate		2425*	426 $_{1}$	341 $_{100}$	339 $_{53}$	183 $_{73}$	76 $_{40}$	1071	4142
3-Bromoquinoline		1490	209 $_{98}$	207 $_{100}$	128 $_{83}$	101 $_{40}$	75 $_{19}$	232	2638
5-Bromosalicylic acid		1530*	216 $_{33}$	198 $_{100}$	170 $_{35}$	142 $_{9}$	63 $_{40}$	258	1996
5-Bromosalicylic acid ME		1465*	230 $_{36}$	198 $_{100}$	170 $_{31}$	143 $_{9}$	63 $_{35}$	305	1997
5-Bromosalicylic acid MEAC		1600*	272 $_{2}$	230 $_{95}$	198 $_{100}$	170 $_{22}$	142 $_{7}$	476	2032
5-Bromosalicylic acid 2ET		1600*	272 $_{100}$	257 $_{33}$	213 $_{97}$	198 $_{32}$	170 $_{16}$	477	1998
5-Bromosalicylic acid 2ME		1500*	244 $_{33}$	213 $_{100}$	183 $_{10}$	170 $_{17}$	155 $_{14}$	354	2031
5-Bromosalicylic acid -CO2		1310*	174 $_{94}$	172 $_{100}$	93 $_{21}$	65 $_{56}$		159	1995
Bromothiophene		<1000*	164 $_{99}$	162 $_{100}$	117 $_{4}$	83 $_{80}$	57 $_{31}$	140	3609
Bromoxynil		1690	277 $_{100}$	275 $_{56}$	168 $_{18}$	117 $_{14}$	88 $_{59}$	487	3630
Bromoxynil ME		1650	289 $_{15}$	248 $_{12}$	202 $_{26}$	86 $_{100}$	72 $_{59}$	555	3631
Bromperidol	G U+UHYAC	3050	419 $_{2}$	281 $_{100}$	268 $_{93}$	250 $_{21}$	123 $_{77}$	1060	2110
Bromperidol TMS		2730	478 $_{2}$	340 $_{40}$	250 $_{35}$	123 $_{50}$	73 $_{50}$	1158	5479
Bromperidol 2TMS		2840	548 $_{6}$	340 $_{84}$	250 $_{31}$	103 $_{33}$	73 $_{90}$	1188	5480
Bromperidol -H2O	U+UHYAC	3020	401 $_{7}$	263 $_{24}$	250 $_{31}$	236 $_{100}$	123 $_{57}$	1019	2115
Bromperidol-M	U+UHYAC	1490*	180 $_{25}$	125 $_{49}$	123 $_{31}$	95 $_{17}$	56 $_{100}$	174	85
Bromperidol-M	UHY	1890	267 $_{15}$	233 $_{96}$	127 $_{38}$	94 $_{44}$	56 $_{100}$	453	141
Bromperidol-M (N-dealkyl-) AC	U+UHYAC	2335	297 $_{35}$	254 $_{36}$	183 $_{22}$	99 $_{28}$	57 $_{100}$	601	166
Bromperidol-M (N-dealkyl-oxo-) -2H2O	U UHY U+UHYAC	1850	233 $_{100}$	154 $_{21}$	127 $_{37}$			314	140
Bromperidol-M 4 AC	U+UHYAC	2260	293 $_{67}$	279 $_{85}$	251 $_{99}$	222 $_{56}$		577	142
Brompheniramine	U UHY U+UHYAC	2105	247 $_{29}$	167 $_{13}$	72 $_{38}$	58 $_{100}$		702	144
Brompheniramine-M (bis-nor-) AC	U+UHYAC	2170	332 $_{4}$	260 $_{100}$	247 $_{38}$	180 $_{22}$	167 $_{35}$	767	2812
Brompheniramine-M (nor-) AC	U+UHYAC	2195	346 $_{12}$	260 $_{48}$	247 $_{100}$	180 $_{20}$	167 $_{50}$	830	145
Brotizolam	G U+UHYAC-I	3090	394 $_{100}$	392 $_{75}$	363 $_{12}$	316 $_{28}$	245 $_{49}$	991	1408
Brotizolam-M (HO-) AC		3140	450 $_{34}$	409 $_{100}$	407 $_{69}$	289 $_{11}$	245 $_{14}$	1113	2052
Brotizolam-M (HO-) -CH2O		3050	380 $_{100}$	378 $_{79}$	299 $_{24}$	245 $_{16}$		944	2051
Brucine	U	3275	394 $_{99}$	379 $_{15}$	355 $_{4}$			999	146
BSTFA		1100	257 $_{3}$	192 $_{35}$	188 $_{47}$	100 $_{58}$	73 $_{200}$	409	5431
Bucetin		2020	223 $_{4}$	179 $_{46}$	137 $_{62}$	109 $_{86}$	108 $_{100}$	285	147
Bucetin AC	UGLUCAC	2095	265 $_{26}$	205 $_{31}$	137 $_{101}$	109 $_{47}$	108 $_{52}$	444	185
Bucetin HYAC	G U+UHYAC	1680	179 $_{66}$	137 $_{51}$	109 $_{97}$	108 $_{100}$	80 $_{18}$	171	186
Bucetin-M	UHY	1240	109 $_{99}$	80 $_{41}$	52 $_{90}$			102	826
Bucetin-M (deethyl-) HYME	UHYME	1100	123 $_{27}$	109 $_{100}$	94 $_{7}$	80 $_{96}$	53 $_{47}$	108	3766
Bucetin-M (HO-) HY2AC	U+UHYAC	1755	237 $_{15}$	195 $_{31}$	153 $_{100}$	124 $_{55}$		333	187
Bucetin-M (O-deethyl-) 2AC	UGLUCAC	2110	279 $_{4}$	237 $_{13}$	177 $_{9}$	151 $_{10}$	109 $_{100}$	510	30
Bucetin-M (p-phenetidine)	UHY	1280	137 $_{68}$	108 $_{99}$	80 $_{38}$	65 $_{9}$		117	844
Bucetin-M HY2AC	U+UHYAC PAC	1765	193 $_{10}$	151 $_{53}$	109 $_{100}$	80 $_{24}$		199	188
Buclizine	G U UHY U+UHYAC	3360	432 $_{5}$	285 $_{37}$	231 $_{100}$	165 $_{35}$	147 $_{53}$	1085	2414
Buclizine artifact-1	G U+UHYAC	1600*	202 $_{30}$	167 $_{100}$	165 $_{52}$	152 $_{17}$	125 $_{7}$	222	2442
Buclizine HY		1830	232 $_{62}$	190 $_{61}$	147 $_{100}$	117 $_{41}$		314	2416
Buclizine HYAC		2020	274 $_{18}$	202 $_{24}$	188 $_{25}$	147 $_{100}$	85 $_{30}$	487	2415
Buclizine-M	U+UHYAC	2210	280 $_{100}$	201 $_{35}$	165 $_{57}$			515	770
Buclizine-M (carbinol)	UHY	1750*	218 $_{17}$	183 $_{7}$	139 $_{39}$	105 $_{100}$	77 $_{87}$	263	2239
Buclizine-M (carbinol) AC	U+UHYAC	1890*	260 $_{8}$	200 $_{40}$	165 $_{100}$	139 $_{10}$	77 $_{29}$	420	1270
Buclizine-M (Cl-benzophenone)	U+UHYAC	1850*	216 $_{43}$	139 $_{58}$	105 $_{100}$	77 $_{44}$		258	1343
Buclizine-M (HO-Cl-benzophenone)	UHY	2300*	232 $_{36}$	197 $_{7}$	139 $_{23}$	121 $_{100}$	111 $_{23}$	311	2240
Buclizine-M (HO-Cl-BPH) iso-1 AC	U+UHYAC	2200*	274 $_{18}$	232 $_{75}$	139 $_{100}$	121 $_{44}$	111 $_{51}$	484	2229
Buclizine-M (HO-Cl-BPH) iso-2 AC	U+UHYAC	2230*	274 $_{7}$	232 $_{43}$	139 $_{25}$	121 $_{100}$	111 $_{27}$	484	2230
Buclizine-M (N-dealkyl-)	UHY	2520	286 $_{13}$	241 $_{48}$	201 $_{50}$	165 $_{65}$	56 $_{100}$	543	2241
Buclizine-M (N-dealkyl-) AC	U+UHYAC	2620	328 $_{7}$	242 $_{19}$	201 $_{48}$	165 $_{66}$	85 $_{100}$	749	1271
Buclizine-M (N-dealkyl-HO-) AC-conj.	U	2580	290 $_{20}$	204 $_{36}$	163 $_{100}$	117 $_{50}$	85 $_{63}$	566	2432
Buclizine-M (N-dealkyl-HO-) 2AC	U+UHYAC	2640	332 $_{12}$	260 $_{37}$	205 $_{100}$	117 $_{34}$	85 $_{59}$	770	2433
Buclizine-M/artifact HYAC	U+UHYAC	2935	280 $_{4}$	201 $_{100}$	165 $_{26}$			515	1272
Budipine		2300	293 $_{2}$	278 $_{100}$	178 $_{5}$	165 $_{5}$	70 $_{28}$	583	6114
Bufexamac ME		1995	237 $_{34}$	222 $_{42}$	166 $_{100}$	122 $_{51}$	107 $_{60}$	334	6086
Bufexamac 2ME		2005	251 $_{23}$	190 $_{19}$	166 $_{51}$	107 $_{100}$	77 $_{10}$	386	6398
Bufexamac artifact (deoxo-)		1970	207 $_{12}$	163 $_{25}$	107 $_{100}$	89 $_{6}$	77 $_{16}$	234	6083
Bufexamac artifact (deoxo-formyl-)		1780	219 $_{3}$	163 $_{36}$	107 $_{100}$	89 $_{4}$	77 $_{11}$	270	6084
Bufexamac-M/artifact (HOOC-) ME		1720*	222 $_{37}$	166 $_{49}$	163 $_{25}$	107 $_{100}$	77 $_{37}$	282	6085
Buflomedil	G P U UHY U+UHYAC	2390	307 $_{2}$	210 $_{5}$	195 $_{31}$	97 $_{73}$	84 $_{100}$	650	2907
Buflomedil TMS		2275	379 $_{5}$	295 $_{28}$	181 $_{29}$	84 $_{170}$	73 $_{72}$	950	6274
Buflomedil-M (O-demethyl-)	UHY	2375	293 $_{13}$	181 $_{49}$	97 $_{83}$	84 $_{100}$	55 $_{10}$	582	3980
Buflomedil-M (O-demethyl-) AC	U+UHYAC	2530	335 $_{3}$	181 $_{16}$	97 $_{83}$	84 $_{100}$	55 $_{15}$	784	3981
Bulbocapnine	UHY	2960	325 $_{100}$	310 $_{98}$	282 $_{27}$	178 $_{30}$	162 $_{44}$	735	4249
Bulbocapnine AC	U+UHYAC	2990	367 $_{100}$	324 $_{32}$	310 $_{83}$	280 $_{28}$	162 $_{23}$	908	4250
Bumadizone		2270	282 $_{10}$	184 $_{100}$	183 $_{69}$	93 $_{31}$	77 $_{80}$	741	5184
Bumadizone ME		2280	296 $_{13}$	184 $_{72}$	183 $_{100}$	77 $_{87}$	57 $_{49}$	808	5185

Table 1-8-1: Compounds in order of names   Butalamine

Name	Detected	RI	Typical ions and intensities					Page	Entry
Bumadizone artifact (azobenzene)		1620	182 $_{14}$	152 $_5$	105 $_{15}$	77 $_{100}$	63 $_2$	180	5186
Bumadizone artifact (hexanilide)		1755	191 $_4$	135 $_6$	93 $_{100}$	77 $_7$	65 $_{71}$	195	5187
Bumatizone artifact AC		2435	366 $_{67}$	184 $_{69}$	183 $_{100}$	105 $_{33}$	77 $_{94}$	903	5188
Bumetanide 2ME		3180	392 $_{100}$	349 $_{76}$	318 $_{44}$	254 $_{42}$	77 $_{25}$	992	2780
Bumetanide 2MEAC		3120	434 $_{100}$	379 $_{11}$	349 $_{23}$	254 $_{21}$	56 $_{57}$	1087	2781
Bumetanide 3ME		2970	406 $_{100}$	363 $_{68}$	318 $_{28}$	298 $_{10}$	254 $_{28}$	1032	2779
Bumetanide 3MEAC		3190	448 $_{24}$	383 $_{100}$	328 $_{48}$			1111	2783
Bumetanide -SO2NH ME		2340	299 $_{38}$	256 $_{100}$	178 $_{11}$	91 $_{22}$	77 $_{23}$	611	2778
Bumetanide -SO2NH MEAC		3150	341 $_{100}$	285 $_7$	254 $_8$	195 $_2$	91 $_4$	812	2782
Bunazosin		3330	373 $_{73}$	260 $_{58}$	247 $_{100}$	233 $_{65}$	221 $_{38}$	932	4690
Bunitrolol		1960	248 $_1$	233 $_{35}$	204 $_6$	86 $_{100}$	57 $_{32}$	372	2608
Bunitrolol AC		2070	275 $_{46}$	119 $_{23}$	98 $_{45}$	86 $_{100}$	56 $_{81}$	565	1351
Bunitrolol TMS		2025	305 $_6$	204 $_3$	176 $_5$	86 $_{100}$	73 $_{15}$	716	6165
Bunitrolol formyl artifact		1980	260 $_3$	245 $_{100}$	86 $_{13}$	70 $_{28}$	57 $_{33}$	422	1350
Bunitrolol-M (deisobutyl-) 2AC	U+UHYAC	2040	276 $_{39}$	233 $_{55}$	158 $_{27}$	96 $_{30}$	86 $_{86}$	496	1586
Bunitrolol-M (HO-) 2AC	U+UHYAC	2300	333 $_{50}$	291 $_{16}$	174 $_{11}$	98 $_{28}$	86 $_{100}$	839	1587
Bunitrolol-M (HO-) artifact AC	U+UHYAC	2370	318 $_1$	303 $_{100}$	261 $_4$	174 $_2$	70 $_{20}$	704	1588
Bunitrolol-M (HO-methoxy-) 2AC	U+UHYAC	2480	363 $_{49}$	321 $_9$	204 $_5$	98 $_{20}$	86 $_{98}$	946	1589
Buphanamine		----	301 $_{100}$	256 $_{35}$	231 $_{45}$	218 $_{39}$	204 $_{41}$	619	4689
Buphenine		2420	176 $_{99}$	121 $_{16}$	91 $_{83}$	71 $_{38}$		612	1409
Bupirimate		2165	316 $_{34}$	273 $_{100}$	208 $_{92}$	166 $_{64}$	108 $_{79}$	694	3319
Bupivacaine	P U	2260	288 $_1$	245 $_1$	140 $_{100}$	98 $_3$	84 $_8$	555	148
Bupranolol		1900	271 $_1$	256 $_{17}$	227 $_6$	86 $_{100}$	57 $_{27}$	475	2609
Bupranolol AC		2370	298 $_{30}$	142 $_9$	112 $_{26}$	86 $_{100}$		680	1346
Bupranolol TMS		2000	343 $_1$	328 $_6$	227 $_6$	86 $_{100}$	73 $_{36}$	822	6147
Bupranolol formyl artifact	P-I	1915	283 $_9$	268 $_{99}$	142 $_{15}$	86 $_{67}$	70 $_{73}$	529	1347
Bupranolol-M (HO-) AC	U+UHYAC	2150	314 $_{36}$	272 $_4$	197 $_4$	112 $_{11}$	86 $_{100}$	754	1590
Bupranolol-M (HO-) 2AC	U+UHYAC	2260	356 $_{60}$	314 $_7$	112 $_{18}$	98 $_{18}$	86 $_{100}$	923	1569
Bupranolol-M (HO-) formyl artifact AC	U+UHYAC	2380	341 $_4$	326 $_{100}$	197 $_7$	86 $_{20}$	70 $_{30}$	812	1591
Bupranolol-M (HO-methoxy-) 2AC	U+UHYAC	2500	386 $_{26}$	112 $_{21}$	86 $_{100}$	70 $_{17}$		1021	1592
Buprenorphine	G	3360	467 $_{10}$	435 $_{23}$	410 $_{41}$	378 $_{86}$	55 $_{71}$	1135	212
Buprenorphine AC	U+UHYAC-I	3410	509 $_6$	452 $_{60}$	420 $_{100}$	408 $_{15}$	55 $_{50}$	1169	211
Buprenorphine HFB		2960	663 $_2$	606 $_{31}$	574 $_{100}$	548 $_{16}$	55 $_{79}$	1201	6336
Buprenorphine ME	UME	3330	481 $_4$	448 $_7$	424 $_{30}$	392 $_{100}$	55 $_{50}$	1149	6318
Buprenorphine PFP		3040	613 $_2$	556 $_{38}$	524 $_{100}$	512 $_{21}$	55 $_{51}$	1195	6123
Buprenorphine TFA		2920	563 $_1$	548 $_2$	506 $_{17}$	474 $_{36}$	55 $_{40}$	1188	6337
Buprenorphine TMS		3890	539 $_4$	506 $_{17}$	482 $_{30}$	450 $_{100}$	55 $_{49}$	1180	5698
Buprenorphine 2HFB		2820	645 $_9$	630 $_{10}$	604 $_8$	562 $_{34}$	55 $_{50}$	1205	6345
Buprenorphine 2PFP		2775	595 $_{12}$	580 $_{27}$	554 $_{24}$	512 $_{71}$	55 $_{100}$	1204	6343
Buprenorphine 2TFA		2800	545 $_{15}$	530 $_{17}$	504 $_{13}$	462 $_{49}$	55 $_{60}$	1199	6341
Buprenorphine -CH3OH HFB		2770	631 $_6$	590 $_7$	547 $_{18}$	273 $_{16}$	55 $_{40}$	1198	6339
Buprenorphine -CH3OH TFA		2785	531 $_{12}$	490 $_6$	447 $_{14}$	273 $_{16}$	55 $_{40}$	1179	6338
Buprenorphine -H2O		3240	449 $_{100}$	434 $_{95}$	408 $_{42}$	392 $_{18}$	55 $_{65}$	1113	3421
Buprenorphine -H2O AC		3320	491 $_{100}$	476 $_{95}$	450 $_{67}$	434 $_{29}$	55 $_{71}$	1159	3418
Buprenorphine -H2O HFB		2800	645 $_{15}$	630 $_{22}$	604 $_{11}$	84 $_8$	55 $_{40}$	1199	6344
Buprenorphine -H2O PFP		2730	595 $_4$	580 $_6$	555 $_1$	498 $_1$	55 $_{40}$	1194	6342
Buprenorphine -H2O TFA		2770	545 $_{16}$	530 $_{22}$	504 $_{13}$	434 $_3$	55 $_{40}$	1183	6340
Buprenorphine-D4 ME	UME	3315	485 $_7$	428 $_{37}$	396 $_{100}$	370 $_{14}$	59 $_7$	1153	6354
Buprenorphine-M (nor-)		3420	413 $_3$	395 $_{19}$	356 $_{54}$	338 $_{100}$	324 $_{91}$	1049	7774
Buprenorphine-M (nor-) ME		3330	427 $_6$	409 $_{19}$	395 $_{33}$	370 $_{46}$	338 $_{100}$	1075	7775
Buprenorphine-M (nor-) 2AC		3870	497 $_1$	482 $_2$	440 $_{100}$	408 $_{45}$	366 $_{26}$	1162	7776
Buprenorphine-M (nor-) 2ME	UGLUCME	3100	441 $_4$	409 $_{11}$	384 $_{70}$	352 $_{100}$	326 $_{15}$	1102	6328
Buprenorphine-M (nor-)-D3		3080	416 $_3$	398 $_{18}$	359 $_{44}$	341 $_{100}$	324 $_{91}$	1055	7301
Buprenorphine-M (nor-)-D3 AC		3670	458 $_2$	443 $_2$	401 $_{83}$	383 $_{84}$	366 $_{100}$	1125	7305
Buprenorphine-M (nor-)-D3 ME		3070	430 $_6$	412 $_{25}$	373 $_{45}$	355 $_{39}$	338 $_{100}$	1082	7302
Buprenorphine-M (nor-)-D3 TMS		3080	488 $_1$	470 $_3$	455 $_{10}$	396 $_{56}$	73 $_{100}$	1156	7308
Buprenorphine-M (nor-)-D3 2AC		3690	500 $_1$	485 $_2$	443 $_{100}$	408 $_{41}$	366 $_{29}$	1164	7304
Buprenorphine-M (nor-)-D3 2ME		3050	444 $_3$	426 $_7$	408 $_{14}$	387 $_{84}$	352 $_{100}$	1107	7303
Buprenorphine-M (nor-)-D3 2TMS		3110	560 $_2$	527 $_9$	503 $_{29}$	468 $_{95}$	73 $_{100}$	1187	7307
Buprenorphine-M (nor-)-D3 -H2O 2TFA		2740	590 $_{10}$	555 $_{17}$	533 $_{100}$	478 $_{18}$	81 $_{55}$	1194	7306
Bupropion		1695	239 $_1$	224 $_{11}$	139 $_{16}$	111 $_{25}$	100 $_{100}$	340	4699
Bupropion AC		2210	264 $_{40}$	225 $_{19}$	208 $_{52}$	183 $_{100}$	57 $_{43}$	521	5700
Bupropion formyl artifact		1755	237 $_5$	222 $_4$	139 $_{19}$	98 $_{38}$	57 $_{100}$	384	4700
Bupropion-M (3-Chlorobenzoic acid)		1430*	156 $_{75}$	139 $_{100}$	111 $_{56}$	75 $_{44}$	65 $_6$	136	6024
Bupropion-M (3-Chlorobenzyl alcohol)		1560*	77 $_{100}$	142 $_{53}$	113 $_{28}$	107 $_{33}$		121	6025
Bupropion-M (HO-)		2040	240 $_3$	224 $_{30}$	166 $_7$	139 $_{29}$	116 $_{100}$	401	7660
Bupropion-M (HO-) AC		2130	224 $_{12}$	166 $_4$	158 $_{100}$	115 $_{44}$	98 $_8$	602	7661
Bupropion-M (HO-) TMS		2075	240 $_{17}$	224 $_{79}$	189 $_{97}$	145 $_{53}$	73 $_{100}$	744	7662
Buspirone	G U+UHYAC	3300	385 $_{43}$	290 $_{21}$	277 $_{100}$	265 $_{95}$	177 $_{78}$	971	1779
Butabarbital	P G U UHY U+UHYAC	1655	183 $_6$	156 $_{100}$	141 $_{84}$			250	149
Butabarbital 2ME		1565	211 $_{10}$	184 $_{75}$	169 $_{100}$			345	646
Butabarbital-M (HO-)	U	1925	213 $_4$	199 $_9$	181 $_6$	156 $_{100}$	141 $_{65}$	300	150
Butabarbital-M (HO-) -H2O	UHY U+UHYAC	1905	210 $_6$	181 $_{51}$	156 $_{40}$	141 $_{48}$	55 $_{100}$	245	2952
Butalamine		2590	273 $_1$	188 $_6$	155 $_7$	142 $_{100}$	100 $_{21}$	694	2285

Butalbital Table 1-8-1: Compounds in order of names

Name	Detected	RI	Typical ions and intensities					Page	Entry
Butalbital	P G U UHY U+UHYAC	1690	209_3	181_{30}	168_{100}	167_{88}	141_{24}	289	151
Butalbital (ME)	P U	1630	223_3	195_{17}	182_{97}	181_{60}	155_{10}	337	153
Butalbital 2ME	PME	1655	237_6	209_{23}	196_{100}	195_{80}	169_{19}	391	154
Butalbital 2TMS		1790	368_1	353_{100}	312_{54}	100_{37}	73_{83}	913	4531
Butalbital-M (HO-)	U UHY U+UHYAC	1940	240_2	168_{100}	167_{41}	141_{18}		344	152
Butallylonal	G P U	1990	223_{39}	167_{100}	124_{21}			622	1916
Butane		400*	58_{13}	43_{100}	41_{40}	29_{42}	27_{42}	91	3808
1,2-Butane diol dibenzoate		2300*	298_1	227_{10}	193_{15}	105_{100}	77_{84}	606	1762
1,2-Butane diol dipivalate		1425*	157_2	143_5	103_{10}	85_{29}	57_{100}	414	6425
1,2-Butane diol phenylboronate		1350*	176_{41}	147_{100}	105_{28}	91_{62}	77_{16}	163	1900
1,3-Butane diol dibenzoate		2300*	269_1	241_4	227_6	176_{25}	105_{100}	606	1763
1,3-Butane diol dipivalate		1420*	173_1	157_{10}	103_{21}	85_{17}	57_{100}	414	6424
1,3-Butane diol phenylboronate		1390*	176_{46}	161_{51}	104_{100}	91_{30}	77_{16}	163	1901
1,4-Butane diol dibenzoate		2400*	298_1	193_8	176_8	105_{100}	77_{51}	606	1764
1,4-Butane diol dipivalate		1520*	156_8	103_{14}	101_{13}	85_{20}	57_{100}	414	1906
1,4-Butane diol phenylboronate		1420*	176_{58}	146_{22}	105_{100}	91_{10}		163	1902
Butanilicaine		2030	254_1	219_7	141_{22}	86_{100}	72_{35}	399	1410
1-Butanol		<1000*	74_1	73_1	56_{76}	41_{71}	31_{100}	94	2448
2-Butanol		<1000*	74_2	59_{20}	45_{100}			94	2447
Butaperazine	G U UHY U+UHYAC	3190	409_6	269_4	141_{20}	113_{50}	70_{100}	1039	155
Butaperazine-M (nor-) AC	U+UHYAC	3800	437_{70}	269_{100}	141_{98}	99_{66}		1093	1254
1-Butene		<1000*	56_{34}	41_{100}	39_{47}	27_{58}		90	3807
2-Butene		<1000*	56_{44}	41_{100}	39_{52}	27_{47}		90	3806
Butethamate		1760	263_{15}	248_{15}	191_{42}	99_{26}	86_{100}	437	156
Butethamate-M/artifact (HOOC-)	U UHY U+UHYAC	1300*	164_{11}	119_{29}	91_{100}	77_{10}		145	2912
Butethamate-M/artifact (HOOC-) ME	UME	1200*	178_{11}	150_3	119_{51}	91_{100}	77_4	168	2911
Butinoline	P G U	2285	291_{53}	290_{51}	115_{70}	105_{96}	70_{100}	570	3237
Butinoline artifact-1	U	1990	221_5	175_{19}	147_{100}	115_{44}	77_{27}	275	3239
Butinoline artifact-2	U	2045*	236_{15}	207_8	165_{100}	115_{47}	77_{34}	327	3238
Butinoline-M (benzophenone)	U+UHYAC	1610*	182_{31}	152_3	105_{100}	77_{70}	51_{39}	180	1624
Butinoline-M/artifact	U UHY U+UHYAC	1850*	209_4	167_{100}	152_{22}	121_8	115_7	239	2081
Butinoline-M/artifact	U	2675	304_{48}	220_{69}	115_{48}	105_{100}	98_{51}	632	3236
Butizide 2ME		3785	381_6	366_{13}	324_{100}	246_2	230_2	955	3094
Butizide 3ME		3455	395_7	380_{12}	338_{100}	313_3	230_4	1000	3095
Butizide 4ME	UEXME	3100	409_2	352_{100}	309_3	244_{10}	145_4	1037	3096
Butobarbital	P G U UHY U+UHYAC	1665	197_2	184_{10}	156_{96}	141_{100}	98_{19}	250	157
Butobarbital 2ME		1585	212_7	184_{61}	169_{100}	112_{24}		345	647
Butobarbital 2TMS		1720	356_2	341_{89}	300_{29}	100_{40}	73_{100}	870	5464
Butobarbital-M (HO-)	U UHY	1920	213_2	199_5	156_{100}	141_{73}	98_{12}	301	159
Butobarbital-M (HO-) AC	U+UHYAC	1940	227_{25}	198_{33}	181_{43}	156_{100}	87_{25}	470	2953
Butobarbital-M (oxo-)	U UHY U+UHYAC	1880	211_4	198_{39}	156_{100}	141_{65}	128_{13}	294	158
Butocarboxim		1595	144_{39}	133_4	87_{100}	75_{47}	55_{56}	193	1327
Butoxycarboxim		1940	165_{16}	108_5	86_{97}	85_{100}	55_{78}	280	4382
Butoxycarboxim artifact		1405	149_1	108_7	86_{100}			125	2271
Buturon		2135	236_{28}	152_9	111_{14}	75_{19}	56_{100}	327	4138
1-Butylamine		<1000	73_{14}	39_9	30_{100}			93	4183
2-Butylamine		<1000	73_7	39_5	30_{100}			93	4190
tert.-Butylamine		<1000	73_1	58_{100}	41_{29}	30_{13}		94	4184
Butyl-2-ethylhexylphthalate		1950*	223_7	205_4	149_{100}	104_4	57_7	781	713
Butylhexadecanoate		2340*	312_{24}	257_{48}	239_{23}	129_{23}	56_{100}	677	160
Bis-tert.-Butylmethylenecyclohexanone		1480*	218_{36}	203_{58}	189_{22}	175_{43}	161_{100}	266	5132
Bis-tert.-Butylmethoxymethylphenol		1710*	250_{26}	235_{100}	219_{20}	193_{13}	91_{11}	382	6367
Butyl-2-methylpropylphthalate		1970*	278_1	223_4	205_4	149_{100}	76_4	507	2995
Butyloctadecanoate		2380*	340_{35}	285_{25}	267_{11}	57_{99}		809	161
Butyloctylphthalate		1950*	223_6	205_4	149_{100}	122_4	104_{11}	781	2361
Butylparaben		1700*	194_{16}	138_{99}	121_{97}			205	162
Bis-tert.-Butylquinone		1465*	220_{53}	177_{100}	149_{36}	135_{44}	67_{39}	274	4949
Butylscopolaminium bromide-M/artifact (scopolamine) -H2O	U+UHYAC	2230	285_{18}	154_{22}	138_{38}	108_{43}	94_{100}	540	960
Butyl stearate		2380*	340_{48}	285_{100}	267_{41}	241_{17}	129_{18}	809	2869
gamma-Butyrolactone		<1000*	86_{11}	56_{16}	42_{100}			97	7275
BZP		1530	176_9	146_5	134_{55}	91_{100}	56_{29}	164	5880
BZP AC		1915	218_4	146_{27}	132_{17}	91_{100}	85_{20}	266	5881
BZP HFB		1730	372_{13}	295_6	281_{15}	175_{13}	91_{100}	926	5884
BZP PFP		1690	322_{18}	245_9	231_{21}	175_{15}	91_{100}	722	5883
BZP TFA		1665	272_3	195_4	181_{10}	146_6	91_{100}	479	5882
BZP TMS		1860	248_{31}	157_{27}	102_{100}	91_{69}	73_{97}	372	5885
BZP-M (piperazine) 2AC		1750	170_{15}	85_{33}	69_{25}	56_{100}		157	879
BZP-M (piperazine) 2HFB		1290	478_3	459_9	309_{100}	281_{22}	252_{10}	1146	6634
BZP-M (piperazine) 2TFA		1005	278_{10}	209_{59}	152_{25}	69_{56}	56_{100}	505	4217
Cafedrine TMS		2815	415_1	250_{100}	207_{32}	73_{40}	70_{20}	1079	6216
Cafedrine -H2O	G P UHY	2960	339_2	277_3	250_{100}	207_{40}	70_{10}	803	1313
Cafedrine -H2O AC	U+UHYAC	3285	381_1	339_2	292_{52}	250_{100}	207_{38}	957	1739
Cafedrine -H2O PFP		2790	485_1	339_6	206_{36}	146_{59}	132_{100}	1152	6118
Cafedrine-M (cathinone) AC		1610	191_2	134_2	105_{35}	86_{100}	77_{48}	195	5901

Table 1-8-1: Compounds in order of names   **Cannabidivarol 2AC**

Name	Detected	RI	Typical ions and intensities					Page	Entry
Cafedrine-M (cathinone) HFB		1395	345 $_1$	240 $_6$	169 $_4$	105 $_{100}$	77 $_{36}$	827	5904
Cafedrine-M (cathinone) PFP		1335	190 $_6$	119 $_7$	105 $_{100}$	77 $_{40}$	69 $_5$	588	5903
Cafedrine-M (cathinone) TFA		1350	245 $_1$	140 $_7$	105 $_{100}$	77 $_{48}$	69 $_{10}$	358	5902
Cafedrine-M (cathinone) TMS		1590	206 $_{14}$	191 $_{15}$	116 $_{100}$	77 $_{27}$	73 $_{74}$	278	5905
Cafedrine-M (etofylline)	UHY	2125	224 $_{47}$	194 $_{17}$	180 $_{100}$	109 $_{35}$	95 $_{74}$	288	771
Cafedrine-M (etofylline) AC	U+UHYAC	2200	266 $_{79}$	206 $_{59}$	180 $_{34}$	122 $_{31}$	87 $_{100}$	448	772
Cafedrine-M (etofylline) TMS	UHYTMS	2160	296 $_{10}$	281 $_{26}$	252 $_6$	180 $_{100}$	73 $_{47}$	599	5696
Cafedrine-M (N-dealkyl-) AC	U+UHYAC	2480	265 $_{35}$	206 $_{100}$	180 $_{35}$	122 $_{18}$	86 $_{27}$	443	1886
Cafedrine-M (norpseudoephedrine)	U UHY	1360	132 $_4$	117 $_9$	105 $_{22}$	79 $_{54}$	77 $_{100}$	129	1154
Cafedrine-M (norpseudoephedrine) TMSTFA		1630	213 $_7$	191 $_7$	179 $_{100}$	149 $_5$	73 $_{80}$	707	6260
Cafedrine-M (norpseudoephedrine) 2AC	U+UHYAC	1740	235 $_2$	176 $_4$	129 $_8$	107 $_9$	86 $_{100}$	324	1155
Cafedrine-M (norpseudoephedrine) 2HFB		1335	330 $_{16}$	303 $_6$	240 $_{100}$	169 $_{19}$	119 $_{12}$	1182	7418
Cafedrine-M (norpseudoephedrine) formyl artifact		1280	117 $_2$	105 $_2$	91 $_4$	77 $_6$	57 $_{100}$	143	4649
Caffeic acid ME2AC		2170*	278 $_3$	236 $_{15}$	194 $_{100}$	163 $_{40}$	134 $_{21}$	505	5967
Caffeic acid ME2HFB		1985*	586 $_{45}$	555 $_{59}$	389 $_{15}$	169 $_{74}$	69 $_{80}$	1192	5971
Caffeic acid ME2PFP		1985*	486 $_{61}$	455 $_{89}$	323 $_{14}$	119 $_{100}$	77 $_{31}$	1153	5970
Caffeic acid ME2TMS		1930*	338 $_{72}$	297 $_9$	219 $_{100}$	191 $_{21}$	73 $_{97}$	797	6013
Caffeic acid 2AC		2240*	264 $_2$	222 $_{12}$	180 $_{100}$	163 $_8$	134 $_{15}$	438	5968
Caffeic acid 2ME		1930*	208 $_{100}$	177 $_{80}$	145 $_{53}$	133 $_{19}$	117 $_{24}$	236	5966
Caffeic acid 3ME	UME	1850*	222 $_{100}$	207 $_{22}$	191 $_{75}$	164 $_{14}$	147 $_{19}$	281	4945
Caffeic acid 3TMS		2115*	396 $_{86}$	381 $_{22}$	219 $_{78}$	191 $_{17}$	73 $_{100}$	1006	6014
Caffeic acid artifact (dihydro-)		2400*	182 $_{17}$	136 $_{13}$	123 $_{100}$	77 $_{55}$	51 $_{67}$	179	5763
Caffeic acid artifact (dihydro-) ME		1870*	196 $_{24}$	136 $_{45}$	123 $_{100}$	91 $_{11}$	77 $_{17}$	210	5764
Caffeic acid artifact (dihydro-) ME2HFB		1720*	588 $_{11}$	528 $_{100}$	349 $_{32}$	169 $_{37}$	69 $_{58}$	1193	5994
Caffeic acid artifact (dihydro-) ME2PFP		1590*	488 $_{19}$	428 $_{100}$	299 $_{26}$	281 $_{26}$	119 $_{73}$	1155	5993
Caffeic acid artifact (dihydro-) ME2TMS		2220*	340 $_{44}$	267 $_{36}$	193 $_{10}$	179 $_{100}$	73 $_{97}$	807	5995
Caffeic acid artifact (dihydro-) MEAC		1980*	280 $_2$	238 $_{15}$	196 $_{100}$	136 $_{65}$	123 $_{68}$	516	5992
Caffeic acid artifact (dihydro-) METFA		1540*	292 $_{88}$	233 $_{100}$	195 $_{27}$	107 $_{46}$	69 $_{62}$	574	5969
Caffeic acid artifact (dihydro-) 3TMS		2250*	398 $_{67}$	280 $_{16}$	267 $_{39}$	179 $_{98}$	73 $_{100}$	1012	5996
Caffeic acid artifact (dihydro-) -CO2		1295*	138 $_{23}$	123 $_{100}$	91 $_{23}$	77 $_{47}$	51 $_{92}$	118	5756
Caffeic acid -CO2		1375*	136 $_{53}$	89 $_{71}$	77 $_{43}$	63 $_{71}$	51 $_{100}$	116	5757
Caffeine	P G U UHY U+UHYAC	1820	194 $_{100}$	109 $_{80}$	82 $_{33}$	67 $_{44}$	55 $_{67}$	204	191
Caffeine-M (1-nor-)	P G U+UHYAC	1980	180 $_{100}$	137 $_{12}$	109 $_{41}$	82 $_{38}$		174	989
Caffeine-M (1-nor-) TMS		2020	252 $_{20}$	237 $_{100}$	109 $_{15}$	100 $_{12}$	73 $_{25}$	390	5452
Caffeine-M (7-nor-)	P G U+UHYAC	2025	180 $_{100}$	95 $_{85}$	68 $_{69}$			174	990
Caffeine-M (7-nor-) TMS		1920	252 $_{61}$	237 $_{100}$	223 $_{14}$	135 $_7$	73 $_{37}$	391	4600
Caffeine-M (HO-) ME	UME	1930	224 $_{100}$	209 $_{66}$	139 $_7$	124 $_{29}$	83 $_{66}$	288	5044
Californine		2615	323 $_{12}$	322 $_{13}$	188 $_{100}$	165 $_4$	130 $_7$	725	5770
Californine-M (bis-(demethylene-methyl-)) iso-1		2860	327 $_{35}$	311 $_5$	190 $_{100}$	175 $_7$		744	6734
Californine-M (bis-(demethylene-methyl-)) iso-1 2AC		2920	411 $_7$	368 $_6$	348 $_8$	232 $_{100}$	190 $_{59}$	1044	6729
Californine-M (bis-(demethylene-methyl-)) iso-2 2AC		3040	411 $_{11}$	368 $_6$	326 $_5$	232 $_{100}$	190 $_{63}$	1045	6730
Californine-M (bis-(demethylene-methyl-)) iso-3 2AC		3055	411 $_{19}$	368 $_{15}$	326 $_{11}$	232 $_{98}$	190 $_{100}$	1045	6731
Californine-M (demethylene-) AC		2960	353 $_{37}$	310 $_{23}$	218 $_{21}$	188 $_{100}$	176 $_{23}$	857	6723
Californine-M (demethylene-) 2AC		3025	395 $_{22}$	310 $_{16}$	218 $_{16}$	188 $_{100}$	176 $_{13}$	1002	6724
Californine-M (demethylene-methyl-) iso-1		2810	325 $_{42}$	310 $_4$	294 $_4$	190 $_{39}$	188 $_{100}$	735	6725
Californine-M (demethylene-methyl-) iso-1 AC		2910	367 $_{33}$	324 $_{13}$	250 $_5$	232 $_{30}$	188 $_{100}$	908	6727
Californine-M (demethylene-methyl-) iso-2		2820	325 $_{67}$	309 $_{13}$	294 $_{10}$	190 $_{91}$	188 $_{100}$	735	6726
Californine-M (demethylene-methyl-) iso-2 AC		2920	367 $_{33}$	324 $_{21}$	310 $_6$	232 $_{18}$	188 $_{100}$	909	6728
Californine-M (nor-)		2625	309 $_{39}$	174 $_{100}$	147 $_{11}$	95 $_6$		659	6732
Californine-M (nor-) AC		3090	351 $_{45}$	308 $_{28}$	292 $_{12}$	216 $_{25}$	174 $_{100}$	847	6733
Californine-M (nor-demethylene-) 3AC		3350	423 $_{15}$	339 $_{16}$	280 $_{22}$	216 $_{37}$	174 $_{100}$	1066	6735
Californine-M (nor-demethylene-methyl-) 2AC		3220	395 $_{37}$	353 $_{37}$	310 $_{41}$	216 $_{36}$	174 $_{100}$	1002	6736
Californine-M/artifact (reframidine)		2735	323 $_{33}$	322 $_{42}$	280 $_{40}$	188 $_{100}$		726	6737
Camazepam	G	2960	371 $_9$	299 $_6$	271 $_{33}$	255 $_{16}$	72 $_{100}$	922	416
Camazepam HY	UHY U+UHYAC	2100	245 $_{95}$	228 $_{38}$	193 $_{29}$	105 $_{38}$	77 $_{100}$	358	272
Camazepam HYAC	U+UHYAC	2260	287 $_{11}$	244 $_{100}$	228 $_{39}$	182 $_{49}$	77 $_{70}$	546	2542
Camazepam-M	P G UGLUC	2320	268 $_{98}$	239 $_{57}$	233 $_{52}$	205 $_{66}$	77 $_{100}$	542	579
Camazepam-M (temazepam)	P UGLUC	2625	300 $_{33}$	271 $_{100}$	256 $_{23}$	228 $_{16}$	77 $_{30}$	614	417
Camazepam-M (temazepam) AC	UGLUCAC	2730	342 $_6$	300 $_{40}$	271 $_{100}$	255 $_{16}$	77 $_{17}$	816	2099
Camazepam-M (temazepam) ME		2600	314 $_{60}$	271 $_{100}$	255 $_{46}$			684	4
Camazepam-M (temazepam) TMS		2665	372 $_{23}$	343 $_{100}$	283 $_{26}$	257 $_{38}$	73 $_{54}$	926	4598
Camazepam-M (temazepam) artifact-1	G	2475	256 $_{19}$	241 $_7$	179 $_{100}$	163 $_7$	77 $_8$	405	5780
Camazepam-M (temazepam) artifact-2	G	2815	270 $_{64}$	269 $_{100}$	254 $_{12}$	228 $_{26}$	191 $_5$	468	5779
Camazepam-M TMS		2635	356 $_9$	341 $_{100}$	312 $_{56}$	239 $_{12}$	135 $_{21}$	876	4577
Camazepam-M 2TMS		2200	430 $_{36}$	429 $_{63}$	340 $_{10}$	313 $_{14}$	73 $_{70}$	1080	5499
Camazepam-M HY	UHY	2050	231 $_{80}$	230 $_{95}$	154 $_{23}$	105 $_{38}$	77 $_{100}$	308	419
Camazepam-M HYAC	U+UHYAC PHYAC	2245	273 $_{31}$	230 $_{100}$	154 $_{13}$	105 $_{23}$	77 $_{50}$	480	273
Camylofine	G U UHY U+UHYAC	2085	320 $_3$	205 $_2$	118 $_6$	86 $_{100}$	58 $_{15}$	716	1411
Cannabidiol	G U-I	2400*	314 $_5$	246 $_{16}$	231 $_{100}$	174 $_{21}$	121 $_{27}$	687	648
Cannabidiol AC		2420*	356 $_{12}$	273 $_{26}$	231 $_{100}$	174 $_{22}$	121 $_{32}$	870	6461
Cannabidiol 2AC		2450*	398 $_{14}$	355 $_{28}$	273 $_{30}$	231 $_{100}$	121 $_{38}$	1013	649
Cannabidiol 2TMS		2330*	458 $_8$	390 $_{93}$	337 $_{38}$	301 $_{20}$	73 $_{81}$	1125	4679
Cannabidivarol		2165*	286 $_7$	218 $_{14}$	203 $_{100}$	174 $_9$	121 $_8$	544	4071
Cannabidivarol 2AC		2630*	370 $_{19}$	327 $_{100}$	285 $_{70}$	257 $_{12}$		921	4072

Table 1-8-1: Compounds in order of names

Name	Detected	RI	Typical ions and intensities					Page	Entry
Cannabielsoic acid -CO2		2405*	330$_{67}$	247$_{76}$	205$_{100}$	148$_{43}$	108$_{20}$	760	4073
Cannabielsoic acid -CO2 2AC		2540*	414$_9$	330$_{51}$	312$_{100}$	247$_{50}$	205$_{70}$	1051	4074
Cannabigerol		2500*	316$_{17}$	247$_{12}$	231$_{31}$	193$_{100}$	123$_{29}$	694	4075
Cannabigerol 2AC		2595*	400$_7$	247$_{73}$	193$_{67}$	123$_{90}$	69$_{100}$	1018	4076
Cannabinol	G UHY	2555*	310$_{13}$	295$_{98}$	238$_{23}$	223$_9$		668	650
Cannabinol AC	U+UHYAC	2540*	352$_{28}$	337$_{100}$	295$_{71}$	238$_{20}$		855	651
Cannabinol TMS		2485*	382$_{10}$	367$_{100}$	310$_6$	238$_5$	73$_{17}$	960	4532
Cannabispirol AC		2350*	290$_{18}$	248$_{19}$	189$_{92}$	176$_{100}$	115$_{16}$	565	6462
Cannabispirone AC		2350*	288$_7$	189$_{10}$	176$_{100}$	115$_9$		553	6463
Canrenoic acid		3100*	358$_{100}$	329$_{35}$	274$_{13}$	201$_{14}$	85$_{68}$	878	2743
Canrenoic acid -H2O	P UHY U+UHYAC	3250*	340$_{100}$	325$_{18}$	267$_{80}$	227$_{15}$		808	2344
Canrenoic acid -H2O ME		3130*	354$_{100}$	339$_{19}$	173$_9$	149$_8$	115$_8$	862	2744
Canrenone	P UHY U+UHYAC	3250*	340$_{100}$	325$_{18}$	267$_{80}$	227$_{15}$		808	2344
Capric acid		1340*	172$_3$	129$_{34}$	87$_{15}$	73$_{99}$	60$_{100}$	160	5629
Capric acid ET		1370*	200$_3$	157$_{23}$	155$_{24}$	101$_{41}$	88$_{100}$	220	5399
Capric acid ME		1360*	186$_3$	155$_8$	143$_{15}$	87$_{45}$	74$_{100}$	189	2665
Caprylic acid ET		1185*	172$_2$	143$_6$	127$_{22}$	101$_{40}$	88$_{100}$	160	5398
Caprylic acid ME		1170*	158$_1$	127$_{12}$	115$_{10}$	87$_{39}$	74$_{100}$	138	2664
Caprylic acid cetylester		2500*	368$_3$	224$_4$	145$_{100}$	88$_{26}$	57$_{65}$	914	6565
Capsaicine		2415	305$_{32}$	195$_{15}$	152$_{43}$	137$_{200}$	122$_{22}$	641	6780
Capsaicine AC		2490	347$_5$	305$_{29}$	195$_{33}$	152$_{45}$	137$_{100}$	836	6782
Captafol		2355	347$_3$	311$_6$	183$_5$	107$_9$	79$_{100}$	833	3320
Captafol artifact-1 (cyclohexenedicarboxylic acid) 2ME		1190*	198$_1$	167$_{18}$	138$_{47}$	107$_{14}$	79$_{100}$	214	4206
Captafol artifact-2 (cyclohexenedicarboximide)		1450	151$_{41}$	123$_9$	80$_{63}$	79$_{100}$		129	3321
Captan		2030	299$_1$	264$_5$	149$_{19}$	117$_{14}$	79$_{100}$	608	2614
Captan artifact-1 (cyclohexenedicarboxylic acid) 2ME		1190*	198$_1$	167$_{18}$	138$_{47}$	107$_{14}$	79$_{100}$	214	4206
Captan artifact-2 (cyclohexenedicarboximide)		1450	151$_{41}$	123$_9$	80$_{63}$	79$_{100}$		129	3321
Captopril		1925	217$_{12}$	199$_{11}$	140$_{17}$	126$_{16}$	70$_{100}$	261	6417
Captopril ME		1730	231$_{13}$	199$_{12}$	172$_{16}$	128$_{15}$	70$_{100}$	309	3005
Captopril 2ME		1810	245$_{25}$	198$_{17}$	128$_{38}$	89$_{39}$	70$_{100}$	358	6418
Captopril artifact (disulfide) 2ME		3200	460$_7$	230$_{100}$	198$_{76}$	128$_{68}$	70$_{99}$	1128	6419
Carazolol	U-I	2810	298$_{17}$	183$_{100}$	154$_{16}$	72$_{59}$		607	1593
Carazolol ME		2815	312$_7$	183$_{21}$	154$_{12}$	86$_{100}$	72$_{12}$	676	1595
Carazolol TMSTFA		2755	466$_{24}$	284$_{100}$	183$_{34}$	129$_{80}$	73$_{58}$	1133	6178
Carazolol 2TMSTFA		2880	538$_{26}$	284$_{100}$	129$_{28}$	73$_{44}$		1180	6179
Carazolol formyl artifact	U-I	2830	310$_{29}$	183$_{100}$	154$_{22}$	127$_{51}$	86$_{26}$	668	1352
Carazolol -H2O AC		3130	322$_{54}$	220$_{100}$	140$_{100}$	98$_{65}$		723	1353
Carazolol-M (deamino-di-HO-) 2AC	UGLUCAC-I	3050	341$_4$	199$_{17}$	159$_{100}$	99$_{16}$		811	1594
Carazolol-M (deamino-tri-HO-) 3AC	UGLUCAC	3290	399$_{27}$	372$_{20}$	199$_{11}$	159$_{63}$		1014	4253
Carbamazepine	P G U+UHYAC	2285	236$_{83}$	193$_{100}$	165$_{31}$			328	420
Carbamazepine TMS		2285	308$_1$	293$_4$	193$_{100}$	165$_{13}$	73$_{15}$	656	4533
Carbamazepine-M (acridine)	U UHY U+UHYAC	1800	179$_{100}$	151$_{14}$				170	421
Carbamazepine-M (formyl-acridine)	U UHY U+UHYAC	2025	207$_{98}$	179$_{100}$	151$_{36}$			232	422
Carbamazepine-M (HO-methoxy-ring)	U UHY	2340	239$_{100}$	224$_{47}$	209$_{42}$	180$_{74}$		339	423
Carbamazepine-M (HO-methoxy-ring) AC	U+UHYAC	2420	281$_{42}$	239$_{100}$	224$_{28}$	196$_{29}$	162$_{16}$	521	2506
Carbamazepine-M (HO-ring)	UHY	2240	209$_{100}$	180$_{16}$	152$_7$			241	2511
Carbamazepine-M (HO-ring) AC	U+UHYAC	2450	251$_{33}$	209$_{100}$	180$_{74}$	152$_{11}$		383	425
Carbamazepine-M (HO-ring) 2AC	U+UHYAC	2490	293$_{21}$	251$_{25}$	209$_{79}$	208$_{100}$	178$_{17}$	579	2672
Carbamazepine-M AC	U+UHYAC	3195	340$_{81}$	298$_{95}$	297$_{100}$	241$_7$	179$_{16}$	805	426
Carbamazepine-M cysteine-conjugate (ME)	U	2715	326$_{34}$	283$_{44}$	180$_{100}$	152$_{33}$		739	428
Carbamazepine-M/artifact (ring)	P U UHY U+UHYAC	1985	193$_{100}$	165$_{19}$	139$_5$	113$_3$	96$_9$	200	309
Carbamazepine-M/artifact AC	U+UHYAC	2040	235$_{27}$	193$_{100}$	192$_{68}$	165$_{17}$		323	2671
Carbaryl		1865	201$_2$	144$_{100}$	115$_{59}$	89$_{10}$	63$_{11}$	221	3751
Carbaryl TFA		1785	297$_{11}$	240$_{54}$	143$_{55}$	115$_{82}$	69$_{100}$	601	4134
Carbaryl-M/artifact (1-naphthol)		1500*	144$_{75}$	115$_{100}$	89$_{17}$	74$_7$	63$_{23}$	123	928
Carbaryl-M/artifact (1-naphthol) AC	U+UHYAC	1555*	186$_{13}$	144$_{100}$	115$_{47}$	89$_8$	63$_7$	188	932
Carbaryl-M/artifact (1-naphthol) HFB		1310*	340$_{46}$	169$_{25}$	143$_{28}$	115$_{100}$	89$_{11}$	806	7476
Carbaryl-M/artifact (1-naphthol) PFP		1510*	290$_{45}$	171$_{100}$	143$_{20}$	115$_{49}$	89$_8$	562	7468
Carbaryl-M/artifact (1-naphthol) TMS		1525*	216$_{100}$	201$_{95}$	185$_{51}$	115$_{39}$	73$_{21}$	260	7460
Carbendazim -C2H2O2		1930	133$_{100}$	105$_{26}$	79$_{13}$	63$_6$		113	4033
Carbetamide		1975	236$_{12}$	165$_3$	119$_{100}$	93$_{35}$	72$_{28}$	329	3172
Carbetamide TFA		1870	332$_{17}$	196$_{73}$	124$_{80}$	119$_{100}$	77$_{34}$	768	4127
Carbetamide 2ME		1965	264$_3$	158$_{30}$	134$_{66}$	86$_{73}$	58$_{100}$	440	4095
Carbidopa 2ME	U+UHYAC	1660	224$_{29}$	164$_3$	137$_{100}$	122$_6$		399	1805
Carbidopa 2MEAC		1990	266$_5$	138$_{37}$	129$_{100}$	97$_{25}$	69$_{27}$	599	1807
Carbidopa 3ME		1680	238$_{19}$	222$_{20}$	162$_{27}$	151$_{100}$	137$_{17}$	461	1806
Carbidopa 3MEAC		2100	294$_6$	280$_{41}$	221$_{22}$	157$_{42}$	143$_{100}$	732	1810
Carbidopa iso-1 3MEAC		2060	280$_6$	221$_6$	152$_{44}$	124$_{100}$	97$_{29}$	667	1808
Carbidopa iso-2 3MEAC		2080	280$_{19}$	221$_9$	143$_{100}$	115$_9$		668	1809
Carbidopa-M (di-HO-phenylacetone) 2AC	U+UHYAC	1735*	250$_3$	208$_{15}$	166$_{45}$	123$_{100}$		379	4210
Carbidopa-M (HO-methoxy-phenylacetone) AC	U+UHYAC	1600*	222$_2$	180$_{22}$	137$_{100}$			280	4211
Carbidopa-M (HO-methoxy-phenylacetone) ME	UHYME	1540*	194$_{25}$	151$_{100}$	135$_4$	107$_{18}$	65$_4$	204	4353
Carbimazole	G U+UHYAC	1705	186$_{74}$	114$_{88}$	109$_{23}$	81$_{35}$	72$_{100}$	187	4705
Carbimazole-M/artifact (thiamazole)	G P-I	1615	114$_{100}$	99$_5$	81$_{16}$	72$_{30}$	69$_{18}$	104	4703

Table 1-8-1: Compounds in order of names

Name	Detected	RI	Typical ions and intensities					Page	Entry
Carbimazole-M/artifact (thiamazole) AC	GAC PAC-I U+UHYAC	1440	156_{40}	114_{100}	86_{10}	81_{18}	72_{20}	136	4704
Carbimazole-M/artifact (thiamazole) ME	GME PME-I	1205	128_{100}	113_8	95_{22}	72_{18}	59_7	111	4687
Carbimazole-M/artifact (thiamazole) TMS	GTMS PTMS-I	1400	186_{51}	171_{100}	116_7	113_8	73_{23}	188	4688
Carbinoxamine	G U+UHYAC	2120	218_1	203_3	167_9	71_{62}	58_{100}	564	1780
Carbinoxamine-M (bis-nor-) AC	U+UHYAC	2430	304_1	218_{100}	203_{65}	167_{97}	86_{50}	634	2171
Carbinoxamine-M (carbinol)	UHY	1670	219_{54}	139_{15}	108_{37}	79_{100}		268	2173
Carbinoxamine-M (carbinol) AC	U+UHYAC	1700	261_4	218_{100}	201_{33}	167_{30}	78_{31}	424	2167
Carbinoxamine-M (Cl-benzoylpyridine)	UHY U+UHYAC	1645	217_{35}	189_{98}	139_{100}	111_{88}	75_{80}	261	2166
Carbinoxamine-M (deamino-HO-) AC	U+UHYAC	2240	305_3	218_{89}	203_{86}	167_{92}	87_{100}	638	2169
Carbinoxamine-M (nor-)	UHY	2150	276_1	220_{40}	203_{100}	167_{61}	139_{13}	496	2174
Carbinoxamine-M (nor-) AC	U+UHYAC	2400	318_1	218_{41}	203_{55}	167_{89}	100_{100}	703	2170
Carbinoxamine-M/artifact	U+UHYAC	2170	239_4	218_{100}	202_{29}	167_{64}	78_{25}	338	2168
Carbinoxamine-M/artifact	UHY	1600	202_{100}	167_{90}	139_{11}			221	2172
Carbochromene	G U UHY U+UHYAC	2850	360_1	316_1	289_3	86_{100}	58_{25}	887	2586
Carbofuran		1660	221_5	164_{100}	149_{56}	123_{17}	58_{16}	276	3899
Carbofuran -C2H3NO		1060*	164_{100}	149_{85}	131_{35}	122_{33}	103_{22}	146	3900
Carbon disulfide		<1000*	76_{100}	44_{17}				95	2754
Carbophenothion		2320*	342_{17}	199_{17}	157_{100}	121_{46}	97_{48}	815	3322
Carboxin		2410	235_{59}	143_{100}	115_9	87_{33}	77_5	322	3884
Carbromal	P G U	1515	208_{41}	191_4	165_{16}	114_{14}	69_{100}	327	652
Carbromal artifact		----	171_3	143_{50}	57_{100}			158	739
Carbromal artifact	P	1115	157_2	129_{97}	114_{14}	87_{19}	57_{100}	137	1026
Carbromal artifact	G P U	1480	223_{29}	191_5	149_{11}	102_{15}	69_{100}	282	1880
Carbromal artifact		1470	191_9	149_{10}	140_{18}	112_9	69_{100}	194	1879
Carbromal artifact		1450	179_{24}	105_{22}	69_{100}			169	1878
Carbromal-M		----	113_{92}	98_{100}	69_{35}	55_{75}		103	658
Carbromal-M (cyamuric acid)		----	129_3	114_{36}	98_{27}	85_{31}	57_{100}	111	657
Carbromal-M (desbromo-)		1380	143_2	130_{59}	113_{95}	87_{99}	71_{62}	138	655
Carbromal-M (desbromo-HO-) -H2O		----	156_9	139_{37}	113_{54}	98_{51}	69_{100}	136	656
Carbromal-M (ethyl-HO-butyric acid) ME		<1000*	117_{10}	87_{100}	69_{20}	57_{73}		124	659
Carbromal-M (HO-carbromide)	U	1340	194_3	181_4	165_{79}	150_{100}	69_{28}	239	654
Carbromal-M/artifact (carbromide)	P G U	1215	165_{67}	150_{18}	114_{52}	69_{100}	55_{38}	199	653
Carisoprodol	P U+UHYAC	2150	260_1	245_5	158_{48}	97_{62}	55_{100}	423	2792
Carisoprodol artifact		1585	202_5	104_{100}	84_{21}	69_{21}	55_{66}	221	5682
Carisoprodol-M (dealkyl-)	P G U+UHYAC	1785	144_{19}	114_{27}	96_{33}	83_{88}	55_{99}	265	1088
Carisoprodol-M (dealkyl-) artifact-1	P G U	1535*	84_{100}	56_{81}				96	1089
Carisoprodol-M (dealkyl-) artifact-2	P U UHY U+UHYAC	1720*	173_2	101_9	84_{99}	56_{90}		160	580
Carphedone		2170	218_{17}	174_{100}	160_{76}	145_{47}	104_{60}	264	5912
Carphedone TMS		2400	290_1	275_5	175_{100}	104_{73}	73_{28}	564	6030
Carphedone 2TMS		2460	362_1	347_4	247_{33}	188_{31}	73_{100}	891	6031
Carprofen		2280	227_{100}	201_5	191_{27}	164_6		481	1999
Carprofen ME		2750	287_{49}	228_{100}	193_{47}	165_4	114_4	546	2001
Carprofen 2ME		2630	301_{69}	242_{100}	227_8	207_{42}	191_8	618	5134
Carprofen -CO2		2250	229_{71}	214_{100}	193_8	178_{13}	152_9	303	2000
Carprofen-M (HO-) iso-1 2ME	UME	2740	317_{58}	258_{73}	216_{100}	181_{16}	129_8	696	6285
Carprofen-M (HO-) iso-1 3ME	UME	2805	331_{100}	316_{81}	272_{81}	256_{25}	222_{16}	763	6288
Carprofen-M (HO-) iso-2 2ME	UME	2810	317_{49}	258_{100}	223_{26}	208_{13}	180_{10}	696	6287
Carprofen-M (HO-) iso-2 3ME	UME	2865	331_{68}	272_{100}	257_6	237_{19}	194_9	763	6289
Carteolol		2670	292_6	277_{13}	202_3	86_{100}	57_{15}	576	2610
Carteolol AC		2700	334_4	319_5	163_6	86_{25}	57_{100}	780	1355
Carteolol formyl artifact		2690	304_{22}	289_{100}	202_{38}	141_{73}	70_{57}	636	1354
Carteolol-M (deisobutyl-) -H2O AC	U+UHYAC	2430	260_{54}	188_{18}	161_{100}	99_{33}	57_{17}	421	1596
Carteolol-M (HO-) 2AC	U+UHYAC	2800	392_{20}	377_{46}	335_{12}	218_{22}	86_{92}	992	1597
Carvedilol TMSTFA		2970	574_{16}	451_{32}	392_{100}	183_{50}	73_{65}	1190	6140
Carzenide 2ME		1920	229_{50}	198_{48}	135_{100}	103_{40}	76_{55}	303	2479
Carzenide 3ME	UME	1850	243_{54}	199_{18}	135_{100}	104_{46}	76_{51}	354	2480
Catechol 2TMS		1245*	254_{13}	239_4	151_6	136_5	73_{80}	399	6021
Cathine	U UHY	1360	132_4	117_9	105_{22}	79_{54}	77_{100}	129	1154
Cathine TMSTFA		1630	213_7	191_7	179_{100}	149_5	73_{80}	707	6260
Cathine 2AC	U+UHYAC	1740	235_2	176_4	129_8	107_9	86_{100}	324	1155
Cathine 2HFB		1335	330_{16}	303_6	240_{100}	169_{19}	119_{12}	1182	7418
Cathine formyl artifact		1280	117_2	105_2	91_4	77_6	57_{100}	143	4649
Cathinone AC		1610	191_2	134_2	105_{35}	86_{100}	77_{48}	195	5901
Cathinone HFB		1395	345_1	240_6	169_4	105_{100}	77_{36}	827	5904
Cathinone PFP		1335	190_6	119_7	105_{40}	77_{100}	69_5	588	5903
Cathinone TFA		1350	245_1	140_7	105_{100}	77_{48}	69_{10}	358	5902
Cathinone TMS		1590	206_{14}	191_{15}	116_{100}	77_{27}	73_{74}	278	5905
Caulophyllin		1995	204_{16}	160_5	146_{10}	117_7	58_{100}	227	5597
2C-B		1785	259_{15}	230_{100}	215_{29}	199_{10}	77_{37}	415	3254
2C-B AC		2180	301_{15}	242_{100}	229_{31}	199_{12}	148_{39}	617	3267
2C-B HFB		2030	455_{32}	242_{98}	229_{79}	199_{28}	148_{92}	1120	6941
2C-B PFP		1995	405_{36}	242_{100}	229_{85}	199_{33}	148_{35}	1029	6936
2C-B TFA		2000	355_{42}	242_{99}	229_{83}	199_{37}	148_{36}	864	6931
2C-B TMS		1935	331_1	272_3	229_2	102_{100}	73_{50}	761	6925

Table 1-8-1: Compounds in order of names

Name	Detected	RI	Typical ions and intensities					Page	Entry
2C-B 2AC		2230	343_7	242_{99}	229_{32}	201_{12}	148_{29}	819	6924
2C-B 2TMS		2195	403_1	388_{14}	272_7	207_{11}	174_{100}	1025	6926
2C-B formyl artifact		1840	271_{12}	240_{65}	229_{27}	199_{11}	77_{24}	473	3245
2C-B formyl artifact		1840	271_{12}	240_{65}	229_{27}	199_{11}	42_{100}	473	5522
2C-B intermediate-1 (2,5-dimethoxyphenyl-2-nitroethene)		1900	209_{100}	162_{51}	147_{52}	133_{61}	77_{62}	240	3286
2C-B intermediate-2 (2,5-dimethoxyphenethylamine)		1630	181_{15}	162_{33}	152_{100}	137_{47}	121_{27}	177	3287
2C-B intermediate-2 (2,5-dimethoxyphenethylamine)		1630	181_{15}	162_{33}	152_{100}	137_{47}	44_{95}	178	5523
2C-B intermediate-2 (2,5-dimethoxyphenethylamine) AC		1935	223_{15}	164_{100}	149_{29}	121_{31}	91_{15}	284	3288
2C-B intermediate-2 (2,5-dimethoxyphenethylamine) formyl artifact		1540	193_{12}	162_{100}	151_{30}	121_{43}	91_{29}	200	3293
2C-B intermediate-2 (2,5-dimethoxyphenethylamine) formyl artifact		1540	193_{12}	162_{100}	151_{30}	121_{43}	42_{64}	200	5524
2C-B precursor (2,5-dimethoxybenzaldehyde)		1345*	166_{100}	151_{39}	120_{36}	95_{61}	63_{62}	150	3278
2C-B-M (deamino-COOH) ME		2030*	288_{100}	273_{12}	241_8	229_{67}	199_{19}	550	7212
2C-B-M (deamino-di-HO-) 2AC	U+UHYAC	2230*	360_{13}	300_{27}	258_{43}	245_{100}	138_{17}	882	7214
2C-B-M (deamino-di-HO-) 2TFA		1790*	468_{100}	354_{14}	341_{77}	311_{14}	276_{76}	1136	7208
2C-B-M (deamino-HO-) AC	U+UHYAC	2300*	302_{20}	242_{100}	227_{25}	183_{33}	148_{42}	622	7198
2C-B-M (deamino-HO-) TFA		1880*	356_{31}	341_{10}	242_{64}	229_{46}	148_{46}	867	7209
2C-B-M (deamino-oxo-)		2020*	258_{35}	229_{100}	215_9	199_{35}	186_8	411	7215
2C-B-M (O-demethyl- N-acetyl-) iso-1 AC	U+UHYAC	2410	329_4	287_{33}	228_{101}	215_{16}	165_5	752	7196
2C-B-M (O-demethyl- N-acetyl-) iso-1 TFA		2090	383_9	324_{100}	255_7	148_{21}	72_{23}	961	7204
2C-B-M (O-demethyl- N-acetyl-) iso-2 AC	U+UHYAC	2440	329_8	287_{21}	228_{100}	215_{17}	72_{13}	752	7197
2C-B-M (O-demethyl- N-acetyl-) iso-2 TFA		2130	383_7	311_{11}	324_{101}	227_{11}	72_{17}	961	7205
2C-B-M (O-demethyl-) iso-1 2AC	U+UHYAC	2410	329_4	287_{33}	228_{101}	215_{16}	165_5	752	7196
2C-B-M (O-demethyl-) iso-1 2TFA		1900	437_{12}	324_{100}	311_{10}	255_7	148_{32}	1092	7206
2C-B-M (O-demethyl-) iso-2 2AC	U+UHYAC	2440	329_8	287_{21}	228_{100}	215_{17}	72_{13}	752	7197
2C-B-M (O-demethyl-) iso-2 2TFA		1950	437_6	324_{100}	311_{32}	253_3	227_{15}	1092	7207
2C-B-M (O-demethyl-deamino-COOH) MEAC		2120*	316_{14}	274_{101}	242_{92}	214_{88}	186_{18}	691	7213
2C-B-M (O-demethyl-deamino-COOH) METFA		1890*	370_{100}	311_{46}	257_{55}	241_{60}	148_{21}	918	7211
2C-B-M (O-demethyl-deamino-COOH) -H2O	U+UHYAC	1980*	242_{56}	214_{70}	186_{26}			349	7203
2C-B-M (O-demethyl-deamino-di-HO-) 3AC	U+UHYAC	2280*	388_6	346_{43}	286_{55}	244_{67}	231_{100}	978	7201
2C-B-M (O-demethyl-deamino-HO-) 2TFA		1800*	438_{31}	341_{16}	324_{100}	311_{35}	227_{55}	1094	7210
2C-B-M (O-demethyl-deamino-HO-) iso-1 2AC	U+UHYAC	2160*	330_5	288_{17}	270_6	228_{100}	213_{12}	757	7199
2C-B-M (O-demethyl-deamino-HO-) iso-2 2AC	U+UHYAC	2180*	330_4	288_{15}	246_{10}	228_{100}	213_{14}	757	7200
2C-B-M (O-demethyl-deamino-HO-oxo-) 2AC	U+UHYAC	2160*	344_{11}	302_{74}	260_{46}	242_{100}	214_{25}	823	7202
2C-D		1605	195_{19}	166_{100}	151_{60}	135_{42}	91_{15}	208	6904
2C-D AC		1940	237_{14}	178_{100}	165_{35}	163_{40}	135_{51}	334	6912
2C-D HFB		1710	391_{49}	226_6	178_{85}	165_{100}	135_{46}	987	6937
2C-D PFP		1680	341_{53}	178_{79}	165_{100}	135_{54}	91_{21}	810	6932
2C-D TFA		1685	291_{43}	178_{73}	165_{100}	135_{57}	91_{22}	568	6927
2C-D TMS		1735	267_7	237_8	166_{20}	102_{43}	73_{91}	456	6914
2C-D 2AC		2010	279_{11}	135_{21}	72_9	178_{100}	163_{34}	511	6913
2C-D 2TMS		2020	339_2	324_{13}	174_{100}	100_{23}	86_{36}	805	6915
2C-D formyl artifact		1530	207_{25}	176_{100}	165_{69}	135_{39}	91_{16}	234	6909
2C-D-M (deamino-COOH) ME		1755*	224_{100}	209_{19}	177_{12}	165_{35}	135_8	288	7229
2C-D-M (deamino-HO-) AC	U+UHYAC	1740*	238_{27}	178_{100}	163_{57}	135_{33}	79_{27}	336	7216
2C-D-M (deamino-oxo-)		1730*	194_{54}	165_{100}	151_{25}	135_{85}	91_{51}	205	7232
2C-D-M (HO-) 2AC	U+UHYAC	2390	295_{33}	236_{100}	223_6	193_{35}	163_{12}	591	7219
2C-D-M (HO-) 2TFA		1950	403_{42}	290_{100}	277_{32}	177_{57}	163_{25}	1025	7228
2C-D-M (HO-) 3AC	U+UHYAC	2400	337_{27}	244_{23}	236_{100}	193_{46}	125_{30}	793	7220
2C-D-M (O-demethyl- N-acetyl-) 2AC	U+UHYAC	2250	307_5	265_7	206_{25}	164_{100}	149_{13}	648	7223
2C-D-M (O-demethyl- N-acetyl-) iso-1 AC	U+UHYAC	2130	265_6	223_{36}	164_{100}	151_{14}	91_4	444	7221
2C-D-M (O-demethyl- N-acetyl-) iso-1 TFA		1990	319_{12}	260_{100}	247_4	191_{18}	163_{26}	707	7224
2C-D-M (O-demethyl- N-acetyl-) iso-2 AC	U+UHYAC	2200	265_7	223_{25}	164_{100}	151_{14}	91_6	444	7222
2C-D-M (O-demethyl- N-acetyl-) iso-2 TFA		2050	319_{10}	260_{100}	247_{13}	245_{22}	163_{39}	707	7225
2C-D-M (O-demethyl-) 3AC	U+UHYAC	2250	307_5	265_7	206_{25}	164_{100}	149_{13}	648	7223
2C-D-M (O-demethyl-) iso-1 2AC	U+UHYAC	2130	265_6	223_{36}	164_{100}	151_{14}	91_4	444	7221
2C-D-M (O-demethyl-) iso-1 2TFA		1780	373_{28}	260_{100}	247_{24}	191_{30}	163_{49}	929	7226
2C-D-M (O-demethyl-) iso-2 2AC	U+UHYAC	2200	265_7	223_{25}	164_{100}	151_{14}	91_6	444	7222
2C-D-M (O-demethyl-) iso-2 2TFA		1850	373_{18}	260_{100}	247_{48}	217_{15}	163_{39}	929	7227
2C-D-M (O-demethyl-deamino-COOH) iso-1 MEAC		1860*	252_{28}	210_{100}	178_{40}	150_{100}	122_{12}	389	7230
2C-D-M (O-demethyl-deamino-COOH) iso-2 MEAC		1900*	252_{21}	210_{100}	193_{10}	163_7	151_{55}	389	7231
2C-D-M (O-demethyl-deamino-HO-) iso-1 2AC	U+UHYAC	1875*	266_5	224_{13}	164_{100}	154_{46}	114_{15}	450	7217
2C-D-M (O-demethyl-deamino-HO-) iso-2 2AC	U+UHYAC	1890*	266_8	224_{12}	164_{100}	121_{10}	206_6	450	7218
2C-E		1660	209_{20}	180_{100}	165_{52}	149_9	91_{17}	243	6905
2C-E AC		2000	251_{12}	192_{100}	177_{25}	149_{13}	91_{13}	386	6916
2C-E HFB		1790	405_{54}	226_7	192_{90}	179_{100}	149_{21}	1031	6938
2C-E PFP		1760	355_{55}	192_{89}	179_{100}	149_{22}	119_{21}	864	6933
2C-E TFA		1765	305_{42}	192_{71}	179_{100}	149_{19}	91_{22}	639	6928
2C-E TMS		1790	281_2	73_{47}	251_3	180_{25}	102_{100}	524	6918
2C-E 2AC		2075	293_9	192_{100}	177_{34}	149_{11}	91_{15}	581	6917
2C-E 2TMS		2065	353_2	338_{15}	174_{100}	100_{18}	86_{24}	859	6919
2C-E formyl artifact		1630	221_{24}	190_{100}	179_{72}	149_{12}	91_{16}	278	6910
2C-E-M (-COOH N-acetyl-) ME		2605	295_{30}	236_{100}	223_{28}	193_{11}	163_{11}	591	7093
2C-E-M (-COOH) MEAC		2605	295_{30}	236_{100}	223_{28}	193_{11}	163_{11}	591	7093
2C-E-M (deamino-COOH) ME		1820*	238_{100}	223_{22}	192_{11}	179_{39}	163_6	337	7091

Table 1-8-1: Compounds in order of names   Cetirizine-M (HO-Cl-BPH) iso-1 AC

Name	Detected	RI	Typical ions and intensities					Page	Entry
2C-E-M (deamino-HO-) AC		1850*	252 $_{18}$	192 $_{100}$	177 $_{55}$	149 $_{19}$	91 $_{23}$	391	7082
2C-E-M (deamino-HO-) TFA		1680*	306 $_{65}$	192 $_{76}$	177 $_{100}$	149 $_{43}$	91 $_{59}$	644	7092
2C-E-M (deamino-oxo-)		1745*	208 $_{57}$	179 $_{100}$	149 $_{24}$	91 $_{89}$	77 $_{66}$	237	7704
2C-E-M (HO- N-acetyl-) 2TFA		2080	459 $_{27}$	345 $_{19}$	304 $_{100}$	276 $_{26}$	69 $_{23}$	1125	7105
2C-E-M (HO- N-acetyl-) -H2O		2175	249 $_{33}$	190 $_{100}$	177 $_{20}$	175 $_{23}$	147 $_{31}$	375	7120
2C-E-M (HO- N-acetyl-) iso-1 TMS		2230	339 $_{26}$	324 $_{10}$	280 $_{49}$	265 $_{100}$	191 $_{18}$	804	7125
2C-E-M (HO- N-acetyl-) iso-1 propionylated		2370	323 $_{17}$	264 $_{100}$	249 $_{34}$	208 $_{43}$	191 $_{120}$	728	7127
2C-E-M (HO- N-acetyl-) iso-2 TMS		2380	339 $_{100}$	294 $_{5}$	251 $_{10}$	249 $_{7}$	73 $_{42}$	804	7126
2C-E-M (HO- N-acetyl-) iso-2 propionylated		2570	323 $_{13}$	252 $_{34}$	208 $_{31}$	196 $_{71}$	57 $_{100}$	728	7128
2C-E-M (HO-) 2TFA		2035	417 $_{43}$	304 $_{87}$	291 $_{18}$	190 $_{53}$	177 $_{100}$	1056	7121
2C-E-M (HO-) -H2O AC		2175	249 $_{33}$	190 $_{100}$	177 $_{20}$	175 $_{23}$	147 $_{31}$	375	7120
2C-E-M (HO-) -H2O TFA		1945	303 $_{64}$	190 $_{100}$	177 $_{85}$	175 $_{21}$	147 $_{65}$	627	7119
2C-E-M (HO-) iso-1 AC		2340	309 $_{22}$	250 $_{100}$	237 $_{6}$	207 $_{50}$	191 $_{77}$	661	7096
2C-E-M (HO-) iso-2 AC		2420	309 $_{8}$	250 $_{8}$	190 $_{52}$	161 $_{9}$	135 $_{5}$	661	7097
2C-E-M (HO-) iso-3 AC		2500	309 $_{22}$	250 $_{32}$	238 $_{22}$	208 $_{28}$	196 $_{100}$	661	7098
2C-E-M (HO-) iso-3 3AC		2595	351 $_{2}$	309 $_{22}$	280 $_{32}$	238 $_{56}$	196 $_{100}$	849	7099
2C-E-M (HO-deamino-COOH) iso-1 AC		2070*	296 $_{88}$	253 $_{38}$	237 $_{100}$	222 $_{21}$	177 $_{33}$	598	7103
2C-E-M (HO-deamino-COOH) iso-2 AC		2150*	296 $_{12}$	236 $_{100}$	177 $_{59}$	161 $_{29}$	147 $_{13}$	598	7104
2C-E-M (O-demethyl- N-acetyl-) iso-1 TFA		1950	333 $_{16}$	274 $_{100}$	259 $_{12}$	205 $_{14}$	177 $_{11}$	772	7108
2C-E-M (O-demethyl- N-acetyl-) iso-1 2TFA		1860	429 $_{15}$	316 $_{6}$	274 $_{100}$	259 $_{24}$	205 $_{39}$	1077	7110
2C-E-M (O-demethyl- N-acetyl-) iso-2 TFA		2020	333 $_{12}$	274 $_{100}$	259 $_{19}$	177 $_{9}$	91 $_{18}$	773	7109
2C-E-M (O-demethyl- N-acetyl-) iso-2 2TFA		1870	429 $_{4}$	274 $_{100}$	261 $_{32}$	259 $_{19}$	231 $_{9}$	1077	7111
2C-E-M (O-demethyl-) iso-1 2AC		2205	279 $_{6}$	237 $_{31}$	178 $_{100}$	165 $_{17}$	122 $_{5}$	511	7083
2C-E-M (O-demethyl-) iso-1 2TFA		1740	387 $_{13}$	274 $_{100}$	259 $_{20}$	205 $_{20}$	177 $_{31}$	975	7106
2C-E-M (O-demethyl-) iso-2 AC		2240	279 $_{6}$	237 $_{28}$	178 $_{100}$	163 $_{15}$	135 $_{6}$	511	7084
2C-E-M (O-demethyl-) iso-2 2TFA		1805	387 $_{11}$	274 $_{100}$	261 $_{41}$	177 $_{27}$	231 $_{7}$	976	7107
2C-E-M (O-demethyl-deamino-COOH) -H2O		1690*	192 $_{78}$	164 $_{100}$	136 $_{32}$	121 $_{27}$	91 $_{17}$	197	7122
2C-E-M (O-demethyl-deamino-COOH) iso-1 MEAC		1940*	266 $_{8}$	224 $_{50}$	192 $_{73}$	164 $_{100}$	136 $_{17}$	450	7100
2C-E-M (O-demethyl-deamino-COOH) iso-1 METFA		1710*	320 $_{93}$	305 $_{14}$	261 $_{53}$	207 $_{68}$	191 $_{100}$	714	7094
2C-E-M (O-demethyl-deamino-COOH) iso-2 MEAC		1980*	266 $_{9}$	224 $_{100}$	207 $_{6}$	165 $_{60}$	135 $_{15}$	450	7101
2C-E-M (O-demethyl-deamino-COOH) iso-2 METFA		1730*	320 $_{51}$	261 $_{100}$	223 $_{17}$	163 $_{13}$	91 $_{50}$	714	7095
2C-E-M (O-demethyl-deamino-HO-) iso-1 2AC		1990*	280 $_{7}$	238 $_{18}$	178 $_{100}$	163 $_{40}$	145 $_{23}$	517	7089
2C-E-M (O-demethyl-deamino-HO-) iso-1 2TFA		1540	388 $_{39}$	274 $_{100}$	259 $_{46}$	205 $_{28}$	177 $_{42}$	979	7116
2C-E-M (O-demethyl-deamino-HO-) iso-2 2AC		2000*	280 $_{6}$	238 $_{9}$	220 $_{8}$	178 $_{100}$	163 $_{26}$	517	7090
2C-E-M (O-demethyl-deamino-HO-) iso-2 2TFA		1580	388 $_{18}$	274 $_{100}$	259 $_{24}$	177 $_{65}$	69 $_{29}$	979	7117
2C-E-M (O-demethyl-HO- N-acetyl-) 2AC		2425	337 $_{2}$	309 $_{2}$	277 $_{22}$	235 $_{41}$	176 $_{100}$	793	7085
2C-E-M (O-demethyl-HO- N-acetyl-) iso-1 -H2O AC		2255	277 $_{10}$	235 $_{37}$	176 $_{100}$	163 $_{11}$	133 $_{8}$	500	7086
2C-E-M (O-demethyl-HO- N-acetyl-) iso-1 -H2O TFA		2015	331 $_{5}$	272 $_{33}$	259 $_{7}$	205 $_{7}$	177 $_{14}$	763	7112
2C-E-M (O-demethyl-HO- N-acetyl-) iso-2 -H2O AC		2280	277 $_{5}$	235 $_{32}$	176 $_{100}$	163 $_{12}$	133 $_{9}$	500	7087
2C-E-M (O-demethyl-HO- N-acetyl-) iso-2 -H2O TFA		2050	331 $_{9}$	272 $_{100}$	259 $_{16}$	203 $_{10}$	192 $_{21}$	763	7113
2C-E-M (O-demethyl-HO-) 3AC		2425	337 $_{2}$	309 $_{2}$	277 $_{22}$	235 $_{41}$	176 $_{100}$	793	7085
2C-E-M (O-demethyl-HO-) 3TFA		1750	499 $_{9}$	386 $_{39}$	373 $_{100}$	343 $_{19}$		1163	7124
2C-E-M (O-demethyl-HO-) -H2O 2TFA		1810	385 $_{29}$	272 $_{100}$	259 $_{16}$	203 $_{13}$	175 $_{21}$	969	7114
2C-E-M (O-demethyl-HO-) iso-1 -H2O 2AC		2255	277 $_{10}$	235 $_{37}$	176 $_{100}$	163 $_{11}$	133 $_{8}$	500	7086
2C-E-M (O-demethyl-HO-) iso-2 -H2O 2AC		2280	277 $_{5}$	235 $_{32}$	176 $_{100}$	163 $_{12}$	133 $_{9}$	500	7087
2C-E-M (O-demethyl-oxo- N-acetyl-)		2320	251 $_{26}$	192 $_{100}$	177 $_{38}$	151 $_{11}$	137 $_{18}$	384	7088
2C-E-M (O-demethyl-oxo- N-acetyl-) AC		2430	293 $_{3}$	251 $_{40}$	192 $_{100}$	176 $_{53}$	137 $_{9}$	579	7118
2C-E-M (O-demethyl-oxo- N-acetyl-) TFA		2115	347 $_{46}$	233 $_{24}$	192 $_{100}$	177 $_{68}$	69 $_{68}$	834	7115
2C-E-M (O-demethyl-oxo-) AC		2320	251 $_{26}$	192 $_{100}$	177 $_{38}$	151 $_{11}$	137 $_{18}$	384	7088
2C-E-M (oxo-deamino-COOH) ME		2025*	252 $_{54}$	237 $_{100}$	193 $_{25}$	177 $_{23}$	163 $_{29}$	390	7102
Cefadroxil-M/artifact ME2AC		1930	265 $_{5}$	233 $_{37}$	164 $_{43}$	122 $_{100}$	120 $_{59}$	443	7653
Cefadroxil-M/artifact ME3AC		2025	307 $_{5}$	265 $_{39}$	233 $_{67}$	180 $_{100}$	120 $_{69}$	648	7654
Cefadroxil-M/artifact 4TMS		1215	455 $_{8}$	440 $_{7}$	216 $_{23}$	172 $_{65}$	73 $_{100}$	1121	7655
Cefalexine artifact MEAC	U+UHYAC	1590	207 $_{1}$	175 $_{7}$	164 $_{11}$	148 $_{27}$	106 $_{100}$	233	5143
Cefazoline artifact		1430	132 $_{100}$	91 $_{5}$	76 $_{15}$	64 $_{26}$	56 $_{61}$	113	7314
Cefazoline artifact ME		1075	146 $_{100}$	105 $_{44}$	91 $_{26}$	76 $_{16}$	59 $_{77}$	123	7315
Celecoxib	P-I G	2770	381 $_{100}$	300 $_{32}$	281 $_{11}$	204 $_{4}$	115 $_{7}$	955	6537
Celiprolol		2610	280 $_{1}$	265 $_{9}$	151 $_{88}$	86 $_{100}$	57 $_{41}$	951	2846
Celiprolol AC		2370	307 $_{64}$	219 $_{5}$	151 $_{83}$	112 $_{36}$	86 $_{100}$	1064	2849
Celiprolol artifact-1		2350	333 $_{100}$	216 $_{9}$	151 $_{7}$	112 $_{19}$	86 $_{96}$	771	2847
Celiprolol artifact-2		2650	291 $_{21}$	277 $_{14}$	151 $_{53}$	114 $_{16}$	86 $_{100}$	567	2850
Celiprolol artifact-3		2740	323 $_{14}$	294 $_{6}$	209 $_{24}$	114 $_{12}$	86 $_{100}$	724	2848
Celiprolol artifact-3 AC		2800	365 $_{60}$	248 $_{6}$	209 $_{18}$	112 $_{28}$	86 $_{100}$	898	2851
Cetirizine ME	PME UME U+UHYAC	2910	402 $_{10}$	229 $_{6}$	201 $_{100}$	165 $_{46}$	146 $_{23}$	1024	4323
Cetirizine artifact		1900*	232 $_{61}$	201 $_{63}$	165 $_{65}$	105 $_{100}$	77 $_{55}$	311	1344
Cetirizine artifact	G U+UHYAC	1600*	202 $_{30}$	167 $_{49}$	165 $_{52}$	152 $_{17}$	125 $_{7}$	222	2442
Cetirizine-M	U+UHYAC	2210	280 $_{100}$	201 $_{35}$	165 $_{57}$			515	770
Cetirizine-M (amino-) AC	U+UHYAC	2310	259 $_{36}$	217 $_{100}$	182 $_{14}$	152 $_{12}$	75 $_{5}$	415	4324
Cetirizine-M (amino-HO-) 2AC	UGLUCAC	2550	317 $_{85}$	275 $_{100}$	216 $_{65}$	181 $_{99}$	121 $_{78}$	696	4325
Cetirizine-M (carbinol)	UHY	1750*	218 $_{17}$	183 $_{7}$	139 $_{39}$	105 $_{100}$	77 $_{87}$	263	2239
Cetirizine-M (carbinol) AC	U+UHYAC	1890*	260 $_{8}$	200 $_{40}$	165 $_{100}$	139 $_{10}$	77 $_{29}$	420	1270
Cetirizine-M (Cl-benzophenone)	U+UHYAC	1850*	216 $_{43}$	139 $_{65}$	105 $_{100}$	77 $_{44}$		258	1343
Cetirizine-M (HO-Cl-benzophenone)	UHY	2300*	232 $_{36}$	197 $_{7}$	139 $_{23}$	121 $_{100}$	111 $_{23}$	311	2240
Cetirizine-M (HO-Cl-BPH) iso-1 AC	U+UHYAC	2200*	274 $_{18}$	232 $_{75}$	139 $_{100}$	121 $_{44}$	111 $_{51}$	484	2229

Cetirizine-M (HO-Cl-BPH) iso-2 AC   Table 1-8-1: Compounds in order of names

Name	Detected	RI	Typical ions and intensities					Page	Entry
Cetirizine-M (HO-Cl-BPH) iso-2 AC	U+UHYAC	2230*	274_7	232_{43}	139_{25}	121_{100}	111_{27}	484	2230
Cetirizine-M (N-dealkyl-)	UHY	2520	286_{13}	241_{48}	201_{50}	165_{65}	56_{100}	543	2241
Cetirizine-M (N-dealkyl-) AC	U+UHYAC	2620	328_7	242_{19}	201_{48}	165_{66}	85_{100}	749	1271
Cetirizine-M (piperazine) 2AC		1750	170_{15}	85_{33}	69_{25}	56_{100}		157	879
Cetirizine-M (piperazine) 2HFB		1290	478_3	459_9	309_{100}	281_{22}	252_{41}	1146	6634
Cetirizine-M (piperazine) 2TFA		1005	278_{10}	209_{59}	152_{25}	69_{56}	56_{100}	505	4129
Cetirizine-M/artifact	P-I U+UHYAC UME	2220	300_{17}	228_{38}	165_{52}	99_{100}	56_{63}	616	670
Cetirizine-M/artifact HYAC	U+UHYAC	2935	280_4	201_{100}	165_{26}			515	1272
Cetobemidone	UHY	2045	247_6	218_1	190_9	119_6	70_{100}	368	429
Cetobemidone AC	U+UHYAC	2095	289_7	247_6	190_7	70_{100}		559	1181
Cetobemidone HFB		1915	443_1	386_2	128_7	96_6	70_{80}	1105	6144
Cetobemidone ME		1950	261_{12}	204_{16}	70_{100}			427	430
Cetobemidone PFP	UHYPFP	1865	393_8	336_{14}	265_4	128_{13}	70_{200}	994	4303
Cetobemidone TFA		1925	343_4	286_{10}	215_3	128_7	70_{100}	821	6210
Cetobemidone TMS	UHYTMS	2070	319_{13}	304_6	262_{20}	71_{81}	70_{100}	711	4302
Cetobemidone-M (methoxy-) AC	U+UHYAC	2265	319_5	220_6	70_{99}			709	1182
Cetobemidone-M (nor-) 2AC	U+UHYAC	2545	317_{11}	261_{99}	218_{31}	70_{54}	58_{99}	698	1183
Chavicine	G P	2900	285_{58}	201_{71}	173_{36}	115_{100}	84_{31}	539	660
Chelerythrine artifact (dihydro-)		2965	349_{100}	348_{88}	332_{15}	318_{12}	290_{12}	842	5772
Chelerythrine artifact (N-demethyl-)		3160	333_{100}	318_{19}	290_{42}	275_{22}	188_{10}	772	5771
Chenodeoxycholic acid ME2AC	U+UHYAC	3435*	430_1	370_{75}	355_{24}	315_{17}	255_{100}	1157	4473
Chenodeoxycholic acid -2H2O ME	U+UHYAC	2680*	370_7	355_{17}	255_{100}	147_{10}	105_{14}	921	4474
Chloral hydrate	G	<1000*	146_9	111_{32}	82_{100}			144	1470
Chloral hydrate-M (trichloroethanol)	P UHY	<1000*	148_3	119_{20}	113_{60}	82_{46}	77_{100}	125	1413
Chloralose 3AC	U+UHYAC-I	2260*	399_2	361_{17}	317_{63}	272_{88}	115_{100}	1087	2128
Chloralose artifact	G U-I UHY-I	2155*	349_1	333_2	279_1	247_{11}	71_{100}	841	2129
Chloralose-M/artifact (destrichloroethylidenyl-) 5HFB		2030*	583_9	269_{25}	169_{91}	72_{86}	69_{100}	1207	5895
Chloralose-M/artifact (destrichloroethylidenyl-) 5PFP		1925*	483_{13}	395_9	273_{12}	119_{100}	72_{78}	1205	5894
Chloralose-M/artifact (destrichloroethylidenyl-) 5TFA		1795*	479_{13}	319_{20}	223_{35}	109_{76}	69_{100}	1200	5893
Chloramben ME		1730	219_{66}	188_{100}	160_{50}	124_{55}	97_{25}	267	4139
Chloramben iso-1 2ME		1795	233_{100}	202_{57}	174_{26}	139_{15}	100_{11}	314	4140
Chloramben iso-2 2ME		1815	233_{75}	205_{38}	188_{100}	161_{49}	124_{41}	314	4141
Chlorambucil		2420	303_7	254_{100}	230_5	132_{18}	118_{63}	626	1414
Chlorambucil ME		2340	317_9	268_{100}	230_6	131_6	118_{30}	696	1781
Chloramphenicol 2AC	U+UHYAC	2630	273_4	212_{50}	170_{36}	153_{100}	118_{30}	1032	1383
Chlorazanil		2650	221_{100}	220_{96}	193_{10}	152_{47}	99_9	275	3081
Chlorazepate artifact	P G U	2430	284_{81}	283_{91}	256_{100}	221_{31}	77_7	532	481
Chlorbenside		2035*	268_{18}	143_4	125_{100}	108_8	89_{16}	457	3512
Chlorbenzoxamine		3350	434_1	218_8	203_{100}	165_{31}	105_{64}	1088	2417
Chlorbenzoxamine artifact-1		2060	276_5	216_{66}	203_{68}	160_{22}	105_{100}	498	2419
Chlorbenzoxamine artifact-2		2580	291_{37}	171_{14}	134_{36}	105_{100}		567	2420
Chlorbenzoxamine artifact-2 HY		1900	234_5	216_{12}	203_{64}	105_{100}	77_{10}	321	2422
Chlorbenzoxamine HY	UHY	1790*	218_{91}	165_{26}	139_{83}	105_{57}	77_{100}	263	2421
Chlorbenzoxamine HYAC	U+UHYAC	1890*	260_6	218_{11}	200_{23}	165_{100}	77_{13}	420	2418
Chlorbenzoxamine-M (HO-phenyl-) HY	UHY	1900*	234_{91}	197_{62}	155_{67}	105_{100}	77_{63}	318	2437
Chlorbenzoxamine-M (HO-phenyl-) HY2AC	U+UHYAC	2170*	318_2	276_9	216_{41}	181_{100}	152_{31}	701	2435
Chlorbenzoxamine-M (N-dealkyl-)	UHY	2150	190_{26}	163_{46}	134_{49}	105_{100}	91_{44}	194	2438
Chlorbenzoxamine-M (N-dealkyl-) AC	U U+UHYAC	2110	232_7	160_{26}	146_{19}	105_{100}	85_{33}	313	2434
Chlorbenzoxamine-M (N-dealkyl-HO-methyl-) AC-conj.	U	2130	248_{20}	160_{39}	146_{23}	105_{100}	85_{24}	372	2439
Chlorbenzoxamine-M (N-dealkyl-HO-methyl-) 2AC	U+UHYAC	2390	290_{17}	218_{69}	163_{95}	121_{100}	85_{73}	565	2436
Chlorbromuron 2ME		1880	306_9	248_{100}	246_{86}	220_{97}	218_{86}	643	3935
Chlorbufam		1720	223_{29}	171_{12}	164_{18}	127_{35}	53_{100}	283	3515
Chlorbufam TFA		1510	319_2	274_4	223_9	154_9	53_{100}	706	4122
Chlorcarvacrol		1505*	184_{61}	169_{100}	134_{22}	133_{21}	105_{31}	184	1979
Chlorcarvacrol AC		1520*	226_{15}	184_{68}	169_{100}	133_{10}	105_{15}	293	1987
Chlorcyclizine	P-I U+UHYAC UME	2220	300_{17}	228_{38}	165_{52}	99_{100}	56_{63}	616	670
Chlorcyclizine-M (nor-)	UHY	2520	286_{13}	241_{48}	201_{50}	165_{65}	56_{100}	543	2241
Chlorcyclizine-M (nor-) AC	U+UHYAC	2620	328_7	242_{19}	201_{48}	165_{66}	85_{100}	749	1271
Chlordecone		2320*	486_1	455_{32}	355_{20}	272_{100}	237_{63}	1153	3324
Chlordiazepoxide	P G	2820	299_{30}	282_{100}	241_{19}	124_9	77_{30}	610	431
Chlordiazepoxide artifact (deoxo-)	P G	2535	283_{83}	282_{100}	247_{14}	220_{13}	124_9	528	432
Chlordiazepoxide HY	UHY	2050	231_{80}	230_{95}	154_{23}	105_{38}	77_{100}	308	419
Chlordiazepoxide HYAC	U+UHYAC PHYAC	2245	273_{31}	230_{100}	154_{13}	105_{23}	77_{50}	480	273
Chlordimeform		1635	196_{80}	181_{67}	152_{49}	117_{90}	89_{61}	211	5196
Chlordimeform artifact-1 (chloromethylbenzamine)		1030	141_{83}	106_{100}	89_{10}	77_{37}	52_{21}	120	5194
Chlordimeform artifact-1 (chloromethylbenzamine) AC		1620	183_{39}	141_{100}	106_{66}	77_{26}	51_{15}	181	5197
Chlordimeform artifact-2		1550	169_{53}	152_6	140_{57}	106_{100}	77_{48}	155	5195
Chlorfenson		2150*	302_{13}	175_{72}	111_{100}	99_{43}	75_{71}	621	3325
Chlorfenvinphos		2080*	358_1	323_{22}	267_{47}	109_{49}	81_{100}	875	3169
Chlorfenvinphos-M/artifact		1495*	222_1	173_{100}	145_{24}	109_{19}	74_{26}	280	3170
Chlorflurenol ME		2095*	274_9	215_{100}	180_4	152_{65}	76_{11}	484	3632
Chlorflurenol impurity (deschloro-) ME		1950*	240_7	181_{100}	152_{31}	126_2	76_6	343	3633
Chloridazone TFA		1170	317_{52}	282_5	105_{19}	77_{100}	69_{24}	695	3749
Chlormadinone AC		3360*	404_3	319_{14}	301_{100}	267_{15}		1029	2477

Table 1-8-1: Compounds in order of names

Name	Detected	RI	Typical ions and intensities					Page	Entry
Chlormadinone -H2O		3340*	344 $_{44}$	234 $_{45}$	175 $_{100}$	147 $_{14}$		825	2478
Chlormephos	G	1385*	234 $_{28}$	154 $_{46}$	121 $_{100}$	97 $_{96}$	65 $_{52}$	317	3299
Chlormezanone	G P U	2210	209 $_9$	152 $_{100}$	98 $_{70}$			480	671
Chlormezanone artifact	G P U	1235	153 $_{57}$	152 $_{100}$	75 $_{35}$			132	672
Chlormezanone-M (chlorobenzoic acid)	G UHY U+UHYAC	1400*	156 $_{61}$	139 $_{100}$	111 $_{54}$	85 $_4$	75 $_{39}$	136	2726
Chlormezanone-M/artifact (N-methyl-4-chlorobenzamide)	U+UHYAC	1555	169 $_{36}$	139 $_{100}$	111 $_{43}$	75 $_{34}$		155	673
3-Chloroaniline AC	U+UHYAC	1580	169 $_{31}$	127 $_{100}$	111 $_2$	99 $_8$		155	6593
3-Chloroaniline HFB		1310	323 $_{60}$	304 $_8$	154 $_{100}$	126 $_{68}$	111 $_{59}$	724	6607
3-Chloroaniline ME		1100	141 $_{74}$	140 $_{100}$	111 $_9$	105 $_{11}$	77 $_{31}$	120	4089
3-Chloroaniline TFA		1125	223 $_{99}$	154 $_{100}$	126 $_{51}$	111 $_{55}$	69 $_{42}$	282	4124
3-Chloroaniline 2ME		1180	155 $_{62}$	154 $_{100}$	140 $_{11}$	118 $_{16}$	75 $_{17}$	134	4090
4-Chlorobenzaldehyde		1105*	140 $_{67}$	139 $_{100}$	111 $_{67}$	75 $_{55}$		119	3171
Chlorobenzilate		2210*	324 $_1$	251 $_{100}$	152 $_5$	139 $_{85}$	111 $_{32}$	730	3511
Chlorobenzilate-M/artifact (HOOC-) ME		2230*	310 $_1$	251 $_{56}$	139 $_{100}$	111 $_{47}$	75 $_{20}$	665	3634
3-Chlorobenzoic acid		1430*	156 $_{75}$	139 $_{100}$	111 $_{56}$	75 $_{44}$	65 $_6$	136	6024
4-Chlorobenzoic acid	G UHY U+UHYAC	1400*	156 $_{61}$	139 $_{100}$	111 $_{54}$	85 $_4$	75 $_{39}$	136	2726
3-Chlorobenzyl alcohol		1560*	77 $_{100}$	142 $_{53}$	113 $_{28}$	107 $_{33}$		121	6025
4-Chlorobenzyl alcohol		1200*	142 $_{40}$	125 $_9$	107 $_{57}$	79 $_{69}$	77 $_{100}$	121	2727
4-Chlorobenzylchloride	U+UHYAC	1150*	160 $_{24}$	125 $_{100}$	99 $_8$	89 $_{25}$	63 $_{19}$	138	5601
4-Chlorobiphenyl	U+UHYAC	1645*	188 $_{100}$	152 $_{47}$	126 $_4$	94 $_4$	76 $_{16}$	190	4702
Chlorocresol	G U UHY	1400*	142 $_{59}$	107 $_{100}$	77 $_{87}$			121	674
Chlorocresol AC	U+UHYAC	1345*	184 $_{12}$	142 $_{100}$	124 $_2$	107 $_{95}$	77 $_{53}$	183	2345
Chlorocresol-M (HO-) 2AC	U+UHYAC	1560*	242 $_2$	200 $_{12}$	158 $_{100}$	123 $_{53}$	65 $_{29}$	349	2346
2-Chloro-4-cyclohexylphenol		1820*	210 $_{87}$	167 $_{45}$	154 $_{70}$	141 $_{100}$	107 $_{34}$	244	5164
2-Chloro-4-cyclohexylphenol AC		1830*	252 $_5$	210 $_{100}$	167 $_{23}$	154 $_{35}$	141 $_{48}$	389	5168
2-Chloro-4-cyclohexylphenol ME		1750*	224 $_{100}$	181 $_{66}$	168 $_{46}$	155 $_{82}$	125 $_{80}$	288	5171
Chloroform		<1000*	118 $_2$	83 $_{100}$	47 $_{40}$	35 $_{36}$		105	675
Chlorophacinone		3280*	374 $_{21}$	201 $_6$	173 $_{100}$	165 $_{22}$	89 $_{13}$	933	2382
2-Chlorophenol		1035*	128 $_{100}$	100 $_{13}$	92 $_{27}$	64 $_{96}$		110	3173
3-Chlorophenol		1750*	128 $_{100}$	100 $_{19}$	92 $_5$	73 $_{10}$	65 $_{41}$	111	2728
4-Chlorophenol	U UHY	1390*	128 $_{100}$	100 $_{21}$	65 $_{54}$			111	676
4-Chlorophenoxyacetic acid		1770*	186 $_{100}$	141 $_{93}$	128 $_{74}$	111 $_{87}$	99 $_{50}$	187	1881
4-Chlorophenoxyacetic acid ME		1510*	200 $_{98}$	141 $_{100}$	111 $_{59}$	99 $_{17}$	75 $_{38}$	218	1077
m-Chlorophenylpiperazine		1910	196 $_{24}$	154 $_{100}$	138 $_{12}$	111 $_9$	75 $_{12}$	211	6885
m-Chlorophenylpiperazine AC	U+UHYAC	2265	238 $_{32}$	195 $_{15}$	166 $_{100}$	154 $_{31}$	111 $_{27}$	336	405
m-Chlorophenylpiperazine HFB	U+UHYHFB	1960	392 $_{100}$	195 $_{38}$	166 $_{41}$	139 $_{36}$	111 $_{25}$	991	6604
m-Chlorophenylpiperazine ME		1820	210 $_{100}$	166 $_{19}$	139 $_{81}$	111 $_{33}$	70 $_{99}$	244	6886
m-Chlorophenylpiperazine TFA	U+UHYTFA	1920	292 $_{100}$	250 $_{12}$	195 $_{77}$	166 $_{79}$	139 $_{66}$	574	6597
m-Chlorophenylpiperazine TMS		2035	268 $_{61}$	253 $_{29}$	128 $_{63}$	101 $_{68}$	86 $_{71}$	461	6888
m-Chlorophenylpiperazine-M (chloroaniline) AC	U+UHYAC	1580	169 $_{31}$	127 $_{100}$	111 $_2$	99 $_8$		155	6593
m-Chlorophenylpiperazine-M (chloroaniline) HFB		1310	323 $_{60}$	304 $_8$	154 $_{100}$	126 $_{68}$	111 $_{59}$	724	6607
m-Chlorophenylpiperazine-M (chloroaniline) ME		1100	141 $_{74}$	140 $_{100}$	111 $_9$	105 $_{11}$	77 $_{31}$	120	4089
m-Chlorophenylpiperazine-M (chloroaniline) TFA		1125	223 $_{99}$	154 $_{100}$	126 $_{51}$	111 $_{55}$	69 $_{42}$	282	4124
m-Chlorophenylpiperazine-M (chloroaniline) 2ME		1180	155 $_{62}$	154 $_{100}$	140 $_{11}$	118 $_{16}$	75 $_{17}$	134	4090
m-Chlorophenylpiperazine-M (deethylene-) 2AC	U+UHYAC	2080	254 $_1$	195 $_{50}$	153 $_{31}$	140 $_{100}$	111 $_{10}$	398	6592
m-Chlorophenylpiperazine-M (deethylene-) 2HFB	U+UHYHFB	1705	562 $_1$	349 $_{42}$	336 $_{73}$	240 $_{26}$	139 $_{100}$	1188	6606
m-Chlorophenylpiperazine-M (deethylene-) 2TFA	U+UHYTFA	1670	362 $_3$	249 $_{77}$	236 $_{100}$	139 $_{66}$	111 $_{28}$	890	6601
m-Chlorophenylpiperazine-M (HO-) TFA	U+UHYTFA	2035	308 $_{100}$	272 $_{19}$	211 $_{46}$	182 $_{36}$	155 $_{46}$	653	6599
m-Chlorophenylpiperazine-M (HO-) iso-1 AC	U+UHYAC	2335	254 $_{65}$	211 $_{23}$	182 $_{100}$	166 $_{71}$	154 $_{33}$	398	5308
m-Chlorophenylpiperazine-M (HO-) iso-1 2AC	U+UHYAC	2515	296 $_{30}$	254 $_{56}$	182 $_{100}$	154 $_{36}$		597	406
m-Chlorophenylpiperazine-M (HO-) iso-1 2TFA	U+UHYTFA	2040	404 $_{28}$	307 $_{74}$	278 $_{34}$	265 $_{13}$	154 $_{40}$	1028	6600
m-Chlorophenylpiperazine-M (HO-) iso-2 AC	U+UHYAC	2345	254 $_{79}$	211 $_5$	182 $_{100}$	169 $_{40}$	154 $_{22}$	398	5307
m-Chlorophenylpiperazine-M (HO-) iso-2 2AC	U+UHYAC	2525	296 $_{20}$	254 $_{74}$	182 $_{100}$	169 $_{72}$	154 $_{24}$	597	32
m-Chlorophenylpiperazine-M (HO-) iso-2 2HFB	U+UHYHFB	2145	604 $_{43}$	585 $_9$	407 $_{100}$	378 $_{22}$	154 $_{22}$	1195	6605
m-Chlorophenylpiperazine-M (HO-) iso-2 2TFA	U+UHYTFA	2045	404 $_{13}$	307 $_{33}$	278 $_{10}$	265 $_4$	154 $_{22}$	1028	6598
m-Chlorophenylpiperazine-M (HO-chloroaniline N-acetyl-) HFB	U+UHYHFB	1820	381 $_{10}$	339 $_{22}$	169 $_{15}$	142 $_{100}$	69 $_{23}$	954	6796
m-Chlorophenylpiperazine-M (HO-chloroaniline N-acetyl-) TFA	U+UHYTFA	1765	281 $_{29}$	239 $_{58}$	142 $_{100}$	114 $_{10}$	69 $_{40}$	520	6797
m-Chlorophenylpiperazine-M (HO-chloroaniline) 2HFB	U+UHYHFB	1540	535 $_{36}$	516 $_8$	338 $_{72}$	169 $_{100}$	143 $_{82}$	1180	6608
m-Chlorophenylpiperazine-M (HO-chloroaniline) iso-1 2AC	U+UHYAC	1980	227 $_{14}$	185 $_{18}$	167 $_{33}$	143 $_{100}$	114 $_7$	297	6594
m-Chlorophenylpiperazine-M (HO-chloroaniline) iso-1 2TFA	U+UHYTFA	1440	335 $_{15}$	238 $_9$	168 $_2$	69 $_{15}$		781	6603
m-Chlorophenylpiperazine-M (HO-chloroaniline) iso-1 3AC	U+UHYAC	1940	269 $_2$	227 $_{13}$	185 $_{17}$	167 $_{33}$	143 $_{100}$	463	6596
m-Chlorophenylpiperazine-M (HO-chloroaniline) iso-2 2AC	U+UHYAC	2020	227 $_4$	185 $_{57}$	143 $_{100}$	79 $_{18}$	114 $_5$	297	404
m-Chlorophenylpiperazine-M (HO-chloroaniline) iso-2 2TFA	U+UHYTFA	1440	335 $_{44}$	266 $_3$	238 $_{52}$	210 $_{14}$	143 $_{42}$	781	6602
m-Chlorophenylpiperazine-M (HO-chloroaniline) iso-2 3AC	U+UHYAC	1900	269 $_4$	227 $_{37}$	185 $_{75}$	143 $_{100}$	79 $_7$	463	6595
Bis-(4-Chlorophenyl-)sulfone		2240*	286 $_{26}$	159 $_{40}$	131 $_9$	111 $_{23}$	75 $_{34}$	541	5739
Chloropicrin		<1000	119 $_{97}$	117 $_{100}$	82 $_{35}$	61 $_{12}$		142	3730
Chloropropham		1620	213 $_{34}$	171 $_{28}$	154 $_{25}$	127 $_{100}$	99 $_{12}$	251	3327
Chloropropylate		2230*	338 $_1$	251 $_{100}$	152 $_5$	139 $_{78}$	111 $_{30}$	796	3513
Chloropropylate-M/artifact (HOOC-) ME		2230*	310 $_1$	251 $_{56}$	139 $_{100}$	111 $_{47}$	75 $_{20}$	665	3634
Chloropyramine	U UHY U+UHYAC	2190	289 $_3$	231 $_5$	125 $_{37}$	72 $_{13}$	58 $_{100}$	558	1416
Chloropyramine-M (bis-nor-) AC	U U+UHYAC	2420	303 $_{10}$	231 $_{43}$	217 $_{11}$	125 $_{100}$	89 $_{16}$	628	2180
Chloropyramine-M (HO-) AC	U+UHYAC	2440	347 $_2$	289 $_6$	234 $_5$	125 $_{31}$	58 $_{100}$	835	2177
Chloropyramine-M (N-dealkyl-)	UHY U+UHYAC	1900	218 $_{78}$	181 $_6$	140 $_{100}$	125 $_{49}$	79 $_{53}$	263	2175
Chloropyramine-M (N-dealkyl-) AC	U+UHYAC	2160	260 $_1$	246 $_{19}$	217 $_{100}$	125 $_{21}$	78 $_{37}$	420	2176

Chloropyramine-M (nor-) Table 1-8-1: Compounds in order of names

Name	Detected	RI	Typical ions and intensities					Page	Entry
Chloropyramine-M (nor-)	U UHY	2210	275_4	232_{25}	219_{32}	125_{100}	107_{69}	490	2179
Chloropyramine-M (nor-) AC	U+UHYAC	2470	317_{13}	231_{36}	217_9	125_{100}	119_{38}	697	2178
Chloroquine	P G U	2595	319_{17}	290_5	245_6	112_8	86_{150}	711	677
Chloroquine-M (deethyl-) AC	U+UHYAC	3010	333_{63}	219_{100}	205_{78}	58_9		774	1759
Chlorothalonil		1775	266_{100}	264_{77}	229_{11}	168_{10}	109_{33}	437	3326
8-Chlorotheophylline	P G U	2500	214_{99}	157_{16}	129_{56}	68_{51}		254	681
8-Chlorotheophylline ET		1910	242_{100}	214_{61}	185_8	157_{13}	129_{50}	350	2399
8-Chlorotheophylline ME	UME	1900	228_{100}	199_6	171_{11}	143_{33}	67_{55}	298	2195
8-Chlorotheophylline TMS		2105	286_{92}	271_{100}	251_{42}	214_7	73_{40}	542	4612
Chlorothiazide artifact 3ME	UEXME	2750	339_{12}	275_{72}	248_{100}	220_{26}	167_{27}	800	6847
Chlorothiazide artifact 5ME	UEXME	2710	355_{100}	263_{17}	248_{26}	220_{59}	139_{39}	864	6846
4-Chlorotoluene		1165*	126_{28}	91_{100}	65_{10}	63_{12}		110	3192
Chlorotrimethoxyhippuric acid ME		2405	317_{39}	286_2	229_{100}	186_5	100_5	695	5181
Chloroxuron		2245	290_{45}	232_{18}	136_5	105_{21}	72_{100}	563	4137
Chloroxuron ME		2430	304_{25}	232_7	168_2	85_{19}	72_{100}	634	4136
Chloroxylenol		1420*	156_{94}	121_{99}	91_{31}	77_{22}		136	678
Chloroxylenol AC	U+UHYAC	1450*	198_{15}	156_{100}	121_{48}	91_{19}		213	121
Chlorphenamine	G P U+UHYAC	2020	274_1	203_{100}	167_{19}	72_{18}	58_{64}	485	679
Chlorphenamine-M (bis-nor-) AC	U+UHYAC	2535	288_4	216_{100}	203_{47}	181_{18}	167_{30}	552	2183
Chlorphenamine-M (deamino-HO-) AC	U+UHYAC	2130	289_{27}	230_{27}	216_{100}	203_{92}	167_{68}	556	2181
Chlorphenamine-M (HO-) AC	U+UHYAC	2405	332_1	261_{61}	219_{100}	72_{11}	58_{55}	768	2182
Chlorphenamine-M (nor-) AC	U+UHYAC	2530	302_8	216_{49}	203_{100}	167_{29}	78_7	624	2040
Chlorphenesin		1690*	202_{10}	153_5	128_{100}	111_9	99_9	222	2768
Chlorphenesin AC		2030*	244_5	141_5	128_{66}	117_{100}	111_{14}	355	2769
Chlorphenesin 2AC		2070*	286_1	159_{100}	128_{37}	99_{35}	75_{14}	542	2770
Chlorphenethazine		2420	304_8	246_3	214_5	152_2	58_{100}	633	4262
Chlorphenoxamine	U	2095	303_1	230_8	178_6	165_{11}	58_{100}	628	1417
Chlorphenoxamine artifact	G U+UHYAC	1700*	214_{49}	179_{100}	152_9	139_3	89_{14}	254	1217
Chlorphenoxamine HY	UHY	1750*	232_{12}	217_{80}	139_{81}	105_{75}	77_{100}	312	1079
Chlorphenoxamine HYAC	U+UHYAC	2180*	274_{18}	232_{75}	197_{14}	139_{35}	121_{100}	485	2185
Chlorphenoxamine-M (HO-)	U	2470	319_1	231_3	195_2	165_4	58_{100}	708	2188
Chlorphenoxamine-M (HO-) -H2O HY	U UHY	2050*	230_{100}	215_{27}	195_{60}	177_{33}	165_{64}	306	2187
Chlorphenoxamine-M (HO-) iso-1 -H2O HYAC	U+UHYAC	2030*	272_{27}	230_{100}	195_{34}	165_{56}	152_{10}	477	2184
Chlorphenoxamine-M (HO-) iso-2 -H2O HYAC	U+UHYAC	2090*	272_{16}	230_{100}	215_{15}	195_{34}	165_{47}	478	2189
Chlorphenoxamine-M (HO-methoxy-) -H2O HYAC	U+UHYAC	2210*	302_{10}	260_{100}	182_{10}	152_{16}	75_4	623	2186
Chlorphenoxamine-M (HO-methoxy-carbinol) -H2O	U UHY	2220*	262_{36}	260_{100}				420	2194
Chlorphenoxamine-M (nor-) AC	U	2580	215_{37}	179_{26}	116_{100}	86_{38}	74_{33}	764	2191
Chlorphenphos-methyl		1540*	232_3	196_{72}	165_{80}	137_{27}	125_{100}	311	4039
Chlorphenphos-methyl -HCl		1455*	196_{53}	165_{10}	137_{100}	101_{39}	75_{26}	210	4040
Chlorphentermine		1355	168_3	125_4	107_8	58_{100}		182	680
Chlorphentermine AC		1730	225_1	166_{13}	100_{53}	86_{31}	58_{100}	291	1418
Chlorphentermine HFB		1560	364_1	254_{100}	214_{11}	166_{15}	125_{23}	947	5048
Chlorphentermine PFP		1515	329_1	204_{100}	166_8	154_7	125_{11}	752	5049
Chlorphentermine TFA		1520	279_1	166_{14}	154_{100}	125_{16}	114_{14}	509	5050
Chlorphentermine TMS		1520	255_1	240_2	130_{100}	114_{30}	73_{88}	402	5447
Chlorpromazine	P-I G U+UHYAC	2500	318_{10}	272_4	232_4	86_{15}	58_{100}	702	310
Chlorpromazine chloro artifact iso-1		2645	352_{19}	86_{24}	268_2	306_{15}	58_{100}	851	7647
Chlorpromazine chloro artifact iso-2		2660	352_{17}	86_{23}	268_5	306_{15}	58_{100}	852	7648
Chlorpromazine-M (bis-nor-) AC	U+UHYAC	2990	332_8	233_{19}	100_{100}			767	1255
Chlorpromazine-M (HO-) ME	UME	2590	348_{20}	302_4	262_8	86_{44}	58_{100}	838	434
Chlorpromazine-M (nor-) AC	U+UHYAC	3070	346_8	232_7	114_{100}	86_9		830	1256
Chlorpromazine-M (ring)	U-I UHY-I U+UHYAC-I	2100	233_{100}	198_{55}				314	311
Chlorpromazine-M/artifact (sulfoxide)	G P U	2900	334_3	318_3	246_{13}	86_6	58_{100}	779	433
Chlorpropamide ME	UME	2250	290_{19}	175_{54}	115_{100}	111_{72}	58_{64}	562	3122
Chlorpropamide TMS		2205	348_1	333_4	173_{19}	73_{47}	58_{100}	838	5024
Chlorpropamide 2ME		2275	304_1	197_{13}	175_{34}	129_{51}	72_{100}	633	4899
Chlorpropamide artifact-1		1685	217_{29}	175_{81}	111_{100}	75_{60}		260	4900
Chlorpropamide artifact-2		1730	191_{63}	175_{41}	128_{43}	111_{100}	75_{81}	194	4901
Chlorpropamide artifact-2 ME	UME	1825	205_{30}	175_{29}	141_{12}	111_{100}	75_{42}	228	3123
Chlorpropamide artifact-2 2ME	UME	1690	219_{43}	175_{27}	111_{100}	75_{37}		268	3124
Chlorpropamide artifact-3 ME		1860	199_{37}	175_{14}	111_{66}	75_{35}	72_{100}	216	3125
Chlorpropamide artifact-4 ME	UME	2135	248_{25}	141_{24}	125_{96}	111_{110}	75_{92}	369	4902
Chlorpropamide artifact-4 2ME	UME	2150	262_{10}	197_7	125_{22}	111_{56}	87_{100}	429	4903
Chlorprothixene	P-I G U+UHYAC	2510	315_3	255_4	221_{17}	58_{100}		688	312
Chlorprothixene artifact (dihydro-)	G UHY U+UHYAC	2490	317_{13}	231_9	152_2	73_7	58_{100}	697	3732
Chlorprothixene-M (bis-nor-) AC	U+UHYAC	2910	329_{15}	270_{87}	255_{22}	235_{77}	221_{100}	753	3736
Chlorprothixene-M (bis-nor-dihydro-) AC	U+UHYAC	2870	331_{12}	231_{100}	195_7	152_6	100_{15}	762	3734
Chlorprothixene-M (bis-nor-HO-) iso-1 2AC	U+UHYAC	3150	387_{16}	328_{100}	286_{56}	269_{45}	238_{78}	975	4167
Chlorprothixene-M (bis-nor-HO-) iso-2 2AC	U+UHYAC	3190	387_{20}	328_{91}	269_{48}	238_{100}	72_{13}	975	4169
Chlorprothixene-M (bis-nor-HO-dihydro-) iso-1 2AC	U+UHYAC	3170	389_{34}	289_{65}	247_{100}	100_{14}	72_{10}	982	3737
Chlorprothixene-M (bis-nor-HO-dihydro-) iso-2 2AC	U+UHYAC	3210	389_{26}	289_{50}	247_{100}	184_2	100_{11}	982	3738
Chlorprothixene-M (bis-nor-HO-methoxy-) 2AC	U+UHYAC	3360	417_{20}	372_{60}	358_{100}	303_{97}	243_{57}	1056	4171
Chlorprothixene-M (bis-nor-HO-methoxy-dihydro-) 2AC	U+UHYAC	3380	419_{28}	319_{37}	277_{100}	234_8	100_{10}	1060	3740
Chlorprothixene-M (HO-) iso-1 AC	U+UHYAC	2750	373_1	273_2	237_7	58_{100}		929	4163

Table 1-8-1: Compounds in order of names 2C-I-M (deamino-HO-O-demethyl-) iso-1 2TFA

Name	Detected	RI	Typical ions and intensities					Page	Entry
Chlorprothixene-M (HO-) iso-2 AC	U+UHYAC	2760	373_1	342_5	237_6	58_{100}		929	4164
Chlorprothixene-M (HO-dihydro-) iso-1	UHY	2750	333_{23}	247_8	58_{100}			772	437
Chlorprothixene-M (HO-dihydro-) iso-1 AC	U+UHYAC	2770	375_{23}	247_8	58_{100}			935	313
Chlorprothixene-M (HO-dihydro-) iso-2	UHY	2790	333_{18}	247_{12}	58_{100}			772	3742
Chlorprothixene-M (HO-dihydro-) iso-2 AC	U+UHYAC	2800	375_{31}	247_8	58_{100}			935	3733
Chlorprothixene-M (HO-methoxy-) AC	U+UHYAC	2870	403_2	358_2	267_5	261_6	58_{100}	1025	4165
Chlorprothixene-M (HO-methoxy-dihydro-)	UHY	2810	363_{12}	277_{10}	58_{100}			894	3743
Chlorprothixene-M (HO-methoxy-dihydro-) AC	U+UHYAC	2890	405_{11}	277_{10}	73_8	58_{100}		1030	3735
Chlorprothixene-M (HO-N-oxide) iso-1 -(CH3)2NOH AC	U+UHYAC	2590*	328_{75}	293_{54}	269_{85}	251_{100}	250_{91}	747	4160
Chlorprothixene-M (HO-N-oxide) iso-2 -(CH3)2NOH AC	U+UHYAC	2620*	328_{49}	285_{47}	268_{100}	251_{50}	221_{29}	747	4161
Chlorprothixene-M (nor-) AC	U+UHYAC	2945	343_{12}	270_{100}	257_{51}	235_{71}	221_{86}	820	1259
Chlorprothixene-M (nor-dihydro-) AC	U+UHYAC	2930	345_{16}	231_{100}	195_8	152_7	114_{11}	827	1258
Chlorprothixene-M (nor-HO-) iso-1 2AC	U+UHYAC	3175	401_{16}	328_{100}	269_{43}	238_{66}	86_{17}	1020	4168
Chlorprothixene-M (nor-HO-) iso-2 2AC	U+UHYAC	3220	401_{20}	328_{100}	273_{91}	237_{60}	86_{11}	1020	4170
Chlorprothixene-M (nor-HO-dihydro-) iso-1 2AC	U+UHYAC	3195	403_{34}	289_{65}	247_{100}	114_{12}	86_8	1025	314
Chlorprothixene-M (nor-HO-dihydro-) iso-2 2AC	U+UHYAC	3240	403_{46}	289_{46}	247_{100}	114_9	86_4	1025	3739
Chlorprothixene-M (nor-HO-methoxy-) 2AC	U+UHYAC	3390	431_{26}	358_{100}	303_{97}	243_{54}		1082	4172
Chlorprothixene-M (nor-HO-methoxy-dihydro-) 2AC	U+UHYAC	3410	433_{28}	319_{37}	277_{100}	234_8	114_4	1086	3741
Chlorprothixene-M (nor-sulfoxide) AC	U+UHYAC	2960	359_{30}	270_{94}	257_{46}	235_{76}	221_{100}	879	4166
Chlorprothixene-M (N-oxide) -(CH3)2NOH	P-I U+UHYAC	2410*	270_{40}	255_{21}	234_{100}	202_{23}	117_{27}	467	438
Chlorprothixene-M (N-oxide-sulfoxide) -(CH3)2NOH	P-I U UGLUC UGLUCAC	2560*	286_{21}	251_{20}	234_{57}	203_{100}	101_{10}	541	436
Chlorprothixene-M / artifact (Cl-thioxanthenone)	U	2260*	246_{10}	218_{46}	183_7	139_{25}	91_{10}	361	2641
Chlorprothixene-M/artifact (sulfoxide)	G P U+UHYAC	2720	331_1	314_1	221_6	189_4	58_{100}	762	4162
Chlorpyrifos	G P-I	1980	349_2	314_{46}	258_{37}	197_{96}	97_{100}	841	1397
Chlorpyrifos HY		1440	197_{100}	169_{67}	161_{15}	134_{21}	107_{39}	212	7439
Chlorpyrifos HYAC		1420	239_3	98_{18}	140_9	197_{106}	169_{14}	338	7440
Chlorpyrifos-methyl		1840	321_2	286_{82}	125_{100}	79_{69}	63_{48}	717	3328
Chlortalidone 3ME		3015	380_1	349_{100}	255_4	176_8		951	3103
Chlortalidone 4ME		2830	394_{39}	379_{100}	363_{95}	285_{53}	176_{62}	998	3104
Chlortalidone artifact 3ME	UME	2950	363_{100}	287_{60}	255_{17}	220_3	176_{23}	893	3105
Chlorthal-methyl		1965*	330_{24}	301_{100}	299_{75}	221_{16}	142_{20}	756	3329
Chlorthiamid		1870	205_{43}	170_{100}	134_{12}	100_{14}	75_{27}	227	3752
Chlorthiamid artifact	U UHY U+UHYAC	1300	171_{94}	136_{20}	100_{33}			157	736
Chlorthiophos iso-1		2210*	360_{16}	289_{21}	257_{37}	222_{61}	97_{100}	882	3300
Chlorthiophos iso-2		2230*	360_3	325_{15}	269_{35}	97_{100}	65_{36}	882	3301
Chlorthiophos iso-3		2250*	360_{26}	325_{42}	269_{77}	208_{20}	97_{100}	882	3302
Chlortoluron ME		1695	226_4	154_3	89_4	72_{100}		293	3973
Chlorzoxazone	U	1800	169_{100}	113_{26}	78_{25}	63_4		155	4372
Chlorzoxazone AC	U+UHYAC	1595	211_{10}	169_{100}	113_{13}	76_{11}	125_7	245	6362
Chlorzoxazone ME		1750	183_{100}	154_{45}	92_{65}	76_{20}	63_{16}	181	2440
Chlorzoxazone artifact Me		1820	201_{43}	142_{100}	78_{36}			220	4373
Chlorzoxazone HY2AC	U+UHYAC	1850	227_{17}	185_{15}	167_{39}	143_{100}	114_{17}	296	6364
Chlorzoxazone HY3AC	U+UHYAC	2160	269_6	227_{20}	185_{100}	129_{10}	86_{15}	463	6363
Cholesta-3,5-dien-7-one		2860*	382_{53}	269_{12}	187_{24}	174_{100}	161_{29}	961	4347
Cholestenone	U UME	3150*	384_{47}	342_{16}	261_{27}	229_{32}	124_{100}	969	6353
Cholesterol	P U UHY	3085*	386_{97}	368_{29}	353_{18}	301_{34}	275_{31}	974	682
Cholesterol TMS		3110*	458_{82}	368_{100}	329_{100}	129_{35}	73_{73}	1125	3209
Cholesterol -H2O	P U UHY U+UHYAC	3050*	368_{100}	353_{19}	260_{18}	147_{40}		914	143
Chrysene		2420*	228_{100}	226_{26}	202_3	113_8	101_4	300	2570
Chrysophanol		2410*	254_{100}	226_{18}	197_{13}	152_{13}	115_{10}	397	3554
Chrysophanol ME		2540*	268_{100}	250_{45}	222_{73}	165_{22}	152_{20}	459	3563
Chrysophanol 2ME		2600*	282_{35}	267_{100}	165_{35}	152_{26}	76_{22}	526	3564
2C-I		2330	307_{16}	278_{100}	263_{20}	247_9	232_3	647	6954
2C-I AC	U+UHYAC	2260	349_{25}	290_{100}	275_{21}	247_{14}	148_{21}	841	6957
2C-I HFB		2110	503_{32}	290_{100}	277_{51}	247_{30}	148_{21}	1166	6948
2C-I PFP		2080	453_{43}	290_{100}	277_{93}	247_{44}	148_{33}	1117	6960
2C-I TFA	UGLUCTFA	2100	403_{20}	290_{100}	277_{69}	247_{49}	148_{31}	1024	6959
2C-I TMS		2070	379_5	320_9	278_7	102_{100}	73_{33}	947	6961
2C-I 2AC	U+UHYAC	2340	391_{16}	290_{100}	275_{19}	247_8	148_{10}	987	6958
2C-I 2ME		2320	335_{18}	290_{100}	275_{16}	247_{11}	148_{10}	782	6962
2C-I deuteroformyl artifact		1850	321_{23}	290_{100}	277_{42}	247_{18}	232_4	717	6956
2C-I formyl artifact		1860	319_{20}	288_{100}	277_{50}	247_{23}	232_6	706	6955
2C-I intermediate-2 (2,5-dimethoxyphenethylamine)		1630	181_{15}	162_{33}	152_{100}	137_{47}	121_{27}	177	3287
2C-I intermediate-2 (2,5-dimethoxyphenethylamine) AC		1935	223_{15}	164_{100}	149_{29}	121_{31}	91_{15}	284	3288
2C-I-M (deamino-HO-)	UGLUC	2020	308_{100}	277_{74}	263_7	247_{16}	150_9	652	6966
2C-I-M (deamino-HO-) AC	UGLUCAC	2150	350_{16}	290_{100}	275_{28}	247_{10}	148_{25}	844	6969
2C-I-M (deamino-HO-) TFA	UGLUCTFA	1980	404_{100}	290_{83}	275_{24}	247_{21}	148_{36}	1027	6978
2C-I-M (deamino-HOOC-) ME	UGLUCMEAC	2115	336_{100}	321_{12}	289_8	277_{57}	247_{18}	786	6982
2C-I-M (deamino-HOOC-O-demethyl-) ME	UGLUCMETFA	2160	322_{62}	290_{87}	262_{100}	234_{21}	191_6	721	6984
2C-I-M (deamino-HOOC-O-demethyl-) MEAC	UGLUCMEAC	2170	364_{26}	322_{100}	290_{53}	262_{44}	234_{10}	896	6981
2C-I-M (deamino-HOOC-O-demethyl-) METFA	UGLUCMETFA	1980	418_{100}	404_{25}	359_{32}	305_{49}	289_{53}	1058	6983
2C-I-M (deamino-HOOC-O-demethyl-) -H2O	UGLUC	2080	290_{100}	262_{69}	234_{42}	191_{11}	127_{16}	561	6965
2C-I-M (deamino-HO-O-demethyl-) iso-1 2AC	UGLUCAC	2240	378_2	336_{16}	276_{100}	261_{10}	134_{16}	944	6970
2C-I-M (deamino-HO-O-demethyl-) iso-1 2TFA	UGLUCTFA	1865	486_{100}	372_{98}	303_{28}	261_9		1153	6979

Table 1-8-1: Compounds in order of names

Name	Detected	RI	Typical ions and intensities					Page	Entry
2C-I-M (deamino-HO-O-demethyl-) iso-2 2AC	UGLUCAC	2275	378_8	336_{33}	276_{100}	261_{34}	150_{24}	944	6971
2C-I-M (deamino-HO-O-demethyl-) iso-2 2TFA	UGLUCTFA	1890	486_{35}	372_{100}	275_{47}	261_{25}	245_{22}	1153	6980
2C-I-M (deamino-oxo-)		1965*	306_{100}	277_{82}	263_{18}	247_{29}	232_{10}	642	7233
2C-I-M (O-demethyl- N-acetyl-) TFA	UGLUCTFA	2270	431_{10}	389_{100}	276_{94}	263_{43}	148_7	1082	6974
2C-I-M (O-demethyl- N-acetyl-) iso-1	UGLUC	2370	335_{15}	276_{100}	263_{11}	233_{11}	134_{14}	782	6963
2C-I-M (O-demethyl- N-acetyl-) iso-1 AC	U+UHYAC	2480	377_5	335_{46}	276_{100}	259_{16}	233_{20}	941	6967
2C-I-M (O-demethyl- N-acetyl-) iso-2	UGLUC	2520	335_{26}	276_{100}	261_{18}	220_3	121_7	782	6964
2C-I-M (O-demethyl- N-acetyl-) iso-2 AC	U+UHYAC	2500	377_8	335_{43}	276_{100}	263_{17}	236_7	941	6968
2C-I-M (O-demethyl-) iso-1 TFA	UGLUCTFA	2100	389_{39}	276_{100}	263_{35}	233_{19}	134_{15}	981	6976
2C-I-M (O-demethyl-) iso-1 2AC	U+UHYAC	2480	377_5	335_{46}	276_{100}	259_{16}	233_{20}	941	6967
2C-I-M (O-demethyl-) iso-1 2TFA	UGLUCTFA	1970	485_{17}	372_{100}	359_{14}	303_{11}	234_9	1152	6972
2C-I-M (O-demethyl-) iso-2 TFA	UGLUCTFA	2275	389_{62}	276_{100}	263_{58}	261_{18}		981	6977
2C-I-M (O-demethyl-) iso-2 2AC	U+UHYAC	2500	377_8	335_{43}	276_{100}	263_{17}	236_7	941	6968
2C-I-M (O-demethyl-) iso-2 2TFA	UGLUCTFA	2010	485_6	372_{100}	359_{23}	275_{11}	126_{10}	1152	6973
2C-I-M (O-demethyl-deamino-di-HO-) 3AC	UGLUCAC	2310*	436_{27}	394_{100}	334_{68}	292_{62}	279_{62}	1092	7130
2C-I-M (O-demethyl-deamino-di-HO-) 3AC	UGLUCAC	2310*	436_{27}	394_{100}	334_{68}	292_{62}	279_{62}	1092	7130
2C-I-M (O-demethyl-deamino-HO-oxo-) 2AC	UGLUCAC	2200*	392_8	350_{61}	308_{37}	290_{100}	262_{87}	991	7129
2C-I-M (O-demethyl-deamino-HO-oxo-) 2AC	UGLUCAC	2200*	392_8	350_{61}	308_{37}	290_{100}	262_{87}	991	7129
Cianidanol 5TMS		2805*	650_2	368_{100}	355_{38}	179_{11}	73_{57}	1199	5817
Cianidanol -H2O 4AC		3025*	440_{34}	398_{87}	356_{73}	314_{100}	272_{80}	1097	5818
Cicloprofen		2305*	238_{37}	193_{100}	178_{46}	165_{44}	96_{10}	336	4275
Cicloprofen ME		2220*	252_{27}	193_{100}	178_{36}	165_{24}	95_9	391	4276
Cilazapril ET	UET	3055	445_2	417_6	311_{75}	239_{43}	171_{100}	1108	4731
Cilazapril ME	PME UME	3010	431_1	358_{13}	297_{41}	225_{49}	157_{100}	1084	4727
Cilazapril METMS		3125	503_1	488_6	369_{51}	215_{100}	73_{52}	1167	4976
Cilazapril TMS		3030	489_1	474_{10}	355_{77}	215_{100}	73_{71}	1157	4975
Cilazapril 2ET	UET	2980	473_2	445_7	274_{90}	239_{90}	171_{100}	1142	4732
Cilazapril 2ME	UME	2945	445_1	417_4	311_{25}	225_{79}	157_{100}	1108	4728
Cilazaprilate 2ET	UET	3055	445_2	417_6	311_{75}	239_{43}	171_{100}	1108	4731
Cilazaprilate 2ME	UME	2945	417_1	389_5	283_{57}	225_{42}	157_{100}	1057	4729
Cilazaprilate 2TMS	UTMS	3055	533_1	518_4	283_{69}	215_{100}	73_{77}	1179	4977
Cilazaprilate 3ET	UET	2980	473_2	445_7	274_{90}	239_{90}	171_{100}	1142	4732
Cilazaprilate 3ME	UME	2960	431_1	372_{22}	297_{45}	225_{93}	157_{100}	1084	4730
Cilazapril-M/artifact (deethyl-) 2ET	UET	3055	445_2	417_6	311_{75}	239_{43}	171_{100}	1108	4731
Cilazapril-M/artifact (deethyl-) 2ME	UME	2945	417_1	389_5	283_{57}	225_{42}	157_{100}	1057	4729
Cilazapril-M/artifact (deethyl-) 2TMS	UTMS	3055	533_1	518_4	283_{69}	215_{100}	73_{77}	1179	4977
Cilazapril-M/artifact (deethyl-) 3ET	UET	2980	473_2	445_7	274_{90}	239_{90}	171_{100}	1142	4732
Cilazapril-M/artifact (deethyl-) 3ME	UME	2960	431_1	372_{22}	297_{45}	225_{93}	157_{100}	1084	4730
Cinchocaine		2890	343_1	326_1	271_2	116_8	86_{100}	822	2126
Cinchonidine		2575	294_1	159_6	136_{100}	95_4	81_7	586	1980
Cinchonidine AC		2740	336_2	294_1	277_7	159_{19}	136_{100}	789	1988
Cinchonine	P-I G U	2590	294_{56}	253_{12}	159_{32}	136_{100}	81_{26}	586	684
Cinchonine AC		2750	336_{34}	277_{34}	253_8	159_{25}	136_{100}	789	2002
Cinnamolaurine	U	2855	297_1	190_{100}	175_2	131_3	91_3	602	5659
Cinnamolaurine-M (nor-)	U	2955	283_1	176_{100}	149_3	118_4	91_6	529	5660
Cinnamolaurine-M (nor-) 2AC	UAC	2930	367_4	324_9	218_{100}	176_{94}	118_4	909	5662
Cinnamoylcocaine iso-1		2345	329_8	238_{10}	182_{60}	96_{62}	82_{100}	754	4402
Cinnamoylcocaine iso-2		2450	329_{10}	238_{13}	182_{54}	96_{58}	82_{100}	755	4403
Cinnarizine	G	3040	368_2	251_{16}	201_{100}	167_{23}	117_{39}	913	1934
Cinnarizine-M (benzophenone)	U+UHYAC	1610*	182_{31}	152_3	105_{100}	77_{70}	51_{39}	180	1624
Cinnarizine-M (carbinol)	UHY	1645*	184_{45}	165_{14}	152_7	105_{100}	77_{63}	184	1333
Cinnarizine-M (carbinol) AC	U+UHYAC	1700*	226_{20}	184_{20}	165_{100}	105_{14}	77_{35}	294	1241
Cinnarizine-M (carbinol) ME	UHY	1655*	198_{70}	167_{94}	121_{100}	105_{56}	77_{71}	215	6779
Cinnarizine-M (HO-BPH) iso-1	UHY	2065*	198_{93}	167_{72}	121_{100}	105_{100}	93_{22}	214	1627
Cinnarizine-M (HO-BPH) iso-1 AC	U+UHYAC	2010*	240_{27}	198_{100}	121_{47}	105_{85}	77_{80}	343	2196
Cinnarizine-M (HO-BPH) iso-2	P-I U UHY	2080*	198_{50}	121_{100}	105_{17}	93_{14}	77_{28}	214	732
Cinnarizine-M (HO-BPH) iso-2 AC	U+UHYAC	2050*	240_{20}	198_{100}	121_{94}	105_{41}	77_{51}	343	2197
Cinnarizine-M (HO-methoxy-BPH)	UHY	2050*	228_{46}	197_6	151_{100}	105_{22}	77_{41}	299	1625
Cinnarizine-M (HO-methoxy-BPH) AC	U+UHYAC	2100*	270_3	228_{92}	151_{100}	105_{25}	77_{40}	468	1622
Cinnarizine-M (HO-methoxy-BPH) AC	U+UHYAC	2090*	284_5	242_{15}	224_{17}	182_{100}	153_{19}	533	2425
Cinnarizine-M (N-dealkyl-)	U+UHYAC	2350	244_{29}	201_{12}	172_{48}	117_{100}	85_{52}	356	2198
Cinnarizine-M (N-dealkyl-HO-) 2AC	U+UHYAC	2580	302_2	243_{20}	141_{100}	99_{21}	56_8	624	2199
Cinnarizine-M (norcyclizine) AC	U+UHYAC	2525	294_{16}	208_{56}	167_{100}	152_{30}	85_{78}	586	1601
Cinnarizine-M (piperazine) 2AC		1750	170_{15}	85_{33}	69_{25}	56_{100}		157	879
Cinnarizine-M (piperazine) 2HFB		1290	478_3	459_9	309_{100}	281_{22}	252_{41}	1146	6634
Cinnarizine-M (piperazine) 2TFA		1005	278_{10}	209_{59}	152_{25}	69_{56}	56_{100}	505	4129
Cinnarizine-M/artifact	UHY	2070*	228_8	186_{100}	157_{10}	128_7	77_4	298	1626
Cinnarizine-M/artifact AC	U+UHYAC	2200*	270_7	228_{17}	186_{100}	157_{10}	77_{18}	467	1623
Cisapride		3895	433_{16}	280_{12}	232_{100}	201_{19}	184_{88}	1133	5607
Cisapride AC		3970	475_8	322_{11}	232_{100}	226_{26}	184_{43}	1168	5608
Cisapride-M (N-dealkyl-) -CH3OH 2AC	U+UHYAC	3195	365_{91}	322_{12}	243_{14}	226_{82}	184_{100}	900	5609
Cisapride-M -CH3OH 2AC	U+UHYAC	3195	365_{91}	322_{12}	243_{14}	226_{82}	184_{100}	900	5609
Citalopram	G P U+UHYAC	2525	324_8	238_6	208_4	190_3	58_{100}	732	4452
Citalopram-M (bis-nor-) AC	U+UHYAC	2780	338_1	320_3	261_7	238_{100}	100_8	797	4454

Table 1-8-1: Compounds in order of names   **Clobenzorex-M (AM) PFP**

Name	Detected	RI	Typical ions and intensities					Page	Entry
Citalopram-M (nor-)	UHY	2500	310_{37}	238_{100}	208_{32}	190_{26}	138_{39}	667	4453
Citalopram-M (nor-) AC	U+UHYAC	2820	352_{1}	261_{8}	238_{100}	114_{16}	86_{11}	854	4455
Citric Acid 3ETAC		1880*	318_{1}	273_{8}	213_{20}	203_{55}	157_{100}	703	4478
Citric Acid 3ME	UME	1410*	175_{14}	143_{100}	101_{86}	69_{18}	59_{44}	319	4451
Citric Acid 4ME		1445*	189_{11}	157_{100}	133_{4}	125_{38}	59_{16}	370	5705
Citric Acid 4TMS	UTMS	1410*	465_{3}	375_{6}	273_{28}	147_{26}	73_{100}	1148	6566
Clemastine	G U	2445	342_{1}	215_{3}	178_{4}	128_{34}	84_{100}	821	1222
Clemastine artifact	G U+UHYAC	1700*	214_{49}	179_{100}	152_{9}	139_{3}	89_{14}	254	1217
Clemastine HY	UHY	1750*	232_{12}	217_{80}	139_{81}	105_{75}	77_{100}	312	1079
Clemastine HYAC	U+UHYAC	2180*	274_{18}	232_{75}	197_{14}	139_{35}	121_{100}	485	2185
Clemastine-M (di-HO-) -H2O HY2AC	U+UHYAC	2440*	330_{9}	288_{23}	246_{100}	211_{6}	152_{11}	758	2190
Clemastine-M (HO-) -H2O HY	U UHY	2050*	230_{100}	215_{27}	195_{60}	177_{33}	165_{64}	306	2187
Clemastine-M (HO-) iso-1 -H2O HYAC	U+UHYAC	2030*	272_{27}	230_{100}	195_{34}	165_{56}	152_{10}	477	2184
Clemastine-M (HO-) iso-2 -H2O HYAC	U+UHYAC	2090*	272_{16}	230_{100}	215_{15}	195_{34}	165_{47}	478	2189
Clemastine-M (HO-methoxy-) -H2O HYAC	U+UHYAC	2210*	302_{10}	260_{100}	182_{10}	152_{16}	75_{4}	623	2186
Clemastine-M (HO-methoxy-carbinol) -H2O	U UHY	2220*	262_{36}	260_{100}				420	2194
Clemizole	G U+UHYAC	2620	325_{1}	256_{100}	255_{90}	131_{65}	125_{35}	735	1613
Clemizole artifact	U+UHYAC	2300	242_{49}	127_{30}	125_{100}	89_{21}		350	1611
Clemizole-M (di-HO-) 2AC	U+UHYAC	3200	371_{35}	286_{17}	204_{60}	162_{100}	125_{60}	1100	5648
Clemizole-M (di-HO-) artifact 2AC	U+UHYAC	2805	358_{22}	316_{30}	274_{100}	146_{68}	125_{65}	876	5652
Clemizole-M (di-HO-methoxy-) 2AC	U+UHYAC	3300	401_{18}	359_{38}	316_{21}	192_{100}	125_{63}	1139	5653
Clemizole-M (di-HO-methoxy-) -H2O AC	U+UHYAC	3750	411_{21}	369_{43}	244_{100}	162_{12}	125_{48}	1043	5655
Clemizole-M (HO-) artifact-1 AC	U+UHYAC	2585	300_{19}	258_{100}	125_{91}	91_{30}		613	5651
Clemizole-M (HO-) artifact-2 AC	U+UHYAC	3080	314_{100}	272_{28}	236_{7}	147_{96}	125_{31}	684	5649
Clemizole-M (HO-deamino-HO-) 2AC	U+UHYAC	2995	372_{33}	330_{83}	287_{100}	245_{29}	125_{93}	926	5654
Clemizole-M (HO-methoxy-deamino-HO-) 2AC	U+UHYAC	2970	402_{13}	360_{73}	299_{14}	175_{22}	125_{100}	1022	5650
Clemizole-M (HO-methoxy-oxo-) AC	U+UHYAC	3190	427_{31}	385_{29}	302_{34}	260_{100}	125_{50}	1073	2861
Clemizole-M (HO-oxo-) AC	U+UHYAC	3120	397_{78}	314_{47}	272_{73}	230_{100}	125_{70}	1009	2860
Clemizole-M (oxo-)	U+UHYAC	2965	339_{75}	255_{100}	214_{48}	131_{70}	125_{58}	801	1612
Clemizole-M/artifact	U+UHYAC	3050	353_{80}	228_{50}	200_{20}	146_{36}	125_{52}	856	5647
Clenbuterol		2100	276_{1}	243_{3}	127_{16}	86_{100}	57_{32}	495	3990
Clenbuterol AC		2090	318_{1}	243_{4}	190_{8}	86_{100}	57_{22}	702	3992
Clenbuterol formyl artifact		2160	288_{1}	243_{16}	188_{9}	99_{100}	57_{45}	551	3989
Clenbuterol -H2O		1895	258_{9}	202_{9}	174_{32}	102_{5}	57_{100}	413	3991
Clenbuterol -H2O AC		2285	300_{1}	244_{21}	202_{100}	166_{29}	57_{48}	614	3993
Climbazole		2205	292_{3}	207_{100}	109_{50}	69_{65}	57_{89}	574	6087
Clindamycin	G P U	2750	388_{1}	341_{1}	126_{100}	82_{5}		1067	4481
Clindamycin 3AC	U+UHYAC	2850	549_{1}	514_{35}	471_{20}	417_{92}	126_{100}	1184	4479
Clindamycin-M (nor-) 4AC	U+UHYAC	2940	531_{11}	452_{30}	428_{28}	154_{84}	112_{100}	1191	4480
Clionasterol		3265*	414_{62}	329_{39}	303_{44}	105_{88}	55_{90}	1052	5622
Clionasterol -H2O		3300*	396_{100}	381_{25}	147_{80}	105_{64}	81_{65}	1008	5626
Clobazam	P-I G U-I	2610	300_{100}	283_{33}	255_{39}	231_{23}	77_{55}	613	439
Clobazam HY	UHY U+UHYAC	2225	274_{50}	257_{100}	231_{25}	215_{24}	77_{66}	485	275
Clobazam-M (HO-)	UGLUC	3000	316_{100}	299_{24}	271_{35}			692	441
Clobazam-M (HO-) AC	UGLUCAC	2900	358_{33}	316_{100}	299_{22}	271_{34}		876	443
Clobazam-M (HO-methoxy-)	UGLUC UGLUCAC	3255	346_{101}	316_{84}	301_{39}	271_{27}	245_{18}	830	442
Clobazam-M (HO-methoxy-) HY	UHY U+UHYAC	2905	320_{100}	240_{23}	206_{25}			715	277
Clobazam-M (nor-)	P-I U-I	2740	286_{100}	244_{33}	215_{45}	77_{60}		541	440
Clobazam-M (nor-) HY	UHY U+UHYAC	2210	242_{100}	206_{8}	166_{9}	77_{48}		350	276
Clobazam-M (nor-HO-) HY	UHY	2650	258_{100}					412	445
Clobazam-M (nor-HO-) HYAC	U+UHYAC	3000	300_{40}	258_{100}				614	279
Clobazam-M (nor-HO-methoxy-) HY	UHY U+UHYAC	2405	288_{101}					551	444
Clobazam-M (nor-HO-methoxy-) HYAC	U+UHYAC	2615	330_{13}	288_{100}				758	278
Clobenzorex	G	1940	259_{1}	244_{1}	168_{100}	125_{88}	91_{24}	416	4409
Clobenzorex AC		2290	301_{1}	266_{3}	210_{60}	168_{100}	125_{70}	618	4410
Clobenzorex HFB		2075	364_{13}	125_{100}	118_{20}	91_{33}		1120	5051
Clobenzorex PFP		2040	405_{1}	314_{21}	125_{100}	118_{21}	91_{18}	1030	5052
Clobenzorex TFA		2075	355_{1}	264_{24}	125_{100}	118_{38}	91_{42}	864	5053
Clobenzorex-M	UHY	1465	181_{9}	138_{100}	122_{18}	94_{24}	77_{16}	178	4351
Clobenzorex-M (4-HO-amfetamine)		1480	151_{10}	107_{69}	91_{10}	77_{42}	56_{100}	129	1802
Clobenzorex-M (4-HO-amfetamine) AC	U+UHYAC	1890	193_{1}	134_{100}	107_{26}	86_{25}	77_{16}	201	1803
Clobenzorex-M (4-HO-amfetamine) TFA		1670	247_{3}	140_{15}	134_{54}	107_{100}	77_{15}	366	6335
Clobenzorex-M (4-HO-amfetamine) 2AC	U+UHYAC	1900	235_{1}	176_{72}	134_{100}	107_{47}	86_{71}	324	1804
Clobenzorex-M (4-HO-amfetamine) 2HFB		<1000	330_{48}	303_{15}	240_{100}	169_{44}	69_{42}	1182	6326
Clobenzorex-M (4-HO-amfetamine) 2PFP		<1000	280_{77}	253_{16}	190_{100}	119_{56}	69_{16}	1104	6325
Clobenzorex-M (4-HO-amfetamine) 2TFA		<1000	230_{94}	203_{14}	140_{130}	92_{15}	69_{76}	819	6324
Clobenzorex-M (4-HO-amfetamine) 2TMS		<1000	280_{7}	179_{9}	149_{8}	116_{100}	73_{78}	594	6327
Clobenzorex-M (4-HO-amfetamine) formyl art.		1220	163_{3}	148_{4}	107_{30}	77_{13}	56_{100}	142	6323
Clobenzorex-M (AM)		1160	134_{1}	120_{1}	91_{6}	77_{1}	65_{4}	115	54
Clobenzorex-M (AM)	U	1160	134_{1}	120_{1}	91_{6}	65_{4}	44_{100}	115	5514
Clobenzorex-M (AM) AC	U+UHYAC	1505	177_{1}	118_{19}	91_{11}	86_{31}	65_{5}	165	55
Clobenzorex-M (AM) AC	UAAC U+UHYAC	1505	177_{1}	118_{19}	91_{11}	86_{31}	44_{100}	165	5515
Clobenzorex-M (AM) HFB		1355	331_{1}	240_{79}	169_{21}	118_{100}	91_{53}	762	5047
Clobenzorex-M (AM) PFP		1330	281_{1}	190_{73}	118_{100}	91_{36}	65_{12}	521	4379

Clobenzorex-M (AM) TFA — Table 1-8-1: Compounds in order of names

Name	Detected	RI	Typical ions and intensities					Page	Entry
Clobenzorex-M (AM) TFA		1095	231_1	140_{100}	118_{92}	91_{45}	69_{19}	309	4000
Clobenzorex-M (AM) TMS		1190	192_6	116_{100}	100_{10}	91_{11}	73_{87}	235	5581
Clobenzorex-M (AM) formyl artifact		1100	147_2	146_6	125_5	91_{12}	56_{100}	125	3261
Clobenzorex-M (AM)-D5 AC	UAAC U+UHYAC	1480	182_3	122_{46}	92_{30}	90_{100}	66_{16}	181	5690
Clobenzorex-M (AM)-D5 HFB		1330	244_{100}	169_{14}	122_{46}	92_{41}	69_{40}	787	6316
Clobenzorex-M (AM)-D5 PFP		1320	194_{100}	123_{42}	119_{32}	92_{46}	69_{11}	543	5566
Clobenzorex-M (AM)-D5 TFA		1085	144_{100}	123_{53}	122_{56}	92_{51}	69_{28}	330	5570
Clobenzorex-M (AM)-D5 TMS		1180	212_1	197_8	120_{100}	92_{11}	73_{57}	251	5582
Clobenzorex-M (AM)-D11 TFA		1615	242_1	144_{100}	128_{82}	98_{43}	70_{14}	353	7283
Clobenzorex-M (AM)-D11PFP		1610	292_1	194_{100}	128_{82}	98_{43}	70_{14}	575	7284
Clobenzorex-M (di-HO-) 3AC	U+UHYAC	2765	417_1	234_8	210_{36}	168_{100}	125_{51}	1056	4415
Clobenzorex-M (HO-) iso-1 2AC	U+UHYAC	2585	359_1	324_6	210_{56}	168_{100}	125_{64}	880	4412
Clobenzorex-M (HO-) iso-2 2AC	U+UHYAC	2630	359_1	324_5	210_{79}	168_{100}	125_{89}	880	4413
Clobenzorex-M (HO-chlorobenzyl-) 2AC	U+UHYAC	2565	324_{10}	268_{23}	226_{100}	183_{30}	141_{29}	880	4411
Clobenzorex-M (HO-HO-alkyl-) 3AC	U+UHYAC	2725	417_1	210_{27}	168_{100}	168_{100}	125_{70}	1057	5106
Clobenzorex-M (HO-HO-chlorobenzyl-) iso-1 3AC	U+UHYAC	2705	417_1	268_{15}	226_{100}	183_{31}	141_{50}	1057	5105
Clobenzorex-M (HO-HO-chlorobenzyl-) iso-2 3AC	U+UHYAC	2725	417_1	268_{38}	226_{77}	183_{23}	141_{100}	1057	5104
Clobenzorex-M (HO-HO-chlorobenzyl-) iso-3 3AC	U+UHYAC	2775	417_1	268_{15}	226_{100}	183_{31}	141_{39}	1057	5103
Clobenzorex-M (HO-HO-chlorobenzyl-) iso-4 3AC	U+UHYAC	2795	417_1	268_{18}	226_{67}	183_{40}	141_{100}	1057	4416
Clobenzorex-M (HO-methoxy-) 2AC	U+UHYAC	2690	389_1	210_{28}	206_{38}	168_{100}	125_{56}	983	4414
Clobenzorex-M (norephedrine) TMSTFA		1890	240_8	198_3	179_{100}	117_5	73_{88}	708	6146
Clobenzorex-M (norephedrine) 2AC		1805	235_1	107_{13}	86_{100}	176_5	134_7	325	2476
Clobenzorex-M (norephedrine) 2HFB	UHYHFB	1455	543_1	330_{14}	240_{100}	169_{44}	69_{57}	1183	5098
Clobenzorex-M (norephedrine) 2PFP	UHYPFP	1380	443_1	280_9	190_{100}	119_{59}	105_{26}	1104	5094
Clobenzorex-M (norephedrine) 2TFA	UTFA	1355	343_1	230_6	203_5	140_{100}	69_{29}	819	5091
Clobenzorex-M 2AC	U+UHYAC	2065	265_3	206_{27}	164_{100}	137_{23}	86_{33}	445	3498
Clobenzorex-M 2AC	U+UHYAC	1820*	266_3	206_9	164_{100}	150_{10}	137_{30}	450	6409
Clobenzorex-M 2HFB	UHFB	1690	360_{82}	333_{15}	240_{100}	169_{42}	69_{39}	1189	6512
Clobutinol	G P U	1895	255_1	240_1	130_{28}	125_{31}	58_{100}	403	2793
Clobutinol AC	U+UHYAC	1980	238_1	222_1	125_{18}	89_9	58_{100}	603	3060
Clofedanol	U UHY	2105	274_1	254_7	111_2	77_6	58_{100}	557	1935
Clofedanol AC	U+UHYAC	2120	296_6	236_3	165_3	58_{100}		764	1936
Clofedanol -H2O	UHY U+UHYAC	2085	271_{12}	270_{21}	236_{100}	160_{41}	58_{68}	474	1639
Clofedanol-M (2-Cl-benzophenone)	U UHY U+UHYAC	1720*	216_{86}	139_{57}	111_{20}	105_{100}	77_{60}	259	1636
Clofedanol-M (aldehyde)	U UHY U+UHYAC	1900*	207_{100}	179_{27}				350	1632
Clofedanol-M (HO-) artifact	U UHY	2040*	230_{100}	195_{70}	177_{56}	165_{65}	152_{20}	306	1637
Clofedanol-M (HO-) -H2O	UHY	2130	287_8	286_{16}	252_{100}	222_{33}	58_{85}	546	1640
Clofedanol-M (HO-) -H2O AC	U+UHYAC	2370	329_7	294_{90}	226_{50}	178_{57}	58_{100}	753	1635
Clofedanol-M (nor-) -H2O	U UHY	2090	257_{13}	256_{17}	222_{100}	163_{85}	134_{72}	410	1641
Clofedanol-M (nor-) -H2O AC	U+UHYAC	2400	299_{34}	256_{16}	226_{65}	191_{54}	98_{100}	610	1633
Clofedanol-M (nor-HO-) -H2O 2AC	U+UHYAC	2800	357_{15}	242_{54}	178_{16}	152_8	98_{100}	872	1634
Clofedanol-M/artifact	U UHY U+UHYAC	1700*	214_{50}	200_5	179_{100}	178_{98}	151_7	255	1631
Clofedanol-M/artifact	UHY	2060*	244_{100}	209_{42}	194_{26}	165_{22}	115_{32}	355	1638
Clofibrate	U	1540*	242_8	169_{16}	128_{100}			351	685
Clofibrate-M (clofibric acid)	U	1640*	214_2	168_9	128_{100}	86_9	65_{17}	254	686
Clofibrate-M (clofibric acid) ME	U	1500*	228_9	169_{16}	128_{100}	99_5	75_9	298	687
Clofibrate-M (clofibric acid) artifact	U+UHYAC	1580*	168_{35}	128_{99}				154	1373
Clofibrate-M/artifact (4-chlorophenol)	U UHY	1390*	128_{100}	100_{21}	65_{54}			111	676
Clofibric acid	U	1640*	214_2	168_9	128_{100}	86_9	65_{17}	254	686
Clofibric acid ME	U	1500*	228_9	169_{16}	128_{100}	99_5	75_9	298	687
Clofibric acid artifact	U+UHYAC	1580*	168_{35}	128_{99}				154	1373
Clofibric acid-M/artifact (4-chlorophenol)	U UHY	1390*	128_{100}	100_{21}	65_{54}			111	676
Clomethiazole	P G U+UHYAC	1230	161_{29}	112_{100}	85_{28}			139	446
Clomethiazole-M (1-HO-ethyl-)	UHY	1560	177_{27}	159_{27}	142_{43}	124_{100}	100_{68}	164	3311
Clomethiazole-M (1-HO-ethyl-) AC	U+UHYAC	1430	219_5	183_{16}	160_{11}	141_{55}	128_{100}	268	452
Clomethiazole-M (1-HO-ethyl-) TMS		1560	249_1	234_8	200_{100}	93_{18}	73_{49}	373	4622
Clomethiazole-M (2-HO-)	P U UHY	1440	177_5	128_{100}	100_7	73_{29}		164	450
Clomethiazole-M (2-HO-) AC	U+UHYAC	1590	219_3	183_{16}	176_{67}	141_{38}	128_{100}	268	3310
Clomethiazole-M (deschloro-2-HO-)	P U UHY	1160	143_{23}	128_{100}	100_{52}	73_{47}		122	449
Clomethiazole-M (deschloro-2-HO-ethyl-)	UHY	1380	143_{31}	113_{42}	112_{100}	85_{30}	71_9	122	448
Clomethiazole-M (deschloro-2-HO-ethyl-) AC	U+UHYAC	1050	185_7	143_5	128_{32}	125_{100}	98_{32}	186	451
Clomethiazole-M (deschloro-di-HO-)	UHY	1685	159_7	128_{100}	100_8	73_{38}		138	3312
Clomethiazole-M (deschloro-di-HO-) -H2O AC	U+UHYAC	1420	183_{19}	170_9	141_{40}	128_{100}		181	1461
Clomethiazole-M (deschloro-HOOC-)	U	1235	157_{50}	128_{17}	112_{100}	85_{24}		137	447
Clomethiazole-M (deschloro-HOOC-2-HO-)	U+UHYAC	1690	155_{50}	125_{100}	97_{61}	70_{32}		134	6560
Clomiphene		2885	405_1	252_2	239_1	100_9	86_{100}	1031	7533
Clomipramine	P G U+UHYAC	2455	314_{17}	269_{50}	227_{11}	85_{42}	58_{100}	686	315
Clomipramine-D3		2440	317_4	88_{49}	130_7	85_{17}	61_{100}	698	5425
Clomipramine-M (bis-nor-) AC	U+UHYAC	2960	328_{42}	242_{100}	227_{38}	100_{45}	72_{20}	749	1177
Clomipramine-M (bis-nor-HO-) 2AC	U+UHYAC	3120	386_{76}	300_{81}	258_{100}	243_{28}	100_{77}	972	3414
Clomipramine-M (HO-) iso-1	U UHY	2540	330_8	285_{16}	245_7	85_{20}	58_{100}	759	453
Clomipramine-M (HO-) iso-1 AC	U+UHYAC	2805	372_5	327_2	285_{17}	85_{17}	58_{100}	927	317
Clomipramine-M (HO-) iso-2	U UHY	2800	330_8	285_{15}	245_7	85_{19}	58_{100}	759	33
Clomipramine-M (HO-) iso-2 AC	U+UHYAC	2905	372_6	327_3	285_{22}	85_{23}	58_{100}	928	122

Table 1-8-1: Compounds in order of names   **Clostebol enol 2TMS**

Name	Detected	RI	Typical ions and intensities					Page	Entry
Clomipramine-M (HO-ring) AC	U+UHYAC	2645	287_{30}	245_{100}	230_{11}	210_{19}	180_{10}	546	4159
Clomipramine-M (nor-)		2620	300_{32}	268_{105}	229_{82}	192_{40}	71_{40}	616	7663
Clomipramine-M (nor-) AC	U+UHYAC	2980	342_{16}	256_{19}	242_{100}	227_{69}	114_{81}	818	1176
Clomipramine-M (nor-) HFB		2650	496_{16}	268_{33}	242_{100}	228_{38}	169_{71}	1161	7666
Clomipramine-M (nor-) PFP		2690	446_{4}	268_{4}	242_{100}	227_{41}	190_{19}	1109	7665
Clomipramine-M (nor-) TFA		2650	396_{14}	242_{100}	227_{53}	191_{22}	69_{74}	1005	7664
Clomipramine-M (nor-) TMS		2575	372_{21}	269_{100}	242_{32}	227_{36}	116_{46}	928	7707
Clomipramine-M (nor-) TMS		2575	372_{21}	269_{100}	242_{18}	227_{21}	73_{80}	928	7785
Clomipramine-M (nor-HO-) 2AC	U+UHYAC	3205	400_{47}	300_{44}	258_{33}	114_{100}	86_{13}	1018	318
Clomipramine-M (N-oxide) -(CH3)2NOH	G P UHY U+UHYAC	2160	269_{42}	228_{100}	193_{60}	165_{14}	89_{15}	464	4346
Clomipramine-M (ring)	U+UHYAC	2230	229_{100}	214_{30}	194_{41}	165_{8}	152_{3}	303	316
Clonazepam	P-I G U-I	2840	315_{89}	314_{97}	286_{55}	280_{100}	234_{53}	688	454
Clonazepam TMS		2795	387_{54}	372_{23}	352_{41}	306_{25}	73_{70}	975	5463
Clonazepam HY	UHY-I U+UHYAC-I	2470	276_{66}	241_{100}	195_{35}	139_{48}	111_{38}	495	280
Clonazepam iso-1 ME		2555	329_{101}	294_{55}	248_{37}			752	460
Clonazepam iso-2 ME		2760	329_{100}	302_{49}	294_{60}	248_{38}		752	461
Clonazepam-M (amino-)	UGLUC-I	2880	285_{100}	256_{62}	250_{24}	222_{18}	111_{44}	538	455
Clonazepam-M (amino-) AC	UGLUCAC-I	3190	327_{100}	299_{40}	292_{36}	256_{30}	220_{16}	743	457
Clonazepam-M (amino-) HY	UHY-I	2285	246_{100}	211_{71}	139_{18}	111_{23}	107_{45}	362	458
Clonazepam-M (amino-) HY2AC	U+UHYAC-I	2845	330_{100}	288_{90}	246_{94}	211_{44}	139_{29}	758	281
Clonazepam-M (amino-HO-)		2935	283_{100}	255_{41}	220_{48}			618	456
Clonazepam-M (amino-HO-) artifact	UHY-I U+UHYAC-I	2325	255_{100}	220_{95}				401	459
Clonidine	G	2090	229_{100}	200_{15}	194_{34}	172_{33}	109_{7}	303	1785
Clonidine AC	U+UHYAC	2060	271_{34}	236_{100}	229_{12}	194_{51}	172_{12}	473	1786
Clonidine TMS		1925	301_{19}	286_{19}	266_{100}	142_{40}	73_{37}	618	6302
Clonidine 2AC	U+UHYAC	2315	313_{17}	278_{57}	236_{100}	194_{66}	85_{17}	678	688
Clonidine 2TMS		2000	373_{18}	358_{16}	338_{100}	214_{44}	73_{62}	930	6303
Clonidine artifact (dehydro-) AC	U+UHYAC	1820	269_{18}	227_{66}	192_{100}	157_{21}	109_{8}	463	1790
Clonidine artifact (dichloroaniline) AC	U+UHYAC	1550	203_{1}	168_{58}	161_{100}	133_{15}	125_{14}	223	1789
Clonidine artifact (dichlorophenylisocyanate)	G P	1350	187_{100}	159_{19}	124_{40}			189	1787
Clonidine artifact (dichlorophenylmethylcarbamate)		1500	219_{10}	184_{100}	174_{24}	160_{14}	133_{18}	266	1788
Clonidine artifact-5		2110	283_{10}	248_{100}	243_{18}	229_{17}	194_{57}	527	1791
Clopamide		2880	345_{1}	330_{12}	218_{30}	127_{71}	111_{100}	827	6879
Clopamide ME		2850	359_{1}	344_{7}	232_{11}	127_{60}	111_{100}	879	6880
Clopamide 2ME		2805	373_{1}	358_{2}	246_{2}	127_{64}	111_{100}	930	3097
Clopamide 3ME	UEXME	2800	372_{1}	246_{7}	141_{90}	112_{100}	83_{20}	976	3098
Clopamide -SO2NH		2195	266_{1}	251_{8}	139_{26}	127_{51}	111_{100}	451	3099
Clopenthixol (cis)	G U	3360	400_{1}	221_{12}	143_{100}	100_{18}	70_{24}	1017	462
Clopenthixol (cis) AC	U+UHYAC	3460	442_{1}	221_{9}	185_{100}	98_{24}	70_{11}	1102	319
Clopenthixol (cis) TMS		3490	472_{1}	457_{6}	221_{19}	215_{100}	98_{23}	1140	4534
Clopenthixol (trans)		3400	400_{1}	221_{15}	143_{100}	100_{21}	70_{30}	1017	4619
Clopenthixol (trans) AC	U+UHYAC	3570	442_{1}	221_{12}	185_{100}	98_{22}	70_{10}	1103	4680
Clopenthixol (trans) TMS		3555	472_{1}	457_{6}	221_{20}	215_{100}	98_{9}	1140	4535
Clopenthixol-M (dealkyl-) AC	U+UHYAC	3490	398_{2}	268_{7}	141_{100}	99_{30}		1011	1261
Clopenthixol-M (dealkyl-dihydro-) AC	U+UHYAC	3450	400_{46}	231_{44}	141_{100}	128_{16}	99_{25}	1018	1260
Clopenthixol-M (N-oxide) -C6H14N2O2	P-I U+UHYAC	2410*	270_{40}	255_{21}	234_{100}	202_{23}	117_{27}	467	438
Clopenthixol-M (N-oxide-sulfoxide) -C6H14N2O2	P-I U UGLUC UGLUCAC	2560*	286_{21}	251_{20}	234_{57}	203_{100}	101_{10}	541	436
Clopenthixol-M / artifact (Cl-thioxanthenone)	U	2260*	246_{44}	218_{46}	183_{9}	139_{25}	91_{10}	361	2641
Clopidogrel		2320	320_{1}	262_{100}	152_{41}	138_{46}	125_{45}	717	5704
Clopyralide ME		1320	205_{15}	174_{27}	147_{100}	110_{50}	75_{31}	227	4119
Clorazepate -H2O -CO2	P G U	2520	270_{86}	269_{97}	242_{100}	241_{82}	77_{17}	468	463
Clorazepate -H2O -CO2 TMS		2300	342_{62}	341_{100}	327_{19}	269_{4}	73_{30}	817	4573
Clorazepate -H2O -CO2 enol AC		2545	312_{55}	270_{34}	241_{100}	227_{8}	205_{9}	674	6102
Clorazepate -H2O -CO2 enol ME		2225	284_{78}	283_{100}	91_{61}			532	464
Clorazepate HY	UHY	2050	231_{80}	230_{95}	154_{23}	105_{38}	77_{100}	308	419
Clorazepate HYAC	U+UHYAC PHYAC	2245	273_{31}	230_{100}	154_{13}	105_{23}	77_{50}	480	273
Clorazepate-M	P G UGLUC	2320	268_{98}	239_{57}	233_{52}	205_{66}	77_{100}	542	579
Clorazepate-M (HO-) artifact AC	U+UHYAC	2515	312_{30}	270_{100}	253_{46}	235_{77}	206_{9}	674	1747
Clorazepate-M (HO-) -H2O -CO2	UGLUC	2750	286_{82}	258_{100}	230_{11}	166_{7}	139_{8}	541	2113
Clorazepate-M (HO-) -H2O -CO2 AC	U+UHYAC	3000	328_{22}	286_{90}	258_{100}	166_{8}	139_{7}	747	2111
Clorazepate-M (HO-) HY	UHY	2400	247_{72}	246_{100}	230_{11}	121_{26}	65_{22}	365	2112
Clorazepate-M (HO-) HYAC	U+UHYAC	2270	289_{18}	247_{86}	246_{100}	105_{77}	77_{35}	556	3143
Clorazepate-M (HO-) iso-1 HY2AC	U+UHYAC	2560	331_{48}	289_{64}	247_{100}	230_{41}	154_{13}	762	2125
Clorazepate-M (HO-) iso-2 HY2AC	U+UHYAC	2610	331_{46}	289_{54}	246_{100}	154_{11}	121_{11}	762	1751
Clorazepate-M (HO-methoxy-) HY2AC	U+UHYAC	2700	361_{7}	319_{27}	276_{38}	260_{5}	246_{4}	886	1752
Clorazepate-M TMS		2635	356_{9}	341_{100}	312_{56}	239_{12}	135_{21}	876	4577
Clorazepate-M 2TMS		2200	430_{36}	429_{63}	340_{10}	313_{14}	73_{70}	1080	5499
Clorazepate-M/artifact AC	U+UHYAC	3000	356_{16}	314_{100}	297_{23}	256_{48}	219_{17}	868	1748
Clorofene	G U UHY	1950*	218_{100}	183_{73}	140_{82}			263	689
Clorofene AC	U+UHYAC	1885*	260_{24}	218_{100}	183_{50}	152_{25}	140_{47}	420	690
Clostebol AC		2965*	364_{25}	328_{100}	287_{26}	269_{23}	147_{32}	898	3945
Clostebol acetate		2965*	364_{25}	328_{100}	287_{26}	269_{23}	147_{32}	898	3945
Clostebol acetate TMS		2870*	436_{62}	401_{12}	230_{6}	133_{12}	73_{100}	1092	3952
Clostebol enol 2TMS		2830*	466_{12}	358_{26}	268_{24}	129_{69}	73_{100}	1134	3953

Table 1-8-1: Compounds in order of names

Name	Detected	RI	Typical ions and intensities					Page	Entry
Clostebol -HCl AC		2700*	328_{20}	286_{27}	253_{18}	133_{100}	91_{34}	750	3951
Clostebol -HCl TMS		2675*	358_{32}	268_{34}	253_{23}	145_{10}	73_{100}	878	3954
Clostebol -HCl enol 2TMS		2640*	430_{38}	415_{8}	231_{6}	207_{10}	73_{100}	1081	3955
Clotiapine	U UHY U+UHYAC	2590	343_{41}	285_{43}	273_{100}	244_{87}	209_{45}	820	2373
Clotiapine artifact (desulfo-)	U UHY U+UHYAC	2600	311_{7}	241_{100}	177_{24}	83_{45}	70_{40}	671	2377
Clotiapine-M (HO-) AC	U+UHYAC	3000	401_{31}	331_{100}	302_{56}	260_{82}	70_{80}	1020	2375
Clotiapine-M (nor-) AC	U U+UHYAC	3030	371_{87}	285_{85}	273_{73}	244_{85}	209_{100}	922	2374
Clotiapine-M (nor-) artifact AC	U UHY U+UHYAC	3070	339_{22}	253_{58}	241_{100}	228_{26}	177_{57}	801	2379
Clotiapine-M (nor-HO-) 2AC	U+UHYAC	3400	429_{86}	344_{46}	302_{91}	260_{100}	112_{23}	1077	2376
Clotiapine-M (oxo-)	U	3030	357_{100}	285_{33}	244_{49}	209_{74}		871	2380
Clotiapine-M (oxo-) artifact	U UHY	3040	325_{23}	253_{30}	241_{100}	213_{23}	177_{27}	734	2378
Clotiazepam	P-I G UGLUC	2540	318_{72}	289_{100}	275_{20}			701	267
Clotiazepam artifact		2280	274_{100}	259_{48}	245_{32}	223_{10}	139_{9}	483	2350
Clotiazepam-M (di-HO-) 2AC	UGLUCAC	2995	434_{36}	374_{81}	332_{100}	319_{61}	291_{52}	1087	271
Clotiazepam-M (HO-)	UGLUC	2705	316_{95}	287_{100}				779	269
Clotiazepam-M (HO-) AC	UGLUCAC	2870	376_{14}	316_{85}	271_{100}	256_{76}		939	270
Clotiazepam-M (oxo-)		2660	332_{75}	303_{100}	297_{28}			767	268
Clotrimazole		2800	277_{100}	239_{15}	199_{5}	165_{30}		824	1753
Clotrimazole artifact-1	U+UHYAC	2240*	278_{38}	243_{33}	201_{9}	165_{100}		506	1756
Clotrimazole artifact-2	U+UHYAC	2530*	294_{100}	217_{60}	183_{54}	139_{50}	105_{55}	585	1757
Clotrimazole artifact-3	U+UHYAC	2550*	308_{100}	277_{76}	231_{86}	165_{60}	139_{60}	654	1758
Cloxazolam		2775	318_{4}	305_{100}	261_{48}	226_{29}	191_{21}	838	2264
Cloxazolam HY	UHY	2180	265_{62}	230_{100}	139_{43}	111_{50}		441	543
Cloxazolam HYAC	U+UHYAC	2300	307_{42}	265_{58}	230_{100}	139_{16}	111_{14}	647	290
Cloxiquine		1565	179_{100}	151_{84}	116_{63}	89_{42}		169	2003
Cloxiquine AC		1790	221_{2}	179_{100}	151_{82}	116_{42}	89_{28}	275	2004
Clozapine	P G U UHY U+UHYAC	2895	326_{27}	256_{75}	243_{100}	192_{37}	70_{31}	739	320
Clozapine AC	U+UHYAC	2870	368_{24}	298_{70}	256_{40}	83_{100}	70_{97}	913	2604
Clozapine TMS		2895	398_{14}	328_{88}	315_{100}	299_{50}	73_{52}	1012	4536
Clozapine-M (HO-) AC	U+UHYAC	3050	384_{38}	314_{70}	301_{83}	259_{100}	70_{35}	967	2605
Clozapine-M (HO-) 2AC	U+UHYAC	2980	426_{24}	356_{71}	314_{47}	83_{100}	70_{93}	1072	2606
Clozapine-M (nor-)	UHY	3105	312_{44}	269_{23}	256_{43}	243_{100}	192_{37}	675	321
Clozapine-M (nor-) AC	U+UHYAC	3650	354_{100}	243_{51}	228_{23}	192_{51}	112_{21}	861	322
Clozapine-M (nor-) 2AC	U+UHYAC	3490	396_{100}	310_{61}	298_{61}	227_{43}	192_{56}	1006	323
Clozapine-M/artifact	U+UHYAC	3875	378_{100}	280_{48}	225_{71}	209_{54}	112_{61}	945	6766
Clozapine-M/artifact AC	U+UHYAC	3855	420_{82}	335_{100}	322_{90}	251_{37}	209_{45}	1062	7802
CN gas (chloroacetophenone)		1020*	154_{3}	105_{100}	91_{4}	77_{61}	51_{26}	133	3731
Cocaethylene	U+UHYAC	2250	317_{58}	272_{24}	196_{100}	82_{74}		698	466
Cocaethylene-M (ethylecgonine) AC		1675	255_{12}	196_{60}	168_{14}	94_{44}	82_{100}	403	6231
Cocaethylene-M (ethylecgonine) TFA		1520	309_{5}	264_{6}	196_{56}	94_{53}	82_{100}	660	6241
Cocaethylene-M (ethylecgonine) TMS		1485	285_{8}	240_{12}	196_{8}	96_{68}	82_{100}	540	6257
Cocaethylene-M (ethylecgonine) TBDMS		1685	327_{1}	270_{5}	204_{10}	196_{9}	82_{100}	746	6249
Cocaethylene-M (nor-)		2115	303_{6}	182_{100}	136_{53}	105_{46}	68_{88}	628	6253
Cocaethylene-M (nor-) AC		2535	345_{9}	182_{89}	136_{48}	109_{88}	105_{90}	829	6233
Cocaethylene-M (nor-) TFA		2245	399_{2}	277_{23}	208_{20}	164_{26}	105_{100}	1014	6245
Cocaine	P-I G UCOME U+UHYAC	2200	303_{6}	272_{3}	198_{8}	182_{66}	82_{100}	628	465
Cocaine-D3		2180	306_{36}	275_{11}	201_{11}	185_{100}	85_{86}	645	5565
Cocaine-M (benzoylecgonine)	U+UHYAC	2570	289_{19}	168_{34}	124_{100}	82_{42}	77_{50}	558	2120
Cocaine-M (benzoylecgonine) ET	U+UHYAC	2250	317_{58}	272_{24}	196_{100}	82_{74}		698	466
Cocaine-M (benzoylecgonine) ME	P-I G UCOME U+UHYAC	2200	303_{6}	272_{3}	198_{8}	182_{66}	82_{100}	628	465
Cocaine-M (benzoylecgonine) PFP		2275	421_{9}	316_{5}	300_{38}	94_{52}	82_{100}	1063	4381
Cocaine-M (benzoylecgonine) TMS		2285	361_{12}	256_{8}	240_{47}	105_{39}	82_{100}	887	5579
Cocaine-M (benzoylecgonine) TBDMS		2465	403_{20}	346_{25}	282_{35}	105_{38}	82_{100}	1027	6236
Cocaine-M (benzoylecgonine)-D3 ME		2180	306_{36}	275_{11}	201_{11}	185_{100}	85_{86}	645	5565
Cocaine-M (benzoylecgonine)-D3 TMS		2275	364_{31}	349_{9}	243_{66}	105_{34}	85_{100}	898	5580
Cocaine-M (cocaethylene)	U+UHYAC	2250	317_{58}	272_{24}	196_{100}	82_{74}		698	466
Cocaine-M (ecgonine) ACTMS		1680	299_{11}	240_{41}	122_{9}	94_{29}	82_{100}	611	6238
Cocaine-M (ecgonine) TMSTFA		1395	353_{15}	267_{35}	240_{82}	94_{38}	82_{100}	858	6255
Cocaine-M (ecgonine) 2TBDMS		1970	398_{1}	356_{12}	275_{4}	96_{28}	82_{100}	1049	6251
Cocaine-M (ecgonine) 2TMS		1680	329_{2}	314_{4}	96_{45}	82_{100}	73_{49}	755	5445
Cocaine-M (ecgonine) TBDMS		1700	299_{5}	242_{21}	205_{5}	96_{34}	82_{100}	612	6250
Cocaine-M (ecgonine)-D3 2TMS		1670	332_{13}	317_{15}	99_{64}	85_{100}	73_{56}	770	5576
Cocaine-M (ethylecgonine) AC		1675	255_{12}	196_{60}	168_{14}	94_{44}	82_{100}	403	6231
Cocaine-M (ethylecgonine) TFA		1520	309_{5}	264_{6}	196_{56}	94_{53}	82_{100}	660	6241
Cocaine-M (ethylecgonine) TMS		1485	285_{8}	240_{12}	196_{8}	96_{68}	82_{100}	540	6257
Cocaine-M (ethylecgonine) TBDMS		1685	327_{1}	270_{5}	204_{10}	196_{9}	82_{100}	746	6249
Cocaine-M (HO-)	UCO	2460	319_{19}	182_{52}	121_{21}	82_{100}		708	468
Cocaine-M (HO-) ME	UCOME	2450	333_{29}	182_{85}	135_{42}	94_{41}	82_{100}	774	470
Cocaine-M (HO-benzoylecgonine) ACTBDMS		2765	461_{13}	404_{9}	282_{18}	121_{14}	82_{100}	1129	6235
Cocaine-M (HO-benzoylecgonine) ACTMS		2565	419_{17}	240_{46}	163_{9}	94_{25}	82_{100}	1061	6239
Cocaine-M (HO-benzoylecgonine) 2TBDMS		2940	533_{37}	476_{25}	282_{50}	235_{14}	82_{100}	1179	6237
Cocaine-M (HO-benzoylecgonine) 2TMS		2505	449_{21}	240_{55}	193_{16}	82_{100}	73_{46}	1112	6258
Cocaine-M (HO-di-methoxy-) AC	UGLUCAC	2750	421_{13}	198_{13}	182_{71}	151_{7}	82_{100}	1063	5945
Cocaine-M (HO-di-methoxy-) HFB	UGLUCHFB	2585	575_{4}	377_{9}	182_{85}	94_{37}	82_{100}	1190	5947

Table 1-8-1: Compounds in order of names   **Codeine-M (O-demethyl-)-D3 2HFB**

Name	Detected	RI	Typical ions and intensities					Page	Entry
Cocaine-M (HO-di-methoxy-) ME		2550	393_{27}	212_{21}	182_{100}	94_{24}	82_{102}	995	5678
Cocaine-M (HO-di-methoxy-) PFP		2555	525_{5}	327_{11}	182_{71}	94_{31}	82_{100}	1177	5948
Cocaine-M (HO-di-methoxy-) TFA		2530	475_{12}	277_{19}	182_{87}	94_{37}	82_{100}	1144	5953
Cocaine-M (HO-di-methoxy-) TMS		2970	451_{15}	198_{7}	182_{61}	94_{29}	82_{100}	1115	5951
Cocaine-M (HO-methoxy-)	UCO	2670	349_{58}	198_{16}	182_{100}	151_{19}	82_{61}	842	469
Cocaine-M (HO-methoxy-) AC	UGLUCAC	2695	391_{13}	198_{10}	182_{67}	151_{19}	82_{100}	988	5944
Cocaine-M (HO-methoxy-) HFB	UGLUCHFB	2500	545_{3}	347_{8}	182_{64}	94_{34}	82_{100}	1183	5946
Cocaine-M (HO-methoxy-) ME	UCOME	2650	363_{18}	198_{8}	182_{86}	94_{37}	82_{100}	894	471
Cocaine-M (HO-methoxy-) PFP	UGLUCPFP	2470	495_{7}	297_{13}	182_{76}	94_{40}	82_{100}	1161	5949
Cocaine-M (HO-methoxy-) TFA	UGLUCTFA	2470	445_{11}	247_{13}	182_{73}	94_{35}	82_{100}	1107	5952
Cocaine-M (HO-methoxy-) TMS	UGLUCTMS	2850	421_{16}	198_{8}	182_{71}	94_{35}	82_{100}	1064	5950
Cocaine-M (HO-methoxy-benzoylecgonine) ACTMS		2505	449_{23}	240_{56}	193_{17}	82_{100}	73_{46}	1112	6240
Cocaine-M (nor-)		2080	289_{5}	168_{100}	136_{45}	77_{87}	68_{89}	558	6252
Cocaine-M (nor-) AC		2495	331_{17}	209_{38}	194_{35}	168_{99}	136_{49}	764	6232
Cocaine-M (nor-) TFA		2185	385_{4}	263_{16}	194_{17}	105_{100}	77_{64}	970	6244
Cocaine-M (nor-benzoylecgonine) ET		2115	303_{6}	182_{100}	136_{53}	105_{46}	68_{88}	628	6253
Cocaine-M (nor-benzoylecgonine) ME		2080	289_{5}	168_{100}	136_{45}	77_{87}	68_{89}	558	6252
Cocaine-M (nor-benzoylecgonine) MEAC		2495	331_{17}	209_{38}	194_{35}	168_{99}	136_{49}	764	6232
Cocaine-M (nor-benzoylecgonine) METFA		2185	385_{4}	263_{16}	194_{17}	105_{100}	77_{64}	970	6244
Cocaine-M (nor-benzoylecgonine) TFATBDMS		2460	428_{24}	306_{5}	179_{20}	105_{100}	77_{39}	1152	6246
Cocaine-M (nor-benzoylecgonine) TBDMS		2375	389_{12}	268_{66}	210_{79}	136_{84}	68_{100}	983	6254
Cocaine-M (nor-cocaethylene)		2115	303_{6}	182_{100}	136_{53}	105_{46}	68_{88}	628	6253
Cocaine-M (nor-cocaethylene) AC		2535	345_{9}	182_{89}	136_{48}	109_{65}	105_{90}	829	6233
Cocaine-M (nor-cocaethylene) TFA		2245	399_{2}	277_{23}	208_{20}	164_{26}	105_{100}	1014	6245
Cocaine-M/artifact (anhydroecgonine) TMS		1345	239_{31}	224_{25}	210_{100}	183_{10}	122_{28}	341	6256
Cocaine-M/artifact (anhydroecgonine) TBDMS		1520	281_{16}	252_{29}	224_{100}	150_{44}	122_{30}	524	6242
Cocaine-M/artifact (anhydromethylecgonine)	UHY-I U+UHYAC-I	1280	181_{31}	152_{100}	138_{9}	122_{15}	82_{18}	178	3574
Cocaine-M/artifact (benzoic acid)	P U UHY	1235*	122_{77}	105_{100}	77_{72}			107	95
Cocaine-M/artifact (benzoic acid) ME	P(ME)	1180*	136_{30}	105_{99}	77_{73}			116	1211
Cocaine-M/artifact (benzoic acid) TBDMS		1295*	221_{1}	179_{100}	135_{30}	105_{55}	77_{51}	330	6247
Cocaine-M/artifact (ecgonine) ACTBDMS		2010	341_{16}	284_{34}	282_{41}	142_{17}	82_{100}	814	6234
Cocaine-M/artifact (ecgonine) ETAC		1675	255_{12}	196_{60}	168_{14}	94_{44}	82_{100}	403	6231
Cocaine-M/artifact (ecgonine) ETPFP		1620	359_{34}	314_{28}	196_{100}	96_{21}	82_{90}	880	5563
Cocaine-M/artifact (ecgonine) MEAC	U+UHYAC	1595	241_{6}	182_{56}	96_{37}	94_{51}	82_{100}	348	472
Cocaine-M/artifact (ecgonine) MEHFB		1620	395_{14}	364_{9}	182_{100}	94_{53}	82_{95}	1001	5676
Cocaine-M/artifact (ecgonine) MEPFP		1530	345_{15}	314_{12}	182_{79}	96_{26}	82_{100}	828	5562
Cocaine-M/artifact (ecgonine) METBDMS		1625	313_{2}	256_{9}	182_{15}	96_{38}	82_{100}	682	6248
Cocaine-M/artifact (ecgonine) METFA		1490	295_{27}	264_{13}	182_{80}	96_{26}	82_{100}	589	5564
Cocaine-M/artifact (ecgonine) METMS		1580	271_{9}	212_{18}	182_{9}	96_{40}	82_{100}	476	5583
Cocaine-M/artifact (ecgonine) TFATBDMS		1585	395_{4}	338_{18}	282_{18}	94_{23}	82_{100}	1003	6243
Cocaine-M/artifact (ecgonine) -H2O TMS		1345	239_{31}	224_{25}	210_{100}	183_{10}	122_{28}	341	6256
Cocaine-M/artifact (ecgonine) -H2O TBDMS		1520	281_{16}	252_{29}	224_{100}	150_{44}	122_{30}	524	6242
Cocaine-M/artifact (methylecgonine)	UCOME	1465	199_{18}	168_{12}	96_{58}	82_{100}		218	467
Cocaine-M/artifact (methylecgonine) AC	U+UHYAC	1595	241_{6}	182_{56}	96_{37}	94_{51}	82_{100}	348	472
Cocaine-M/artifact (methylecgonine) HFB		1620	395_{14}	364_{9}	182_{100}	94_{53}	82_{95}	1001	5676
Cocaine-M/artifact (methylecgonine) PFP		1530	345_{15}	314_{12}	182_{79}	96_{26}	82_{100}	828	5562
Cocaine-M/artifact (methylecgonine) TFA		1490	295_{27}	264_{13}	182_{80}	96_{26}	82_{100}	589	5564
Cocaine-M/artifact (methylecgonine) TMS		1580	271_{9}	212_{18}	182_{9}	96_{40}	82_{100}	476	5583
Cocaine-M/artifact (methylecgonine) -H2O	UHY-I U+UHYAC-I	1280	181_{31}	152_{100}	138_{9}	122_{15}	82_{18}	178	3574
Cocaine-M/artifact (methylecgonine) TBDMS		1625	313_{2}	256_{9}	182_{15}	96_{38}	82_{100}	682	6248
Codeine	P G U UHY	2375	299_{100}	229_{26}	162_{46}	124_{23}		611	473
Codeine AC	U+UHYAC PAC	2500	341_{100}	282_{40}	229_{20}	204_{17}	162_{5}	812	224
Codeine HFB		2320	495_{8}	438_{3}	282_{100}	266_{9}	225_{11}	1161	6142
Codeine PFP		2430	445_{100}	388_{7}	282_{73}	266_{6}	119_{10}	1107	2252
Codeine TFA		2280	395_{64}	338_{5}	282_{100}	115_{20}	69_{34}	1001	4011
Codeine TMS		2520	371_{50}	196_{34}	178_{51}	146_{36}	73_{100}	924	2464
Codeine Cl-artifact AC	U+UHYAC	2630	375_{86}	316_{100}	263_{12}	204_{42}	162_{14}	936	2991
Codeine-D3		2495	344_{100}	285_{91}	232_{52}	193_{36}	156_{37}	825	7300
Codeine-D3		2370	302_{100}	232_{26}	165_{47}	127_{24}		625	7295
Codeine-M (hydrocodone)	G UHY U+UHYAC	2440	299_{100}	242_{51}	185_{23}	96_{24}	59_{23}	611	238
Codeine-M (hydrocodone) Cl-artifact		2630	375_{100}	340_{47}	318_{28}	146_{13}	115_{10}	936	4401
Codeine-M (hydrocodone) enol AC		2500	341_{100}	298_{65}	242_{32}	162_{26}		813	258
Codeine-M (hydrocodone) enol TMS		2475	371_{31}	356_{14}	313_{9}	234_{30}	73_{100}	925	6215
Codeine-M (nor-) 2AC	U+UHYAC	2945	369_{14}	327_{3}	223_{37}	87_{100}	72_{36}	916	226
Codeine-M (O-demethyl-)	G UHY	2455	285_{100}	268_{15}	162_{58}	124_{20}		539	474
Codeine-M (O-demethyl-) TFA		2285	381_{55}	268_{100}	146_{14}	115_{13}	69_{23}	955	5569
Codeine-M (O-demethyl-) 2AC	G PHYAC U+UHYAC	2620	369_{59}	327_{100}	310_{36}	268_{47}	162_{11}	915	225
Codeine-M (O-demethyl-) 2HFB		2375	677_{10}	480_{11}	464_{100}	407_{9}	169_{8}	1201	6120
Codeine-M (O-demethyl-) 2PFP		2360	577_{51}	558_{7}	430_{8}	414_{100}	119_{22}	1191	2251
Codeine-M (O-demethyl-) 2TFA		2250	477_{71}	364_{100}	307_{6}	115_{8}	69_{31}	1145	4008
Codeine-M (O-demethyl-) 2TMS	UHYTMS	2560	429_{19}	236_{21}	196_{15}	146_{21}	73_{100}	1079	2463
Codeine-M (O-demethyl-) Cl-artifact 2AC	U+UHYAC	2680	403_{59}	361_{100}	344_{63}	302_{90}	204_{55}	1026	2992
Codeine-M (O-demethyl-)-D3 TFA		2275	384_{39}	271_{100}	211_{8}	165_{6}	152_{7}	967	5572
Codeine-M (O-demethyl-)-D3 2HFB		2375	680_{4}	483_{9}	467_{100}	169_{23}	414_{7}	1202	6126

Codeine-M (O-demethyl-)-D3 2PFP   Table 1-8-1: Compounds in order of names

Name	Detected	RI	Typical ions and intensities					Page	Entry
Codeine-M (O-demethyl-)-D3 2PFP		2350	580 $_{16}$	433 $_7$	417 $_{100}$	269 $_5$	119 $_8$	1192	5567
Codeine-M (O-demethyl-)-D3 2TFA		2240	480 $_{32}$	383 $_6$	367 $_{100}$	314 $_6$	307 $_6$	1148	5571
Codeine-M (O-demethyl-)-D3 2TMS		2550	432 $_{100}$	290 $_{30}$	239 $_{66}$	199 $_{44}$	73 $_{110}$	1085	5578
Codeine-M 2PFP		2440	563 $_{100}$	400 $_{10}$	355 $_{38}$	327 $_7$	209 $_{15}$	1188	3534
Codeine-M 3AC	U+UHYAC	2955	397 $_8$	355 $_9$	209 $_{41}$	87 $_{100}$	72 $_{33}$	1010	1194
Codeine-M 3PFP	UHYPFP	2405	709 $_{80}$	533 $_{28}$	388 $_{29}$	367 $_{51}$	355 $_{100}$	1203	3533
Codeine-M 3TMS	UHYTMS	2605	487 $_9$	416 $_9$	222 $_{18}$	131 $_{10}$	73 $_{50}$	1155	3525
Colchicine		3200	399 $_{51}$	371 $_{20}$	312 $_{100}$	297 $_{36}$	281 $_{31}$	1014	2852
Colecalciferol		3150*	384 $_{43}$	351 $_{100}$	325 $_{38}$	143 $_{37}$	57 $_{35}$	969	2794
Colecalciferol AC		3300*	426 $_{16}$	398 $_{84}$	382 $_{100}$	351 $_{44}$	145 $_{78}$	1072	2796
Colecalciferol -H2O	P	3130*	366 $_{100}$	351 $_{16}$	271 $_9$	158 $_{11}$	143 $_{19}$	906	2795
Coniine		1610	127 $_2$	98 $_2$	84 $_{100}$	70 $_4$	56 $_{13}$	110	4459
Coniine AC	U+UHYAC	1405	169 $_5$	154 $_3$	126 $_{43}$	98 $_3$	84 $_{100}$	156	4460
Cotinine	P U+UHYAC	1715	176 $_{36}$	118 $_{62}$	98 $_{100}$			163	692
Coumachlor ET		2780*	370 $_{35}$	327 $_{100}$	299 $_{78}$	187 $_{38}$	139 $_{45}$	919	4812
Coumachlor ME	UME UGLUCME	2770*	356 $_{21}$	313 $_{100}$	201 $_9$	189 $_8$	125 $_{19}$	868	4143
Coumachlor TMS		2870*	414 $_{38}$	371 $_{100}$	261 $_{21}$	75 $_{41}$	73 $_{56}$	1049	4962
Coumachlor artifact	UME	1575*	180 $_{32}$	165 $_{100}$	137 $_{49}$	102 $_{51}$	145 $_{41}$	173	4427
Coumachlor enol 2TMS		2990*	486 $_9$	443 $_{77}$	247 $_{14}$	193 $_{11}$	73 $_{100}$	1154	4963
Coumachlor iso-1 AC		2810*	384 $_{14}$	342 $_{34}$	299 $_{100}$	187 $_{20}$	121 $_{44}$	965	4816
Coumachlor iso-2 AC		2810*	384 $_{40}$	342 $_{25}$	299 $_{100}$	187 $_8$	121 $_{27}$	966	4817
Coumachlor-M (di-HO-) 3ME	UME	3195*	416 $_{27}$	373 $_{100}$	359 $_{11}$	180 $_3$	125 $_7$	1054	4425
Coumachlor-M (HO-) 2TMS	UTMS	3150*	502 $_{26}$	459 $_{100}$	446 $_{34}$	281 $_{31}$	73 $_{85}$	1165	4964
Coumachlor-M (HO-) enol 3TMS	UTMS	3240*	574 $_{19}$	531 $_{100}$	335 $_{36}$	281 $_{19}$	73 $_{78}$	1190	4965
Coumachlor-M (HO-) iso-1 2ET	UET	3020*	414 $_{33}$	371 $_{100}$	343 $_{69}$	231 $_{16}$	139 $_{22}$	1049	4813
Coumachlor-M (HO-) iso-1 2ME	UME UHYME	2990*	386 $_{31}$	343 $_{100}$	231 $_6$	151 $_{20}$	125 $_{13}$	972	4422
Coumachlor-M (HO-) iso-2 2ET	UET	3095*	414 $_{29}$	371 $_{100}$	343 $_{56}$	231 $_{12}$	139 $_{21}$	1050	4814
Coumachlor-M (HO-) iso-2 2ME	UME UHYME	3035*	386 $_{42}$	343 $_{100}$	231 $_8$	151 $_{14}$	125 $_{19}$	972	4423
Coumachlor-M (HO-dihydro-) 2ME	UME	3095*	388 $_{100}$	343 $_{94}$	329 $_{69}$	245 $_{48}$	125 $_{57}$	980	4426
Coumachlor-M (HO-dihydro-) 3TMS	UME	3170*	576 $_8$	459 $_{77}$	446 $_{100}$	281 $_{61}$	73 $_{73}$	1190	4966
Coumachlor-M (HO-methoxy-) 2ET	UET	3320*	444 $_{29}$	401 $_{100}$	373 $_{37}$	263 $_{11}$	139 $_{16}$	1105	4815
Coumachlor-M (HO-methoxy-) 2ME	UME	3195*	416 $_{27}$	373 $_{100}$	359 $_{11}$	180 $_3$	125 $_7$	1054	4425
Coumaphos		2575*	362 $_{73}$	226 $_{58}$	210 $_{36}$	109 $_{100}$	97 $_{100}$	889	3330
m-Coumaric acid		1940*	164 $_{100}$	147 $_{20}$	118 $_{32}$	91 $_{49}$	65 $_{39}$	145	5765
m-Coumaric acid AC		1970*	206 $_{14}$	164 $_{100}$	147 $_{29}$	118 $_{20}$	91 $_{21}$	230	5998
m-Coumaric acid HFB		1820*	360 $_{100}$	169 $_{36}$	147 $_{43}$	91 $_{35}$	69 $_{93}$	883	6003
m-Coumaric acid ME		1720*	178 $_{61}$	147 $_{100}$	119 $_{28}$	91 $_{49}$	65 $_{33}$	167	5997
m-Coumaric acid MEAC		1760*	220 $_{13}$	178 $_{96}$	147 $_{100}$	119 $_{19}$	91 $_{25}$	273	5999
m-Coumaric acid MEHFB		1665*	374 $_{45}$	343 $_{100}$	169 $_{32}$	101 $_{36}$	69 $_{44}$	933	6002
m-Coumaric acid MEPFP		1580*	324 $_{49}$	293 $_{100}$	119 $_{33}$	101 $_{36}$	69 $_{26}$	730	6001
m-Coumaric acid METMS		1750*	250 $_{76}$	235 $_{76}$	203 $_{100}$	89 $_{45}$	73 $_{64}$	380	6005
m-Coumaric acid PFP		1670*	310 $_{100}$	293 $_{13}$	146 $_{25}$	119 $_{58}$	69 $_{33}$	665	6000
m-Coumaric acid 2TMS		1910*	308 $_{40}$	293 $_{46}$	249 $_{25}$	203 $_{39}$	73 $_{51}$	656	6004
p-Coumaric acid		2225*	164 $_{34}$	118 $_{32}$	91 $_{82}$	65 $_{100}$	63 $_{96}$	145	5760
p-Coumaric acid AC		1910*	206 $_6$	164 $_{100}$	147 $_{29}$	118 $_{17}$	89 $_{12}$	231	5981
p-Coumaric acid HFB		1855*	360 $_{100}$	343 $_{15}$	169 $_{29}$	147 $_{20}$	69 $_{81}$	883	5986
p-Coumaric acid ME		1800*	178 $_{56}$	147 $_{100}$	119 $_{39}$	91 $_{37}$	65 $_{31}$	167	5979
p-Coumaric acid MEAC		1785*	220 $_5$	178 $_{89}$	147 $_{100}$	119 $_{21}$	89 $_{18}$	273	5980
p-Coumaric acid MEHFB		1695*	374 $_{57}$	343 $_{100}$	315 $_{13}$	129 $_{28}$	69 $_{51}$	933	5985
p-Coumaric acid METFA		1540*	274 $_{65}$	243 $_{100}$	215 $_{17}$	99 $_{47}$	69 $_{61}$	484	5982
p-Coumaric acid METMS		2750*	250 $_{71}$	235 $_{13}$	203 $_{13}$	179 $_9$	73 $_{70}$	380	6020
p-Coumaric acid PFP		1720*	310 $_{100}$	293 $_{13}$	163 $_{20}$	119 $_{55}$	69 $_{61}$	665	5984
p-Coumaric acid TFA		1665*	260 $_{76}$	243 $_{14}$	101 $_{23}$	89 $_{27}$	69 $_{80}$	419	5983
p-Coumaric acid 2TMS		2040*	308 $_{35}$	293 $_{48}$	249 $_{29}$	219 $_{63}$	73 $_{100}$	656	6019
p-Coumaric acid -CO2		1045*	120 $_{51}$	91 $_{100}$	65 $_{62}$	63 $_{56}$	51 $_{52}$	106	5761
Coumarin	G	1550*	146 $_{66}$	118 $_{100}$	90 $_{31}$	63 $_{23}$		124	4365
Coumarin-M (HO-)	UHY	1780*	162 $_{94}$	134 $_{100}$	105 $_{23}$	78 $_{28}$	63 $_9$	140	4366
Coumarin-M (HO-) AC	U+UHYAC	1840*	204 $_{16}$	162 $_{100}$	134 $_{85}$	105 $_9$	77 $_{11}$	225	4367
Coumarin-M (HO-) HFB		1685*	358 $_{100}$	330 $_{51}$	169 $_{25}$	133 $_{80}$	105 $_{23}$	875	7614
Coumarin-M (HO-) ME		1750*	176 $_{100}$	148 $_{76}$	133 $_{86}$	105 $_{12}$	77 $_{16}$	162	7611
Coumarin-M (HO-) PFP		1550*	308 $_{50}$	280 $_{50}$	261 $_5$	161 $_3$	133 $_{49}$	653	7613
Coumarin-M (HO-) TFA		1540*	258 $_{100}$	230 $_{50}$	133 $_{49}$	119 $_{14}$	105 $_8$	412	7615
Coumarin-M (HO-) TMS		1925*	234 $_{87}$	219 $_{100}$	191 $_{20}$	163 $_{58}$	73 $_{25}$	319	7612
Coumatetralyl	G	2660*	292 $_{100}$	188 $_{68}$	130 $_{41}$	121 $_{69}$	91 $_{24}$	575	1431
Coumatetralyl AC	U+UHYAC	2725*	334 $_{11}$	292 $_{100}$	188 $_{47}$	175 $_{23}$	121 $_{44}$	780	4789
Coumatetralyl TMS		2765*	364 $_{76}$	349 $_{16}$	260 $_{43}$	193 $_{16}$	73 $_{100}$	897	5026
Coumatetralyl HY		2250*	266 $_3$	248 $_{13}$	220 $_{15}$	130 $_{100}$	121 $_{39}$	452	4809
Coumatetralyl HYAC		2350*	308 $_1$	265 $_7$	248 $_{12}$	130 $_{64}$	121 $_{100}$	657	4811
Coumatetralyl HYME		2300*	280 $_4$	135 $_{70}$	130 $_{100}$	115 $_{17}$	77 $_{28}$	518	4810
Coumatetralyl iso-1 ET		2680*	320 $_{54}$	291 $_{62}$	175 $_{100}$	129 $_{21}$	121 $_{32}$	715	4800
Coumatetralyl iso-1 ME	UME	2655*	306 $_{100}$	291 $_{47}$	175 $_{72}$	121 $_{34}$	115 $_{41}$	645	4790
Coumatetralyl iso-2 ET		2705*	320 $_{58}$	291 $_{69}$	175 $_{100}$	129 $_{26}$	121 $_{33}$	715	4801
Coumatetralyl iso-2 ME	UME	2690*	306 $_{100}$	291 $_{28}$	202 $_{23}$	175 $_{27}$	115 $_{25}$	645	2084
Coumatetralyl-M (di-HO-) 3ET	UET	3290*	408 $_{44}$	379 $_{100}$	219 $_{14}$			1037	4807

Table 1-8-1: Compounds in order of names

CS gas (o-chlorobenzylidenemalonitrile)

Name	Detected	RI	Typical ions and intensities					Page	Entry
Coumatetralyl-M (di-HO-) iso-1 2ME	UME	3005*	352_1	333_{100}	319_9	205_{11}	151_6	853	4798
Coumatetralyl-M (di-HO-) iso-1 3TMS	UTMS	2955*	540_6	348_{100}	333_{33}	73_{36}		1181	5030
Coumatetralyl-M (di-HO-) iso-2 2ME	UME	3085*	352_2	333_{100}	205_{15}	177_{14}	151_{20}	853	4797
Coumatetralyl-M (di-HO-) iso-2 3TMS	UTMS	3230*	540_1	525_3	449_{39}	348_{100}	73_{82}	1181	5031
Coumatetralyl-M (di-HO-) iso-3 3ME	UME	3105*	366_{100}	351_{92}	232_{33}	193_{24}	159_{45}	904	4794
Coumatetralyl-M (HO-) iso-1 ET	UET	2905*	336_{13}	318_{50}	289_{79}	217_{97}	121_{60}	788	4802
Coumatetralyl-M (HO-) iso-1 ME	UME	2910*	322_{11}	203_{100}	303_{91}	187_{28}	121_{18}	722	4795
Coumatetralyl-M (HO-) iso-1 2TMS	UTMS	2835*	452_{100}	437_{19}	348_{20}	193_{17}	131_{45}	1116	5028
Coumatetralyl-M (HO-) iso-2 2ET	UET	2910*	364_{51}	335_{100}	219_{80}	161_{55}	91_{60}	897	4803
Coumatetralyl-M (HO-) iso-2 2ME	UME	2925*	336_{100}	321_{12}	232_{34}	217_{33}	205_{62}	788	4791
Coumatetralyl-M (HO-) iso-2 2TMS	UTMS	2880*	452_8	362_{100}	361_{99}	233_{33}	73_{98}	1116	5029
Coumatetralyl-M (HO-) iso-3 2ET	UET	2920*	364_{54}	335_{100}	187_{16}	175_{50}	121_{28}	897	4804
Coumatetralyl-M (HO-) iso-3 2ME	UME	2935*	336_{62}	321_{120}	305_{34}	175_{60}	121_{48}	788	4793
Coumatetralyl-M (HO-) iso-3 2TMS	UTMS	3015*	452_{68}	437_{15}	348_{29}	333_{25}	73_{100}	1116	5027
Coumatetralyl-M (HO-) iso-4 2ET	UET	3000*	364_{95}	335_{100}	245_{37}	219_{69}	165_{17}	897	4805
Coumatetralyl-M (HO-) iso-4 2ME	UME	2990*	336_{100}	321_{26}	232_{50}	217_{29}	205_{61}	788	4792
Coumatetralyl-M (HO-methoxy-) 2ET	UET	3070*	394_{100}	378_{37}	365_{70}	349_{51}	275_{16}	999	4806
Coumatetralyl-M (HO-methoxy-) 2ME	UME	3070*	366_{100}	351_{37}	262_{42}	235_{37}	181_{14}	904	4796
Coumatetralyl-M (tri-HO-) -H2O 2ET	UET	3320*	378_{82}	349_{73}	206_{100}	165_{84}	137_{71}	946	4808
Coumatetralyl-M (tri-HO-) -H2O 2ME	UME	3175*	350_{100}	335_{11}	205_{28}	177_{30}	151_{21}	845	4799
2C-P		1720	223_{21}	194_{100}	179_{13}	165_{51}	135_{15}	286	6906
2C-P AC		2090	265_{14}	206_{100}	193_{28}	177_{60}	135_{13}	446	6920
2C-P HFB		1895	419_{49}	206_{76}	193_{100}	177_{42}	163_{13}	1061	6940
2C-P PFP		1865	369_{54}	206_{77}	193_{100}	177_{42}	119_{15}	915	6935
2C-P TFA		1870	319_{36}	206_{69}	193_{100}	177_{39}	149_{38}	708	6930
2C-P TMS		1860	295_5	265_5	194_{20}	102_{100}	73_{64}	595	6922
2C-P 2AC		2160	307_{20}	206_{100}	193_{24}	177_{42}	135_8	650	6921
2C-P 2TMS		2130	367_1	352_5	174_{100}	100_{11}	86_{28}	912	6923
2C-P formyl artifact		1755	235_{23}	204_{100}	193_{62}	163_9	135_{11}	325	6908
m-Cresol TMS		1040*	180_{38}	165_{100}	149_7	135_7	91_{31}	175	5674
p-Cresol	UHY	1060*	108_{76}	107_{100}	77_{32}	53_{17}		102	4220
p-Cresol AC	U+UHYAC	1110*	150_{10}	108_{100}	77_{18}			127	4225
Crimidine	G	1560	171_{76}	156_{73}	142_{100}	120_{31}	93_{36}	158	693
Crinosterol		3135*	398_{100}	300_{62}	271_{71}	255_{85}	69_{96}	1013	5619
Crinosterol -H2O		3210*	380_{100}	255_{39}	81_{53}	69_{53}	55_{67}	954	5623
Croconazole		2390	310_{60}	243_{13}	185_{59}	125_{100}	89_{75}	665	5686
Cropropamide		1725	240_1	195_4	168_{14}	100_{100}	69_{55}	346	694
Crotamiton (cis)	P G U	1560	203_{12}	188_{16}	135_{14}	120_{40}	69_{100}	224	5347
Crotamiton (trans)	P G U	1600	203_9	188_{14}	135_{13}	120_{27}	69_{100}	224	695
Crotamiton-M (4-HO-crotyl-) (cis)	UGLUC	1790	219_{14}	135_{21}	120_{100}	91_{39}	85_{29}	270	5357
Crotamiton-M (4-HO-crotyl-) (trans)	UGLUC	1865	219_{31}	201_{38}	133_{64}	120_{100}	85_{78}	270	5356
Crotamiton-M (4-HO-crotyl-) (trans) AC	UGLUC	1940	261_{76}	219_{87}	133_{86}	118_{65}	85_{108}	426	5358
Crotamiton-M (4-HO-crotyl-) (trans) TMS	UGLUCTMS	1800	291_{62}	276_{39}	162_{21}	91_{31}	73_{80}	571	5359
Crotamiton-M (di-HO-) 2AC	UGLUCAC	2215	319_{32}	260_{36}	246_{86}	118_{100}	85_{86}	708	5362
Crotamiton-M (di-HO-) 2TMS	UGLUCTMS	2050	379_1	364_{10}	276_{31}	132_{44}	73_{60}	949	5363
Crotamiton-M (di-HO-dihydro-)	UGLUC	1900	237_6	219_{19}	206_{47}	162_{82}	120_{100}	334	5360
Crotamiton-M (di-HO-dihydro-) 2AC	UGLUCAC	2105	321_{33}	219_{52}	162_{84}	135_{90}	120_{92}	718	5361
Crotamiton-M (HO-ethyl-) (cis)	UGLUC	1805	219_2	188_{25}	150_{15}	118_{85}	69_{100}	271	5354
Crotamiton-M (HO-ethyl-) (trans)	P U UGLUC	1830	219_2	188_{25}	150_{17}	118_{80}	69_{100}	271	5353
Crotamiton-M (HO-ethyl-) (trans) AC	UGLUC	1905	261_{18}	188_{80}	150_{33}	118_{100}	69_{92}	426	5355
Crotamiton-M (HO-ethyl-HOOC-) MEAC	UGLUCAC	2135	305_6	245_{15}	232_{100}	132_{90}	118_{79}	639	5374
Crotamiton-M (HO-methyl-disulfide)	UGLUC	2235	299_{13}	253_{100}	174_{77}	162_{62}	134_{81}	610	5370
Crotamiton-M (HO-methyl-disulfide) AC	UGLUCAC	2315	341_{28}	295_{85}	202_{100}	162_{79}	134_{75}	811	5371
Crotamiton-M (HO-methylthio-)	UGLUC	2025	267_2	221_{90}	190_{62}	162_{92}	134_{100}	455	5351
Crotamiton-M (HO-methylthio-) AC	UGLUCAC	2115	309_{19}	263_{100}	190_{89}	162_{95}	134_{99}	661	5352
Crotamiton-M (HOOC-)	U	1940	233_5	188_{39}	134_{100}	120_{67}	99_{49}	315	697
Crotamiton-M (HOOC-) (cis) TMS	UTMS	1855	305_9	290_{42}	187_{60}	134_{84}	73_{100}	639	5350
Crotamiton-M (HOOC-) (trans) TMS	UTMS	1875	305_{13}	290_{53}	187_{58}	134_{71}	73_{100}	639	5349
Crotamiton-M (HOOC-) ME	UME	1865	247_{27}	216_{30}	188_{83}	134_{100}	113_{73}	367	5348
Crotamiton-M (HOOC-dihydro-)	UGLUC	1900	235_{40}	202_{26}	190_{82}	162_{40}	134_{100}	324	5364
Crotamiton-M (HOOC-dihydro-) ME	UGLUCME	1845	249_{42}	218_{63}	162_{61}	120_{96}	115_{100}	375	5365
Crotamiton-M (HOOC-methyl-thio-) ME	UGLUCME	2010	295_1	249_{75}	234_{27}	162_{100}	134_{95}	590	5376
Crotamiton-M (HOOC-thio-)	UGLUC	2150	267_{96}	208_{25}	174_{73}	162_{120}	135_{41}	454	5375
Crotamiton-M (HO-thio-)	UGLUC	1970	253_7	209_{66}	134_{59}	120_{100}	91_{45}	393	5367
Crotamiton-M (HO-thio-) AC	UGLUCAC	2070	295_{69}	262_{42}	209_{71}	176_{83}	162_{100}	590	5368
Crotamiton-M (HO-thio-) 2AC	UGLUCAC	2210	337_{27}	295_{89}	262_{37}	209_{59}	162_{100}	792	5369
Crotamiton-M (N-deethyl-)	UHY	1415	175_{27}	107_{100}	96_{24}	83_{32}	69_{77}	161	5373
Crotamiton-M (N-deethyl-HO-methyl-)	UGLUC	1995	191_{61}	123_{100}	94_{10}	69_{72}		195	696
Crotamiton-M (N-deethyl-HO-methyl-) AC	UGLUCAC	2055	233_{53}	191_{91}	123_{100}	69_{83}		315	5372
Crotamiton-M/artifact (methyl-thio-chloro-)	UGLUC	1985	285_{12}	239_{69}	190_{40}	162_{95}	134_{100}	538	5366
Crotethamide		1675	226_1	181_5	154_{20}	86_{100}	69_{48}	296	698
Crotylbarbital	P G U UHY U+UHYAC	1620	210_7	181_{85}	156_{93}	141_{39}	55_{100}	245	699
Crotylbarbital-M (HO-) -H2O	U UHY U+UHYAC	1600	208_3	179_{32}	157_{100}	141_{65}		237	700
CS gas (o-chlorobenzylidenemalonitrile)		1500	188_{43}	161_{15}	153_{100}	99_{17}	75_{29}	190	3539

2C-T-2

Table 1-8-1: Compounds in order of names

Name	Detected	RI	Typical ions and intensities					Page	Entry
2C-T-2		1980	241 $_{22}$	212 $_{100}$	197 $_{20}$	183 $_{35}$	153 $_{22}$	347	5035
2C-T-2 AC	U+UHYAC	2310	283 $_{28}$	224 $_{100}$	211 $_{34}$	181 $_{12}$	153 $_{12}$	529	5037
2C-T-2 HFB		2040	437 $_{24}$	224 $_{21}$	211 $_{100}$	181 $_{19}$	169 $_{17}$	1092	6816
2C-T-2 PFP		2090	387 $_{34}$	224 $_{29}$	211 $_{100}$	181 $_{18}$	153 $_{15}$	976	6817
2C-T-2 TFA	UGLUCTFA	2210	337 $_{26}$	224 $_{17}$	211 $_{100}$	181 $_{13}$	153 $_{10}$	791	6818
2C-T-2 TMS		2405	313 $_{1}$	299 $_{3}$	174 $_{100}$	147 $_{4}$	86 $_{7}$	681	6814
2C-T-2 2AC	U+UHYAC	2395	325 $_{17}$	224 $_{100}$	211 $_{50}$	181 $_{16}$	153 $_{14}$	735	5038
2C-T-2 2TMS		2405	385 $_{1}$	370 $_{3}$	254 $_{4}$	211 $_{4}$	174 $_{100}$	970	6815
2C-T-2 deuteroformyl artifact		1935	255 $_{37}$	224 $_{31}$	211 $_{100}$	181 $_{14}$	153 $_{16}$	402	5036
2C-T-2-M (aryl-HOOC-)	UGLUC	1970	242 $_{100}$	227 $_{20}$	183 $_{68}$	153 $_{14}$		350	6893
2C-T-2-M (aryl-HOOC-) ME	USPEME	1960*	256 $_{100}$	241 $_{10}$	197 $_{49}$	181 $_{12}$	167 $_{15}$	405	6842
2C-T-2-M (deamino-HO-)	UGLUC	1905*	242 $_{71}$	211 $_{100}$	181 $_{15}$	153 $_{11}$		352	6839
2C-T-2-M (deamino-HO-) AC	U+UHYAC	2050*	284 $_{40}$	224 $_{100}$	209 $_{30}$	167 $_{20}$	150 $_{10}$	534	6892
2C-T-2-M (deamino-HOOC-)		2130*	256 $_{100}$	242 $_{31}$	211 $_{60}$	195 $_{30}$	181 $_{32}$	405	6840
2C-T-2-M (deamino-HOOC-) ME	UHYME	1910*	270 $_{100}$	255 $_{24}$	211 $_{82}$	195 $_{44}$	181 $_{46}$	469	6838
2C-T-2-M (deamino-HOOC-) TMS	USPETMS	2075*	328 $_{100}$	313 $_{29}$	298 $_{25}$	255 $_{61}$	211 $_{57}$	749	6841
2C-T-2-M (deamino-oxo-)		2130*	240 $_{33}$	211 $_{100}$	181 $_{11}$	153 $_{19}$	122 $_{11}$	343	7234
2C-T-2-M (HO- N-acetyl-) TFA	UGLUCTFA	2270	427 $_{33}$	367 $_{5}$	259 $_{140}$	167 $_{8}$		1073	6834
2C-T-2-M (HO- sulfone) AC	U+UHYAC	2730	331 $_{7}$	272 $_{100}$	259 $_{74}$	238 $_{26}$	165 $_{47}$	763	6828
2C-T-2-M (HO- sulfone) 2AC	U+UHYAC	2780	373 $_{35}$	314 $_{84}$	302 $_{75}$	272 $_{54}$	259 $_{100}$	930	6833
2C-T-2-M (O-demethyl- N-acetyl-) TFA	U+UHYTFA	2250	365 $_{11}$	323 $_{19}$	306 $_{100}$	293 $_{18}$	197 $_{9}$	899	6942
2C-T-2-M (O-demethyl- N-acetyl-) 2TFA	U+UHYTFA	2180	461 $_{16}$	306 $_{100}$	293 $_{43}$	209 $_{25}$		1129	6894
2C-T-2-M (O-demethyl- sulfone) 2AC	U+UHYAC	2510	343 $_{3}$	301 $_{90}$	242 $_{100}$	230 $_{48}$	153 $_{9}$	820	6835
2C-T-2-M (O-demethyl-) 2AC	U+UHYAC	2120	311 $_{48}$	297 $_{31}$	269 $_{100}$	252 $_{46}$	210 $_{78}$	671	6837
2C-T-2-M (O-demethyl-) 2TFA	UGLUCTFA	1980	419 $_{28}$	306 $_{100}$	293 $_{92}$	209 $_{20}$	69 $_{32}$	1060	6821
2C-T-2-M (O-demethyl-) 3AC	U+UHYAC	2290	353 $_{22}$	311 $_{32}$	252 $_{33}$	210 $_{100}$	197 $_{20}$	858	6836
2C-T-2-M (O-demethyl-sulfone N-acetyl-) TFA	UGLUCTFA	2450	397 $_{4}$	355 $_{81}$	242 $_{120}$	153 $_{16}$		1008	6820
2C-T-2-M (S-deethyl-) AC	U+UHYAC	2170	255 $_{18}$	196 $_{100}$	183 $_{41}$	181 $_{34}$	153 $_{21}$	401	6831
2C-T-2-M (S-deethyl-) 3AC	U+UHYAC	2420	339 $_{22}$	297 $_{5}$	238 $_{21}$	196 $_{100}$	183 $_{28}$	802	6827
2C-T-2-M (S-deethyl-) iso-1 2AC		2240	297 $_{29}$	210 $_{14}$	196 $_{100}$	183 $_{35}$	181 $_{29}$	602	6823
2C-T-2-M (S-deethyl-) iso-2 2AC	U+UHYAC	2360	297 $_{16}$	255 $_{11}$	238 $_{20}$	196 $_{100}$	183 $_{37}$	602	6826
2C-T-2-M (S-deethyl-methyl- N-acetyl-)	U+UHYAC	2230	269 $_{19}$	210 $_{100}$	197 $_{35}$	195 $_{21}$	167 $_{27}$	465	6832
2C-T-2-M (S-deethyl-methyl- sulfone) AC	U+UHYAC	2580	301 $_{7}$	242 $_{100}$	230 $_{4}$	196 $_{7}$	124 $_{7}$	618	6829
2C-T-2-M (S-deethyl-methyl- sulfoxide) AC	U+UHYAC	2460	285 $_{16}$	268 $_{23}$	226 $_{33}$	211 $_{100}$	197 $_{31}$	538	6830
2C-T-2-M (sulfone N-acetyl-) TFA	UGLUCTFA	2400	411 $_{23}$	256 $_{100}$	242 $_{4}$	181 $_{10}$	167 $_{11}$	1043	6822
2C-T-2-M (sulfone) AC	U+UHYAC	2600	315 $_{15}$	256 $_{100}$	244 $_{9}$	167 $_{12}$	91 $_{8}$	689	6825
2C-T-2-M (sulfone) TFA	UGLUCTFA	2310	369 $_{44}$	256 $_{100}$	243 $_{7}$	211 $_{4}$	167 $_{23}$	915	6819
2C-T-2-M (sulfone) 2AC	U+UHYAC	2640	357 $_{10}$	256 $_{100}$	244 $_{6}$	167 $_{7}$	91 $_{7}$	873	6824
2C-T-7		2470	255 $_{26}$	226 $_{100}$	183 $_{63}$	169 $_{31}$	153 $_{34}$	403	6855
2C-T-7 AC		2410	297 $_{22}$	238 $_{100}$	225 $_{30}$	181 $_{23}$	153 $_{14}$	603	6858
2C-T-7 HFB		2175	451 $_{14}$	238 $_{24}$	225 $_{100}$	181 $_{23}$	153 $_{21}$	1114	6861
2C-T-7 PFP		2160	401 $_{25}$	238 $_{17}$	225 $_{100}$	181 $_{17}$	153 $_{19}$	1020	6862
2C-T-7 TFA		2170	351 $_{26}$	238 $_{17}$	225 $_{100}$	181 $_{24}$	153 $_{23}$	847	6863
2C-T-7 2AC		2470	339 $_{14}$	238 $_{100}$	225 $_{30}$	181 $_{22}$	153 $_{17}$	803	6859
2C-T-7 2TMS		2395	399 $_{1}$	384 $_{4}$	369 $_{4}$	225 $_{7}$	174 $_{100}$	1016	6860
2C-T-7 deuteroformyl artifact		2060	269 $_{27}$	238 $_{25}$	225 $_{100}$	183 $_{13}$	153 $_{24}$	466	6857
2C-T-7 formyl artifact		2050	267 $_{35}$	236 $_{27}$	225 $_{100}$	183 $_{26}$	153 $_{46}$	455	6856
2C-T-7-M (deamino-HO-)	UGLUC	2000*	256 $_{69}$	225 $_{100}$	183 $_{23}$	150 $_{56}$	135 $_{23}$	407	6864
2C-T-7-M (deamino-HO-) AC	U+UHYAC	2080*	298 $_{73}$	238 $_{100}$	181 $_{45}$	147 $_{22}$		606	6869
2C-T-7-M (deamino-HOOC-)		2110*	270 $_{100}$	225 $_{55}$	213 $_{46}$	181 $_{34}$	153 $_{21}$	469	6872
2C-T-7-M (deamino-HOOC-) ME		1950*	284 $_{100}$	227 $_{50}$	225 $_{74}$	183 $_{24}$	153 $_{25}$	534	6873
2C-T-7-M (deamino-oxo-)		2190*	254 $_{42}$	225 $_{100}$	183 $_{14}$	153 $_{24}$	137 $_{8}$	399	7235
2C-T-7-M (HO- N-acetyl-)	UGLUC	2525	313 $_{40}$	254 $_{100}$	242 $_{44}$	210 $_{38}$	183 $_{21}$	680	6866
2C-T-7-M (HO- N-acetyl-) TFA		2345	409 $_{27}$	350 $_{100}$	337 $_{9}$	236 $_{5}$	181 $_{13}$	1038	6871
2C-T-7-M (HO- sulfone N-acetyl-)	UGLUC	2740	286 $_{73}$	345 $_{31}$	164 $_{100}$	151 $_{27}$	120 $_{18}$	828	6865
2C-T-7-M (HO- sulfone) 2AC	U+UHYAC	2760	387 $_{31}$	340 $_{42}$	328 $_{100}$	268 $_{36}$	108 $_{33}$	976	6868
2C-T-7-M (HO-) 2AC	U+UHYAC	2585	355 $_{51}$	296 $_{72}$	283 $_{10}$	236 $_{92}$	101 $_{100}$	865	6867
2C-T-7-M (HO-) 2TFA		2110	463 $_{80}$	434 $_{43}$	350 $_{60}$	337 $_{100}$	231 $_{67}$	1130	6870
2C-T-7-M (HO-) 3AC	U+UHYAC	2630	397 $_{69}$	296 $_{99}$	283 $_{12}$	236 $_{100}$	101 $_{64}$	1010	6875
2C-T-7-M (S-depropyl-) AC	U+UHYAC	2170	255 $_{18}$	196 $_{100}$	183 $_{41}$	181 $_{34}$	153 $_{21}$	401	6831
2C-T-7-M (S-depropyl-) iso-1 2AC		2240	297 $_{29}$	210 $_{14}$	196 $_{100}$	183 $_{35}$	181 $_{29}$	602	6823
2C-T-7-M (S-depropyl-) iso-2 2AC	U+UHYAC	2360	297 $_{16}$	255 $_{11}$	238 $_{20}$	196 $_{100}$	183 $_{37}$	602	6826
2C-T-7-M (S-depropyl-methyl- N-acetyl-)	U+UHYAC	2230	269 $_{19}$	210 $_{100}$	197 $_{35}$	195 $_{21}$	167 $_{27}$	465	6832
2C-T-7-M (S-depropyl-methyl- sulfone) AC	U+UHYAC	2580	301 $_{7}$	242 $_{100}$	230 $_{4}$	196 $_{7}$	124 $_{7}$	618	6829
2C-T-7-M (S-depropyl-methyl- sulfoxide) AC	U+UHYAC	2460	285 $_{16}$	268 $_{23}$	226 $_{33}$	211 $_{100}$	197 $_{31}$	538	6830
Cyamemazine		2565	323 $_{2}$	277 $_{1}$	223 $_{1}$	100 $_{2}$	58 $_{100}$	727	4248
Cyamemazine-M (bis-nor-) AC	U+UHYAC	3035	337 $_{65}$	237 $_{93}$	205 $_{49}$	114 $_{100}$	72 $_{25}$	791	4393
Cyamemazine-M (bis-nor-HO-) 2AC	U+UHYAC	3300	395 $_{86}$	295 $_{25}$	253 $_{59}$	114 $_{100}$	72 $_{48}$	1001	4396
Cyamemazine-M (HO-) AC	U+UHYAC	3000	381 $_{20}$	294 $_{5}$	239 $_{6}$	100 $_{4}$	58 $_{100}$	956	4391
Cyamemazine-M (HO-methoxy-) AC	U+UHYAC	3110	411 $_{35}$	324 $_{5}$	269 $_{27}$	100 $_{10}$	58 $_{100}$	1044	4392
Cyamemazine-M (nor-) AC	U+UHYAC	3080	351 $_{45}$	237 $_{47}$	205 $_{31}$	128 $_{100}$	86 $_{45}$	847	4394
Cyamemazine-M (nor-HO-) 2AC	U+UHYAC	3320	409 $_{48}$	295 $_{13}$	253 $_{27}$	128 $_{100}$	86 $_{35}$	1038	4395
Cyamemazine-M (nor-HO-methoxy-) 2AC	U+UHYAC	3500	439 $_{27}$	269 $_{19}$	128 $_{100}$	86 $_{44}$		1096	4397
Cyamemazine-M (nor-sulfoxide) AC	U+UHYAC	3285	367 $_{4}$	350 $_{32}$	277 $_{100}$	237 $_{87}$	128 $_{27}$	908	4398

Table 1-8-1: Compounds in order of names Cyclopentolate

Name	Detected	RI	Typical ions and intensities					Page	Entry
Cyamemazine-M (sulfoxide)		2960	339 $_1$	322 $_{22}$	237 $_{68}$	224 $_{10}$	58 $_{100}$	802	4399
Cyamemazine-M/artifact (ring)	U UHY U+UHYAC	2555	224 $_{100}$	192 $_{32}$				286	1281
Cyamemazine-M/artifact (ring) TMS		2310	296 $_{34}$	281 $_3$	223 $_6$	73 $_{100}$		596	5437
Cyamemazine-M/artifact (ring-COOH) METMS		2430	329 $_{19}$	314 $_5$	197 $_{39}$	73 $_{100}$		753	5438
Cyanazine		1960	240 $_{24}$	225 $_{45}$	198 $_{30}$	172 $_{35}$	68 $_{100}$	343	3175
Cyanophenphos		2310	303 $_{11}$	185 $_{22}$	169 $_{48}$	157 $_{100}$	63 $_{68}$	626	3331
Cyanophos		1720	243 $_{35}$	125 $_{53}$	109 $_{100}$	79 $_{38}$	63 $_{35}$	353	3332
Cyanuric acid	U UHY U+UHYAC	2880	129 $_{100}$	86 $_{20}$	70 $_{13}$			112	4424
Cyclamate 2TMS		1680	323 $_7$	280 $_{66}$	210 $_9$	147 $_{51}$	73 $_{100}$	727	4537
Cyclamate-M AC	U+UHYAC	1290	141 $_{26}$	98 $_{19}$	67 $_{15}$	60 $_{100}$	56 $_{94}$	120	1229
Cyclandelate		1975	276 $_1$	125 $_{20}$	107 $_{80}$	83 $_{31}$	69 $_{100}$	497	7524
Cyclandelate AC		2080	149 $_8$	125 $_{15}$	107 $_{29}$	83 $_{26}$	69 $_{100}$	704	7525
Cyclandelate-M/artifact (mandelic acid)		1485*	166 $_7$	107 $_{100}$	79 $_{59}$	77 $_{40}$		149	1071
Cyclandelate-M/artifact (mandelic acid)		1890*	152 $_{10}$	107 $_{100}$	79 $_{75}$	77 $_{55}$	51 $_{63}$	131	5759
Cyclizine	G U UHY U+UHYAC	2045	266 $_{46}$	207 $_{50}$	194 $_{54}$	165 $_{38}$	99 $_{100}$	453	1782
Cyclizine-M (benzophenone)	U+UHYAC	1610*	182 $_{31}$	152 $_3$	105 $_{100}$	77 $_{70}$	51 $_{39}$	180	1624
Cyclizine-M (carbinol)	UHY	1645*	184 $_{45}$	165 $_{14}$	152 $_7$	105 $_{100}$	77 $_{63}$	184	1333
Cyclizine-M (carbinol) AC	U+UHYAC	1700*	226 $_{20}$	184 $_{20}$	165 $_{100}$	105 $_{14}$	77 $_{35}$	294	1241
Cyclizine-M (carbinol) ME	UHY	1655*	198 $_{70}$	167 $_{94}$	121 $_{100}$	105 $_{56}$	77 $_{71}$	215	6779
Cyclizine-M (HO-BPH) iso-1	UHY	2065*	198 $_{93}$	121 $_{72}$	105 $_{100}$	93 $_{22}$	77 $_{66}$	214	1627
Cyclizine-M (HO-BPH) iso-1 AC	U+UHYAC	2010*	240 $_{27}$	198 $_{100}$	121 $_{47}$	105 $_{85}$	77 $_{80}$	343	2196
Cyclizine-M (HO-BPH) iso-2	P-I U UHY	2080*	198 $_{50}$	121 $_{100}$	105 $_{17}$	93 $_{14}$	77 $_{28}$	214	732
Cyclizine-M (HO-BPH) iso-2 AC	U+UHYAC	2050*	240 $_{22}$	198 $_{100}$	121 $_{94}$	105 $_{41}$	77 $_{51}$	343	2197
Cyclizine-M (HO-methoxy-BPH)	UHY	2050*	228 $_{46}$	197 $_6$	151 $_{100}$	105 $_{22}$	77 $_{41}$	299	1625
Cyclizine-M (HO-methoxy-BPH) AC	U+UHYAC	2100*	270 $_3$	228 $_{92}$	151 $_{100}$	105 $_{25}$	77 $_{40}$	468	1622
Cyclizine-M (nor-)	U UHY	2120	252 $_{12}$	207 $_{58}$	167 $_{100}$	152 $_{33}$	85 $_{49}$	392	1602
Cyclizine-M (nor-) AC	U+UHYAC	2525	294 $_{16}$	208 $_{56}$	167 $_{100}$	152 $_{30}$	85 $_{78}$	586	1601
Cyclizine-M/artifact	UHY	2070*	228 $_8$	186 $_{100}$	157 $_{10}$	128 $_7$	77 $_4$	298	1626
Cyclizine-M/artifact AC	U+UHYAC	2200*	270 $_7$	228 $_{17}$	186 $_{100}$	157 $_{10}$	128 $_7$	467	1623
Cycloate		1610	215 $_5$	154 $_{54}$	83 $_{100}$	72 $_{26}$	55 $_{93}$	257	3174
Cyclobarbital	P G U+UHYAC	1970	236 $_1$	207 $_{100}$	157 $_4$	141 $_{22}$	79 $_{16}$	329	701
Cyclobarbital (ME)	P	1940	221 $_{100}$	155 $_{44}$	143 $_9$	87 $_{18}$		380	2288
Cyclobarbital 2ME	PME	1845	264 $_1$	235 $_{100}$	178 $_7$	169 $_{45}$	79 $_{12}$	440	705
Cyclobarbital 2TMS		1890	380 $_{12}$	365 $_{19}$	351 $_{63}$	150 $_{49}$	73 $_{70}$	953	5496
Cyclobarbital-M (di-HO-) -2H2O	P G U+UHYAC	1965	232 $_{14}$	204 $_{100}$	161 $_{18}$	146 $_{12}$	117 $_{37}$	312	854
Cyclobarbital-M (di-HO-) -2H2O ME	P G U UHY U+UHYAC	1895	246 $_{10}$	218 $_{100}$	146 $_{23}$	117 $_{39}$		363	1120
Cyclobarbital-M (di-HO-) -2H2O 2ME	PME UME	1860	260 $_2$	232 $_{100}$	175 $_{20}$	146 $_{24}$	117 $_{34}$	421	1121
Cyclobarbital-M (di-HO-) -2H2O 2TMS		2015	376 $_2$	361 $_{34}$	261 $_{15}$	146 $_{100}$	73 $_{46}$	939	4582
Cyclobarbital-M (di-oxo-)	U UHY U+UHYAC	1980	264 $_4$	235 $_{100}$	207 $_{26}$	193 $_{12}$	79 $_{11}$	439	4461
Cyclobarbital-M (di-oxo-) 2ME	UME UHYME	2100	292 $_{35}$	263 $_{100}$	235 $_{50}$	207 $_{18}$	178 $_{25}$	574	4462
Cyclobarbital-M (HO-) 3TMS	UTMS	2600	468 $_1$	453 $_{32}$	439 $_{100}$	349 $_{57}$	73 $_{100}$	1137	4463
Cyclobarbital-M (HO-) -H2O	U UHY U+UHYAC	2170	234 $_{57}$	205 $_{100}$	156 $_{45}$	141 $_{76}$	79 $_{21}$	319	702
Cyclobarbital-M (oxo-)	U+UHYAC	2190	250 $_{11}$	221 $_{100}$	193 $_{35}$	179 $_{25}$	150 $_{15}$	379	703
Cyclobarbital-M (oxo-) 2ME	U UHY	2050	278 $_1$	249 $_{100}$	221 $_{23}$	164 $_{11}$	79 $_8$	506	706
Cyclobarbital-M (oxo-) 2TMS	UTMS	2570	394 $_{11}$	379 $_{103}$	264 $_{49}$	164 $_{15}$	73 $_{95}$	999	4464
Cyclobenzaprine	UHY U+UHYAC	2235	275 $_3$	215 $_{41}$	202 $_{23}$	189 $_{14}$	58 $_{100}$	492	40
Cyclobenzaprine-M (bis-nor-) AC	U+UHYAC	2710	289 $_{15}$	230 $_{100}$	215 $_{70}$	202 $_{31}$	189 $_5$	558	1873
Cyclobenzaprine-M (HO-N-oxide) -(CH3)2NOH	UHY	2280*	246 $_{100}$	228 $_{35}$	215 $_{70}$	202 $_{36}$	178 $_{33}$	363	2698
Cyclobenzaprine-M (HO-N-oxide) -(CH3)2NOH AC	U+UHYAC	2530*	288 $_{35}$	246 $_{100}$	229 $_{58}$	215 $_{89}$	202 $_{37}$	552	2541
Cyclobenzaprine-M (nor-)	UHY	2600	261 $_{14}$	218 $_{99}$	215 $_{100}$	202 $_{66}$	189 $_{23}$	427	2270
Cyclobenzaprine-M (nor-) AC	U+UHYAC	2670	303 $_{20}$	230 $_{100}$	215 $_{74}$	202 $_{35}$	86 $_{18}$	629	42
Cyclobenzaprine-M (N-oxide) -(CH3)2NOH	U UHY U+UHYAC	2000*	230 $_{100}$	215 $_{40}$				307	46
Cyclocumarol		2670	322 $_{100}$	265 $_{70}$	249 $_{13}$	148 $_{17}$	72 $_{87}$	722	4047
Cyclofenil		2710	364 $_{71}$	322 $_{76}$	280 $_{100}$	263 $_{25}$	199 $_{11}$	897	2282
Cyclofenil artifact (deacetyl-)		2680	322 $_{76}$	280 $_{100}$	263 $_{25}$	199 $_{11}$	107 $_{15}$	723	3210
Cyclofenil HY		2700*	280 $_{100}$	237 $_{17}$	199 $_{32}$			518	2278
Cyclohexadecane		1950*	224 $_1$	196 $_3$	97 $_{59}$	83 $_{73}$	55 $_{100}$	290	2355
Cyclohexane		<1000*	84 $_{10}$	69 $_{42}$	56 $_{100}$	41 $_{80}$	27 $_{41}$	96	3774
Cyclohexanol		<1000*	100 $_4$	82 $_{57}$	71 $_{17}$	67 $_{38}$	57 $_{100}$	100	707
Cyclohexanol -H2O		<1000*	82 $_{66}$	67 $_{58}$	43 $_{100}$			96	1629
Cyclohexanone		<1000*	98 $_{30}$	83 $_7$	69 $_{27}$	55 $_{100}$	42 $_{86}$	99	3610
Cyclohexene		<1000*	82 $_{66}$	67 $_{58}$	43 $_{100}$			96	1629
2-Cyclohexylphenol		1580*	176 $_{61}$	133 $_{87}$	120 $_{70}$	107 $_{100}$	91 $_{27}$	164	5162
2-Cyclohexylphenol AC		1615*	218 $_9$	176 $_{100}$	133 $_{47}$	120 $_{40}$	107 $_{48}$	265	5166
2-Cyclohexylphenol ME		1565*	190 $_{39}$	147 $_{100}$	134 $_{23}$	121 $_{60}$	91 $_{45}$	194	5170
4-Cyclohexylphenol		1595*	176 $_{45}$	133 $_{100}$	120 $_{38}$	107 $_{52}$	91 $_{12}$	164	5163
4-Cyclohexylphenol AC		1720*	218 $_6$	176 $_{100}$	133 $_{88}$	120 $_{30}$	107 $_{36}$	265	5167
Cyclopentamine		1230	141 $_1$	126 $_2$	67 $_2$	58 $_{100}$		121	2771
Cyclopentamine AC		1680	183 $_1$	168 $_{17}$	100 $_{60}$	58 $_{100}$		183	2284
Cyclopentaphenanthrene		2000*	190 $_{100}$	189 $_{82}$	163 $_4$	161 $_1$	95 $_{21}$	193	2565
Cyclopenthiazide 4ME	UEXME	3660	435 $_1$	352 $_{100}$	309 $_3$	244 $_{10}$	145 $_2$	1090	6849
Cyclopentobarbital	P G U UHY U+UHYAC	1865	193 $_{51}$	169 $_{46}$	67 $_{100}$			319	708
Cyclopentobarbital 2ME		1775	221 $_{75}$	196 $_{34}$	67 $_{100}$			430	709
Cyclopentolate		2025	175 $_1$	163 $_1$	91 $_{15}$	71 $_{13}$	58 $_{100}$	571	2760

Cyclopentolate -H2O   **Table 1-8-1:** Compounds in order of names

Name	Detected	RI	Typical ions and intensities					Page	Entry
Cyclopentolate -H2O		2000	273_1	129_2	91_6	71_{34}	58_{100}	482	2772
Cyclophosphamide		2065	260_1	211_{15}	175_{100}	147_{25}	69_{18}	419	1496
Cyclophosphamide -HCl	P	1975	224_{13}	175_{100}	147_{25}	69_{18}		287	1489
Cyclotetradecane		1860*	196_1	111_{25}	97_{51}	83_{66}	55_{100}	211	2354
Cyclothiazide 4ME	UEXME	3730	445_3	352_{100}	244_{15}	145_8		1107	6850
Cycloxydim		2580	279_7	251_4	178_{100}	149_8	108_{21}	736	3635
Cycloxydim ME		2380	293_9	192_{100}	164_3	123_{15}	95_{27}	804	3636
Cycluron		1760	198_3	127_{10}	99_7	89_{17}	72_{100}	216	3936
Cycluron ME		1720	212_2	141_{13}	113_{22}	102_8	72_{100}	251	3937
Cyfluthrin		2755	433_1	226_{35}	206_{58}	163_{100}	127_{29}	1086	3514
Cypermethrin		2815	415_3	209_{15}	181_{78}	163_{100}	91_{72}	1052	3176
Cypermethrin-M/artifact (deacyl-) ME		2590	239_2	197_{100}	141_{30}	115_{30}	77_{19}	340	2819
Cypermethrin-M/artifact (deacyl-) -HCN		1700*	198_{100}	169_{64}	141_{74}	115_{50}	77_{68}	214	2797
Cypermethrin-M/artifact (HOOC-) ME		1170*	222_4	187_{64}	163_{70}	127_{65}	91_{100}	280	4207
Cyphenothrin		2960	375_6	181_{27}	167_{13}	123_{100}	81_{30}	937	3881
Cyprazepam artifact		2505	265_{70}	264_{100}	230_{79}	177_{24}	75_{33}	441	4010
Cyprazepam artifact (deoxo-)		2730	323_{38}	294_{100}	241_{21}	91_{68}	55_{91}	726	4012
Cyprazepam HY	UHY	2050	231_{80}	230_{95}	154_{23}	105_{38}	77_{100}	308	419
Cyprazepam HYAC	U+UHYAC PHYAC	2245	273_{31}	230_{100}	154_{13}	105_{23}	77_{50}	480	273
Cyproheptadine	G U+UHYAC	2340	287_{100}	215_{52}	96_{80}	70_{27}		549	710
Cyproheptadine-M (HO-)	UHY-I	3060	303_{100}	243_6	217_{26}	202_{41}	178_9	629	1620
Cyproheptadine-M (nor-)	U-I UHY-I	2400	273_{100}	229_{29}	215_{84}	165_3	82_{16}	481	1619
Cyproheptadine-M (nor-) AC	U+UHYAC	2920	315_{100}	300_{41}	243_{39}	229_{48}	215_{57}	689	1614
Cyproheptadine-M (nor-HO-) AC	U+UHYAC-I	2980	331_{100}	241_{19}	229_{35}	215_{54}	202_{58}	764	1616
Cyproheptadine-M (nor-HO-) 2AC	U+UHYAC-I	3000	373_2	331_{29}	303_{100}	202_{61}	82_{24}	931	1615
Cyproheptadine-M (nor-HO-) -H2O	U-I UHY-I	2450	271_{100}	241_{11}	213_7	193_{23}	165_{25}	475	1618
Cyproheptadine-M (nor-HO-) -H2O AC	U+UHYAC-I	2940	313_{59}	243_{47}	229_{80}	215_{100}	202_{29}	680	1617
Cyproheptadine-M (nor-HO-aryl-) 2AC	U+UHYAC-I	3060	373_{100}	358_{36}	316_{38}	259_{66}	72_7	931	2691
Cyproheptadine-M (oxo-)	U-I UHY-I U+UHYAC-I	2960	301_{56}	258_{25}	229_{100}	215_{82}	202_{53}	620	1621
Cyproterone AC		3340*	416_8	356_{51}	313_{100}	246_{34}	175_{70}	1055	1415
Cyproterone -H2O	U+UHYAC	3310*	356_{43}	246_{44}	175_{100}			869	1208
Cyproterone-M/artifact-1 AC	U+UHYAC	3320*	374_8	356_{33}	339_{100}	175_{44}		932	1209
Cyproterone-M/artifact-2 AC	U+UHYAC	3330*	372_{15}	354_{39}	339_{100}			926	1210
Cytisine		2100	190_{73}	160_{24}	146_{100}	134_{25}		194	1630
Cytisine AC		2480	232_{54}	189_{15}	160_{19}	146_{150}	134_{20}	312	7442
Cytisine HFB		2255	386_{26}	240_{10}	217_6	189_{17}	146_{100}	972	7445
Cytisine PFP		2245	336_{59}	292_6	189_{23}	146_{200}	217_8	787	7444
Cytisine TFA		2230	286_{47}	242_6	189_{17}	146_{150}	69_{59}	542	7443
Cytisine TMS		2110	262_{39}	218_{34}	146_{22}	116_{96}	73_{100}	431	7446
Cytosine 2TMS		1480	255_{42}	254_{100}	240_{110}	170_{25}	73_{68}	402	7555
Danazole		2880	337_{25}	270_{100}	121_{81}	105_{71}	79_{59}	795	6112
Danazole AC		2820	379_2	337_{45}	173_{53}	146_{57}	91_{100}	950	6113
Danthron		2330*	240_{100}	212_{16}	184_{18}	138_{10}	92_{14}	341	3555
Danthron AC		2460*	282_{11}	240_{100}	212_9	184_9	127_7	526	3678
Danthron ET		2500*	268_{32}	253_{100}	236_9	152_{14}	139_{13}	459	3695
Danthron ME		2435*	254_{100}	236_{29}	208_{45}	168_{15}	139_{20}	397	3693
Danthron TMS		2465*	312_2	297_{100}	253_5	240_4	127_6	675	3697
Danthron 2AC		2595*	324_1	282_{27}	240_{100}	212_7	155_8	730	3679
Danthron 2ET		2560*	296_{100}	253_{60}	237_{12}	165_{10}	139_9	597	3696
Danthron 2ME		2475*	268_{100}	239_{39}	180_{10}	139_{12}	126_{23}	459	3694
Danthron 2TMS		2530*	369_{100}	297_8	268_2	210_2	73_{50}	966	3698
Dantrolene		1900	214_{100}	184_{36}	156_{28}	140_{47}	113_{36}	683	2033
Dantrolene artifact		1880	184_{93}	155_{100}	130_{20}	102_8	92_{11}	183	2034
Dapsone	P-I	2865	248_{100}	184_{13}	140_{58}	108_{86}	92_{40}	370	6534
Dapsone 2AC	U+UHYAC-I	3960	332_{100}	290_{61}	248_{51}	140_{19}	108_{31}	768	6535
Dapsone 2HFB		2695	640_{44}	336_{100}	304_{53}	141_{55}	118_{46}	1198	6563
Dapsone 2PFP		2670	540_{71}	286_{100}	254_{51}	141_{49}	119_{64}	1181	6562
Dapsone 2TFA		2700	440_{70}	236_{73}	204_{34}	188_{21}	109_{35}	1097	6564
Dazomet		1660	162_{54}	129_3	89_{100}	72_{16}	57_{35}	140	3915
o,p'-DDD	G P U	2230*	318_6	235_{100}	199_{12}	165_{25}		700	1783
o,p'-DDD -HCl	P U	1800*	282_{49}	247_{18}	212_{100}	176_{34}		525	1888
o,p'-DDD-M (dichlorophenylmethane)	P U	1900*	236_{54}	201_{100}	165_{82}	82_{31}		327	1743
o,p'-DDD-M (HO-) -2HCl	P U	1790*	264_8	235_{100}	199_{19}	165_{46}		438	1884
o,p'-DDD-M (HO-HOOC-)	P U	2040*	296_1	251_{100}	139_{88}	111_{28}		595	1893
o,p'-DDD-M (HOOC-) ME	P U	2530*	294_{16}	259_{15}	235_{100}	199_{26}	165_{66}	584	1889
o,p'-DDD-M/artifact (dehydro-)	G P U	2100*	316_{46}	281_7	246_{100}	210_7	176_{11}	691	1784
p,p'-DDD		2240*	318_2	235_{100}	199_{16}	165_{74}	75_{32}	700	1954
p,p'-DDD -HCl		2390*	282_{61}	247_{17}	212_{100}	176_{49}	75_{31}	525	3177
o,p'-DDE	G P U	2100*	316_{46}	281_7	246_{100}	210_7	176_{11}	691	1784
p,p'-DDE	U	2150*	316_{54}	246_{100}	210_{17}	176_{52}	75_{38}	691	1931
o,p'-DDT		2275*	352_1	235_{100}	199_{23}	165_{65}	75_{31}	851	3178
p,p'-DDT	U	2320*	352_1	235_{100}	199_{17}	165_{66}	75_{31}	851	1932
Decamethrin		2900	503_1	253_{57}	181_{100}	93_{40}	77_{55}	1166	2818
Decamethrin-M/artifact (deacyl-) ME		2590	239_2	197_{100}	141_{30}	115_{30}	77_{19}	340	2819

Table 1-8-1: Compounds in order of names   Dextromoramide-M (HO-)

Name	Detected	RI	Typical ions and intensities					Page	Entry
Decamethrin-M/artifact (deacyl-) -HCN		1700*	198 100	169 64	141 74	115 50	77 68	214	2797
Decamethrin-M/artifact (HOOC-) ME		1540*	310 3	253 46	231 44	172 37	91 100	664	2798
Decamethyltetrasiloxane		1300*	295 19	207 100	191 17	147 20	73 37	666	5429
Decane		1000*	142 3	120 11	105 24	71 25	57 77	122	3776
Decyldodecylphthalate		2990*	474 1	335 6	307 8	149 100	57 11	1143	3542
Decylhexylphthalate		2665*	390 1	307 5	251 10	233 2	149 100	986	6402
Decyloctylphthalate		2675*	418 1	307 9	279 12	149 100	57 28	1059	3544
Decyltetradecylphthalate		3250*	502 1	363 4	307 7	149 100	57 14	1166	3543
DEET		1550	190 40	162 4	119 100	91 48	65 24	196	4501
Dehydroabietic acid	P	2590*	300 20	285 73	239 100	197 39	141 24	616	4493
Dehydroepiandrosterone		2530*	288 100	270 43	255 56	203 31	91 55	555	3760
Dehydroepiandrosterone enol 2TMS		2580*	432 44	417 36	327 27	169 25	73 100	1085	3800
Dehydroepiandrosterone -H2O	U UHY U+UHYAC	2595*	270 100	255 23	121 74	91 46	79 39	472	3770
1-Dehydrotestosterone		2610*	286 4	147 13	122 100	91 28	55 17	544	3892
1-Dehydrotestosterone AC		2690*	328 4	147 23	122 100	91 26	55 15	750	3922
1-Dehydrotestosterone TMS		2640*	358 5	268 7	147 39	122 100	73 93	878	3926
1-Dehydrotestosterone enol 2TMS		2600*	430 38	415 12	325 14	206 58	73 100	1081	3965
Deiquate artifact		1460	156 100	128 34	102 6	78 31	51 54	136	105
Delorazepam HY	UHY	2180	265 62	230 100	139 43	111 50		441	543
Delorazepam HYAC	U+UHYAC	2300	307 42	265 58	230 100	139 16	111 14	647	290
Deltamethrin		2900	503 1	253 57	181 100	93 40	77 55	1166	2818
Deltamethrin-M/artifact (deacyl-) ME		2590	239 2	197 100	141 30	115 30	77 19	340	2819
Deltamethrin-M/artifact (deacyl-) -HCN		1700*	198 100	169 64	141 74	115 50	77 68	214	2797
Deltamethrin-M/artifact (HOOC-) ME		1540*	310 3	253 46	231 44	172 37	91 100	664	2798
Demedipham TFA		2460	396 52	277 100	218 59	205 99	119 73	1005	4125
Demedipham-M/artifact (phenol)		1740	181 60	122 62	109 100	81 43	53 28	177	3750
Demedipham-M/artifact (phenol) TFA		1540	277 100	218 70	205 92	91 25	69 34	499	4126
Demedipham-M/artifact (phenol) 2ME		1640	209 100	150 53	136 72	108 57	77 35	241	4099
Demedipham-M/artifact (phenol) 3ME		1560	195 100	164 8	136 47	108 34	72 57	207	4093
Demedipham-M/artifact (phenylcarbamic acid) ME	G	1320	151 100	119 56	106 76	92 36	65 66	129	3909
Demedipham-M/artifact (phenylcarbamic acid) 2ME		1190	165 100	134 13	120 40	106 56	77 72	147	4100
Demeton-S-methyl	G P-I U-I	1635*	230 3	142 12	109 23	88 100	60 64	305	1112
Demeton-S-methylsulfone	G	1865*	262 1	169 100	125 45	109 87	79 21	429	3428
Demeton-S-methylsulfoxide	G P-I	1860*	218 1	169 60	125 47	109 100	79 26	361	1500
Demetryn		1800	213 89	198 64	171 44	82 80	58 100	252	3829
Deoxycholic acid -H2O ME		3630*	388 5	370 27	273 76	255 100	55 28	981	3126
Deoxycortone		2785*	330 52	288 60	245 38	147 60	124 100	760	6069
Deoxycortone AC		3175*	372 6	299 100	271 54	253 54	147 38	928	6068
Deoxycortone acetate		3175*	372 6	299 100	271 54	253 54	147 38	928	6068
Desipramine	UHY	2225	266 28	235 61	208 61	195 100	71 59	453	324
Desipramine AC	U+UHYAC PAC	2670	308 60	208 150	193 82	114 93		657	325
Desipramine HFB		2450	462 23	268 13	240 20	208 100	193 54	1130	7706
Desipramine PFP		2450	412 68	234 28	218 35	208 180	193 85	1047	7667
Desipramine TFA		2430	208 100	140 17	69 21	193 51	362 14	891	7786
Desipramine TMS		2470	338 5	235 62	143 28	116 50	73 70	800	5461
Desipramine-M (di-HO-) 3AC	U+UHYAC	3380	424 44	324 35	282 34	240 27	114 100	1068	3315
Desipramine-M (di-HO-ring)	UHY	2600	227 100	196 7				297	2296
Desipramine-M (di-HO-ring) 2AC	U+UHYAC	2750	311 28	269 23	227 100	196 7		670	2292
Desipramine-M (HO-) 2AC	U+UHYAC	3065	366 27	266 39	114 100			906	1175
Desipramine-M (HO-methoxy-ring)	UHY	2390	241 100	226 17	210 12	180 14		347	2315
Desipramine-M (HO-methoxy-ring) AC	U+UHYAC	2370	283 10	241 100	226 17	210 12	180 14	529	2867
Desipramine-M (HO-ring)	UHY	2240	211 100	196 15	180 10	152 4		246	2295
Desipramine-M (HO-ring) AC	U+UHYAC	2535	253 26	211 100	196 19	180 11	152 4	393	1218
Desipramine-M (nor-) AC	U+UHYAC	2640	294 23	208 100	193 43	152 10	100 17	586	3313
Desipramine-M (nor-HO-) 2AC	U+UHYAC	2980	352 60	266 88	224 100	180 15	100 48	855	3314
Desipramine-M (ring)	U U+UHYAC	1930	195 100	180 40	167 9	96 33	83 22	207	308
Desipramine-M (ring) ME		1915	209 70	194 100	178 13	165 11		242	6352
Desloratadine AC	U+UHYAC	3120	352 100	294 32	280 33	266 60	245 29	853	5610
Detajmium bitartrate artifact -H2O		3700	437 1	365 24	196 8	112 30	86 100	1094	4263
Detajmium bitartrate artifact -H2O AC		3680	479 1	407 31	144 20	112 79	86 100	1147	4272
Dextro-Methorphan	G P-I U+UHYAC	2145	271 31	214 15	171 14	150 29	59 100	476	227
Dextro-Methorphan-M (bis-demethyl-) 2AC	U+UHYAC	2710	327 11	240 8	199 12	87 100	72 62	745	228
Dextro-Methorphan-M (nor-) AC	U+UHYAC	2590	299 13	213 42	171 22	87 100	72 52	612	4477
Dextro-Methorphan-M (O-demethyl-)	UHY	2255	257 38	200 17	150 28	59 100		411	475
Dextro-Methorphan-M (O-demethyl-) AC	U+UHYAC	2280	299 100	231 42	200 20	150 48	59 15	612	230
Dextro-Methorphan-M (O-demethyl-) HFB		2100	453 100	396 18	385 91	169 19	150 27	1118	6151
Dextro-Methorphan-M (O-demethyl-) PFP	UHYPFP	2060	403 93	335 78	303 14	150 100	119 58	1027	4305
Dextro-Methorphan-M (O-demethyl-) TFA		2015	353 69	285 80	150 100	115 26	69 72	858	4006
Dextro-Methorphan-M (O-demethyl-) TMS	UHYTMS	2230	329 31	272 20	150 39	73 26	59 100	756	4304
Dextro-Methorphan-M (O-demethyl-HO-) 2AC	U+UHYAC	2580	357 68	247 22	215 17	150 100	59 30	874	1187
Dextro-Methorphan-M (O-demethyl-methoxy-) AC	U+UHYAC	2520	329 48	261 23	229 23	150 100	59 28	756	4476
Dextro-Methorphan-M (O-demethyl-oxo-) AC	U+UHYAC	2695	313 16	240 11	199 98	157 12	73 100	681	4475
Dextromoramide	G P-I U UHY U+UHYAC	2920	306 1	265 35	165 6	128 42	100 100	993	229
Dextromoramide-M (HO-)	UHY	3095	322 2	281 54	165 6	128 43	100 100	1037	1185

77

Dextromoramide-M (HO-) AC Table 1-8-1: Compounds in order of names

Name	Detected	RI	Typical ions and intensities					Page	Entry
Dextromoramide-M (HO-) AC	U+UHYAC	3210	364_1	323_{33}	194_{13}	128_{41}	100_{100}	1113	1184
Dextropropoxyphene	G P	2205	250_2	193_3	178_2	91_{15}	58_{100}	805	476
Dextropropoxyphene artifact		1755*	208_{56}	193_{41}	130_{38}	115_{100}	91_{42}	238	477
Dextropropoxyphene-M (HY)	UHY	2395	281_9	190_{76}	119_{96}	105_{100}	56_{96}	520	480
Dextropropoxyphene-M (nor-) -H2O	UHY	2240	251_{30}	217_{95}	119_{99}			387	479
Dextropropoxyphene-M (nor-) -H2O AC	U+UHYAC	2365	293_{18}	220_{99}	205_{38}			583	232
Dextropropoxyphene-M (nor-) -H2O N-prop.	U UHY U+UHYAC	2555	307_8	234_{75}	105_{100}	100_{74}	91_{67}	651	231
Dextropropoxyphene-M (nor-) N-prop.	P U	2400	307_{16}	220_{68}	100_{100}	57_{83}		736	478
Dextrorphan	UHY	2255	257_{38}	200_{17}	150_{28}	59_{100}		411	475
Dextrorphan AC	U+UHYAC	2280	299_{100}	231_{42}	200_{20}	150_{48}	59_{15}	612	230
Dextrorphan HFB		2100	453_{100}	396_{18}	385_{91}	169_{19}	150_{27}	1118	6151
Dextrorphan PFP	UHYPFP	2060	403_{93}	335_{78}	303_{14}	150_{100}	119_{58}	1027	4305
Dextrorphan TFA		2015	353_{69}	285_{80}	150_{100}	115_{26}	69_{72}	858	4006
Dextrorphan TMS	UHYTMS	2230	329_{31}	272_{20}	150_{39}	73_{26}	59_{100}	756	4304
Dextrorphan-M (methoxy-) AC	U+UHYAC	2520	329_{48}	261_{23}	229_{23}	150_{100}	59_{28}	756	4476
Dextrorphan-M (nor-) 2AC	U+UHYAC	2710	327_{11}	240_8	199_{12}	87_{100}	72_{62}	745	228
Dextrorphan-M (oxo-) AC	U+UHYAC	2695	313_{16}	240_{11}	199_{98}	157_{12}	73_{100}	681	4475
Dialifos		2545	357_4	208_{100}	129_{14}	97_{14}	76_{21}	993	3833
Diallate		1670	254_1	234_{35}	152_4	128_{18}	86_{100}	463	3429
Diazepam	P G U	2430	284_{81}	283_{91}	256_{100}	221_{31}	77_7	532	481
Diazepam HY	UHY U+UHYAC	2100	245_{95}	228_{38}	193_{29}	105_{38}	77_{100}	358	272
Diazepam HYAC	U+UHYAC	2260	287_{11}	244_{100}	228_{39}	182_{49}	77_{70}	546	2542
Diazepam-D5		2425	289_{81}	287_{89}	261_{100}	226_{18}		557	6848
Diazepam-M	P G U	2520	270_{86}	269_{97}	242_{100}	241_{82}	77_{17}	468	463
Diazepam-M	P G UGLUC	2320	268_{98}	239_{57}	233_{52}	205_{66}	77_{100}	542	579
Diazepam-M (3-HO-)	P UGLUC	2625	300_{33}	271_{100}	256_{23}	228_{16}	77_{30}	614	417
Diazepam-M (3-HO-) AC	UGLUCAC	2730	342_6	300_{40}	271_{100}	255_{16}	77_{17}	816	2099
Diazepam-M (3-HO-) ME		2600	314_{60}	271_{100}	255_{46}			684	418
Diazepam-M (3-HO-) TMS		2665	372_{23}	343_{100}	283_{26}	257_{38}	73_{54}	926	4598
Diazepam-M (3-HO-) artifact-1	G	2475	256_{19}	241_7	179_{100}	163_7	77_8	405	5780
Diazepam-M (3-HO-) artifact-2	G	2815	270_{64}	269_{100}	254_{12}	228_{26}	191_5	468	5779
Diazepam-M (HO-)	UGLUC	2670	300_{67}	272_{100}	237_{10}			614	619
Diazepam-M (HO-) AC	UGLUCAC	2790	342_{16}	300_{61}	272_{100}	237_9		816	621
Diazepam-M (HO-) HY	UHY	2580	261_{91}	260_{100}	244_{42}	209_{21}	121_{17}	424	2048
Diazepam-M (HO-) HYAC	U+UHYAC	2600	303_{77}	260_{100}	244_{47}	121_{11}		626	2060
Diazepam-M (nor-HO-)	UGLUC	2750	286_{82}	258_{100}	230_{11}	166_7	139_8	541	2113
Diazepam-M (nor-HO-) AC	U+UHYAC	3000	328_{22}	286_{90}	258_{100}	166_8	139_7	747	2111
Diazepam-M (nor-HO-) HY	UHY	2400	247_{72}	246_{100}	230_{11}	121_{26}	65_{22}	365	2112
Diazepam-M (nor-HO-) HYAC	U+UHYAC	2270	289_{18}	247_{86}	246_{100}	105_7	77_{35}	556	3143
Diazepam-M (nor-HO-) iso-1 HY2AC	U+UHYAC	2560	331_{48}	289_{64}	247_{100}	230_{41}	154_{13}	762	2125
Diazepam-M (nor-HO-) iso-2 HY2AC	U+UHYAC	2610	331_{46}	289_{54}	246_{100}	154_{11}	121_{11}	762	1751
Diazepam-M (nor-HO-methoxy-) HY2AC	U+UHYAC	2700	361_7	319_{27}	276_{38}	260_5	246_4	886	1752
Diazepam-M TMS		2635	356_9	341_{100}	312_{56}	239_{12}	135_{21}	876	4577
Diazepam-M TMS		2300	342_{62}	341_{100}	327_{19}	269_4	73_{30}	817	4573
Diazepam-M 2TMS		2200	430_{36}	429_{63}	340_{10}	313_{14}	73_{70}	1080	5499
Diazepam-M artifact-3	P-I UHY U+UHYAC	2060	240_{59}	239_{100}	205_{81}	177_{16}	151_9	342	300
Diazepam-M artifact-4	UHY U+UHYAC	2070	254_{77}	253_{100}	219_{98}			397	301
Diazepam-M HY	UHY	2050	231_{80}	230_{95}	154_{23}	105_{38}	77_{100}	308	419
Diazepam-M HYAC	U+UHYAC PHYAC	2245	273_{31}	230_{100}	154_{13}	105_{23}	77_{50}	480	273
Diazinon	P G	1760	304_{27}	199_{50}	179_{81}	152_{64}	137_{100}	634	2784
Diazinon artifact-1		1140	166_{41}	151_{99}	138_{50}	109_{41}	93_{38}	151	1399
Diazinon artifact-2	P-I U	1400*	198_{26}	170_{38}	138_{95}	111_{100}	81_{80}	213	1442
Diazinon artifact-3		1685	152_{49}	137_{99}	124_{17}	109_{18}	84_{38}	132	1375
Dibenzepin	P-I G U UHY U+UHYAC	2465	295_1	224_{21}	180_2	72_6	58_{100}	593	326
Dibenzepin-M (bis-nor-)	UHY	2700	267_1	235_{100}	207_{20}	179_{10}	103_8	455	2221
Dibenzepin-M (bis-nor-) AC	U+UHYAC	2870	309_{13}	236_{65}	223_{98}	195_{62}	100_{16}	661	327
Dibenzepin-M (HO-) iso-1 AC	PAC U+UHYAC	2600	353_1	282_{11}	240_5	71_8	58_{100}	859	3335
Dibenzepin-M (HO-) iso-2 AC	PAC U+UHYAC	2770	353_3	282_{18}	240_{12}	209_{18}	58_{100}	859	3337
Dibenzepin-M (N5-demethyl-)	U+UHYAC	2460	281_1	237_1	210_{51}	72_5	58_{100}	523	482
Dibenzepin-M (N5-demethyl-HO-) iso-1 AC	U+UHYAC	2680	339_1	268_{15}	226_8	71_{13}	58_{100}	803	3336
Dibenzepin-M (N5-demethyl-HO-) iso-2 AC	U+UHYAC	2825	339_1	268_{14}	226_{12}	71_5	58_{100}	803	3338
Dibenzepin-M (nor-) AC	PAC U+UHYAC	2800	323_{30}	250_8	237_{47}	209_{100}	100_{15}	728	1165
Dibenzepin-M (nor-HO-) iso-1 2AC	U+UHYAC	3110	381_{31}	308_{76}	266_{87}	253_{100}	100_{45}	956	3309
Dibenzepin-M (nor-HO-) iso-2 2AC	U+UHYAC	3290	381_{41}	308_{61}	266_{100}	225_{69}	100_{32}	956	3339
Dibenzepin-M (ter-nor-)	UHY	2680	235_{100}	207_{23}	179_{12}	117_7	103_7	394	2222
Dibenzepin-M (ter-nor-) AC	U+UHYAC	2825	295_{29}	236_{45}	223_{100}	195_{65}	167_{27}	591	328
Dibenzo[a,h]anthracene		3055*	278_{100}	250_3	139_{35}	125_{11}	113_6	506	3705
Dibenzofuran		1520*	168_{100}	139_{27}	113_3	84_6	70_4	154	2559
Dibutyladipate		2385*	258_3	185_{93}	129_{100}	111_{64}		414	722
Dibutylpentylpyridine		1930	261_1	232_{20}	190_{35}	163_{100}	120_{13}	428	5133
Dicamba		1795*	220_{62}	191_{35}	173_{100}	113_{26}	73_{41}	272	3637
Dicamba ME	G P-I	1525*	234_{22}	203_{100}	188_{23}	97_{13}	75_{10}	318	3639
Dicamba TMS	UTMS	1735*	292_{22}	277_{54}	203_{100}	188_{37}	73_{61}	573	6464
Dicamba -CO2		1200*	176_{100}	161_{25}	133_{72}	75_{24}	63_{37}	162	3638

Table 1-8-1: Compounds in order of names Dicloxacillin artifact-4

Name	Detected	RI	Typical ions and intensities				Page	Entry
4,4'-Dicarbonitrile-1,1'-biphenyl	U+UHYAC	1960	204$_{100}$	177$_8$	150$_4$	102$_6$	226	2408
Dichlobenil	U UHY U+UHYAC	1300	171$_{94}$	136$_{20}$	100$_{33}$		157	736
Dichlobenil-M (HO-)	U UHY	1540	187$_{100}$	159$_{62}$	88$_{53}$	86$_{57}$	189	2986
Dichlobenil-M (HO-) AC	U+UHYAC	1660	229$_{24}$	187$_{100}$	159$_{26}$	120$_{44}$ 88$_{40}$	302	2987
Dichlofenthion		1870*	314$_1$	279$_{72}$	223$_{81}$	162$_{43}$ 97$_{100}$	682	3431
Dichlofluanid		1950	332$_7$	224$_{24}$	167$_{34}$	123$_{100}$ 77$_{27}$	766	2999
Dichloran		1730	206$_{100}$	176$_{75}$	160$_{44}$	124$_{89}$ 62$_{33}$	230	3432
2,3-Dichloroaniline	G P U UHY	1400	161$_{100}$	126$_{15}$	99$_{18}$	90$_{24}$ 63$_{25}$	139	3427
3,4-Dichloroaniline	P-I U UHY U+UHYAC	1420	161$_{100}$	126$_{14}$	99$_{20}$	90$_{18}$ 63$_{18}$	139	4234
3,4-Dichloroaniline AC	U+UHYAC	1990	203$_{28}$	161$_{100}$	133$_{13}$	90$_9$ 63$_{26}$	223	4235
1,2-Dichlorobenzene		1040*	146$_{100}$	111$_{51}$	84$_7$	75$_{51}$	123	3179
1,3-Dichlorobenzene		1040*	146$_{100}$	128$_{53}$	111$_{43}$	75$_{38}$ 64$_{33}$	123	3180
p,p'-Dichlorobenzophenone (DCBP)		2340*	250$_{25}$	215$_9$	139$_{100}$	111$_{38}$ 75$_{26}$	378	1953
Dichlorodifluoromethane		<1000*	120$_1$	101$_{20}$	85$_{100}$	66$_{10}$ 50$_{36}$	106	3793
Dichloromethane		<1000*	84$_{74}$	49$_{100}$			96	1543
2,5-Dichloromethoxybenzene		1200*	176$_{100}$	161$_{25}$	133$_{72}$	75$_{24}$ 63$_{37}$	162	3638
Dichlorophen 2AC		2250*	352$_{11}$	310$_{32}$	268$_{100}$	233$_{12}$ 128$_{32}$	851	2035
Dichlorophen 2ET		2225*	324$_{100}$	309$_{44}$	289$_{30}$	273$_7$ 215$_8$	730	2005
Dichlorophen 2ME		2245*	296$_{51}$	261$_{19}$	155$_{47}$	141$_{40}$ 121$_{100}$	596	2721
2,4-Dichlorophenol	U	1320*	164$_{59}$	162$_{100}$	126$_{15}$	98$_{37}$ 63$_{65}$	140	712
2,4-Dichlorophenoxyacetic acid (D)	P U	1800*	220$_{55}$	175$_{22}$	162$_{100}$	133$_{29}$ 111$_{22}$	272	711
2,4-Dichlorophenoxyacetic acid (D) ME	P U PME UME	1580*	234$_{57}$	199$_{100}$	175$_{51}$	145$_{26}$ 111$_{32}$	317	2370
2,4-Dichlorophenoxyacetic acid (D)-M (dichlorophenol)	U	1320*	164$_{59}$	162$_{100}$	126$_{15}$	98$_{37}$ 63$_{65}$	140	712
2,4-Dichlorophenoxybutyric acid ME		1835*	262$_2$	231$_9$	162$_{32}$	101$_{100}$ 59$_{43}$	429	4118
p,p'-Dichlorophenylacetate ME		2160*	294$_{19}$	235$_{100}$	199$_{20}$	165$_{70}$ 82$_{16}$	584	3184
p,p'-Dichlorophenylethanol		2185*	266$_4$	235$_{55}$	199$_{19}$	165$_{100}$ 75$_{19}$	447	3181
o,p'-Dichlorophenylmethane	P U	1900*	236$_{54}$	201$_{100}$	165$_{82}$	82$_{31}$	327	1743
p,p'-Dichlorophenylmethane		1855*	236$_{35}$	201$_{98}$	165$_{100}$	125$_{20}$ 82$_{50}$	327	3182
p,p'-Dichlorophenylmethanol		2080*	252$_7$	217$_6$	139$_{100}$	111$_{15}$ 77$_{40}$	388	3183
Dichloroquinolinol	P G UHY U+UHYAC	1850	213$_{100}$	185$_9$	150$_{11}$		251	714
Dichlorprop	G P-I U-I	1840*	234$_{19}$	220$_5$	162$_{100}$	133$_{11}$ 109$_{11}$	318	2371
Dichlorprop ME		1630*	248$_{35}$	189$_{43}$	162$_{100}$	133$_{11}$ 109$_{12}$	369	2372
Dichlorprop-M (2,4-dichlorophenol)	U	1320*	164$_{59}$	162$_{100}$	126$_{15}$	98$_{37}$ 63$_{65}$	140	712
Dichlorvos		1275*	220$_4$	185$_{19}$	145$_7$	109$_{100}$ 79$_{26}$	272	1423
Diclofenac	G P	2205	295$_{17}$	242$_{54}$	214$_{100}$	179$_9$ 108$_{30}$	588	4469
Diclofenac ET		2240	323$_{33}$	277$_{12}$	242$_{52}$	214$_{100}$ 179$_{11}$	725	6488
Diclofenac ME	P(ME) G(ME)	2195	309$_{48}$	277$_{10}$	242$_{48}$	214$_{100}$ 179$_{20}$	658	717
Diclofenac TMS		2170	367$_{19}$	352$_{11}$	242$_{38}$	214$_{100}$ 73$_{86}$	907	5467
Diclofenac 2ME		2220	323$_{53}$	264$_{10}$	228$_{100}$	214$_{21}$	725	2323
Diclofenac -H2O	P G U+UHYAC	2135	277$_{72}$	242$_{62}$	214$_{99}$	179$_{29}$ 89$_{25}$	498	716
Diclofenac -H2O ET		2130	305$_{100}$	290$_{69}$	270$_{97}$	242$_{65}$ 227$_{42}$	637	6390
Diclofenac -H2O ME	G P	2300	291$_{19}$	263$_{37}$	228$_{100}$	200$_{33}$ 109$_8$	567	2324
Diclofenac -H2O TMS		2180	349$_{100}$	314$_{27}$	241$_{11}$	190$_{48}$ 73$_{81}$	841	4538
Diclofenac-M (di-HO-) 3ME	UME	2490	369$_{100}$	337$_{18}$	322$_{56}$	274$_{59}$ 231$_{16}$	914	6388
Diclofenac-M (di-HO-) -H2O 2AC	U+UHYAC	2880	393$_{16}$	351$_{100}$	309$_{89}$	274$_{21}$ 246$_{30}$	993	4467
Diclofenac-M (glycine conjugate) ME	P-I	2550	366$_{15}$	331$_{28}$	242$_6$	214$_{100}$ 179$_9$	903	6411
Diclofenac-M (HO-) ME	P	2540	325$_{23}$	258$_{29}$	230$_{100}$	201$_5$ 166$_{12}$	733	5958
Diclofenac-M (HO-) 2ME	UHYME	2460	339$_{78}$	272$_{62}$	244$_{100}$	201$_{30}$ 166$_{16}$	801	2325
Diclofenac-M (HO-) -H2O	P U+UHYAC	2400	293$_{39}$	258$_{100}$	230$_{46}$	195$_{34}$ 166$_{40}$	578	6467
Diclofenac-M (HO-) -H2O ME	G P	2365	307$_{65}$	209$_{15}$	201$_{47}$	272$_{97}$ 244$_{100}$	647	6490
Diclofenac-M (HO-) -H2O iso-1 AC	U+UHYAC	2520	335$_{88}$	293$_{100}$	258$_{93}$	230$_{93}$ 166$_{23}$	782	2321
Diclofenac-M (HO-) -H2O iso-2 AC	U+UHYAC	2540	335$_{22}$	293$_{100}$	258$_{29}$	230$_{48}$ 195$_{17}$	782	1212
Diclofenac-M (HO-methoxy-) 2ME	UME	2550	369$_{96}$	337$_{30}$	302$_{42}$	274$_{100}$ 260$_{86}$	915	6389
Diclofenac-M (HO-methoxy-) 2ME	UME	2490	369$_{100}$	337$_{18}$	322$_{56}$	274$_{59}$ 231$_{16}$	914	6388
Diclofenac-M (HO-methoxy-) -H2O	U+UHYAC	2505	323$_{43}$	288$_{100}$	260$_{33}$	245$_{21}$ 89$_{29}$	724	6466
Diclofenac-M (HO-methoxy-) iso-1 -H2O AC	U+UHYAC	2595	365$_{15}$	323$_{32}$	288$_{100}$	260$_{25}$ 180$_{31}$	899	4468
Diclofenac-M (HO-methoxy-) iso-2 -H2O AC	U+UHYAC	2640	365$_{17}$	323$_{46}$	288$_{100}$	260$_{37}$ 180$_{27}$	899	6465
Diclofenac-M/artifact	U+UHYAC	2980	355$_{88}$	320$_{100}$	292$_{10}$	228$_{16}$ 75$_6$	863	2322
Diclofenac-M/artifact AC	U+UHYAC	3225	413$_{30}$	371$_{100}$	336$_{50}$	214$_8$	1047	26
Diclofenac-M/artifact AC	U+UHYAC	2680	347$_{84}$	305$_{100}$	270$_{53}$	258$_{52}$ 89$_{61}$	833	6468
Diclofenamide 4ME		2540	360$_{22}$	316$_6$	253$_{100}$	144$_{13}$ 108$_{37}$	882	3127
Diclofop-methyl		2360*	340$_{48}$	281$_{40}$	253$_{100}$	120$_{59}$ 59$_{67}$	806	3832
Dicloxacillin artifact-1	U UHY U+UHYAC	1300	171$_{94}$	136$_{20}$	100$_{33}$		157	736
Dicloxacillin artifact-10 HYAC	U+UHYAC	2030	294$_8$	259$_{100}$	252$_{44}$	197$_{28}$ 102$_{48}$	584	3015
Dicloxacillin artifact-11 HYAC	U+UHYAC	2220	310$_9$	275$_{100}$	97$_{43}$	70$_{46}$ 58$_{53}$	664	3016
Dicloxacillin artifact-12 HYAC	U+UHYAC	2640	335$_{100}$	293$_{30}$	247$_7$		781	3017
Dicloxacillin artifact-13 HYAC	U+UHYAC	2460	354$_6$	312$_{89}$	277$_{42}$	254$_{29}$ 212$_{100}$	860	3018
Dicloxacillin artifact-14 HYAC	U+UHYAC	2560	368$_8$	333$_{100}$	326$_{20}$	266$_{10}$ 70$_{10}$	912	3019
Dicloxacillin artifact-15 HYAC	U+UHYAC	2785	386$_{57}$	351$_{12}$	254$_{19}$	214$_{66}$ 212$_{100}$	971	3020
Dicloxacillin artifact-16 HYAC	U+UHYAC	3370	398$_2$	216$_{28}$	174$_{100}$	114$_{11}$	1011	3021
Dicloxacillin artifact-17 HYAC	U+UHYAC	3340	467$_{35}$	393$_{35}$	212$_{100}$	139$_{56}$ 97$_{66}$	1134	3022
Dicloxacillin artifact-2	G U UHY U+UHYAC	1800	229$_4$	214$_{18}$	194$_{100}$	171$_{23}$ 123$_8$	302	2978
Dicloxacillin artifact-3	G U UHY U+UHYAC	1845	250$_{100}$	212$_{12}$	183$_7$		378	3006
Dicloxacillin artifact-4	U UHY U+UHYAC	2060	266$_{47}$	254$_{26}$	214$_{64}$	212$_{100}$ 75$_{11}$	447	3007

Table 1-8-1: Compounds in order of names

Name	Detected	RI	Typical ions and intensities					Page	Entry
Dicloxacillin artifact-5	G P U UHY U+UHYAC	2095	235_{100}	212_{13}	75_6			322	3008
Dicloxacillin artifact-5 AC	U+UHYAC	2105	277_{50}	235_{100}	212_{17}	98_7		498	3013
Dicloxacillin artifact-6	G U UHY U+UHYAC	2295	307_{100}	254_8	247_{18}	212_{30}		646	3009
Dicloxacillin artifact-7		2340	364_{29}	321_{91}	247_{24}	212_{100}	100_{19}	896	3010
Dicloxacillin artifact-8 HY	UHY	2710	382_{38}	254_{13}	212_{40}	127_{98}	100_{100}	958	3011
Dicloxacillin artifact-8 HYAC	U+UHYAC	3500	424_{12}	249_{16}	212_{19}	155_{34}	142_{100}	1067	3014
Dicloxacillin artifact-9 HY	UHY	2905	407_{100}	372_{15}	254_{38}	212_{99}	153_{60}	1033	3012
Dicloxacillin-M (HO-) artifact-1 AC	U+UHYAC	2090	308_{100}	270_7	211_{11}	172_5	148_5	653	3023
Dicloxacillin-M (HO-) artifact-2 AC	U+UHYAC	2210	324_{26}	291_{45}	289_{25}	254_{24}	212_{100}	730	3024
Dicloxacillin-M/artifact-1 HY	UHY U+UHYAC	1795	220_{100}	185_7	102_9			272	3025
Dicloxacillin-M/artifact-10 HYAC	U+UHYAC	2830	423_{100}	310_8	254_{11}	212_{65}	169_{52}	1065	3034
Dicloxacillin-M/artifact-2 HY	UHY U+UHYAC	1970	252_{100}	220_{53}	172_{32}	152_{24}		387	3026
Dicloxacillin-M/artifact-3 HY	UHY U+UHYAC	2155	274_{100}	241_{24}	192_{18}	148_{40}	94_{37}	483	3027
Dicloxacillin-M/artifact-4 HYAC	U+UHYAC	2015	219_{100}	172_9	141_5			267	3028
Dicloxacillin-M/artifact-5 HYAC	U+UHYAC	2110	287_3	252_{14}	214_{100}	171_{24}	123_8	545	3029
Dicloxacillin-M/artifact-6 HYAC	U+UHYAC	2295	350_{38}	293_{100}	251_{41}	212_{60}	184_{11}	844	3030
Dicloxacillin-M/artifact-7 HYAC	U+UHYAC	2300	354_9	319_{57}	254_{10}	212_{100}	183_9	860	3031
Dicloxacillin-M/artifact-8 HYAC	U+UHYAC	2520	397_{10}	369_{98}	254_{16}	212_{100}	59_{70}	1008	3032
Dicloxacillin-M/artifact-9 HYAC	U+UHYAC	2790	449_4	391_{100}	356_{16}	254_{19}	212_{50}	1111	3033
Dicofol		2485*	368_1	251_{59}	199_4	139_{100}	111_{36}	912	4147
Dicofol artifact (DCBP)		2340*	250_{25}	215_9	139_{100}	111_{38}	75_{26}	378	1953
Dicrotophos		1645	237_7	193_{10}	127_{100}	109_{10}	67_{46}	332	3433
Dicycloverine		2120	309_3	294_3	165_9	99_{14}	86_{100}	663	718
Diethylallylacetamide	P G U	1285	155_1	140_{25}	126_{100}	69_{94}	55_{86}	135	719
Diethylallylacetamide-M	U	1510*	144_{40}	113_{41}	95_{52}	69_{85}	55_{100}	160	720
Diethylallylacetamide-M AC	U+UHYAC-I	1725*	186_{18}	141_{56}	126_{54}	95_{91}	69_{100}	256	4245
Diethylamine		<1000	73_{17}	58_{94}	44_{25}	30_{100}		93	4188
Diethyldithiocarbamic acid ME	P	1340	163_{82}	116_{60}	91_{49}	88_{100}	60_{76}	142	6458
Diethylene glycol dibenzoate		2445*	227_1	149_{71}	105_{100}	77_{46}		684	1755
Diethylene glycol dipivalate	PPIV	1520*	159_1	129_{73}	113_4	85_{16}	57_{100}	487	1904
Diethylene glycol monoethylether pivalate	PPIV	1345*	218_1	129_{26}	85_{13}	72_{66}	57_{100}	266	6422
Diethylether		<1000*	74_{58}	59_{100}				94	2755
Diethylphthalate		1495*	222_3	177_{18}	149_{100}			281	721
Diethylstilbestrol		2295*	268_{99}	239_{45}	159_{15}	145_{42}	107_{60}	461	1419
Diethylstilbestrol 2AC		2450*	352_{27}	310_{70}	268_{100}	239_{39}	107_{48}	854	1420
Diethylstilbestrol 2ME		2190*	296_{100}	267_{26}	159_{32}	121_{43}		599	1421
Difenzoquate -C2H6SO4		1665	234_{100}	189_{11}	165_6	118_{16}	77_{48}	320	3958
Diflubenzuron 2ME		2290	338_8	154_8	141_{100}	113_{19}	63_{10}	796	3974
Diflufenicam		2670	394_9	266_{100}	246_{12}	169_9	101_9	998	3891
Diflunisal		2095*	250_{57}	232_{100}	204_{30}	175_{34}	151_{19}	379	1478
Diflunisal ME		2050*	264_{61}	232_{100}	204_{30}	175_{35}	151_8	438	2223
Diflunisal MEAC		2060*	306_{100}	247_{74}	199_4	175_8	143_9	644	2224
Diflunisal 2ME		1990*	278_{99}	247_{81}	204_{25}	188_{27}	175_{33}	505	1432
Diflunisal -CO2		1950*	206_{100}	177_{29}	151_{18}	115_{18}		230	2225
Digitoxigenin -2H2O		2840*	338_{100}	323_{43}	282_{37}	228_{74}	91_{92}	800	5243
Digitoxigenin -H2O AC		3180*	398_5	338_{52}	323_{100}	145_{81}	91_{81}	1013	5242
Digitoxin -2H2O HY		2840*	338_{100}	323_{43}	282_{37}	228_{74}	91_{92}	800	5243
Digitoxin -H2O HYAC		3180*	398_5	338_{52}	323_{100}	145_{81}	91_{81}	1013	5242
Dihexylamine		1380	185_4	114_{100}	100_3	79_5	57_8	186	4947
Dihydrobrassicasterol		3190*	400_{60}	315_{33}	289_{41}	105_{84}	55_{100}	1019	5620
Dihydrobrassicasterol -H2O		3270*	382_{100}	213_{22}	147_{78}	105_{62}	55_{56}	961	5624
Dihydrocapsaicine		2430	307_{10}	195_9	151_{13}	137_{100}	122_9	652	5927
Dihydrocapsaicine AC		2540	349_3	308_7	195_{17}	151_{15}	137_{100}	844	5928
Dihydrocapsaicine HFB		2490	503_{31}	404_{22}	391_{40}	347_{100}	333_{93}	1167	5931
Dihydrocapsaicine ME		2470	321_1	195_{23}	151_{25}	137_{100}	122_9	720	6781
Dihydrocapsaicine MEAC		2510	363_2	321_{27}	195_{17}	151_{19}	137_{100}	895	6783
Dihydrocapsaicine PFP		2410	453_{34}	354_{22}	341_{45}	297_{100}	283_{92}	1118	5930
Dihydrocapsaicine TFA		2410	403_6	304_{22}	291_{41}	247_{100}	233_{93}	1027	5929
Dihydrocapsaicine TMS		2700	379_{17}	364_{10}	209_{100}	179_{36}	73_{63}	951	6034
Dihydrocapsaicine 2TMS		2700	451_5	436_4	339_{23}	209_{87}	73_{100}	1115	6035
Dihydrocodeine	G P UHY	2410	301_{100}	244_{10}	164_{28}	115_{25}	70_{37}	620	483
Dihydrocodeine AC	U+UHYAC	2435	343_{100}	300_{33}	284_{30}	226_{14}	70_{10}	822	233
Dihydrocodeine HFB		2315	497_{100}	440_8	300_{24}	284_{79}	227_{18}	1162	6143
Dihydrocodeine PFP		2360	447_{100}	390_{16}	300_{17}	284_{55}	119_{45}	1109	2248
Dihydrocodeine TFA		2265	397_{100}	340_8	284_{29}	70_{28}	59_{47}	1010	4001
Dihydrocodeine TMS		2480	373_{59}	236_{15}	178_{14}	146_{30}	73_{100}	932	2468
Dihydrocodeine Br-artifact	U+UHYAC	2485	379_{100}	362_{21}	322_{30}	265_{22}	164_{73}	947	2988
Dihydrocodeine Cl-artifact AC		2500	377_{100}	334_{31}	318_{41}	260_{17}	164_{10}	942	2989
Dihydrocodeine-M (dehydro-)	G UHY U+UHYAC	2440	299_{100}	242_{15}	185_{23}	96_{24}	59_{23}	611	238
Dihydrocodeine-M (dehydro-) enol AC		2500	341_{100}	298_{65}	242_{32}	162_{26}		813	258
Dihydrocodeine-M (dehydro-) enol TMS		2475	371_{31}	356_{14}	313_9	234_{30}	73_{100}	925	6215
Dihydrocodeine-M (dehydro-) enol Cl-artifact AC		2630	375_{100}	340_{47}	318_{28}	146_{13}	115_{10}	936	4401
Dihydrocodeine-M (N,O-bis-demethyl-) 3AC	U+UHYAC	2790	399_{20}	357_{50}	229_{19}	87_{100}	72_{22}	1014	3050
Dihydrocodeine-M (nor-)	UHY	2440	287_{100}	244_{24}	242_{22}	150_{32}	115_{24}	547	4368

Table 1-8-1: Compounds in order of names   Dimefuron ME

Name	Detected	RI	Typical ions and intensities					Page	Entry
Dihydrocodeine-M (nor-) AC	U+UHYAC	2700	329 $_{40}$	243 $_{42}$	183 $_{26}$	87 $_{100}$	72 $_{44}$	755	3054
Dihydrocodeine-M (nor-) 2AC	U+UHYAC	2750	371 $_{20}$	285 $_{7}$	243 $_{26}$	87 $_{100}$	72 $_{33}$	924	235
Dihydrocodeine-M (nor-) Cl-artifact 2AC	U+UHYAC	2820	405 $_{8}$	320 $_{2}$	259 $_{4}$	87 $_{100}$	72 $_{26}$	1031	2990
Dihydrocodeine-M (O-demethyl-)	UHY	2400	287 $_{100}$	230 $_{14}$	164 $_{35}$	115 $_{28}$	70 $_{54}$	547	484
Dihydrocodeine-M (O-demethyl-) AC	U+UHYAC	2490	329 $_{100}$	287 $_{56}$	230 $_{10}$	164 $_{20}$	70 $_{21}$	755	3055
Dihydrocodeine-M (O-demethyl-) TFA		2250	383 $_{100}$	286 $_{19}$	270 $_{44}$	213 $_{19}$	69 $_{50}$	963	6199
Dihydrocodeine-M (O-demethyl-) 2AC	U+UHYAC	2545	371 $_{83}$	329 $_{100}$	286 $_{34}$	212 $_{21}$	70 $_{33}$	924	234
Dihydrocodeine-M (O-demethyl-) 2HFB		2260	679 $_{41}$	482 $_{53}$	466 $_{100}$	360 $_{13}$	169 $_{21}$	1202	6197
Dihydrocodeine-M (O-demethyl-) 2PFP		2330	579 $_{60}$	432 $_{21}$	416 $_{49}$	310 $_{4}$	119 $_{100}$	1191	2460
Dihydrocodeine-M (O-demethyl-) 2TFA		2190	479 $_{91}$	382 $_{25}$	366 $_{61}$	260 $_{7}$	69 $_{100}$	1147	6198
Dihydrocodeine-M (O-demethyl-) 2TMS		2520	431 $_{33}$	373 $_{7}$	236 $_{23}$	146 $_{15}$	73 $_{100}$	1083	2469
Dihydrocodeine-M (O-demethyl-dehydro-)	UHY	2445	285 $_{100}$	228 $_{18}$	214 $_{10}$	171 $_{10}$	96 $_{30}$	539	527
Dihydrocodeine-M (O-demethyl-dehydro-) AC	U+UHYAC	2595	327 $_{34}$	285 $_{100}$	229 $_{36}$	200 $_{14}$	171 $_{13}$	745	240
Dihydroergotamine artifact-1	U+UHYAC P-I	2375	244 $_{36}$	153 $_{50}$	125 $_{100}$	91 $_{38}$	70 $_{50}$	356	4495
Dihydroergotamine artifact-1 AC		2360	286 $_{29}$	153 $_{62}$	125 $_{100}$	91 $_{57}$	70 $_{73}$	544	5217
Dihydroergotamine artifact-2		2440	314 $_{45}$	244 $_{35}$	153 $_{84}$	125 $_{75}$	70 $_{100}$	683	4494
Dihydromorphine	UHY	2400	287 $_{100}$	230 $_{14}$	164 $_{35}$	115 $_{28}$	70 $_{54}$	547	484
Dihydromorphine AC	U+UHYAC	2490	329 $_{100}$	287 $_{56}$	230 $_{10}$	164 $_{20}$	70 $_{21}$	755	3055
Dihydromorphine TFA		2250	383 $_{100}$	286 $_{19}$	270 $_{44}$	213 $_{19}$	69 $_{50}$	963	6199
Dihydromorphine 2AC	U+UHYAC	2545	371 $_{83}$	329 $_{100}$	286 $_{34}$	212 $_{21}$	70 $_{33}$	924	234
Dihydromorphine 2HFB		2260	679 $_{41}$	482 $_{53}$	466 $_{100}$	360 $_{13}$	169 $_{21}$	1202	6197
Dihydromorphine 2PFP		2330	579 $_{60}$	432 $_{21}$	416 $_{49}$	310 $_{4}$	119 $_{100}$	1191	2460
Dihydromorphine 2TFA		2190	479 $_{91}$	382 $_{25}$	366 $_{61}$	260 $_{7}$	69 $_{100}$	1147	6198
Dihydromorphine 2TMS		2520	431 $_{33}$	373 $_{7}$	236 $_{23}$	146 $_{15}$	73 $_{100}$	1083	2469
Dihydromorphine-M (nor-) 3AC	U+UHYAC	2790	399 $_{20}$	357 $_{50}$	229 $_{19}$	87 $_{100}$	72 $_{22}$	1014	3050
Dihydrotestosterone		2510*	290 $_{95}$	247 $_{49}$	220 $_{86}$	161 $_{44}$	55 $_{100}$	566	3896
Dihydrotestosterone AC		2620*	332 $_{73}$	272 $_{100}$	257 $_{44}$	201 $_{46}$	79 $_{67}$	771	3918
Dihydrotestosterone TMS		2485*	362 $_{15}$	347 $_{10}$	246 $_{24}$	129 $_{70}$	73 $_{100}$	893	3963
Dihydrotestosterone enol 2TMS		2450*	434 $_{100}$	405 $_{12}$	202 $_{7}$	143 $_{41}$	73 $_{90}$	1089	3964
3,4-Dihydroxybenzoic acid ME2AC		1750*	252 $_{1}$	210 $_{19}$	168 $_{100}$	137 $_{60}$	109 $_{8}$	388	5254
3,4-Dihydroxybenzoic acid 3ME		1600*	196 $_{100}$	165 $_{97}$	137 $_{12}$	125 $_{18}$	79 $_{26}$	210	4942
3,4-Dihydroxybenzylamine 3AC		2100	265 $_{14}$	223 $_{23}$	181 $_{100}$	138 $_{24}$	122 $_{37}$	442	5692
3,4-Dihydroxycinnamic acid ME2AC		2170*	278 $_{3}$	236 $_{15}$	194 $_{100}$	163 $_{40}$	134 $_{21}$	505	5967
3,4-Dihydroxycinnamic acid ME2HFB		1985*	586 $_{45}$	555 $_{59}$	389 $_{15}$	169 $_{74}$	69 $_{80}$	1192	5971
3,4-Dihydroxycinnamic acid ME2PFP		1985*	486 $_{61}$	455 $_{89}$	323 $_{14}$	119 $_{100}$	77 $_{31}$	1153	5970
3,4-Dihydroxycinnamic acid ME2TMS		1930*	338 $_{72}$	297 $_{9}$	219 $_{100}$	191 $_{21}$	73 $_{97}$	797	6013
3,4-Dihydroxycinnamic acid 2AC		2240*	264 $_{2}$	222 $_{12}$	180 $_{100}$	163 $_{8}$	134 $_{15}$	438	5968
3,4-Dihydroxycinnamic acid 3TMS		2115*	396 $_{86}$	381 $_{22}$	219 $_{78}$	191 $_{17}$	73 $_{100}$	1006	6014
3,4-Dihydroxycinnamic acid -CO2		1375*	136 $_{53}$	89 $_{71}$	77 $_{48}$	63 $_{71}$	51 $_{100}$	116	5757
Dihydroxynorcholanoic acid -H2O MEAC	U+UHYAC	2980*	416 $_{1}$	356 $_{21}$	343 $_{53}$	255 $_{100}$	145 $_{36}$	1055	2455
3,4-Dihydroxyphenethylamine 3AC	U+UHYAC	2150	279 $_{3}$	237 $_{22}$	220 $_{21}$	178 $_{30}$	136 $_{100}$	510	5284
3,4-Dihydroxyphenethylamine 4AC		2245	321 $_{1}$	220 $_{33}$	178 $_{36}$	136 $_{100}$	123 $_{32}$	718	5285
3,4-Dihydroxyphenylacetic acid		2440*	168 $_{14}$	123 $_{100}$	105 $_{6}$	77 $_{48}$	51 $_{47}$	154	5754
3,4-Dihydroxyphenylacetic acid ME		1870*	182 $_{21}$	123 $_{100}$	105 $_{4}$	77 $_{20}$	51 $_{59}$	179	5755
3,4-Dihydroxyphenylacetic acid ME2AC		2105*	266 $_{2}$	224 $_{13}$	182 $_{89}$	123 $_{85}$	94 $_{9}$	448	5960
3,4-Dihydroxyphenylacetic acid ME2HFB		1680*	574 $_{25}$	515 $_{33}$	302 $_{39}$	69 $_{87}$	59 $_{100}$	1189	5964
3,4-Dihydroxyphenylacetic acid ME2PFP		1590*	474 $_{42}$	415 $_{52}$	252 $_{47}$	119 $_{86}$	59 $_{100}$	1142	5962
3,4-Dihydroxyphenylacetic acid ME2TFA		1560*	374 $_{30}$	315 $_{32}$	202 $_{38}$	69 $_{65}$	59 $_{67}$	932	5961
3,4-Dihydroxyphenylacetic acid ME2TMS		1695*	326 $_{57}$	267 $_{35}$	179 $_{110}$	149 $_{14}$	73 $_{99}$	740	6011
3,4-Dihydroxyphenylacetic acid MEHFB		1905*	378 $_{61}$	319 $_{100}$	169 $_{32}$	94 $_{39}$	69 $_{65}$	945	5965
3,4-Dihydroxyphenylacetic acid MEPFP		1680*	328 $_{49}$	269 $_{100}$	137 $_{17}$	119 $_{46}$	59 $_{47}$	747	5963
3,4-Dihydroxyphenylacetic acid 3TMS		1880*	384 $_{17}$	297 $_{100}$	237 $_{13}$	209 $_{16}$	73 $_{53}$	968	6012
Diisodecylphthalate		2800*	446 $_{1}$	307 $_{24}$	167 $_{27}$	149 $_{100}$	57 $_{7}$	1109	3541
Diisohexylphthalate		2380*	334 $_{1}$	251 $_{10}$	233 $_{2}$	149 $_{100}$	104 $_{3}$	781	6397
Diisononylphthalate		2700	418 $_{1}$	293 $_{35}$	167 $_{19}$	149 $_{100}$	71 $_{54}$	1059	1232
Diisooctylphthalate		2520*	390 $_{1}$	279 $_{15}$	167 $_{41}$	149 $_{100}$	57 $_{29}$	986	723
Diisopropylidene-fructopyranose		1680*	245 $_{100}$	229 $_{22}$	171 $_{46}$	127 $_{54}$	69 $_{81}$	421	5707
Diisopropylidene-fructopyranose TMS		1900*	317 $_{31}$	257 $_{13}$	229 $_{65}$	199 $_{20}$	171 $_{100}$	769	5709
Dilaurylthiodipropionate		3970*	514 $_{8}$	329 $_{19}$	178 $_{36}$	143 $_{42}$	55 $_{100}$	1172	3532
Dilazep-M/artifact (trimethoxybenzoic acid)		1780*	212 $_{100}$	197 $_{57}$	169 $_{13}$	141 $_{27}$		248	1949
Dilazep-M/artifact (trimethoxybenzoic acid) ET		1770*	240 $_{100}$	225 $_{44}$	212 $_{17}$	195 $_{45}$	141 $_{24}$	344	5219
Dilazep-M/artifact (trimethoxybenzoic acid) ME		1740*	226 $_{100}$	211 $_{49}$	195 $_{23}$	155 $_{21}$		293	1950
Diltiazem	G P U+UHYAC	2960	414 $_{1}$	150 $_{2}$	121 $_{4}$	71 $_{27}$	58 $_{100}$	1050	2504
Diltiazem-M (deacetyl-)	P UHY	2990	372 $_{1}$	178 $_{1}$	150 $_{2}$	71 $_{24}$	58 $_{100}$	927	2505
Diltiazem-M (deacetyl-) TMS		2835	444 $_{1}$	429 $_{1}$	374 $_{9}$	222 $_{27}$	58 $_{100}$	1106	4539
Diltiazem-M (deamino-HO-) AC	U+UHYAC	3060	372 $_{1}$	341 $_{25}$	240 $_{15}$	150 $_{100}$	121 $_{48}$	1078	2705
Diltiazem-M (deamino-HO-) -H2O	U+UHYAC	3310	369 $_{1}$	309 $_{100}$	150 $_{71}$	121 $_{44}$	100 $_{19}$	915	2703
Diltiazem-M (deamino-HO-) HY	UHY	3020	345 $_{11}$	316 $_{88}$	208 $_{35}$	150 $_{31}$	121 $_{100}$	828	2706
Diltiazem-M (O-demethyl-) AC	U+UHYAC	3080	442 $_{1}$	178 $_{1}$	136 $_{7}$	71 $_{34}$	58 $_{100}$	1103	2701
Diltiazem-M (O-demethyl-) HY	UHY	3050	358 $_{1}$	136 $_{7}$	107 $_{9}$	71 $_{25}$	58 $_{100}$	877	2707
Diltiazem-M (O-demethyl-deamino-HO-) 2AC	U+UHYAC	3170	457 $_{2}$	369 $_{15}$	178 $_{38}$	136 $_{100}$	87 $_{20}$	1124	2702
Diltiazem-M (O-demethyl-deamino-HO-) -H2O AC	U+UHYAC	3540	397 $_{2}$	337 $_{93}$	178 $_{34}$	136 $_{100}$	100 $_{37}$	1009	2704
Dimefuron +H2O 3ME		2600	398 $_{10}$	314 $_{3}$	255 $_{8}$	72 $_{100}$	57 $_{55}$	1012	3939
Dimefuron ME		2520	352 $_{5}$	269 $_{3}$	225 $_{4}$	127 $_{2}$	72 $_{100}$	852	3938

Dimetacrine   Table 1-8-1: Compounds in order of names

Name	Detected	RI	Typical ions and intensities					Page	Entry
Dimetacrine	G U	2315	294 $_{18}$	279 $_{28}$	86 $_{100}$	58 $_{80}$		587	329
Dimetacrine-M (N-oxide) -(CH3)2NOH	U	2020	249 $_{18}$	234 $_{100}$	194 $_{72}$			377	1170
Dimetacrine-M (ring)	U	1905	209 $_{12}$	194 $_{101}$				242	1169
Dimetamfetamine		1250	163 $_1$	148 $_1$	117 $_1$	91 $_6$	72 $_{60}$	143	1427
Dimetamfetamine-M (nor-)	U	1195	148 $_1$	134 $_2$	115 $_1$	91 $_9$	58 $_{100}$	127	1093
Dimetamfetamine-M (nor-) AC	U+UHYAC	1575	191 $_1$	117 $_2$	100 $_{42}$	91 $_6$	58 $_{100}$	196	1094
Dimetamfetamine-M (nor-) HFB		1460	254 $_{100}$	210 $_{44}$	169 $_{15}$	118 $_{41}$	91 $_{38}$	827	5069
Dimetamfetamine-M (nor-) PFP		1415	204 $_{100}$	160 $_{46}$	118 $_{35}$	91 $_{25}$	69 $_4$	589	5070
Dimetamfetamine-M (nor-) TFA		1300	245 $_1$	154 $_{100}$	118 $_{48}$	110 $_{55}$	91 $_{23}$	358	3998
Dimetamfetamine-M (nor-) TMS		1325	206 $_4$	130 $_{100}$	91 $_{17}$	73 $_{83}$	59 $_{13}$	279	6214
Dimethachlor		1565	255 $_1$	210 $_{10}$	197 $_{32}$	134 $_{100}$	77 $_{18}$	401	3830
Dimethoate	P G U	1725	229 $_7$	125 $_{45}$	93 $_{62}$	87 $_{100}$		302	724
Dimethoate-M (HO-)	U	1430	245 $_{19}$	218 $_{20}$	125 $_{45}$	93 $_{100}$		357	2119
Dimethoate-M (HOOC-) ME	U	1400*	230 $_{49}$	198 $_{66}$	125 $_{67}$	93 $_{100}$	79 $_{43}$	305	2118
Dimethoate-M (oxo-)	G P-I	1585	213 $_7$	156 $_{90}$	110 $_{99}$	79 $_{52}$	58 $_{70}$	251	1501
2,5-Dimethoxy-4-iodophenethylamine		2330	307 $_{16}$	278 $_{100}$	263 $_{20}$	247 $_9$	232 $_3$	647	6954
2,5-Dimethoxy-4-iodophenethylamine	UGLUCAC	2310*	436 $_{27}$	394 $_{100}$	334 $_{68}$	292 $_{62}$	279 $_{62}$	1092	7130
2,5-Dimethoxy-4-iodophenethylamine	UGLUCAC	2200*	392 $_8$	350 $_{61}$	308 $_{37}$	290 $_{100}$	262 $_{87}$	991	7129
2,5-Dimethoxy-4-iodophenethylamine (O-demethyl- N-acetyl-) iso-1	UGLUC	2370	335 $_{15}$	276 $_{100}$	263 $_{11}$	233 $_{11}$	134 $_{14}$	782	6963
2,5-Dimethoxy-4-iodophenethylamine (O-demethyl- N-acetyl-) iso-1 AC	U+UHYAC	2480	377 $_5$	335 $_{46}$	276 $_{100}$	259 $_{16}$	233 $_{20}$	941	6967
2,5-Dimethoxy-4-iodophenethylamine (O-demethyl- N-acetyl-) iso-2 AC	U+UHYAC	2500	377 $_8$	335 $_{43}$	276 $_{100}$	263 $_{17}$	236 $_7$	941	6968
2,5-Dimethoxy-4-iodophenethylamine (O-demethyl-) iso-1 2AC	U+UHYAC	2480	377 $_5$	335 $_{46}$	276 $_{100}$	259 $_{16}$	233 $_{20}$	941	6967
2,5-Dimethoxy-4-iodophenethylamine (O-demethyl-) iso-2 2AC	U+UHYAC	2500	377 $_8$	335 $_{43}$	276 $_{100}$	263 $_{17}$	236 $_7$	941	6968
2,5-Dimethoxy-4-iodophenethylamine AC	U+UHYAC	2260	349 $_{25}$	290 $_{100}$	275 $_{21}$	247 $_{14}$	148 $_{21}$	841	6957
2,5-Dimethoxy-4-Iodophenethylamine HFB		2110	503 $_{32}$	290 $_{100}$	277 $_{51}$	247 $_{30}$	148 $_{21}$	1166	6948
2,5-Dimethoxy-4-iodophenethylamine PFP		2080	453 $_{43}$	290 $_{100}$	277 $_{93}$	247 $_{44}$	148 $_{33}$	1117	6960
2,5-Dimethoxy-4-iodophenethylamine TFA	UGLUCTFA	2100	403 $_{20}$	290 $_{100}$	277 $_{69}$	247 $_{49}$	148 $_{31}$	1024	6959
2,5-Dimethoxy-4-iodophenethylamine TMS		2070	379 $_5$	320 $_9$	278 $_7$	102 $_{100}$	73 $_{33}$	947	6961
2,5-Dimethoxy-4-iodophenethylamine 2AC	U+UHYAC	2340	391 $_{16}$	290 $_{100}$	275 $_{19}$	247 $_8$	148 $_{10}$	987	6958
2,5-Dimethoxy-4-iodophenethylamine 2ME		2320	335 $_{18}$	290 $_{100}$	275 $_{16}$	247 $_{21}$	148 $_{10}$	782	6962
2,5-Dimethoxy-4-iodophenethylamine deuteroformyl artifact		1850	321 $_{23}$	290 $_{100}$	277 $_{42}$	247 $_{18}$	232 $_4$	717	6956
2,5-Dimethoxy-4-iodophenethylamine formyl artifact		1860	319 $_{20}$	288 $_{100}$	277 $_{50}$	247 $_{23}$	232 $_6$	706	6955
2,5-Dimethoxy-4-iodophenethylamine-M (deamino-HO-)	UGLUC	2020	308 $_{100}$	277 $_{74}$	263 $_7$	247 $_{16}$	150 $_9$	652	6966
2,5-Dimethoxy-4-iodophenethylamine-M (deamino-HO-) AC	UGLUCAC	2150	350 $_{16}$	290 $_{100}$	275 $_{28}$	247 $_{10}$	148 $_{25}$	844	6969
2,5-Dimethoxy-4-iodophenethylamine-M (deamino-HO-) TFA	UGLUCTFA	1980	404 $_{100}$	290 $_{83}$	275 $_{24}$	247 $_{12}$	148 $_{36}$	1027	6978
2,5-Dimethoxy-4-iodophenethylamine-M (deamino-HOOC-) ME	UGLUCMEAC	2115	336 $_{100}$	321 $_{12}$	289 $_8$	277 $_{57}$	247 $_{18}$	786	6982
2,5-Dimethoxy-4-iodophenethylamine-M (deamino-HOOC-O-demethyl-) ME	UGLUCMETFA	2160	322 $_{62}$	290 $_{87}$	262 $_{100}$	234 $_{21}$	191 $_6$	721	6984
2,5-Dimethoxy-4-iodophenethylamine-M (deamino-HOOC-O-demethyl-) MEAC	UGLUCMEAC	2170	364 $_{26}$	322 $_{100}$	290 $_{53}$	262 $_{44}$	234 $_{10}$	896	6981
2,5-Dimethoxy-4-iodophenethylamine-M (deamino-HOOC-O-demethyl-) METFA	UGLUCMETFA	1980	418 $_{100}$	404 $_{25}$	359 $_{32}$	305 $_{49}$	289 $_{53}$	1058	6983
2,5-Dimethoxy-4-iodophenethylamine-M (deamino-HOOC-O-demethyl-) -H2O	UGLUC	2080	290 $_{100}$	262 $_{69}$	234 $_{42}$	191 $_{11}$	127 $_{16}$	561	6965
2,5-Dimethoxy-4-iodophenethylamine-M (deamino-HO-O-demethyl-) iso-1 2AC	UGLUCAC	2240	378 $_2$	336 $_{16}$	276 $_{100}$	261 $_{10}$	134 $_{16}$	944	6970
2,5-Dimethoxy-4-iodophenethylamine-M (deamino-HO-O-demethyl-) iso-1 2TFA	UGLUCTFA	1865	486 $_{100}$	372 $_{98}$	303 $_{28}$	261 $_9$		1153	6979
2,5-Dimethoxy-4-iodophenethylamine-M (deamino-HO-O-demethyl-) iso-2 2AC	UGLUCAC	2275	378 $_8$	336 $_{33}$	276 $_{100}$	261 $_{34}$	150 $_{24}$	944	6971
2,5-Dimethoxy-4-iodophenethylamine-M (deamino-HO-O-demethyl-) iso-2 2TFA	UGLUCTFA	1890	486 $_{35}$	372 $_{100}$	275 $_{47}$	261 $_{25}$	245 $_{22}$	1153	6980
2,5-Dimethoxy-4-iodophenethylamine-M (deamino-oxo-)		1965*	306 $_{100}$	277 $_{82}$	263 $_{18}$	247 $_{29}$	232 $_{10}$	642	7233
2,5-Dimethoxy-4-iodophenethylamine-M (O-demethyl- N-acetyl-) TFA	UGLUCTFA	2270	431 $_{10}$	389 $_{100}$	276 $_{94}$	263 $_{43}$	148 $_7$	1082	6974
2,5-Dimethoxy-4-iodophenethylamine-M (O-demethyl- N-acetyl-) iso-2	UGLUC	2520	335 $_{26}$	276 $_{100}$	261 $_{18}$	220 $_3$	121 $_7$	782	6964
2,5-Dimethoxy-4-iodophenethylamine-M (O-demethyl-) iso-1 TFA	UGLUCTFA	2100	389 $_{39}$	276 $_{100}$	263 $_{35}$	233 $_{19}$	134 $_9$	981	6976
2,5-Dimethoxy-4-iodophenethylamine-M (O-demethyl-) iso-1 2TFA	UGLUCTFA	1970	485 $_{17}$	372 $_{100}$	359 $_{14}$	303 $_{11}$	234 $_9$	1152	6972
2,5-Dimethoxy-4-iodophenethylamine-M (O-demethyl-) iso-2 TFA	UGLUCTFA	2275	389 $_{62}$	276 $_{100}$	263 $_{58}$	261 $_{18}$		981	6977
2,5-Dimethoxy-4-iodophenethylamine-M (O-demethyl-) iso-2 2TFA	UGLUCTFA	2010	485 $_6$	372 $_{100}$	359 $_{23}$	275 $_{11}$	126 $_{10}$	1152	6973
2,5-Dimethoxybenzaldehyde		1615*	166 $_{200}$	151 $_{60}$	123 $_{45}$	95 $_{72}$	63 $_{82}$	149	7705
Dimethoxyethane		<1000*	89 $_1$	75 $_{38}$	59 $_{100}$	31 $_{73}$	29 $_{100}$	98	3778
3,4-Dimethoxyhydrocinnamic acid ME	UME	1705*	224 $_{38}$	164 $_{15}$	151 $_{100}$	121 $_9$	107 $_{10}$	288	4943
2,5-Dimethoxyphenethylamine-M (O-demethyl- N-acetyl-)	UGLUCTFA	2270*	209 $_{20}$	150 $_{100}$	135 $_{31}$	107 $_{25}$	77 $_{12}$	240	6975
3,4-Dimethoxyphenethylamine		1530	181 $_9$	152 $_{100}$	137 $_{21}$	107 $_{17}$	91 $_8$	178	7350
3,4-Dimethoxyphenethylamine AC		1900	223 $_6$	164 $_{100}$	151 $_{57}$	107 $_{10}$	91 $_9$	284	7352
3,4-Dimethoxyphenethylamine HFB		1665	377 $_{13}$	164 $_{75}$	151 $_{100}$	107 $_{15}$	91 $_{11}$	942	7356
3,4-Dimethoxyphenethylamine PFP		1630	327 $_{22}$	164 $_{44}$	151 $_{100}$	107 $_7$	91 $_5$	743	7355
3,4-Dimethoxyphenethylamine TFA		1645	277 $_{34}$	164 $_{64}$	151 $_{100}$	107 $_{21}$	91 $_9$	500	7354
3,4-Dimethoxyphenethylamine TMS		1650	253 $_1$	238 $_7$	151 $_{11}$	102 $_{100}$	73 $_{57}$	395	7357
3,4-Dimethoxyphenethylamine 2AC		1995	265 $_5$	164 $_{100}$	151 $_{62}$	107 $_{13}$	91 $_7$	444	7353
3,4-Dimethoxyphenethylamine 2TMS		1945	325 $_1$	310 $_4$	174 $_{100}$	86 $_{34}$	73 $_{62}$	736	7358
3,4-Dimethoxyphenethylamine formyl artifact		1510	193 $_{17}$	164 $_3$	151 $_{100}$	107 $_{11}$	91 $_8$	200	7351
Dimethoxyphthalic acid 2ME		1870*	254 $_{37}$	223 $_{100}$	207 $_{16}$	191 $_{62}$	77 $_{30}$	398	5152
Dimethylamine		<1000	45 $_{50}$	44 $_{100}$	28 $_{70}$			90	3618
N,N-Dimethyl-4-aminophenol	UHY	1220	137 $_{68}$	136 $_{100}$	121 $_{23}$	94 $_9$	65 $_{14}$	117	3415
N,N-Dimethyl-4-aminophenol AC	U+UHYAC	1370	179 $_{20}$	137 $_{96}$	136 $_{100}$	121 $_7$	65 $_{13}$	171	3416
N,N-Dimethyl-4-aminophenol-M	UHY	1240	109 $_{99}$	80 $_{41}$	52 $_{90}$			102	826
N,N-Dimethyl-4-aminophenol-M (nor-) 2AC	U+UHYAC	1615	207 $_9$	193 $_3$	165 $_{40}$	123 $_{100}$	94 $_{10}$	233	3417
N,N-Dimethyl-4-aminophenol-M 2AC	U+UHYAC PAC	1765	193 $_{10}$	151 $_{53}$	109 $_{100}$	80 $_{24}$		199	188
2,6-Dimethylaniline		1180	121 $_{100}$	106 $_{77}$				107	725
2,6-Dimethylaniline AC	U+UHYAC	1470	163 $_{33}$	121 $_{100}$	106 $_{66}$	91 $_{17}$	77 $_{30}$	142	57
Dimethylbromophenol		1470*	200 $_{82}$	121 $_{99}$	91 $_{60}$	77 $_{57}$		218	1424

Table 1-8-1: Compounds in order of names

2,2-Diphenylethylamine 2TMS

Name	Detected	RI	Typical ions and intensities					Page	Entry
2,2-Dimethylbutane		<1000*	71_{45}	57_{61}	43_{100}	41_{71}	29_{51}	97	3815
1,3-Dimethylcyclopentane		<1000*	98_9	83_{17}	70_{83}	56_{100}	41_{95}	99	3821
1,2-Dimethylcyclopropane		<1000*	70_{29}	55_{100}	42_{73}	39_{73}	29_{53}	92	3813
Dimethylformamide		<1000	73_{81}	58_7	44_{100}	42_{60}	28_{53}	93	3781
N,N-Dimethyl-5-methoxy-tryptamine	G U UHY U+UHYAC	2040	218_{16}	160_{10}	145_7	117_{10}	58_{100}	266	4059
N,N-Dimethyl-5-methoxy-tryptamine-M (HO-)	U UHY	2335	234_{16}	175_4	163_6	72_{12}	58_{100}	321	4060
N,N-Dimethyl-5-methoxy-tryptamine-M (O-demethyl-HO-) 2AC	U+UHYAC	2400	304_3	234_5	175_{10}	149_{12}	58_{100}	635	4061
1,5-Dimethylnaphthalene		1340*	156_{100}	153_{34}	141_{63}	128_{10}	115_{10}	136	2557
2,6-Dimethylphenol		1155*	122_{99}	107_{89}				107	726
2,6-Dimethylphenol AC		1130*	164_{13}	122_{100}	107_{48}	91_{12}	77_{17}	145	857
1,2-Dimethyl-3-phenyl-aziridine		1145	147_1	121_2	105_2	77_5	58_{100}	125	7526
Dimethylphenylthiazolanimin		1760	206_{42}	191_8	132_{15}	118_{27}	58_{100}	231	1426
Dimethylphthalate		1450*	194_7	163_{100}	133_9	104_7	77_{15}	203	4948
Dimethylsulfoxide		<1000*	78_{72}	63_{100}	45_{24}			95	1469
Dimetindene		2290	292_3	218_3	58_{100}			576	727
Dimetindene-M (nor-) AC	U+UHYAC	2775	320_8	218_8	100_{100}	86_{19}	58_{37}	716	1331
Dimetindene-M (nor-HO-) 2AC	U+UHYAC	3090	378_{12}	276_3	234_6	100_{100}		946	1332
Dimetotiazine	G U UHY U+UHYAC	3060	391_1	320_2	276_1	179_1	72_{100}	988	1937
Dimetotiazine-M (bis-nor-) AC	U+UHYAC	3380	405_{48}	346_{39}	319_{100}	210_{29}	100_{18}	1031	1644
Dimetotiazine-M (HO-) AC	U+UHYAC	3200	449_3	398_5	245_{18}	198_8	72_{100}	1112	1645
Dimetotiazine-M (nor-)	U UHY	3150	377_3	320_{25}	306_{23}	198_{15}	72_{100}	942	1642
Dimetotiazine-M (nor-) AC	U+UHYAC	3360	419_{41}	346_{55}	319_{75}	114_{92}	58_{100}	1061	1643
Dimpylate	P G	1760	304_{27}	199_{50}	179_{81}	152_{64}	137_{100}	634	2784
Dimpylate artifact-1		1140	166_{41}	151_{99}	138_{50}	109_{41}	93_{38}	151	1399
Dimpylate artifact-2	P-I U	1400*	198_{26}	170_{38}	138_{95}	111_{100}	81_{80}	213	1442
Dimpylate artifact-3		1685	152_{49}	137_{99}	124_{17}	109_{18}	84_{38}	132	1375
2,4-Dinitrophenol		1520	184_{100}	154_{28}	107_{41}	91_{45}	63_{88}	183	728
Dinobuton		2060	267_4	240_{12}	211_{100}	163_{26}	147_{19}	739	3516
Dinocap		2460	364_1	197_1	130_1	103_1	69_{100}	897	3828
Dinoseb		1780	240_{16}	211_{100}	163_{48}	147_{35}	117_{33}	342	3640
Dinoterb		1760	240_{12}	225_{100}	177_{40}	131_{22}	77_{22}	342	3641
Dioctylphthalate		2655*	390_1	279_{12}	261_2	167_2	149_{100}	986	6401
Dioctylsebacate	U UHY U+UHYAC	2705*	426_1	315_2	297_4	185_{100}	112_{19}	1072	5408
Dioxacarb		1825	193_1	166_{63}	149_{22}	121_{100}	73_{35}	284	3914
Dioxacarb -C2H3NO	U	1325*	166_{30}	149_3	121_{100}	104_{15}	73_{28}	149	729
Dioxane		<1000*	88_{100}	58_{89}	43_{33}	31_{45}		97	730
Dioxathion		1705*	270_{11}	197_6	125_{39}	97_{100}	73_{37}	1122	3831
Dioxethedrine ME3AC		2060	351_1	222_2	153_9	114_{70}	72_{100}	849	1793
Dioxethedrine 4AC		2090	319_8	150_9	114_{52}	72_{100}	70_{27}	948	1795
Dioxethedrine -H2O 2AC		1950	277_{27}	235_9	193_{20}	114_{13}	70_{100}	501	1792
Dioxethedrine -H2O 3AC		2075	319_{37}	277_{52}	235_{49}	193_{32}	70_{100}	708	1794
Diphenhydramine	P G U	1870	227_1	165_9	152_3	73_{14}	58_{50}	403	731
Diphenhydramine HY	UHY	1645*	184_{45}	165_{14}	152_7	105_{100}	77_{63}	184	1333
Diphenhydramine HYAC	U+UHYAC	1700*	226_{20}	184_{20}	165_{100}	105_{14}	77_{35}	294	1241
Diphenhydramine HYME	UHY	1655*	198_{70}	167_{94}	121_{100}	105_{56}	77_{71}	215	6779
Diphenhydramine-M (benzophenone)	U+UHYAC	1610*	182_{31}	152_3	105_{100}	77_{70}	51_{39}	180	1624
Diphenhydramine-M (bis-nor-) AC	U+UHYAC	2240	183_{30}	167_{100}	87_{45}	72_{26}		465	2080
Diphenhydramine-M (deamino-HO-)	P U	1760	228_9	183_{46}	167_{100}	152_{26}	105_{27}	301	2049
Diphenhydramine-M (deamino-HO-) AC	U+UHYAC	1820	270_1	183_{98}	167_{100}	152_{31}	87_{67}	470	2079
Diphenhydramine-M (di-HO-)	U	1895*	244_{61}	213_{100}	167_{78}			356	733
Diphenhydramine-M (HO-)	P U	1890	213_{11}	183_{25}	167_{19}	58_{100}		475	734
Diphenhydramine-M (HO-) HY2AC	U+UHYAC	2090*	284_5	242_{15}	224_{17}	182_{100}	153_{19}	533	2425
Diphenhydramine-M (HO-BPH) iso-1	UHY	2065*	198_{93}	121_{72}	105_{100}	93_{22}	77_{66}	214	1627
Diphenhydramine-M (HO-BPH) iso-1 AC	U+UHYAC	2010*	240_{27}	198_{100}	121_{47}	105_{85}	77_{80}	343	2196
Diphenhydramine-M (HO-BPH) iso-2	P-I U UHY	2080*	198_{50}	121_{100}	105_{17}	93_{14}	77_{28}	214	732
Diphenhydramine-M (HO-BPH) iso-2 AC	U+UHYAC	2050*	240_{20}	198_{100}	121_{94}	105_{41}	77_{51}	343	2197
Diphenhydramine-M (HO-methoxy-BPH)	UHY	2050*	228_{46}	197_6	151_{100}	105_{22}	77_{41}	299	1625
Diphenhydramine-M (HO-methoxy-BPH) AC	U+UHYAC	2100*	270_3	228_{92}	151_{100}	105_{25}	77_{40}	468	1622
Diphenhydramine-M (methoxy-)	U	2010	285_1	183_4	165_3	73_{12}	58_{100}	540	2078
Diphenhydramine-M (methoxy-) HY	UHY	1875*	214_{41}	183_{100}	137_{80}	121_{39}	105_{30}	255	4483
Diphenhydramine-M (methoxy-) HYAC	U+UHYAC	1780	256_{11}	214_{58}	183_{100}	105_{42}	77_{47}	406	2077
Diphenhydramine-M (nor-)	P U	1520	167_{100}	165_{41}	152_{21}			348	2047
Diphenhydramine-M (nor-) AC	P U+UHYAC	2265	241_1	167_{100}	152_{30}	101_{85}	86_{27}	529	735
Diphenhydramine-M/artifact	UHY	2070*	228_8	186_{100}	157_{10}	128_7	77_4	298	1626
Diphenhydramine-M/artifact	U UHY U+UHYAC	1850*	209_4	167_{100}	152_{22}	121_8	115_7	239	2081
Diphenhydramine-M/artifact AC	U+UHYAC	2200*	270_7	228_{17}	186_{100}	157_{10}	128_7	467	1623
Diphenoxylate		3415	452_8	377_{21}	246_{100}	193_{19}	165_{18}	1117	236
Diphenylamine		1595	169_{100}	168_{74}	141_4	84_{24}	77_{15}	155	3434
1,1-Diphenyl-1-butene	U+UHYAC	1900*	208_{100}	193_{68}	178_{31}	165_{33}	130_{49}	238	5294
2,2-Diphenylethylamine HFB		1720	226_3	180_{62}	167_{100}	165_{52}	152_{23}	993	7626
2,2-Diphenylethylamine PFP		1650	224_1	180_{60}	167_{100}	165_{49}	152_{24}	820	7627
2,2-Diphenylethylamine TFA		1665	224_1	180_{54}	167_{100}	165_{59}	152_{28}	578	7628
2,2-Diphenylethylamine TMS		1650	269_1	254_{13}	165_{26}	102_{100}	73_{77}	466	7624
2,2-Diphenylethylamine 2TMS		1950	341_1	326_3	174_{100}	86_{15}	73_{28}	814	7625

Table 1-8-1: Compounds in order of names

Name	Detected	RI	Typical ions and intensities					Page	Entry
2,2-Diphenylethylamine formyl artifact		1510	209_1	178_3	167_{100}	152_{18}	105_{12}	242	7623
Diphenyloctylamine		2330	281_{11}	210_{40}	194_7	180_4	92_9	524	5145
Diphenylprolinol		2120	181_7	77_{31}	70_{100}	165_5	105_{22}	395	7804
Diphenylprolinol AC		2405	181_9	165_6	113_{89}	77_{53}	70_{200}	593	7805
Diphenylprolinol HFB		2185	266_{36}	239_{14}	183_{100}	105_{83}	77_{54}	1111	7813
Diphenylprolinol ME		2070	181_8	165_5	152_6	105_{19}	84_{100}	456	7806
Diphenylprolinol PFP		2160	216_{49}	183_{100}	119_{60}	105_{89}	77_{69}	1014	7811
Diphenylprolinol TFA		2185	183_{77}	166_{31}	139_{17}	105_{100}	77_{81}	842	7807
Diphenylprolinol 2TMS		2160	382_1	255_4	239_{12}	142_{100}	73_{66}	1011	7814
Diphenylprolinol -H2O		2095	206_2	165_{14}	105_{25}	83_{100}	55_{10}	325	7803
Diphenylprolinol -H2O AC		2265	277_{85}	234_{100}	167_{52}	165_{55}	152_{20}	501	7809
Diphenylprolinol -H2O HFB		2065	431_{100}	354_{10}	262_{13}	234_{57}	206_{82}	1082	7812
Diphenylprolinol -H2O PFP		2050	381_{100}	262_{17}	234_{52}	206_{80}	119_{57}	955	7810
Diphenylprolinol -H2O TFA		2075	331_{100}	262_{34}	234_{41}	206_{86}	69_{83}	763	7808
Diphenylprolinol-M/artif. (benzophenone)	U+UHYAC	1610*	182_{31}	152_3	105_{100}	77_{70}	51_{39}	180	1624
Diphenylpyraline	G U+UHYAC	2115	281_1	167_{15}	114_{30}	99_{100}		524	737
Diphenylpyraline HY	UHY	1645*	184_{45}	165_{14}	152_7	105_{100}	77_{63}	184	1333
Diphenylpyraline HYAC	U+UHYAC	1700*	226_{20}	184_{20}	165_{100}	105_{14}	77_{35}	294	1241
Diphenylpyraline HYME	UHY	1655*	198_{70}	167_{94}	121_{100}	105_{56}	77_{71}	215	6779
Diphenylpyraline-M (benzophenone)	U+UHYAC	1610*	182_{31}	152_3	105_{100}	77_{70}	51_{39}	180	1624
Diphenylpyraline-M (HO-BPH) iosmer-2 AC	U+UHYAC	2050*	240_{20}	198_{100}	121_{94}	105_{41}	77_{51}	343	2197
Diphenylpyraline-M (HO-BPH) iso-1	UHY	2065*	198_{93}	121_{72}	105_{100}	93_{22}	77_{66}	214	1627
Diphenylpyraline-M (HO-BPH) iso-1 AC	U+UHYAC	2010*	240_{27}	198_{100}	121_{47}	105_{85}	77_{80}	343	2196
Diphenylpyraline-M (HO-BPH) iso-2	P-I U UHY	2080*	198_{50}	121_{100}	105_{17}	93_{14}	77_{28}	214	732
Diphenylpyraline-M (HO-methoxy-BPH)	UHY	2050*	228_{46}	197_6	151_{100}	105_{22}	77_{41}	299	1625
Diphenylpyraline-M (HO-methoxy-BPH) AC	U+UHYAC	2100*	270_3	228_{92}	151_{100}	105_{25}	77_{40}	468	1622
Diphenylpyraline-M/artifact	UHY	2070*	228_8	186_{100}	157_{10}	128_7	77_4	298	1626
Diphenylpyraline-M/artifact AC	U+UHYAC	2200*	270_7	228_{17}	186_{100}	157_{10}	128_7	467	1623
Dipivefrin TFATMS		2400	519_1	379_{56}	295_{41}	211_{22}	57_{100}	1174	6333
Dipivefrin 2AC		2760	435_1	362_{10}	307_{17}	86_{41}	57_{100}	1090	2747
Dipivefrin 2TMS		2410	495_1	480_1	116_{100}	73_{37}	57_{24}	1161	6332
Dipivefrin -H2O		2505	333_1	249_2	205_1	85_8	57_{100}	775	2745
Dipivefrin -H2O AC		2720	375_1	362_9	307_{18}	115_{20}	57_{100}	937	2746
Diprophylline 2AC		2455	338_{52}	236_{14}	194_{31}	180_{100}	159_{26}	797	1433
Dipropylbarbital	P G U UHY U+UHYAC	1650	170_{95}	141_{99}	98_{16}			250	1428
Dipropylbarbital 2ME	UME PME	1580	198_{44}	183_4	169_{100}	140_5	112_{19}	345	6406
Dipropylbarbital-M (HO-) iso-1	U UHY	1930	213_3	171_{22}	141_{100}	112_{12}	98_{31}	301	2955
Dipropylbarbital-M (HO-) iso-1 AC	U+UHYAC	1950	226_9	184_{38}	168_{97}	141_{100}	101_{74}	470	2957
Dipropylbarbital-M (HO-) iso-2	U UHY	1980	210_1	186_{17}	168_{54}	141_{100}	98_{35}	301	2956
Dipropylbarbital-M (HO-) iso-2 AC	U+UHYAC	2000	227_5	210_3	168_{100}	141_{26}	97_{10}	470	2958
Dipropylbarbital-M (oxo-)	U UHY U+UHYAC	1870	226_2	184_{24}	169_{100}	141_{39}	98_{20}	294	2954
Dipyrone	G P U	1995	215_{34}	123_{99}	91_{18}	56_{40}		670	197
Dipyrone-M (bis-dealkyl-)	P U UHY	1955	203_{23}	93_{14}	84_{59}	56_{100}		224	219
Dipyrone-M (bis-dealkyl-) AC	P U U+UHYAC	2270	245_{30}	203_{13}	84_{50}	56_{100}		359	183
Dipyrone-M (bis-dealkyl-) 2AC	U+UHYAC	2280	287_8	245_{31}	203_{15}	84_{56}	56_{100}	547	3333
Dipyrone-M (bis-dealkyl-) artifact	U UHY	1945	180_{13}	119_{99}	91_{45}			173	424
Dipyrone-M (dealkyl-)	P U UHY	1980	217_{17}	123_{14}	97_7	83_{17}	56_{100}	261	220
Dipyrone-M (dealkyl-) AC	P U+UHYAC	2395	259_{20}	217_7	123_8	56_{99}		416	184
Dipyrone-M (dealkyl-) ME artifact	P G U-I	1895	231_{36}	123_7	111_{17}	97_{56}	56_{100}	310	189
Disopyramide	P G U UHY U+UHYAC	2490	239_{14}	212_{65}	195_{100}	167_{42}	114_{52}	805	2872
Disopyramide artifact	P G U UHY U+UHYAC	1980	193_{97}	165_6				198	330
Disopyramide -CHNO	UHY U+UHYAC	2030	296_1	253_6	196_{32}	169_{100}	128_{38}	600	2873
Disopyramide-M (bis-dealkyl-) -NH3	U+UHYAC	2245	238_{80}	194_{23}	182_{70}	167_{100}	152_8	336	2874
Disopyramide-M (N-dealkyl-) AC	U+UHYAC	2640	339_1	296_4	212_{70}	195_{100}	167_{61}	804	2876
Disopyramide-M (N-dealkyl-) ME		2345	280_8	224_{60}	194_8	167_{55}	98_{100}	673	7581
Disopyramide-M (N-dealkyl-) TMS		2155	369_1	354_5	284_{100}	195_{53}	167_{51}	917	2155
Disopyramide-M (N-dealkyl-) TMS		2155	369_1	354_6	284_{100}	195_{52}	167_{51}	917	7583
Disopyramide-M (N-dealkyl-) 2TMS		2200	426_1	336_2	284_{100}	195_{34}	73_{54}	1102	7582
Disopyramide-M (N-dealkyl-) -CHNO AC	U+UHYAC	2330	296_1	196_7	182_{13}	169_{100}	72_{11}	600	2875
Disopyramide-M (N-dealkyl-) -H2O		2075	279_{51}	236_{19}	193_{72}	182_{100}	167_{64}	514	1926
Disopyramide-M (N-dealkyl-) -H2O AC		2300	321_1	278_1	221_{15}	194_{100}	167_3	719	1929
Disopyramide-M (N-dealkyl-) -H2O HFB		2075	475_1	226_4	221_{14}	207_{37}	194_{100}	1144	7586
Disopyramide-M (N-dealkyl-) -H2O PFP		2080	425_4	382_{32}	306_{46}	278_{31}	193_{100}	1069	7584
Disopyramide-M (N-dealkyl-) -H2O TFA		2120	375_1	332_{26}	306_{27}	278_{26}	193_{100}	936	7585
Disopyramide-M (N-dealkyl-) -NH3		2100	280_{76}	209_{29}	194_{100}	180_{43}	167_{58}	518	1925
Disugram	G P-I	1525*	234_{22}	203_{100}	188_{23}	97_{13}	75_{10}	318	3639
Disulfiram	G P-I	2470	296_1	148_{19}	116_{100}	88_{76}	60_{60}	596	1494
Disulfiram-M/artifact (di-oxo-)		2215	264_8	116_{100}	88_{86}	76_{33}	60_{47}	439	4471
Disulfoton		1780*	274_{15}	186_{11}	125_{12}	97_{28}	88_{100}	483	1429
Ditalimfos		2095	299_{39}	243_{31}	209_{26}	148_{45}	130_{100}	609	3435
Ditazol		2900	324_{100}	293_{37}	249_{77}	165_{53}	77_{82}	732	1430
Ditazol 2AC	PAC U+UHYAC	2985	408_{100}	365_4	322_{17}	262_{15}	87_{71}	1036	738
Ditazol-M (benzil)	U UHY U+UHYAC	1825*	210_3	105_{100}	77_{45}	51_{26}		244	1233
Ditazol-M (bis-dealkyl-)	UHY	2280	236_{100}	165_{30}	105_{78}	104_{77}	77_{44}	328	2544

Table 1-8-1: Compounds in order of names   DOET

Name	Detected	RI	Typical ions and intensities					Page	Entry
Ditazol-M (bis-dealkyl-) AC	U+UHYAC	2560	278_{36}	236_{100}	165_9	105_{72}	77_{39}	506	1234
Ditazol-M (bis-dealkyl-HO-) MEAC	UHYMEAC	2960	308_{42}	266_{100}	135_{54}	134_{45}	77_{24}	654	1205
Ditazol-M (bis-dealkyl-HO-) 2AC	U+UHYAC	2845	336_{26}	294_{22}	252_{100}	121_{35}	105_{20}	787	1202
Ditazol-M (dealkyl-) 2AC	U+UHYAC	2620	364_{17}	322_{100}	249_{44}	105_{10}	87_{25}	896	2547
Ditazol-M (dealkyl-HO-) ME2AC	UHYMEAC	2970	394_{35}	352_{100}	279_{13}	135_{20}	87_{25}	999	1206
Ditazol-M (dealkyl-HO-) 3AC	U+UHYAC	3020	422_{21}	380_{78}	338_{75}	252_{100}	87_{33}	1064	1203
Ditazol-M (deamino-HO-)	UHY U+UHYAC	2580	237_{60}	165_{10}	105_{44}	104_{100}	77_{53}	332	2543
Ditazol-M (HO-) ME2AC	UHYMEAC	3200	438_{100}	352_{15}	279_{10}	135_{20}	87_{75}	1094	1207
Ditazol-M (HO-) 3AC	U+UHYAC	3250	466_{100}	424_{37}	338_{16}	278_{13}	87_{76}	1133	1204
Ditazol-M (HO-benzil) AC	U+UHYAC	2160*	268_1	226_1	163_{31}	121_{100}	105_{30}	459	2546
Ditazol-M (HO-benzil) ME	UHYME	2290*	240_3	135_{100}	105_{18}	77_{10}		342	2545
Diuron ME		1880	246_{14}	174_5	145_4	109_5	72_{100}	362	4092
Diuron-M (3,4-dichloroaniline)	P-I U UHY U+UHYAC	1420	161_{100}	126_{14}	99_{20}	90_{18}	63_{18}	139	4234
Diuron-M (3,4-dichloroaniline) AC	U+UHYAC	1990	203_{28}	161_{100}	90_9	63_{26}		223	4235
Diuron-M/artifact (3,4-dichlorcarbanilic acid) ME	G P-I U UHY U+UHYAC	1850	219_{100}	187_{77}	174_{86}	160_{47}	133_{56}	267	850
Diuron-M/artifact (dichlorophenylisocyanate)	P U	1960	187_{100}	159_{30}	124_{74}	62_{21}		189	4508
Dixyrazine	UHY	3220	427_{73}	352_{33}	212_{100}	187_{63}		1075	485
Dixyrazine AC	U+UHYAC	3530	469_{44}	366_{28}	229_{77}	212_{100}	180_{26}	1138	331
Dixyrazine-M (amino-) AC	U+UHYAC	2765	312_{60}	212_{100}	114_{57}			675	1240
Dixyrazine-M (N-dealkyl-) AC	U+UHYAC	3355	381_{16}	339_1	199_{20}	141_{100}	99_{21}	957	1263
Dixyrazine-M (O-dealkyl-) AC	U+UHYAC	3350	425_{56}	365_9	199_{67}	185_{101}	98_{15}	1070	1262
Dixyrazine-M (ring)	P G U UHY U+UHYAC	2010	199_{100}	167_{45}				216	10
Dixyrazine-M AC	U+UHYAC	2550	257_{23}	215_{100}	183_7			409	12
Dixyrazine-M 2AC	U+UHYAC	2865	315_{54}	273_{34}	231_{100}	202_{11}		688	2618
DMA		1535	195_1	152_{18}	137_5	121_2	77_3	208	3255
DMA		1535	195_1	152_{18}	137_5	121_2	44_{100}	208	5525
DMA AC		1870	237_6	178_{39}	152_5	121_6	86_{12}	334	3268
DMA AC		1870	237_6	178_{39}	121_6	86_{12}	44_{100}	334	5526
DMA formyl artifact		1550	207_5	176_{45}	151_{11}	121_{12}	56_{100}	234	3243
DMA intermediate (2,5-dimethoxyphenyl-2-nitropropene)		1860	223_{98}	176_{54}	161_{100}	147_{78}	91_{97}	284	3284
DMA precursor (2,5-dimethoxybenzaldehyde)		1345*	166_{100}	151_{39}	120_{36}	95_{61}	63_{62}	150	3278
DMCC		<1000	152_7	151_{13}	137_{18}	109_{100}	84_{12}	132	3580
DNOC		1660	198_{100}	168_{39}	121_{60}	105_{63}	53_{60}	213	2508
DOB		1800	273_1	232_6	230_6	199_1	77_7	480	2548
DOB		1800	273_1	230_6	105_3	77_7	44_{100}	480	5527
DOB AC		2150	315_3	256_{20}	229_1	162_4	86_{15}	688	2549
DOB AC		2150	315_3	256_{20}	162_4	86_{22}	44_{100}	688	5528
DOB HFB		1945	469_{24}	256_{88}	240_{55}	229_{100}	199_{29}	1137	6008
DOB PFP		1905	419_{21}	256_{69}	229_{94}	190_{55}	119_{87}	1059	6007
DOB TFA		1935	369_{28}	256_{81}	229_{100}	199_{40}	69_{81}	914	6006
DOB TMS		1920	345_1	272_2	229_1	116_{80}	73_{63}	827	6009
DOB formyl artifact		1790	285_3	254_{15}	229_5	199_3	56_{100}	537	3242
DOB precursor (2,5-dimethoxybenzaldehyde)		1345*	166_{100}	151_{39}	120_{36}	95_{61}	63_{62}	150	3278
DOB-M (bis-O-demethyl-) 3AC	U+UHYAC	2325	371_2	329_{35}	287_{54}	228_{100}	86_{44}	921	7075
DOB-M (bis-O-demethyl-) artifact 2AC	U+UHYAC	2225	311_{23}	269_{98}	227_{59}	212_{39}	133_{18}	670	7184
DOB-M (deamino-HO-) AC	U+UHYAC	1950*	316_7	274_{22}	214_{96}	186_{17}		692	7061
DOB-M (deamino-oxo-)	U+UHYAC	1835*	272_7	229_{11}				477	7062
DOB-M (HO-) 2AC	U+UHYAC	2270	373_3	86_{100}	313_{14}	271_{37}		929	7081
DOB-M (HO-) -H2O	U+UHYAC	1960*	273_{31}	271_{29}	258_{43}	256_{42}		473	7073
DOB-M (HO-) -H2O AC	U+UHYAC	2130*	313_{24}	271_{79}	256_{100}			678	7074
DOB-M (O-demethyl-) iso-1 AC	U+UHYAC	2120	301_{18}	242_{100}	215_{11}	185_{13}	86_{20}	617	7070
DOB-M (O-demethyl-) iso-1 2AC	U+UHYAC	2235	343_{12}	301_{23}	284_{57}	242_{100}	86_{81}	819	7065
DOB-M (O-demethyl-) iso-2 AC	U+UHYAC	2180	301_{29}	242_{100}	215_{14}	86_{36}		618	7071
DOB-M (O-demethyl-) iso-2 2AC	U+UHYAC	2275	343_2	284_{56}	242_{100}	215_{13}	86_{23}	819	7066
DOB-M (O-demethyl-deamino-oxo-) AC	U+UHYAC	1930*	300_8	258_{94}	215_{100}			612	7063
DOB-M (O-demethyl-deamino-oxo-) iso-1	U+UHYAC	1870*	260_{66}	258_{72}	217_{99}	215_{100}		411	7068
DOB-M (O-demethyl-deamino-oxo-) iso-2	U+UHYAC	1885*	260_{99}	258_{100}	217_{97}	215_{93}		411	7069
DOB-M (O-demethyl-HO-) 3AC	U+UHYAC	2385	401_1	359_5	317_7	258_{10}	86_7	1019	7067
DOB-M (O-demethyl-HO-) -H2O 2AC	U+UHYAC	2280	341_{32}	299_{62}	257_{100}	242_{72}		809	7072
DOB-M (O-demethyl-HO-deamino-HO-) 3AC	U+UHYAC	2145*	402_2	360_9	315_{13}	300_8	231_{24}	1022	7064
Dobutamine 3TMS		2875	517_1	502_6	250_{26}	73_{60}	58_{100}	1174	4540
Dobutamine 3TMSTFA		2780	613_7	280_{100}	267_{56}	179_{64}	73_{71}	1195	6182
Dobutamine 4AC	U+UHYAC	3495	469_2	262_{100}	220_{57}	107_{51}	58_{68}	1138	3531
Dobutamine 4TMS		3025	589_1	574_2	322_{52}	130_{100}	73_{75}	1194	4541
Dobutamine-M (N-dealkyl-O-methyl-) AC	U+UHYAC	2330	209_{100}	180_{16}	150_{43}	138_{17}	58_{10}	241	2980
Dobutamine-M (N-dealkyl-O-methyl-) 2AC	U+UHYAC	2070	251_2	209_7	150_{100}	137_{15}		384	1273
Dobutamine-M (O-methyl-)	UHY	3200	315_1	178_{24}	151_7	107_{30}	58_{100}	690	2979
Dobutamine-M (O-methyl-) 2AC	U+UHYAC	3100	399_2	250_{87}	220_{29}	150_{58}	58_{58}	1015	2981
Dobutamine-M (O-methyl-) 3AC	U+UHYAC	3350	441_1	262_{51}	220_{34}	150_{82}	58_{100}	1101	2484
Docosane		2200*	310_1	99_{11}	85_{45}	71_{64}	57_{100}	669	4946
Dodecane		1200*	170_6	85_{39}	99_7	71_{64}	57_{100}	157	4701
Dodemorph		2020	281_{10}	238_9	210_4	154_{100}	55_{30}	525	4034
DOET		1610	223_1	180_{33}	165_{10}	91_7	77_4	286	3260
DOET		1610	223_1	180_{33}	165_{10}	91_7	44_{100}	286	5529

85

DOET AC Table 1-8-1: Compounds in order of names

Name	Detected	RI	Typical ions and intensities					Page	Entry
DOET AC		1990	265_9	206_{60}	179_{12}	165_7	86_{13}	446	3269
DOET AC		1990	265_9	206_{60}	179_{12}	86_{13}	44_{100}	446	5530
DOET formyl artifact		1600	235_{12}	204_{73}	179_{51}	91_{23}	56_{100}	326	3247
DOET precursor (2,5-dimethoxyacetophenone)		1280*	180_{43}	165_{100}	150_{12}	107_{32}	77_{36}	175	3283
DOET precursor (2,5-dimethoxyacetophenone)		1280*	180_{43}	165_{100}	107_{32}	77_{36}	43_{61}	175	5531
DOI		2025	321_2	278_{100}	263_{13}	247_9	77_{32}	717	7172
DOI AC	U+UHYAC	2295	363_{12}	304_{150}	277_9	247_9	86_{64}	893	7174
DOI HFB		2070	517_{10}	304_{58}	277_{100}	247_{36}	69_{71}	1173	7179
DOI PFP		2055	467_{46}	304_{79}	277_{150}	247_{35}	190_{30}	1135	7178
DOI TFA		2075	417_{49}	304_{63}	277_{200}	247_{25}	140_{20}	1056	7176
DOI 2AC		2360	405_{12}	304_{100}	277_{15}	247_5	86_{39}	1030	7175
DOI 2ME		2305	349_{23}	304_{100}	277_{18}	162_{11}	72_{65}	841	7569
DOI 2TFA		1940	513_7	304_{25}	277_{100}	247_{18}	69_{53}	1171	7177
DOI formyl artifact		1960	333_1	302_{10}	277_5	247_2	56_{100}	772	7173
DOI-M (bis-O-demethyl-) 3AC		2480	419_4	377_{32}	360_{36}	276_{78}	86_{100}	1059	7837
DOI-M (bis-O-demethyl-) artifact 2AC		2425	359_{20}	317_{100}	275_{68}	260_{43}	133_{28}	879	7182
DOI-M (O-demethyl-) iso-1 2AC	U+UHYAC	2395	391_2	349_9	332_{36}	290_{100}	86_{55}	987	7180
DOI-M (O-demethyl-) iso-2 2AC	U+UHYAC	2410	391_3	349_{40}	332_{11}	290_{100}	86_{30}	987	7181
DOM		1660	209_3	166_{42}	151_{24}	135_5	91_8	243	2573
DOM		1660	209_3	166_{42}	151_{24}	135_5	44_{100}	243	5532
DOM AC	UAAC	2020	251_4	192_{50}	165_9	135_9	86_{15}	386	2574
DOM AC	UAAC	2020	251_4	192_{50}	165_9	86_{15}	44_{100}	386	5533
DOM PFP	UPFP	1730	355_{19}	192_{60}	165_{100}	135_{34}	119_{29}	865	2591
DOM 2AC		2090	293_6	192_{100}	165_{30}	135_{12}	86_{36}	582	2575
DOM formyl artifact		1565	221_{10}	190_{63}	165_{51}	135_{33}	56_{100}	278	3248
DOM intermediate (2,5-dimethoxytoluene)		1020*	152_{61}	137_{100}	109_{17}	77_{17}	65_{11}	131	3289
DOM precursor-1 (hydroquinone dimethylether)		<1000*	138_{56}	123_{100}	95_{54}	63_{22}		118	3282
DOM precursor-2 (2-methylhydroquinone)		1210*	124_{100}	107_{12}	95_{33}	77_{23}	67_{30}	109	3280
DOM precursor-2 (2-methylhydroquinone) 2AC		1440*	208_3	166_{13}	124_{100}	95_3	77_4	236	3281
DOM precursor-2 (2-methylhydroquinone) 2AC		1440*	208_3	166_{13}	124_{100}	95_3	43_{58}	236	5534
DOM-M (deamino-oxo-HO-) 2AC	UAAC	2560*	308_{17}	249_{65}	223_{29}	206_{30}	164_{100}	655	2587
DOM-M (deamino-oxo-HO-) 2PFP	UPFP	2045*	516_{37}	353_{100}	326_{57}	233_{38}	206_{35}	1173	2590
DOM-M (HO-) 2AC	UAAC	2260	309_{18}	250_{100}	191_1	164_{14}	86_{15}	662	2588
DOM-M (HO-) 2PFP	UPFP	1830	517_{20}	354_{100}	327_6	190_{20}	119_{15}	1173	2589
DOM-M (O-demethyl-) 2PFP	UGLUCPFP	1780	487_4	324_{100}	297_{55}	190_{24}	119_{15}	1155	2592
Donepezil	U+UHYAC	3150	379_{46}	288_{44}	188_{25}	175_{58}	91_{100}	950	6548
Donepezil-M (O-demethyl-)	U+UHYAC	3180	365_{31}	274_{44}	175_{38}	146_{30}	91_{100}	902	6549
Dopamine 3AC	U+UHYAC	2150	279_3	237_{22}	220_{21}	178_{30}	136_{100}	510	5284
Dopamine 4AC		2245	321_1	220_{33}	178_{36}	136_{100}	123_{32}	718	5285
Dopamine-M (O-methyl-) AC	U+UHYAC	2330	209_{100}	180_{16}	150_{43}	138_{17}	58_{10}	241	2980
Dopamine-M (O-methyl-) 2AC	U+UHYAC	2070	251_2	209_7	150_{100}	137_{15}		384	1273
Dorzolamide		2715	307_4	282_{28}	218_{61}	203_{56}	138_{300}	730	7427
Dorzolamide ME		2670	321_4	296_{21}	232_{57}	217_{35}	138_{200}	796	7425
Dorzolamide 2TMS		2695	453_8	381_{29}	290_{92}	275_{69}	138_{250}	1136	7428
Dorzolamide iso-1 2ME		2640	335_3	310_{21}	246_{64}	231_{26}	138_{300}	852	7426
Dorzolamide iso-2 2ME		2660	352_1	310_8	246_{10}	199_5	138_{250}	852	7424
Dosulepin	P G U UHY U+UHYAC	2385	295_3	234_2	221_5	202_7	58_{100}	591	435
Dosulepin-M (bis-nor-) AC	U+UHYAC	2800	309_{53}	250_{38}	235_{43}	217_{100}	202_{43}	660	2943
Dosulepin-M (HO-)	U UHY	2500	311_2	217_{27}	202_{23}	165_2	58_{100}	671	2939
Dosulepin-M (HO-) iso-1 AC	U+UHYAC	2660	353_1	272_1	219_1	202_1	58_{100}	858	2942
Dosulepin-M (HO-) iso-2 AC	U+UHYAC	2690	353_1	266_1	219_1	150_4	58_{100}	858	2944
Dosulepin-M (HO-N-oxide) -(CH3)2NOH	U UHY	2130*	266_{100}	251_{28}	233_{28}	206_{28}	165_{77}	448	2937
Dosulepin-M (HO-N-oxide) -(CH3)2NOH AC	U+UHYAC	2480*	308_{70}	266_{100}	233_{28}	206_{28}	165_{40}	654	2941
Dosulepin-M (nor-)	U UHY	2370	281_{21}	238_{10}	204_{10}	178_8	165_9	522	2940
Dosulepin-M (nor-) AC	U+UHYAC	2820	323_{24}	250_{72}	217_{100}	202_{36}	86_{47}	726	2934
Dosulepin-M (nor-HO-) iso-1 2AC	U+UHYAC	3110	381_6	308_{14}	266_{17}	203_{16}	86_{100}	955	2935
Dosulepin-M (nor-HO-) iso-2 2AC	U+UHYAC	3150	381_7	308_{64}	266_{35}	235_{100}	86_{57}	955	2945
Dosulepin-M (N-oxide) -(CH3)2NOH	U+UHYAC	2100*	250_{76}	235_{67}	217_{100}	202_{53}	165_{10}	379	2938
Doxepin	P-I G U+UHYAC	2240	279_1	178_4	165_5	58_{100}	219_3	513	332
Doxepin artifact	G U+UHYAC	1905*	210_{100}	181_{74}	165_6	152_{26}	89_8	244	4470
Doxepin-M (HO-) iso-1	U UHY	2535	295_5	178_4	165_8	58_{100}		593	488
Doxepin-M (HO-) iso-1 AC	U+UHYAC	2540	337_1	178_1	165_1	58_{100}		794	336
Doxepin-M (HO-) iso-2	U UHY	2560	295_5	178_4	165_8	58_{100}		593	920
Doxepin-M (HO-) iso-2 AC	U+UHYAC	2585	337_1	165_1	152_1	58_{100}		794	883
Doxepin-M (HO-dihydro-)	UHY	2530	297_4	71_6	58_{100}			603	487
Doxepin-M (HO-dihydro-) AC	U+UHYAC	2340	339_4	211_7	165_2	58_{100}		804	334
Doxepin-M (HO-methoxy-) iso-1 AC	U+UHYAC	2735	337_1	178_1	165_1	58_{100}		910	6777
Doxepin-M (HO-methoxy-) iso-2 AC	U+UHYAC	2780	337_1	178_1	165_2	58_{100}		910	6778
Doxepin-M (HO-N-oxide) -(CH3)2NOH	UHY	2120*	250_{100}	231_{45}	203_{37}			380	557
Doxepin-M (HO-N-oxide) -(CH3)2NOH AC	U+UHYAC	2360*	292_{79}	250_{84}	233_{100}	165_{66}		575	335
Doxepin-M (nor-)	UHY	2270	265_{11}	222_{73}	204_{100}	178_{64}	115_{43}	445	486
Doxepin-M (nor-) AC	U+UHYAC	2700	307_{22}	234_{100}	219_{44}	86_{49}		649	337
Doxepin-M (nor-) HFB		2395	461_6	240_{43}	234_{100}	219_{76}	178_{56}	1129	7709
Doxepin-M (nor-) PFP		2580	411_{15}	234_{100}	219_{55}	190_{40}	178_{56}	1043	7669

Table 1-8-1: Compounds in order of names   Duloxetine 2PFP

Name	Detected	RI	Typical ions and intensities					Page	Entry
Doxepin-M (nor-) TFA		2495	361 $_{28}$	234 $_{100}$	219 $_{86}$	202 $_{52}$	178 $_{69}$	887	7668
Doxepin-M (nor-) TMS		2340	337 $_{1}$	322 $_{2}$	219 $_{14}$	178 $_{21}$	116 $_{100}$	794	7708
Doxepin-M (nor-HO-)	U UHY	2540	281 $_{11}$	238 $_{27}$	220 $_{45}$	165 $_{21}$	152 $_{21}$	523	489
Doxepin-M (nor-HO-) iso-1 2AC	U+UHYAC	2995	365 $_{17}$	292 $_{27}$	250 $_{100}$	237 $_{39}$	86 $_{69}$	901	338
Doxepin-M (nor-HO-) iso-2 2AC	U+UHYAC	3035	365 $_{20}$	292 $_{69}$	250 $_{85}$	233 $_{91}$	86 $_{79}$	901	31
Doxepin-M (N-oxide) -(CH3)2NOH	P U UHY U+UHYAC	1970*	234 $_{100}$	219 $_{71}$	165 $_{42}$			320	333
Doxylamine	P-I G U+UHYAC	1920	270 $_{1}$	182 $_{4}$	167 $_{8}$	71 $_{62}$	58 $_{100}$	471	740
Doxylamine HY	U+UHYAC	1630	199 $_{100}$	184 $_{48}$				218	743
Doxylamine-M	UHY U+UHYAC	1520	183 $_{56}$	182 $_{99}$	167 $_{43}$			182	741
Doxylamine-M (bis-nor-) AC	U+UHYAC	2280	284 $_{1}$	198 $_{32}$	182 $_{100}$	167 $_{82}$	86 $_{84}$	535	746
Doxylamine-M (bis-nor-HO-) 2AC	U+UHYAC	2720	284 $_{6}$	241 $_{51}$	198 $_{100}$	183 $_{41}$	86 $_{61}$	818	2696
Doxylamine-M (carbinol) -H2O	U+UHYAC	1560	181 $_{35}$	180 $_{100}$	152 $_{11}$	77 $_{16}$		177	742
Doxylamine-M (deamino-HO-) AC	U+UHYAC	1960	285 $_{6}$	198 $_{100}$	182 $_{50}$	167 $_{41}$	87 $_{81}$	539	2692
Doxylamine-M (HO-) AC	U+UHYAC	2300	258 $_{1}$	198 $_{11}$	183 $_{18}$	71 $_{61}$	58 $_{100}$	750	744
Doxylamine-M (HO-carbinol) AC	U+UHYAC	2980	257 $_{98}$	242 $_{55}$	200 $_{79}$	137 $_{100}$	78 $_{93}$	410	2693
Doxylamine-M (HO-carbinol) -H2O	UHY	1800	197 $_{44}$	196 $_{100}$	167 $_{16}$	139 $_{4}$	89 $_{3}$	212	2688
Doxylamine-M (HO-carbinol) -H2O AC	U+UHYAC	1940	239 $_{14}$	197 $_{44}$	196 $_{100}$	167 $_{16}$	139 $_{4}$	340	2689
Doxylamine-M (HO-methoxy-) AC	U+UHYAC	2320	258 $_{1}$	198 $_{4}$	183 $_{5}$	71 $_{59}$	58 $_{100}$	878	745
Doxylamine-M (HO-methoxy-carbinol) AC	U+UHYAC	2030	287 $_{47}$	245 $_{86}$	230 $_{100}$	167 $_{48}$	106 $_{39}$	546	2695
Doxylamine-M (HO-methoxy-carbinol) -H2O AC	U+UHYAC	2010	269 $_{7}$	227 $_{58}$	226 $_{100}$	211 $_{25}$	154 $_{21}$	464	2694
Doxylamine-M (nor-) AC	U U+UHYAC	2340	298 $_{1}$	212 $_{2}$	182 $_{100}$	167 $_{75}$	100 $_{68}$	607	2690
Doxylamine-M (nor-HO-) 2AC	U+UHYAC	2760	257 $_{4}$	241 $_{3}$	198 $_{53}$	183 $_{35}$	100 $_{100}$	870	2697
Drofenine		2180	317 $_{1}$	173 $_{5}$	99 $_{17}$	86 $_{100}$		700	747
Dronabinol	G	2470*	314 $_{85}$	299 $_{100}$	271 $_{41}$	243 $_{29}$	231 $_{45}$	687	981
Dronabinol AC	U+UHYAC-I	2450*	356 $_{13}$	313 $_{35}$	297 $_{100}$	243 $_{12}$	231 $_{24}$	870	982
Dronabinol ET		2390*	342 $_{90}$	327 $_{100}$	313 $_{39}$	271 $_{27}$	259 $_{30}$	818	2531
Dronabinol ME		2360*	328 $_{82}$	313 $_{100}$	285 $_{28}$	257 $_{24}$	245 $_{27}$	751	2530
Dronabinol TMS		2405*	386 $_{100}$	371 $_{86}$	315 $_{37}$	303 $_{31}$	73 $_{80}$	974	4599
Dronabinol iso-1 PFP		2150*	460 $_{100}$	445 $_{14}$	417 $_{70}$	392 $_{30}$	377 $_{100}$	1128	5669
Dronabinol iso-2 PFP		2170*	460 $_{100}$	445 $_{65}$	417 $_{75}$	389 $_{87}$	297 $_{80}$	1128	5668
Dronabinol-D3		2450*	317 $_{96}$	302 $_{100}$	274 $_{43}$	258 $_{26}$	234 $_{55}$	700	5663
Dronabinol-D3 AC		2750*	359 $_{9}$	316 $_{18}$	300 $_{100}$	274 $_{9}$	234 $_{17}$	882	7309
Dronabinol-D3 ME		2355*	331 $_{81}$	316 $_{100}$	288 $_{34}$	257 $_{29}$	248 $_{39}$	766	6040
Dronabinol-D3 TMS		2385*	389 $_{100}$	374 $_{96}$	346 $_{26}$	315 $_{59}$	306 $_{41}$	984	5670
Dronabinol-D3 iso-1 PFP		2130*	463 $_{60}$	420 $_{54}$	395 $_{25}$	380 $_{100}$	342 $_{13}$	1131	5665
Dronabinol-D3 iso-1 TFA		2160*	413 $_{51}$	370 $_{31}$	345 $_{14}$	330 $_{100}$	232 $_{10}$	1048	5667
Dronabinol-D3 iso-2 PFP		2150*	463 $_{100}$	448 $_{65}$	420 $_{70}$	389 $_{85}$	300 $_{81}$	1132	5664
Dronabinol-D3 iso-2 TFA		2180*	413 $_{100}$	398 $_{74}$	370 $_{71}$	339 $_{78}$	300 $_{73}$	1049	5666
Dronabinol-M (11-HO-)		2775*	330 $_{12}$	299 $_{100}$	231 $_{10}$	217 $_{9}$	193 $_{7}$	760	4661
Dronabinol-M (11-HO-) 2ME		2580*	358 $_{13}$	313 $_{100}$	257 $_{3}$	231 $_{3}$		879	4659
Dronabinol-M (11-HO-) 2PFP		2350*	622 $_{15}$	607 $_{5}$	551 $_{9}$	458 $_{100}$	415 $_{24}$	1197	4658
Dronabinol-M (11-HO-) 2TFA		2450*	522 $_{13}$	451 $_{8}$	408 $_{100}$	395 $_{13}$	365 $_{24}$	1175	4657
Dronabinol-M (11-HO-) 2TMS		2630*	474 $_{5}$	459 $_{4}$	403 $_{2}$	371 $_{100}$	73 $_{14}$	1143	4656
Dronabinol-M (11-HO-) -H2O AC		2740*	354 $_{48}$	312 $_{100}$	297 $_{19}$	269 $_{31}$	91 $_{21}$	863	4660
Dronabinol-M (HO-nor-delta-9-HOOC-) 2ME	UTHCME-I	2840*	388 $_{42}$	373 $_{65}$	329 $_{100}$	201 $_{24}$	189 $_{28}$	981	3466
Dronabinol-M (nor-delta-9-HOOC-) 2ME	UTHCME UGLUCEXME	2620*	372 $_{52}$	357 $_{79}$	341 $_{9}$	313 $_{100}$	245 $_{6}$	929	1439
Dronabinol-M (nor-delta-9-HOOC-) 2PFP		2440*	622 $_{35}$	607 $_{49}$	459 $_{100}$	445 $_{76}$	69 $_{82}$	1197	4380
Dronabinol-M (nor-delta-9-HOOC-) 2TMS		2470*	488 $_{40}$	473 $_{5}$	398 $_{12}$	371 $_{100}$	73 $_{66}$	1156	5671
Dronabinol-M (nor-delta-9-HOOC-)-D3 2ME		2590*	375 $_{39}$	360 $_{68}$	356 $_{19}$	316 $_{100}$	301 $_{9}$	938	6187
Dronabinol-M (nor-delta-9-HOOC-)-D3 2PFP		2425*	625 $_{40}$	610 $_{57}$	462 $_{100}$	448 $_{62}$	432 $_{43}$	1197	6039
Dronabinol-M (nor-delta-9-HOOC-)-D3 2TMS		2660*	491 $_{44}$	476 $_{55}$	374 $_{100}$	300 $_{15}$	73 $_{28}$	1159	5672
Dronabinol-M (oxo-nor-delta-9-HOOC-) 2ME	UTHCME-I	2860*	386 $_{55}$	371 $_{73}$	327 $_{100}$	314 $_{22}$	189 $_{11}$	974	3467
Droperidol		9999	379 $_{7}$	246 $_{100}$	165 $_{41}$	134 $_{57}$	123 $_{97}$	949	1495
Droperidol ME		3370	393 $_{6}$	246 $_{83}$	165 $_{81}$	123 $_{100}$		995	490
Droperidol 2TMS		3485	523 $_{6}$	300 $_{73}$	271 $_{45}$	255 $_{29}$	73 $_{100}$	1176	4542
Droperidol-M	U+UHYAC	1490*	180 $_{25}$	125 $_{49}$	123 $_{35}$	95 $_{17}$	56 $_{100}$	174	85
Droperidol-M (benzimidazolone)	UHY-I	1950	134 $_{100}$	106 $_{36}$	79 $_{62}$	67 $_{20}$		114	491
Droperidol-M (benzimidazolone) 2AC	U+UHYAC-I	1730	218 $_{11}$	176 $_{19}$	134 $_{98}$	106 $_{11}$		263	171
Dropropizine		2205	236 $_{14}$	175 $_{100}$	132 $_{41}$	104 $_{46}$	70 $_{83}$	331	2775
Dropropizine AC		2390	278 $_{14}$	175 $_{100}$	132 $_{35}$	104 $_{33}$	70 $_{53}$	508	2776
Dropropizine 2AC		2330	320 $_{6}$	260 $_{6}$	175 $_{100}$	132 $_{44}$	70 $_{51}$	716	2777
Dropropizine-M (HO-) 3AC		2675	378 $_{17}$	259 $_{17}$	233 $_{100}$	216 $_{20}$	191 $_{35}$	946	6805
Dropropizine-M (HO-phenylpiperazine) 2AC	U+UHYAC	2350	262 $_{39}$	220 $_{60}$	177 $_{21}$	148 $_{100}$	135 $_{59}$	430	6610
Dropropizine-M (phenylpiperazine) AC	U+UHYAC	1920	204 $_{48}$	161 $_{21}$	132 $_{99}$	56 $_{77}$		227	1276
Drostanolone		2555*	304 $_{49}$	245 $_{58}$	177 $_{21}$	95 $_{57}$	55 $_{100}$	637	2773
Drostanolone AC		2700*	346 $_{26}$	286 $_{48}$	271 $_{48}$	149 $_{64}$	55 $_{100}$	833	2774
Drostanolone TMS		2575*	376 $_{8}$	361 $_{18}$	286 $_{26}$	129 $_{81}$	73 $_{100}$	941	3956
Drostanolone enol 2TMS		2625*	448 $_{100}$	405 $_{14}$	157 $_{12}$	141 $_{25}$	73 $_{74}$	1111	3957
Drostanolone propionate		2985*	360 $_{3}$	286 $_{13}$	271 $_{12}$	149 $_{16}$	57 $_{100}$	885	2761
Duloxetine		2500	297 $_{100}$	265 $_{37}$	239 $_{30}$	221 $_{41}$	181 $_{26}$	602	7461
Duloxetine ME		2490	311 $_{14}$	237 $_{13}$	209 $_{11}$	141 $_{14}$	58 $_{100}$	671	7462
Duloxetine 2AC		3160	381 $_{62}$	266 $_{100}$	239 $_{74}$	221 $_{45}$	87 $_{75}$	956	7464
Duloxetine 2HFB		2435	689 $_{3}$	476 $_{13}$	435 $_{50}$	249 $_{100}$	237 $_{94}$	1202	7477
Duloxetine 2PFP		2425	589 $_{25}$	385 $_{72}$	249 $_{89}$	237 $_{91}$	119 $_{100}$	1193	7470

Duloxetine iosmer-1 TFA Table 1-8-1: Compounds in order of names

Name	Detected	RI	Typical ions and intensities					Page	Entry
Duloxetine iosmer-1 TFA		2690	393 $_{71}$	266 $_{17}$	239 $_{100}$	140 $_{50}$	69 $_{51}$	993	7473
Duloxetine iosmer-2 TFA		2700	393 $_{45}$	265 $_{7}$	239 $_{100}$	221 $_{41}$	69 $_{30}$	993	7467
Duloxetine iso-1 AC		3050	339 $_{68}$	266 $_{52}$	237 $_{100}$	182 $_{27}$	87 $_{62}$	802	7463
Duloxetine iso-1 HFB		2650	493 $_{46}$	266 $_{14}$	239 $_{100}$	182 $_{44}$	169 $_{49}$	1159	7478
Duloxetine iso-1 PFP		2300	443 $_{44}$	239 $_{100}$	190 $_{49}$	182 $_{43}$	119 $_{55}$	1105	7471
Duloxetine iso-1 TMS		2510	369 $_{46}$	338 $_{11}$	311 $_{24}$	249 $_{33}$	73 $_{100}$	916	7480
Duloxetine iso-1 2TMS		2545	441 $_{19}$	369 $_{14}$	338 $_{9}$	116 $_{79}$	73 $_{100}$	1101	7482
Duloxetine iso-2 AC		3150	339 $_{57}$	266 $_{80}$	239 $_{100}$	221 $_{82}$	87 $_{93}$	802	7474
Duloxetine iso-2 HFB		2725	493 $_{39}$	266 $_{6}$	239 $_{100}$	169 $_{33}$	221 $_{34}$	1159	7479
Duloxetine iso-2 PFP		2700	443 $_{44}$	239 $_{100}$	190 $_{49}$	182 $_{43}$	119 $_{55}$	1105	7472
Duloxetine iso-2 TMS		2550	369 $_{56}$	337 $_{26}$	311 $_{28}$	249 $_{19}$	73 $_{100}$	916	7481
Duloxetine iso-2 2TMS		2620	441 $_{19}$	369 $_{4}$	337 $_{10}$	116 $_{63}$	73 $_{100}$	1101	7483
Duloxetine-M (1-naphthol)		1500*	144 $_{75}$	115 $_{100}$	89 $_{17}$	74 $_{7}$	63 $_{21}$	123	928
Duloxetine-M (1-naphthol) AC	U+UHYAC	1555*	186 $_{13}$	144 $_{100}$	115 $_{47}$	89 $_{6}$	63 $_{7}$	188	932
Duloxetine-M (1-naphthol) HFB		1310*	340 $_{46}$	169 $_{25}$	143 $_{28}$	115 $_{100}$	89 $_{11}$	806	7476
Duloxetine-M (1-naphthol) PFP		1510*	290 $_{45}$	171 $_{100}$	143 $_{20}$	115 $_{49}$	89 $_{8}$	562	7468
Duloxetine-M (1-naphthol) TMS		1525*	216 $_{100}$	201 $_{95}$	185 $_{51}$	115 $_{39}$	73 $_{21}$	260	7460
Duloxetine-M (4-HO-1-naphthol) 2AC	U+UHYAC	1900*	244 $_{14}$	202 $_{19}$	160 $_{100}$	131 $_{21}$	103 $_{8}$	355	933
Duloxetine-M/artifact -H2O AC		1760	195 $_{20}$	152 $_{27}$	138 $_{52}$	123 $_{27}$	98 $_{100}$	207	7465
Duloxetine-M/artifact -H2O HFB		1560	349 $_{23}$	252 $_{62}$	180 $_{56}$	123 $_{100}$	69 $_{73}$	841	7475
Duloxetine-M/artifact -H2O PFP		1535	299 $_{47}$	202 $_{65}$	180 $_{72}$	123 $_{100}$	119 $_{79}$	609	7469
Duloxetine-M/artifact -H2O TFA		1545	249 $_{43}$	180 $_{25}$	152 $_{60}$	123 $_{63}$	69 $_{100}$	373	7466
2,3-EBDB		1670	192 $_{5}$	135 $_{5}$	86 $_{100}$	77 $_{6}$	58 $_{11}$	278	5417
2,3-EBDB AC		2000	263 $_{3}$	192 $_{4}$	176 $_{18}$	128 $_{42}$	86 $_{100}$	435	5511
2,3-EBDB HFB		1790	417 $_{1}$	282 $_{100}$	176 $_{25}$	135 $_{24}$	77 $_{17}$	1056	5594
2,3-EBDB PFP		1755	367 $_{3}$	232 $_{100}$	176 $_{16}$	119 $_{30}$	69 $_{10}$	908	5595
2,3-EBDB TFA		1780	317 $_{5}$	182 $_{100}$	176 $_{37}$	154 $_{16}$	135 $_{14}$	697	5512
2,3-EBDB TMS		1825	278 $_{2}$	264 $_{5}$	158 $_{100}$	135 $_{16}$	73 $_{48}$	583	5596
ECC -HCN		<1000	125 $_{4}$	110 $_{12}$	96 $_{44}$	82 $_{33}$	56 $_{97}$	110	3598
ECC -HCN		<1000	125 $_{4}$	110 $_{12}$	96 $_{44}$	56 $_{97}$	41 $_{100}$	109	5535
Ecgonidine ME	UHY-I U+UHYAC-I	1280	181 $_{31}$	152 $_{100}$	138 $_{9}$	122 $_{15}$	82 $_{18}$	178	3574
Ecgonidine TMS		1345	239 $_{31}$	224 $_{25}$	210 $_{100}$	183 $_{10}$	122 $_{28}$	341	6256
Ecgonidine TBDMS		1520	281 $_{16}$	252 $_{29}$	224 $_{100}$	150 $_{44}$	122 $_{30}$	524	6242
Ecgonine ACTMS		1680	299 $_{11}$	240 $_{41}$	122 $_{9}$	94 $_{29}$	82 $_{100}$	611	6238
Ecgonine TMSTFA		1395	353 $_{15}$	267 $_{35}$	240 $_{82}$	94 $_{38}$	82 $_{100}$	858	6255
Ecgonine 2TMS		1680	329 $_{2}$	314 $_{4}$	96 $_{45}$	82 $_{100}$	73 $_{49}$	755	5445
Ecgonine-D3 2TMS		1670	332 $_{13}$	317 $_{15}$	99 $_{64}$	85 $_{100}$	73 $_{56}$	770	5576
Econazole	U	3550	380 $_{3}$	299 $_{9}$	206 $_{4}$	125 $_{100}$	81 $_{27}$	951	2550
EDDP	U UHY U+UHYAC	2040	277 $_{105}$	276 $_{100}$	262 $_{43}$	220 $_{23}$	165 $_{10}$	503	242
EDDP-M (HO-) AC	U+UHYAC	2350	335 $_{100}$	304 $_{27}$	292 $_{26}$	276 $_{16}$	234 $_{18}$	785	5297
EDDP-M (nor-) AC	U+UHYAC	2220	305 $_{100}$	290 $_{54}$	262 $_{29}$	236 $_{29}$	220 $_{25}$	640	5292
EDDP-M (nor-HO-) 2AC	U+UHYAC	2645	363 $_{100}$	348 $_{43}$	320 $_{63}$	278 $_{22}$	149 $_{15}$	895	5299
EDTA 3ME1ET		2125*	362 $_{14}$	303 $_{28}$	289 $_{18}$	188 $_{100}$	174 $_{64}$	891	6452
EDTA 4ME		2105*	348 $_{8}$	289 $_{25}$	188 $_{23}$	174 $_{100}$	146 $_{40}$	839	6451
Eicosane		2000*	282 $_{6}$	99 $_{15}$	85 $_{38}$	71 $_{65}$	57 $_{100}$	527	2352
Eicosanoic acid ME		2275*	326 $_{14}$	283 $_{6}$	143 $_{14}$	87 $_{61}$	74 $_{100}$	742	3035
Elemicin		1435*	208 $_{100}$	193 $_{60}$	133 $_{25}$	118 $_{22}$	77 $_{32}$	238	7136
Elemicin-M (1-HO-) AC	U+UHYAC	2035*	266 $_{100}$	223 $_{9}$	207 $_{27}$	195 $_{21}$	176 $_{40}$	450	7142
Elemicin-M (1-HO-) ME	UME	2085*	238 $_{100}$	223 $_{50}$	207 $_{15}$	195 $_{13}$	163 $_{29}$	337	7151
Elemicin-M (bisdemethyl-) 2AC	U+UHYAC	1880*	264 $_{1}$	222 $_{30}$	180 $_{100}$	147 $_{15}$	91 $_{34}$	439	7148
Elemicin-M (demethyl-) iso-1 AC	U+UHYAC	1755*	236 $_{35}$	194 $_{100}$	179 $_{43}$	119 $_{15}$	91 $_{12}$	328	7140
Elemicin-M (demethyl-) iso-2 AC	U+UHYAC	1790*	236 $_{9}$	194 $_{100}$	179 $_{9}$	133 $_{14}$	119 $_{7}$	328	7141
Elemicin-M (demethyl-dihydroxy-) iso-1 3AC	U+UHYAC	2275*	354 $_{2}$	294 $_{27}$	280 $_{25}$	252 $_{100}$	210 $_{30}$	861	7138
Elemicin-M (demethyl-dihydroxy-) iso-2 3AC	U+UHYAC	2300*	354 $_{5}$	312 $_{20}$	252 $_{100}$	210 $_{8}$	167 $_{18}$	861	7139
Elemicin-M (dihydroxy-) 2AC	U+UHYAC	2195*	326 $_{28}$	266 $_{100}$	223 $_{28}$	207 $_{23}$	181 $_{59}$	740	7137
Eletriptan		3650	380 $_{1}$	156 $_{5}$	129 $_{1}$	84 $_{100}$	82 $_{4}$	959	7491
Eletriptan HFB		3370	576 $_{1}$	352 $_{1}$	156 $_{2}$	129 $_{1}$	84 $_{100}$	1191	7494
Eletriptan TFA		3650	476 $_{1}$	252 $_{2}$	156 $_{3}$	129 $_{1}$	84 $_{100}$	1146	7493
Eletriptan TMS		3580	452 $_{1}$	371 $_{1}$	228 $_{1}$	156 $_{3}$	84 $_{100}$	1120	7492
Emetine	G	4055	480 $_{12}$	288 $_{21}$	272 $_{33}$	206 $_{32}$	192 $_{100}$	1148	5611
Emetine ET		3320	508 $_{2}$	302 $_{3}$	206 $_{100}$	190 $_{7}$	150 $_{6}$	1169	5614
Emetine ME		4010	494 $_{2}$	288 $_{4}$	272 $_{3}$	206 $_{100}$	190 $_{9}$	1160	5612
Emtricitabine		2555	229 $_{3}$	190 $_{24}$	130 $_{55}$	100 $_{47}$	87 $_{100}$	365	7485
Emtricitabine 2AC		2580	188 $_{5}$	154 $_{12}$	130 $_{18}$	100 $_{100}$	87 $_{26}$	762	7486
Emtricitabine 2TFA		2350	250 $_{3}$	200 $_{7}$	182 $_{26}$	154 $_{69}$	100 $_{100}$	1095	7487
Emtricitabine 2TMS		2455	376 $_{4}$	190 $_{40}$	100 $_{65}$	87 $_{95}$	73 $_{100}$	988	7484
Enalapril ET		2715	404 $_{1}$	331 $_{11}$	234 $_{100}$	160 $_{17}$	91 $_{31}$	1029	4738
Enalapril ME	PME UME	2675	390 $_{1}$	317 $_{12}$	234 $_{100}$	91 $_{50}$	70 $_{41}$	986	3200
Enalapril METMS		2800	462 $_{1}$	447 $_{2}$	375 $_{10}$	248 $_{100}$	91 $_{18}$	1130	4984
Enalapril TMS		2740	448 $_{1}$	433 $_{3}$	375 $_{20}$	234 $_{100}$	73 $_{14}$	1111	4608
Enalapril 2ET		2745	432 $_{1}$	359 $_{13}$	262 $_{100}$	188 $_{9}$	91 $_{13}$	1085	4739
Enalapril 2ME	UME	2690	404 $_{1}$	331 $_{11}$	248 $_{100}$	174 $_{14}$	91 $_{36}$	1029	3201
Enalapril 2TMS		2790	520 $_{1}$	505 $_{20}$	447 $_{14}$	306 $_{100}$	73 $_{8}$	1175	4979
Enalapril -H2O	G UHY U+UHYAC UME	2770	358 $_{22}$	254 $_{90}$	208 $_{100}$	160 $_{25}$	91 $_{59}$	878	3199

Table 1-8-1: Compounds in order of names Ephedrine TMSTFA

Name	Detected	RI	Typical ions and intensities					Page	Entry
Enalapril-M/artifact (deethyl-) METMS		2730	434_1	419_5	375_{10}	220_{100}	91_{23}	1089	4609
Enalapril-M/artifact (deethyl-) 2ET		2715	404_1	331_{12}	234_{100}	160_{17}	91_{31}	1029	4738
Enalapril-M/artifact (deethyl-) 2ME	UME	2620	376_1	317_8	220_{100}	116_{21}	91_{48}	940	3198
Enalapril-M/artifact (deethyl-) 2TMS	UME	2780	492_1	477_6	375_{39}	278_{100}	234_{23}	1159	4978
Enalapril-M/artifact (deethyl-) 3ET		2745	432_1	359_{13}	262_{100}	188_9	91_{13}	1085	4739
Enalapril-M/artifact (deethyl-) 3ME	UME	2680	390_1	331_{11}	234_{100}	174_{13}	130_{24}	986	4733
Enalapril-M/artifact (deethyl-) -H2O ME	UME	2735	344_{17}	240_{62}	208_{71}	91_{60}	70_{50}	825	3202
Enalapril-M/artifact (deethyl-HOOC-) 2ME	UME	1870	279_2	220_{100}	160_{10}	117_{28}	91_{57}	512	4734
Enalapril-M/artifact (deethyl-HOOC-) 3ET	UET	2095	335_1	262_{100}	234_7	188_7	91_{27}	785	4741
Enalapril-M/artifact (deethyl-HOOC-) 3ME	UME	1935	293_2	234_{100}	174_8	130_{16}	91_{50}	582	4735
Enalapril-M/artifact (HOOC-) ET	UET	2025	307_2	234_{100}	160_{12}	117_{17}	91_{32}	651	4740
Enalapril-M/artifact (HOOC-) ME	UME	1930	293_3	234_{36}	220_{100}	160_{11}	91_{23}	582	4736
Enalapril-M/artifact (HOOC-) 2ET	UET	2095	335_1	262_{100}	234_7	188_7	91_{27}	785	4741
Enalapril-M/artifact (HOOC-) 2ME	UME	1985	307_2	248_{34}	234_{100}	174_5	91_6	651	4737
Enalaprilate METMS		2730	434_1	419_5	375_{10}	220_{100}	91_{23}	1089	4609
Enalaprilate 2ET		2715	404_1	331_{12}	234_{100}	160_{17}	91_{31}	1029	4738
Enalaprilate 2ME	UME	2620	376_1	317_8	220_{100}	116_{21}	91_{48}	940	3198
Enalaprilate 2TMS	UME	2780	492_1	477_6	375_{39}	278_{100}	234_{23}	1159	4978
Enalaprilate 3ET		2745	432_1	359_{13}	262_{100}	188_9	91_{13}	1085	4739
Enalaprilate 3ME	UME	2680	390_1	331_{11}	234_{100}	174_{13}	130_{24}	986	4733
Enalaprilate -H2O ME	UME	2735	344_{17}	240_{62}	208_{71}	91_{60}	70_{50}	825	3202
Enalaprilate-M/artifact (HOOC-) 2ET	UET	2025	307_2	234_{100}	160_{12}	117_{17}	91_{32}	651	4740
Enalaprilate-M/artifact (HOOC-) 2ME	UME	1870	279_2	220_{100}	160_{10}	117_{28}	91_{57}	512	4734
Enalaprilate-M/artifact (HOOC-) 3ET	UET	2095	335_1	262_{100}	234_7	188_7	91_{27}	785	4741
Enalaprilate-M/artifact (HOOC-) 3ME	UME	1935	293_2	234_{100}	174_8	130_{16}	91_{50}	582	4735
Endogenous biomolecule	U+UHYAC	1550	200_3	97_{16}	86_{100}	71_{28}	55_{50}	218	492
Endogenous biomolecule	UHY	2520*	288_{65}	255_{100}	197_{42}	134_{57}	91_{74}	550	715
Endogenous biomolecule	UHY U+UHYAC	1790*	192_6	153_{13}	137_{100}	121_{78}	82_{95}	197	1947
Endogenous biomolecule	UHY U+UHYAC	2545*	310_{51}	267_8	197_{100}	153_9		664	2368
Endogenous biomolecule	UME	2050*	278_{74}	246_{17}	203_{16}	151_{100}	150_{91}	504	4954
Endogenous biomolecule	UME	1750*	268_{15}	208_{91}	195_{40}	179_{100}	165_{10}	458	4952
Endogenous biomolecule	UME	1640*	179_{11}	171_{88}	130_{20}	115_{100}	99_{36}	169	4951
Endogenous biomolecule	UME	2140*	294_{27}	238_{56}	235_{47}	206_{100}	195_{18}	584	4958
Endogenous biomolecule	UME	2510*	320_{16}	265_{20}	239_{100}	210_{17}	83_{29}	713	4957
Endogenous biomolecule (ME)	UME	2100	252_{32}	192_{100}	179_{39}	147_{14}	84_{17}	388	5040
Endogenous biomolecule AC	U+UHYAC	2620*	370_{30}	310_{100}	254_{12}	239_{14}	161_{29}	918	43
Endogenous biomolecule AC	U+UHYAC	2400	315_{100}	255_{77}	214_4	161_7	147_{13}	687	622
Endogenous biomolecule AC	U+UHYAC	2575*	330_{62}	270_{14}	255_{100}	197_{42}	117_{45}	757	984
Endogenous biomolecule AC	U+UHYAC	1350*	155_{49}	140_{10}	112_{36}	69_{11}		134	2367
Endogenous biomolecule AC	U+UHYAC	3040*	474_{16}	414_{10}	294_{55}	269_{100}	173_{55}	1142	2454
Endogenous biomolecule AC	U+UHYAC	2240*	264_1	222_{30}	137_{100}	122_6	85_{23}	438	2452
Endogenous biomolecule AC	U+UHYAC	1640*	208_6	166_{49}	151_{100}	123_{12}		236	2483
Endogenous biomolecule ME	UME	2235*	306_{52}	259_{10}	217_9	179_{100}	91_{21}	643	4956
Endogenous biomolecule ME	UME	2160*	292_{31}	203_8	165_{100}	121_{10}	91_{24}	573	4955
Endogenous biomolecule ME	UME	1945*	296_{12}	236_{85}	223_{33}	179_{100}	147_{21}	595	4953
Endogenous biomolecule 2AC	U+UHYAC	2910*	400_{14}	340_{100}	265_{33}	172_{20}	157_{29}	1016	985
Endogenous biomolecule 2AC	U+UHYAC	2280*	292_3	250_{19}	208_{100}	123_{39}	85_{13}	573	1002
Endogenous biomolecule 2AC	U+UHYAC	1920	247_8	205_{27}	163_{100}	135_{25}		365	1135
Endogenous biomolecule 2AC	U+UHYAC	2000	263_{29}	221_{100}	177_{57}	162_{45}	133_{11}	433	1508
Endogenous biomolecule 2AC	U+UHYAC	1875	235_{10}	193_{56}	151_{85}	136_{44}	108_{16}	322	1566
Endogenous biomolecule 2AC	U+UHYAC	2650*	394_3	352_9	310_{100}	197_{43}		998	2369
Endogenous biomolecule 2AC	U+UHYAC	1800*	296_5	236_9	193_{100}	149_{46}	135_{77}	595	2482
Endogenous biomolecule 2AC	U+UHYAC	1695	193_1	151_8	133_{19}	109_{100}	80_{14}	199	3212
Endogenous biomolecule 2AC	U+UHYAC	2800	361_{22}	319_{12}	277_{100}	127_{91}	84_{33}	886	3744
Endogenous biomolecule 2AC	U+UHYAC	1820*	266_3	206_9	164_{100}	150_{10}	137_{30}	450	6409
Endogenous biomolecule 3AC	U+UHYAC	2060*	276_7	234_{15}	192_{32}	150_{100}	122_5	494	493
Endogenous biomolecule 3AC	U+UHYAC	1950	305_{23}	263_{65}	221_{100}	179_{41}	161_{66}	637	1481
Endogenous biomolecule 3AC	U+UHYAC	1760	239_2	197_7	155_{58}	140_{22}	113_{100}	338	2453
Endogenous biomolecule 3AC	U+UHYAC	1710	235_1	193_6	151_{34}	109_{100}	80_{11}	322	3213
Endogenous biomolecule -H2O AC	U+UHYAC	2830*	340_{100}	265_{37}	172_{22}	157_{33}	145_{17}	806	802
Endogenous biomolecule iso-1 AC	U+UHYAC	2750	340_1	265_6	144_{31}	102_{23}	84_{100}	806	3664
Endogenous biomolecule iso-1 2AC	U+UHYAC	2700*	326_{31}	284_{90}	242_{64}	123_{100}	120_{70}	737	2428
Endogenous biomolecule iso-2 AC	U+UHYAC	2825	352_2	144_{52}	102_{35}	84_{100}	60_{31}	851	3665
Endogenous biomolecule iso-2 2AC	U+UHYAC	2750*	326_{29}	284_{65}	242_{36}	123_{55}	120_{100}	737	2429
Endosulfan		2080*	339_{17}	265_{51}	237_{84}	195_{100}	159_{77}	1027	3834
Endosulfan sulfate		2260*	420_3	387_{33}	272_{100}	227_{76}	85_{48}	1062	3835
Endothal		1370*	140_9	100_{29}	81_{13}	68_{100}	53_{18}	188	4154
Endrin		2175*	345_6	281_{24}	263_{53}	113_{27}	81_{100}	944	3836
Enilconazole		2140	296_{10}	240_{12}	215_{100}	173_{74}	81_{32}	596	2054
Enoximone	U+UHYAC	2770	248_{100}	247_{92}	201_{43}	151_{72}	124_{22}	370	5212
Enoximone AC	U+UHYAC	2600	290_{13}	248_{100}	201_{43}	151_{44}	108_{18}	563	5211
Enoximone 2AC		2560	332_8	290_{25}	248_{100}	201_{70}	151_{60}	768	5210
Ephedrine	G UHY	1375	146_1	131_1	105_3	77_{12}	58_{100}	148	748
Ephedrine TMSTFA		1620	318_1	227_9	179_{100}	110_8	73_{79}	773	6038

Ephedrine 2AC Table 1-8-1: Compounds in order of names

Name	Detected	RI	Typical ions and intensities					Page	Entry
Ephedrine 2AC	PAC U+UHYAC	1795	249_1	100_{57}	58_{100}	148_2	117_2	375	749
Ephedrine 2HFB	UHYHFB	1500	344_{10}	254_{100}	210_{27}	169_{18}	69_{36}	1186	5097
Ephedrine 2PFP		1370	338_1	294_3	204_{100}	160_{25}	119_{14}	1123	2577
Ephedrine 2TFA		1345	338_1	244_4	154_{100}	110_{72}	69_{47}	872	3997
Ephedrine 2TMS		1620	294_3	163_4	147_8	130_{100}	73_{84}	663	4543
Ephedrine formyl artifact	G U	1430	177_1	121_{10}	107_6	71_{100}	56_{76}	165	4500
Ephedrine -H2O AC	U+UHYAC	1560	189_1	148_3	121_6	100_{49}	58_{100}	192	5646
Ephedrine-M (HO-)		1875	148_1	95_1	77_4	71_6	58_{100}	178	1971
Ephedrine-M (HO-) ME2AC		2000	279_1	247_1	206_3	100_{60}	58_{100}	513	2348
Ephedrine-M (HO-) 3AC	U+UHYAC	2145	307_1	247_1	205_1	100_{72}	58_{100}	649	750
Ephedrine-M (HO-) formyl artifact		1790	133_2	121_{10}	107_6	71_{100}	56_{76}	201	4499
Ephedrine-M (HO-) -H2O 2AC	U+UHYAC	1990	247_9	205_8	163_{24}	107_8	56_{100}	367	1972
Ephedrine-M (nor-)	P U	1370	132_2	118_8	107_{11}	91_{10}	77_{100}	130	2475
Ephedrine-M (nor-) TMSTFA		1890	240_8	198_3	179_{100}	117_5	73_{88}	708	6146
Ephedrine-M (nor-) 2AC	U+UHYAC	1805	235_1	107_{13}	86_{100}	176_5	134_7	325	2476
Ephedrine-M (nor-) 2HFB	UHYHFB	1455	543_1	330_{14}	240_{100}	169_{44}	69_{57}	1183	5098
Ephedrine-M (nor-) 2PFP	UHYPFP	1380	443_1	280_9	190_{100}	119_{59}	105_{26}	1104	5094
Ephedrine-M (nor-) 2TFA	UTFA	1355	343_1	230_6	203_5	140_{100}	69_{29}	819	5091
Ephedrine-M (nor-) 2TMS		1555	280_5	163_4	147_{10}	116_{100}	73_{83}	594	4574
Ephedrine-M (nor-) formyl artifact		1240	117_2	105_4	91_4	77_{10}	57_{100}	143	4650
Ephedrine-M (nor-HO-) 3AC	U+UHYAC	2135	234_6	165_8	123_{11}	86_{100}	58_{45}	580	4961
Epiandrosterone		2520*	290_{100}	246_{28}	147_{25}	107_{71}	67_{57}	567	3898
Epiandrosterone AC		2630*	332_{19}	272_{100}	218_{41}	201_{57}	107_{80}	771	3919
Epiandrosterone TMS		2500*	362_{11}	347_{60}	272_{11}	155_{11}	75_{100}	893	3959
Epiandrosterone enol 2TMS		2570*	434_{78}	419_{100}	329_9	239_3	73_{50}	1090	3960
Epinastine		2430	249_{44}	194_{100}	178_{34}	165_{30}	116_{22}	375	7262
Epinastine AC		2600	291_{36}	276_{48}	248_{43}	194_{100}	178_{42}	569	7264
Epinastine HFB		2530	445_7	276_{100}	248_7	178_{27}	165_{28}	1107	7267
Epinastine ME		2380	263_{74}	262_{100}	194_{46}	178_{44}	165_{51}	435	7263
Epinastine PFP		2520	395_{31}	276_{100}	249_{27}	194_{33}	165_{34}	1001	7266
Epinastine TFA		2580	276_{100}	178_{28}	165_{29}	69_{72}	345_{32}	828	7265
Epinastine TMS		2450	321_{13}	249_{28}	194_{100}	178_{29}	165_{23}	719	7268
Epinastine 2TMS		2470	393_{100}	378_{94}	279_{74}	171_{88}	73_{97}	996	7269
Epinephrine artifact (3,4-dihydroxybenzoic acid) ME2AC		1750*	252_1	210_{19}	168_{100}	137_{60}	109_8	388	5254
Epitestosterone enol 2TMS		2620*	432_{91}	417_8	327_8	209_{16}	73_{100}	1085	3802
Eplerenone		3455*	414_{11}	399_5	355_{51}	194_{23}	55_{100}	1050	7270
Eplerenone HFB		3015*	610_1	551_5	533_2	169_{60}	69_{100}	1195	7273
Eplerenone PFP		2985*	560_1	501_6	483_2	119_{100}	55_{52}	1187	7272
Eplerenone TFA		2995*	510_1	451_{15}	433_6	111_{44}	69_{100}	1170	7271
Eplerenone TMS		3430*	486_4	427_{21}	291_3	111_{17}	73_{100}	1154	7274
Eprazinone		2820	380_1	245_{100}	139_9	111_{30}	105_{60}	954	1938
Eprosartan 2ME		3335	452_{100}	410_{48}	351_{43}	149_{59}	121_{77}	1116	7592
Eprosartan 2TMS		3480	568_{42}	451_{23}	409_{40}	271_{38}	73_{80}	1189	7593
EPTC		1350	189_{14}	160_5	132_{25}	128_{78}	86_{100}	192	3188
Ergometrine		----	325_{20}	307_{40}	221_{100}	196_{82}		736	751
Ergost-3,5-ene		3270*	382_{100}	213_{22}	147_{78}	105_{62}	55_{56}	961	5624
Ergost-5-en-3-ol		3190*	400_{63}	315_{33}	289_{41}	105_{84}	55_{100}	1019	5620
Ergost-5-en-3-ol -H2O		3270*	382_{100}	213_{22}	147_{78}	105_{62}	55_{56}	961	5624
Ergosta-3,5,22-triene		3210*	380_{100}	255_{39}	81_{53}	69_{53}	55_{67}	954	5623
Ergosta-5,22-dien-3-ol		3135*	398_{100}	300_{62}	271_{71}	255_{85}	69_{96}	1013	5619
Ergosta-5,22-dien-3-ol -H2O		3210*	380_{100}	255_{39}	81_{53}	69_{53}	55_{67}	954	5623
Ergosterol	G	3130*	396_{65}	363_{100}	337_{34}	253_{30}	143_{36}	1008	5137
Ergotamine artifact-1	U+UHYAC P-I	2375	244_{36}	153_{50}	125_{100}	91_{38}	70_{50}	356	4495
Ergotamine artifact-1 AC		2360	286_{29}	153_{100}	125_{100}	91_{57}	70_{73}	544	5217
Ergotamine artifact-2		2440	314_{45}	244_{35}	153_{84}	125_{75}	70_{100}	683	4494
Erucic acid ME		2490*	352_5	320_{45}	97_{44}	69_{58}	55_{100}	856	2670
Erythritol 4AC		1595*	217_{11}	145_{94}	128_{35}	115_{100}	103_{88}	563	5605
Esmolol		2225	294_1	251_2	116_5	107_5	72_{100}	594	6266
Esmolol TMSTFA		2130	463_1	448_2	284_{100}	270_{10}	131_{19}	1131	6270
Esmolol 2AC		2400	291_2	200_{100}	140_{26}	98_{40}	72_{82}	949	5136
Esmolol 2HFB		2005	687_1	508_{56}	466_{100}	252_{39}	226_{26}	1202	6269
Esmolol 2PFP		2115	587_1	408_{28}	366_{100}	202_{40}	176_{35}	1193	6268
Esmolol 2TFA		1990	487_1	308_{72}	266_{100}	152_{19}	126_9	1155	6271
Esmolol formyl artifact	G	2290	307_{11}	292_{37}	127_{62}	112_{48}	56_{70}	651	5135
Esmolol -H2O AC		2575	319_3	200_{100}	140_{19}	72_{48}	98_{18}	709	6267
Estazolam		3070	294_{82}	259_{100}	239_{43}	205_{61}	101_{54}	584	2392
Estradiol		2550*	272_{99}	213_{29}	172_{24}	160_{32}	146_{24}	479	1434
Estradiol 2AC	U+UHYAC	2780*	356_{12}	314_{100}	172_{19}	146_{19}		870	1435
Estradiol undecylate		3900*	440_{60}	255_{19}	159_{77}	133_{52}	57_{100}	1099	5244
Estriol		2940*	288_{99}	213_{25}	172_{21}	160_{32}	146_{25}	554	1436
Estriol 3AC	U+UHYAC	3010*	414_{14}	372_{100}	330_4	270_8	160_{30}	1051	1476
Estriol-M (HO-) 4AC	U+UHYAC	3280*	472_8	430_{100}	268_{69}	250_{20}	107_{11}	1140	4290
Estrone		2580*	270_{100}	213_{19}	185_{40}	172_{36}	146_{55}	471	5178
Estrone AC	U+UHYAC	2630*	312_5	270_{100}	213_{19}	185_{40}	146_{39}	676	5207

Table 1-8-1: Compounds in order of names   Ethylamine

Name	Detected	RI	Typical ions and intensities					Page	Entry
Estrone ME		2530*	284_{100}	227_{18}	199_{82}	160_{72}	115_{40}	536	5206
Etacrinic acid ET		2230*	330_{27}	315_{37}	263_{67}	261_{100}	203_{6}	758	2631
Etacrinic acid ME		2195*	316_{11}	281_{5}	263_{66}	261_{100}	243_{40}	691	2630
Etafenone		2680	325_{1}	310_{1}	99_{20}	86_{100}	58_{15}	736	2503
Etafenone-M (di-HO-) 2AC	U+UHYAC	3070	441_{1}	426_{5}	99_{9}	86_{100}	58_{6}	1101	3358
Etafenone-M (HO-) iso-1	UHY	2800	341_{1}	326_{1}	99_{10}	86_{100}	58_{7}	814	3347
Etafenone-M (HO-) iso-1 AC	U+UHYAC	2775	383_{1}	368_{2}	99_{16}	86_{100}	58_{10}	964	3355
Etafenone-M (HO-) iso-2	UHY	2820	341_{2}	326_{1}	99_{15}	86_{100}	58_{7}	814	3348
Etafenone-M (HO-) iso-2 AC	U+UHYAC	2810	383_{1}	368_{2}	99_{15}	86_{100}	58_{11}	965	3356
Etafenone-M (HO-methoxy-)	U UHY	2830	371_{9}	137_{5}	99_{8}	86_{100}	58_{6}	925	3349
Etafenone-M (HO-methoxy-) AC	U+UHYAC	2955	413_{1}	398_{4}	137_{7}	99_{15}	86_{100}	1048	3357
Etafenone-M (O-dealkyl-)	G P-I U+UHYAC	1830*	226_{35}	207_{14}	121_{100}	91_{28}	65_{32}	294	896
Etafenone-M (O-dealkyl-) AC	U+UHYAC	2130*	268_{6}	225_{38}	208_{32}	121_{100}	91_{15}	460	3726
Etafenone-M (O-dealkyl-di-HO-) 2AC	U+UHYAC	2620*	342_{6}	300_{40}	258_{87}	136_{37}	121_{100}	817	3354
Etafenone-M (O-dealkyl-HO-) 2AC	U+UHYAC	2515*	326_{15}	284_{26}	242_{20}	224_{27}	121_{100}	739	3352
Etafenone-M (O-dealkyl-HO-) iso-1	UHY	2345*	242_{35}	223_{22}	121_{100}	107_{67}	65_{26}	351	3344
Etafenone-M (O-dealkyl-HO-) iso-1 AC	U+UHYAC	2215*	284_{30}	242_{96}	137_{100}	91_{79}		533	899
Etafenone-M (O-dealkyl-HO-) iso-2	UHY	2355*	242_{43}	223_{27}	121_{100}	107_{78}	65_{30}	351	3345
Etafenone-M (O-dealkyl-HO-) iso-2 AC	U+UHYAC	2370*	284_{17}	242_{26}	224_{38}	121_{100}	65_{17}	533	3350
Etafenone-M (O-dealkyl-HO-) iso-3 AC	U+UHYAC	2410*	284_{8}	242_{31}	224_{26}	121_{100}	107_{81}	533	3351
Etafenone-M (O-dealkyl-HO-methoxy-)	UHY	2400*	272_{36}	151_{9}	137_{100}	121_{52}	65_{31}	479	3346
Etafenone-M (O-dealkyl-HO-methoxy-) iso-1 AC	U+UHYAC	2525*	314_{6}	272_{58}	137_{100}	121_{61}	65_{32}	685	3353
Etafenone-M (O-dealkyl-HO-methoxy-) iso-2 AC	U+UHYAC	2580*	314_{19}	272_{89}	167_{99}	137_{16}	91_{52}	685	900
Etambutol 4AC	U+UHYAC	2455	329_{1}	299_{2}	199_{37}	144_{100}	84_{33}	878	6440
Etamiphylline	G U UHY U+UHYAC	2210	279_{2}	99_{6}	86_{100}	58_{9}		514	1201
Etamiphylline-M (deethyl-) AC	U+UHYAC	2560	293_{10}	250_{10}	206_{50}	114_{26}	58_{100}	581	1723
Etamivan	G UHY	1900	223_{35}	151_{100}	72_{25}			285	752
Etamivan AC	U+UHYAC	1970	265_{11}	222_{38}	194_{12}	151_{100}	72_{26}	444	753
Ethacridine	G	3000	253_{84}	224_{100}	196_{33}	179_{3}	169_{7}	394	6376
Ethadione		1120	157_{50}	70_{37}	58_{99}			137	221
Ethanol		<1000*	46_{24}	45_{38}	31_{100}	28_{50}		90	1545
Ethaverine	P G U+UHYAC	2940	395_{57}	366_{100}	352_{11}	252_{5}	236_{7}	1003	754
Ethaverine-M (bis-deethyl-) iso-1 2AC	U+UHYAC	3050	423_{88}	381_{96}	352_{79}	310_{100}	133_{13}	1066	3668
Ethaverine-M (bis-deethyl-) iso-2 2AC	U+UHYAC	3085	423_{45}	380_{86}	352_{17}	310_{100}	133_{13}	1066	3669
Ethaverine-M (HO-) AC	U+UHYAC	3160	453_{31}	424_{22}	410_{33}	394_{100}	382_{69}	1118	3670
Ethaverine-M (HO-) ME	UHYME	2905	425_{51}	396_{100}	368_{5}	228_{3}	213_{8}	1070	3713
Ethaverine-M (O-deethyl-) iso-1	UHY	2900	367_{63}	338_{100}	310_{14}	236_{8}	196_{5}	910	3666
Ethaverine-M (O-deethyl-) iso-1 AC	U+UHYAC	2980	409_{94}	380_{42}	366_{41}	338_{100}	310_{13}	1039	3074
Ethaverine-M (O-deethyl-) iso-1 ME	UHYME	2850	381_{69}	352_{100}	324_{7}	236_{6}	196_{4}	957	3715
Ethaverine-M (O-deethyl-) iso-2	UHY	2930	367_{60}	338_{100}	310_{12}	236_{6}	208_{4}	910	3667
Ethaverine-M (O-deethyl-) iso-2 AC	U+UHYAC	3020	409_{46}	380_{37}	366_{51}	338_{100}	310_{6}	1039	3075
Ethaverine-M (O-deethyl-) iso-2 ME	UHYME	2880	381_{94}	352_{100}	324_{8}	236_{5}	196_{3}	957	3716
Ethaverine-M (O-deethyl-HO-) 2AC	U+UHYAC	3210	467_{13}	424_{36}	408_{100}	382_{28}	366_{46}	1135	3671
Ethaverine-M (O-deethyl-HO-) 2ME	UHYME	2980	411_{100}	396_{75}	382_{51}	358_{21}	233_{4}	1046	3714
Ethchlorvynol		<1000*	115_{100}	109_{20}	89_{20}	53_{34}		122	2407
Ethenzamide	G P	1575	165_{1}	150_{38}	120_{100}	105_{53}	92_{67}	147	192
Ethenzamide-M (deethyl-)	P G UHY	1460	137_{90}	120_{100}	92_{80}	65_{56}		117	755
Ethenzamide-M (deethyl-) AC	U+UHYAC	1660	179_{39}	137_{65}	120_{100}	92_{39}	63_{20}	170	193
Ethenzamide-M (deethyl-) 2ME		1480	165_{11}	135_{100}	105_{6}	92_{15}	77_{30}	148	6395
Ethenzamide-M (deethyl-) 2TMS		1725	281_{3}	266_{100}	250_{80}	176_{40}	73_{88}	522	4596
Ethinamate	P G U	1395	167_{1}	124_{64}	95_{99}	91_{84}	81_{77}	153	756
Ethinylestradiol		2525*	296_{40}	228_{13}	213_{100}	160_{40}	133_{24}	599	5177
Ethinylestradiol AC		2610*	338_{15}	296_{89}	228_{25}	213_{100}	160_{40}	799	5180
Ethinylestradiol -HCCH		2580*	270_{100}	213_{19}	185_{40}	172_{36}	146_{55}	471	5178
Ethinylestradiol -HCCH AC	U+UHYAC	2630*	312_{5}	270_{100}	213_{19}	185_{40}	146_{39}	676	5207
Ethinylestradiol -HCCH ME		2530*	284_{100}	227_{18}	199_{82}	160_{72}	115_{40}	536	5206
Ethiofencarb		1835	225_{4}	168_{62}	139_{7}	107_{100}	77_{44}	291	3444
Ethiofencarb-M/artifact (decarbamoyl-)		1390*	168_{54}	137_{2}	107_{100}	77_{55}		154	3445
Ethion	G	2235*	384_{5}	231_{67}	153_{55}	125_{51}	97_{100}	965	3837
Ethirimol		2080	209_{13}	194_{3}	166_{100}	96_{37}	55_{17}	243	3642
Ethofumesate		1985*	286_{23}	207_{100}	161_{87}	137_{52}	79_{49}	542	4080
Ethoprofos		1700*	242_{12}	200_{27}	158_{83}	139_{45}	97_{100}	350	4081
Ethosuximide	P G U+UHYAC	1225	141_{8}	113_{95}	70_{86}	55_{100}		120	757
Ethosuximide ME		1130	155_{1}	127_{89}	112_{12}	70_{69}	55_{100}	135	2922
Ethosuximide-M (3-HO-)	U UHY	1325	157_{12}	129_{76}	86_{78}	71_{100}		137	758
Ethosuximide-M (3-HO-) AC	U+UHYAC	1350	199_{3}	171_{100}	129_{54}	86_{95}	84_{73}	217	760
Ethosuximide-M (HO-ethyl-)	U UHY	1370	157_{1}	142_{4}	113_{100}	85_{46}	69_{40}	137	759
Ethosuximide-M (HO-ethyl-) AC	U+UHYAC	1390	171_{20}	155_{34}	139_{32}	113_{99}		217	761
Ethosuximide-M (oxo-)	U+UHYAC	1270	155_{82}	113_{46}	98_{31}	70_{100}	55_{77}	134	2913
7-Ethoxycoumarin		----*	190_{100}	162_{50}	134_{94}			193	762
Ethoxyphenyldiethylphenyl butyramine		2350	325_{1}	252_{4}	206_{100}	178_{8}	105_{34}	737	763
Ethoxyquin		1720	217_{15}	202_{100}	174_{56}	145_{25}	115_{7}	261	3851
Ethylacetate		<1000*	88_{5}	70_{14}	61_{16}	43_{100}	29_{33}	98	60
Ethylamine		<1000	45_{19}	30_{100}				90	3617

Table 1-8-1: Compounds in order of names

Name	Detected	RI	Typical ions and intensities					Page	Entry
N-Ethylcarboxamido-adenosine 2AC		2735	392 $_5$	333 $_{13}$	262 $_{66}$	136 $_{75}$	85 $_{100}$	992	3092
N-Ethylcarboxamido-adenosine -2H2O		2930	272 $_{27}$	228 $_{100}$	172 $_{30}$	136 $_{28}$	66 $_{30}$	478	3093
N-Ethylcarboxamido-adenosine 3AC		3265	434 $_5$	375 $_{15}$	363 $_{16}$	304 $_{50}$	85 $_{100}$	1088	3091
4-Ethyl-2,5-dimethoxyphenethylamine		1660	209 $_{20}$	180 $_{100}$	165 $_{52}$	149 $_9$	91 $_{17}$	243	6905
4-Ethyl-2,5-dimethoxyphenethylamine AC		2000	251 $_{12}$	192 $_{100}$	177 $_{25}$	149 $_{13}$	91 $_{13}$	386	6916
4-Ethyl-2,5-dimethoxyphenethylamine HFB		1790	405 $_{54}$	226 $_7$	192 $_{90}$	179 $_{100}$	149 $_{21}$	1031	6938
4-Ethyl-2,5-dimethoxyphenethylamine PFP		1760	355 $_{55}$	192 $_{89}$	179 $_{100}$	149 $_{22}$	119 $_{21}$	864	6933
4-Ethyl-2,5-dimethoxyphenethylamine TFA		1765	305 $_{42}$	192 $_{71}$	179 $_{100}$	149 $_{19}$	91 $_{22}$	639	6928
4-Ethyl-2,5-dimethoxyphenethylamine TMS		1790	281 $_2$	73 $_{47}$	251 $_3$	180 $_{25}$	102 $_{100}$	524	6918
4-Ethyl-2,5-dimethoxyphenethylamine 2AC		2075	293 $_9$	192 $_{100}$	177 $_{34}$	149 $_{11}$	91 $_{15}$	581	6917
4-Ethyl-2,5-dimethoxyphenethylamine 2TMS		2065	353 $_2$	338 $_{15}$	174 $_{100}$	100 $_{18}$	86 $_{24}$	859	6919
4-Ethyl-2,5-dimethoxyphenethylamine formyl artifact		1630	221 $_{24}$	190 $_{100}$	179 $_{72}$	149 $_{12}$	91 $_{18}$	278	6910
4-Ethyl-2,5-dimethoxyphenethylamine-M (-COOH) MEAC		2605	295 $_{30}$	236 $_{100}$	223 $_{28}$	193 $_{11}$	163 $_{11}$	591	7093
4-Ethyl-2,5-dimethoxyphenethylamine-M (deamino-COOH) ME		1820*	238 $_{100}$	223 $_{22}$	192 $_{11}$	179 $_{39}$	163 $_6$	337	7091
4-Ethyl-2,5-dimethoxyphenethylamine-M (deamino-HO-) AC		1850*	252 $_{18}$	192 $_{100}$	177 $_{55}$	149 $_{19}$	91 $_{23}$	391	7082
4-Ethyl-2,5-dimethoxyphenethylamine-M (deamino-HO-) TFA		1680*	306 $_{65}$	192 $_{76}$	177 $_{100}$	149 $_{43}$	91 $_{59}$	644	7092
4-Ethyl-2,5-dimethoxyphenethylamine-M (deamino-oxo-)		1745*	208 $_{57}$	179 $_{100}$	149 $_{24}$	91 $_{89}$	77 $_{66}$	237	7704
4-Ethyl-2,5-dimethoxyphenethylamine-M (HO- N-acetyl-) 2TFA		2080	459 $_{27}$	345 $_{19}$	304 $_{100}$	276 $_{26}$	69 $_{23}$	1125	7105
4-Ethyl-2,5-dimethoxyphenethylamine-M (HO- N-acetyl-) iso-1		2370	323 $_{17}$	264 $_{100}$	249 $_{34}$	208 $_{43}$	191 $_{120}$	728	7127
4-Ethyl-2,5-dimethoxyphenethylamine-M (HO- N-acetyl-) iso-1 TMS		2230	339 $_{26}$	324 $_{10}$	280 $_{49}$	265 $_{100}$	191 $_{18}$	804	7125
4-Ethyl-2,5-dimethoxyphenethylamine-M (HO- N-acetyl-) iso-2		2570	323 $_{13}$	252 $_{34}$	208 $_{31}$	196 $_{71}$	57 $_{100}$	728	7128
4-Ethyl-2,5-dimethoxyphenethylamine-M (HO- N-acetyl-) iso-2 TMS		2380	339 $_{100}$	294 $_5$	251 $_{10}$	249 $_7$	73 $_{42}$	804	7126
4-Ethyl-2,5-dimethoxyphenethylamine-M (HO-) 2TFA		2035	417 $_{43}$	304 $_{87}$	291 $_{18}$	190 $_{53}$	177 $_{100}$	1056	7121
4-Ethyl-2,5-dimethoxyphenethylamine-M (HO-) -H2O AC		2175	249 $_{33}$	190 $_{100}$	177 $_{20}$	175 $_{23}$	147 $_{31}$	375	7120
4-Ethyl-2,5-dimethoxyphenethylamine-M (HO-) -H2O TFA		1945	303 $_{64}$	190 $_{100}$	177 $_{85}$	175 $_{21}$	147 $_{65}$	627	7119
4-Ethyl-2,5-dimethoxyphenethylamine-M (HO-) iso-1 AC		2340	309 $_{22}$	250 $_{100}$	237 $_6$	207 $_{50}$	191 $_{77}$	661	7096
4-Ethyl-2,5-dimethoxyphenethylamine-M (HO-) iso-2 AC		2420	309 $_8$	250 $_8$	190 $_{52}$	161 $_9$	135 $_5$	661	7097
4-Ethyl-2,5-dimethoxyphenethylamine-M (HO-) iso-3 AC		2500	309 $_{22}$	250 $_{32}$	238 $_{32}$	208 $_{28}$	196 $_{100}$	661	7098
4-Ethyl-2,5-dimethoxyphenethylamine-M (HO-) iso-3 3AC		2595	351 $_2$	309 $_{22}$	280 $_{32}$	238 $_{56}$	196 $_{100}$	849	7099
4-Ethyl-2,5-dimethoxyphenethylamine-M (HO-deamino-COOH) iso-1 AC		2070*	296 $_{88}$	253 $_{38}$	237 $_{100}$	222 $_{21}$	177 $_{33}$	598	7103
4-Ethyl-2,5-dimethoxyphenethylamine-M (HO-deamino-COOH) iso-2 AC		2150*	296 $_{12}$	236 $_{100}$	177 $_{59}$	161 $_{29}$	147 $_{13}$	598	7104
4-Ethyl-2,5-dimethoxyphenethyl-M (O-demethyl- N-acetyl-) iso-1 TFA		1950	333 $_{16}$	274 $_{100}$	259 $_{12}$	205 $_{14}$	177 $_{11}$	772	7108
4-Ethyl-2,5-dimethoxyphenethylamine-M (O-demethyl- N-acetyl-) iso-1 2TFA		1860	429 $_{15}$	316 $_6$	274 $_{100}$	259 $_{24}$	205 $_{39}$	1077	7110
4-Ethyl-2,5-dimethoxyphenethylamine-M (O-demethyl- N-acetyl-) iso-2 TFA		2020	333 $_{12}$	274 $_{100}$	259 $_{19}$	177 $_9$	91 $_{18}$	773	7109
4-Ethyl-2,5-dimethoxyphenethylamine-M (O-demethyl- N-acetyl-) iso-2 2TFA		1870	429 $_4$	274 $_{100}$	261 $_{32}$	259 $_{19}$	231 $_9$	1077	7111
4-Ethyl-2,5-dimethoxyphenethylamine-M (O-demethyl-) iso-1 2AC		2205	279 $_6$	237 $_{31}$	178 $_{100}$	165 $_{17}$	122 $_5$	511	7083
4-Ethyl-2,5-dimethoxyphenethylamine-M (O-demethyl-) iso-1 2TFA		1740	387 $_{13}$	274 $_{100}$	259 $_{20}$	205 $_{20}$	177 $_{31}$	975	7106
4-Ethyl-2,5-dimethoxyphenethylamine-M (O-demethyl-) iso-2 AC		2240	279 $_6$	237 $_{28}$	178 $_{100}$	163 $_{15}$	135 $_6$	511	7084
4-Ethyl-2,5-dimethoxyphenethylamine-M (O-demethyl-) iso-2 2TFA		1805	387 $_{11}$	274 $_{100}$	261 $_{41}$	177 $_{27}$	231 $_7$	976	7107
4-Ethyl-2,5-dimethoxyphenethylamine-M (O-demethyl-deamino-COOH) -H2O		1690*	192 $_{78}$	164 $_{100}$	136 $_{32}$	121 $_{27}$	91 $_{17}$	197	7122
4-Ethyl-2,5-dimethoxyphenethylamine-M (O-demethyl-deamino-COOH) iso-1 MEAC		1940*	266 $_8$	224 $_{50}$	192 $_{73}$	164 $_{100}$	136 $_{17}$	450	7100
4-Ethyl-2,5-dimethoxyphenethylamine-M (O-demethyl-deamino-COOH) iso-1 METFA		1710*	320 $_{93}$	305 $_{14}$	261 $_{53}$	207 $_{68}$	191 $_{100}$	714	7094
4-Ethyl-2,5-dimethoxyphenethylamine-M (O-demethyl-deamino-COOH) iso-2 MEAC		1980*	266 $_9$	224 $_{100}$	207 $_6$	165 $_{60}$	135 $_{15}$	460	7101
4-Ethyl-2,5-dimethoxyphenethylamine-M (O-demethyl-deamino-COOH) iso-2 METFA		1730*	320 $_{51}$	261 $_{100}$	223 $_{17}$	163 $_{13}$	91 $_{50}$	714	7095
4-Ethyl-2,5-dimethoxyphenethylamine-M (O-demethyl-deamino-HO-) iso-1 2AC		1990*	280 $_7$	238 $_{18}$	178 $_{100}$	163 $_{40}$	145 $_{23}$	517	7089
4-Ethyl-2,5-dimethoxyphenethylamine-M (O-demethyl-deamino-HO-) iso-1 2TFA		1540	388 $_{39}$	274 $_{100}$	259 $_{46}$	205 $_{28}$	177 $_{42}$	979	7116
4-Ethyl-2,5-dimethoxyphenethylamine-M (O-demethyl-deamino-HO-) iso-2 2AC		2000*	280 $_6$	238 $_9$	220 $_8$	178 $_{100}$	163 $_{26}$	517	7090
4-Ethyl-2,5-dimethoxyphenethylamine-M (O-demethyl-deamino-HO-) iso-2 2TFA		1580	388 $_{18}$	274 $_{100}$	259 $_{24}$	177 $_{65}$	69 $_{29}$	979	7117
4-Ethyl-2,5-dimethoxyphenethylamine-M (O-demethyl-HO- N-acetyl-) iso-1 -H2O TFA		2015	331 $_5$	272 $_{33}$	259 $_7$	205 $_7$	177 $_{14}$	763	7112
4-Ethyl-2,5-dimethoxyphenethylamine-M (O-demethyl-HO- N-acetyl-) iso-2 -H2O TFA		2050	331 $_9$	272 $_{100}$	259 $_{16}$	203 $_{10}$	192 $_{21}$	763	7113
4-Ethyl-2,5-dimethoxyphenethylamine-M (O-demethyl-HO-) 3AC		2425	337 $_2$	309 $_2$	277 $_{22}$	235 $_{41}$	176 $_{100}$	793	7085
4-Ethyl-2,5-dimethoxyphenethylamine-M (O-demethyl-HO-) 3TFA		1750	499 $_9$	386 $_{39}$	373 $_{100}$	343 $_{11}$		1163	7124
4-Ethyl-2,5-dimethoxyphenethylamine-M (O-demethyl-HO-) -H2O 2TFA		1810	385 $_{29}$	272 $_{100}$	259 $_{16}$	203 $_{13}$	175 $_{21}$	969	7114
4-Ethyl-2,5-dimethoxyphenethylamine-M (O-demethyl-HO-) iso-1 -H2O 2AC		2255	277 $_{10}$	235 $_{37}$	176 $_{100}$	163 $_{11}$	133 $_8$	500	7086
4-Ethyl-2,5-dimethoxyphenethylamine-M (O-demethyl-HO-) iso-2 -H2O 2AC		2280	277 $_5$	235 $_{32}$	176 $_{100}$	163 $_{12}$	133 $_9$	500	7087
4-Ethyl-2,5-dimethoxyphenethylamine-M (O-demethyl-oxo- N-acetyl-) AC		2430	293 $_3$	251 $_{40}$	192 $_{100}$	176 $_{53}$	137 $_9$	579	7118
4-Ethyl-2,5-dimethoxyphenethylamine-M (O-demethyl-oxo- N-acetyl-) TFA		2115	347 $_{46}$	233 $_{24}$	192 $_{100}$	177 $_{68}$	69 $_{68}$	834	7115
4-Ethyl-2,5-dimethoxyphenethylamine-M (O-demethyl-oxo-) AC		2320	251 $_{26}$	192 $_{100}$	177 $_{38}$	151 $_{11}$	137 $_{18}$	384	7088
4-Ethyl-2,5-dimethoxyphenethylamine-M (oxo-deamino-COOH) ME		2025*	252 $_{54}$	237 $_{100}$	193 $_{25}$	177 $_{23}$	163 $_{29}$	390	7102
Ethyldimethylbenzene		1065*	134 $_{32}$	119 $_{100}$	105 $_{26}$	91 $_{44}$	77 $_{20}$	114	3790
Ethylenediaminetetraacetic acid 3ME1ET		2125*	362 $_{14}$	303 $_{28}$	289 $_{18}$	188 $_{100}$	174 $_{64}$	891	6452
Ethylenediaminetetraacetic acid 4ME		2105*	348 $_8$	289 $_{25}$	188 $_{23}$	174 $_{100}$	146 $_{40}$	839	6451
Ethylene glycol		<1000*	62 $_{33}$	43 $_{64}$	33 $_{100}$	31 $_{100}$		92	765
Ethylene glycol 2AC		<1000*	116 $_{25}$	103 $_{23}$	86 $_{99}$	73 $_{41}$		124	766
Ethylene glycol dibenzoate		2120*	270 $_1$	227 $_{14}$	162 $_{10}$	105 $_{100}$	77 $_{87}$	469	1741
Ethylene glycol dipivalate		1320*	185 $_1$	143 $_2$	129 $_9$	85 $_{28}$	57 $_{100}$	308	1903
Ethylene glycol monomethylether		<1000*	76 $_5$	58 $_4$	45 $_{100}$	31 $_{30}$	29 $_{66}$	95	3779
Ethylene glycol phenylboronate		1210*	148 $_{85}$	118 $_{34}$	91 $_{100}$	77 $_{14}$		125	1896
Ethylene oxide		<1000*	44 $_{57}$	29 $_{100}$				90	4195
Ethylene thiourea		2080	102 $_{100}$	73 $_{13}$	60 $_4$			100	3910
2-Ethylhexyldiphenylphosphate	P G U UHY U+UHYAC	2450*	362 $_{10}$	251 $_{100}$	170 $_5$	94 $_{14}$	77 $_{14}$	891	3053
Ethylhexylmethylphthalate		2010*	181 $_{24}$	163 $_{100}$	149 $_{48}$	83 $_{11}$	70 $_{34}$	576	5319
Ethylloflazepate artifact	U+UHYAC	2050	272 $_{70}$	271 $_{100}$	237 $_{78}$	151 $_{11}$	110 $_8$	477	2409
Ethylloflazepate -C3H4O2	G P-I UGLUC	2470	288 $_{100}$	287 $_{64}$	260 $_{93}$	259 $_{75}$		551	508

Table 1-8-1: Compounds in order of names — 4-Ethylthio-2,5-dimethoxyphenethylamine-M (sulfone) TFA

Name	Detected	RI	Typical ions and intensities					Page	Entry
Ethylloflazepate -C3H4O2 TMS		2470	360 $_{44}$	359 $_{48}$	341 $_{30}$	197 $_{12}$	73 $_{70}$	883	4621
Ethylloflazepate HY	UHY	2030	249 $_{97}$	154 $_{33}$	123 $_{45}$	95 $_{41}$		373	512
Ethylloflazepate HYAC	U+UHYAC	2195	291 $_{52}$	249 $_{100}$	123 $_{57}$	95 $_{61}$		567	286
Ethylloflazepate-M (HO-) artifact-1	U+UHYAC	2380	330 $_{26}$	288 $_{100}$	287 $_{91}$	271 $_{39}$	253 $_{82}$	758	2410
Ethylloflazepate-M (HO-) artifact-2	U+UHYAC	2420	316 $_{24}$	258 $_{100}$	221 $_{15}$	95 $_{19}$	75 $_{16}$	692	2412
Ethylloflazepate-M (HO-) HY2AC	U+UHYAC	2500	349 $_{47}$	307 $_{82}$	265 $_{100}$	264 $_{81}$	139 $_{46}$	841	2411
1-Ethyl-2-methylbenzene		<1000*	120 $_{27}$	105 $_{100}$	91 $_{13}$	77 $_{15}$	63 $_{7}$	106	3787
2-Ethyl-3-methyl-1-butene		<1000*	98 $_{32}$	83 $_{64}$	69 $_{100}$	55 $_{80}$	41 $_{92}$	99	3824
1-Ethyl-4-methylbenzene		<1000*	120 $_{40}$	105 $_{100}$	91 $_{12}$	77 $_{17}$	65 $_{7}$	106	3827
2-Ethyl-5-methyl-3,3-diphenyl-1-pyrroline (EMDP)	U+UHYAC	1940	263 $_{1}$	208 $_{100}$	193 $_{78}$	179 $_{38}$	130 $_{50}$	436	5295
Ethylmethylphthalate		1520*	208 $_{2}$	176 $_{11}$	163 $_{100}$	149 $_{58}$	77 $_{25}$	237	4940
Ethylmorphine	U UHY	2420	313 $_{100}$	284 $_{19}$	162 $_{54}$	124 $_{29}$		681	494
Ethylmorphine AC	U+UHYAC	2530	355 $_{100}$	327 $_{24}$	204 $_{18}$	162 $_{10}$	124 $_{14}$	866	237
Ethylmorphine PFP		2430	459 $_{63}$	430 $_{8}$	402 $_{2}$	296 $_{100}$	119 $_{21}$	1126	2461
Ethylmorphine TFA		2320	409 $_{100}$	380 $_{15}$	296 $_{97}$	115 $_{21}$	59 $_{39}$	1038	4014
Ethylmorphine TMS		2540	385 $_{45}$	234 $_{17}$	192 $_{31}$	146 $_{31}$	73 $_{100}$	971	2467
Ethylmorphine-M (nor-) 2AC	U+UHYAC	2930	383 $_{23}$	237 $_{29}$	209 $_{47}$	87 $_{100}$	72 $_{34}$	964	1193
Ethylmorphine-M (O-deethyl-)	G UHY	2455	285 $_{100}$	268 $_{15}$	162 $_{58}$	124 $_{20}$		539	474
Ethylmorphine-M (O-deethyl-) TFA		2285	381 $_{55}$	268 $_{100}$	146 $_{14}$	115 $_{13}$	69 $_{23}$	955	5569
Ethylmorphine-M (O-deethyl-) 2AC	G PHYAC U+UHYAC	2620	369 $_{59}$	327 $_{100}$	310 $_{36}$	268 $_{47}$	162 $_{11}$	915	225
Ethylmorphine-M (O-deethyl-) 2HFB		2375	677 $_{10}$	480 $_{11}$	464 $_{100}$	407 $_{9}$	169 $_{8}$	1201	6120
Ethylmorphine-M (O-deethyl-) 2PFP		2360	577 $_{51}$	558 $_{7}$	430 $_{8}$	414 $_{100}$	119 $_{22}$	1191	2251
Ethylmorphine-M (O-deethyl-) 2TFA		2250	477 $_{71}$	364 $_{100}$	307 $_{6}$	115 $_{8}$	69 $_{31}$	1145	4008
Ethylmorphine-M (O-deethyl-) 2TMS	UHYTMS	2560	429 $_{19}$	236 $_{21}$	196 $_{15}$	146 $_{21}$	73 $_{100}$	1079	2463
Ethylmorphine-M (O-deethyl-) Cl-artifact 2AC	U+UHYAC	2680	403 $_{59}$	361 $_{100}$	344 $_{63}$	302 $_{90}$	204 $_{55}$	1026	2992
Ethylmorphine-M (O-deethyl-)-D3 TFA		2275	384 $_{39}$	271 $_{100}$	211 $_{8}$	165 $_{6}$	152 $_{7}$	967	5572
Ethylmorphine-M (O-deethyl-)-D3 2HFB		2375	680 $_{4}$	483 $_{9}$	467 $_{100}$	169 $_{23}$	414 $_{7}$	1202	6126
Ethylmorphine-M (O-deethyl-)-D3 2PFP		2350	580 $_{16}$	433 $_{7}$	417 $_{100}$	269 $_{5}$	119 $_{8}$	1192	5567
Ethylmorphine-M (O-deethyl-)-D3 2TFA		2240	480 $_{32}$	383 $_{6}$	367 $_{100}$	314 $_{6}$	307 $_{6}$	1148	5571
Ethylmorphine-M (O-deethyl-)-D3 2TMS		2550	432 $_{100}$	290 $_{30}$	239 $_{66}$	199 $_{44}$	73 $_{110}$	1085	5578
Ethylmorphine-M 2PFP		2440	563 $_{100}$	400 $_{10}$	355 $_{38}$	327 $_{7}$	209 $_{15}$	1188	3534
Ethylmorphine-M 3AC	U+UHYAC	2955	397 $_{8}$	355 $_{9}$	209 $_{41}$	87 $_{100}$	72 $_{33}$	1010	1194
Ethylmorphine-M 3PFP	UHYPFP	2405	709 $_{80}$	533 $_{28}$	388 $_{29}$	367 $_{51}$	355 $_{100}$	1203	3533
Ethylmorphine-M 3TMS	UHYTMS	2605	487 $_{9}$	416 $_{9}$	222 $_{18}$	131 $_{10}$	73 $_{50}$	1155	3525
Ethylparaben		1580*	166 $_{32}$	138 $_{29}$	121 $_{100}$			150	767
Ethylparaben-M (4-hydroxyhippuric acid) ME	U	1820	209 $_{100}$	177 $_{32}$	149 $_{34}$	121 $_{87}$		240	817
1-Ethylpiperidine		<1000	113 $_{12}$	98 $_{100}$	70 $_{10}$	58 $_{39}$	42 $_{76}$	104	3613
4-Ethylthio-2,5-dimethoxyphenethylamine		1980	241 $_{22}$	212 $_{100}$	197 $_{20}$	183 $_{35}$	153 $_{22}$	347	5035
4-Ethylthio-2,5-dimethoxyphenethylamine AC	U+UHYAC	2310	283 $_{28}$	224 $_{100}$	211 $_{34}$	181 $_{12}$	153 $_{12}$	529	5037
4-Ethylthio-2,5-dimethoxyphenethylamine HFB		2040	437 $_{24}$	224 $_{21}$	211 $_{100}$	181 $_{18}$	169 $_{17}$	1092	6816
4-Ethylthio-2,5-dimethoxyphenethylamine PFP		2090	387 $_{34}$	224 $_{29}$	211 $_{100}$	181 $_{18}$	153 $_{15}$	976	6817
4-Ethylthio-2,5-dimethoxyphenethylamine TFA	UGLUCTFA	2210	337 $_{26}$	224 $_{17}$	211 $_{100}$	181 $_{13}$	153 $_{10}$	791	6818
4-Ethylthio-2,5-dimethoxyphenethylamine TMS		2405	313 $_{1}$	299 $_{3}$	174 $_{100}$	147 $_{4}$	86 $_{7}$	681	6814
4-Ethylthio-2,5-dimethoxyphenethylamine 2AC	U+UHYAC	2395	325 $_{17}$	224 $_{100}$	211 $_{50}$	181 $_{16}$	153 $_{14}$	735	5038
4-Ethylthio-2,5-dimethoxyphenethylamine 2TMS		2405	385 $_{1}$	370 $_{3}$	254 $_{4}$	211 $_{4}$	174 $_{100}$	970	6815
4-Ethylthio-2,5-dimethoxyphenethylamine deuteroformyl artifact		1935	255 $_{37}$	224 $_{31}$	211 $_{100}$	181 $_{14}$	153 $_{16}$	402	5036
4-Ethylthio-2,5-dimethoxyphenethylamine-M (aryl-HOOC-)	UGLUC	1970	242 $_{100}$	227 $_{20}$	183 $_{68}$	153 $_{11}$		350	6893
4-Ethylthio-2,5-dimethoxyphenethylamine-M (aryl-HOOC-) ME	USPEME	1960*	256 $_{100}$	241 $_{10}$	197 $_{49}$	181 $_{12}$	167 $_{15}$	405	6842
4-Ethylthio-2,5-dimethoxyphenethylamine-M (deamino-HO-)	UGLUC	1905*	242 $_{71}$	211 $_{100}$	181 $_{15}$	153 $_{11}$		352	6839
4-Ethylthio-2,5-dimethoxyphenethylamine-M (deamino-HO-) AC	U+UHYAC	2050*	284 $_{40}$	224 $_{100}$	209 $_{30}$	167 $_{20}$	150 $_{10}$	534	6892
4-Ethylthio-2,5-dimethoxyphenethylamine-M (deamino-HOOC-)		2130*	256 $_{100}$	242 $_{31}$	211 $_{60}$	195 $_{30}$	181 $_{32}$	405	6840
4-Ethylthio-2,5-dimethoxyphenethylamine-M (deamino-HOOC-) ME	UHYME	1910*	270 $_{100}$	255 $_{74}$	211 $_{82}$	195 $_{44}$	181 $_{46}$	469	6838
4-Ethylthio-2,5-dimethoxyphenethylamine-M (deamino-HOOC-) TMS	USPETMS	2075*	328 $_{100}$	313 $_{29}$	298 $_{25}$	255 $_{61}$	211 $_{57}$	749	6841
4-Ethylthio-2,5-dimethoxyphenethylamine-M (deamino-oxo-)		2130*	240 $_{33}$	211 $_{100}$	181 $_{11}$	153 $_{19}$	122 $_{11}$	343	7234
4-Ethylthio-2,5-dimethoxyphenethylamine-M (HO- N-acetyl-) TFA	UGLUCTFA	2270	427 $_{33}$	367 $_{5}$	259 $_{140}$	167 $_{8}$		1073	6834
4-Ethylthio-2,5-dimethoxyphenethylamine-M (HO- sulfone) AC	U+UHYAC	2730	331 $_{7}$	272 $_{100}$	259 $_{74}$	238 $_{26}$	165 $_{47}$	763	6828
4-Ethylthio-2,5-dimethoxyphenethylamine-M (HO- sulfone) 2AC	U+UHYAC	2780	373 $_{35}$	314 $_{84}$	302 $_{75}$	272 $_{54}$	259 $_{100}$	930	6833
4-Ethylthio-2,5-dimethoxyphenethylamine-M (O-demethyl- N-acetyl-) TFA	U+UHYTFA	2250	365 $_{11}$	323 $_{19}$	306 $_{100}$	293 $_{18}$	197 $_{9}$	899	6942
4-Ethylthio-2,5-dimethoxyphenethylamine-M (O-demethyl- N-acetyl-) 2TFA	U+UHYTFA	2180	461 $_{16}$	306 $_{100}$	293 $_{43}$	209 $_{25}$		1129	6894
4-Ethylthio-2,5-dimethoxyphenethylamine-M (O-demethyl- sulfone) 2AC	U+UHYAC	2510	343 $_{3}$	301 $_{90}$	242 $_{100}$	230 $_{48}$	153 $_{9}$	820	6835
4-Ethylthio-2,5-dimethoxyphenethylamine-M (O-demethyl-) 2AC	U+UHYAC	2120	311 $_{48}$	297 $_{31}$	269 $_{100}$	252 $_{46}$	210 $_{78}$	671	6837
4-Ethylthio-2,5-dimethoxyphenethylamine-M (O-demethyl-) 2TFA	UGLUCTFA	1980	419 $_{28}$	306 $_{100}$	293 $_{92}$	209 $_{20}$	69 $_{32}$	1060	6821
4-Ethylthio-2,5-dimethoxyphenethylamine-M (O-demethyl-) 3AC	U+UHYAC	2290	353 $_{22}$	311 $_{32}$	252 $_{33}$	210 $_{100}$	197 $_{20}$	858	6836
4-Ethylthio-2,5-dimethoxyphenethylamine-M (O-demethyl-sulfone N-acetyl-) TFA	UGLUCTFA	2450	397 $_{4}$	355 $_{81}$	242 $_{120}$	153 $_{16}$		1008	6820
4-Ethylthio-2,5-dimethoxyphenethylamine-M (S-deethyl-) AC	U+UHYAC	2170	255 $_{18}$	196 $_{100}$	183 $_{41}$	181 $_{34}$	153 $_{21}$	401	6831
4-Ethylthio-2,5-dimethoxyphenethylamine-M (S-deethyl-) 3AC	U+UHYAC	2420	339 $_{22}$	297 $_{5}$	238 $_{21}$	196 $_{100}$	183 $_{28}$	802	6827
4-Ethylthio-2,5-dimethoxyphenethylamine-M (S-deethyl-) iso-1 2AC		2240	297 $_{29}$	210 $_{14}$	196 $_{100}$	183 $_{35}$	181 $_{29}$	602	6823
4-Ethylthio-2,5-dimethoxyphenethylamine-M (S-deethyl-) iso-2 2AC	U+UHYAC	2360	297 $_{16}$	255 $_{11}$	238 $_{20}$	196 $_{100}$	183 $_{37}$	602	6826
4-Ethylthio-2,5-dimethoxyphenethylamine-M (S-deethyl-methyl- N-acetyl-)	U+UHYAC	2230	269 $_{19}$	210 $_{100}$	197 $_{35}$	195 $_{21}$	167 $_{27}$	465	6832
4-Ethylthio-2,5-dimethoxyphenethylamine-M (S-deethyl-methyl- sulfone) AC	U+UHYAC	2580	301 $_{7}$	242 $_{100}$	230 $_{4}$	196 $_{7}$	124 $_{7}$	618	6829
4-Ethylthio-2,5-dimethoxyphenethylamine-M (S-deethyl-methyl- sulfoxide) AC	U+UHYAC	2460	285 $_{16}$	268 $_{23}$	226 $_{33}$	211 $_{100}$	197 $_{31}$	538	6830
4-Ethylthio-2,5-dimethoxyphenethylamine-M (sulfone N-acetyl-) TFA	UGLUCTFA	2400	411 $_{23}$	256 $_{100}$	242 $_{4}$	181 $_{10}$	167 $_{11}$	1043	6822
4-Ethylthio-2,5-dimethoxyphenethylamine-M (sulfone) AC	U+UHYAC	2600	315 $_{15}$	256 $_{100}$	244 $_{9}$	167 $_{12}$	91 $_{8}$	689	6825
4-Ethylthio-2,5-dimethcoxyphenethylamine-M (sulfone) TFA	UGLUCTFA	2310	369 $_{44}$	256 $_{100}$	243 $_{7}$	211 $_{4}$	167 $_{23}$	915	6819

4-Ethylthio-2,5-dimethoxyphenethylamine-M (sulfone) 2AC   Table 1-8-1: Compounds in order of names

Name	Detected	RI	Typical ions and intensities					Page	Entry
4-Ethylthio-2,5-dimethoxyphenethylamine-M (sulfone) 2AC	U+UHYAC	2640	357 $_{10}$	256 $_{100}$	244 $_6$	167 $_7$	91 $_7$	873	6824
Ethyltolylbarbital 2ET		2010	302 $_{15}$	274 $_{100}$	246 $_{11}$	160 $_{15}$	117 $_{18}$	624	2597
Eticyclidine		1545	203 $_{18}$	160 $_{100}$	146 $_{18}$	117 $_{16}$	91 $_{22}$	225	3602
Eticyclidine intermediate (ECC) -HCN		<1000	125 $_4$	110 $_{12}$	96 $_{44}$	82 $_{33}$	56 $_{97}$	110	3598
Eticyclidine intermediate (ECC) -HCN		<1000	125 $_4$	110 $_{12}$	96 $_{44}$	56 $_{97}$	41 $_{100}$	109	5535
Eticyclidine precursor (ethylamine)		<1000	45 $_{19}$	30 $_{100}$				90	3617
Etidocaine		2040	276 $_1$	259 $_3$	245 $_8$	128 $_{100}$	86 $_{12}$	498	1437
Etifelmin		1880	237 $_6$	208 $_{100}$	191 $_{41}$	165 $_{17}$	91 $_{29}$	335	1796
Etifelmin AC		2220	279 $_{37}$	220 $_{53}$	205 $_{73}$	191 $_{84}$	112 $_{100}$	513	1441
Etilamfetamine	U	1230	162 $_1$	148 $_1$	117 $_1$	91 $_{10}$	72 $_{40}$	143	764
Etilamfetamine AC	U+UHYAC	1675	205 $_1$	114 $_{53}$	91 $_{12}$	72 $_{100}$		229	1438
Etilamfetamine HFB		1485	359 $_1$	268 $_{100}$	240 $_{46}$	118 $_{18}$	91 $_{25}$	880	5085
Etilamfetamine PFP		1450	309 $_1$	218 $_{100}$	190 $_{48}$	118 $_{34}$	91 $_{35}$	659	5082
Etilamfetamine TFA		1450	213 $_1$	168 $_{100}$	140 $_{38}$	118 $_{36}$	69 $_{56}$	416	4004
Etilamfetamine-M	UHY	1465	181 $_9$	138 $_{100}$	122 $_{18}$	94 $_{24}$	77 $_{16}$	178	4351
Etilamfetamine-M (AM)		1160	134 $_1$	120 $_1$	91 $_6$	77 $_1$	65 $_4$	115	54
Etilamfetamine-M (AM)	U	1160	134 $_1$	120 $_1$	91 $_6$	65 $_4$	44 $_{100}$	115	5514
Etilamfetamine-M (AM) AC	U+UHYAC	1505	177 $_1$	118 $_{19}$	91 $_{11}$	86 $_{31}$	65 $_5$	165	55
Etilamfetamine-M (AM) AC	UAAC U+UHYAC	1505	177 $_1$	118 $_{19}$	91 $_{11}$	86 $_{31}$	44 $_{100}$	165	5515
Etilamfetamine-M (AM) HFB		1355	331 $_1$	240 $_{73}$	169 $_{21}$	118 $_{100}$	91 $_{53}$	762	5047
Etilamfetamine-M (AM) PFP		1330	281 $_1$	190 $_{73}$	118 $_{100}$	91 $_{36}$	65 $_{12}$	521	4379
Etilamfetamine-M (AM) TFA		1095	231 $_1$	140 $_{100}$	118 $_{92}$	91 $_{45}$	69 $_{19}$	309	4000
Etilamfetamine-M (AM) TMS		1190	192 $_6$	116 $_{100}$	100 $_{10}$	91 $_{11}$	73 $_{87}$	235	5581
Etilamfetamine-M (AM) formyl artifact		1100	147 $_2$	146 $_6$	125 $_5$	91 $_{12}$	56 $_{100}$	125	3261
Etilamfetamine-M (AM)-D5 AC	UAAC U+UHYAC	1480	182 $_3$	122 $_{46}$	92 $_{30}$	90 $_{100}$	66 $_{16}$	181	5690
Etilamfetamine-M (AM)-D5 HFB		1330	244 $_{100}$	169 $_{14}$	122 $_{46}$	92 $_{41}$	69 $_{40}$	787	6316
Etilamfetamine-M (AM)-D5 PFP		1320	194 $_{100}$	123 $_{42}$	119 $_{32}$	92 $_{46}$	69 $_{11}$	543	5566
Etilamfetamine-M (AM)-D5 TFA		1085	144 $_{100}$	123 $_{53}$	122 $_{56}$	92 $_{51}$	69 $_{28}$	330	5570
Etilamfetamine-M (AM)-D5 TMS		1180	212 $_1$	197 $_8$	120 $_{100}$	92 $_{11}$	73 $_{57}$	251	5582
Etilamfetamine-M (AM)-D11 PFP		1610	292 $_1$	194 $_{100}$	128 $_{82}$	98 $_{43}$	70 $_{14}$	575	7284
Etilamfetamine-M (AM)-D11 TFA		1615	242 $_1$	144 $_{100}$	128 $_{82}$	98 $_{43}$	70 $_{14}$	353	7283
Etilamfetamine-M (AM-4-HO-)		1480	151 $_{10}$	107 $_{69}$	91 $_{10}$	77 $_{42}$	56 $_{100}$	129	1802
Etilamfetamine-M (AM-4-HO-) AC	U+UHYAC	1890	193 $_1$	134 $_{100}$	107 $_{26}$	86 $_{25}$	77 $_{16}$	201	1803
Etilamfetamine-M (AM-4-HO-) TFA		1670	247 $_4$	140 $_{15}$	134 $_{54}$	107 $_{100}$	77 $_{15}$	366	6335
Etilamfetamine-M (AM-4-HO-) 2AC	U+UHYAC	1900	235 $_1$	176 $_{72}$	134 $_{100}$	107 $_{47}$	86 $_{71}$	324	1804
Etilamfetamine-M (AM-4-HO-) 2HFB		<1000	330 $_{48}$	303 $_{15}$	240 $_{100}$	169 $_{44}$	69 $_{42}$	1182	6326
Etilamfetamine-M (AM-4-HO-) 2PFP		<1000	280 $_{77}$	253 $_{16}$	190 $_{100}$	119 $_{56}$	69 $_{16}$	1104	6325
Etilamfetamine-M (AM-4-HO-) 2TFA		<1000	230 $_{94}$	203 $_{14}$	140 $_{130}$	92 $_{15}$	69 $_{76}$	819	6324
Etilamfetamine-M (AM-4-HO-) 2TMS		<1000	280 $_7$	179 $_9$	149 $_8$	116 $_{100}$	73 $_{78}$	594	6327
Etilamfetamine-M (AM-4-HO-) formyl art.		1220	163 $_3$	148 $_4$	107 $_{30}$	77 $_{13}$	56 $_{100}$	142	6323
Etilamfetamine-M (AM-HO-methoxy-) ME	UHYME	1550	195 $_1$	152 $_{100}$	137 $_{17}$	107 $_{16}$	77 $_{14}$	208	4352
Etilamfetamine-M (deamino-oxo-HO-methoxy-)	UHY	1510*	180 $_{19}$	137 $_{100}$	122 $_{19}$	107 $_2$	94 $_{16}$	175	4247
Etilamfetamine-M (di-HO-) 3AC	U+UHYAC	2200	321 $_1$	234 $_4$	150 $_{10}$	114 $_{46}$	72 $_{100}$	718	4208
Etilamfetamine-M (HO-) ME		1660	192 $_1$	149 $_1$	121 $_5$	91 $_2$	72 $_{100}$	202	5831
Etilamfetamine-M (HO-) MEAC	U+UHYAC	1855	235 $_1$	148 $_{32}$	121 $_7$	114 $_{26}$	72 $_{100}$	326	5322
Etilamfetamine-M (HO-) MEHFB	UHFB	1785	389 $_2$	268 $_{100}$	240 $_{46}$	148 $_{62}$	121 $_{43}$	983	5834
Etilamfetamine-M (HO-) MEPFP	UPFP	1765	339 $_1$	218 $_{100}$	190 $_{44}$	148 $_{59}$	121 $_{49}$	802	5833
Etilamfetamine-M (HO-) METFA	UTFA	1775	289 $_1$	168 $_{100}$	148 $_{63}$	140 $_{41}$	121 $_{62}$	557	5832
Etilamfetamine-M (HO-) METMS		2065	264 $_1$	250 $_{14}$	144 $_{100}$	121 $_{17}$	73 $_{84}$	447	5836
Etilamfetamine-M (HO-) 2AC	U+UHYAC	1995	263 $_1$	176 $_9$	134 $_{15}$	114 $_{44}$	72 $_{100}$	435	5323
Etilamfetamine-M (HO-) 2ME		1780	206 $_1$	121 $_5$	86 $_{100}$	72 $_3$	58 $_{20}$	235	5835
Etilamfetamine-M (HO-methoxy-)	UHY	1640	209 $_1$	137 $_{12}$	122 $_7$	94 $_9$	72 $_{100}$	242	4364
Etilamfetamine-M (HO-methoxy-) AC	U+UHYAC	2000	251 $_1$	164 $_{46}$	137 $_7$	114 $_{33}$	72 $_{100}$	386	4274
Etilamfetamine-M (HO-methoxy-) ME	UHYME	1930	223 $_{17}$	194 $_7$	151 $_{36}$	94 $_{12}$	72 $_{100}$	286	4350
Etilamfetamine-M (HO-methoxy-) 2AC	U+UHYAC	2080	293 $_1$	206 $_{20}$	164 $_{38}$	114 $_{77}$	72 $_{100}$	582	4209
Etilamfetamine-M AC	U+UHYAC	1600*	222 $_2$	180 $_{22}$	137 $_{100}$			280	4211
Etilamfetamine-M ME	UHYME	1540*	194 $_{25}$	151 $_{100}$	135 $_4$	107 $_{18}$	65 $_4$	204	4353
Etilamfetamine-M 2AC	U+UHYAC	2065	265 $_3$	206 $_{27}$	164 $_{100}$	137 $_{23}$	86 $_{33}$	445	3498
Etilamfetamine-M 2AC	U+UHYAC	1735*	250 $_3$	208 $_{15}$	166 $_{45}$	123 $_{100}$		379	4210
Etilamfetamine-M 2AC	U+UHYAC	1820*	266 $_3$	206 $_9$	164 $_{100}$	150 $_{10}$	137 $_{30}$	450	6409
Etilamfetamine-M 2HFB	UHFB	1690	360 $_{82}$	333 $_{15}$	240 $_{100}$	169 $_{42}$	69 $_{39}$	1189	6512
Etilefrine		1690	181 $_1$	121 $_1$	95 $_3$	77 $_5$	58 $_{100}$	178	4667
Etilefrine ME2AC		2000	279 $_1$	247 $_4$	192 $_{25}$	100 $_{23}$	58 $_{100}$	512	1970
Etilefrine 3AC	U+UHYAC	2150	307 $_2$	264 $_8$	247 $_4$	100 $_{76}$	58 $_{100}$	649	768
Etilefrine 3TMS		1885	397 $_1$	382 $_5$	147 $_{10}$	130 $_{100}$	73 $_{89}$	1011	4544
Etilefrine formyl artifact		1860	193 $_{19}$	178 $_9$	135 $_{36}$	107 $_{16}$	58 $_{100}$	201	1969
3-alpha-Etiocholanolone		2515*	290 $_{100}$	257 $_{16}$	246 $_{18}$	107 $_{27}$	79 $_{28}$	566	3759
3-alpha-Etiocholanolone AC		2585*	332 $_{10}$	272 $_{100}$	257 $_{47}$	108 $_{60}$	67 $_{56}$	770	3769
3-alpha-Etiocholanolone 2TMS		2520*	434 $_{54}$	419 $_{52}$	329 $_{41}$	169 $_{33}$	73 $_{100}$	1089	3799
3-beta-Etiocholanolone		2465*	290 $_{63}$	244 $_{50}$	201 $_{27}$	93 $_{66}$	67 $_{100}$	566	3897
3-beta-Etiocholanolone AC		2540*	332 $_{18}$	272 $_{100}$	257 $_{57}$	79 $_{89}$	67 $_{56}$	770	3921
3-beta-Etiocholanolone TMS		2430*	362 $_9$	347 $_{15}$	272 $_{100}$	244 $_{61}$	75 $_{99}$	893	3961
3-beta-Etiocholanolone 2TMS		2485*	434 $_{46}$	419 $_{43}$	329 $_{29}$	169 $_{19}$	73 $_{100}$	1089	3962
Etiroxate artifact ME		3700	490 $_1$	448 $_1$	387 $_3$	130 $_{100}$	102 $_{11}$	1017	2750

Table 1-8-1: Compounds in order of names   Famciclovir TFA

Name	Detected	RI	Typical ions and intensities					Page	Entry
Etiroxate artifact-1		2285	416_{10}	288_1	132_3	116_{100}	88_{17}	1053	2749
Etiroxate artifact-1 2AC		2690	559_3	500_4	458_{42}	158_{56}	116_{100}	1187	2763
Etiroxate artifact-2 AC		3300	651_{12}	550_{100}	158_{38}	116_{49}	609_{21}	1199	2764
Etiroxate artifact-3		3360	506_1	451_1	337_2	116_{100}	88_{14}	1168	2748
Etiroxate artifact-4 AC		3800	777_{10}	735_{57}	676_{100}	158_{95}	116_{97}	1205	2765
Etizolam		2980	342_{100}	313_{37}	266_{32}	224_{24}	137_{14}	816	4022
Etodolac ME		2225	301_{21}	272_{63}	228_{100}	198_{32}	115_9	620	6128
Etodolac TMS		2350	359_{26}	330_{62}	309_{27}	228_{100}	73_{49}	881	6129
Etodroxizine	G UHY	3155	418_8	299_{14}	201_{100}	165_{17}		1059	769
Etodroxizine AC	U+UHYAC	3180	460_{12}	299_{37}	201_{100}	165_{18}	87_{12}	1128	1797
Etodroxizine artifact		1900*	232_{61}	201_{63}	165_{65}	105_{100}	77_{55}	311	1344
Etodroxizine artifact-1	G U+UHYAC	1600*	202_{30}	167_{100}	165_{52}	152_{17}	125_7	222	2442
Etodroxizine-M	U+UHYAC	2210	280_{100}	201_{35}	165_{57}			515	770
Etodroxizine-M (carbinol)	UHY	1750*	218_{17}	183_7	139_{39}	105_{100}	77_{87}	263	2239
Etodroxizine-M (carbinol) AC	U+UHYAC	1890*	260_8	200_{40}	165_{100}	139_{10}	77_{29}	420	1270
Etodroxizine-M (Cl-benzophenone)	U+UHYAC	1850*	216_{43}	139_{58}	105_{100}	77_{44}		258	1343
Etodroxizine-M (HO-Cl-benzophenone)	UHY	2300*	232_{36}	197_7	139_{23}	121_{100}	111_{23}	311	2240
Etodroxizine-M (HO-Cl-BPH) iso-1 AC	U+UHYAC	2200*	274_{18}	232_{75}	139_{100}	121_{44}	111_{51}	484	2229
Etodroxizine-M (HO-Cl-BPH) iso-2 AC	U+UHYAC	2230*	274_7	232_{43}	139_{25}	121_{100}	111_{27}	484	2230
Etodroxizine-M (N-dealkyl-)	UHY	2520	286_{13}	241_{48}	201_{50}	165_{65}	56_{100}	543	2241
Etodroxizine-M (N-dealkyl-) AC	U+UHYAC	2620	328_7	242_{19}	201_{48}	165_{66}	85_{100}	749	1271
Etodroxizine-M/artifact	P-I U+UHYAC UME	2220	300_{17}	228_{38}	165_{52}	99_{100}	56_{63}	616	670
Etodroxizine-M/artifact 2AC	U+UHYAC	2300	302_1	199_1	154_4	141_{100}	99_{24}	625	2445
Etodroxizine-M/artifact HYAC	U+UHYAC	2935	280_4	201_{100}	165_{26}			515	1272
Etofenamate		2510	369_{32}	263_{100}	243_5	235_{13}	167_{11}	915	6093
Etofenamate AC		2590	411_{15}	263_{100}	235_{17}	167_{13}	87_{20}	1043	6094
Etofenamate-M/artifact (flufenamic acid) ME	PME	1880	295_{51}	263_{100}	235_{15}	166_{11}	92_7	589	5147
Etofenamate-M/artifact (HO-flufenamic acid) 2ME	PME	2115	325_{100}	293_{91}	278_{88}	250_{67}	202_{19}	734	6377
Etofenamate-M/artifact (oxoethyl-)		2125	323_{36}	263_{100}	243_6	235_{12}	167_8	725	6092
Etofibrate		2520	363_4	236_{64}	128_{100}	106_{78}	78_{54}	893	2762
Etofibrate-M (clofibric acid)	U	1640*	214_2	168_9	128_{100}	86_9	65_{17}	254	686
Etofibrate-M (clofibric acid) ME	U	1500*	228_9	169_{16}	128_{100}	99_5	75_9	298	687
Etofibrate-M artifact	U+UHYAC	1580*	168_{35}	128_{99}				154	1373
Etofibrate-M/artifact (denicotinyl-)		2030*	258_5	169_{13}	128_{100}	111_8	69_{19}	413	2751
Etofylline	UHY	2125	224_{47}	194_{17}	180_{100}	109_{35}	95_{74}	288	771
Etofylline AC	U+UHYAC	2200	266_{79}	206_{59}	180_{34}	122_{31}	87_{100}	448	772
Etofylline TMS	UHYTMS	2160	296_{10}	281_{26}	252_6	180_{100}	73_{47}	599	5696
Etofylline clofibrate	G	3125	420_9	293_{43}	206_{25}	113_{35}	69_{100}	1062	1939
Etofylline clofibrate-M (clofibric acid)	U	1640*	214_2	168_9	128_{100}	86_9	65_{17}	254	686
Etofylline clofibrate-M (clofibric acid) ME	U	1500*	228_9	169_{16}	128_{100}	99_5	75_9	298	687
Etofylline clofibrate-M (etofylline)	UHY	2125	224_{47}	194_{17}	180_{100}	109_{35}	95_{74}	288	771
Etofylline clofibrate-M (etofylline) AC	U+UHYAC	2200	266_{79}	206_{59}	180_{34}	122_{31}	87_{100}	448	772
Etofylline clofibrate-M (etofylline) TMS	UHYTMS	2160	296_{10}	281_{26}	252_6	180_{100}	73_{47}	599	5696
Etofylline clofibrate-M artifact	U+UHYAC	1580*	168_{35}	128_{99}				154	1373
Etoloxamine		2120	283_2	268_1	181_2	165_4	86_{100}	530	4264
Etomidate	G P U	1870	244_{16}	199_3	105_{100}	77_{22}		356	1924
Etomidate-M (HOOC-) ME	UME	1840	230_{11}	199_1	105_{100}	77_{18}		307	3371
Etonitazene		3375	396_1	135_4	107_4	86_{100}	58_{30}	1007	3655
Etonitazene intermediate-1		2515	267_1	196_1	117_1	86_{100}	58_{15}	527	2843
Etonitazene intermediate-2		2540	252_3	164_4	118_4	86_{100}	58_{38}	392	2844
Etonitazene intermediate-2 2AC		2745	336_1	321_1	118_1	86_{100}	58_8	789	2845
Etoricoxib	G	2750	357_{100}	278_{29}	263_3	243_4	202_3	876	7447
Etozoline		2390	284_{14}	251_{11}	211_{44}	154_{20}	84_{100}	534	3107
Etridiazole		1480	246_{11}	211_{100}	183_{83}	140_{48}	108_{34}	361	4051
Etridiazole artifact (deschloro-)		1320	212_{25}	184_{79}	149_{100}	141_{66}	106_{44}	247	4052
Etrimfos		1850	292_{40}	181_{52}	153_{59}	125_{61}	56_{100}	574	2509
Etryptamine		1860	188_7	131_{100}	130_{78}	103_{13}	58_{110}	191	5552
Etryptamine AC		2380	230_{10}	171_{82}	156_{24}	130_{100}	58_{74}	308	4694
Etryptamine ACPFP		2150	376_{18}	213_{24}	184_{42}	172_{100}	130_{32}	939	5556
Etryptamine HFB		1945	384_{17}	171_9	130_{100}	103_4	77_4	966	6196
Etryptamine PFP		1880	334_{17}	171_{10}	130_{100}	103_6	77_7	780	5555
Etryptamine TFA		1950	284_{19}	171_9	156_4	130_{100}	103_7	534	5558
Etryptamine 2HFB		1830	580_{10}	367_{95}	326_{100}	254_{62}	129_{17}	1191	6195
Etryptamine 2PFP		1840	480_{15}	317_{92}	276_{100}	204_{61}	129_{45}	1147	5554
Etryptamine 2TFA		1860	380_{27}	267_{98}	226_{100}	154_{64}	129_{32}	952	5557
Etryptamine 2TMS		1880	332_1	317_3	203_{17}	130_{100}	73_{60}	770	5559
Etryptamine formyl artifact		1890	200_{20}	169_7	143_{100}	115_3	58_6	219	5553
Exemestane		2580*	296_4	268_{11}	211_9	148_{100}	133_{18}	599	7621
Exemestane TMS		2590*	368_{29}	353_{19}	221_{25}	148_{67}	73_{80}	913	7622
Famciclovir		2430	321_{27}	278_{21}	262_{100}	202_{55}	136_{62}	718	7739
Famciclovir AC		2645	363_{16}	304_{90}	262_{100}	202_{73}	135_{77}	894	7741
Famciclovir HFB		2405	517_{11}	458_{100}	412_{45}	398_{42}	332_{38}	1173	7746
Famciclovir PFP		2380	467_2	407_{10}	348_{100}	334_{82}	308_{39}	1135	7743
Famciclovir TFA		2400	417_5	348_{76}	298_{76}	284_{100}	162_{68}	1056	7742

Famciclovir TMS   Table 1-8-1: Compounds in order of names

Name	Detected	RI	Typical ions and intensities					Page	Entry
Famciclovir TMS		2485	393 $_{47}$	378 $_{26}$	334 $_{62}$	318 $_{100}$	276 $_{63}$	995	7748
Famciclovir artifact (deacetyl) HFB		2299	475 $_{68}$	432 $_{41}$	416 $_{100}$	262 $_{53}$	202 $_{31}$	1144	7747
Famciclovir artifact (deacetyl) ME		2280	293 $_{78}$	278 $_{32}$	251 $_{41}$	163 $_{74}$	135 $_{100}$	581	7740
Famciclovir artifact (deacetyl) PFP		2340	425 $_{71}$	366 $_{150}$	262 $_{87}$	202 $_{72}$	135 $_{80}$	1069	7744
Famciclovir artifact (deacetyl) TFA		2350	375 $_{68}$	332 $_{52}$	316 $_{100}$	262 $_{47}$	202 $_{40}$	935	7745
Famciclovir artifact (deacetyl) TMS		2375	351 $_{83}$	292 $_{120}$	202 $_{88}$	163 $_{88}$	135 $_{83}$	850	7749
Famciclovir artifact (deacetyl) 2TMS		2430	423 $_{64}$	364 $_{67}$	348 $_{71}$	220 $_{49}$	73 $_{100}$	1067	7750
Famotidine artifact (sulfurylamine)		1625	96 $_{63}$	82 $_{7}$	80 $_{100}$	64 $_{29}$		98	6055
Famotidine artifact (sulfurylamine) ME		1345	110 $_{82}$	109 $_{69}$	94 $_{100}$	80 $_{4}$	64 $_{54}$	103	6057
Famotidine artifact (sulfurylamine) 2ME		1140	124 $_{64}$	94 $_{100}$	78 $_{13}$	60 $_{33}$		109	6056
Famprofazone		2965	377 $_{2}$	286 $_{100}$	229 $_{22}$	136 $_{7}$	91 $_{13}$	944	1968
Famprofazone-M (AM)		1160	134 $_{1}$	120 $_{1}$	91 $_{6}$	77 $_{1}$	65 $_{4}$	115	54
Famprofazone-M (AM)	U	1160	134 $_{1}$	120 $_{1}$	91 $_{6}$	65 $_{4}$	44 $_{100}$	115	5514
Famprofazone-M (AM) AC	U+UHYAC	1505	177 $_{1}$	118 $_{19}$	91 $_{11}$	86 $_{31}$	65 $_{5}$	165	55
Famprofazone-M (AM) AC	UAAC U+UHYAC	1505	177 $_{1}$	118 $_{19}$	91 $_{11}$	86 $_{31}$	44 $_{100}$	165	5515
Famprofazone-M (AM) HFB		1355	331 $_{1}$	240 $_{79}$	169 $_{21}$	118 $_{100}$	91 $_{53}$	762	5047
Famprofazone-M (AM) PFP		1330	281 $_{1}$	190 $_{73}$	118 $_{100}$	91 $_{36}$	65 $_{12}$	521	4379
Famprofazone-M (AM) TFA		1095	231 $_{1}$	140 $_{100}$	118 $_{92}$	91 $_{45}$	69 $_{19}$	309	4000
Famprofazone-M (AM) TMS		1190	192 $_{6}$	116 $_{100}$	100 $_{10}$	91 $_{11}$	73 $_{87}$	235	5581
Famprofazone-M (HO-metamfetamine)		1885	150 $_{1}$	135 $_{1}$	107 $_{5}$	77 $_{5}$	58 $_{100}$	148	1766
Famprofazone-M (HO-metamfetamine) TFA		1770	261 $_{1}$	154 $_{100}$	134 $_{68}$	110 $_{42}$	107 $_{41}$	425	6180
Famprofazone-M (HO-metamfetamine) TMSTFA		1690	333 $_{3}$	206 $_{72}$	179 $_{100}$	154 $_{53}$	73 $_{50}$	773	6228
Famprofazone-M (HO-metamfetamine) 2AC	U+UHYAC	1995	249 $_{1}$	176 $_{6}$	134 $_{7}$	100 $_{43}$	58 $_{100}$	376	1767
Famprofazone-M (HO-metamfetamine) 2HFB		1670	538 $_{1}$	330 $_{17}$	254 $_{100}$	210 $_{32}$	169 $_{22}$	1186	5076
Famprofazone-M (HO-metamfetamine) 2PFP		1605	295 $_{1}$	280 $_{18}$	204 $_{100}$	160 $_{47}$	119 $_{39}$	1123	5077
Famprofazone-M (HO-metamfetamine) 2TFA		1585	357 $_{1}$	230 $_{22}$	154 $_{100}$	110 $_{42}$	69 $_{29}$	872	5078
Famprofazone-M (HO-metamfetamine) 2TMS		1620	309 $_{10}$	206 $_{70}$	179 $_{100}$	154 $_{32}$	73 $_{40}$	663	6190
Famprofazone-M (HO-propyphenazone)	UHY	2410	246 $_{51}$	231 $_{100}$	215 $_{9}$	77 $_{16}$		364	912
Famprofazone-M (HO-propyphenazone) AC	U+UHYAC	2240	288 $_{62}$	273 $_{82}$	245 $_{99}$	232 $_{94}$	190 $_{39}$	553	206
Famprofazone-M (metamfetamine)	U	1195	148 $_{1}$	134 $_{2}$	115 $_{9}$	91 $_{9}$	58 $_{100}$	127	1093
Famprofazone-M (metamfetamine) AC	U+UHYAC	1575	191 $_{1}$	117 $_{2}$	100 $_{42}$	91 $_{6}$	58 $_{100}$	196	1094
Famprofazone-M (metamfetamine) HFB		1460	254 $_{100}$	210 $_{44}$	169 $_{15}$	118 $_{41}$	91 $_{38}$	827	5069
Famprofazone-M (metamfetamine) PFP		1415	204 $_{100}$	160 $_{46}$	118 $_{35}$	91 $_{25}$	69 $_{4}$	589	5070
Famprofazone-M (metamfetamine) TFA		1300	245 $_{1}$	154 $_{100}$	118 $_{48}$	110 $_{55}$	91 $_{23}$	358	3998
Famprofazone-M (metamfetamine) TMS		1325	206 $_{4}$	130 $_{100}$	91 $_{17}$	73 $_{83}$	59 $_{13}$	279	6214
Felbamate -C2H3NO2		1890	165 $_{2}$	134 $_{9}$	104 $_{100}$	91 $_{23}$	77 $_{21}$	147	4696
Felbamate -CH3NO2		2210	177 $_{9}$	134 $_{100}$	104 $_{63}$	91 $_{26}$	77 $_{22}$	165	4695
Felbamate-M/artifact (bis-decarbamoyl-) -H2O		1450*	134 $_{1}$	121 $_{11}$	104 $_{100}$	91 $_{26}$	77 $_{25}$	114	4698
Felbamate-M/artifact (bis-decarbamoyl-) -H2O AC		2010*	176 $_{11}$	134 $_{100}$	104 $_{75}$	91 $_{19}$	77 $_{20}$	163	4697
Felbinac ET		1980*	240 $_{46}$	167 $_{100}$	165 $_{29}$	152 $_{14}$	83 $_{4}$	344	6075
Felbinac ME		1960*	226 $_{33}$	167 $_{100}$	165 $_{23}$	152 $_{9}$	83 $_{5}$	294	6074
Felodipine	UME	2670	383 $_{7}$	354 $_{11}$	238 $_{100}$	210 $_{36}$	150 $_{7}$	962	4627
Felodipine ME		2725	397 $_{7}$	338 $_{16}$	324 $_{51}$	252 $_{100}$	224 $_{59}$	1008	4853
Felodipine HY		2240*	254 $_{6}$	210 $_{4}$	101 $_{3}$	82 $_{100}$	54 $_{15}$	396	6064
Felodipine-M (dehydro-COOH) ET	UET	2665	439 $_{1}$	404 $_{100}$	376 $_{40}$	344 $_{25}$	309 $_{18}$	1096	4862
Felodipine-M (dehydro-COOH) ME	UME	2570	425 $_{1}$	390 $_{100}$	362 $_{53}$	309 $_{8}$	245 $_{9}$	1068	4854
Felodipine-M (dehydro-COOH) TMS	UTMS	2840	448 $_{33}$	434 $_{46}$	343 $_{79}$	287 $_{100}$	117 $_{26}$	1150	5007
Felodipine-M (dehydro-deethyl-COOH) 2ME	UME	2520	411 $_{1}$	376 $_{100}$	352 $_{8}$	295 $_{3}$	172 $_{8}$	1043	4857
Felodipine-M (dehydro-deethyl-HO-) -H2O	UME	2235	351 $_{1}$	316 $_{100}$	301 $_{8}$	284 $_{4}$		846	4858
Felodipine-M (dehydro-demethyl-COOH) 2ET	UET	2600	453 $_{1}$	418 $_{100}$	390 $_{40}$	344 $_{78}$	244 $_{21}$	1118	4864
Felodipine-M (dehydro-demethyl-HO-) -H2O	UME	2560	365 $_{1}$	330 $_{81}$	302 $_{100}$	267 $_{9}$	164 $_{7}$	899	4859
Felodipine-M (dehydro-HO-)	UET	2430	397 $_{5}$	362 $_{96}$	334 $_{100}$	295 $_{22}$	260 $_{22}$	1008	4863
Felodipine-M/artifact (dehydro-)	UME	2280	381 $_{1}$	346 $_{100}$	318 $_{76}$	286 $_{18}$	173 $_{32}$	954	4855
Felodipine-M/artifact (dehydro-deethyl-) ME	UME	2235	367 $_{1}$	332 $_{100}$	300 $_{7}$	258 $_{6}$	173 $_{7}$	907	4856
Felodipine-M/artifact (dehydro-deethyl-) TMS	UTMS	2610	390 $_{100}$	380 $_{8}$	362 $_{41}$	164 $_{5}$	139 $_{2}$	1069	5005
Felodipine-M/artifact (dehydro-deethyl-) -CO2	UME	2235	309 $_{1}$	274 $_{100}$	259 $_{11}$	215 $_{4}$	139 $_{5}$	658	4860
Felodipine-M/artifact (dehydro-demethyl-) ET	UET	2375	395 $_{1}$	360 $_{96}$	332 $_{100}$	286 $_{21}$	173 $_{10}$	1000	4861
Felodipine-M/artifact (dehydro-demethyl-deethyl-) -CO2 TMS	UTMS	2250	332 $_{100}$	300 $_{4}$	257 $_{2}$	173 $_{3}$	137 $_{2}$	907	5006
Fenamiphos		2020	303 $_{100}$	260 $_{26}$	217 $_{27}$	154 $_{43}$	80 $_{27}$	627	3436
Fenarimol		2605	330 $_{38}$	251 $_{44}$	219 $_{49}$	139 $_{100}$	107 $_{80}$	757	3437
Fenazepam		2440	350 $_{42}$	348 $_{72}$	321 $_{100}$	313 $_{71}$	177 $_{31}$	837	5850
Fenazepam TMS		2790	422 $_{100}$	420 $_{72}$	405 $_{48}$	385 $_{80}$	73 $_{68}$	1062	5853
Fenazepam artifact-1	U UHY U+UHYAC	2230	320 $_{100}$	318 $_{79}$	283 $_{47}$	239 $_{66}$	75 $_{34}$	701	2152
Fenazepam artifact-2	U UHY U+UHYAC	2250	334 $_{98}$	332 $_{79}$	297 $_{50}$	253 $_{100}$	75 $_{86}$	766	2153
Fenazepam HY	UHY	2270	311 $_{100}$	309 $_{71}$	276 $_{88}$	274 $_{90}$	195 $_{41}$	658	2151
Fenazepam HYAC	U+UHYAC	2500	353 $_{44}$	351 $_{32}$	311 $_{85}$	276 $_{100}$	274 $_{100}$	846	2149
Fenazepam iso-1 ME		2395	364 $_{100}$	362 $_{82}$	327 $_{81}$	212 $_{21}$	125 $_{24}$	889	5851
Fenazepam iso-2 ME		2530	364 $_{100}$	362 $_{85}$	336 $_{86}$	327 $_{95}$	299 $_{45}$	889	5852
Fenazepam-M HY	UHY	2270	311 $_{100}$	309 $_{71}$	276 $_{88}$	274 $_{90}$	195 $_{41}$	658	2151
Fenazepam-M HYAC	U+UHYAC	2500	353 $_{44}$	351 $_{32}$	311 $_{85}$	276 $_{100}$	274 $_{100}$	846	2149
Fenbendazole ME		2965	313 $_{100}$	281 $_{12}$	254 $_{31}$	225 $_{7}$		679	7407
Fenbendazole 2ME		2935	327 $_{100}$	239 $_{14}$	59 $_{26}$	268 $_{75}$	254 $_{9}$	743	7409
Fenbendazole artifact (decarbamoyl-) AC		2910	297 $_{59}$	255 $_{100}$	225 $_{6}$	208 $_{10}$	195 $_{5}$	601	7412
Fenbendazole artifact (decarbamoyl-) AC		2930	283 $_{79}$	241 $_{100}$	209 $_{11}$	199 $_{11}$	171 $_{17}$	527	7411

Table 1-8-1: Compounds in order of names Fenoprop

Name	Detected	RI	Typical ions and intensities					Page	Entry
Fenbendazole artifact (decarbamoyl-) ME		2985	255 $_{100}$	239 $_{12}$	225 $_{11}$	199 $_{6}$	171 $_{9}$	401	7408
Fenbendazole artifact (decarbamoyl-) 2ME		2700	269 $_{100}$	254 $_{13}$	241 $_{7}$	227 $_{4}$	184 $_{9}$	464	7410
Fenbuconazole		2665	336 $_{1}$	211 $_{6}$	198 $_{48}$	129 $_{100}$	125 $_{39}$	787	6089
Fenbufen		2010*	254 $_{12}$	181 $_{100}$	152 $_{38}$	127 $_{4}$	76 $_{10}$	398	5245
Fenbufen ME		1975*	268 $_{14}$	237 $_{9}$	181 $_{100}$	152 $_{43}$	76 $_{10}$	460	5246
Fenbufen-M (acetic acid HO-) 2ME	UME	2200*	256 $_{50}$	197 $_{100}$	182 $_{13}$	154 $_{30}$	128 $_{20}$	406	6292
Fenbufen-M (dihydro-) ME	UME	1995*	270 $_{50}$	211 $_{100}$	178 $_{21}$	165 $_{23}$	152 $_{16}$	470	6291
Fenbutrazate	U	2680	367 $_{11}$	261 $_{34}$	190 $_{28}$	91 $_{80}$	69 $_{100}$	911	773
Fencamfamine	G U UA UHY	1685	215 $_{43}$	186 $_{18}$	98 $_{100}$	84 $_{49}$	58 $_{68}$	258	774
Fencamfamine AC	U+UHYAC	2085	257 $_{4}$	170 $_{100}$	142 $_{97}$	58 $_{74}$		411	775
Fencamfamine HFB		1795	342 $_{2}$	280 $_{8}$	170 $_{100}$	142 $_{95}$	91 $_{47}$	1044	6305
Fencamfamine PFP		1755	292 $_{2}$	230 $_{11}$	170 $_{100}$	142 $_{88}$	91 $_{34}$	887	6304
Fencamfamine TFA		1970	242 $_{2}$	180 $_{9}$	170 $_{95}$	142 $_{100}$	91 $_{39}$	671	3699
Fencamfamine TMS		1780	287 $_{33}$	272 $_{21}$	258 $_{43}$	170 $_{44}$	73 $_{46}$	549	6306
Fencamfamine-M (deethyl-) AC	U+UHYAC	2005	229 $_{5}$	170 $_{99}$	142 $_{79}$			304	776
Fencamfamine-M (deethyl-HO-) 2AC	U+UHYAC	2305	287 $_{6}$	228 $_{25}$	168 $_{36}$	142 $_{100}$		547	777
Fencarbamide		2470	326 $_{1}$	196 $_{7}$	169 $_{37}$	99 $_{23}$	86 $_{100}$	750	1444
Fenchlorphos		1905*	320 $_{12}$	285 $_{94}$	167 $_{6}$	125 $_{100}$	79 $_{19}$	713	3438
Fendiline	U UHY	2450	315 $_{26}$	181 $_{13}$	167 $_{18}$	132 $_{33}$	105 $_{99}$	691	1445
Fendiline AC	U+UHYAC	2825	357 $_{46}$	162 $_{25}$	110 $_{25}$	105 $_{99}$	72 $_{79}$	874	1446
Fendiline-M (deamino-HO-) -H2O	UHY U+UHYAC	1940*	194 $_{49}$	167 $_{100}$	165 $_{67}$	152 $_{34}$	116 $_{17}$	205	3388
Fendiline-M (HO-)	UHY	2785	331 $_{53}$	316 $_{14}$	197 $_{35}$	120 $_{28}$	105 $_{100}$	766	3389
Fendiline-M (HO-) 2AC	U+UHYAC	3275	415 $_{35}$	251 $_{13}$	177 $_{52}$	105 $_{100}$	72 $_{95}$	1053	3394
Fendiline-M (HO-methoxy-)	UHY	2820	361 $_{32}$	227 $_{45}$	120 $_{35}$	105 $_{100}$	91 $_{21}$	888	3390
Fendiline-M (HO-methoxy-) 2AC	U+UHYAC	3410	445 $_{54}$	239 $_{59}$	177 $_{76}$	105 $_{100}$	72 $_{84}$	1108	3395
Fendiline-M (N-dealkyl-) AC	U+UHYAC	2320	253 $_{15}$	193 $_{12}$	165 $_{19}$	152 $_{10}$	73 $_{100}$	395	3391
Fendiline-M (N-dealkyl-HO-) 2AC	U+UHYAC	2635	311 $_{54}$	269 $_{12}$	239 $_{21}$	183 $_{63}$	73 $_{100}$	672	3392
Fendiline-M (N-dealkyl-HO-methoxy-) 2AC	U+UHYAC	2700	341 $_{10}$	299 $_{54}$	213 $_{74}$	152 $_{14}$	73 $_{100}$	812	3393
Fenetylline	G P-I U UHY	2830	326 $_{1}$	250 $_{100}$	207 $_{34}$	91 $_{23}$	70 $_{28}$	814	778
Fenetylline AC	U+UHYAC	3110	383 $_{9}$	292 $_{25}$	250 $_{99}$	207 $_{50}$		964	779
Fenetylline HFB		2815	537 $_{7}$	446 $_{54}$	266 $_{35}$	207 $_{22}$	91 $_{100}$	1180	5054
Fenetylline PFP		2790	487 $_{9}$	396 $_{90}$	369 $_{59}$	207 $_{43}$	91 $_{100}$	1155	5055
Fenetylline TFA		2840	437 $_{9}$	346 $_{52}$	319 $_{40}$	166 $_{65}$	91 $_{100}$	1093	5056
Fenetylline-M (AM)		1160	134 $_{1}$	120 $_{1}$	91 $_{6}$	77 $_{1}$	65 $_{4}$	115	54
Fenetylline-M (AM)	U	1160	134 $_{1}$	120 $_{1}$	91 $_{6}$	65 $_{4}$	44 $_{100}$	115	5514
Fenetylline-M (AM) AC	U+UHYAC	1505	177 $_{1}$	118 $_{19}$	91 $_{11}$	86 $_{31}$	65 $_{5}$	165	55
Fenetylline-M (AM) AC	UAAC U+UHYAC	1505	177 $_{1}$	118 $_{19}$	91 $_{11}$	86 $_{31}$	44 $_{100}$	165	5515
Fenetylline-M (AM) HFB		1355	331 $_{1}$	240 $_{79}$	169 $_{21}$	118 $_{100}$	91 $_{53}$	762	5047
Fenetylline-M (AM) PFP		1330	281 $_{1}$	190 $_{73}$	118 $_{100}$	91 $_{36}$	65 $_{12}$	521	4379
Fenetylline-M (AM) TFA		1095	231 $_{1}$	140 $_{49}$	118 $_{92}$	91 $_{45}$	69 $_{19}$	309	4000
Fenetylline-M (AM) TMS		1190	192 $_{6}$	116 $_{100}$	100 $_{10}$	91 $_{11}$	73 $_{87}$	235	5581
Fenetylline-M (AM)-D5 AC	UAAC U+UHYAC	1480	182 $_{3}$	122 $_{46}$	92 $_{30}$	90 $_{100}$	66 $_{16}$	181	5690
Fenetylline-M (AM)-D5 HFB		1330	244 $_{100}$	169 $_{14}$	122 $_{46}$	92 $_{41}$	69 $_{40}$	787	6316
Fenetylline-M (AM)-D5 PFP		1320	194 $_{100}$	123 $_{42}$	119 $_{32}$	92 $_{46}$	69 $_{11}$	543	5566
Fenetylline-M (AM)-D5 TFA		1085	144 $_{100}$	123 $_{53}$	122 $_{56}$	92 $_{51}$	69 $_{28}$	330	5570
Fenetylline-M (AM)-D5 TMS		1180	212 $_{1}$	197 $_{8}$	120 $_{100}$	92 $_{11}$	73 $_{57}$	251	5582
Fenetylline-M (AM)-D11 PFP		1610	292 $_{1}$	194 $_{100}$	128 $_{82}$	98 $_{43}$	70 $_{14}$	575	7284
Fenetylline-M (AM)-D11 TFA		1615	242 $_{1}$	144 $_{100}$	128 $_{82}$	98 $_{43}$	70 $_{14}$	353	7283
Fenetylline-M (etofylline)	UHY	2125	224 $_{47}$	194 $_{17}$	180 $_{100}$	109 $_{35}$	95 $_{74}$	288	771
Fenetylline-M (etofylline) AC	U+UHYAC	2200	266 $_{79}$	206 $_{59}$	180 $_{34}$	122 $_{31}$	87 $_{100}$	448	772
Fenetylline-M (etofylline) TMS	UHYTMS	2160	296 $_{10}$	281 $_{26}$	252 $_{6}$	180 $_{100}$	73 $_{47}$	599	5696
Fenetylline-M (N-dealkyl-) AC	U+UHYAC	2480	265 $_{35}$	206 $_{100}$	180 $_{35}$	122 $_{18}$	86 $_{27}$	443	1886
Fenfluramine	G P U	1250	230 $_{1}$	216 $_{2}$	159 $_{7}$	72 $_{100}$		309	780
Fenfluramine AC	U+UHYAC	1580	254 $_{1}$	216 $_{1}$	159 $_{6}$	114 $_{33}$	72 $_{100}$	481	781
Fenfluramine HFB		1495	427 $_{1}$	408 $_{6}$	268 $_{100}$	240 $_{54}$	159 $_{45}$	1073	5057
Fenfluramine PFP		1455	377 $_{1}$	358 $_{3}$	218 $_{100}$	190 $_{59}$	159 $_{35}$	942	5058
Fenfluramine TFA		1455	327 $_{1}$	308 $_{3}$	186 $_{7}$	168 $_{100}$	140 $_{48}$	743	5059
Fenfluramine-M (deethyl-) AC	U+UHYAC	1510	245 $_{2}$	226 $_{5}$	186 $_{4}$	159 $_{13}$	86 $_{100}$	358	782
Fenfluramine-M (deethyl-HO-) 2AC	U+UHYAC	1980	176 $_{4}$	159 $_{12}$	133 $_{10}$	107 $_{14}$	86 $_{100}$	627	5657
Fenfluramine-M (di-HO-) 3AC	U+UHYAC	2585	191 $_{1}$	150 $_{2}$	114 $_{49}$	72 $_{100}$		983	5656
Fenfluramine-M (HO-) 2AC	U+UHYAC	1895	175 $_{1}$	162 $_{1}$	134 $_{2}$	114 $_{47}$	72 $_{100}$	764	4472
Fenfuram		1900	201 $_{26}$	184 $_{1}$	144 $_{1}$	109 $_{100}$	65 $_{7}$	221	2532
Fenitrothion		1925	277 $_{39}$	260 $_{24}$	125 $_{100}$	109 $_{90}$	79 $_{53}$	498	2510
Fenitrothion-M/artifact (3-methyl-4-nitrophenol)		1560	153 $_{34}$	136 $_{68}$	108 $_{11}$	77 $_{100}$	53 $_{77}$	132	7537
Fenitrothion-M/artifact (3-methyl-4-nitrophenol) AC		1455	195 $_{20}$	153 $_{83}$	136 $_{200}$	108 $_{18}$	77 $_{59}$	206	7538
Fenofibrate		2515*	360 $_{29}$	273 $_{62}$	232 $_{60}$	139 $_{61}$	121 $_{100}$	884	1940
Fenofibrate-M (HOOC-) ME	P UME U UHY U+UHYAC	2430*	332 $_{19}$	273 $_{13}$	232 $_{39}$	139 $_{29}$	121 $_{100}$	767	3039
Fenofibrate-M (O-dealkyl-)	UHY	2300*	232 $_{36}$	197 $_{7}$	139 $_{23}$	121 $_{100}$	111 $_{23}$	311	2240
Fenofibrate-M (O-dealkyl-) AC	U+UHYAC	2230*	274 $_{7}$	232 $_{43}$	139 $_{25}$	121 $_{100}$	111 $_{27}$	484	2230
Fenoprofen		2035*	242 $_{85}$	197 $_{100}$	104 $_{25}$	91 $_{44}$	77 $_{46}$	352	5112
Fenoprofen ME		1970*	256 $_{81}$	197 $_{100}$	181 $_{25}$	103 $_{29}$	91 $_{36}$	406	5111
Fenoprofen artifact		1765*	212 $_{64}$	197 $_{100}$	169 $_{19}$	141 $_{40}$	115 $_{23}$	247	5113
Fenoprofen-M (HO-) 2ME	UME	2130*	286 $_{100}$	227 $_{51}$	123 $_{17}$	91 $_{8}$	152 $_{7}$	543	6290
Fenoprop	P-I G U	1760*	268 $_{17}$	223 $_{8}$	196 $_{100}$	167 $_{13}$	97 $_{33}$	457	783

Fenoprop ME Table 1-8-1: Compounds in order of names

Name	Detected	RI	Typical ions and intensities					Page	Entry
Fenoprop ME		1720*	282_{31}	223_{35}	196_{100}	87_{31}	59_{92}	525	2397
Fenoprop-M (2,4,5-trichlorophenol)	U	1440*	198_{93}	196_{100}	132_{34}	97_{64}	73_{14}	209	784
Fenoterol -H2O 4AC	U+UHYAC	3440	453_{7}	304_{19}	262_{100}	220_{19}	107_{11}	1118	3146
Fenoxaprop-ethyl	U+UHYAC	2615	361_{77}	288_{100}	261_{28}	119_{35}	76_{58}	886	4120
Fenoxaprop-ethyl-M/artifact (phenol)		1630*	210_{30}	137_{64}	110_{100}	81_{35}	65_{27}	244	4121
Fenpipramide		2690	322_{1}	238_{2}	211_{6}	112_{40}	98_{100}	724	785
Fenpipramide TMS		2690	394_{1}	283_{38}	112_{56}	98_{100}	73_{66}	1000	4614
Fenpropathrin		2450	349_{3}	265_{15}	208_{13}	181_{48}	97_{100}	843	3843
Fenpropemorph		2010	303_{9}	147_{2}	128_{10}	91_{5}	70_{9}	632	3439
Fenproporex	U	1585	173_{1}	132_{2}	97_{100}	91_{18}	56_{31}	191	786
Fenproporex AC	U+UHYAC	1915	139_{18}	118_{5}	97_{100}	91_{13}	56_{23}	308	787
Fenproporex HFB		1730	293_{100}	240_{33}	118_{61}	91_{48}	56_{54}	966	5060
Fenproporex PFP		1685	243_{100}	190_{50}	118_{77}	91_{44}	56_{52}	780	5061
Fenproporex TFA		1705	193_{100}	140_{47}	118_{62}	91_{42}	56_{37}	534	5062
Fenproporex-M	UHY	1465	181_{9}	138_{100}	122_{18}	94_{24}	77_{16}	178	4351
Fenproporex-M (AM)		1160	134_{1}	120_{1}	91_{6}	77_{1}	65_{4}	115	54
Fenproporex-M (AM)	U	1160	134_{1}	120_{1}	91_{6}	65_{4}	44_{100}	115	5514
Fenproporex-M (AM) AC	U+UHYAC	1505	177_{1}	118_{19}	91_{11}	86_{31}	65_{5}	165	55
Fenproporex-M (AM) AC	UAAC U+UHYAC	1505	177_{1}	118_{19}	91_{11}	86_{31}	44_{100}	165	5515
Fenproporex-M (AM) HFB		1355	331_{1}	240_{79}	169_{21}	118_{100}	91_{53}	762	5047
Fenproporex-M (AM) PFP		1330	281_{1}	190_{73}	118_{100}	91_{36}	65_{12}	521	4379
Fenproporex-M (AM) TFA		1095	231_{1}	140_{100}	118_{92}	91_{45}	69_{19}	309	4000
Fenproporex-M (AM) TMS		1190	192_{6}	116_{100}	100_{10}	91_{11}	73_{87}	235	5581
Fenproporex-M (AM) formyl artifact		1100	147_{2}	146_{6}	125_{5}	91_{12}	56_{100}	125	3261
Fenproporex-M (di-HO-) 3AC	U+UHYAC	2575	346_{1}	234_{7}	192_{7}	150_{11}	97_{100}	832	4386
Fenproporex-M (HO-) iso-1 2AC	U+UHYAC	2260	288_{1}	176_{7}	139_{13}	134_{9}	97_{100}	553	4383
Fenproporex-M (HO-) iso-2 2AC	U+UHYAC	2350	288_{1}	176_{13}	139_{13}	134_{18}	97_{100}	553	4384
Fenproporex-M (HO-methoxy-) 2AC	U+UHYAC	2495	318_{1}	206_{12}	164_{31}	137_{11}	97_{100}	703	4385
Fenproporex-M (N-dealkyl-3-HO-) TMSTFA		1630	319_{8}	206_{86}	191_{32}	140_{100}	73_{58}	708	6141
Fenproporex-M (N-dealkyl-3-HO-) 2AC	U+UHYAC	1930	235_{2}	176_{48}	134_{52}	107_{21}	86_{100}	324	4387
Fenproporex-M (N-dealkyl-3-HO-) 2HFB		1620	330_{30}	303_{11}	240_{100}	169_{15}	69_{29}	1182	5737
Fenproporex-M (N-dealkyl-3-HO-) 2PFP		1520	280_{36}	253_{9}	190_{100}	119_{19}	69_{8}	1104	5738
Fenproporex-M (N-dealkyl-3-HO-) 2TFA		<1000	230_{4}	203_{1}	140_{14}	115_{1}		819	6224
Fenproporex-M (N-dealkyl-3-HO-) 2TMS	UHYTMS	1850	280_{11}	179_{3}	116_{100}	100_{12}	73_{72}	594	5693
Fenproporex-M (N-dealkyl-4-HO-)		1480	151_{10}	107_{69}	91_{10}	77_{42}	56_{100}	129	1802
Fenproporex-M (N-dealkyl-4-HO-) AC	U+UHYAC	1890	193_{1}	134_{100}	107_{26}	86_{25}	77_{16}	201	1803
Fenproporex-M (N-dealkyl-4-HO-) TFA		1670	247_{4}	140_{15}	134_{54}	107_{100}	77_{15}	366	6335
Fenproporex-M (N-dealkyl-4-HO-) 2AC	U+UHYAC	1900	235_{1}	176_{72}	134_{100}	107_{47}	86_{71}	324	1804
Fenproporex-M (N-dealkyl-4-HO-) 2HFB		<1000	330_{48}	303_{15}	240_{100}	169_{44}	69_{42}	1182	6326
Fenproporex-M (N-dealkyl-4-HO-) 2PFP		<1000	280_{77}	253_{16}	190_{100}	119_{56}	69_{16}	1104	6325
Fenproporex-M (N-dealkyl-4-HO-) 2TFA		<1000	230_{94}	203_{14}	140_{130}	92_{15}	69_{76}	819	6324
Fenproporex-M (N-dealkyl-4-HO-) 2TMS		<1000	280_{7}	179_{9}	149_{8}	116_{100}	73_{78}	594	6327
Fenproporex-M (N-dealkyl-di-HO-) 3AC	U+UHYAC	2150	293_{2}	234_{54}	192_{48}	150_{99}	86_{100}	579	3725
Fenproporex-M (N-dealkyl-HO-methoxy-) 2HFB	UHFB	1690	360_{82}	333_{15}	240_{100}	169_{42}	69_{39}	1189	6512
Fenproporex-M (norephedrine) TMSTFA		1890	240_{8}	198_{3}	179_{100}	117_{5}	73_{88}	708	6146
Fenproporex-M (norephedrine) 2AC	U+UHYAC	1805	235_{1}	107_{13}	86_{100}	176_{5}	134_{7}	325	2476
Fenproporex-M (norephedrine) 2HFB	UHYHFB	1455	543_{1}	330_{14}	240_{100}	169_{44}	69_{57}	1183	5098
Fenproporex-M (norephedrine) 2PFP	UHYPFP	1380	443_{1}	280_{7}	190_{100}	119_{59}	105_{26}	1104	5094
Fenproporex-M (norephedrine) 2TFA	UTFA	1355	343_{1}	230_{6}	203_{5}	140_{100}	69_{29}	819	5091
Fenproporex-M 2AC	U+UHYAC	2065	265_{3}	206_{27}	164_{100}	137_{23}	86_{33}	445	3498
Fenproporex-M 2AC	U+UHYAC	1820*	266_{3}	206_{9}	164_{100}	150_{10}	137_{30}	450	6409
Fenproporex-M formyl art.		1220	163_{3}	148_{4}	107_{30}	77_{13}	56_{100}	142	6323
Fenproporex-M-D5 AC	UAAC U+UHYAC	1480	182_{3}	122_{46}	92_{30}	90_{100}	66_{16}	181	5690
Fenproporex-M-D5 HFB		1330	244_{100}	169_{14}	122_{46}	92_{46}	69_{40}	787	6316
Fenproporex-M-D5 PFP		1320	194_{100}	123_{42}	119_{32}	92_{46}	69_{11}	543	5566
Fenproporex-M-D5 TFA		1085	144_{100}	123_{53}	122_{56}	92_{51}	69_{28}	330	5570
Fenproporex-M-D5 TMS		1180	212_{1}	197_{8}	120_{100}	92_{11}	73_{57}	251	5582
Fenproporex-M-D11 PFP		1610	292_{1}	194_{100}	128_{82}	98_{43}	70_{14}	575	7284
Fenproporex-M-D11 TFA		1615	242_{1}	144_{100}	128_{82}	98_{43}	70_{14}	353	7283
Fenson		1980*	268_{28}	141_{76}	99_{14}	77_{100}	51_{35}	458	3440
Fensulfothion		2250*	308_{59}	141_{48}	125_{68}	97_{81}		653	1447
Fensulfothion impurity		1910*	292_{100}	156_{53}	140_{46}	125_{31}	97_{47}	574	1452
Fentanyl	P-I U+UHYAC	2720	336_{1}	245_{100}	189_{62}	146_{94}		790	788
Fentanyl-D5		2710	341_{1}	250_{100}	207_{13}	194_{29}	151_{53}	815	7368
Fenthion	G	1930*	278_{100}	169_{35}	125_{77}	109_{63}	79_{33}	505	3838
Fenticonazole		3410	454_{5}	209_{8}	199_{100}	185_{7}	81_{10}	1119	6088
Fenuron ME		1405	178_{13}	133_{10}	106_{29}	72_{100}		168	3967
Fenvalerate iso-1		2890	419_{5}	225_{25}	181_{44}	167_{72}	125_{100}	1060	3839
Fenvalerate iso-2		3839	419_{5}	225_{25}	181_{43}	167_{73}	125_{100}	1061	3840
Ferulic acid ME		1930*	208_{100}	177_{80}	145_{53}	133_{19}	117_{24}	236	5966
Ferulic acid MEAC		1950*	250_{3}	208_{100}	177_{47}	145_{26}	117_{9}	379	5814
Ferulic acid 2ME	UME	1850*	222_{100}	207_{22}	191_{75}	164_{14}	147_{19}	281	4945
Ferulic acid 2TMS		2160*	338_{92}	323_{56}	308_{48}	249_{45}	73_{100}	797	5815
Ferulic acid -CO2		1195*	150_{34}	135_{33}	107_{68}	77_{100}	51_{68}	127	5752

Table 1-8-1: Compounds in order of names

Flunitrazepam-D7

Name	Detected	RI	Typical ions and intensities					Page	Entry
Ferulic acid glycine conjugate ME		2380	265 $_{75}$	204 $_7$	177 $_{100}$	145 $_{47}$	117 $_{13}$	443	5766
Ferulic acid glycine conjugate 2ME		2450	279 $_{41}$	191 $_{100}$	163 $_{12}$	148 $_5$	133 $_5$	510	5825
Ferulic acid glycine conjugate 3TMS		2540	467 $_{61}$	453 $_8$	336 $_{22}$	249 $_{67}$	73 $_{100}$	1135	5826
Fexofenadine 2TMS		3950	645 $_{18}$	280 $_{67}$	183 $_{18}$	105 $_{25}$	73 $_{100}$	1199	7731
Fexofenadine -H2O 2TMS		3690	627 $_{11}$	262 $_{82}$	248 $_{25}$	129 $_8$	73 $_{100}$	1197	7732
Fexofenadine -H2O -CO2	U+UHYAC	3650	439 $_6$	280 $_{100}$	262 $_{12}$	131 $_{10}$	105 $_{17}$	1097	5223
Fexofenadine-M (benzophenone)	U+UHYAC	1610*	182 $_{31}$	152 $_3$	105 $_{100}$	77 $_{70}$	51 $_{39}$	180	1624
Fexofenadine-M (N-dealkyl-oxo-) -2H2O	U+UHYAC	2190	245 $_{82}$	167 $_{100}$	152 $_{17}$	139 $_{21}$	115 $_{16}$	359	2218
Fipexide		3090	388 $_7$	261 $_8$	253 $_8$	176 $_{17}$	135 $_{100}$	980	6718
Fipexide Cl-artifact		1295*	170 $_{25}$	135 $_{100}$	105 $_8$	77 $_{41}$		156	6635
Fipexide-M (deethylene-MDBP) 2AC	U+UHYAC	2320	278 $_4$	235 $_{64}$	177 $_{11}$	150 $_{25}$	135 $_{100}$	506	6626
Fipexide-M (deethylene-MDBP) 2HFB	U+UHYHFB	2080	586 $_3$	389 $_{24}$	346 $_7$	240 $_7$	135 $_{100}$	1193	6632
Fipexide-M (deethylene-MDBP) 2TFA	U+UHYTFA	2230	386 $_8$	289 $_{19}$	246 $_{10}$	150 $_4$	135 $_{100}$	972	6629
Fipexide-M (HO-methoxy-BZP) AC	U+UHYAC	2410	264 $_9$	192 $_8$	137 $_{100}$	122 $_{18}$	85 $_{42}$	440	6509
Fipexide-M (HO-methoxy-BZP) HFB	U+UHYHFB	2135	418 $_{22}$	295 $_{15}$	281 $_{27}$	138 $_{67}$	137 $_{100}$	1058	6575
Fipexide-M (HO-methoxy-BZP) TFA	U+UHYTFA	2120	318 $_8$	181 $_{10}$	137 $_{100}$	122 $_{21}$	69 $_{12}$	703	6570
Fipexide-M (HO-methoxy-BZP) 2AC	U+UHYAC	2380	306 $_2$	234 $_9$	179 $_{13}$	137 $_{100}$	85 $_{64}$	645	6508
Fipexide-M (N-dealkyl-) AC		2460	296 $_{20}$	169 $_{23}$	155 $_{62}$	113 $_{100}$	85 $_{77}$	597	6809
Fipexide-M (N-dealkyl-deethylene-) AC		2280	270 $_{18}$	211 $_{45}$	185 $_{48}$	111 $_{51}$	87 $_{100}$	468	6810
Fipexide-M (piperazine) 2AC		1750	170 $_{15}$	85 $_{33}$	69 $_{25}$	56 $_{100}$		157	879
Fipexide-M (piperazine) 2HFB		1290	478 $_3$	459 $_9$	309 $_{100}$	281 $_{22}$	252 $_{41}$	1146	6634
Fipexide-M (piperazine) 2TFA		1005	278 $_{14}$	209 $_{59}$	152 $_{25}$	69 $_{56}$	56 $_{100}$	505	4129
Fipexide-M/artifact (HOOC-)		1770*	186 $_{100}$	141 $_{93}$	128 $_{74}$	111 $_{87}$	99 $_{50}$	187	1881
Fipexide-M/artifact (HOOC-) ME		1510*	200 $_{98}$	141 $_{100}$	111 $_{59}$	99 $_{17}$	75 $_{38}$	218	1077
Fipexide-M/artifcat (MDBP)		1890	220 $_{21}$	178 $_{14}$	164 $_{13}$	135 $_{100}$	85 $_{36}$	274	6624
Fipexide-M/artifcat (MDBP) AC	U+UHYAC	2350	262 $_{16}$	190 $_{11}$	176 $_{21}$	135 $_{100}$	85 $_{45}$	431	6625
Fipexide-M/artifcat (MDBP) TMS		2080	292 $_{87}$	157 $_{53}$	135 $_{100}$	102 $_{57}$	73 $_{85}$	576	6887
Flamprop-isopropyl		2225	363 $_1$	276 $_4$	156 $_3$	105 $_{100}$	77 $_{27}$	894	3844
Flamprop-methyl		2155	335 $_1$	276 $_1$	156 $_3$	105 $_{100}$	77 $_{48}$	782	3845
Flecainide	P-I G U UHY	2520	395 $_1$	301 $_6$	209 $_3$	97 $_{10}$	84 $_{100}$	1050	2822
Flecainide AC	U+UHYAC	2515	456 $_3$	301 $_8$	218 $_2$	126 $_{75}$	84 $_{100}$	1122	1449
Flecainide 2TMS		2520	558 $_1$	543 $_3$	301 $_8$	156 $_{100}$	73 $_{58}$	1187	4545
Flecainide formyl artifact	P G U UHY	2500	426 $_{71}$	301 $_{79}$	218 $_{19}$	125 $_{100}$	97 $_{37}$	1071	1448
Flecainide-M (HO-) 2AC	U+UHYAC	2680	514 $_2$	301 $_{23}$	184 $_{29}$	142 $_{100}$	100 $_{29}$	1172	2868
Flecainide-M (O-dealkyl-) 2AC	U+UHYAC	2780	416 $_2$	301 $_7$	219 $_{16}$	126 $_{84}$	84 $_{100}$	1054	2390
Fluanisone	U UHY U+UHYAC	2795	356 $_{79}$	218 $_{100}$	205 $_{100}$	162 $_{50}$	123 $_{69}$	870	172
Fluanisone-M	U+UHYAC	1490*	180 $_{25}$	125 $_{49}$	123 $_{35}$	95 $_{17}$	56 $_{100}$	174	85
Fluanisone-M (N,O-bis-dealkyl-) 2AC	U+UHYAC	2140	262 $_{10}$	220 $_{18}$	148 $_{100}$	120 $_{51}$	86 $_{12}$	431	170
Fluanisone-M (O-demethyl-)	UHY	2715	342 $_{52}$	194 $_{100}$	165 $_{84}$	123 $_{83}$		818	495
Fluanisone-M (O-demethyl-) AC	U+UHYAC	2830	384 $_{33}$	246 $_{88}$	233 $_{100}$	123 $_{100}$		968	173
Fluanisone-M/artifact AC	U+UHYAC	2445	292 $_{100}$	250 $_{45}$	178 $_{82}$	154 $_{92}$	123 $_{44}$	576	496
Fluazifop-butyl		2200	383 $_{11}$	282 $_{78}$	146 $_{97}$	91 $_{100}$	57 $_{98}$	963	3846
Flubenzimine		2430	416 $_{100}$	212 $_{27}$	186 $_{64}$	135 $_{66}$	77 $_{51}$	1053	3847
Fluchloralin		1800	355 $_1$	326 $_{30}$	306 $_{34}$	264 $_{25}$	63 $_{100}$	864	3841
Fluconazole	P U+UHYAC	2210	224 $_{100}$	155 $_6$	141 $_{22}$	127 $_{64}$	82 $_{61}$	644	4349
Fludiazepam		2530	302 $_{100}$	301 $_{90}$	274 $_{99}$	239 $_{18}$	183 $_{16}$	623	3069
Fludiazepam HY		2180	263 $_{100}$	246 $_{51}$	211 $_{33}$	95 $_{30}$	75 $_{26}$	433	3070
Fludiazepam-M (nor-)	G P-I UGLUC	2470	288 $_{100}$	287 $_{64}$	260 $_{93}$	259 $_{75}$		551	508
Fludiazepam-M (nor-) TMS		2470	360 $_{44}$	359 $_{48}$	341 $_{30}$	197 $_{12}$	73 $_{70}$	883	4621
Fludiazepam-M (nor-) artifact	U+UHYAC	2050	272 $_{70}$	271 $_{100}$	237 $_{78}$	151 $_{11}$	110 $_8$	477	2409
Fludiazepam-M (nor-) HY	UHY	2030	249 $_{97}$	154 $_{33}$	123 $_{45}$	95 $_{41}$		373	512
Fludiazepam-M (nor-) HYAC	U+UHYAC	2195	291 $_{52}$	249 $_{100}$	123 $_{57}$	95 $_{61}$		567	286
Flufenamic acid		1935	281 $_{69}$	263 $_{100}$	235 $_{29}$	166 $_{26}$	92 $_{29}$	520	5149
Flufenamic acid ME	PME	1880	295 $_{51}$	263 $_{100}$	235 $_{15}$	166 $_{11}$	92 $_7$	589	5147
Flufenamic acid MEAC		1950	337 $_6$	306 $_{11}$	295 $_{100}$	263 $_{89}$	235 $_{18}$	790	5150
Flufenamic acid TMS		2095	353 $_{28}$	263 $_{100}$	235 $_9$	167 $_{11}$	75 $_{11}$	857	6331
Flufenamic acid 2ME		1785	309 $_{100}$	276 $_{70}$	248 $_{58}$	180 $_{32}$	77 $_{35}$	659	5148
Flufenamic acid-M (HO-) 2ME	PME	2115	325 $_{100}$	293 $_{91}$	278 $_{88}$	250 $_{67}$	202 $_{19}$	734	6377
Flumazenil		2660	303 $_{23}$	257 $_{43}$	229 $_{100}$	201 $_{22}$	94 $_{13}$	626	3674
Flumazenil-M (HOOC-) ME		2555	289 $_{42}$	257 $_{75}$	229 $_{100}$	201 $_{38}$	94 $_9$	556	3675
Flumazenil-M (HOOC-) -CO2		2245	231 $_{100}$	203 $_{70}$	189 $_{27}$	147 $_{26}$	94 $_{19}$	309	3676
Flunarizine	G U+UHYAC	3135	404 $_2$	287 $_{18}$	201 $_{100}$	183 $_{19}$	117 $_{67}$	1029	789
Flunarizine-M (carbinol)	UHY	1690*	220 $_8$	123 $_{100}$	97 $_{26}$	95 $_{23}$	75 $_{13}$	273	3378
Flunarizine-M (carbinol) AC	U+UHYAC	1740*	262 $_5$	202 $_{100}$	201 $_{98}$	158 $_{67}$	116 $_{44}$	430	3374
Flunarizine-M (difluoro-benzophenone)	U UHY U+UHYAC	1595*	218 $_{42}$	123 $_{100}$	109 $_{23}$	95 $_{57}$		263	3373
Flunarizine-M (HO-difluoro-benzophenone)	UHY	1965*	234 $_{64}$	139 $_{52}$	123 $_{100}$	95 $_{53}$	75 $_{17}$	318	3379
Flunarizine-M (HO-difluoro-benzophenone) AC	U+UHYAC	1995*	276 $_{18}$	234 $_{42}$	139 $_{45}$	123 $_{100}$	95 $_{47}$	495	3375
Flunarizine-M (HO-methoxy-difluoro-benzophenone) AC	U+UHYAC	2565*	306 $_{29}$	264 $_{60}$	185 $_{100}$	143 $_{38}$		644	3377
Flunarizine-M (N-dealkyl-) AC	U+UHYAC	2350	244 $_{29}$	201 $_{12}$	172 $_{48}$	117 $_{100}$	85 $_{52}$	356	2198
Flunarizine-M (N-dealkyl-HO-) 2AC	U+UHYAC	2580	302 $_3$	243 $_{20}$	141 $_{100}$	99 $_{27}$	56 $_8$	624	2199
Flunarizine-M (N-desciannamyl-) AC	U+UHYAC	2545	330 $_4$	244 $_{29}$	203 $_{59}$	146 $_{47}$	85 $_{100}$	759	3376
Flunitrazepam	G P-I	2610	313 $_{89}$	312 $_{98}$	285 $_{100}$	266 $_{46}$	238 $_{37}$	679	497
Flunitrazepam HY	UHY-I U+UHYAC-I	2370	274 $_{100}$	257 $_{22}$	211 $_{22}$	123 $_{15}$	95 $_9$	484	282
Flunitrazepam-D7		2600	320 $_{48}$	318 $_{69}$	301 $_{37}$	292 $_{100}$	272 $_{40}$	715	7777

Flunitrazepam-D7 HY **Table 1-8-1:** Compounds in order of names

Name	Detected	RI	Typical ions and intensities					Page	Entry
Flunitrazepam-D7 HY		2360	281 $_{100}$	263 $_{21}$	217 $_{37}$	127 $_{40}$	99 $_{29}$	522	7778
Flunitrazepam-M (amino-)	P-I U-I UGLUC-I	2615	283 $_{100}$	255 $_{64}$	254 $_{52}$	240 $_{12}$		528	498
Flunitrazepam-M (amino-) AC	UGLUCAC	2950	325 $_{100}$	306 $_{22}$	297 $_{66}$	255 $_{24}$		734	501
Flunitrazepam-M (amino-) TMS		2585	355 $_{100}$	312 $_{10}$	73 $_{70}$	327 $_{48}$	336 $_{12}$	865	7502
Flunitrazepam-M (amino-) formyl artifact		2580	295 $_{97}$	276 $_{24}$	267 $_{100}$	239 $_{11}$	183 $_{11}$	590	6322
Flunitrazepam-M (amino-) HY	UHY	2795	244 $_{100}$	227 $_{44}$				355	504
Flunitrazepam-M (amino-) HY2AC	U+UHYAC-I	2870	328 $_{64}$	286 $_{80}$	244 $_{43}$	205 $_{100}$		749	285
Flunitrazepam-M (nor-)		2705	299 $_{65}$	298 $_{100}$	272 $_{76}$	252 $_{40}$	224 $_{60}$	610	500
Flunitrazepam-M (nor-) TMS		2450	371 $_{78}$	356 $_{26}$	352 $_{38}$	324 $_{18}$	73 $_{75}$	922	7501
Flunitrazepam-M (nor-) HY	UHY	2335	260 $_{100}$	241 $_{12}$	123 $_{64}$	95 $_{44}$		420	283
Flunitrazepam-M (nor-) HYAC	U+UHYAC	2380	302 $_{21}$	260 $_{87}$	259 $_{100}$	241 $_{18}$	123 $_{31}$	623	6321
Flunitrazepam-M (nor-amino-)		2690	269 $_{100}$	241 $_{52}$	240 $_{64}$			464	499
Flunitrazepam-M (nor-amino-) AC		3035	311 $_{100}$	283 $_{59}$	241 $_{33}$			670	502
Flunitrazepam-M (nor-amino-) HY		2165	230 $_{100}$	211 $_{34}$				307	503
Flunitrazepam-M (nor-amino-) HY2AC	U+UHYAC-I	2715	314 $_{100}$	272 $_{80}$	230 $_{98}$	123 $_{23}$		684	284
Fluocortolone		3225*	345 $_{100}$	299 $_{38}$	279 $_{7}$	171 $_{12}$	139 $_{14}$	940	1798
Fluocortolone AC		3420*	418 $_{10}$	398 $_{64}$	345 $_{100}$	299 $_{63}$	279 $_{58}$	1059	1800
Fluocortolone 2AC		3400*	460 $_{12}$	418 $_{24}$	387 $_{100}$	299 $_{63}$		1128	1799
Fluoranthene		1970*	202 $_{100}$	200 $_{20}$	174 $_{2}$	150 $_{1}$	101 $_{7}$	222	2566
Fluorene		1570*	166 $_{100}$	165 $_{94}$	139 $_{5}$	115 $_{3}$	82 $_{12}$	151	2560
4-Fluorophenylacetic acid		<1000*	154 $_{22}$	109 $_{100}$	83 $_{17}$	63 $_{4}$	57 $_{7}$	133	5156
4-Fluorophenylacetic acid ME		1005*	168 $_{19}$	109 $_{100}$	89 $_{5}$	83 $_{8}$	63 $_{3}$	154	5157
Fluorouracil		2090	130 $_{100}$	87 $_{64}$	60 $_{64}$			112	4174
Fluoxetine	G	1920	309 $_{22}$	183 $_{17}$	162 $_{36}$	104 $_{100}$	91 $_{67}$	660	4277
Fluoxetine	G	1950	309 $_{2}$	183 $_{2}$	162 $_{4}$	104 $_{10}$	44 $_{100}$	660	7249
Fluoxetine AC	U+UHYAC	2250	351 $_{1}$	190 $_{45}$	145 $_{2}$	117 $_{35}$	86 $_{100}$	848	4278
Fluoxetine HFB		1980	344 $_{66}$	252 $_{9}$	240 $_{100}$	169 $_{56}$	117 $_{85}$	1167	7672
Fluoxetine ME		1920	323 $_{1}$	183 $_{1}$	104 $_{5}$	58 $_{60}$		727	7248
Fluoxetine PFP		2080	294 $_{21}$	202 $_{4}$	190 $_{85}$	174 $_{5}$	117 $_{100}$	1121	7671
Fluoxetine TFA		1950	244 $_{64}$	183 $_{9}$	162 $_{24}$	140 $_{100}$	117 $_{77}$	1030	7670
Fluoxetine TMS		2060	381 $_{5}$	262 $_{7}$	219 $_{13}$	116 $_{100}$	73 $_{58}$	957	4546
Fluoxetine -H2O HYAC	U+UHYAC-I	1680	189 $_{6}$	146 $_{30}$	115 $_{56}$	98 $_{100}$	70 $_{8}$	192	4339
Fluoxetine -H2O HYHFB		1470	343 $_{9}$	252 $_{100}$	174 $_{63}$	146 $_{5}$	115 $_{25}$	820	7240
Fluoxetine -H2O HYPFP		1450	293 $_{17}$	202 $_{87}$	174 $_{100}$	117 $_{51}$	115 $_{90}$	578	7242
Fluoxetine -H2O HYTMS		1580	219 $_{27}$	204 $_{44}$	161 $_{30}$	103 $_{49}$	75 $_{100}$	271	7246
Fluoxetine HY2AC	U+UHYAC	1890	249 $_{2}$	206 $_{100}$	146 $_{36}$	98 $_{78}$	86 $_{34}$	375	4340
Fluoxetine HY2HFB		1490	557 $_{1}$	434 $_{4}$	360 $_{7}$	343 $_{31}$	241 $_{100}$	1186	7241
Fluoxetine HY2PFP		1430	457 $_{2}$	334 $_{4}$	310 $_{16}$	239 $_{28}$	190 $_{100}$	1123	7243
Fluoxetine HY2TFA		1435	357 $_{4}$	243 $_{20}$	174 $_{32}$	140 $_{100}$	117 $_{42}$	871	7244
Fluoxetine-D6		1890	315 $_{52}$	257 $_{8}$	162 $_{72}$	110 $_{100}$	83 $_{44}$	690	7788
Fluoxetine-D6 AC		1900	357 $_{1}$	196 $_{100}$	123 $_{52}$	110 $_{32}$	86 $_{44}$	874	7789
Fluoxetine-D6 HFB		1750	350 $_{52}$	123 $_{70}$	252 $_{6}$	240 $_{100}$	169 $_{38}$	1170	7790
Fluoxetine-D6 TFA		1730	250 $_{58}$	162 $_{8}$	140 $_{100}$	123 $_{63}$	110 $_{18}$	1044	7792
Fluoxetine-D6 TMS		1670	387 $_{9}$	262 $_{15}$	219 $_{21}$	116 $_{100}$	73 $_{45}$	977	7793
Fluoxetine-D6 HY2PFP		1420	463 $_{3}$	334 $_{9}$	298 $_{41}$	190 $_{100}$	119 $_{87}$	1131	7791
Fluoxetine-M (nor-) AC	U+UHYAC	2190	251 $_{1}$	176 $_{51}$	117 $_{100}$	104 $_{20}$	72 $_{91}$	792	4338
Fluoxetine-M (nor-) HFB		1895	330 $_{62}$	226 $_{48}$	169 $_{61}$	162 $_{50}$	117 $_{100}$	1157	7672
Fluoxetine-M (nor-) TFA		1900	230 $_{61}$	183 $_{10}$	162 $_{45}$	126 $_{25}$	117 $_{100}$	987	7673
Fluoxetine-M (nor-) TMS		1830	367 $_{14}$	248 $_{33}$	219 $_{74}$	102 $_{200}$	73 $_{63}$	909	7712
Fluoxetine-M (nor-) 2TMS		2010	439 $_{6}$	320 $_{19}$	219 $_{11}$	174 $_{100}$	104 $_{16}$	1096	7713
Fluoxetine-M (nor-) formyl artifact		1750	307 $_{5}$	183 $_{23}$	162 $_{25}$	146 $_{100}$	119 $_{33}$	648	7710
Fluoxetine-M (nor-) HY2AC		1870	235 $_{1}$	192 $_{100}$	133 $_{67}$	84 $_{41}$	72 $_{42}$	324	5342
Fluoxetine-M (nor-) HY2PFP		1400	443 $_{8}$	296 $_{40}$	280 $_{7}$	239 $_{100}$	177 $_{14}$	1104	7711
Fluoxymesterone		2835*	336 $_{51}$	279 $_{33}$	175 $_{27}$	123 $_{49}$	71 $_{100}$	790	3893
Fluoxymesterone AC		2850*	378 $_{1}$	336 $_{53}$	279 $_{33}$	175 $_{24}$	71 $_{100}$	947	3923
Fluoxymesterone TMS		2785*	408 $_{100}$	335 $_{4}$	207 $_{7}$	111 $_{9}$	73 $_{94}$	1037	3928
Fluoxymesterone enol 3TMS		2840*	552 $_{80}$	462 $_{14}$	407 $_{7}$	319 $_{4}$	73 $_{100}$	1185	3966
Flupentixol		3055	434 $_{2}$	403 $_{4}$	289 $_{8}$	143 $_{100}$	100 $_{14}$	1088	1314
Flupentixol AC	U+UHYAC	3045	476 $_{2}$	457 $_{2}$	291 $_{7}$	221 $_{10}$	185 $_{100}$	1145	1315
Flupentixol TMS		3360	506 $_{1}$	491 $_{8}$	403 $_{4}$	215 $_{100}$	98 $_{49}$	1168	5697
Flupentixol-M (dealkyl-dihydro-) AC	U+UHYAC	3055	434 $_{18}$	265 $_{32}$	185 $_{98}$	141 $_{66}$	99 $_{26}$	1088	1265
Flupentixol-M (dihydro-) AC	U+UHYAC	3005	478 $_{47}$	265 $_{100}$	185 $_{70}$	125 $_{59}$	98 $_{32}$	1146	1264
Flupentixol-M/artifact (N-oxide) -C6H14N2O2		2120*	304 $_{100}$	303 $_{93}$	289 $_{41}$	234 $_{72}$	202 $_{16}$	633	1891
Fluphenazine	G UHY	3050	437 $_{18}$	280 $_{99}$	143 $_{47}$	113 $_{42}$	70 $_{100}$	1093	505
Fluphenazine AC	G U UHY U+UHYAC	3170	479 $_{48}$	419 $_{41}$	280 $_{100}$	185 $_{52}$	125 $_{56}$	1147	339
Fluphenazine TMS		3155	509 $_{5}$	494 $_{4}$	406 $_{25}$	280 $_{100}$	73 $_{18}$	1169	4547
Fluphenazine-M (amino-) AC	U+UHYAC	2765	366 $_{47}$	280 $_{17}$	266 $_{40}$	248 $_{18}$	100 $_{300}$	903	1267
Fluphenazine-M (dealkyl-) AC	U+UHYAC	3145	435 $_{49}$	267 $_{90}$	141 $_{100}$	99 $_{59}$		1090	1266
Fluphenazine-M (ring)	U+UHYAC	2190	267 $_{100}$	235 $_{10}$				453	1266
Flupirtine	P-I	2880	304 $_{89}$	258 $_{17}$	231 $_{33}$	124 $_{28}$	109 $_{100}$	635	1811
Flupirtine 2AC	U+UHYAC	2900	388 $_{3}$	346 $_{41}$	303 $_{21}$	258 $_{91}$	109 $_{100}$	980	1815
Flupirtine -C2H5OH	G	2930	258 $_{95}$	163 $_{6}$	135 $_{14}$	109 $_{100}$		413	1812
Flupirtine -C2H5OH AC	U+UHYAC	2840	300 $_{10}$	258 $_{76}$	163 $_{7}$	124 $_{15}$	109 $_{100}$	614	1813
Flupirtine -C2H5OH 2AC	U+UHYAC	2860	342 $_{20}$	300 $_{35}$	257 $_{69}$	135 $_{13}$	109 $_{100}$	817	1814

Table 1-8-1: Compounds in order of names

Name	Detected	RI	Typical ions and intensities					Page	Entry
Flupirtine -C2H5OH 2TMS		2640	402 $_{100}$	387 $_{13}$	293 $_{20}$	109 $_{47}$	73 $_{69}$	1024	4548
Flupirtine -C2H5OH 3TMS		2600	474 $_{41}$	459 $_{99}$	401 $_{55}$	109 $_{29}$	73 $_{100}$	1143	4673
Flupirtine-M (decarbamoyl-) 3AC	U+UHYAC	2700	358 $_{2}$	315 $_{65}$	273 $_{75}$	231 $_{59}$	109 $_{100}$	877	4342
Flupirtine-M (decarbamoyl-) formyl artifact 3AC	U+UHYAC	2570	370 $_{29}$	328 $_{29}$	285 $_{82}$	271 $_{33}$	109 $_{100}$	920	4341
Flupirtine-M (decarbamoyl-) -H2O 2AC	U+UHYAC	2780	340 $_{7}$	298 $_{35}$	256 $_{47}$	133 $_{31}$	109 $_{100}$	807	4343
Flurazepam	G P-I	2780	387 $_{4}$	315 $_{1}$	245 $_{1}$	99 $_{7}$	86 $_{100}$	976	506
Flurazepam HY		2555	348 $_{7}$	86 $_{100}$				838	511
Flurazepam-M (bis-deethyl-) AC	U+UHYAC	3025	373 $_{74}$	314 $_{66}$	286 $_{53}$	273 $_{99}$	246 $_{65}$	930	1451
Flurazepam-M (bis-deethyl-) -H2O	P-I U+UHYAC-I	2650	313 $_{100}$					679	1450
Flurazepam-M (bis-deethyl-) -H2O HY	UHY	2295	274 $_{62}$	246 $_{80}$	211 $_{101}$			484	513
Flurazepam-M (bis-deethyl-) -H2O HYAC	U+UHYAC	2460	316 $_{44}$	315 $_{50}$	274 $_{53}$	246 $_{100}$	211 $_{57}$	692	287
Flurazepam-M (dealkyl-)	G P-I UGLUC	2470	288 $_{100}$	287 $_{64}$	260 $_{93}$	259 $_{75}$		551	508
Flurazepam-M (dealkyl-) ME		2530	302 $_{100}$	301 $_{90}$	274 $_{99}$	239 $_{18}$	183 $_{16}$	623	3069
Flurazepam-M (dealkyl-) TMS		2470	360 $_{44}$	359 $_{48}$	341 $_{30}$	197 $_{12}$	73 $_{70}$	883	4621
Flurazepam-M (dealkyl-) artifact	U+UHYAC	2050	272 $_{70}$	271 $_{100}$	237 $_{78}$	151 $_{11}$	110 $_{8}$	477	2409
Flurazepam-M (dealkyl-) HY	UHY	2030	249 $_{97}$	154 $_{33}$	123 $_{45}$	95 $_{41}$		373	512
Flurazepam-M (dealkyl-) HYAC	U+UHYAC	2195	291 $_{52}$	249 $_{100}$	123 $_{57}$	95 $_{61}$		567	286
Flurazepam-M (dealkyl-HO-)		2265	286 $_{100}$	258 $_{92}$	223 $_{97}$	75 $_{92}$		633	507
Flurazepam-M (deethyl-) AC	U+UHYAC	2990	401 $_{23}$	358 $_{5}$	314 $_{20}$	100 $_{15}$	58 $_{100}$	1020	1845
Flurazepam-M (HO-ethyl-)	UGLUC	2660	332 $_{39}$	288 $_{100}$	273 $_{61}$	211 $_{22}$		767	509
Flurazepam-M (HO-ethyl-) AC	UGLUCAC	2725	374 $_{54}$	346 $_{32}$	314 $_{100}$	287 $_{60}$	87 $_{12}$	933	510
Flurazepam-M (HO-ethyl-) HY	UHY	2385	293 $_{34}$	262 $_{100}$	166 $_{63}$	109 $_{86}$		578	514
Flurazepam-M (HO-ethyl-) HYAC	U+UHYAC	2470	335 $_{32}$	275 $_{54}$	262 $_{100}$	166 $_{75}$	109 $_{89}$	782	288
Flurazepam-M/artifact	G P-I	2510	314 $_{79}$	285 $_{100}$	258 $_{23}$	223 $_{10}$	75 $_{13}$	683	3545
Flurazepam-M/artifact AC	U+UHYAC	2430	287 $_{46}$	245 $_{100}$	217 $_{3}$	210 $_{8}$	181 $_{9}$	545	5735
Flurbiprofen	G	1900*	244 $_{51}$	199 $_{99}$	183 $_{17}$	170 $_{14}$		355	1453
Flurbiprofen ME	UME	1880*	258 $_{22}$	199 $_{100}$	183 $_{16}$	178 $_{23}$	170 $_{10}$	413	1456
Flurbiprofen-M (di-HO-) 3ME		2310*	318 $_{100}$	259 $_{68}$	215 $_{8}$			703	1455
Flurbiprofen-M (HO-) 2ME	UME	2180*	288 $_{78}$	229 $_{100}$				553	1454
Flurbiprofen-M (HO-methoxy-) 2ME		2310*	318 $_{100}$	259 $_{68}$	215 $_{8}$			703	1455
Flurenol ME		1950*	240 $_{7}$	181 $_{100}$	152 $_{31}$	126 $_{2}$	76 $_{6}$	343	3633
Flurenol artifact		1790*	180 $_{100}$	152 $_{60}$	126 $_{7}$	76 $_{31}$	63 $_{16}$	174	3186
Flurochloridone		2005	311 $_{58}$	187 $_{87}$	145 $_{83}$	103 $_{60}$	75 $_{100}$	669	3187
Flurodifen		2120	328 $_{6}$	309 $_{6}$	190 $_{100}$	126 $_{40}$	75 $_{28}$	747	3842
Fluroxypyr ME		1830	268 $_{40}$	237 $_{7}$	209 $_{100}$	181 $_{51}$	152 $_{17}$	457	4149
Fluroxypyr 2ME		1890	282 $_{68}$	237 $_{10}$	209 $_{100}$	181 $_{52}$	152 $_{12}$	525	4150
Flusilazole		2150	315 $_{20}$	300 $_{9}$	246 $_{7}$	233 $_{100}$	206 $_{46}$	688	7523
Fluspirilene		9999	475 $_{64}$	418 $_{13}$	244 $_{100}$	187 $_{35}$	72 $_{50}$	1144	1499
Fluspirilene AC		3340	517 $_{31}$	475 $_{10}$	286 $_{72}$	109 $_{39}$	72 $_{100}$	1174	519
Fluspirilene-M (deamino-carboxy-)	P-I UHY U+UHYAC	2230*	276 $_{7}$	216 $_{17}$	203 $_{100}$	183 $_{22}$		496	169
Fluspirilene-M (deamino-carboxy-) ME	P-I UHYME U+UHYAC	2125*	290 $_{12}$	258 $_{13}$	216 $_{30}$	203 $_{100}$	183 $_{22}$	563	3372
Fluspirilene-M (deamino-HO-)	UHY-I	2120*	262 $_{16}$	203 $_{100}$	201 $_{16}$	183 $_{12}$		430	515
Fluspirilene-M (deamino-HO-) AC	U+UHYAC	2150*	304 $_{7}$	244 $_{22}$	216 $_{41}$	203 $_{100}$	183 $_{17}$	635	307
Fluspirilene-M (N-dealkyl-) ME	UHY-I	2500	245 $_{17}$	71 $_{100}$	57 $_{60}$			360	518
Fluspirilene-M (N-dealkyl-oxo-)	UHY-I	2405	245 $_{49}$	57 $_{100}$				359	517
Fluspirilene-M (N-dealkyl-oxo-) AC	U+UHYAC-I	2730	287 $_{8}$	245 $_{16}$	57 $_{100}$			547	180
Fluspirilene-M (N-dealkyl-oxo-) ME	UHYME-I	2350	259 $_{12}$	71 $_{100}$				417	516
Flutazolam		2460	289 $_{25}$	259 $_{38}$	245 $_{72}$	210 $_{55}$	183 $_{34}$	939	4026
Flutazolam AC		2475	331 $_{11}$	289 $_{7}$	245 $_{9}$	210 $_{6}$	87 $_{100}$	1058	4027
Flutazolam artifact		2185	259 $_{100}$	209 $_{24}$	183 $_{16}$	130 $_{14}$	111 $_{30}$	415	4028
Flutazolam HY	UHY	2385	293 $_{34}$	262 $_{100}$	166 $_{63}$	109 $_{86}$		578	514
Flutazolam HYAC	U+UHYAC	2470	335 $_{32}$	275 $_{54}$	262 $_{100}$	166 $_{75}$	109 $_{89}$	782	288
Fluvoxamine	G	1890	299 $_{10}$	276 $_{57}$	187 $_{34}$	172 $_{23}$	71 $_{100}$	703	1819
Fluvoxamine AC	U+UHYAC	2240	360 $_{8}$	341 $_{9}$	258 $_{27}$	102 $_{60}$	86 $_{100}$	885	1820
Fluvoxamine HFB		1990	514 $_{1}$	495 $_{6}$	258 $_{40}$	240 $_{94}$	226 $_{88}$	1172	7677
Fluvoxamine PFP		1930	464 $_{1}$	445 $_{6}$	258 $_{90}$	226 $_{87}$	190 $_{100}$	1132	7676
Fluvoxamine TFA		1950	414 $_{1}$	395 $_{8}$	258 $_{78}$	140 $_{96}$	71 $_{96}$	1050	7675
Fluvoxamine TMS		1925	390 $_{1}$	185 $_{10}$	145 $_{21}$	102 $_{100}$	73 $_{62}$	985	7678
Fluvoxamine artifact	G U+UHYAC	1895*	329 $_{13}$	311 $_{18}$	258 $_{72}$	226 $_{49}$	71 $_{100}$	751	1818
Fluvoxamine artifact (imine)		1560	259 $_{6}$	244 $_{12}$	200 $_{60}$	187 $_{100}$	172 $_{88}$	416	1817
Fluvoxamine artifact (ketone)	G U+UHYAC	1525*	260 $_{1}$	242 $_{3}$	228 $_{38}$	173 $_{100}$	145 $_{33}$	421	1816
Fluvoxamine-M (HO-HOOC-) (ME)2AC	U+UHYAC-I	2655	330 $_{2}$	198 $_{20}$	102 $_{75}$	86 $_{100}$	60 $_{71}$	1084	5341
Fluvoxamine-M (HOOC-) (ME)AC	U+UHYAC	2355	355 $_{1}$	272 $_{3}$	102 $_{77}$	86 $_{100}$	60 $_{83}$	933	5338
Fluvoxamine-M (HOOC-) artifact (ketone)	U+UHYAC	1545*	260 $_{3}$	241 $_{25}$	214 $_{10}$	173 $_{100}$	145 $_{43}$	420	5339
Fluvoxamine-M (HOOC-) artifact (ketone) (ME)	U+UHYAC	1550*	274 $_{1}$	255 $_{3}$	242 $_{6}$	173 $_{100}$	145 $_{51}$	485	5336
Fluvoxamine-M (O-demethyl-) 2AC	U+UHYAC-I	2355	244 $_{7}$	187 $_{16}$	102 $_{72}$	86 $_{53}$	60 $_{80}$	980	5300
Fluvoxamine-M (O-demethyl-) artifact (ketone) AC	U+UHYAC-I	2010*	243 $_{23}$	214 $_{23}$	173 $_{100}$	145 $_{34}$	101 $_{23}$	552	5340
Fluvoxate	G	3210	391 $_{1}$	234 $_{5}$	147 $_{24}$	111 $_{39}$	98 $_{100}$	989	4520
Fluvoxate artifact (dehydro-)		3230	389 $_{20}$	234 $_{5}$	207 $_{7}$	109 $_{15}$	96 $_{100}$	983	4647
Fluvoxate artifact (dihydro-)	G	2940	393 $_{2}$	111 $_{14}$	98 $_{100}$	70 $_{4}$	55 $_{7}$	995	4645
Fluvoxate-M/artifact (alcohol)	UHY	1270	129 $_{3}$	98 $_{100}$	84 $_{3}$	70 $_{8}$	55 $_{12}$	112	4516
Fluvoxate-M/artifact (alcohol) AC	U+UHYAC	1530	171 $_{1}$	138 $_{3}$	111 $_{5}$	98 $_{100}$	70 $_{8}$	158	4517
Fluvoxate-M/artifact (HOOC-)	G UHY U+UHYAC	2770*	280 $_{46}$	279 $_{100}$	205 $_{4}$	147 $_{7}$	115 $_{32}$	516	4519
Fluvoxate-M/artifact (HOOC-) ET		2615*	308 $_{39}$	307 $_{100}$	279 $_{44}$	263 $_{16}$	147 $_{16}$	654	4646

Fluvoxate-M/artifact (HOOC-) ME   Table 1-8-1: Compounds in order of names

Name	Detected	RI	Typical ions and intensities					Page	Entry
Fluvoxate-M/artifact (HOOC-) ME	G P UHY U+UHYAC	2580*	294_{49}	293_{100}	263_9	147_{15}	115_{17}	585	4518
Fluvoxate-M/artifact (HOOC-) isopropylester	G UHY U+UHYAC	2625*	322_{56}	321_{77}	279_{100}	147_{26}	115_{22}	722	4521
Folpet		2000	295_{19}	260_{91}	130_{62}	104_{94}	76_{100}	588	3441
Fonofos		1750*	246_{34}	137_{48}	109_{100}	81_{17}	63_{19}	361	3442
Formaldehyde		<1000*	30_{90}	29_{100}				89	4192
Formetanate		2100	221_{100}	163_{84}	149_{79}	122_{66}	92_{66}	277	3901
Formetanate -C2HNO		1660	166_9	121_4	109_{100}	80_{28}	65_5	152	3902
Formoterol HY		1225	165_1	122_{16}	107_2	91_3	77_7	148	3249
Formoterol HY		1225	165_1	122_{16}	91_3	77_7	44_{100}	148	5517
Formoterol HY AC	U+UHYAC	1720	207_1	148_{41}	121_{16}	91_5	86_{10}	235	3265
Formoterol HY AC	U+UHYAC	1720	207_1	148_{41}	121_{16}	86_{10}	44_{100}	234	5537
Formoterol HY -2H		1320	163_4	148_4	107_{27}	77_{11}	56_{100}	143	4079
Formoterol HY formyl artifact		1255	177_6	162_4	121_{60}	77_{12}	56_{100}	166	3250
Formoterol HYHFB		1560	361_2	240_5	169_5	148_{53}	121_{100}	886	6769
Formoterol HYPFP		1460	311_7	190_4	148_{43}	121_{100}	91_7	670	6775
Formoterol HYTFA		1460	261_6	148_{33}	140_4	121_{100}	91_8	425	6774
Formothion		1820	257_9	224_{11}	170_{20}	125_{89}	93_{100}	408	3443
Formothion -CO	P G U	1725	229_7	125_{45}	93_{62}	87_{100}		302	724
4-Formyl-phenazone		2285	216_{22}	188_{60}	121_{100}	77_{37}	56_{54}	259	4214
Fosazepam		3070	360_{25}	283_{17}	255_{100}	227_{63}	91_{84}	883	4025
Frangula-emodin		2620*	270_{100}	242_8	213_5	185_6	139_6	468	3565
Frangula-emodin AC		2740*	312_{19}	270_{100}	242_8	213_6	139_5	674	3566
Frangula-emodin 2ME		2775*	298_{100}	280_{48}	252_{55}	135_{22}	115_{12}	605	3567
Frangula-emodin 3ME		2845*	312_{65}	297_{100}	295_{27}	267_7	142_{13}	675	3568
Frigen 11		<1000*	136_1	117_2	101_{77}	82_5	66_{13}	116	3794
Frigen 12		<1000*	120_1	101_{20}	85_{100}	66_{10}	50_{36}	106	3793
Frovatriptan		2960	243_{43}	212_9	186_{100}	170_{36}	142_{15}	354	7751
Frovatriptan ME		2785	257_{26}	212_4	186_{100}	170_{24}	71_{29}	411	7641
Frovatriptan TMS		2800	315_{61}	258_{180}	243_{89}	200_{47}	75_{44}	690	7644
Frovatriptan iso-1 2TMS		2985	387_{31}	372_{14}	330_{120}	214_{41}	73_{93}	977	7643
Frovatriptan iso-1 3TMS		2745	459_{58}	401_{57}	330_{72}	287_{49}	147_{100}	1127	7645
Frovatriptan iso-2 2TMS		3000	387_{20}	258_{100}	243_{49}	129_{61}	73_{62}	978	7646
Frovatriptan iso-2 3TMS		3075	459_{12}	444_9	330_{100}	214_{22}	73_{39}	1127	7642
Fructose 5AC		1995*	331_4	317_{11}	275_{68}	187_{24}	101_{100}	984	1958
Fructose 5HFB		1620*	750_1	322_{52}	309_{75}	169_{100}	69_{80}	1207	5793
Fructose 5PFP		1250*	419_{13}	405_7	203_{16}	147_{23}	119_{100}	1205	5792
Fructose 5TFA		1470*	450_2	222_{40}	209_{75}	125_{24}	69_{100}	1200	5791
Fuberidazole		1940	184_{100}	155_{24}	129_{13}	92_{10}	63_{10}	184	3643
Furalaxyl		1960	301_{11}	242_{30}	152_{14}	95_{100}	77_9	619	4044
Furmecyclox		1850	251_{13}	138_2	123_{100}	81_{29}	53_{33}	387	2998
Furosemide ME	P PME UME	2890	344_{26}	311_9	283_4	96_9	81_{100}	823	2329
Furosemide 2ME	PME UME	2850	358_{46}	325_8	297_4	96_8	81_{100}	876	2330
Furosemide 2TMS		2895	474_{19}	459_7	355_{12}	81_{100}	73_{30}	1142	4549
Furosemide 3ME	PME UME	2800	372_{35}	339_7	311_4	96_8	81_{100}	926	2331
Furosemide 3TMS		2805	546_3	531_{13}	147_{33}	81_{100}	73_{36}	1184	4550
Furosemide -SO2NH	P U	2040	251_{23}	233_3	96_{12}	81_{100}	53_{19}	382	3367
Furosemide -SO2NH ME	PME-I UME UHYME	2020	265_{33}	232_{10}	204_7	96_{14}	81_{100}	441	2332
Furosemide -SO2NH 2ME	UME	2050	279_{32}	250_{24}	232_{36}	204_{12}	81_{100}	509	2333
Furosemide-M (N-dealkyl-) ME	UME	2750	264_{100}	232_{86}	200_9	169_{11}	141_8	437	2334
Furosemide-M (N-dealkyl-) MEAC		2440	306_1	263_1	200_1	169_2	56_{100}	643	2336
Furosemide-M (N-dealkyl-) 2ME	UME	2450	278_{100}	248_{32}	200_{60}	185_{41}	169_{32}	505	2335
Furosemide-M (N-dealkyl-) 2MEAC		2375	320_{18}	278_{42}	200_{35}	169_{22}	56_{100}	714	2337
Furosemide-M (N-dealkyl-) -SO2NH ME		1470	185_{52}	153_{100}	126_{43}	99_{13}	63_{20}	186	2338
Furosemide-M (N-dealkyl-) -SO2NH MEAC		1650	227_{19}	185_{60}	153_{100}	126_{33}	63_{50}	297	2340
Furosemide-M (N-dealkyl-) -SO2NH 2ME		1500	199_{19}	185_{64}	153_{100}	126_{38}	90_{22}	216	2339
Gabapentin -H2O	G U+UHYAC	1750	153_{82}	110_{39}	96_{35}	81_{100}	67_{74}	133	3112
Gabapentin -H2O AC	U+UHYAC	1730	195_{100}	167_{49}	153_{43}	81_{96}	67_{47}	208	6555
Gabapentin -H2O ME	UME U+UHYAC	1560	167_{79}	124_{42}	110_{21}	81_{100}	67_{81}	153	3113
Galactose 5AC	U+UHYAC	1995*	331_{14}	245_{42}	168_{77}	143_{100}	103_{83}	984	1959
Galactose 5HFB		1505*	519_{25}	465_8	277_{12}	249_8	169_{100}	1207	5796
Galactose 5PFP		1200*	419_{13}	365_5	227_8	147_{22}	119_{100}	1205	5795
Galactose 5TFA		1190*	547_1	407_6	319_{31}	265_{26}	69_{100}	1200	5794
Galantamine		2340	287_{83}	286_{100}	270_{13}	244_{21}	216_{23}	547	6710
Galantamine AC		2450	329_{31}	328_{34}	270_{100}	216_{27}	165_{14}	755	6712
Galantamine HFB		2330	483_{36}	482_{32}	270_{100}	216_{22}	174_{24}	1150	6717
Galantamine PFP		2295	433_{43}	432_{43}	270_{100}	216_{19}	174_{21}	1086	6716
Galantamine TFA		2300	383_{89}	382_{67}	270_{100}	216_{25}	174_{26}	963	6715
Galantamine TMS		2420	359_{100}	358_{99}	344_{16}	316_{23}	216_{18}	937	6714
Galantamine -H2O		2180	269_{100}	268_{46}	226_{25}	211_{57}	165_{59}	465	6711
Galantamine HYAC	U+UHYAC	2280	311_{36}	268_{10}	226_{100}	211_{19}	165_{32}	672	6713
Galantamine-M (nor-) HYAC	U+UHYAC	2410	339_{42}	296_{17}	280_{18}	212_{100}	165_{25}	802	7385
Gallopamil	G P-I U UHY U+UHYAC	3190	484_1	483_1	333_{100}	151_{15}	58_{50}	1151	2520
Gallopamil-M (N-bis-dealkyl-) AC	U+UHYAC	2500	348_{54}	305_{70}	263_{100}	194_{47}	100_{17}	840	2908
Gallopamil-M (N-dealkyl-)		2180	320_7	289_{12}	194_{85}	70_{70}	57_{100}	716	2522

Table 1-8-1: Compounds in order of names Glimepiride artifact-5 ME

Name	Detected	RI	Typical ions and intensities					Page	Entry
Gallopamil-M (N-dealkyl-) AC	U+UHYAC	2520	362_{58}	319_{83}	277_{100}	114_{68}	86_{56}	892	2524
Gallopamil-M (N-dealkyl-bis-O-demethyl-) 2AC	U+UHYAC	2650	376_{19}	334_{100}	291_{44}	114_{92}	86_{40}	940	2910
Gallopamil-M (N-dealkyl-O-demethyl-) 2AC	U+UHYAC	2600	390_{5}	348_{80}	305_{100}	263_{58}	114_{87}	986	2909
Gallopamil-M (nor-)		3260	470_{1}	319_{100}	290_{10}	151_{32}	57_{9}	1139	2521
Gallopamil-M (nor-) AC	U+UHYAC	3520	512_{3}	348_{11}	319_{23}	164_{100}	151_{16}	1171	2523
Gallopamil-M (O-demethyl-) AC	U+UHYAC	3300	511_{1}	361_{100}	319_{85}	276_{11}	58_{79}	1171	1927
GC septum bleed		----	503_{25}	355_{83}	281_{54}	147_{83}	73_{100}	1166	2220
GC stationary phase (methylsilicone)		----	429_{10}	355_{31}	281_{67}	207_{100}	73_{27}	1076	2627
GC stationary phase (OV-101)		----	355_{10}	281_{45}	207_{99}	73_{35}		863	1016
GC stationary phase (OV-17)		----	452_{20}	394_{34}	315_{100}	198_{51}	135_{46}	1115	1017
GC stationary phase (UCC-W-982)		----	429_{5}	355_{9}	281_{36}	207_{100}	73_{45}	1077	1018
Gemfibrozil ME		1855*	264_{6}	143_{100}	122_{55}	83_{92}	59_{57}	441	2799
Gepefrine TMSTFA		1630	319_{8}	206_{86}	191_{32}	140_{100}	73_{58}	708	6141
Gepefrine 2AC	U+UHYAC	1930	235_{2}	176_{48}	134_{52}	107_{21}	86_{100}	324	4387
Gepefrine 2HFB		1620	330_{30}	303_{11}	240_{100}	169_{15}	69_{29}	1182	5737
Gepefrine 2PFP		1520	280_{36}	253_{9}	190_{100}	119_{19}	69_{8}	1104	5738
Gepefrine 2TFA		<1000	230_{4}	203_{1}	140_{14}	115_{1}		819	6224
Gepefrine 2TMS	UHYTMS	1850	280_{11}	179_{3}	116_{100}	100_{12}	73_{72}	594	5693
Gepefrine formyl artifact ME		1290	177_{2}	162_{4}	121_{4}	77_{5}	56_{100}	165	5129
Gestonorone caproate		3440*	414_{1}	371_{17}	316_{9}	273_{100}	99_{59}	1051	2279
Gestonorone -H2O		3410*	298_{64}	283_{41}	255_{100}	161_{17}	91_{30}	608	2075
GHB 2TMS		1520*	248_{1}	233_{26}	147_{100}	117_{30}	73_{32}	371	5430
GHB -H2O		<1000*	86_{11}	56_{16}	42_{100}			97	7275
Glibenclamide artifact-1		2035	224_{25}	143_{22}	99_{27}	83_{7}	56_{100}	290	4904
Glibenclamide artifact-2		2480	289_{25}	198_{15}	169_{100}	126_{16}	111_{10}	556	4905
Glibenclamide artifact-3 ME	UME	3445	382_{1}	353_{3}	287_{13}	198_{13}	169_{100}	958	3128
Glibenclamide artifact-3 2ME	UME	3355	353_{8}	289_{42}	198_{5}	169_{100}	111_{8}	1004	4906
Glibenclamide artifact-4 ME		3460	468_{2}	287_{15}	198_{21}	169_{100}	126_{6}	1136	4907
Glibornuride AC		1923	408_{14}	393_{35}	315_{100}	229_{72}	91_{76}	1036	2013
Glibornuride 2TMS		2855	510_{2}	495_{22}	355_{64}	155_{49}	73_{100}	1170	5020
Glibornuride artifact-1		1390	169_{4}	154_{9}	140_{21}	98_{42}	84_{100}	156	2006
Glibornuride artifact-1 2AC	U+UHYAC	1800	253_{2}	238_{4}	193_{41}	168_{40}	95_{100}	395	2012
Glibornuride artifact-1 2TMS		1555	313_{4}	298_{10}	170_{28}	156_{66}	73_{100}	682	5021
Glibornuride artifact-1 -H2O AC		1370	193_{6}	150_{11}	137_{50}	134_{71}	95_{100}	202	2010
Glibornuride artifact-2		1405	181_{3}	152_{100}	123_{7}	95_{18}	70_{25}	176	2007
Glibornuride artifact-3	UME	1620	197_{33}	155_{50}	91_{100}	65_{36}		212	4910
Glibornuride artifact-4	G P-I U+UHYAC UME	1700	171_{25}	155_{19}	107_{6}	91_{100}	65_{25}	158	2008
Glibornuride artifact-4 AC		1550	237_{1}	195_{28}	151_{23}	134_{27}	95_{100}	332	2011
Glibornuride artifact-4 ME	UME	1740	185_{25}	155_{19}	121_{6}	91_{100}	65_{25}	186	3131
Glibornuride artifact-4 TMS	UTMS	1875	243_{3}	228_{100}	167_{6}	149_{12}	91_{9}	354	5022
Glibornuride artifact-4 2ME		1690	199_{34}	155_{15}	91_{100}	65_{23}		217	3130
Glibornuride artifact-5	U+UHYAC UME	1840	195_{17}	164_{12}	134_{14}	109_{45}	95_{100}	208	2009
Glibornuride artifact-5 ME	UME	1715	209_{53}	139_{46}	109_{49}	100_{62}	95_{100}	243	4916
Glibornuride artifact-6 ME		1845	212_{1}	179_{46}	122_{21}	91_{100}	72_{53}	247	3132
Glibornuride -H2O ME		2670	362_{5}	207_{66}	150_{65}	134_{100}	91_{100}	891	3129
Glibornuride-M (HO-) artifact AC	U+UHYAC	2180	229_{16}	187_{80}	107_{80}	89_{99}	77_{44}	303	4915
Glibornuride-M (HO-) artifact ME	UME	2265	201_{40}	172_{45}	107_{100}	89_{81}	77_{91}	220	4913
Glibornuride-M (HO-) artifact 2ME	UME	2030	215_{60}	171_{29}	107_{100}	89_{80}	77_{72}	257	4914
Glibornuride-M (HO-) artifact 3TMS	UTMS	2000	403_{20}	388_{79}	272_{35}	258_{55}	73_{80}	1026	5018
Glibornuride-M (HO-bornyl-) artifact	UME	2305	211_{3}	181_{7}	125_{13}	108_{22}	95_{100}	247	4917
Glibornuride-M (HO-bornyl-) artifact 2TMS	UTMS	1955	355_{11}	340_{36}	265_{67}	107_{56}	73_{100}	867	5023
Glibornuride-M (HOOC-) artifact 2ME		1920	229_{50}	198_{48}	135_{100}	103_{40}	76_{55}	303	2479
Glibornuride-M (HOOC-) artifact 3ME	UME	1850	243_{54}	199_{18}	135_{100}	104_{46}	76_{51}	354	2480
Gliclazide		2440	278_{14}	125_{62}	110_{100}	81_{91}		726	4908
Gliclazide artifact-1 ME	UME	1545	184_{13}	125_{44}	110_{100}	81_{70}	67_{32}	185	4909
Gliclazide artifact-2	UME	1620	197_{33}	155_{50}	91_{100}	65_{36}		212	4910
Gliclazide artifact-3	U+UHYAC UME	1670	169_{1}	125_{68}	110_{79}	81_{100}	67_{46}	156	4911
Gliclazide artifact-4	G P-I U+UHYAC UME	1700	171_{25}	155_{19}	107_{6}	91_{100}	65_{25}	158	2008
Gliclazide artifact-4 ME	UME	1740	185_{25}	155_{19}	121_{6}	91_{100}	65_{25}	186	3131
Gliclazide artifact-4 TMS	UTMS	1875	243_{3}	228_{100}	167_{6}	149_{12}	91_{9}	354	5022
Gliclazide artifact-4 2ME		1690	199_{34}	155_{15}	91_{100}	65_{23}		217	3130
Gliclazide artifact-5 AC	U+UHYAC	1535	210_{7}	168_{69}	125_{61}	110_{100}	81_{78}	243	4912
Gliclazide-M (HO-) artifact AC	U+UHYAC	2180	229_{16}	187_{80}	107_{80}	89_{99}	77_{44}	303	4915
Gliclazide-M (HO-) artifact ME	UME	2265	201_{40}	172_{45}	107_{100}	89_{81}	77_{91}	220	4913
Gliclazide-M (HO-) artifact 2ME	UME	2030	215_{60}	171_{29}	107_{100}	89_{80}	77_{72}	257	4914
Gliclazide-M (HO-) artifact 3TMS	UTMS	2000	403_{20}	388_{79}	272_{35}	258_{55}	73_{80}	1026	5018
Gliclazide-M (HOOC-) artifact 2ME		1920	229_{50}	198_{48}	135_{100}	103_{40}	76_{55}	303	2479
Gliclazide-M (HOOC-) artifact 3ME	UME	1850	243_{54}	199_{18}	135_{100}	104_{46}	76_{51}	354	2480
Glimepiride artifact-1 ME		1195	171_{11}	156_{3}	114_{100}	101_{9}	76_{21}	158	4918
Glimepiride artifact-2 ME		1265	183_{6}	151_{41}	124_{64}	96_{87}	81_{100}	182	4925
Glimepiride artifact-3		1275	125_{100}	110_{46}	96_{67}	82_{22}	67_{27}	109	4919
Glimepiride artifact-3 TMS		1360	197_{67}	182_{100}	166_{35}	126_{19}	73_{46}	213	5025
Glimepiride artifact-4		2130	252_{32}	157_{33}	113_{15}	95_{7}	56_{100}	392	4922
Glimepiride artifact-5 ME		2325	240_{41}	184_{32}	146_{24}	120_{100}	89_{28}	342	4920

Glimepiride artifact-5 2ME — Table 1-8-1: Compounds in order of names

Name	Detected	RI	Typical ions and intensities					Page	Entry
Glimepiride artifact-5 2ME		2690	254_4	240_{45}	184_{34}	125_{38}	120_{100}	398	4921
Glimepiride-M (HO-) artifact ME		1630	187_4	155_8	128_{16}	114_{100}	76_{34}	190	4923
Glimepiride-M (HOOC-) artifact 2ME		1670	215_6	184_{10}	156_{60}	114_{100}	101_{32}	257	4924
Glipizide artifact-1		2035	224_{25}	143_{22}	99_{27}	83_7	56_{100}	290	4904
Glipizide artifact-2 ME	UME	3020	334_{35}	239_{10}	150_{100}	121_{82}	93_{67}	779	3133
Glipizide artifact-2 TMS	UTMS	3195	392_{19}	377_8	240_{37}	150_{100}	121_{43}	991	5019
Glipizide artifact-2 TMS	UTMS	3195	392_{19}	377_8	240_{37}	150_{100}	121_{43}	992	4926
Glipizide artifact-2 2ME	UME	3005	348_3	241_{92}	150_{100}	121_{96}	93_{81}	838	4927
Gliquidone artifact-1		1845	219_{54}	204_{100}	176_{21}	158_{25}	133_{39}	269	4928
Gliquidone artifact-2		2035	224_{25}	143_{22}	99_{27}	83_7	56_{100}	290	4904
Gliquidone artifact-3	U+UHYAC UME	2555	323_{25}	219_{94}	204_{100}	191_{60}	176_{96}	728	4929
Gliquidone artifact-4		3440	402_{26}	321_{38}	219_{74}	204_{100}	175_{76}	1023	4930
Gliquidone artifact-4 ME	UME	3460	416_{30}	321_{28}	219_{87}	204_{100}	175_{70}	1054	4931
Gliquidone artifact-4 TMS	UTMS	3585	474_{11}	459_{11}	219_{61}	204_{100}	176_{63}	1143	5016
Gliquidone artifact-4 2ME	UME	3415	430_9	323_{72}	220_{100}	204_{63}	175_{47}	1080	3134
Gliquidone artifact-5 ME		3415*	502_{12}	321_{100}	219_{65}	204_{77}	175_{52}	1165	4932
Glisoxepide artifact-1 ME	UME	1315	172_{12}	113_{76}	98_{80}	68_{70}	59_{100}	160	3140
Glisoxepide artifact-3 ME	UME	2855	323_1	228_{42}	197_{43}	139_{87}	110_{100}	725	4933
Glisoxepide artifact-3 2ME	UME	2840	337_1	294_{16}	230_{100}	199_{30}	110_{59}	791	4934
Gluconic acid ME5AC		1975*	361_3	289_{14}	173_{58}	145_{72}	115_{100}	1062	5227
Glucose 5AC	U+UHYAC	2010*	331_1	242_{25}	157_{64}	115_{83}	98_{78}	985	790
Glucose 5HFB		1460*	519_{21}	321_6	277_{11}	197_{14}	169_{100}	1207	5784
Glucose 5PFP		1180*	747_1	419_{26}	227_{15}	147_{30}	119_{100}	1206	5783
Glucose 5TFA		1200*	547_1	413_4	319_{33}	265_{25}	69_{100}	1200	5782
Glucose 5TMS		2050*	435_1	217_{17}	204_{100}	191_{67}	73_{91}	1181	4333
Glutethimide	P G U UHY U+UHYAC	1830	217_{20}	189_{100}	160_{36}	132_{64}	117_{87}	261	791
Glutethimide TMS		1800	274_{13}	245_{10}	174_{26}	132_{100}	117_{39}	559	5481
Glutethimide 2TMS		1845	361_8	346_{20}	332_{25}	147_{14}	73_{40}	888	5482
Glutethimide-M (HO-ethyl-)	U UHY	1865	233_{69}	205_{14}	189_{32}	146_{100}	104_{78}	315	792
Glutethimide-M (HO-ethyl-) AC	U+UHYAC	2060	275_{20}	247_{37}	233_{47}	189_{100}	187_{96}	489	794
Glutethimide-M (HO-phenyl-)	U UHY	1875	233_{89}	204_{98}	176_{79}	148_{43}	133_{95}	315	793
Glutethimide-M (HO-phenyl-) AC	U+UHYAC	2250	275_{17}	233_{100}	204_{84}	189_{83}	176_{46}	489	795
Glycerol 3AC	U+UHYAC	1485*	158_3	145_{86}	116_{55}	103_{100}	86_{30}	264	2014
Glycerol 3TMS		1125*	293_2	218_{13}	205_{37}	147_{65}	73_{100}	657	7451
Glyceryl dimyristate -H2O	G	3830*	494_6	285_{29}	98_{49}	71_{65}	57_{100}	1161	5631
Glyceryl monomyristate	G	2260*	285_6	271_{10}	211_{66}	98_{83}	55_{100}	625	5587
Glyceryl monooleate 2AC	U+UHYAC	2790*	380_{34}	264_{71}	159_{100}	81_{49}	69_{85}	1099	5602
Glyceryl monopalmitate	G	2420*	313_3	299_{11}	239_{49}	98_{100}	57_{99}	760	5588
Glyceryl monopalmitate 2AC	U+UHYAC	2645*	354_4	239_{34}	159_{100}	98_{29}	84_{18}	1051	5412
Glyceryl monopalmitate 2TMS	G	2620*	460_3	372_{77}	205_{25}	147_{62}	73_{170}	1143	7449
Glyceryl monostearate 2AC	U+UHYAC	2790*	382_6	267_{26}	159_{100}	98_{30}	84_{19}	1104	5413
Glyceryl monostearate 2TMS		2780*	488_2	400_{67}	205_{23}	147_{63}	73_{150}	1166	7450
Glyceryl triacetate	U+UHYAC	1485*	158_3	145_{86}	116_{55}	103_{100}	86_{30}	264	2014
Glyceryl tridecanoate	G	3280*	383_{16}	355_{49}	155_{57}	127_{42}	57_{100}	1186	4466
Glyceryl trioctanoate	G	2850*	452_1	327_{24}	201_8	127_{73}	57_{100}	1139	4465
Glycophen		2470	329_5	187_{18}	142_{58}	127_{84}	56_{100}	752	3848
Glyphosate 3ME		1410	211_5	179_{15}	152_{75}	102_{100}	74_{70}	246	4153
Glyphosate 4ME		1390	225_6	166_{67}	116_{100}	93_{71}	58_{39}	290	4152
Granisetron	P U UHY U+UHYAC	2880	312_{24}	159_{80}	136_{40}	110_{70}	96_{100}	677	3185
Grepafloxacin ME		3540	373_{37}	317_{100}	242_{16}	85_{11}	70_{30}	931	7733
Grepafloxacin TMS		3570	431_{69}	416_{34}	375_{100}	331_{22}	214_8	1083	7735
Grepafloxacin 2ME		3520	387_{72}	317_{100}	242_{20}	85_{17}	70_{40}	977	7734
Grepafloxacin -CO2		3120	315_{25}	259_{100}	245_5	215_3	174_3	690	7738
Grepafloxacin -CO2 ME		3130	329_{21}	294_4	259_{100}	215_8	189_5	756	7737
Grepafloxacin -CO2 TMS		3120	387_{34}	372_{22}	331_{100}	301_{16}	273_7	977	7736
Guaifenesin	P G	1610*	198_{15}	124_{100}	109_{75}			215	796
Guaifenesin AC		2000*	240_{10}	124_{43}	117_{100}	109_{42}	77_{15}	344	1992
Guaifenesin 2AC	U+UHYAC	1865*	282_3	159_{99}	124_{13}	99_{18}		526	799
Guaifenesin 2TMS		1850*	342_{16}	196_{33}	149_{32}	103_{47}	73_{100}	818	4551
Guaifenesin-M (HO-) 3AC	U+UHYAC	2235*	340_2	298_1	159_{100}	140_{10}	99_8	807	797
Guaifenesin-M (HO-methoxy-) 2AC	U+UHYAC	2290*	328_2	245_5	170_{15}	159_{100}	99_9	748	801
Guaifenesin-M (HO-methoxy-) 3AC	U+UHYAC	2265*	370_9	230_7	212_6	170_{20}	159_{100}	919	798
Guaifenesin-M (O-demethyl-)	UHY	1700*	184_{26}	135_4	110_{100}	81_8	64_5	184	2683
Guaifenesin-M (O-demethyl-) 3AC	U+UHYAC	1920*	310_1	268_{13}	159_{100}	117_{12}	110_{19}	666	800
Guanfacine		1890	225_2	159_9	123_7	101_{100}	89_7	357	7561
Guanfacine AC		2020	287_1	267_1	159_{17}	143_{45}	101_{100}	545	7567
Guanfacine HFB		1985	406_1	282_{15}	272_{58}	159_{61}	86_{80}	1099	7572
Guanfacine PFP		1965	391_1	356_4	272_{100}	159_{87}	86_{110}	986	7571
Guanfacine TFA		1995	341_1	306_8	306_8	272_{59}	159_{72}	809	7570
Guanfacine artifact (-COONH2)		1680	203_1	168_{100}	159_{29}	125_{70}	89_{29}	223	7562
Guanfacine artifact (-COONH2) TMS		1685	275_1	260_{17}	160_{25}	116_{50}	73_{100}	488	7564
Guanfacine artifact (-COONH2) 2AC		2150	272_3	252_4	159_{21}	128_{87}	86_{100}	545	7568
Guanfacine artifact (-COONH2) 2TMS		1635	347_2	332_4	188_{64}	147_{43}	73_{100}	834	7563
Guanfacine artifact (HOOC-) ME		1390*	218_6	183_{69}	159_{100}	123_{23}	89_{25}	262	7560

Table 1-8-1: Compounds in order of names   Heroin-M (6-acetyl-morphine) HFB

Name	Detected	RI	Typical ions and intensities					Page	Entry
Guanfacine artifact (HOOC-) TMS		1510	261 $_{27}$	241 $_{16}$	232 $_8$	159 $_{13}$	73 $_{100}$	494	7565
Guanfacine artifact (-NH3) TMS		1880	285 $_6$	265 $_{30}$	237 $_{27}$	141 $_{77}$	73 $_{100}$	613	7566
Halazepam	P U+UHYAC	2335	352 $_{60}$	324 $_{100}$	289 $_{14}$	241 $_{11}$		852	2083
Halazepam HY	UHY U+UHYAC	2380	313 $_{100}$	296 $_6$	244 $_{11}$	105 $_{15}$	77 $_{16}$	678	2091
Halazepam-M (HO-) iso-1 HYAC	U+UHYAC	2350	371 $_{100}$	328 $_{73}$	312 $_{40}$	260 $_9$	208 $_7$	921	2121
Halazepam-M (HO-) iso-2 HYAC	U+UHYAC	2370	371 $_{100}$	328 $_{76}$	312 $_{28}$	260 $_{16}$		921	2122
Halazepam-M (HO-methoxy-) HYAC	U+UHYAC	2500	401 $_{47}$	358 $_{100}$	342 $_{40}$	273 $_{21}$	85 $_{37}$	1019	2123
Halazepam-M (N-dealkyl-HO-)	UGLUC	2750	286 $_{82}$	258 $_{100}$	230 $_{11}$	166 $_7$	139 $_8$	541	2113
Halazepam-M (N-dealkyl-HO-) AC	U+UHYAC	3000	328 $_{22}$	286 $_{90}$	258 $_{100}$	166 $_8$	139 $_7$	747	2111
Halazepam-M (N-dealkyl-HO-) HY	UHY	2400	247 $_{72}$	246 $_{100}$	230 $_{11}$	121 $_{26}$	65 $_{22}$	365	2112
Halazepam-M (N-dealkyl-HO-) HYAC	U+UHYAC	2270	289 $_{18}$	247 $_{86}$	246 $_{100}$	105 $_7$	77 $_{35}$	556	3143
Halazepam-M (N-dealkyl-HO-) iso-1 HY2AC	U+UHYAC	2560	331 $_{48}$	289 $_{64}$	247 $_{100}$	230 $_{41}$	154 $_{13}$	762	2125
Halazepam-M (N-dealkyl-HO-) iso-2 HY2AC	U+UHYAC	2610	331 $_{46}$	289 $_{54}$	246 $_{100}$	154 $_{11}$	121 $_{11}$	762	1751
Halazepam-M (N-dealkyl-HO-methoxy-) HY2AC	U+UHYAC	2700	361 $_7$	319 $_{27}$	276 $_{38}$	260 $_5$	246 $_4$	886	1752
Halazepam-M artifact	P-I UHY U+UHYAC	2060	240 $_{59}$	239 $_{100}$	205 $_{81}$	177 $_{16}$	151 $_9$	342	300
Halazepam-M HYAC	U+UHYAC PHYAC	2245	273 $_{31}$	230 $_{100}$	154 $_{13}$	105 $_{23}$	77 $_{50}$	480	273
Haloperidol	G P-I U UHY	2940	375 $_2$	237 $_{92}$	224 $_{100}$	123 $_{50}$	95 $_{31}$	936	340
Haloperidol TMS		2965	447 $_2$	432 $_7$	296 $_{100}$	206 $_{97}$	123 $_{98}$	1110	4552
Haloperidol 2TMS		3055	519 $_1$	504 $_3$	296 $_{100}$	206 $_{55}$	73 $_{75}$	1174	4553
Haloperidol -H2O	U+UHYAC	2915	357 $_{13}$	206 $_{39}$	192 $_{100}$	123 $_{36}$	95 $_{23}$	873	523
Haloperidol-D4		2930	379 $_1$	316 $_7$	237 $_{89}$	224 $_{100}$	127 $_{53}$	948	5427
Haloperidol-D4 TMS		2960	451 $_1$	436 $_7$	309 $_{75}$	296 $_{100}$	127 $_{61}$	1115	7286
Haloperidol-D4 2TMS		3050	523 $_1$	508 $_2$	296 $_{100}$	206 $_{77}$	73 $_{74}$	1176	7285
Haloperidol-D4 -H2O		2900	361 $_6$	206 $_{36}$	192 $_{100}$	127 $_{36}$		887	5428
Haloperidol-M	U+UHYAC	1490*	180 $_{25}$	125 $_{49}$	123 $_{35}$	95 $_{17}$	56 $_{100}$	174	85
Haloperidol-M	U	2250	239 $_9$	189 $_{20}$	139 $_{24}$	100 $_{26}$	56 $_{100}$	339	522
Haloperidol-M	U UHY	1750	223 $_{20}$	189 $_{19}$	139 $_{23}$	84 $_{40}$	56 $_{100}$	282	520
Haloperidol-M (N-dealkyl-)	UHY	1800	211 $_6$	139 $_{20}$	84 $_{42}$	56 $_{100}$		246	521
Haloperidol-M (N-dealkyl-) AC	U U+UHYAC	2235	253 $_{31}$	210 $_{53}$	193 $_{14}$	139 $_{42}$	57 $_{100}$	393	524
Haloperidol-M (N-dealkyl-) -H2O AC	U+UHYAC	2155	235 $_{100}$	192 $_{58}$	158 $_{51}$	129 $_{53}$	82 $_{33}$	322	182
Haloperidol-M (N-dealkyl-oxo-) -2H2O	U UHY U+UHYAC	1650	189 $_{100}$	154 $_{32}$	127 $_{33}$	101 $_8$	75 $_{17}$	191	181
Halothane		<1000*	196 $_{50}$	177 $_{16}$	117 $_{100}$	98 $_{35}$	67 $_{47}$	209	2996
Harmaline	G	2430	214 $_{91}$	213 $_{100}$	198 $_{29}$	170 $_{25}$	115 $_{12}$	256	4062
Harmaline AC		2670	256 $_{73}$	213 $_{100}$	186 $_{13}$	170 $_{23}$	115 $_9$	407	4063
Harmaline HFB		2590	410 $_{28}$	241 $_{100}$	226 $_7$	198 $_6$	170 $_5$	1040	5925
Harmaline PFP		2540	360 $_{40}$	241 $_{100}$	198 $_8$	184 $_7$	121 $_{14}$	884	5917
Harmaline TFA		2525	310 $_{58}$	241 $_{100}$	198 $_6$	184 $_7$	169 $_7$	665	5919
Harmaline 2AC		2800	298 $_{33}$	255 $_{100}$	241 $_{17}$	212 $_{11}$	141 $_5$	606	4064
Harmaline -2H	G U UHY U+UHYAC	2460	212 $_{100}$	197 $_{29}$	169 $_{67}$	140 $_6$	115 $_8$	249	4066
Harmaline -2H AC		2545	254 $_{42}$	212 $_{100}$	197 $_{18}$	169 $_{26}$	140 $_4$	399	4067
Harmaline artifact (dihydro-)		2375	216 $_{39}$	201 $_{100}$	186 $_{14}$	172 $_{18}$	144 $_5$	260	4065
Harmaline-M (O-demethyl-) -2H	UHY	2550	198 $_{100}$	170 $_{17}$	140 $_2$	99 $_{11}$	75 $_5$	214	4068
Harmaline-M (O-demethyl-) -2H AC	U+UHYAC	2600	240 $_{22}$	198 $_{100}$	169 $_{11}$	140 $_3$	115 $_5$	343	4069
Harmine	G U UHY U+UHYAC	2460	212 $_{100}$	197 $_{29}$	169 $_{67}$	140 $_6$	115 $_8$	249	4066
Harmine AC		2545	254 $_{42}$	212 $_{100}$	197 $_{18}$	169 $_{26}$	140 $_4$	399	4067
Harmine-M (O-demethyl-)	UHY	2550	198 $_{100}$	170 $_{17}$	140 $_2$	99 $_{11}$	75 $_5$	214	4068
Harmine-M (O-demethyl-) AC	U+UHYAC	2600	240 $_{22}$	198 $_{100}$	169 $_{11}$	140 $_3$	115 $_5$	343	4069
Heptabarbital	P G U UHY U+UHYAC	2070	221 $_{100}$	141 $_{32}$				381	803
Heptabarbital (ME)	G P	1800	235 $_{100}$	221 $_{32}$	169 $_8$	155 $_{48}$	141 $_{10}$	440	1885
Heptabarbital 2ME		1915	249 $_{100}$	169 $_{42}$	133 $_{13}$			508	806
Heptabarbital 2TMS		1980	394 $_2$	379 $_{15}$	365 $_{87}$	100 $_{32}$	73 $_{90}$	999	5492
Heptabarbital-M (HO-)	U	2275	266 $_6$	237 $_{46}$	219 $_{97}$	141 $_{51}$	93 $_{100}$	451	804
Heptabarbital-M (HO-) -H2O	U+UHYAC	2300	248 $_8$	219 $_{53}$	157 $_{43}$	141 $_{39}$	93 $_{100}$	371	805
Heptachlor		1860*	370 $_1$	337 $_1$	272 $_{15}$	135 $_{14}$	100 $_{100}$	918	3849
Heptachlorepoxide		2015*	351 $_2$	253 $_{19}$	183 $_{49}$	135 $_{49}$	81 $_{100}$	971	3850
2,2',3,4,4',5,5'-Heptachlorobiphenyl		2460*	394 $_{100}$	392 $_{44}$	324 $_{48}$	252 $_{14}$	162 $_{15}$	990	885
Heptadecane		1700*	240 $_4$	127 $_6$	85 $_{43}$	71 $_{60}$	57 $_{100}$	346	2977
Heptadecane		1700*	240 $_{20}$	85 $_{60}$	71 $_{73}$	57 $_{100}$	43 $_{85}$	346	7687
Heptadecanoic acid ET		2035*	298 $_5$	255 $_8$	157 $_{18}$	101 $_{54}$	88 $_{100}$	608	5404
Heptadecanoic acid ME		2025*	284 $_{20}$	241 $_{14}$	143 $_{19}$	87 $_{66}$	74 $_{100}$	537	3037
Heptafluorobutanoic acid		<1000*	197 $_3$	169 $_{34}$	119 $_{90}$	69 $_{100}$	45 $_{10}$	253	5548
Heptafluorobutanoic acid		<1000*	197 $_3$	169 $_{34}$	150 $_{78}$	119 $_{90}$	69 $_{100}$	253	5545
Heptaminol		1125	127 $_{17}$	113 $_{43}$	110 $_{26}$	69 $_{54}$	59 $_{100}$	123	1459
Heptaminol 2AC		1530	172 $_{14}$	169 $_{30}$	114 $_{22}$	95 $_{29}$	86 $_{100}$	305	1460
Heptane		700*	100 $_{11}$	71 $_{36}$	57 $_{53}$	43 $_{100}$	29 $_{69}$	100	3823
Heptenophos		1570*	250 $_2$	141 $_6$	124 $_{100}$	109 $_{48}$	89 $_{91}$	379	3852
Heroin	G PHYAC U+UHYAC	2620	369 $_{59}$	327 $_{100}$	310 $_{36}$	268 $_{47}$	162 $_{11}$	915	225
Heroin Cl-artifact	U+UHYAC	2680	403 $_{59}$	361 $_{100}$	344 $_{63}$	302 $_{90}$	204 $_{55}$	1026	2992
Heroin-D3		2510	372 $_{87}$	330 $_{100}$	313 $_{57}$	271 $_{87}$	218 $_{56}$	928	7294
Heroin-M (3-acetyl-morphine)		2500	327 $_{100}$	310 $_{10}$	285 $_{73}$	215 $_{19}$	162 $_{22}$	744	2341
Heroin-M (3-acetyl-morphine) PFP		2490	473 $_{25}$	431 $_{45}$	310 $_{60}$	268 $_{100}$	119 $_{16}$	1141	2462
Heroin-M (3-acetyl-morphine) TMS		2570	399 $_{51}$	357 $_{45}$	234 $_{31}$	164 $_{34}$	73 $_{100}$	1015	2466
Heroin-M (6-acetyl-morphine)	U-I UMAM	2535	327 $_{100}$	268 $_{53}$	162 $_{11}$	124 $_{13}$		744	525
Heroin-M (6-acetyl-morphine) HFB		2425	523 $_{42}$	464 $_{100}$	411 $_{39}$	204 $_{56}$	69 $_{45}$	1176	6121

Heroin-M (6-acetyl-morphine) PFP — Table 1-8-1: Compounds in order of names

Name	Detected	RI	Typical ions and intensities					Page	Entry
Heroin-M (6-acetyl-morphine) PFP	UMAMPFP	2650	473$_{90}$	430$_{11}$	414$_{100}$	361$_{27}$	204$_{24}$	1141	2253
Heroin-M (6-acetyl-morphine) TFA		2630	423$_{65}$	380$_{15}$	364$_{100}$	311$_{36}$	204$_{30}$	1065	5575
Heroin-M (6-acetyl-morphine) TMS	UMAMTMS	2590	399$_{92}$	340$_{50}$	287$_{32}$	204$_{25}$	73$_{100}$	1015	2465
Heroin-M (6-acetyl-morphine)-D3		2515	330$_{100}$	287$_{13}$	271$_{99}$	218$_{37}$	165$_{24}$	760	5574
Heroin-M (6-acetyl-morphine)-D3 HFB		2415	526$_{44}$	467$_{100}$	414$_{36}$	207$_{45}$	69$_{45}$	1177	6122
Heroin-M (6-acetyl-morphine)-D3 PFP		2640	476$_{63}$	417$_{100}$	364$_{31}$	207$_{46}$	165$_{15}$	1145	5568
Heroin-M (6-acetyl-morphine)-D3 TFA		2630	426$_{53}$	367$_{100}$	314$_{28}$	207$_{43}$	69$_{25}$	1072	5573
Heroin-M (6-acetyl-morphine)-D3 TMS		2580	402$_{100}$	343$_{75}$	290$_{43}$	207$_{22}$	73$_{90}$	1024	5577
Heroin-M (morphine)	G UHY	2455	285$_{100}$	268$_{15}$	162$_{58}$	124$_{20}$		539	474
Heroin-M (morphine) TFA		2285	381$_{55}$	268$_{100}$	146$_{14}$	115$_{13}$	69$_{23}$	955	5569
Heroin-M (morphine) 2HFB		2375	677$_{10}$	480$_{11}$	464$_{100}$	407$_{9}$	169$_{8}$	1201	6120
Heroin-M (morphine) 2PFP		2360	577$_{51}$	558$_{7}$	430$_{8}$	414$_{100}$	119$_{22}$	1191	2251
Heroin-M (morphine) 2TFA		2250	477$_{71}$	364$_{100}$	307$_{6}$	115$_{8}$	69$_{31}$	1145	4008
Heroin-M (morphine) 2TMS	UHYTMS	2560	429$_{19}$	236$_{21}$	196$_{15}$	146$_{21}$	73$_{100}$	1079	2463
Heroin-M (morphine)-D3 TFA		2275	384$_{39}$	271$_{100}$	211$_{8}$	165$_{6}$	152$_{7}$	967	5572
Heroin-M (morphine)-D3 2HFB		2375	680$_{4}$	483$_{9}$	467$_{100}$	169$_{23}$	414$_{7}$	1202	6126
Heroin-M (morphine)-D3 2PFP		2350	580$_{16}$	433$_{7}$	417$_{100}$	269$_{5}$	119$_{8}$	1192	5567
Heroin-M (morphine)-D3 2TFA		2240	480$_{32}$	383$_{6}$	367$_{100}$	314$_{6}$	307$_{6}$	1148	5571
Heroin-M (morphine)-D3 2TMS		2550	432$_{100}$	290$_{30}$	239$_{66}$	199$_{44}$	73$_{110}$	1085	5578
Heroin-M 2PFP		2440	563$_{100}$	400$_{10}$	355$_{38}$	327$_{7}$	209$_{15}$	1188	3534
Heroin-M 3AC	U+UHYAC	2955	397$_{8}$	355$_{9}$	209$_{41}$	87$_{100}$	72$_{33}$	1010	1194
Heroin-M 3PFP	UHYPFP	2405	709$_{80}$	533$_{28}$	388$_{29}$	367$_{51}$	355$_{100}$	1203	3533
Heroin-M 3TMS	UHYTMS	2605	487$_{9}$	416$_{9}$	222$_{18}$	131$_{10}$	73$_{50}$	1155	3525
Hexachlorobenzene		1690*	284$_{100}$	282$_{53}$	247$_{11}$	212$_{7}$	142$_{15}$	525	1462
2,2',3,4,4',5'-Hexachlorobiphenyl		2290*	362$_{80}$	360$_{100}$	358$_{52}$	288$_{34}$	218$_{12}$	875	884
2,2',4,4',5,5'-Hexachlorobiphenyl		2330*	362$_{79}$	360$_{100}$	358$_{45}$	288$_{27}$	218$_{9}$	875	2633
alpha-Hexachlorocyclohexane (HCH)		1690*	252$_{1}$	217$_{43}$	181$_{95}$	109$_{100}$	51$_{97}$	549	3853
gamma-Hexachlorocyclohexane (HCH)	P-I	1740*	288$_{2}$	252$_{8}$	217$_{67}$	181$_{100}$	109$_{48}$	550	1067
delta-Hexachlorocyclohexane (HCH)		1710*	252$_{2}$	217$_{32}$	181$_{64}$	109$_{100}$	51$_{81}$	549	3854
1,2,3,4,7,8-Hexachlorodibenzofuran (HXCDF)		----*	374$_{100}$	372$_{56}$	309$_{19}$	239$_{34}$	156$_{39}$	925	3497
1,2,3,6,7,8-Hexachlorodibenzofuran (HXCDF)		----*	374$_{100}$	372$_{50}$	309$_{15}$	239$_{31}$	187$_{28}$	925	3495
2,3,4,6,7,8-Hexachlorodibenzofuran (HXCDF)		----*	374$_{100}$	372$_{58}$	309$_{19}$	239$_{34}$	156$_{41}$	926	3496
Hexachlorophene		2790*	404$_{37}$	369$_{15}$	335$_{17}$	209$_{52}$	196$_{100}$	1027	3644
Hexacosane		2600*	366$_{3}$	99$_{13}$	85$_{36}$	71$_{56}$	57$_{100}$	907	2365
Hexadecane		1600*	226$_{4}$	99$_{10}$	85$_{31}$	71$_{59}$	57$_{100}$	296	2353
Hexamid	U	2380	331$_{3}$	259$_{4}$	117$_{5}$	86$_{100}$	58$_{7}$	766	1908
Hexamid-M (bis-deethyl-) AC	U UAAC	2570	317$_{2}$	258$_{12}$	117$_{21}$	85$_{100}$	72$_{34}$	697	1910
Hexamid-M (deethyl-)	UAAC	2200	303$_{1}$	117$_{15}$	71$_{51}$	58$_{100}$		629	1912
Hexamid-M (deethyl-) AC	UAAC	2780	345$_{11}$	302$_{61}$	113$_{44}$	100$_{45}$	58$_{100}$	829	1909
Hexamid-M (deethyl-HO-) 2AC	UAAC	3140	403$_{16}$	360$_{66}$	318$_{27}$	100$_{37}$	58$_{100}$	1027	1911
Hexamid-M (phenobarbital)	P G U+UHYAC	1965	232$_{14}$	204$_{100}$	161$_{8}$	146$_{12}$	117$_{37}$	312	854
Hexamid-M (phenobarbital) 2TMS		2015	376$_{2}$	361$_{4}$	261$_{15}$	146$_{100}$	73$_{46}$	939	4582
Hexane		600*	86$_{14}$	57$_{94}$	43$_{81}$	41$_{70}$	29$_{71}$	97	3775
Hexazinone		2295	252$_{4}$	171$_{100}$	128$_{24}$	83$_{57}$	71$_{47}$	392	4053
Hexethal	P G U	1835	211$_{6}$	156$_{100}$	141$_{93}$	55$_{36}$		345	807
Hexethal 2ME		1745	210$_{6}$	184$_{88}$	169$_{100}$	112$_{18}$	55$_{20}$	462	808
Hexobarbital	P G U+UHYAC	1855	236$_{10}$	221$_{100}$	157$_{43}$	155$_{26}$	81$_{70}$	330	809
Hexobarbital ME		1805	250$_{3}$	235$_{100}$	169$_{27}$	81$_{59}$		381	811
Hexobarbital-M (HO-) -H2O	U+UHYAC	1970	234$_{36}$	219$_{59}$	156$_{96}$	91$_{73}$	79$_{100}$	320	2265
Hexobarbital-M (nor-)		1980	222$_{7}$	207$_{41}$	143$_{50}$	81$_{100}$		281	1917
Hexobarbital-M (oxo-)	U+UHYAC	2055	250$_{60}$	235$_{80}$	193$_{27}$	156$_{28}$	95$_{100}$	380	810
Hexobarbital-M (oxo-) ME	PME UME	2020	264$_{28}$	249$_{100}$	221$_{20}$	207$_{17}$	95$_{37}$	439	2759
Hexobendine-M/artifact (trimethoxybenzoic acid)		1780*	212$_{100}$	197$_{57}$	169$_{13}$	141$_{27}$		248	1949
Hexobendine-M/artifact (trimethoxybenzoic acid) ET		1770*	240$_{100}$	225$_{44}$	212$_{17}$	195$_{45}$	141$_{24}$	344	5219
Hexobendine-M/artifact (trimethoxybenzoic acid) ME		1740*	226$_{100}$	211$_{49}$	195$_{21}$	155$_{21}$		293	1950
Hexyloctylphthalate		2500*	362$_{1}$	279$_{4}$	251$_{5}$	149$_{100}$	85$_{5}$	893	6053
Hexylresorcinol	P	1830*	194$_{41}$	123$_{100}$	95$_{9}$	77$_{10}$		206	1981
Hexylresorcinol AC		1875*	236$_{6}$	194$_{30}$	123$_{100}$			330	1989
Hexylresorcinol 2AC		1935*	278$_{7}$	236$_{22}$	194$_{90}$	123$_{100}$		507	1990
Hippuric acid	U	1745	179$_{1}$	161$_{2}$	135$_{22}$	105$_{100}$	77$_{9}$	169	96
Hippuric acid ME	UME	1660	193$_{5}$	161$_{7}$	134$_{19}$	105$_{100}$	77$_{45}$	199	97
Hippuric acid TMS	UTMS	1925	251$_{1}$	236$_{8}$	206$_{71}$	105$_{100}$	73$_{85}$	383	5813
Hippuric acid 2TMS	UTMS	2070	323$_{10}$	308$_{16}$	280$_{11}$	206$_{50}$	105$_{100}$	727	5812
Histapyrrodine	G U UHY U+UHYAC	2240	280$_{8}$	196$_{27}$	91$_{62}$	84$_{100}$	65$_{9}$	519	1646
Histapyrrodine-M (HO-)	UHY	2500	296$_{23}$	212$_{81}$	120$_{7}$	91$_{84}$	84$_{100}$	600	1650
Histapyrrodine-M (HO-) AC	U+UHYAC	2630	338$_{8}$	254$_{23}$	120$_{5}$	91$_{56}$	84$_{100}$	799	1652
Histapyrrodine-M (N-dealkyl-)	UHY	1930	183$_{25}$	106$_{16}$	91$_{100}$	77$_{25}$	65$_{25}$	182	2065
Histapyrrodine-M (N-dealkyl-) AC	U+UHYAC	2080	225$_{22}$	183$_{23}$	106$_{10}$	91$_{100}$	77$_{25}$	291	2066
Histapyrrodine-M (N-debenzyl-)	UHY	1800	190$_{12}$	120$_{3}$	106$_{6}$	84$_{100}$	77$_{10}$	194	1654
Histapyrrodine-M (N-debenzyl-oxo-)	UHY	2120	204$_{19}$	119$_{54}$	106$_{100}$	98$_{17}$	77$_{23}$	227	1653
Histapyrrodine-M (N-debenzyl-oxo-) AC	U+UHYAC	2160	246$_{4}$	161$_{59}$	119$_{49}$	106$_{100}$	98$_{29}$	363	1649
Histapyrrodine-M (N-dephenyl-) AC	U+UHYAC	2120	246$_{2}$	176$_{3}$	91$_{19}$	84$_{100}$	65$_{5}$	364	1647
Histapyrrodine-M (N-dephenyl-HO-) -H2O	UHY	2100	216$_{29}$	159$_{29}$	97$_{10}$	91$_{100}$	69$_{23}$	260	1655
Histapyrrodine-M (N-dephenyl-oxo-) AC	U+UHYAC	2260	260$_{3}$	217$_{45}$	175$_{12}$	120$_{79}$	91$_{100}$	422	1648

Table 1-8-1: Compounds in order of names   Hydromorphone enol 2TFA

Name	Detected	RI	Typical ions and intensities					Page	Entry
Histapyrrodine-M (oxo-)	U UHY U+UHYAC	2570	294_7	209_9	196_{44}	120_5	91_{100}	587	1651
Homatropine		2340	275_6	142_4	124_{100}	94_{21}	82_{23}	490	6259
Homatropine AC		2250	317_3	245_2	124_{100}	94_{18}	82_{22}	698	6264
Homatropine TMS		2090	347_4	179_{23}	124_{100}	94_{15}	73_{30}	836	6307
Homatropine-M (mandelic acid)		1890*	152_{10}	107_{100}	79_{75}	77_{55}	51_{63}	131	5759
Homatropine-M (mandelic acid) ME		1485*	166_7	107_{100}	79_{59}	77_{40}		149	1071
Homatropine-M (nor-) 2AC		2565	345_{11}	302_{10}	152_{39}	110_{100}	68_{24}	829	6265
Homatropine-M/artifact (tropine) AC		1240	183_{25}	140_{11}	124_{100}	94_{42}	82_{63}	183	5125
Homofenazine	G	3165	451_1	433_9	280_{51}	167_{63}	58_{100}	1115	526
Homofenazine AC		3260	493_3	433_{29}	280_{65}	167_{100}	87_{83}	1159	341
Homofenazine-M (amino-) AC	U+UHYAC	2765	366_{47}	280_{17}	266_{40}	248_{18}	100_{300}	903	1267
Homofenazine-M (dealkyl-) AC	U+UHYAC	3240	449_{80}	267_{100}	112_{27}			1112	1269
Homofenazine-M (ring)	U+UHYAC	2190	267_{100}	235_{19}				453	1266
Homovanillic acid	U	1610*	182_{35}	137_{100}	122_{12}	94_7	65_9	179	3368
Homovanillic acid HFB		1770*	378_{79}	333_{39}	181_{60}	107_{100}	69_{72}	945	5975
Homovanillic acid ME	UME	1750*	196_{25}	137_{100}	122_{16}	107_2	94_{12}	210	812
Homovanillic acid MEAC	U+UHYAC	1700*	238_2	196_{47}	137_{100}	122_6	107_6	336	2973
Homovanillic acid MEHFB		1570*	392_{100}	333_{84}	169_{24}	107_{78}	69_{60}	991	5974
Homovanillic acid MEPFP		1570*	342_{96}	283_{100}	195_{35}	119_{37}	107_{51}	815	5972
Homovanillic acid METMS		1670*	268_{41}	238_{69}	209_{52}	179_{100}	73_{95}	461	6016
Homovanillic acid PFP		1685*	328_{85}	283_{45}	181_{58}	119_{33}	107_{100}	747	5973
Homovanillic acid 2ME	UME	1720*	210_{31}	151_{100}	123_4	107_6		244	5959
Homovanillic acid 2TMS		1760*	326_{32}	267_{20}	209_{46}	179_{34}	73_{100}	740	6015
Hydrocaffeic acid		2400*	182_{17}	136_{13}	123_{100}	77_{55}	51_{67}	179	5763
Hydrocaffeic acid ME		1870*	196_{24}	136_{45}	123_{100}	91_{11}	77_{17}	210	5764
Hydrocaffeic acid ME2AC		1980*	280_2	238_{15}	196_{100}	136_{65}	123_{68}	516	5992
Hydrocaffeic acid ME2HFB		1720*	588_{11}	528_{100}	349_{32}	169_{37}	69_{58}	1193	5994
Hydrocaffeic acid ME2PFP		1590*	488_{19}	428_{100}	299_{26}	281_{26}	119_{73}	1155	5993
Hydrocaffeic acid ME2TMS		2220*	340_{44}	267_{36}	193_{10}	179_{100}	73_{97}	807	5995
Hydrocaffeic acid METFA		1540*	292_{88}	233_{100}	195_{27}	107_{46}	69_{62}	574	5969
Hydrocaffeic acid 3TMS		2250*	398_{67}	280_{16}	267_{39}	179_{98}	73_{100}	1012	5996
Hydrocaffeic acid -CO2		1295*	138_{23}	123_{100}	91_{28}	77_{47}	51_{92}	118	5756
Hydrochlorothiazide		9999	297_{53}	269_{100}	221_{18}			600	813
Hydrochlorothiazide 4ME	UME	2905	353_{100}	310_{90}	288_{45}	218_{38}	138_{44}	856	6536
Hydrochlorothiazide artifact ME	UME	1980	220_{31}	191_{10}	142_{25}	127_{100}	99_{21}	272	3003
Hydrochlorothiazide -SO2NH ME	UME	2170	232_{94}	167_{48}	139_{100}	127_{85}	125_{70}	311	3002
Hydrocodone	G UHY U+UHYAC	2440	299_{100}	242_{51}	185_{23}	96_{24}	59_{23}	611	238
Hydrocodone enol AC		2500	341_{100}	298_{65}	242_{32}	162_{26}		813	258
Hydrocodone enol TMS		2475	371_{31}	356_{14}	313_9	234_{30}	73_{100}	925	6215
Hydrocodone enol Cl-artifact AC		2630	375_{100}	340_{47}	318_{28}	146_{13}	115_{10}	936	4401
Hydrocodone-M (dihydro-) 6-beta isomer TMS		2495	373_{100}	316_{27}	236_6	146_7	73_{200}	932	6762
Hydrocodone-M (N,O-bisdemethyl-) enol 3TMS		2635	487_{100}	472_{45}	357_{65}	292_{32}	73_{200}	1155	6764
Hydrocodone-M (N,O-bis-demethyl-dihydro-) 3AC	U+UHYAC	2790	399_{20}	357_{50}	229_{19}	87_{100}	72_{22}	1014	3050
Hydrocodone-M (N,O-bisdemethyl-dihydro-) 6-beta isomer 3TMS		2600	489_{100}	474_{20}	374_{53}	294_{23}	73_{200}	1157	6765
Hydrocodone-M (N-demethyl-) enol 2TMS		2610	429_{88}	414_{26}	314_{37}	292_{30}	73_{150}	1079	6763
Hydrocodone-M (N-demethyl-dihydro-) 6-beta isomer 2TMS		2560	431_{100}	316_{33}	234_{12}	144_{14}	73_{200}	1083	6761
Hydrocodone-M (nor-) AC	U+UHYAC	2760	327_{18}	241_{16}	87_{100}	72_{37}		744	239
Hydrocodone-M (nor-dihydro-)	UHY	2440	287_{100}	244_{24}	242_{22}	150_{32}	115_{24}	547	4368
Hydrocodone-M (nor-dihydro-) AC	U+UHYAC	2700	329_{40}	243_{42}	183_{26}	87_{100}	72_{44}	755	3054
Hydrocodone-M (nor-dihydro-) 2AC	U+UHYAC	2750	371_{20}	285_7	243_{26}	87_{100}	72_{33}	924	235
Hydrocodone-M (O-demethyl-) TMS		2475	357_{51}	342_{18}	300_{38}	96_{29}	73_{100}	873	6209
Hydrocodone-M (O-demethyl-) enol 2AC	U+UHYAC	2625	369_{39}	327_{100}	284_{54}	228_{37}	162_{40}	916	1186
Hydrocodone-M (O-demethyl-) enol 2TFA		2230	477_{100}	380_{32}	258_{18}	223_7	117_{10}	1145	4009
Hydrocodone-M (O-demethyl-) enol 2TMS		2520	429_{22}	414_{22}	357_{10}	234_{23}	73_{100}	1079	6208
Hydrocodone-M (O-demethyl-dihydro-)	UHY	2400	287_{100}	230_{14}	164_{35}	115_{28}	70_{54}	547	484
Hydrocodone-M (O-demethyl-dihydro-) AC	U+UHYAC	2490	329_{100}	287_{56}	230_{10}	164_{20}	70_{21}	755	3055
Hydrocodone-M (O-demethyl-dihydro-) TFA		2250	383_{100}	286_{19}	270_{44}	213_{19}	69_{50}	963	6199
Hydrocodone-M (O-demethyl-dihydro-) 2AC	U+UHYAC	2545	371_{83}	329_{100}	286_{34}	212_{21}	70_{33}	924	234
Hydrocodone-M (O-demethyl-dihydro-) 2HFB		2260	679_{41}	482_{53}	466_{100}	360_{13}	169_{21}	1202	6197
Hydrocodone-M (O-demethyl-dihydro-) 2PFP		2330	579_{60}	432_{21}	416_{49}	310_4	119_{100}	1191	2460
Hydrocodone-M (O-demethyl-dihydro-) 2TFA		2190	479_{91}	382_{25}	366_{61}	260_7	69_{100}	1147	6198
Hydrocodone-M (O-demethyl-dihydro-) 6-alpha isomer 2TMS		2520	431_{33}	373_7	236_{23}	146_{15}	73_{100}	1083	2469
Hydrocodone-M (O-demethyl-dihydro-) 6-beta isomer 2TMS		2540	431_{87}	374_{28}	236_7	146_8	73_{200}	1083	6760
Hydrocortisone	UME	2740*	302_{53}	189_{31}	163_{100}	123_{62}	91_{59}	892	3295
Hydrocotarnine		1790	221_{60}	220_{100}	205_{23}	178_{66}	163_{21}	276	2862
Hydromorphone	UHY	2445	285_{100}	228_{14}	214_{10}	171_{10}	96_{30}	539	527
Hydromorphone AC	U+UHYAC	2595	327_{34}	285_{100}	229_{36}	200_{14}	171_{13}	745	240
Hydromorphone HFB		2385	481_{100}	452_{17}	425_{77}	410_{22}	284_8	1149	6137
Hydromorphone PFP		2250	431_{100}	375_{78}	360_{33}	346_{25}	119_{50}	1082	2662
Hydromorphone TMS		2475	357_{51}	342_{18}	300_{38}	96_{29}	73_{100}	873	6209
Hydromorphone 2HFB		2325	677_{100}	480_{31}	464_{18}	358_{10}	169_7	1201	6138
Hydromorphone enol 2AC	U+UHYAC	2625	369_{39}	327_{100}	284_{54}	228_{37}	162_{40}	916	1186
Hydromorphone enol 2PFP		2320	577_{48}	430_{100}	414_{62}	372_{30}	308_{68}	1191	2663
Hydromorphone enol 2TFA		2230	477_{100}	380_{32}	258_{18}	223_7	117_{10}	1145	4009

Hydromorphone enol 2TMS

Table 1-8-1: Compounds in order of names

Name	Detected	RI	Typical ions and intensities					Page	Entry
Hydromorphone enol 2TMS		2520	429 $_{22}$	414 $_{22}$	357 $_{10}$	234 $_{23}$	73 $_{100}$	1079	6208
Hydromorphone-M (dihydro-)	UHY	2400	287 $_{100}$	230 $_{14}$	164 $_{35}$	115 $_{28}$	70 $_{54}$	547	484
Hydromorphone-M (dihydro-) AC	U+UHYAC	2490	329 $_{100}$	287 $_{56}$	230 $_{10}$	164 $_{20}$	70 $_{21}$	755	3055
Hydromorphone-M (dihydro-) TFA		2250	383 $_{100}$	286 $_{19}$	270 $_{44}$	213 $_{19}$	69 $_{50}$	963	6199
Hydromorphone-M (dihydro-) 2AC	U+UHYAC	2545	371 $_{83}$	329 $_{100}$	286 $_{34}$	212 $_{21}$	70 $_{33}$	924	234
Hydromorphone-M (dihydro-) 2HFB		2260	679 $_{41}$	482 $_{53}$	466 $_{100}$	360 $_{13}$	169 $_{21}$	1202	6197
Hydromorphone-M (dihydro-) 2PFP		2330	579 $_{60}$	432 $_{21}$	416 $_{49}$	310 $_{4}$	119 $_{100}$	1191	2460
Hydromorphone-M (dihydro-) 2TFA		2190	479 $_{91}$	382 $_{25}$	366 $_{61}$	260 $_{7}$	69 $_{100}$	1147	6198
Hydromorphone-M (dihydro-) 6-alpha isomer 2TMS		2520	431 $_{33}$	373 $_{7}$	236 $_{23}$	146 $_{15}$	73 $_{100}$	1083	2469
Hydromorphone-M (dihydro-) 6-beta isomer 2TMS		2540	431 $_{87}$	374 $_{28}$	236 $_{7}$	146 $_{8}$	73 $_{200}$	1083	6760
Hydromorphone-M (N-demethyl-) enol 3TMS		2635	487 $_{100}$	472 $_{45}$	357 $_{65}$	292 $_{32}$	73 $_{200}$	1155	6764
Hydromorphone-M (N-demethyl-dihydro-) 6-beta isomer 3TMS		2600	489 $_{100}$	474 $_{20}$	374 $_{53}$	294 $_{23}$	73 $_{200}$	1157	6765
Hydroquinone	UHY	<1000*	110 $_{100}$	81 $_{27}$				103	814
Hydroquinone 2AC	U+UHYAC	1395*	194 $_{8}$	152 $_{26}$	110 $_{100}$			203	815
Hydroquinone 2ME		<1000*	138 $_{56}$	123 $_{100}$	95 $_{54}$	63 $_{22}$		118	3282
Hydroquinone-M (2-HO-)	UHY	1460*	126 $_{100}$	109 $_{18}$	79 $_{26}$	53 $_{9}$		110	3163
Hydroquinone-M (2-HO-) 3AC	U+UHYAC	1710*	252 $_{1}$	210 $_{7}$	168 $_{46}$	126 $_{100}$	97 $_{7}$	388	4336
Hydroquinone-M (2-methoxy-) 2AC	U+UHYAC	1450*	224 $_{3}$	182 $_{23}$	140 $_{100}$	125 $_{71}$	97 $_{9}$	287	4337
N-Hydroxy-Amfetamine		1180	149 $_{2}$	116 $_{3}$	91 $_{31}$	65 $_{12}$	60 $_{100}$	129	5906
N-Hydroxy-Amfetamine AC		1300	149 $_{7}$	107 $_{12}$	102 $_{74}$	91 $_{62}$	60 $_{100}$	201	5907
N-Hydroxy-Amfetamine TFA		1195	156 $_{88}$	118 $_{63}$	91 $_{100}$	69 $_{41}$	65 $_{33}$	366	5909
N-Hydroxy-Amfetamine 2AC		1720	144 $_{27}$	118 $_{26}$	102 $_{100}$	91 $_{56}$	60 $_{39}$	325	5908
Hydroxyandrostanedione	UHY	2530*	304 $_{74}$	286 $_{72}$	232 $_{99}$	191 $_{54}$		636	816
Hydroxyandrostanedione AC	U+UHYAC	2630*	346 $_{25}$	286 $_{76}$	271 $_{100}$	232 $_{61}$	191 $_{89}$	833	2699
Hydroxyandrostene	U	2300*	274 $_{100}$	259 $_{75}$	241 $_{66}$	148 $_{86}$	94 $_{85}$	487	614
Hydroxyandrostene AC	UGLUCAC	2860*	316 $_{100}$	256 $_{59}$	241 $_{24}$	215 $_{25}$		695	266
11-Hydroxyandrosterone		2640*	306 $_{100}$	288 $_{52}$	270 $_{41}$	147 $_{45}$	55 $_{86}$	646	3763
11-Hydroxyandrosterone AC		2760*	348 $_{45}$	273 $_{44}$	270 $_{39}$	105 $_{68}$	55 $_{100}$	840	3771
11-Hydroxyandrosterone enol 3TMS		2705*	522 $_{24}$	417 $_{23}$	256 $_{25}$	168 $_{48}$	73 $_{100}$	1175	3805
3-Hydroxybenzoic acid		1620*	138 $_{65}$	121 $_{63}$	93 $_{42}$	65 $_{100}$		118	5758
3-Hydroxybenzoic acid AC		1560*	180 $_{10}$	138 $_{100}$	121 $_{50}$	93 $_{11}$	63 $_{17}$	173	5978
3-Hydroxybenzoic acid ME		1330*	152 $_{46}$	121 $_{100}$	93 $_{43}$	65 $_{33}$	53 $_{7}$	130	5976
3-Hydroxybenzoic acid MEAC		1375*	194 $_{6}$	163 $_{8}$	152 $_{72}$	121 $_{100}$	93 $_{21}$	203	5977
3-Hydroxybenzoic acid 2ME		1490*	166 $_{58}$	135 $_{97}$	107 $_{43}$	92 $_{23}$	77 $_{40}$	149	1110
3-Hydroxybenzoic acid 2TMS		1535*	282 $_{38}$	267 $_{100}$	223 $_{42}$	193 $_{40}$	73 $_{86}$	526	6017
4-Hydroxybenzoic acid 2ME		1270*	166 $_{34}$	135 $_{100}$	107 $_{15}$	92 $_{16}$	77 $_{26}$	150	6446
3-Hydroxybenzylalcohol		1310*	124 $_{100}$	107 $_{15}$	95 $_{60}$	77 $_{66}$	65 $_{22}$	109	4663
Bis-(2-Hydroxy-3-tert-butyl-5-ethylphenyl)methane		2450*	368 $_{49}$	312 $_{17}$	191 $_{100}$	175 $_{61}$	163 $_{46}$	914	2870
4-Hydroxybutyric acid 2TMS		1520*	248 $_{1}$	233 $_{26}$	147 $_{100}$	117 $_{30}$	73 $_{32}$	371	5430
gamma-Hydroxybutyric acid 2TMS		1520*	248 $_{1}$	233 $_{26}$	147 $_{100}$	117 $_{30}$	73 $_{32}$	371	5430
1,1-Bis-(2-Hydroxy-3,5-dimethylphenyl-)-2-methylpropane		2050*	298 $_{21}$	255 $_{100}$	237 $_{7}$	209 $_{4}$	179 $_{4}$	608	5658
Hydroxyethylsalicylate		1540*	182 $_{13}$	164 $_{32}$	120 $_{100}$	92 $_{45}$	65 $_{21}$	179	5224
Hydroxyethylsalicylate 2AC		1800*	266 $_{1}$	224 $_{7}$	164 $_{6}$	120 $_{32}$	87 $_{100}$	448	5225
Hydroxyethylurea		9999	104 $_{1}$	86 $_{6}$	74 $_{100}$	61 $_{7}$		101	1551
11-Hydroxyetiocholanolone		2675*	306 $_{100}$	273 $_{64}$	147 $_{30}$	107 $_{40}$	55 $_{53}$	646	3764
11-Hydroxyetiocholanolone AC		2770*	348 $_{20}$	288 $_{80}$	270 $_{71}$	255 $_{62}$	55 $_{100}$	840	3772
11-Hydroxyetiocholanolone enol 3TMS		2735*	522 $_{22}$	417 $_{12}$	327 $_{23}$	168 $_{72}$	73 $_{100}$	1176	3798
4-Hydroxyhippuric acid ME	U	1820	209 $_{100}$	177 $_{32}$	149 $_{34}$	121 $_{87}$		240	817
5-Hydroxyindole	UHY	1340	133 $_{100}$	105 $_{25}$	78 $_{16}$	51 $_{13}$		113	3285
5-Hydroxyindole AC	U+UHYAC	1370	175 $_{8}$	133 $_{100}$	106 $_{13}$	78 $_{17}$	51 $_{11}$	161	4273
5-Hydroxyindoleacetic acid 2ME		1995	219 $_{55}$	160 $_{100}$	145 $_{23}$	117 $_{16}$	89 $_{7}$	269	5042
5-Hydroxyindolepropanoic acid 2ME		1695	233 $_{35}$	174 $_{16}$	160 $_{100}$	149 $_{12}$	130 $_{19}$	315	5041
N-Hydroxy-MDA 2AC		2010	279 $_{1}$	162 $_{100}$	135 $_{76}$	102 $_{58}$	60 $_{23}$	510	5910
N-Hydroxy-MDA TFA		1665	291 $_{12}$	162 $_{37}$	135 $_{100}$	105 $_{3}$	77 $_{22}$	568	5911
Hydroxy-methoxy-acetophenone AC	U+UHYAC	1640*	208 $_{6}$	166 $_{49}$	151 $_{100}$	123 $_{12}$		236	2483
4-Hydroxy-3-methoxy-benzylamine 2AC		1995	237 $_{9}$	195 $_{100}$	152 $_{48}$	137 $_{43}$	122 $_{33}$	333	5691
4-Hydroxy-3-methoxy-cinnamic acid ME		1930*	208 $_{100}$	177 $_{80}$	145 $_{53}$	133 $_{19}$	117 $_{24}$	236	5966
4-Hydroxy-3-methoxy-cinnamic acid MEAC		1950*	250 $_{3}$	208 $_{100}$	177 $_{47}$	145 $_{26}$	117 $_{9}$	379	5814
4-Hydroxy-3-methoxy-cinnamic acid 2ME	UME	1850*	222 $_{100}$	207 $_{22}$	191 $_{75}$	164 $_{14}$	147 $_{19}$	281	4945
4-Hydroxy-3-methoxy-cinnamic acid 2TMS		2160*	338 $_{92}$	323 $_{56}$	308 $_{48}$	249 $_{45}$	73 $_{100}$	797	5815
4-Hydroxy-3-methoxy-cinnamic acid -CO2		1195*	150 $_{34}$	135 $_{33}$	107 $_{68}$	77 $_{100}$	51 $_{68}$	127	5752
4-Hydroxy-3-methoxy-cinnamic acid glycine conjugate		2380	265 $_{75}$	204 $_{7}$	177 $_{100}$	145 $_{47}$	117 $_{13}$	443	5766
4-Hydroxy-3-methoxy-cinnamic acid glycine conjugate 2ME		2450	279 $_{41}$	191 $_{100}$	163 $_{12}$	148 $_{6}$	133 $_{5}$	510	5825
4-Hydroxy-3-methoxy-cinnamic acid glycine conjugate 3TMS		2540	467 $_{61}$	453 $_{8}$	336 $_{22}$	249 $_{67}$	73 $_{100}$	1135	5826
Hydroxymethoxyflavone AC		2610*	310 $_{64}$	268 $_{100}$	253 $_{17}$	132 $_{40}$	117 $_{9}$	665	5599
Hydroxymethoxyflavone ME		2600*	282 $_{100}$	267 $_{15}$	150 $_{15}$	132 $_{64}$	89 $_{14}$	526	5600
4-Hydroxy-3-methoxyhydrocinnamic acid ME		1670*	210 $_{26}$	150 $_{20}$	137 $_{100}$	122 $_{9}$	107 $_{8}$	244	5822
4-Hydroxy-3-methoxyhydrocinnamic acid MEAC		1860*	252 $_{3}$	210 $_{59}$	179 $_{8}$	150 $_{29}$	137 $_{100}$	390	5823
4-Hydroxy-3-methoxyhydrocinnamic acid 2TMS		1940*	340 $_{71}$	310 $_{33}$	209 $_{100}$	192 $_{56}$	73 $_{89}$	807	5824
4-Hydroxy-3-methoxy-phenethylamine		1410	167 $_{10}$	138 $_{100}$	123 $_{34}$	94 $_{19}$	77 $_{10}$	152	5615
15-Hydroxy-5,8,11,13-eicosatetraenoic acid METFA		2390*	430 $_{1}$	316 $_{6}$	131 $_{43}$	117 $_{66}$	91 $_{100}$	1080	4354
15-Hydroxy-5,8,11,13-eicosatetraenoic acid -H2O ME		2360*	316 $_{4}$	189 $_{37}$	119 $_{54}$	105 $_{100}$	91 $_{92}$	694	4355
Hydroxypethidine AC	U+UHYAC	2205	305 $_{9}$	230 $_{6}$	188 $_{10}$	71 $_{100}$		640	1195
3-Hydroxyphenylacetic acid 2TMS		1695*	296 $_{20}$	281 $_{19}$	73 $_{100}$	164 $_{50}$	147 $_{39}$	598	6010

Table 1-8-1: Compounds in order of names   Imipramine-M (bis-nor-HO-) 2AC

Name	Detected	RI	Typical ions and intensities					Page	Entry
4-Hydroxyphenylacetic acid	U	1565*	152_{26}	107_{100}	77_{18}			130	818
4-Hydroxyphenylacetic acid AC		1565*	194_{4}	152_{46}	107_{100}	77_{18}		203	5819
4-Hydroxyphenylacetic acid HFB		1495*	348_{40}	303_{100}	275_{35}	169_{30}	69_{70}	837	5956
4-Hydroxyphenylacetic acid ME	UME	1570*	166_{18}	121_{3}	107_{100}	77_{19}	59_{3}	150	4224
4-Hydroxyphenylacetic acid MEAC		1550*	149_{3}	208_{3}	166_{37}	107_{100}	77_{12}	236	5820
4-Hydroxyphenylacetic acid MEHFB		1405*	362_{36}	303_{100}	275_{33}	78_{49}	59_{56}	890	5957
4-Hydroxyphenylacetic acid MEPFP		1220*	312_{34}	253_{100}	225_{28}	119_{39}	78_{43}	674	5955
4-Hydroxyphenylacetic acid METFA		1120*	262_{33}	203_{100}	175_{21}	69_{56}	59_{35}	429	5750
4-Hydroxyphenylacetic acid METMS		1485*	238_{48}	223_{17}	179_{100}	163_{68}	73_{78}	336	6018
4-Hydroxyphenylacetic acid TFA		1450*	248_{36}	203_{100}	175_{29}	77_{37}	69_{57}	369	5954
4-Hydroxyphenylacetic acid 2ME	UME	1420*	180_{17}	121_{100}	91_{9}	77_{10}		175	4228
4-Hydroxyphenylacetic acid 2PFP		1340*	430_{22}	253_{100}	225_{13}	119_{45}	69_{52}	1079	5675
4-Hydroxyphenylacetic acid 2TMS		1675*	296_{16}	281_{18}	252_{18}	179_{35}	73_{100}	599	5821
Hydroxyprogesterone -H2O		2650*	312_{84}	297_{35}	269_{100}	227_{34}	91_{39}	677	5182
Hydroxyproline ME2AC		1690	229_{1}	198_{1}	169_{9}	110_{37}	68_{100}	303	2709
Hydroxyproline MEAC		1635	187_{7}	128_{54}	86_{100}	68_{28}		189	2283
2-Hydroxyquinoxaline		2020	146_{100}	118_{91}	91_{42}	76_{3}	64_{23}	124	7413
2-Hydroxyquinoxaline ME		1750	160_{100}	131_{71}	104_{24}	90_{11}	77_{17}	138	7414
3-Hydroxytyramine 3AC	U+UHYAC	2150	279_{3}	237_{22}	220_{21}	178_{30}	136_{100}	510	5284
3-Hydroxytyramine 4AC		2245	321_{1}	220_{33}	178_{36}	136_{100}	123_{32}	718	5285
Hydroxyzine	G P-I U	2900	374_{13}	299_{17}	201_{100}	165_{32}		934	820
Hydroxyzine AC	U+UHYAC	3000	416_{5}	299_{16}	201_{100}	165_{49}	87_{19}	1055	1463
Hydroxyzine artifact		1900*	232_{61}	201_{63}	165_{100}	105_{100}	77_{55}	311	1344
Hydroxyzine artifact	G U+UHYAC	1600*	202_{30}	167_{100}	165_{52}	152_{17}	125_{7}	222	2442
Hydroxyzine-M	U+UHYAC	2210	280_{100}	201_{35}	165_{57}			515	770
Hydroxyzine-M (carbinol)	UHY	1750*	218_{17}	183_{7}	139_{39}	105_{100}	77_{87}	263	2239
Hydroxyzine-M (carbinol) AC	U+UHYAC	1890*	260_{8}	200_{40}	165_{100}	139_{10}	77_{29}	420	1270
Hydroxyzine-M (Cl-benzophenone)	U+UHYAC	1850*	216_{43}	139_{58}	105_{100}	77_{44}		258	1343
Hydroxyzine-M (HO-Cl-benzophenone)	UHY	2300*	232_{36}	197_{7}	139_{23}	121_{100}	111_{23}	311	2240
Hydroxyzine-M (HO-Cl-BPH) iso-1 AC	U+UHYAC	2200*	274_{18}	232_{75}	139_{100}	121_{44}	111_{51}	484	2229
Hydroxyzine-M (HO-Cl-BPH) iso-2 AC	U+UHYAC	2230*	274_{7}	232_{43}	139_{25}	121_{100}	111_{27}	484	2230
Hydroxyzine-M (HOOC-) ME	G PME UME U+UHYAC	2910	402_{10}	229_{6}	201_{100}	165_{46}	146_{23}	1024	4323
Hydroxyzine-M (N-dealkyl-)	UHY	2520	286_{13}	241_{48}	201_{50}	165_{65}	56_{100}	543	2241
Hydroxyzine-M (N-dealkyl-) AC	U+UHYAC	2620	328_{7}	242_{19}	201_{48}	165_{66}	85_{100}	749	1271
Hydroxyzine-M AC	U+UHYAC	2380	299_{4}	285_{49}	226_{100}	191_{58}	84_{45}	609	821
Hydroxyzine-M/artifact	P-I U+UHYAC UME	2220	300_{17}	228_{38}	165_{52}	99_{100}	56_{63}	616	670
Hydroxyzine-M/artifact 2AC		2005	258_{2}	199_{6}	141_{100}	112_{9}	99_{20}	414	2443
Hydroxyzine-M/artifact HYAC	U+UHYAC	2935	280_{4}	201_{100}	165_{26}			515	1272
Hymecromone	G UHY	2015*	176_{74}	148_{100}	147_{71}	120_{12}	91_{27}	162	2571
Hymecromone AC	U+UHYAC	2005*	218_{11}	176_{92}	148_{100}	120_{9}	91_{29}	263	2572
Hymexazol		1300	99_{100}	71_{18}	54_{14}			99	3645
Hyoscyamine	P G U	2215	289_{9}	272_{1}	140_{5}	124_{100}	94_{6}	559	69
Hyoscyamine AC	U+UHYAC	2275	331_{5}	140_{8}	124_{100}	94_{22}	82_{35}	765	71
Hyoscyamine TMS		2295	361_{5}	140_{5}	124_{100}	82_{20}	73_{50}	888	4526
Hyoscyamine -H2O	P G U UHY U+UHYAC	2085	271_{33}	140_{8}	124_{100}	96_{21}	82_{13}	475	70
Ibuprofen	G P U+UHYAC	1615*	206_{44}	163_{92}	161_{100}	119_{44}	91_{75}	231	1941
Ibuprofen ME	PME UME UHYME	1505*	220_{25}	177_{28}	161_{100}	119_{18}	91_{19}	274	1942
Ibuprofen TMS		1665*	278_{5}	263_{17}	160_{31}	117_{21}	73_{100}	508	4554
Ibuprofen-M (HO-) MEAC	U+UHYAC	1880*	278_{1}	218_{41}	177_{12}	159_{100}	117_{30}	508	3385
Ibuprofen-M (HO-) -H2O	U+UHYAC	1700*	204_{43}	159_{100}	128_{13}	117_{30}	91_{14}	226	3382
Ibuprofen-M (HO-) -H2O ME	UHYME	1585*	218_{23}	159_{100}	128_{11}	117_{25}	91_{11}	265	3380
Ibuprofen-M (HO-) iso-1 ME	UME	1680*	236_{1}	178_{52}	119_{100}	118_{99}	91_{72}	330	3381
Ibuprofen-M (HO-) iso-2 ME	UME	1770*	236_{2}	193_{100}	133_{32}	105_{66}	77_{16}	331	6386
Ibuprofen-M (HO-) iso-3 ME	PME UME	1830*	236_{13}	205_{6}	177_{100}	159_{33}	117_{54}	331	3383
Ibuprofen-M (HO-) iso-4 ME	UME	1925*	236_{20}	161_{12}	118_{100}	91_{9}	59_{24}	331	6387
Ibuprofen-M (HO-HOOC-) -H2O 2ME	UME	1900*	262_{17}	203_{100}	157_{15}	143_{43}	128_{59}	430	3386
Ibuprofen-M (HOOC-) 2ME	UME	1810*	264_{24}	205_{100}	177_{65}	145_{95}	117_{42}	439	3384
Idobutal	P G U	1700	181_{14}	167_{100}	124_{36}			289	1036
Idobutal 2ME		1610	223_{6}	195_{100}	181_{21}	169_{20}	138_{41}	391	1037
Imazalil		2140	296_{10}	240_{12}	215_{100}	173_{74}	81_{32}	596	2054
Imidapril ME		2700	419_{2}	346_{28}	234_{100}	159_{21}	91_{41}	1061	6279
Imidapril TMS		2700	477_{2}	404_{31}	234_{100}	160_{32}	91_{70}	1146	6281
Imidapril artifact		2000	234_{19}	220_{100}	160_{13}	117_{27}	91_{79}	318	6280
Imidaprilate 2ME		2695	405_{2}	346_{14}	220_{100}	159_{20}	91_{56}	1031	6282
Imidaprilate 2TMS		2770	160_{37}	506_{3}	521_{1}	404_{72}	278_{100}	1175	6284
Imidaprilate 3ME		2710	419_{1}	360_{15}	234_{100}	159_{15}	91_{33}	1061	6283
Imidapril-M (deethyl-) 2ME		2695	405_{2}	346_{14}	220_{100}	159_{20}	91_{56}	1031	6282
Imidapril-M (deethyl-) 2TMS		2770	160_{37}	506_{3}	521_{1}	404_{72}	278_{100}	1175	6284
Imidapril-M (deethyl-) 3ME		2710	419_{1}	360_{15}	234_{100}	159_{15}	91_{33}	1061	6283
2,2'-Iminodibenzyl	U U+UHYAC	1930	195_{100}	180_{40}	167_{9}	96_{33}	83_{22}	207	308
2,2'-Iminodibenzyl ME		1915	209_{70}	194_{100}	178_{13}	165_{11}		242	6352
Imipramine	P-I G U+UHYAC	2215	280_{11}	234_{39}	193_{24}	85_{56}	58_{100}	519	342
Imipramine-M (bis-nor-) AC	U+UHYAC	2640	294_{23}	208_{100}	193_{43}	152_{6}	100_{17}	586	3313
Imipramine-M (bis-nor-HO-) 2AC	U+UHYAC	2980	352_{60}	266_{88}	224_{100}	180_{15}	100_{48}	855	3314

Imipramine-M (di-HO-ring)

Table 1-8-1: Compounds in order of names

Name	Detected	RI	Typical ions and intensities	Page	Entry
Imipramine-M (di-HO-ring)	UHY	2600	227_{100} 196_7	297	2296
Imipramine-M (di-HO-ring) 2AC	U+UHYAC	2750	311_{28} 269_{23} 227_{100} 196_7	670	2292
Imipramine-M (HO-)	UHY	2565	296_{34} 251_{43} 85_{39} 58_{99}	600	528
Imipramine-M (HO-) AC	U+UHYAC	2610	338_3 251_{15} 211_{18} 85_{30} 58_{100}	799	343
Imipramine-M (HO-) ME		2480	310_{16} 265_{13} 225_9 85_8 58_{100}	668	529
Imipramine-M (HO-methoxy-ring)	UHY	2390	241_{100} 226_{17} 210_{12} 180_{14}	347	2315
Imipramine-M (HO-methoxy-ring) AC	U+UHYAC	2370	283_{10} 241_{100} 226_{17} 210_{12} 180_{14}	529	2867
Imipramine-M (HO-ring)	UHY	2240	211_{100} 196_{15} 180_{10} 152_4	246	2295
Imipramine-M (HO-ring) AC	U+UHYAC	2535	253_{26} 211_{100} 196_{19} 180_{11} 152_4	393	1218
Imipramine-M (nor-)	UHY	2225	266_{28} 235_{61} 208_{61} 195_{100} 71_{59}	453	324
Imipramine-M (nor-) AC	U+UHYAC PAC	2670	308_{60} 208_{150} 193_{82} 114_{93}	657	325
Imipramine-M (nor-) HFB		2450	462_{23} 268_{13} 240_{20} 208_{100} 193_{54}	1130	7706
Imipramine-M (nor-) PFP		2450	412_{68} 234_{28} 218_{35} 208_{180} 193_{85}	1047	7667
Imipramine-M (nor-) TFA		2430	208_{100} 140_{17} 69_{21} 193_{51} 362_{14}	891	7786
Imipramine-M (nor-) TMS		2470	338_5 235_{62} 143_{28} 116_{50} 73_{70}	800	5461
Imipramine-M (nor-di-HO-) 3AC	U+UHYAC	3380	424_{44} 324_{35} 282_{34} 240_{27} 114_{100}	1068	3315
Imipramine-M (nor-HO-) 2AC	U+UHYAC	3065	366_{27} 266_{39} 114_{100}	906	1175
Imipramine-M (ring)	U U+UHYAC	1930	195_{100} 180_{40} 167_9 96_{33} 83_{22}	207	308
Imipramine-M (ring) ME		1915	209_{70} 194_{100} 178_{13} 165_{11}	242	6352
Impurity		1730*	249_1 180_{12} 123_{27} 97_{100} 57_{92}	373	116
Impurity		3580	441_{100} 385_1 308_6 147_{14} 57_{91}	1099	3573
Impurity	P	1490*	178_{18} 163_{100} 135_{28} 115_7 91_{25}	166	6953
Impurity AC		1625*	181_3 122_{18} 92_6 80_{100}	176	2359
Impurity AC	U+UHYAC	1430*	195_1 152_6 122_{20} 92_6 80_{100}	206	2358
Impurity AC	U+UHYAC	2780*	219_1 175_3 131_7 87_{100} 73_{13}	267	2499
Impurity AC	U+UHYAC	2570*	219_1 175_3 131_6 87_{100} 73_{13}	268	2498
Impurity AC	U+UHYAC	2340*	219_1 175_1 131_4 87_{100} 73_8	267	2497
Impurity AC	U+UHYAC	2095*	240_1 179_1 131_1 87_{100} 73_3	341	2496
Impurity AC	U+UHYAC	1800*	194_1 179_2 117_1 87_{100} 58_4	203	2495
Impurity AC	U+UHYAC	3020*	219_1 175_2 131_3 87_{100} 73_6	267	2501
Impurity AC	U+UHYAC	2950*	219_1 175_2 131_5 87_{100} 73_{10}	267	2500
Impurity TMS		2555	484_{51} 469_8 427_{39} 233_{26} 73_{100}	1151	4613
Indanavir	U+UHYAC	3435	464_1 364_{100} 272_{29} 174_{33} 92_{73}	1196	7316
Indanavir TFA		3170	459_{55} 367_{93} 284_{14} 193_{19} 91_{100}	1203	7320
Indanavir artifact		1300	161_{23} 144_{72} 132_{100} 115_{69} 104_{74}	139	7317
Indanavir artifact -H2O AC	U+UHYAC	1780	173_{40} 148_{10} 131_{100} 118_{11} 103_{26}	161	7321
Indanavir artifact -H2O HFB		1450	327_{100} 169_{17} 130_{36} 115_{38} 103_{17}	742	7324
Indanavir artifact -H2O PFP		1450	277_{100} 130_{52} 119_{40} 103_{54} 77_{36}	499	7323
Indanavir artifact -H2O TFA		1485	227_{100} 146_{11} 130_{44} 103_{54} 69_{51}	297	7322
Indanazoline AC		2415	243_{49} 200_{100} 130_{16} 115_{45} 86_{33}	354	2800
Indapamide -2H 3ME		2940	405_{88} 298_{100} 246_5 130_{41} 77_{10}	1030	3115
Indapamide 3ME		3035	407_{27} 246_1 161_{100} 132_{35} 91_7	1033	3114
Indapamide artifact (ME)		2215	234_{44} 199_{11} 127_{100} 99_{46} 90_{39}	318	3116
Indapamide-M/artifact (H2N-)		1100	147_{30} 132_{100} 117_{44} 91_{12} 77_8	125	3117
Indapamide-M/artifact (HOOC-) 3ME		2130	277_{71} 169_{100} 138_{61} 110_{67} 75_{93}	499	3118
Indeloxazine		2085	231_{44} 132_{33} 115_{27} 100_{50} 56_{70}	309	6109
Indeloxazine AC		2400	273_{48} 142_{44} 132_7 100_{100} 86_{52}	481	6111
Indeloxazine ME		2030	245_{38} 131_9 114_{100} 84_{28} 70_{46}	359	6110
Indeloxazine TFA		2080	327_{82} 196_{31} 140_{48} 132_{200} 103_{58}	743	7755
Indeloxazine TMS		2080	303_{23} 172_{58} 157_{46} 73_{100} 56_{96}	630	7754
Indene		1050*	116_{100} 115_{82} 89_{12} 63_{10}	105	2553
Indeno[1,2,3-c,d]pyrene		3075*	276_{100} 248_2 138_9 125_7	496	3706
Indole		1350	117_{99} 90_{37} 63_{18}	105	1466
Indole acetic acid ME	UME	1900	189_{25} 130_{100} 103_5 77_8	192	1011
Indole propionic acid ME	P	1910	203_{18} 143_{16} 130_{100} 115_{17} 91_{12}	224	6375
Indometacin	G P-I	2550	313_{30} 139_{101} 111_{24}	871	1038
Indometacin ET		2820	385_{40} 312_{19} 158_6 139_{100} 111_{20}	969	3168
Indometacin ME	P(ME) G(ME) U(ME)	2770	371_9 312_6 139_{100} 111_{17}	922	1039
Indometacin TMS		2650	429_{23} 370_4 312_{17} 139_{100} 73_{22}	1078	5462
Indometacin artifact ME	PME UME	2130	233_{38} 174_{100}	315	1230
Indometacin artifact 2ME	PME UME	2090	247_{38} 188_{100} 173_{11} 145_8	367	6294
Indometacin-M (chlorobenzoic acid)	G UHY U+UHYAC	1400*	156_{61} 139_{100} 111_{54} 85_4 75_{39}	136	2726
Indometacin-M (HO-) 2ME	UME	2880	401_{27} 262_4 139_{100} 111_{23}	1020	6293
Inositol 6AC		2060*	373_2 270_6 210_{68} 168_{100} 126_{58}	1084	5677
Instillagel (TM) ingredient	G U UHY U+UHYAC	1830	232_{14} 218_{37} 132_{19} 85_{33} 71_{101}	313	1040
4-Iodo-2,5-dimethoxy-amfetamine		2025	321_2 278_{100} 263_{13} 247_9 77_{32}	717	7172
4-Iodo-2,5-dimethoxy-amfetamine AC	U+UHYAC	2295	363_{12} 304_{150} 277_9 247_9 86_{64}	893	7174
4-Iodo-2,5-dimethoxy-amfetamine HFB		2070	517_{10} 304_{58} 277_{100} 247_{36} 69_{71}	1173	7179
4-Iodo-2,5-dimethoxy-amfetamine PFP		2055	467_{46} 304_{79} 277_{150} 247_{35} 190_{30}	1135	7178
4-Iodo-2,5-dimethoxy-amfetamine TFA		2075	417_{49} 304_{63} 277_{200} 247_{25} 140_{20}	1056	7176
4-Iodo-2,5-dimethoxy-amfetamine 2AC		2360	405_{12} 304_{100} 277_{15} 247_5 86_{39}	1030	7175
4-Iodo-2,5-dimethoxy-amfetamine 2ME		2305	349_{23} 304_{100} 277_{18} 162_{11} 72_{65}	841	7569
4-Iodo-2,5-dimethoxy-amfetamine 2TFA		1940	513_7 304_{25} 277_{100} 247_{18} 69_{53}	1171	7177
4-Iodo-2,5-dimethoxy-amfetamine formyl artifact		1960	333_1 302_{10} 277_5 247_2 56_{100}	772	7173

Table 1-8-1: Compounds in order of names — Karbutilate -C3H5NO

Name	Detected	RI	Typical ions and intensities					Page	Entry
4-Iodo-2,5-dimethoxy-amfetamine-M (bis-O-demethyl-) 3AC		2480	419 $_4$	377 $_{32}$	360 $_{36}$	276 $_{78}$	86 $_{100}$	1059	7837
4-Iodo-2,5-dimethoxy-amfetamine-M (bis-O-demethyl-) artifact 2AC		2425	359 $_{20}$	317 $_{100}$	275 $_{68}$	260 $_{43}$	133 $_{28}$	879	7182
4-Iodo-2,5-dimethoxy-amfetamine-M (O-demethyl-) iso-1 2AC	U+UHYAC	2395	391 $_2$	349 $_9$	332 $_{35}$	290 $_{100}$	86 $_{55}$	987	7180
4-Iodo-2,5-dimethoxy-amfetamine-M (O-demethyl-) iso-2 2AC	U+UHYAC	2410	391 $_3$	349 $_{40}$	332 $_{11}$	290 $_{100}$	86 $_{30}$	987	7181
Iodofenphos		2150*	412 $_2$	377 $_{100}$	362 $_9$	125 $_{37}$	79 $_{24}$	1046	3448
Ionol		1515*	220 $_{16}$	205 $_{100}$	177 $_{15}$	145 $_{16}$	57 $_{22}$	275	1041
Ionol-4		1710*	250 $_{26}$	235 $_{100}$	219 $_{20}$	193 $_{13}$	91 $_{11}$	382	6367
Ionol-acetamide		2070	277 $_{49}$	262 $_{40}$	220 $_{18}$	203 $_{100}$	178 $_{45}$	503	5751
Ioxynil ME		1885	385 $_{100}$	370 $_{43}$	243 $_{23}$	127 $_7$	88 $_{26}$	969	4145
IPCC		<1000	166 $_1$	151 $_{62}$	123 $_{38}$	81 $_{28}$	54 $_{31}$	152	3584
IPCC		<1000	166 $_1$	151 $_{62}$	123 $_{38}$	54 $_{31}$	44 $_{100}$	152	5536
IPCC -HCN		<1000	139 $_{19}$	124 $_{59}$	96 $_{32}$	82 $_{26}$	54 $_{100}$	119	3586
IPCC -HCN		<1000	139 $_{19}$	124 $_{59}$	96 $_{32}$	54 $_{100}$	41 $_{83}$	119	5538
Irbesartan ME	UME	3500	442 $_2$	413 $_{12}$	400 $_{100}$	192 $_{30}$	165 $_{23}$	1104	5039
Irganox	G P U UHY U+UHYAC	3390*	530 $_{100}$	515 $_{51}$	219 $_{57}$	203 $_{22}$	57 $_{67}$	1178	4648
Isoaminile	U UHY U+UHYAC	1705	244 $_1$	229 $_3$	158 $_1$	115 $_1$	72 $_{100}$	357	4389
Isoaminile-M (nor-)	U UHY U+UHYAC	1725	229 $_2$	215 $_{11}$	188 $_{100}$	173 $_{27}$	91 $_7$	308	4390
Isobutylbenzene		1050*	134 $_{21}$	119 $_8$	105 $_{100}$	91 $_{13}$	77 $_{17}$	114	3789
Isocarbamide 2ME		1685	213 $_1$	170 $_{32}$	127 $_{100}$	86 $_{69}$	70 $_{54}$	253	4157
Isocitric acid 3ME		1495*	175 $_{33}$	143 $_{76}$	115 $_{100}$	83 $_{24}$	129 $_{23}$	319	6453
Isoconazole	U+UHYAC	3150	414 $_4$	333 $_{19}$	159 $_{100}$	123 $_6$	81 $_{14}$	1049	2055
Isofenphos		2005	345 $_2$	255 $_{18}$	213 $_{45}$	121 $_{34}$	58 $_{100}$	828	3446
Isofenphos-M/artifact (HOOC-) ME		1980	317 $_2$	259 $_5$	227 $_{57}$	121 $_{30}$	58 $_{100}$	696	3447
Iso-LSD TMS		3515	395 $_{72}$	293 $_{35}$	279 $_{19}$	253 $_{31}$	73 $_{80}$	1075	6222
Iso-Lysergide (iso-LSD) TMS		3515	395 $_{72}$	293 $_{35}$	279 $_{19}$	253 $_{31}$	73 $_{80}$	1075	6222
Isoniazid	P-I G U	1650	137 $_{38}$	122 $_5$	106 $_{77}$	78 $_{100}$	51 $_{73}$	117	1043
Isoniazid AC	U+UHYAC	1950	179 $_{96}$	137 $_{67}$	106 $_{100}$	78 $_{66}$	51 $_{57}$	170	1044
Isoniazid 2AC	U+UHYAC	1825	221 $_7$	179 $_{26}$	161 $_{44}$	137 $_{55}$	106 $_{100}$	276	1045
Isoniazid acetone derivate	U+UHYAC	1840	177 $_{11}$	162 $_{73}$	106 $_{79}$	78 $_{100}$		165	1046
Isoniazid artifact (HOOC-) TMS		1295	195 $_4$	180 $_{100}$	136 $_{36}$	106 $_{30}$	78 $_{37}$	207	4555
Isoniazid formyl artifact	P G U+UHYAC	1510	149 $_5$	122 $_{55}$	106 $_{76}$	78 $_{100}$	51 $_{69}$	126	4057
Isoniazid formyl artifact AC		1785	191 $_{17}$	149 $_{100}$	106 $_{51}$	78 $_{39}$	51 $_{30}$	195	4058
Isoniazid-M glycine conjugate	U	----	180 $_{100}$	165 $_{34}$	137 $_{39}$	106 $_{41}$	78 $_{46}$	174	1047
Isonicotinic acid TMS		1295	195 $_4$	180 $_{100}$	136 $_{36}$	106 $_{30}$	78 $_{37}$	207	4555
Isooctane		<1000*	99 $_6$	57 $_{100}$				104	2753
Isoprenaline 4AC		2460	365 $_6$	319 $_{24}$	277 $_{46}$	193 $_{50}$	84 $_{100}$	948	1468
Isopropanol		<1000*	60 $_3$	45 $_{100}$				92	1546
N-Isopropyl-BDB		1720	206 $_2$	135 $_{10}$	100 $_{100}$	77 $_7$	58 $_{77}$	326	5419
N-Isopropyl-BDB AC		2095	206 $_1$	176 $_{36}$	142 $_{27}$	100 $_{100}$	58 $_{35}$	502	5509
N-Isopropyl-BDB TFA		1895	331 $_3$	196 $_{30}$	176 $_{38}$	154 $_{100}$	135 $_{22}$	764	5510
Isopropylbenzene		<1000*	120 $_{30}$	105 $_{100}$	77 $_{16}$	65 $_8$	51 $_{11}$	107	3785
Isoproturon ME		1685	220 $_8$	205 $_5$	148 $_8$	132 $_5$	72 $_{100}$	274	3968
Isopyrin	G	2045	245 $_{28}$	230 $_{12}$	137 $_{31}$	83 $_{25}$	56 $_{100}$	360	530
Isopyrin AC		2400	287 $_{35}$	244 $_{37}$	137 $_{31}$	56 $_{99}$		548	194
Isopyrin-M (nor-) 2AC	U+UHYAC	2365	315 $_5$	273 $_{45}$	231 $_{48}$	123 $_{100}$	70 $_{94}$	689	195
Isopyrin-M (nor-HO-) -H2O 2AC	U+UHYAC	2160	313 $_3$	271 $_9$	229 $_{28}$	77 $_{16}$	56 $_{100}$	680	531
Isosteviol		2620*	318 $_{25}$	300 $_{21}$	121 $_{66}$	109 $_{71}$	55 $_{100}$	705	3680
Isosteviol ME		2520*	332 $_{73}$	300 $_{71}$	273 $_{28}$	121 $_{81}$	55 $_{71}$	771	3681
Isothipendyl	P-I G U+UHYAC	2245	285 $_1$	214 $_1$	200 $_2$	181 $_1$	72 $_{100}$	538	1467
Isothipendyl-M (bis-nor-)	UHY	2230	257 $_{75}$	214 $_{100}$	181 $_{24}$	58 $_{58}$		410	1666
Isothipendyl-M (bis-nor-) AC	U+UHYAC	2520	299 $_{41}$	213 $_{100}$	200 $_{25}$	181 $_{43}$	100 $_{10}$	610	1662
Isothipendyl-M (HO-)	UHY	2450	301 $_{22}$	218 $_{11}$	197 $_{10}$	72 $_{100}$		619	1665
Isothipendyl-M (HO-) AC	U+UHYAC	2640	343 $_2$	272 $_3$	229 $_3$	197 $_4$	72 $_{100}$	821	1663
Isothipendyl-M (HO-ring)	U UHY	2800	216 $_{100}$	187 $_{68}$	168 $_{50}$	140 $_9$		259	2272
Isothipendyl-M (HO-ring) AC	U+UHYAC	2575	258 $_{30}$	216 $_{100}$	187 $_{13}$	183 $_{13}$	155 $_{10}$	412	2275
Isothipendyl-M (nor-)	UHY	2220	271 $_3$	214 $_{43}$	199 $_3$	181 $_6$	58 $_{100}$	474	1664
Isothipendyl-M (nor-) AC	U+UHYAC	2600	313 $_{16}$	213 $_{100}$	181 $_{48}$	114 $_{58}$	58 $_{89}$	679	1661
Isothipendyl-M (nor-HO-) 2AC	U+UHYAC	2940	371 $_{31}$	271 $_{45}$	229 $_{78}$	114 $_{100}$	58 $_{58}$	923	2441
Isothipendyl-M (nor-sulfone) AC	U+UHYAC	2900	345 $_1$	272 $_{26}$	257 $_{15}$	100 $_{55}$	58 $_{100}$	828	2687
Isothipendyl-M (nor-sulfoxide) AC	U+UHYAC	2880	329 $_3$	312 $_{26}$	213 $_{99}$	100 $_{28}$	58 $_{100}$	753	2686
Isothipendyl-M (ring)	U+UHYAC	2045	200 $_{100}$	168 $_{36}$	156 $_{13}$			219	386
Isovanillic acid MEAC		1630*	224 $_6$	193 $_2$	182 $_{92}$	151 $_{100}$	123 $_{13}$	287	5329
Isoxaben		2910	332 $_1$	250 $_1$	165 $_{100}$	150 $_8$	107 $_5$	769	3885
Isradipine	UME	2680	371 $_9$	284 $_{14}$	252 $_{42}$	210 $_{100}$	150 $_{17}$	923	4628
Isradipine ME	UME	2670	385 $_{38}$	326 $_{27}$	298 $_{100}$	268 $_{51}$	224 $_{70}$	970	4852
Isradipine-M (dehydro-demethyl-HO-) -H2O	UME	2635	353 $_8$	311 $_{100}$	294 $_{49}$	267 $_{49}$	237 $_{61}$	856	4869
Isradipine-M/artifact (dehydro-)	UME	2360	369 $_{21}$	327 $_{85}$	295 $_{100}$	265 $_{46}$	221 $_{45}$	915	4865
Isradipine-M/artifact (dehydro-deisopropyl-) ME	UME	2270	341 $_{51}$	309 $_{100}$	294 $_{36}$	279 $_{34}$	264 $_{63}$	810	4866
Isradipine-M/artifact (dehydro-deisopropyl-) TMS	UTMS	2395	399 $_{46}$	384 $_{100}$	309 $_{77}$	264 $_{18}$	164 $_{13}$	1014	5009
Isradipine-M/artifact (dehydro-demethyl-) TMS	UTMS	2535	427 $_{20}$	385 $_{28}$	370 $_{100}$	295 $_{62}$	251 $_{23}$	1073	5010
Isradipine-M/artifact (deisopropyl-) ME	UME	2610	343 $_{10}$	284 $_{18}$	254 $_{11}$	224 $_{100}$	192 $_9$	821	4866
Isradipine-M/artifact (deisopropyl-) 2ME	UME	2655	357 $_{40}$	326 $_{14}$	298 $_{100}$	268 $_{51}$	238 $_{62}$	873	4867
Kadethrin		3190*	396 $_1$	170 $_{100}$	143 $_{56}$	128 $_{66}$	91 $_{41}$	1006	2801
Karbutilate -C3H5NO		1640	208 $_{31}$	164 $_5$	136 $_{13}$	92 $_9$	72 $_{100}$	238	4151

Karbutilate -C5H9NO **Table 1-8-1:** Compounds in order of names

Name	Detected	RI	Typical ions and intensities					Page	Entry
Karbutilate -C5H9NO		1890	167_{63}	135_{100}	122_{31}	81_{28}	52_{36}	175	3907
Kavain	G P	2235*	230_{32}	202_{40}	104_{31}	98_{98}	68_{100}	307	1048
Kavain -CO2	P	1705*	186_{100}	155_{23}	128_{51}	95_{74}	77_{18}	188	1049
Kavain-M (O-demethyl-) -CO2	U UHY	1680*	172_{84}	157_{58}	128_{100}	95_{39}	77_{60}	159	2936
Kebuzone		2525	332_{1}	264_{6}	183_{20}	93_{53}	77_{100}	723	4265
Kebuzone artifact		2150	266_{37}	183_{70}	118_{8}	105_{28}	77_{100}	449	4266
Kebuzone enol ME	UME	2510	336_{58}	266_{16}	183_{51}	105_{52}	77_{100}	789	6378
Kebuzone-M (HO-) enol 2ME	UME	2690	366_{77}	296_{2}	213_{8}	107_{74}	77_{100}	905	6379
Kelevan		2895*	488_{1}	455_{15}	357_{23}	272_{100}	77_{49}	1197	4045
Kelevan artifact		2320*	486_{1}	455_{22}	355_{20}	272_{100}	237_{42}	1153	3324
Ketamine	P U UHY	1835	237_{1}	209_{22}	180_{1}	152_{14}	102_{15}	333	1050
Ketamine AC	U+UHYAC	2170	279_{2}	216_{100}	208_{91}	180_{93}	152_{80}	510	1056
Ketamine TMS		1800	309_{1}	294_{11}	157_{100}	152_{63}	73_{81}	660	4556
Ketamine isomer	G P	1735	237_{1}	180_{8}	152_{100}	138_{19}	102_{11}	333	5561
Ketamine-D4		1825	241_{1}	213_{26}	184_{100}	156_{14}	142_{15}	348	7779
Ketamine-D4 AC		2165	283_{1}	255_{11}	220_{100}	212_{95}	184_{63}	529	7780
Ketamine-D4 HFB		1895	437_{1}	402_{4}	374_{29}	366_{34}	210_{100}	1093	7784
Ketamine-D4 ME		1840	255_{1}	236_{4}	227_{20}	198_{100}	156_{12}	403	7781
Ketamine-D4 TFA		1900	337_{2}	309_{12}	274_{81}	240_{51}	110_{100}	791	7783
Ketamine-D4 TMS		1795	313_{3}	298_{11}	229_{6}	157_{100}	73_{48}	681	7782
Ketamine-M (nor-)	P U	1810	223_{3}	195_{34}	166_{100}	138_{10}	131_{14}	283	1055
Ketamine-M (nor-) AC	U+UHYAC	2035	265_{1}	230_{89}	202_{100}	166_{74}	138_{29}	442	7826
Ketamine-M (nor-di-HO-) -2H2O	U	1920	219_{90}	190_{100}	184_{22}	156_{43}	129_{25}	268	1054
Ketamine-M (nor-di-HO-) -2H2O AC	U+UHYAC	1970	261_{5}	219_{85}	190_{100}	184_{32}	157_{46}	424	3672
Ketamine-M (nor-HO-) -H2O	U UHY	1960	221_{28}	193_{15}	166_{100}	131_{44}	102_{27}	275	1051
Ketamine-M (nor-HO-) -H2O AC	U+UHYAC	2080	263_{3}	228_{41}	160_{100}	153_{23}	102_{11}	433	3673
Ketamine-M (nor-HO-) -NH3	P U UHY	1740*	222_{24}	187_{100}	159_{11}	115_{18}	77_{21}	280	1053
Ketamine-M (nor-HO-) -NH3 -H2O	U UHY	1620*	204_{49}	169_{100}	139_{32}	115_{18}	70_{37}	225	1052
Ketamine-M (nor-HO-) -NH3 -H2O AC	U+UHYAC	1670*	246_{17}	204_{100}	169_{63}	139_{19}	107_{10}	362	1231
Ketamine-M/artifact	U UHY U+UHYAC	1630	189_{50}	154_{100}	127_{17}	75_{8}		191	3683
Ketanserin-M/artifact		2470	201_{31}	123_{100}	95_{41}	75_{16}	51_{21}	221	4232
Ketazolam artifact	P G U	2430	284_{81}	283_{91}	256_{100}	221_{31}	77_{7}	532	481
Ketazolam HY	UHY U+UHYAC	2100	245_{95}	228_{38}	193_{29}	105_{38}	77_{100}	358	272
Ketazolam HYAC	U+UHYAC	2260	287_{11}	244_{100}	228_{39}	182_{49}	77_{70}	546	2542
Ketazolam-M	P G U	2520	270_{86}	269_{97}	242_{100}	241_{82}	77_{17}	468	463
Ketazolam-M	P G U	2430	284_{81}	283_{91}	256_{100}	221_{31}	77_{7}	532	481
Ketazolam-M	P G UGLUC	2320	268_{98}	239_{57}	233_{52}	205_{66}	77_{100}	542	579
Ketazolam-M TMS		2635	356_{9}	341_{100}	312_{56}	239_{12}	135_{21}	876	4577
Ketazolam-M TMS		2300	342_{62}	341_{100}	327_{19}	269_{4}	73_{30}	817	4573
Ketazolam-M 2TMS		2200	430_{36}	429_{63}	340_{10}	313_{14}	73_{70}	1080	5499
Ketazolam-M artifact-1	P-I UHY U+UHYAC	2060	240_{59}	239_{100}	205_{81}	177_{16}	151_{9}	342	300
Ketazolam-M artifact-2	UHY U+UHYAC	2070	254_{77}	253_{100}	219_{98}			397	301
Ketazolam-M HY	UHY	2050	231_{80}	230_{95}	154_{28}	105_{38}	77_{100}	308	419
Ketazolam-M HYAC	U+UHYAC PHYAC	2245	273_{31}	230_{100}	154_{13}	105_{23}	77_{50}	480	273
Ketoprofen		2245*	254_{40}	209_{50}	177_{62}	105_{100}	77_{56}	398	1425
Ketoprofen ME	PME UME	2090*	268_{50}	209_{100}	191_{22}	105_{76}	77_{63}	460	1471
Ketoprofen-M (HO-) ME	UME	2345*	284_{59}	225_{71}	191_{6}	121_{100}	65_{26}	533	5215
Ketoprofen-M (HO-) iso-1 2ME	UME	2250*	298_{80}	239_{100}	191_{34}	135_{54}	77_{30}	606	5213
Ketoprofen-M (HO-) iso-2 2ME	UME	2295*	298_{70}	239_{50}	191_{7}	135_{100}	77_{18}	606	5214
Ketorolac ME		2265	269_{51}	210_{84}	132_{9}	105_{100}	77_{57}	464	4625
Ketotifen	G U+UHYAC	2600	309_{100}	237_{10}	208_{12}	96_{29}	70_{6}	660	1472
Ketotifen-M (dihydro-) -H2O	UHY U+UHYAC	2480	293_{100}	249_{27}	221_{44}	96_{19}	202_{9}	579	4482
Ketotifen-M (nor-)	U-I UHY-I	2700	295_{100}	253_{84}	208_{43}	165_{24}	152_{24}	589	2202
Ketotifen-M (nor-) AC	U+UHYAC-I	3180	337_{100}	277_{24}	265_{47}	221_{36}	208_{39}	791	2203
LAAM	G P U+UHYAC	2230	353_{2}	338_{1}	225_{4}	91_{6}	72_{100}	859	5616
Labetalol 2TMS		2530	472_{1}	439_{4}	162_{62}	73_{63}	58_{100}	1140	5489
Labetalol 3AC	U+UHYAC	3400	376_{33}	335_{100}	293_{24}	133_{32}	91_{50}	1119	1357
Labetalol 3TMS		2620	511_{2}	365_{4}	234_{100}	130_{48}	91_{35}	1183	5490
Labetalol artifact		1320	149_{14}	132_{59}	117_{42}	91_{99}	57_{34}	126	1356
Labetalol artifact AC	U+UHYAC	1780	191_{21}	132_{15}	117_{43}	87_{100}	72_{33}	196	1701
Labetalol-M (HO-) iso-1 artifact 2AC	U+UHYAC	1940	249_{3}	206_{40}	147_{58}	104_{37}	86_{38}	375	1702
Labetalol-M (HO-) iso-2 artifact 2AC	U+UHYAC	2000	249_{9}	207_{14}	148_{34}	133_{31}	87_{80}	376	1703
Lacidipine		2955	455_{1}	382_{5}	326_{10}	252_{28}	57_{100}	1121	5749
Lactose 8AC	U+UHYAC	3100*	331_{79}	169_{100}	127_{23}	109_{46}	81_{28}	1202	1960
Lactose 8HFB		2070*	537_{4}	519_{19}	293_{16}	169_{100}	81_{73}	1208	5787
Lactose 8PFP		1950*	601_{1}	437_{6}	419_{55}	273_{12}	119_{100}	1208	5786
Lactose 8TFA		1980*	337_{11}	319_{87}	223_{30}	193_{43}	69_{100}	1207	5785
Lactose 8TMS		2730*	361_{8}	217_{31}	204_{83}	191_{86}	73_{100}	1206	4334
Lactylphenetidine	UGLUC	1885	209_{59}	137_{74}	109_{85}	108_{100}		241	532
Lactylphenetidine AC	UGLUCAC	1960	251_{51}	137_{100}	109_{85}	108_{45}		384	196
Lactylphenetidine HYAC	G U+UHYAC	1680	179_{66}	137_{51}	109_{97}	108_{100}	80_{18}	171	186
Lactylphenetidine-M	UHY	1240	109_{99}	80_{41}	52_{90}			102	826
Lactylphenetidine-M (deethyl-) HYME	UHYME	1100	123_{27}	109_{100}	94_{7}	80_{96}	53_{47}	108	3766
Lactylphenetidine-M (HO-) HY2AC	U+UHYAC	1755	237_{15}	195_{31}	153_{100}	124_{55}		333	187

Table 1-8-1: Compounds in order of names　　　　　　　　　　　　　　　　　　　　　　　Levodopa-M (homovanillic acid) 2ME

Name	Detected	RI	Typical ions and intensities					Page	Entry
Lactylphenetidine-M (O-deethyl-) 2AC	UGLUCAC	1975	265 $_5$	223 $_{20}$	151 $_9$	109 $_{100}$		443	533
Lactylphenetidine-M (p-phenetidine)	UHY	1280	137 $_{68}$	108 $_{99}$	80 $_{38}$	65 $_9$		117	844
Lactylphenetidine-M HY2AC	U+UHYAC PAC	1765	193 $_{10}$	151 $_{53}$	109 $_{100}$	80 $_{24}$		199	188
Lamotrigine	P	2635	255 $_{45}$	185 $_{100}$	157 $_{16}$	123 $_{30}$	114 $_{26}$	400	4636
Lamotrigine AC	PAC U+UHYAC	2665	297 $_7$	268 $_{14}$	185 $_{100}$	157 $_{21}$	114 $_{12}$	601	4637
Lamotrigine 2AC	PAC U+UHYAC	2855	339 $_6$	297 $_{14}$	255 $_{11}$	185 $_{100}$	114 $_{16}$	800	4638
LAMPA TMS		3740	395 $_{57}$	337 $_{10}$	293 $_{26}$	253 $_{36}$	73 $_{65}$	1003	6263
Laudanosine	P U+UHYAC	2575	357 $_1$	206 $_{100}$	190 $_{23}$	162 $_8$	151 $_7$	874	6106
Laudanosine-M (O-bisdemethyl-) 2AC	U+UHYAC	3370	427 $_{50}$	385 $_9$	354 $_{55}$	312 $_{100}$	137 $_{76}$	1074	6789
Laudanosine-M (O-bisdemethyl-) 2AC	U+UHYAC	3020	413 $_1$	178 $_{29}$	151 $_7$	262 $_{34}$	220 $_{43}$	1048	6788
Laudanosine-M (O-demethyl-) AC	U+UHYAC	3210	399 $_{59}$	326 $_{47}$	313 $_{19}$	295 $_{25}$	151 $_{70}$	1015	6790
Laudanosine-M (O-demethyl-) AC	U+UHYAC	2595	385 $_1$	234 $_{88}$	192 $_{100}$	177 $_{16}$	151 $_{17}$	970	6787
Lauric acid		1670*	200 $_7$	157 $_{16}$	129 $_{25}$	73 $_{100}$	60 $_{87}$	220	5630
Lauric acid ET		1570*	228 $_5$	185 $_{13}$	157 $_{16}$	101 $_{51}$	88 $_{100}$	302	5400
Lauric acid ME		1550*	214 $_3$	183 $_6$	143 $_{12}$	87 $_{52}$	74 $_{100}$	256	2666
Lauric acid TMS		1670*	272 $_6$	257 $_{95}$	117 $_{75}$	75 $_{90}$	73 $_{110}$	480	5716
Lauroscholtzine		2665	341 $_{89}$	340 $_{100}$	326 $_{42}$	310 $_{22}$	267 $_{24}$	812	5773
Lauroscholtzine AC		2750	383 $_{100}$	382 $_{92}$	368 $_{56}$	340 $_{64}$	326 $_{34}$	964	5774
Lauroscholtzine ME		2680	355 $_{94}$	354 $_{100}$	340 $_{57}$	324 $_{31}$	281 $_{29}$	866	5775
Lauroscholtzine artifact (dehydro-)		3180	339 $_{100}$	324 $_{77}$	296 $_{22}$	281 $_{17}$	238 $_{14}$	803	6742
Lauroscholtzine artifact (dehydro-) AC		3285	381 $_{100}$	339 $_{16}$	324 $_{90}$	296 $_{17}$	280 $_{12}$	956	6743
Lauroscholtzine artifact (dehydro-) ME		3235	353 $_{100}$	338 $_{71}$	307 $_{12}$	280 $_{26}$	176 $_{13}$	858	6744
Lauroscholtzine-M (bis-O-demethyl-) 3AC		3170	439 $_{78}$	424 $_{39}$	396 $_{100}$	354 $_{46}$	312 $_{85}$	1096	6754
Lauroscholtzine-M (O-demethyl-) iso-1 2AC		3095	411 $_{91}$	396 $_{100}$	368 $_{62}$	326 $_{47}$	295 $_{32}$	1045	6752
Lauroscholtzine-M (O-demethyl-) iso-2 2AC		3055	411 $_{95}$	368 $_{100}$	354 $_{37}$	326 $_{73}$	224 $_{36}$	1045	6753
Lauroscholtzine-M (seco-O-demethyl-) 3AC		3650	453 $_{38}$	380 $_{63}$	338 $_{100}$	296 $_{85}$	283 $_{69}$	1118	6755
Lauroscholtzine-M/artifact (nor-seco-) AC		3230	369 $_{60}$	310 $_{45}$	297 $_{100}$	283 $_{14}$	263 $_9$	916	6750
Lauroscholtzine-M/artifact (nor-seco-) 2AC		3315	411 $_{100}$	352 $_{40}$	339 $_{88}$	310 $_{74}$	297 $_{77}$	1045	6751
Lauroscholtzine-M/artifact (seco-) AC		3405	383 $_{52}$	310 $_{20}$	297 $_{65}$	263 $_{20}$	251 $_{11}$	964	6745
Lauroscholtzine-M/artifact (seco-) ME		3035	355 $_5$	297 $_3$	165 $_1$	152 $_2$	58 $_{100}$	866	6747
Lauroscholtzine-M/artifact (seco-) MEAC		3120	397 $_4$	339 $_3$	297 $_6$	251 $_5$	58 $_{100}$	1011	6748
Lauroscholtzine-M/artifact (seco-) 2AC		3470	425 $_{75}$	352 $_{100}$	339 $_{91}$	310 $_{95}$	297 $_{45}$	1070	6746
Lauroscholtzine-M/artifact (seco-) 2ME		3030	369 $_{17}$	311 $_6$	265 $_7$	165 $_3$	58 $_{100}$	917	6749
Laurylmethylthiodipropionate		2550*	360 $_{13}$	192 $_{58}$	175 $_{27}$	146 $_{69}$	55 $_{100}$	885	4400
Leflunomide HYAC	U+UHYAC	1420	203 $_{22}$	184 $_4$	161 $_{100}$	142 $_{20}$	111 $_{12}$	223	7372
Lenacil		2275	153 $_{100}$	136 $_8$	110 $_{13}$	67 $_9$	53 $_{24}$	321	3855
Lenacil ME		2260	248 $_2$	167 $_{100}$	124 $_{14}$	95 $_{13}$	67 $_{22}$	372	3969
Lenacil 2ME		2280	262 $_3$	181 $_{100}$	165 $_{13}$	138 $_{12}$	67 $_{21}$	432	3970
Lercanidipine-M (N-dealkyl-) AC	U+UHYAC	2320	253 $_{15}$	193 $_{12}$	165 $_{19}$	152 $_{10}$	73 $_{100}$	395	3391
Lercanidipine-M/artifact (alcohol)		2010	238 $_{16}$	165 $_{100}$	152 $_7$	91 $_{14}$	58 $_{100}$	603	7596
Lercanidipine-M/artifact (alcohol) AC		2080	339 $_1$	282 $_1$	238 $_{34}$	165 $_{15}$	58 $_{100}$	805	7595
Lercanidipine-M/artifact (alcohol) -H2O		1845	279 $_3$	165 $_8$	115 $_5$	98 $_{15}$	58 $_{15}$	515	7594
Lercanidipine-M/artifact -CO2	U UHY U+UHYAC UME	2175	286 $_{30}$	269 $_{100}$	239 $_{48}$	180 $_8$	139 $_{13}$	542	3656
Letrozole		2630	285 $_{49}$	217 $_{100}$	190 $_{61}$	156 $_{15}$	102 $_{22}$	538	7510
Levacetylmethadol	G P U+UHYAC	2230	353 $_1$	338 $_1$	225 $_4$	91 $_6$	72 $_{100}$	859	5616
Levallorphan	UHY	2355	283 $_{101}$	256 $_{55}$	176 $_{64}$	157 $_{70}$	85 $_{81}$	530	534
Levallorphan AC	U+UHYAC	2390	325 $_{72}$	299 $_{37}$	257 $_{20}$	176 $_{30}$	85 $_{100}$	737	1473
Levallorphan HFB		2205	479 $_{100}$	452 $_{79}$	411 $_{46}$	353 $_{14}$	176 $_{14}$	1147	6227
Levallorphan PFP		2120	429 $_{95}$	402 $_{82}$	361 $_{57}$	176 $_{38}$	119 $_{100}$	1078	6226
Levallorphan TFA		2110	379 $_{100}$	352 $_{79}$	311 $_{52}$	176 $_{35}$	69 $_{100}$	949	6225
Levallorphan TMS		2375	355 $_{16}$	272 $_{16}$	176 $_{32}$	85 $_{68}$	73 $_{100}$	867	6213
Levetiracetam	P G U+UHYAC	1740	170 $_3$	126 $_{100}$	112 $_2$	98 $_{12}$	69 $_{33}$	157	6876
Levetiracetam AC	U+UHYAC	1780	212 $_1$	153 $_2$	126 $_{100}$	98 $_9$	69 $_{21}$	250	6877
Levetiracetam HFB		1590	366 $_1$	169 $_7$	126 $_4$	126 $_{100}$	69 $_{33}$	903	7362
Levetiracetam PFP		1540	316 $_1$	153 $_3$	126 $_{100}$	98 $_6$	69 $_{24}$	693	7361
Levetiracetam TFA		1500	266 $_1$	153 $_4$	126 $_{100}$	98 $_{11}$	69 $_{32}$	448	7359
Levetiracetam TMS		1655	242 $_1$	227 $_8$	126 $_{100}$	112 $_{81}$	98 $_{44}$	353	7365
Levetiracetam 2HFB		1670	562 $_1$	349 $_4$	322 $_{100}$	265 $_{18}$	152 $_{27}$	1188	7363
Levetiracetam 2TFA		1190	319 $_4$	250 $_3$	222 $_{100}$	165 $_{39}$	69 $_{33}$	890	7360
Levetiracetam 2TMS		1700	314 $_1$	299 $_7$	199 $_{44}$	184 $_{46}$	73 $_{100}$	686	7364
Levobunolol		2430	291 $_1$	276 $_{17}$	115 $_5$	86 $_{100}$	57 $_{11}$	571	2611
Levobunolol AC		2460	333 $_8$	318 $_{27}$	259 $_{15}$	200 $_9$	86 $_{100}$	775	1540
Levobunolol formyl artifact		2450	303 $_{10}$	288 $_{100}$	201 $_{11}$	141 $_{13}$	70 $_{42}$	630	1539
Levobunolol -H2O AC		2570	315 $_2$	259 $_{44}$	200 $_{29}$	160 $_{20}$	57 $_{100}$	690	1541
Levodopa 3ME	UME	1870	239 $_7$	194 $_6$	180 $_6$	151 $_{100}$	102 $_{36}$	341	2903
Levodopa-M (homovanillic acid)	U	1610*	182 $_{35}$	137 $_{100}$	122 $_{12}$	94 $_7$	65 $_9$	179	3368
Levodopa-M (homovanillic acid) HFB		1770*	378 $_{79}$	333 $_{39}$	181 $_{60}$	107 $_{100}$	69 $_{72}$	945	5975
Levodopa-M (homovanillic acid) ME	UME	1750*	196 $_{25}$	137 $_{100}$	122 $_{16}$	107 $_2$	94 $_{12}$	210	812
Levodopa-M (homovanillic acid) MEAC	U+UHYAC	1700*	238 $_2$	196 $_{47}$	137 $_{100}$	122 $_6$	107 $_6$	336	2973
Levodopa-M (homovanillic acid) MEHFB		1570*	392 $_{100}$	333 $_{84}$	169 $_{24}$	107 $_{78}$	69 $_{60}$	991	5974
Levodopa-M (homovanillic acid) MEPFP		1570*	342 $_{96}$	283 $_{100}$	195 $_{35}$	119 $_{37}$	107 $_{51}$	815	5972
Levodopa-M (homovanillic acid) METMS		1670*	268 $_{41}$	238 $_{69}$	209 $_{52}$	179 $_{100}$	73 $_{95}$	461	6016
Levodopa-M (homovanillic acid) PFP		1685*	328 $_{85}$	283 $_{45}$	181 $_{58}$	119 $_{33}$	107 $_{100}$	747	5973
Levodopa-M (homovanillic acid) 2ME	UME	1720*	210 $_{31}$	151 $_{100}$	123 $_4$	107 $_6$		244	5959

Levodopa-M (homovanillic acid) 2TMS   Table 1-8-1: Compounds in order of names

Name	Detected	RI	Typical ions and intensities					Page	Entry
Levodopa-M (homovanillic acid) 2TMS		1760*	326$_{32}$	267$_{20}$	209$_{46}$	179$_{34}$	73$_{100}$	740	6015
Levodopa-M (O-methyl-dopamine) AC	U+UHYAC	2330	209$_{100}$	180$_{16}$	150$_{43}$	138$_{17}$	58$_{10}$	241	2980
Levodopa-M (O-methyl-dopamine) 2AC	U+UHYAC	2070	251$_{2}$	209$_{7}$	150$_{100}$	137$_{15}$		384	1273
Levomepromazine	P-I G U UHY U+UHYAC	2540	328$_{10}$	228$_{7}$	185$_{7}$	100$_{7}$	58$_{100}$	750	344
Levomepromazine-M (di-HO-) 2AC	U+UHYAC	3185	444$_{31}$	356$_{14}$	258$_{9}$	100$_{23}$	58$_{100}$	1106	3052
Levomepromazine-M (HO-)	UHY	2735	344$_{6}$	100$_{7}$	58$_{100}$			825	537
Levomepromazine-M (HO-) AC	U+UHYAC	2745	386$_{15}$	299$_{2}$	244$_{3}$	100$_{6}$	58$_{100}$	973	345
Levomepromazine-M (nor-)	UHY	2600	314$_{24}$	229$_{69}$	213$_{100}$	72$_{89}$		686	536
Levomepromazine-M (nor-) AC	U+UHYAC	2970	356$_{6}$	242$_{5}$	228$_{4}$	128$_{100}$		869	346
Levomepromazine-M (nor-HO-)	UHY	2750	330$_{45}$	258$_{37}$	245$_{80}$	86$_{49}$	72$_{100}$	759	538
Levomepromazine-M (nor-HO-) AC	U+UHYAC	3140	372$_{46}$	258$_{28}$	244$_{44}$	128$_{100}$	86$_{28}$	927	6415
Levomepromazine-M (nor-HO-) 2AC	U+UHYAC	3220	414$_{19}$	372$_{7}$	300$_{4}$	244$_{11}$	128$_{100}$	1050	347
Levomepromazine-M (nor-O-demethyl-) 2AC	U+UHYAC	2930	384$_{74}$	270$_{17}$	228$_{29}$	214$_{22}$	128$_{100}$	967	15
Levomepromazine-M (O-demethyl-)	UHY	2650	314$_{19}$	228$_{4}$	214$_{5}$	100$_{10}$	58$_{100}$	685	11
Levomepromazine-M (O-demethyl-) AC	U+UHYAC	2600	356$_{15}$	228$_{3}$	214$_{7}$	100$_{15}$	58$_{100}$	869	13
Levomepromazine-M/artifact (sulfoxide)	G P U+UHYAC	2940	344$_{4}$	242$_{41}$	229$_{14}$	210$_{7}$	58$_{100}$	825	535
Levorphanol	UHY	2255	257$_{38}$	200$_{17}$	150$_{47}$	59$_{100}$		411	475
Levorphanol AC	U+UHYAC	2280	299$_{100}$	231$_{42}$	200$_{20}$	150$_{48}$	59$_{15}$	612	230
Levorphanol HFB		2100	453$_{100}$	396$_{18}$	385$_{91}$	169$_{19}$	150$_{27}$	1118	6151
Levorphanol PFP	UHYPFP	2060	403$_{93}$	335$_{78}$	303$_{14}$	150$_{100}$	119$_{58}$	1027	4305
Levorphanol TFA		2015	353$_{69}$	285$_{80}$	150$_{100}$	115$_{26}$	69$_{72}$	858	4006
Levorphanol TMS	UHYTMS	2230	329$_{31}$	272$_{20}$	150$_{39}$	73$_{26}$	59$_{100}$	756	4304
Levorphanol-M (HO-) 2AC	U+UHYAC	2580	357$_{68}$	247$_{22}$	215$_{17}$	150$_{100}$	59$_{30}$	874	1187
Levorphanol-M (methoxy-) AC	U+UHYAC	2520	329$_{48}$	261$_{23}$	229$_{23}$	150$_{100}$	59$_{28}$	756	4476
Levorphanol-M (nor-) 2AC	U+UHYAC	2710	327$_{11}$	240$_{8}$	199$_{12}$	87$_{100}$	72$_{62}$	745	228
Levorphanol-M (oxo-) AC	U+UHYAC	2695	313$_{16}$	240$_{11}$	199$_{98}$	157$_{12}$	73$_{100}$	681	4475
Lidocaine	P G U+UHYAC	1875	234$_{6}$	120$_{6}$	86$_{100}$	72$_{10}$	58$_{26}$	321	1061
Lidocaine AC	U+UHYAC	1860	276$_{5}$	204$_{1}$	120$_{2}$	86$_{100}$	58$_{17}$	498	2585
Lidocaine TMS		1785	306$_{1}$	291$_{7}$	235$_{27}$	220$_{51}$	86$_{100}$	646	4557
Lidocaine artifact		1855	232$_{5}$	217$_{23}$	132$_{11}$	85$_{100}$	70$_{21}$	313	6784
Lidocaine-M (deethyl-)	U UHY	1790	206$_{3}$	163$_{12}$	121$_{10}$	58$_{100}$		231	1063
Lidocaine-M (deethyl-) AC	U+UHYAC	2115	248$_{4}$	128$_{26}$	100$_{15}$	58$_{100}$		372	1066
Lidocaine-M (dimethylaniline)		1180	121$_{100}$	106$_{77}$				107	725
Lidocaine-M (dimethylaniline) AC	U+UHYAC	1470	163$_{33}$	121$_{100}$	106$_{66}$	91$_{17}$	77$_{30}$	142	57
Lidocaine-M (dimethylhydroxyaniline)	UHY	1460	137$_{100}$	122$_{30}$	107$_{26}$			117	1062
Lidocaine-M (dimethylhydroxyaniline) 2AC	U+UHYAC	1885	221$_{9}$	179$_{39}$	137$_{100}$			277	1064
Lidocaine-M (dimethylhydroxyaniline) 3AC	U+UHYAC	1900	263$_{26}$	221$_{40}$	179$_{100}$	137$_{80}$	108$_{4}$	434	1065
Lidocaine-M (HO-)	UHY	2350	250$_{3}$	194$_{3}$	86$_{100}$	58$_{19}$		382	4070
Lidocaine-M (HO-) AC	U+UHYAC	2300	292$_{2}$	204$_{3}$	127$_{4}$	86$_{100}$	58$_{8}$	576	3361
Lidoflazine		3870	491$_{18}$	343$_{100}$	288$_{6}$	260$_{4}$	109$_{6}$	1158	2725
Lidoflazine-M (deamino-carboxy-)	P-I UHY U+UHYAC	2230*	276$_{7}$	216$_{17}$	203$_{100}$	183$_{22}$		496	169
Lidoflazine-M (deamino-carboxy-) ME	P-I UHYME U+UHYAC	2125*	290$_{12}$	258$_{13}$	216$_{30}$	203$_{100}$	183$_{22}$	563	3372
Lidoflazine-M (deamino-HO-) AC	U+UHYAC	2150*	304$_{7}$	244$_{22}$	216$_{41}$	203$_{100}$	183$_{17}$	635	307
Lidoflazine-M (N-dealkyl-) AC	U+UHYAC	2970	141$_{100}$	109$_{9}$	372$_{29}$	300$_{13}$	201$_{11}$	928	3370
Lignoceric acid ME		2745*	382$_{25}$	339$_{6}$	143$_{15}$	87$_{66}$	74$_{100}$	961	3796
Lincomycin -H2O (4)AC	U	2725	416$_{1}$	126$_{100}$	82$_{3}$			1190	5128
Lincomycin -H2O (4)AC	U	2695	527$_{1}$	126$_{100}$	82$_{2}$			1190	5127
Lincomycin -H2O (4)AC	U	2660	126$_{100}$	82$_{4}$				1190	5126
Lindane	P-I	1740*	288$_{2}$	252$_{8}$	217$_{67}$	181$_{100}$	109$_{48}$	550	1067
Lindane-M (dichloro-HO-thiophenol)	U	1470*	194$_{100}$	159$_{84}$	131$_{27}$	95$_{34}$		202	3365
Lindane-M (dichlorothiophenol)	U	1250*	178$_{100}$	143$_{88}$	107$_{26}$	69$_{19}$		166	3362
Lindane-M (tetrachlorocyclohexene)	U	1470*	218$_{1}$	183$_{9}$	147$_{100}$	111$_{56}$	77$_{39}$	262	3369
Lindane-M (2,3,4,5-tetrachlorophenol)	U	1500*	232$_{100}$	230$_{79}$	194$_{15}$	166$_{20}$	131$_{26}$	305	3366
Lindane-M (2,4,5-trichlorophenol)	U	1440*	198$_{93}$	196$_{100}$	132$_{34}$	97$_{64}$	73$_{14}$	209	784
Lindane-M (2,4,6-trichlorophenol)	U	1420*	198$_{98}$	196$_{100}$	160$_{11}$	132$_{37}$	97$_{31}$	209	3363
Lindane-M (trichlorothiophenol)	U	1450*	212$_{100}$	177$_{89}$	142$_{23}$	106$_{16}$		247	3364
Linezolide		2770	337$_{100}$	293$_{30}$	234$_{43}$	209$_{60}$	164$_{48}$	792	7318
Linezolide TMS		2380	409$_{98}$	312$_{8}$	281$_{32}$	150$_{21}$	73$_{100}$	1039	7325
Linezolide artifact		2270	293$_{100}$	234$_{18}$	209$_{82}$	151$_{26}$	85$_{15}$	579	7319
Linezolide artifact (deacetyl-) HFB		2580	491$_{100}$	433$_{33}$	175$_{25}$	164$_{28}$	149$_{36}$	1158	7328
Linezolide artifact (deacetyl-) PFP		2550	441$_{100}$	383$_{43}$	176$_{39}$	149$_{42}$	119$_{31}$	1100	7326
Linezolide artifact (deacetyl-) TFA		2575	391$_{100}$	333$_{35}$	175$_{25}$	164$_{28}$	149$_{36}$	988	7327
Linoleic acid	G	2140*	280$_{3}$	95$_{33}$	81$_{46}$	67$_{64}$	55$_{70}$	519	2551
Linoleic acid ET		2150*	308$_{10}$	263$_{6}$	95$_{51}$	81$_{85}$	67$_{100}$	658	5642
Linoleic acid ME		2110*	294$_{10}$	263$_{6}$	95$_{43}$	81$_{67}$	67$_{100}$	587	1068
Linolenic acid ME		2130*	292$_{16}$	191$_{6}$	121$_{26}$	95$_{74}$	79$_{100}$	577	2668
Linuron ME		1785	262$_{11}$	231$_{10}$	202$_{87}$	174$_{100}$	109$_{22}$	429	3940
Lisinopril 3TMS		3165	621$_{1}$	606$_{6}$	361$_{95}$	324$_{100}$	73$_{84}$	1197	4982
Lisinopril 4TMS		3260	693$_{1}$	678$_{13}$	479$_{23}$	318$_{38}$	73$_{53}$	1203	4983
Lisofylline	G P UHY	2505	280$_{25}$	236$_{20}$	193$_{32}$	180$_{100}$	109$_{31}$	518	1213
Lisofylline AC	U+UHYAC	2560	322$_{32}$	262$_{15}$	193$_{35}$	180$_{100}$		723	1214
Lisofylline -H2O	U+UHYAC	2300	262$_{25}$	193$_{35}$	181$_{100}$	137$_{24}$	109$_{45}$	431	1732
Lobeline		1820	337$_{1}$	216$_{5}$	120$_{42}$	105$_{99}$	77$_{79}$	795	1474
Lobeline artifact		1880	217$_{16}$	97$_{84}$	96$_{100}$	77$_{14}$		262	1821

114

Table 1-8-1: Compounds in order of names   **Lovastatin -H2O -C5H10O2**

Name	Detected	RI	Typical ions and intensities					Page	Entry
Lobeline artifact AC		1900	259 $_{21}$	200 $_4$	97 $_{42}$	96 $_{100}$		417	1822
Lodoxamide artifact		1790	167 $_{100}$	139 $_{11}$	132 $_7$	105 $_{25}$	77 $_8$	152	7519
Lodoxamide artifact AC		2080	209 $_{17}$	174 $_{83}$	167 $_{100}$	139 $_{11}$	105 $_{14}$	239	7520
Lodoxamide artifact 2AC		2325	251 $_7$	216 $_{100}$	174 $_{70}$	167 $_{80}$	139 $_8$	382	7522
Lodoxamide artifact 3AC		2235	293 $_1$	258 $_{16}$	251 $_{19}$	216 $_{100}$	167 $_{26}$	578	7521
Lofepramine-M (dealkyl-)	UHY	2225	266 $_{28}$	235 $_{61}$	208 $_{61}$	195 $_{100}$	71 $_{59}$	453	324
Lofepramine-M (dealkyl-) AC	U+UHYAC PAC	2670	308 $_{60}$	208 $_{150}$	193 $_{82}$	114 $_{93}$		657	325
Lofepramine-M (dealkyl-) HFB		2450	462 $_{23}$	268 $_{13}$	240 $_{20}$	208 $_{100}$	193 $_{54}$	1130	7706
Lofepramine-M (dealkyl-) PFP		2450	412 $_{68}$	234 $_{28}$	218 $_{35}$	208 $_{180}$	193 $_{85}$	1047	7667
Lofepramine-M (dealkyl-) TFA		2430	208 $_{100}$	140 $_{17}$	69 $_{21}$	193 $_{51}$	362 $_{14}$	891	7786
Lofepramine-M (dealkyl-) TMS		2470	338 $_5$	235 $_{62}$	143 $_{28}$	116 $_{50}$	73 $_{70}$	800	5461
Lofepramine-M (dealkyl-HO-) 2AC	U+UHYAC	3065	366 $_{27}$	266 $_{39}$	114 $_{100}$			906	1175
Lofepramine-M (di-HO-ring)	UHY	2600	227 $_{100}$	196 $_7$				297	2296
Lofepramine-M (di-HO-ring) 2AC	U+UHYAC	2750	311 $_{28}$	269 $_{23}$	227 $_{100}$	196 $_7$		670	2292
Lofepramine-M (HO-methoxy-ring)	UHY	2390	241 $_{100}$	226 $_{17}$	210 $_{12}$	180 $_{14}$		347	2315
Lofepramine-M (HO-methoxy-ring) AC	U+UHYAC	2370	283 $_{10}$	241 $_{100}$	226 $_{17}$	210 $_{12}$	180 $_{14}$	529	2867
Lofepramine-M (HO-ring)	UHY	2240	211 $_{100}$	196 $_{15}$	180 $_{10}$	152 $_4$		246	2295
Lofepramine-M (HO-ring) AC	U+UHYAC	2535	253 $_{26}$	211 $_{100}$	196 $_{19}$	180 $_{11}$	152 $_4$	393	1218
Lofepramine-M (ring)	U U+UHYAC	1930	195 $_{100}$	180 $_{40}$	167 $_9$	96 $_{33}$	83 $_{22}$	207	308
Lofepramine-M (ring) ME		1915	209 $_{70}$	194 $_{100}$	178 $_{13}$	165 $_{11}$		242	6352
Lofexidine		1910	258 $_1$	243 $_{100}$	223 $_{71}$	95 $_{69}$	67 $_{91}$	412	5208
Lofexidine AC		2200	300 $_1$	285 $_7$	257 $_{20}$	139 $_{58}$	86 $_{100}$	613	5209
Lonazolac	G	3000	312 $_{82}$	267 $_{100}$	232 $_{30}$	104 $_{20}$	77 $_{69}$	674	1913
Lonazolac ET		2950	340 $_{52}$	267 $_{100}$	232 $_{20}$	164 $_7$	77 $_{25}$	806	1994
Lonazolac ME	PME UME U+UHYAC	2685	326 $_{88}$	267 $_{99}$	232 $_{21}$	164 $_9$	77 $_{51}$	738	1377
Lonazolac -CO2		2400	268 $_{100}$	232 $_5$	164 $_7$	130 $_{38}$	77 $_{32}$	460	1975
Lonazolac-M (HO-) 2ME	UME	2880	356 $_{100}$	297 $_{79}$	262 $_{78}$	247 $_{13}$	117 $_8$	868	6296
Loperamide AC		3370	476 $_3$	432 $_1$	266 $_{10}$	238 $_{100}$	224 $_{44}$	1174	1824
Loperamide artifact	G U+UHYAC	3380	401 $_{90}$	250 $_{100}$	238 $_{77}$	222 $_{39}$	115 $_{37}$	1021	1825
Loperamide -H2O	U+UHYAC	3000	458 $_{13}$	266 $_{100}$	239 $_{24}$	192 $_{81}$	72 $_{17}$	1124	1823
Loperamide-M (N-dealkyl-)	UHY	1800	211 $_6$	139 $_{20}$	84 $_{42}$	56 $_{100}$		246	521
Loperamide-M (N-dealkyl-) AC	U U+UHYAC	2235	253 $_{31}$	210 $_{53}$	193 $_{14}$	139 $_{42}$	57 $_{100}$	393	524
Loperamide-M (N-dealkyl-oxo-) -2H2O	U UHY U+UHYAC	1650	189 $_{100}$	154 $_{32}$	127 $_{33}$	101 $_8$	75 $_{17}$	191	181
Loprazolam HY	UHY-I U+UHYAC-I	2470	276 $_{66}$	241 $_{100}$	195 $_{35}$	139 $_{48}$	111 $_{38}$	495	280
Loratadine	G U+UHYAC	3050	382 $_{100}$	292 $_{33}$	280 $_{34}$	266 $_{50}$	244 $_{33}$	959	5283
Loratadine-M/artifact (-COOCHCH3) AC	U+UHYAC	3120	352 $_{100}$	294 $_{32}$	280 $_{33}$	266 $_{60}$	245 $_{29}$	853	5610
Lorazepam	P-I G UGLUC	2440	302 $_{36}$	274 $_{49}$	239 $_{73}$	75 $_{80}$		714	539
Lorazepam 2AC		2730	345 $_{18}$	307 $_{42}$	265 $_{53}$	230 $_{101}$		1028	540
Lorazepam 2TMS		2380	464 $_2$	449 $_6$	429 $_{100}$	347 $_{13}$	73 $_{34}$	1132	4607
Lorazepam artifact-1	UHY U+UHYAC	2140	288 $_{57}$	287 $_{54}$	253 $_{100}$	177 $_{16}$	150 $_{13}$	550	2526
Lorazepam artifact-2	UHY-I U+UHYAC-I	2170	274 $_{44}$	273 $_{37}$	239 $_{100}$	177 $_7$	110 $_{10}$	483	289
Lorazepam artifact-3	U+UHYAC	2325	290 $_{76}$	255 $_{100}$				562	544
Lorazepam HY	UHY	2180	265 $_{62}$	230 $_{100}$	139 $_{43}$	111 $_{50}$		441	543
Lorazepam HYAC	U+UHYAC	2300	307 $_{42}$	265 $_{58}$	230 $_{100}$	139 $_{16}$	111 $_{14}$	647	290
Lorazepam iso-1 2ME		2485	348 $_{15}$	305 $_{43}$	75 $_{100}$			838	541
Lorazepam iso-2 2ME		2525	330 $_{10}$	316 $_{46}$	274 $_{38}$	239 $_{40}$	75 $_{100}$	757	542
Lorazepam-M (HO-) artifact	UHY-I	2400	304 $_{98}$	303 $_{98}$	277 $_{62}$	275 $_{100}$	203 $_{31}$	632	2529
Lorazepam-M (HO-) artifact AC	U+UHYAC-I	2550	346 $_9$	305 $_{79}$	303 $_{100}$	275 $_{25}$	239 $_8$	830	2527
Lorazepam-M (HO-) HY	UHY-I	2360	281 $_{62}$	246 $_{100}$	228 $_{26}$	154 $_{44}$	126 $_{43}$	520	545
Lorazepam-M (HO-) HY2AC	U+UHYAC	2600	365 $_{28}$	323 $_{30}$	281 $_{39}$	246 $_{100}$		899	2528
Lorazepam-M (HO-methoxy-) HY	UHY-I	2780	311 $_{23}$	281 $_{47}$	246 $_{100}$			669	546
Lorcainide	G U UHY U+UHYAC	2815	370 $_5$	355 $_{33}$	110 $_{75}$	91 $_{69}$	82 $_{100}$	920	1477
Lorcainide-M (deacyl-)	UHY U+UHYAC	2100	252 $_{18}$	237 $_9$	125 $_{23}$	110 $_{100}$	58 $_{47}$	391	2890
Lorcainide-M (deacyl-) AC	U+UHYAC	2200	294 $_5$	279 $_{55}$	110 $_{64}$	82 $_{94}$	56 $_{100}$	586	2891
Lorcainide-M (HO-) AC	U+UHYAC	2880	428 $_3$	413 $_{30}$	251 $_{21}$	110 $_{82}$	82 $_{100}$	1076	2893
Lorcainide-M (HO-di-methoxy-) AC	U+UHYAC	3010	488 $_{10}$	473 $_{65}$	251 $_{32}$	110 $_{66}$	82 $_{100}$	1156	2895
Lorcainide-M (HO-methoxy-) AC	U+UHYAC	2940	458 $_7$	443 $_{62}$	251 $_{30}$	110 $_{66}$	82 $_{100}$	1124	2894
Lorcainide-M (N-dealkyl-deacyl-) 2AC	U+UHYAC	2490	294 $_7$	251 $_{35}$	192 $_{24}$	125 $_{37}$	83 $_{100}$	585	2892
Lormetazepam	P-I G UGLUC	2735	334 $_8$	307 $_{66}$	305 $_{100}$	111 $_6$	75 $_{10}$	778	547
Lormetazepam AC		2740	376 $_6$	334 $_{22}$	305 $_{100}$	291 $_{20}$	255 $_9$	939	5604
Lormetazepam artifact-1	UHY-I U+UHYAC-I	2170	274 $_{44}$	273 $_{37}$	239 $_{100}$	177 $_7$	110 $_{10}$	483	289
Lormetazepam artifact-2		2585	291 $_{72}$	289 $_{100}$	253 $_6$	179 $_{44}$	109 $_6$	556	2381
Lormetazepam artifact-3	G	2850	304 $_{98}$	269 $_{100}$	262 $_{87}$	254 $_{22}$	75 $_{31}$	632	5640
Lormetazepam artifact-4	G	3120	334 $_{79}$	262 $_{47}$	228 $_{100}$	195 $_{84}$	75 $_{39}$	778	5641
Lormetazepam HY	UHY U+UHYAC	2220	279 $_{91}$	244 $_{100}$	229 $_{66}$	111 $_{53}$	75 $_{54}$	509	291
Lormetazepam iso-1 TMS		2735	406 $_{29}$	363 $_{66}$	267 $_{100}$	228 $_{51}$	73 $_{66}$	1032	4606
Lormetazepam iso-2 TMS		2735	406 $_{10}$	391 $_{22}$	377 $_{100}$	291 $_{25}$	73 $_{34}$	1032	4558
Lormetazepam-M (HO-) HY	UHY	2470	295 $_{86}$	260 $_{100}$	245 $_{70}$			588	548
Lormetazepam-M (nor-)	P-I G UGLUC	2440	302 $_{36}$	274 $_{49}$	239 $_{73}$	75 $_{80}$		714	539
Lormetazepam-M (nor-) 2TMS		2380	464 $_2$	449 $_6$	429 $_{100}$	347 $_{13}$	73 $_{34}$	1132	4607
Lormetazepam-M (nor-) HY	UHY	2180	265 $_{62}$	230 $_{100}$	139 $_{43}$	111 $_{50}$		441	543
Lormetazepam-M (nor-) HYAC	U+UHYAC	2300	307 $_{42}$	265 $_{58}$	230 $_{100}$	139 $_{16}$	111 $_{14}$	647	290
Losartan 2ME	UME	3555	450 $_{18}$	249 $_{92}$	201 $_{93}$	192 $_{100}$	165 $_{72}$	1113	4841
Lovastatin -H2O -C5H10O2	G	2775*	284 $_{23}$	199 $_{100}$	198 $_{90}$	172 $_{61}$	157 $_{80}$	536	6449

Table 1-8-1: Compounds in order of names

Name	Detected	RI	Typical ions and intensities					Page	Entry
Loxapine	G U+UHYAC	2555	327_6	257_{23}	83_{49}	70_{100}		744	549
Loxapine-M (HO-) AC	U+UHYAC	2935	385_5	315_{10}	83_{65}	70_{101}		970	1274
Loxapine-M (nor-HO-) 2AC	U+UHYAC	3450	413_{31}	207_{101}	112_{20}			1047	1275
LSD		3445	323_{100}	221_{73}	207_{53}	181_{48}	72_{50}	729	1069
LSD TMS	ULSDTMS	3595	395_{100}	293_{53}	253_{84}	73_{97}		1003	1070
LSD-M (2-oxo-3-HO-) 2TMS		3430	499_{53}	325_9	309_{75}	235_{16}	73_{70}	1163	6223
LSD-M (nor-) TMS		3705	381_{64}	279_{46}	254_{22}	100_6	73_{70}	958	6262
LSD-M (nor-) 2TMS		3515	453_{64}	351_{23}	279_{13}	253_{36}	73_{70}	1119	6261
Lupanine	U UHY U+UHYAC	2230	248_{42}	149_{51}	136_{100}	98_{26}	84_{18}	373	2877
Lynestrenol	G	2260*	284_{26}	201_{50}	159_{32}	91_{100}	79_{96}	536	2242
Lynestrenol AC		2280*	326_{29}	266_{39}	201_{74}	159_{49}	91_{100}	742	2263
Lysergic acid N,N-methylpropylamine TMS		3740	395_{57}	337_{10}	293_{26}	253_{36}	73_{65}	1003	6263
Lysergide		3445	323_{100}	221_{73}	207_{53}	181_{48}	72_{50}	729	1069
Lysergide TMS	ULSDTMS	3595	395_{100}	293_{53}	253_{84}	73_{97}		1003	1070
Lysergide alpha isomer (iso-LSD) TMS		3515	395_{72}	293_{35}	279_{19}	253_{31}	73_{80}	1075	6222
Lysergide-M (2-oxo-3-HO-) 2TMS		3430	499_{53}	325_9	309_{75}	235_{16}	73_{70}	1163	6223
Lysergide-M (nor-) TMS		3705	381_{64}	279_{46}	254_{22}	100_6	73_{70}	958	6262
Lysergide-M (nor-) 2TMS		3515	453_{64}	351_{23}	279_{13}	253_{36}	73_{70}	1119	6261
Mafenide		2340	186_3	185_{21}	141_{11}	106_{100}	77_{36}	187	5228
Mafenide AC		2425	228_{57}	185_{22}	147_{60}	106_{74}	105_{100}	299	5232
Mafenide MEAC		2300	242_{42}	199_{26}	185_{50}	161_{58}	119_{100}	351	5233
Mafenide 2ME		1920	214_{17}	133_3	89_{10}	74_{11}	58_{100}	255	5230
Mafenide 3ME		1900	228_{13}	214_4	133_4	89_{13}	58_{100}	300	5229
Mafenide 4ME		1870	242_{18}	134_7	107_{11}	89_{19}	58_{100}	352	5231
Malaoxon		1890*	314_1	268_{16}	195_{17}	127_{100}	99_{51}	683	3449
Malathion		1940*	330_2	285_8	173_{86}	127_{99}	93_{56}	758	1401
Malathion-M (malaoxon)		1890*	314_1	268_{16}	195_{17}	127_{100}	99_{51}	683	3449
Maleic acid 2TMS		1080*	245_{15}	147_{10}	133_7	83_{10}	73_{49}	421	4674
Maleic hydrazide (MH)		1735	112_{100}	97_5	82_{80}	68_4	55_{75}	103	3916
Mandelic acid		1890*	152_{10}	107_{100}	79_{75}	77_{55}	51_{63}	131	5759
Mandelic acid ME		1485*	166_7	107_{100}	79_{59}	77_{40}		149	1071
Mannitol 6AC	U+UHYAC	2080*	361_{26}	289_{34}	187_{53}	139_{70}	115_{100}	1087	1965
Mannitol 6HFB		1510*	521_3	478_5	307_{52}	240_{29}	169_{100}	1208	5802
Mannitol 6PFP		1510*	378_3	257_{17}	219_{34}	190_{24}	119_{100}	1206	5801
Mannitol 6TFA		1370*	321_3	278_{13}	265_{12}	140_{56}	69_{100}	1204	5800
Mannose 5AC	PAC U+UHYAC	2000*	331_{10}	242_{23}	157_{66}	115_{100}	98_{66}	985	1964
Mannose 5HFB		1805*	519_1	465_{11}	321_{21}	257_{24}	169_{100}	1208	5805
Mannose 5PFP		1285*	419_{13}	365_7	227_{11}	147_{29}	119_{100}	1206	5804
Mannose 5TFA		1650*	290_{10}	265_{34}	221_{32}	157_{36}	69_{100}	1200	5803
Mannose iso-1 5TMS		1885*	435_1	217_{26}	204_{100}	191_{57}	73_{96}	1182	4559
Mannose iso-2 5TMS		1990*	435_2	217_{32}	204_{100}	191_{66}	73_{98}	1182	4560
Maprotiline	P-I G UHY	2390	277_{24}	204_{100}	189_{79}	70_{56}	59_{63}	503	550
Maprotiline (ME)		2360	291_2	203_{19}	189_{14}	178_9	58_{80}	572	2254
Maprotiline AC	U+UHYAC	2800	319_5	291_{100}	218_{75}	203_{74}	191_{66}	711	349
Maprotiline HFB		2525	473_1	445_{50}	240_{51}	203_{59}	191_{80}	1142	7681
Maprotiline PFP		2530	423_2	395_{77}	203_{61}	191_{100}	119_{71}	1066	7680
Maprotiline TFA		2430	373_3	345_{100}	203_{67}	191_{100}	140_{62}	931	7679
Maprotiline TMS		2565	349_{11}	203_9	191_{15}	116_{100}	73_{46}	844	4561
Maprotiline-M (deamino-di-HO-)	UHY	2570*	280_{15}	252_{100}	207_{64}			518	551
Maprotiline-M (deamino-di-HO-) 2AC	U+UHYAC	2820*	364_3	336_{67}	294_{100}	234_{11}	207_{26}	897	351
Maprotiline-M (deamino-HO-propyl-) AC	U+UHYAC	2425*	306_2	278_{100}	218_{12}	203_{17}	191_{49}	645	350
Maprotiline-M (deamino-tri-HO-) 3AC	U+UHYAC	3200*	422_2	394_{84}	352_{56}	310_{100}	223_{21}	1065	355
Maprotiline-M (HO-anthryl-) AC	U+UHYAC	2995	335_{56}	307_{100}	234_{32}	207_{41}	100_9	784	6478
Maprotiline-M (HO-anthryl-) 2AC	U+UHYAC	3095	377_8	349_{100}	307_{45}	234_{35}	207_{16}	943	353
Maprotiline-M (HO-ethanediyl-) 2AC	U+UHYAC	2995	377_1	291_{100}	218_{44}	191_{25}	100_9	943	352
Maprotiline-M (nor-) AC	P U+UHYAC	2760	305_4	277_{100}	218_{35}	191_{36}		640	348
Maprotiline-M (nor-di-HO-anthryl-) 3AC	U+UHYAC	3100	421_2	393_{100}	351_{38}	309_{49}	223_{15}	1063	3359
Maprotiline-M (nor-HO-anthryl-) AC	U+UHYAC	3020	321_{42}	293_{100}	234_{20}	222_{30}	207_{25}	719	6479
Maprotiline-M (nor-HO-anthryl-) 2AC	U+UHYAC	3150	363_3	335_{100}	293_{77}	234_{20}	207_{25}	894	354
Maprotiline-M (nor-HO-ethanediyl-) 2AC	U+UHYAC	2970	363_1	277_{100}	218_{32}	191_{38}	179_{10}	895	6477
Mazindol AC		2705	326_{73}	284_{28}	256_{101}	220_{78}		738	1073
Mazindol -H2O		2345	266_{100}	231_{22}	102_{27}			448	1072
2,3-MBDB		1610	178_3	135_3	89_4	72_{100}	57_7	234	5416
2,3-MBDB AC		1965	249_4	176_{17}	135_7	114_{35}	72_{100}	375	5507
2,3-MBDB HFB		1735	403_3	268_{100}	210_{17}	176_8	135_5	1025	5591
2,3-MBDB PFP		1710	353_8	218_{100}	176_{23}	160_{14}	135_{12}	857	5592
2,3-MBDB TFA		1725	303_4	176_{43}	168_{13}	135_{14}	110_{23}	627	5508
2,3-MBDB TMS		1730	264_{12}	176_{13}	144_{100}	135_{13}	73_{54}	514	5593
2,3-MBDB-M (nor-)		1550	193_1	164_2	135_2	77_5	58_{100}	200	5414
2,3-MBDB-M (nor-) AC		1895	235_7	176_{43}	135_{10}	100_{21}	58_{100}	323	5504
2,3-MBDB-M (nor-) HFB		1660	389_{23}	345_8	254_{59}	176_{100}	135_{57}	982	5505
2,3-MBDB-M (nor-) PFP		1615	339_4	204_7	176_{43}	135_{100}	119_{14}	801	5544
2,3-MBDB-M (nor-) TFA		1705	289_{30}	176_{100}	154_{74}	135_{62}	77_{24}	556	5506
2,3-MBDB-M (nor-) formyl artifact		1575	205_9	176_3	135_9	77_9	70_{100}	228	5415

Table 1-8-1: Compounds in order of names   MDA TMS

Name	Detected	RI	Typical ions and intensities					Page	Entry
MBDB		1630	207_1	178_2	135_5	72_{100}	57_7	234	3256
MBDB AC		1995	249_1	176_{33}	135_6	114_{24}	72_{100}	376	3270
MBDB HFB		1815	403_{13}	268_{100}	210_{54}	176_{81}	135_{57}	1026	5088
MBDB PFP		1785	353_{17}	218_{100}	176_{80}	160_{55}	135_{61}	857	5084
MBDB TFA		1800	303_{21}	176_{84}	168_{100}	135_{67}	110_{48}	627	5081
MBDB intermediate-1		1385*	176_{59}	131_{100}	103_{80}	77_{41}	63_{26}	163	3291
MBDB intermediate-2 (1-(1,3-benzodioxol-5-yl)-butan-2-one)		1525*	192_{20}	135_{100}	105_6	77_{22}	57_{21}	197	3292
MBDB intermediate-3 (1-(1,3-benzodioxol-5-yl)-butan-1-ol)		1560*	194_{17}	151_{100}	123_{21}	93_{72}	65_{37}	205	3290
MBDB intermediate-3 AC		1670*	236_{21}	193_{10}	151_{100}	135_{30}	93_{16}	328	3294
MBDB precursor (piperonal)		1160*	150_{80}	149_{100}	121_{34}	91_{12}	63_{49}	127	3275
MBDB-M (demethylenyl-) 3AC		2295	321_1	248_7	164_8	114_{46}	72_{100}	718	5110
MBDB-M (demethylenyl-methyl-) 2AC		2170	293_1	220_{16}	178_{19}	114_{41}	72_{100}	582	5109
MBDB-M (nor-)	U+UHYAC	1570	193_1	164_2	136_{12}	77_6	58_{100}	201	3253
MBDB-M (nor-) AC	U+UHYAC	1950	235_3	176_{23}	162_{14}	100_{13}	58_{100}	324	3262
MBDB-M (nor-) HFB		1690	389_4	176_{42}	135_{100}	77_{11}		982	5288
MBDB-M (nor-) PFP		1700	339_4	204_7	176_{43}	135_{100}	119_{14}	801	5287
MBDB-M (nor-) TFA		1705	289_4	176_{33}	154_{11}	135_{100}	77_{12}	557	5286
MBDB-M (nor-) formyl artifact		1585	205_8	176_3	135_{25}	77_{16}	70_{100}	228	3246
MBDB-M (nor-demethylenyl-) 3AC	U+UHYAC	2235	307_1	248_2	164_6	100_{29}	58_{100}	649	5551
MBDB-M (nor-demethylenyl-methyl-) 2AC	U+UHYAC	2140	279_4	220_{32}	178_{74}	100_{24}	58_{100}	512	5550
MCC		1560	194_9	164_9	151_{100}	124_{41}	56_{21}	206	3578
MCC -HCN		1260	167_{81}	152_{52}	108_{100}	94_{36}	81_{77}	153	3579
MCPA	P U	1580*	200_{99}	155_{27}	141_{100}	125_{32}	77_{58}	218	1074
MCPA ME	P U PME UME	1525*	214_{100}	182_{10}	155_{57}	141_{85}	125_{49}	254	2266
MCPB	U	1845*	228_{13}	142_{100}	107_{35}	77_{19}		299	1075
MCPB ME		1760*	242_8	211_7	142_9	101_{100}	59_{76}	351	2267
mCPP		1910	196_{24}	154_{100}	138_{12}	111_9	75_{12}	211	6885
mCPP AC	U+UHYAC	2265	238_{32}	195_{15}	166_{100}	154_{31}	111_{27}	336	405
mCPP HFB	U+UHYHFB	1960	392_{100}	195_{38}	166_{41}	139_{36}	111_{25}	991	6604
mCPP ME		1820	210_{100}	166_{19}	139_{81}	111_{33}	70_{99}	244	6886
mCPP TFA	U+UHYTFA	1920	292_{100}	250_{12}	195_{77}	166_{79}	139_{66}	574	6597
mCPP TMS		2035	268_{61}	253_{29}	128_{63}	101_{68}	86_{71}	461	6888
mCPP-M (chloroaniline) AC	U+UHYAC	1580	169_{31}	127_{100}	111_2	99_8		155	6593
mCPP-M (chloroaniline) HFB		1310	323_{60}	304_8	154_{100}	126_{68}	111_{59}	724	6607
mCPP-M (chloroaniline) ME		1100	141_{74}	140_{100}	111_9	105_{11}	77_{31}	120	4089
mCPP-M (chloroaniline) TFA		1125	223_{99}	154_{100}	126_{51}	111_{55}	69_{42}	282	4124
mCPP-M (chloroaniline) 2ME		1180	155_{62}	154_{100}	140_{11}	118_{16}	75_{17}	134	4090
mCPP-M (deethylene-) 2AC	U+UHYAC	2080	254_1	195_{50}	153_{31}	140_{100}	111_{10}	398	6592
mCPP-M (deethylene-) 2HFB	U+UHYHFB	1705	562_1	349_{42}	336_{73}	240_{26}	139_{100}	1188	6606
mCPP-M (deethylene-) 2TFA	U+UHYTFA	1670	362_3	249_{77}	236_{100}	139_{66}	111_{28}	890	6601
mCPP-M (HO-) TFA	U+UHYTFA	2035	308_{100}	272_{19}	211_{46}	182_{30}	155_{46}	653	6599
mCPP-M (HO-) iso-1 AC	U+UHYAC	2335	254_{65}	211_{23}	182_{100}	166_{71}	154_{33}	398	5308
mCPP-M (HO-) iso-1 2AC	U+UHYAC	2515	296_{30}	254_{56}	182_{100}	154_{36}		597	406
mCPP-M (HO-) iso-1 2TFA	U+UHYTFA	2040	404_{28}	307_{74}	278_{34}	265_{13}	154_{40}	1028	6600
mCPP-M (HO-) iso-2 AC	U+UHYAC	2345	254_{79}	211_{15}	182_{100}	169_{40}	154_{40}	398	5307
mCPP-M (HO-) iso-2 2AC	U+UHYAC	2525	296_{20}	254_{74}	182_{100}	169_{72}	154_{24}	597	32
mCPP-M (HO-) iso-2 2HFB	U+UHYHFB	2145	604_{43}	585_9	407_{100}	378_{22}	154_{22}	1195	6605
mCPP-M (HO-) iso-2 2TFA	U+UHYTFA	2045	404_{13}	307_{33}	278_{10}	265_4	154_{22}	1028	6598
mCPP-M (HO-chloroaniline N-acetyl-) HFB	U+UHYHFB	1820	381_{10}	339_{22}	169_{15}	142_{100}	69_{23}	954	6796
mCPP-M (HO-chloroaniline N-acetyl-) TFA	U+UHYTFA	1765	281_{29}	239_{58}	142_{100}	114_{10}	69_{40}	520	6797
mCPP-M (HO-chloroaniline) 2HFB	U+UHYHFB	1540	535_{36}	516_8	338_{72}	169_{100}	143_{82}	1180	6608
mCPP-M (HO-chloroaniline) iso-1 2AC	U+UHYAC	1980	227_{14}	185_{18}	167_{33}	143_{100}	114_7	297	6594
mCPP-M (HO-chloroaniline) iso-1 2TFA	U+UHYTFA	1440	335_{15}	238_9	168_2	69_{15}		781	6603
mCPP-M (HO-chloroaniline) iso-1 3AC	U+UHYAC	1940	269_2	227_{13}	185_{17}	167_{33}	143_{100}	463	6596
mCPP-M (HO-chloroaniline) iso-2 2AC	U+UHYAC	2020	227_4	185_{57}	143_{100}	79_{18}	114_5	297	404
mCPP-M (HO-chloroaniline) iso-2 2TFA	U+UHYTFA	1440	335_{44}	266_3	238_{52}	210_{14}	143_{42}	781	6602
mCPP-M (HO-chloroaniline) iso-2 3AC	U+UHYAC	1900	269_4	227_{37}	185_{75}	143_{100}	79_7	463	6595
2,3-MDA		1470	179_{34}	164_{13}	135_{96}	77_{100}	51_{91}	170	5420
2,3-MDA		1470	179_3	135_7	77_{11}	51_8	44_{100}	170	5513
2,3-MDA AC		1770	221_{31}	162_{100}	135_{35}	86_8	77_{40}	276	5589
2,3-MDA AC		1770	221_{18}	162_{71}	135_{35}	77_{64}	44_{100}	276	6310
2,3-MDA HFB		1595	375_{12}	240_{42}	169_{11}	162_{100}	135_{61}	935	5502
2,3-MDA PFP		1545	325_8	190_{42}	162_{100}	135_{71}	119_{46}	733	5542
2,3-MDA TFA		1585	275_{32}	162_{100}	140_{56}	135_{77}	77_{26}	488	5503
2,3-MDA TMS		1655	251_1	236_7	135_{19}	116_{100}	73_{71}	385	5590
2,3-MDA formyl artifact		1490	191_{11}	135_6	105_4	77_{11}	56_{100}	195	5421
MDA	U UHY	1495	179_1	136_{18}	105_1	77_5		170	3241
MDA	U UHY	1495	179_1	136_{18}	77_5	51_5	44_{100}	171	5518
MDA AC	U+UHYAC	1860	221_3	162_{46}	135_{12}	86_8	77_{12}	277	3263
MDA AC	U+UHYAC	1860	221_3	162_{46}	135_{12}	86_8	44_{100}	277	5519
MDA HFB	UHFB	1650	375_4	240_6	169_{10}	162_{50}	135_{100}	935	5291
MDA PFP	UPFP	1605	325_5	190_{10}	162_{46}	135_{100}	119_{18}	733	5290
MDA TFA	UTFA	1615	275_4	162_{39}	135_{100}	105_6	77_{12}	488	5289
MDA TMS		1735	251_1	236_4	135_8	116_{100}	73_{47}	385	6334

MDA formyl artifact   Table 1-8-1: Compounds in order of names

Name	Detected	RI	Typical ions and intensities					Page	Entry
MDA formyl artifact	P-I	1520	191_9	135_{27}	105_4	77_{17}	56_{100}	195	3252
MDA precursor-1 (piperonal)		1160*	150_{80}	149_{100}	121_{34}	91_{12}	63_{49}	127	3275
MDA precursor-2 (isosafrole)		1215*	162_{100}	131_{51}	104_{71}	77_{53}	63_{28}	141	3276
MDA precursor-3 (piperonylacetone)		1365*	178_{20}	135_{100}	105_7	77_{41}	51_{38}	167	3274
MDA R-(-)-enantiomer HFBP		2280	472_2	294_{15}	266_{100}	162_{66}	135_8	1140	6640
MDA S-(+)-enantiomer HFBP		2290	472_2	294_{16}	266_{90}	162_{69}	135_9	1140	6641
MDA-D5 AC		1840	226_{13}	167_{100}	166_{93}	136_{48}	78_{28}	295	5688
MDA-D5 HFB		1630	380_{10}	244_{10}	167_{31}	166_{29}	136_{100}	952	6773
MDA-D5 2AC		1910	268_3	167_{100}	166_{81}	136_{47}	90_{59}	462	5689
MDA-D5 R-(-)-enantiomer HFBP		2275	477_{18}	341_6	294_{27}	266_{100}	167_{59}	1145	6798
MDA-D5 S-(+)-enantiomer HFBP		2285	477_9	341_6	294_{26}	266_{100}	167_{51}	1145	6799
MDA-M	UHY	1510*	180_{19}	137_{100}	122_{19}	107_2	94_{16}	175	4247
MDA-M (deamino-HO-) AC	U+UHYAC	1620*	222_9	162_{100}	135_{56}	104_{10}	77_{17}	281	6410
MDA-M (deamino-oxo-)		1365*	178_{20}	135_{100}	105_7	77_{41}	51_{38}	167	3274
MDA-M (deamino-oxo-demethylenyl-) 2AC	U+UHYAC	1735*	250_3	208_{15}	166_{45}	123_{100}		379	4210
MDA-M (deamino-oxo-demethylenyl-methyl-) AC	U+UHYAC	1600*	222_2	180_{22}	137_{100}			280	4211
MDA-M (deamino-oxo-demethylenyl-methyl-) ME	UHYME	1540*	194_{25}	151_{100}	135_4	107_{18}	65_4	204	4353
MDA-M (demethylenyl-) 3AC	U+UHYAC	2150	293_2	234_{54}	192_{48}	150_{99}	86_{100}	579	3725
MDA-M (demethylenyl-methyl-)	UHY	1465	181_9	138_{100}	122_{18}	94_{24}	77_{16}	178	4351
MDA-M (demethylenyl-methyl-) ME	UHYME	1550	195_1	152_{100}	137_{17}	107_{16}	77_{14}	208	4352
MDA-M (demethylenyl-methyl-) 2AC	U+UHYAC	2065	265_3	206_{27}	164_{100}	137_{23}	86_{33}	445	3498
MDA-M (demethylenyl-methyl-) 2HFB	UHFB	1690	360_{82}	333_{15}	240_{100}	169_{42}	69_{39}	1189	6512
MDA-M (HO-methoxy-hippuric acid) ME	UHYME	2165	239_9	151_{100}				339	4213
MDA-M (methylenedioxy-hippuric acid) ME	UME UHYME	2065	237_{16}	178_4	149_{100}	121_{15}	65_{13}	332	4212
MDA-M 2AC	U+UHYAC	1820*	266_3	206_9	164_{100}	150_{10}	137_{30}	450	6409
MDBP		1890	220_{21}	178_{14}	164_{13}	135_{100}	85_{36}	274	6624
MDBP AC	U+UHYAC	2350	262_{16}	190_{11}	176_{21}	135_{100}	85_{45}	431	6625
MDBP HFB	U+UHYHFB	2190	416_2	281_{15}	148_9	135_{100}	105_{11}	1053	6631
MDBP TFA	U+UHYTFA	2350	316_9	181_{14}	148_6	135_{100}	105_6	693	6628
MDBP TMS		2080	292_{87}	157_{53}	135_{100}	102_{57}	73_{85}	576	6887
MDBP artifact (piperonylacetate)		1530*	194_{63}	152_{66}	135_{100}	122_{17}	104_{18}	204	6637
MDBP Cl-artifact		1295*	170_{25}	135_{100}	105_8	77_{41}		156	6635
MDBP-M (deethylene-) 2AC	U+UHYAC	2320	278_4	235_{64}	177_{11}	150_{25}	135_{100}	506	6626
MDBP-M (deethylene-) 2HFB	U+UHYHFB	2080	586_3	389_{24}	346_7	240_7	135_{100}	1193	6632
MDBP-M (deethylene-) 2TFA	U+UHYTFA	2230	386_3	289_{19}	246_{10}	150_4	135_{100}	972	6629
MDBP-M (demethylene-methyl-) Cl-artifact		1625*	172_{41}	137_{100}	122_6	108_{13}	77_7	159	6636
MDBP-M (demethylenyl-methyl-) AC	U+UHYAC	2410	264_9	192_8	137_{100}	122_{18}	85_{42}	440	6509
MDBP-M (demethylenyl-methyl-) HFB	U+UHYHFB	2135	418_{22}	295_{15}	281_{27}	138_{87}	137_{100}	1058	6575
MDBP-M (demethylenyl-methyl-) TFA	U+UHYTFA	2120	318_8	181_{10}	137_{100}	122_{21}	69_{12}	703	6570
MDBP-M (demethylenyl-methyl-) 2AC	U+UHYAC	2380	306_2	234_9	179_{13}	137_{100}	85_{64}	645	6508
MDBP-M (piperonylamine) AC	U+UHYAC	2015	193_{100}	150_{80}	135_{54}	121_{37}	93_{27}	199	6627
MDBP-M (piperonylamine) HFB	U+UHYHFB	1640	347_{100}	317_6	289_{10}	178_6	135_{75}	834	6633
MDBP-M (piperonylamine) PFP		1755	297_{100}	267_8	239_{13}	178_7	135_{51}	601	7632
MDBP-M (piperonylamine) TFA	U+UHYTFA	1775	247_{100}	217_9	189_{16}	148_{12}	135_{80}	365	6630
MDBP-M (piperonylamine) 2AC		2230	235_{23}	192_{35}	150_{100}	135_{17}	93_9	323	7631
MDBP-M (piperonylamine) 2TMS		2130	295_{27}	280_{100}	206_{15}	179_{50}	73_{51}	592	7630
MDBP-M (piperonylamine) formyl artifact		1560	163_{22}	135_{100}	121_3	105_{10}	77_{37}	142	7629
MDBP-M/artifact (piperazine) 2AC		1750	170_{15}	85_{33}	69_{25}	56_{100}		157	879
MDBP-M/artifact (piperazine) 2HFB		1290	478_3	459_9	309_{100}	281_{22}	252_{41}	1146	6634
MDBP-M/artifact (piperazine) 2TFA		1005	278_{10}	209_{59}	152_{25}	69_{56}	56_{100}	505	4129
2,3-MDE-M (deethyl-)		1470	179_{34}	164_{13}	135_{96}	77_{100}	51_{91}	170	5420
2,3-MDE-M (deethyl-)		1470	179_3	135_7	77_{11}	51_8	44_{100}	170	5513
2,3-MDE-M (deethyl-) AC		1770	221_{31}	162_{100}	135_{35}	86_8	77_{40}	276	5589
2,3-MDE-M (deethyl-) AC		1770	221_{18}	162_{71}	135_{35}	77_{64}	44_{100}	276	6310
2,3-MDE-M (deethyl-) HFB		1595	375_{12}	240_{42}	169_{11}	162_{100}	135_{61}	935	5502
2,3-MDE-M (deethyl-) PFP		1545	325_8	190_{42}	162_{100}	135_{71}	119_{46}	733	5542
2,3-MDE-M (deethyl-) TFA		1585	275_{32}	162_{100}	140_{56}	135_{77}	77_{26}	488	5503
2,3-MDE-M (deethyl-) TMS		1655	251_1	236_7	135_{19}	116_{100}	73_{71}	385	5590
2,3-MDE-M (deethyl-) formyl artifact		1490	191_{11}	135_6	105_4	77_{11}	56_{100}	195	5421
MDE	G	1560	207_1	163_1	135_4	77_4	72_{100}	234	3257
MDE AC	U+UHYAC	1985	249_1	162_{41}	135_6	114_{23}	72_{100}	376	3271
MDE HFB		1790	403_5	268_{100}	240_{53}	162_{62}	135_{47}	1026	5087
MDE PFP		1755	353_{16}	218_{100}	190_{73}	162_{71}	135_{38}	857	5083
MDE TFA		1770	303_{14}	168_{100}	162_{76}	140_{63}	135_{49}	627	5080
MDE TMS		1825	264_{14}	144_{100}	135_{23}	100_4	73_{79}	514	4604
MDE precursor-1 (piperonal)		1160*	150_{80}	149_{100}	121_{34}	91_{12}	63_{49}	127	3275
MDE precursor-2 (isosafrole)		1215*	162_{100}	131_{51}	104_{71}	77_{53}	63_{28}	141	3276
MDE precursor-3 (piperonylacetone)		1365*	178_{20}	135_{100}	105_7	77_{41}	51_{38}	167	3274
MDE R-(-)-enantiomer HFBP		2460	500_1	365_6	266_{100}	162_{46}	135_8	1164	6644
MDE S-(+)-enantiomer HFBP		2470	500_1	365_6	266_{100}	162_{54}	135_6	1164	6645
MDE-D5		1555	212_1	135_8	105_2	77_{100}		250	7287
MDE-D5 HFB		1770	408_3	273_{100}	241_{35}	162_{47}	135_{20}	1036	6772
MDE-D5 PFP		1750	358_4	223_{93}	191_{53}	162_{100}	135_{44}	877	7289
MDE-D5 TFA		1765	308_9	173_{100}	162_{71}	141_{35}	135_{29}	656	7288

Table 1-8-1: Compounds in order of names

MDMA-M (methylenedioxy-hippuric acid) ME

Name	Detected	RI	Typical ions and intensities					Page	Entry
MDE-D5 TMS		1820	284_1	269_3	149_{100}	135_6	73_{32}	536	7290
MDE-D5 R-(-)-enantiomer HFBP		2455	505_1	162_{31}	370_{17}	266_{100}		1167	6802
MDE-D5 S-(+)-enantiomer HFBP		2465	505_1	162_{38}	370_8	266_{100}		1168	6803
MDE-M	UHY	1510*	180_{19}	137_{100}	122_{19}	107_2	94_{16}	175	4247
MDE-M	UHY	1465	181_9	138_{100}	122_{18}	94_{24}	77_{16}	178	4351
MDE-M (deamino-HO-) AC	U+UHYAC	1620*	222_9	162_{100}	135_{56}	104_{10}	77_{17}	281	6410
MDE-M (deamino-oxo-)		1365*	178_{20}	135_{100}	105_7	77_{41}	51_{38}	167	3274
MDE-M (deethyl-)	U UHY	1495	179_1	136_{18}	105_1	77_5		170	3241
MDE-M (deethyl-)	U UHY	1495	179_1	136_{18}	77_5	51_5	44_{100}	171	5518
MDE-M (deethyl-) AC	U+UHYAC	1860	221_3	162_{46}	135_{12}	86_8	77_{12}	277	3263
MDE-M (deethyl-) AC	U+UHYAC	1860	221_3	162_{46}	135_{12}	86_8	44_{100}	277	5519
MDE-M (deethyl-) HFB	UHFB	1650	375_4	240_6	169_{10}	162_{50}	135_{100}	935	5291
MDE-M (deethyl-) PFP	UPFP	1605	325_5	190_{10}	162_{46}	135_{100}	119_{18}	733	5290
MDE-M (deethyl-) TFA	UTFA	1615	275_4	162_{39}	135_{100}	105_6	77_{12}	488	5289
MDE-M (deethyl-) TMS		1735	251_1	236_4	135_8	116_{100}	73_{47}	385	6334
MDE-M (deethyl-) R-(-)-enantiomer HFBP		2280	472_2	294_{15}	266_{100}	162_{66}	135_8	1140	6640
MDE-M (deethyl-) S-(+)-enantiomer HFBP		2290	472_2	294_{16}	266_{90}	162_{69}	135_9	1140	6641
MDE-M (deethyl-)-D5 AC		1840	226_{13}	167_{100}	166_{93}	136_{48}	78_{28}	295	5688
MDE-M (deethyl-)-D5 2AC		1910	268_3	167_{100}	166_{81}	136_{47}	90_{59}	462	5689
MDE-M (deethyl-demethylenyl-) 3AC	U+UHYAC	2150	293_2	234_{54}	192_{48}	150_{99}	86_{100}	579	3725
MDE-M (deethyl-demethylenyl-methyl-) 2AC	U+UHYAC	2065	265_3	206_{27}	164_{100}	137_{23}	86_{33}	445	3498
MDE-M (demethylenyl-) 3AC	U+UHYAC	2200	321_1	234_4	150_{10}	114_{46}	72_{100}	718	4208
MDE-M (demethylenyl-methyl-)	UHY	1640	209_1	137_{12}	122_7	94_9	72_{100}	242	4364
MDE-M (demethylenyl-methyl-) AC	U+UHYAC	2000	251_1	164_{46}	137_7	114_{33}	72_{100}	386	4274
MDE-M (demethylenyl-methyl-) ME	UHYME	1930	223_{17}	194_7	151_{36}	94_{12}	72_{100}	286	4350
MDE-M (demethylenyl-methyl-) 2AC	U+UHYAC	2080	293_1	206_{20}	164_{38}	114_{77}	72_{100}	582	4209
MDE-M (HO-methoxy-hippuric acid) ME	UHYME	2165	239_9	151_{100}				339	4213
MDE-M (methylenedioxy-hippuric acid) ME	UME UHYME	2065	237_{16}	178_4	149_{100}	121_{15}	65_{13}	332	4212
MDE-M AC	U+UHYAC	1600*	222_2	180_{22}	137_{100}			280	4211
MDE-M ME	UHYME	1540*	194_{25}	151_{100}	135_4	107_{18}	65_4	204	4353
MDE-M ME	UHYME	1550	195_1	152_{100}	137_{17}	107_{16}	77_{14}	208	4352
MDE-M 2AC	U+UHYAC	1735*	250_3	208_{15}	166_{45}	123_{100}		379	4210
MDE-M 2AC	U+UHYAC	1820*	266_3	206_9	164_{100}	150_{10}	137_{30}	450	6409
MDE-M 2HFB	UHFB	1690	360_{82}	333_{15}	240_{100}	169_{42}	69_{39}	1189	6512
2,3-MDMA-M (nor-)		1470	179_{34}	164_{13}	135_{96}	77_{100}	51_{91}	170	5420
2,3-MDMA-M (nor-)		1470	179_3	135_7	77_{11}	51_8	44_{100}	170	5513
2,3-MDMA-M (nor-) AC		1770	221_{31}	162_{100}	135_{35}	86_8	77_{40}	276	5589
2,3-MDMA-M (nor-) AC		1770	221_{18}	162_{71}	135_{35}	77_{64}	44_{100}	276	6310
2,3-MDMA-M (nor-) HFB		1595	375_{12}	240_{42}	169_{11}	162_{100}	135_{61}	935	5502
2,3-MDMA-M (nor-) PFP		1545	325_8	190_{42}	162_{100}	135_{71}	119_{46}	733	5542
2,3-MDMA-M (nor-) TFA		1585	275_{32}	162_{100}	140_{56}	135_{77}	77_{26}	488	5503
2,3-MDMA-M (nor-) TMS		1655	251_1	236_7	135_{19}	116_{100}	73_{71}	385	5590
2,3-MDMA-M (nor-) formyl artifact		1490	191_{11}	135_6	105_4	77_{11}	56_{100}	195	5421
MDMA	G P-I	1790	193_4	177_3	135_{19}	77_{10}	58_{100}	201	2599
MDMA AC	U+UHYAC	2140	235_1	162_{20}	100_{14}	77_7	58_{100}	324	2600
MDMA HFB	PHFB	1740	389_7	254_{100}	210_{43}	162_{82}	135_{56}	982	5086
MDMA PFP		1750	339_{23}	204_{100}	162_{64}	135_{38}	119_{17}	801	2601
MDMA TFA		1720	289_5	162_{75}	154_{100}	135_{67}	110_{44}	557	5079
MDMA TMS		1710	250_7	135_{16}	130_{100}	77_{15}	73_{67}	445	4562
MDMA intermediate		2025	207_{35}	160_{50}	131_{11}	103_{100}	77_{69}	232	2842
MDMA precursor-1 (piperonal)		1160*	150_{80}	149_{100}	121_{34}	91_{12}	63_{49}	127	3275
MDMA precursor-2 (isosafrole)		1215*	162_{100}	131_{51}	104_{71}	77_{53}	63_{28}	141	3276
MDMA precursor-3 (piperonylacetone)		1365*	178_{20}	135_{100}	105_7	77_{41}	51_{38}	167	3274
MDMA R-(-)-enantiomer HFBP		2450	486_1	351_5	266_{100}	162_{31}	135_4	1153	6642
MDMA S-(+)-enantiomer HFBP		2460	486_1	351_5	266_{100}	162_{60}	135_7	1153	6643
MDMA-D5		1770	198_1	136_5	78_4	62_{100}		216	6356
MDMA-D5 AC		2130	240_2	164_{35}	136_9	104_{26}	62_{100}	345	6355
MDMA-D5 HFB		1750	394_{13}	258_{100}	213_{32}	164_{45}	136_{27}	998	6359
MDMA-D5 PFP		1740	344_{15}	208_{100}	163_{62}	136_{33}	119_{10}	824	6358
MDMA-D5 TFA		1700	294_{13}	164_{49}	158_{100}	136_{55}	113_{28}	585	6357
MDMA-D5 TMS		1700	255_6	134_{100}	73_{30}			471	6360
MDMA-D5 R-(-)-enantiomer HFBP		2445	491_1	355_6	266_{100}	164_{38}		1158	6800
MDMA-D5 S-(+)-enantiomer HFBP		2455	491_3	355_{11}	266_{100}	164_{43}		1158	6801
MDMA-M	UHY	1510*	180_{19}	137_{100}	122_{19}	107_2	94_{16}	175	4247
MDMA-M	UHY	1465	181_9	138_{100}	122_{18}	94_{24}	77_{16}	178	4351
MDMA-M (deamino-HO-) AC	U+UHYAC	1620*	222_9	162_{100}	135_{56}	104_{10}	77_{17}	281	6410
MDMA-M (deamino-oxo-)		1365*	178_{20}	135_{100}	105_7	77_{41}	51_{38}	167	3274
MDMA-M (demethylenyl-) 3AC	U+UHYAC	2190	307_1	234_7	150_{12}	100_{59}	58_{100}	649	4244
MDMA-M (demethylenyl-methyl-)	UHY	1810	195_1	137_4	122_2	94_3	58_{100}	208	4246
MDMA-M (demethylenyl-methyl-) 2HFB	UHFB	1760	360_{28}	333_6	254_{100}	210_{23}	169_{18}	1193	6492
MDMA-M (demethylenyl-methyl-) iso-1 2AC	U+UHYAC	2095	279_1	206_{15}	164_{23}	100_{39}	58_{90}	512	6757
MDMA-M (demethylenyl-methyl-) iso-2 2AC	U+UHYAC	2115	279_1	206_{19}	164_{26}	100_{48}	58_{100}	512	4243
MDMA-M (HO-methoxy-hippuric acid) ME	UHYME	2165	239_9	151_{100}				339	4213
MDMA-M (methylenedioxy-hippuric acid) ME	UME UHYME	2065	237_{16}	178_4	149_{100}	121_{15}	65_{13}	332	4212

MDMA-M (nor-)   Table 1-8-1: Compounds in order of names

Name	Detected	RI	Typical ions and intensities					Page	Entry
MDMA-M (nor-)	U UHY	1495	179_1	136_{18}	105_1	77_5		170	3241
MDMA-M (nor-)	U UHY	1495	179_1	136_{18}	77_5	51_5	44_{100}	171	5518
MDMA-M (nor-) AC	U+UHYAC	1860	221_3	162_{46}	135_{12}	86_8	77_{12}	277	3263
MDMA-M (nor-) AC	U+UHYAC	1860	221_3	162_{46}	135_{12}	86_8	44_{100}	277	5519
MDMA-M (nor-) HFB	UHFB	1650	375_4	240_6	169_{10}	162_{50}	135_{100}	935	5291
MDMA-M (nor-) PFP	UPFP	1605	325_5	190_{10}	162_{46}	135_{100}	119_{18}	733	5290
MDMA-M (nor-) TFA	UTFA	1615	275_4	162_{39}	135_{100}	105_6	77_{12}	488	5289
MDMA-M (nor-) TMS		1735	251_1	236_4	135_8	116_{100}	73_{47}	385	6334
MDMA-M (nor-) R-(-)-enantiomer HFBP		2280	472_2	294_{15}	266_{100}	162_{66}	135_8	1140	6640
MDMA-M (nor-) S-(+)-enantiomer HFBP		2290	472_2	294_{16}	266_{90}	162_{69}	135_9	1140	6641
MDMA-M (nor-)-D5 AC		1840	226_{13}	167_{100}	166_{93}	136_{48}	78_{28}	295	5688
MDMA-M (nor-)-D5 2AC		1910	268_3	167_{100}	166_{81}	136_{47}	90_{59}	462	5689
MDMA-M (nor-demethylenyl-) 3AC	U+UHYAC	2150	293_2	234_{54}	192_{48}	150_{99}	86_{100}	579	3725
MDMA-M (nor-demethylenyl-methyl-) 2AC	U+UHYAC	2065	265_3	206_{27}	164_{100}	137_{23}	86_{33}	445	3498
MDMA-M (nor-demethylenyl-methyl-) 2HFB	UHFB	1690	360_{82}	333_{15}	240_{100}	169_{42}	69_{39}	1189	6512
MDMA-M AC	U+UHYAC	1600*	222_2	180_{22}	137_{100}			280	4211
MDMA-M ME	UHYME	1540*	194_{25}	151_{100}	135_4	107_{18}	65_4	204	4353
MDMA-M ME	UHYME	1550	195_1	152_{100}	137_{17}	107_{16}	77_{14}	208	4352
MDMA-M 2AC	U+UHYAC	1735*	250_3	208_{15}	166_{45}	123_{100}		379	4210
MDMA-M 2AC	U+UHYAC	1820	266_3	206_9	164_{100}	150_{10}	137_{30}	450	6409
MDPPP		1995	178_1	149_2	121_2	98_{100}	56_{13}	367	5422
MDPPP-M (4-HO-3-methoxy-benzoic acid) 2ET		1675*	224_{44}	196_{17}	179_{68}	151_{100}		288	6531
MDPPP-M (deamino-oxo-)		1525*	192_{24}	149_{100}	121_{24}	65_{10}		197	6529
MDPPP-M (demethylene-) 2ET		2165	290_1	193_1	165_2	137_3	98_{100}	571	6525
MDPPP-M (demethylene-deamino-oxo-) 2ET		1720*	236_4	193_{100}	165_{48}	137_{45}	109_{12}	328	6530
MDPPP-M (demethylene-methyl-) ET		2135	277_1	208_1	179_2	151_7	98_{100}	502	6524
MDPPP-M (demethylene-methyl-) HFB		1960	445_1	347_{100}	69_{100}			1107	6532
MDPPP-M (demethylene-methyl-) ME		2070	165_1	98_{100}	79_3	69_4	56_{11}	436	6538
MDPPP-M (demethylene-methyl-) TMS		1960	321_1	306_3	223_3	165_1	98_{100}	719	6533
MDPPP-M (demethylene-methyl-deamino-oxo-) ET		1680*	222_4	179_{100}	151_{81}	123_{19}		281	6523
MDPPP-M (demethylene-methyl-oxo-) ET		2290	290_1	208_{12}	179_{19}	151_{11}	112_{100}	569	6527
MDPPP-M (demethylene-oxo-) 2ET		2325	305_1	222_3	193_4	151_{11}	112_{100}	640	6526
MDPPP-M (dihydro-)		2040	248_1	149_3	121_2	98_{100}	56_{19}	376	6698
MDPPP-M (dihydro-) AC		2065	232_3	149_6	121_2	98_{100}	56_{15}	569	6703
MDPPP-M (dihydro-) TMS		1965	306_2	232_1	149_3	98_{100}	73_{24}	719	6708
MDPPP-M (oxo-)		2290	261_2	178_{26}	149_{19}	121_{10}	112_{100}	425	6528
Mebendazole ME		2950	309_{100}	250_{50}	232_{96}	200_{42}	77_{67}	659	7540
Mebendazole artifact (amine) 3ME		2930	279_{100}	264_{70}	249_{22}	173_{15}	77_{23}	511	7543
Mebendazole artifact (amine) iso-1 2ME		2930	265_{100}	250_{10}	188_{58}	160_{11}	77_{10}	444	7542
Mebendazole artifact (amine) iso-2 2ME		2950	265_{100}	250_{12}	188_{60}	160_{24}	77_{30}	444	7545
Mebendazole iso-1 2ME		2785	323_{62}	264_{100}	246_{16}	159_{13}	77_{35}	726	7544
Mebendazole iso-2 2ME		2930	323_{62}	264_{100}	246_{16}	159_{13}	77_{35}	726	7541
Mebeverine		3045	428_1	308_{100}	165_{18}	121_7	84_2	1079	4404
Mebeverine-M (3,4-dihydroxybenzoic acid) ME2AC		1750*	252_1	210_{19}	168_{100}	137_{60}	109_8	388	5254
Mebeverine-M (HO-phenyl-alcohol) 2AC		2415	186_{100}	165_4	137_6	98_3	72_4	902	5326
Mebeverine-M (HO-phenyl-O-demethyl-alcohol) 3AC		2525	193_2	186_{100}	151_4	123_4	72_4	996	5327
Mebeverine-M (isovanillic acid) MEAC		1630*	224_6	193_2	182_{92}	151_{100}	123_{13}	287	5329
Mebeverine-M (N-dealkyl-)		1660	192_1	149_1	121_5	91_2	72_{100}	202	5831
Mebeverine-M (N-dealkyl-) AC	U+UHYAC	1855	235_1	148_{32}	121_7	114_{26}	72_{100}	326	5322
Mebeverine-M (N-dealkyl-) HFB	UHFB	1785	389_2	268_{100}	240_{46}	148_{62}	121_{43}	983	5834
Mebeverine-M (N-dealkyl-) ME		1780	206_1	121_5	86_{100}	72_3	58_{20}	235	5835
Mebeverine-M (N-dealkyl-) PFP	UPFP	1765	339_1	218_{100}	190_{44}	148_{59}	121_{49}	802	5833
Mebeverine-M (N-dealkyl-) TFA	UTFA	1775	289_1	168_{100}	148_{63}	140_{41}	121_{62}	557	5832
Mebeverine-M (N-dealkyl-) TMS		2065	264_1	250_{14}	144_{100}	121_{17}	73_{84}	447	5836
Mebeverine-M (N-dealkyl-N-deethyl-) AC	U+UHYAC	1720	207_1	148_{41}	121_{16}	91_5	86_{10}	235	3265
Mebeverine-M (N-dealkyl-N-deethyl-) AC	U+UHYAC	1720	207_1	148_{41}	121_{16}	86_{10}	44_{100}	234	5537
Mebeverine-M (N-dealkyl-O-demethyl-) 2AC	U+UHYAC	1995	263_1	176_9	134_{15}	114_{44}	72_{100}	435	5323
Mebeverine-M (N-deethyl-alcohol) 2AC	U+UHYAC	2390	321_1	200_{25}	158_{100}	148_{68}	98_{65}	720	5321
Mebeverine-M (N-deethyl-O-demethyl-alcohol) 3AC	U+UHYAC	2535	200_{37}	158_{100}	134_{21}	107_{20}	98_{44}	843	5320
Mebeverine-M (O-demethyl-alcohol) AC		2245	186_{100}	135_3	107_{17}	72_8		583	5325
Mebeverine-M (O-demethyl-alcohol) 2AC		2305	234_3	186_{100}	135_4	107_9	72_7	785	5324
Mebeverine-M (vanillic acid) ME		1455*	182_{50}	151_{100}	123_{14}	108_{16}	77_{16}	180	5216
Mebeverine-M (vanillic acid) MEAC		1640*	224_6	193_2	182_{96}	151_{100}	123_{11}	287	5328
Mebeverine-M/artifact (alcohol)		2110	264_1	144_{100}	121_{15}	72_{52}	55_{19}	447	4405
Mebeverine-M/artifact (alcohol) AC	U+UHYAC	2210	292_1	186_{100}	121_{12}	72_{16}	55_{20}	652	4406
Mebeverine-M/artifact (veratric acid)	P UHY	1730*	182_{100}	167_{29}	121_{20}	111_{29}	77_{38}	179	4407
Mebeverine-M/artifact (veratric acid) ME	UHYME	1585*	196_{97}	181_{100}	165_{100}	137_{15}	137_{15}	211	4408
Mebhydroline	U UHY U+UHYAC	2445	276_{43}	233_{89}	115_{22}	91_{100}	65_{29}	497	1667
Mebhydroline-M (nor-) AC	U+UHYAC	2820	304_{75}	232_{35}	213_{28}	91_{100}	65_{20}	635	1668
Mebhydroline-M (nor-HO-) 2AC	U+UHYAC	3130	362_{56}	320_{35}	249_{49}	187_{12}	91_{100}	891	1669
MECC		<1000	138_3	137_{10}	123_{13}	95_{100}	82_{15}	118	3593
MECC -HCN		<1000	111_{25}	91_{13}	82_{27}	68_{100}	55_{67}	103	3597
Meclofenamic acid		2350	295_{32}	242_{100}	214_{13}	179_{11}	151_8	588	5768
Meclofenamic acid AC		2320	337_{33}	295_{13}	242_{100}	214_{14}	180_{10}	790	5769

Table 1-8-1: Compounds in order of names Mefenorex-M (AM) HFB

Name	Detected	RI	Typical ions and intensities					Page	Entry
Meclofenamic acid ME		2240	309_{36}	277_{15}	242_{100}	214_{14}	179_{11}	658	5701
Meclofenamic acid TMS		2750	367_{22}	352_6	277_{14}	242_{100}	214_9	908	5703
Meclofenamic acid 2ME		2275	323_{24}	277_{10}	242_{100}	214_{10}	180_7	725	5702
Meclofenamic acid -CO2		2035	251_{34}	216_{13}	181_{100}	152_9	77_{13}	382	5767
Meclofenoxate	G	1790	257_5	141_7	111_{22}	71_{37}	58_{100}	409	1076
Meclofenoxate-M (HOOC-)		1770*	186_{100}	141_{93}	128_{74}	111_{87}	99_{50}	187	1881
Meclofenoxate-M (HOOC-) ME		1510*	200_{98}	141_{100}	111_{59}	99_{17}	75_{38}	218	1077
Mecloxamine	G	2180	317_1	215_1	179_3	165_4	72_{100}	698	1078
Mecloxamine artifact	G U+UHYAC	1700*	214_{49}	179_{100}	152_9	139_3	89_{14}	254	1217
Mecloxamine HY	UHY	1750*	232_{12}	217_{80}	139_{81}	105_{75}	77_{100}	312	1079
Mecloxamine HYAC	U+UHYAC	2180*	274_{18}	232_{75}	197_{14}	139_{35}	121_{100}	485	2185
Mecloxamine-M (HO-) -H2O HY	U UHY	2050*	230_{100}	215_{27}	195_{60}	177_{33}	165_{64}	306	2187
Mecloxamine-M (HO-) iso-1 -H2O HYAC	U+UHYAC	2030*	272_{27}	230_{100}	195_{34}	165_{56}	152_{10}	477	2184
Mecloxamine-M (HO-) iso-2 -H2O HYAC	U+UHYAC	2090*	272_{16}	230_{100}	215_{15}	195_{34}	165_{47}	478	2189
Mecloxamine-M (HO-methoxy-) -H2O HYAC	U+UHYAC	2210*	302_{10}	260_{100}	182_{10}	152_{16}	75_4	623	2186
Mecloxamine-M (HO-methoxy-carbinol) -H2O	U UHY	2220*	262_{36}	260_{100}				420	2194
Mecloxamine-M (nor-)	U	2440	303_1	233_7	179_4	165_7	58_{100}	628	2192
Mecloxamine-M (nor-) AC	U	2580	310_{10}	215_{19}	130_{29}	100_{67}	58_{100}	829	2193
Meclozine	G	3040	390_{13}	285_{12}	189_{89}	105_{100}		985	1080
Meclozine artifact		1900*	232_{61}	201_{63}	165_{65}	105_{100}	77_{55}	311	1344
Meclozine artifact	G U+UHYAC	1600*	202_{30}	167_{100}	165_{52}	152_{17}	125_7	222	2442
Meclozine-M (carbinol)	UHY	1750*	218_{17}	183_7	139_{39}	105_{100}	77_{87}	263	2239
Meclozine-M (carbinol) AC	U+UHYAC	1890*	260_8	200_{40}	165_{100}	139_{10}	77_{29}	420	1270
Meclozine-M (Cl-benzophenone)	U+UHYAC	1850*	216_{43}	139_{58}	105_{100}	77_{44}		258	1343
Meclozine-M (HO-Cl-benzophenone)	UHY	2300*	232_{36}	197_7	139_{23}	121_{100}	111_{23}	311	2240
Meclozine-M (HO-Cl-BPH) iso-1 AC	U+UHYAC	2200*	274_{18}	232_{75}	139_{100}	121_{44}	111_{51}	484	2229
Meclozine-M (HO-Cl-BPH) iso-2 AC	U+UHYAC	2230*	274_7	232_{43}	139_{25}	121_{100}	111_{27}	484	2230
Meclozine-M (N-dealkyl-)	UHY	2520	286_{13}	241_{48}	201_{50}	165_{65}	56_{100}	543	2241
Meclozine-M (N-dealkyl-) AC	U+UHYAC	2620	328_7	242_{100}	201_{48}	165_{60}	85_{100}	749	1271
Meclozine-M AC	U+UHYAC	2380	299_4	285_{61}	226_{100}	191_{58}	84_{45}	609	821
Meclozine-M/artifact	P-I U+UHYAC UME	2220	300_{17}	228_{38}	165_{52}	99_{100}	56_{63}	616	670
Meclozine-M/artifact AC		2010	232_7	160_{26}	146_{17}	105_{100}	85_{32}	313	2444
Meconin	U+UHYAC	1780*	194_{92}	176_{52}	165_{100}	147_{69}	77_{29}	204	2326
Mecoprop	U	1540*	214_{38}	169_{22}	142_{100}	107_{36}	77_{31}	254	1081
Mecoprop ME		1500*	228_{40}	169_{81}	142_{64}	107_{58}	77_{32}	299	2268
Medazepam	G P-I U+UHYAC-I	2235	270_{38}	242_{98}	207_{100}	165_{24}		469	292
Medazepam-M	P G U	2520	270_{86}	269_{97}	242_{100}	241_{82}	77_{17}	468	463
Medazepam-M (nor-)	P-I U UHY	2280	256_{73}	228_{100}	193_{85}	165_{41}		405	293
Medazepam-M (nor-) AC	U+UHYAC	2470	298_{42}	297_{80}	256_{51}	228_{101}	193_{52}	605	1324
Medazepam-M (nor-HO-) 2AC	U+UHYAC	2590	356_{59}	355_{100}	314_{44}	271_{38}	244_{77}	868	3046
Medazepam-M (oxo-)	P G U	2430	284_{81}	283_{91}	256_{100}	221_{31}	77_7	532	481
Medazepam-M (oxo-) HY	UHY U+UHYAC	2100	245_{95}	228_{38}	193_{29}	105_{38}	77_{100}	358	272
Medazepam-M (oxo-) HYAC	U+UHYAC	2260	287_{11}	244_{100}	228_{39}	182_{49}	77_{70}	546	2542
Medazepam-M TMS		2300	342_{62}	341_{100}	327_{19}	269_4	73_{30}	817	4573
Medazepam-M HY	UHY	2050	231_{80}	230_{95}	154_{23}	105_{38}	77_{100}	308	419
Medazepam-M HYAC	U+UHYAC PHYAC	2245	273_{31}	230_{100}	154_{13}	105_{23}	77_{50}	480	273
Medroxyprogesterone AC		3050*	386_2	344_{76}	301_{29}	283_{100}	243_{13}	974	2803
Medroxyprogesterone -H2O		3010*	326_{100}	311_{35}	283_{70}	138_{54}	91_{65}	742	2802
Medrylamine	G U	2230	285_1	257_2	213_{13}	73_{41}	58_{100}	540	2423
Medrylamine HY	UHY	1930*	214_{34}	135_{46}	109_{100}	105_{56}	77_{71}	255	2426
Medrylamine HYAC	U+UHYAC	1980*	256_{13}	196_{100}	181_{46}	153_{46}	77_{26}	406	2424
Medrylamine-M (HO-benzophenone) AC	U+UHYAC	2050*	240_{20}	198_{100}	121_{94}	105_{41}	77_{51}	343	2197
Medrylamine-M (methoxy-benzophenone)	UHY U+UHYAC	1930*	212_{30}	135_{100}	105_{16}	92_{16}	77_{47}	249	2431
Medrylamine-M (nor-) AC	U	2450	313_1	213_{40}	197_{60}	101_{100}	86_{51}	681	2430
Medrylamine-M (O-demethyl-) HY2AC	U+UHYAC	2090*	284_5	242_{15}	224_{17}	182_{100}	153_{19}	533	2425
Mefenamic acid		2195	241_{96}	223_{100}	208_{69}	194_{33}	180_{36}	347	5189
Mefenamic acid ET		2160	269_{85}	223_{100}	208_{46}	194_{24}	180_{29}	465	5192
Mefenamic acid ME		2115	255_{69}	223_{100}	208_{70}	194_{33}	180_{42}	402	5190
Mefenamic acid MEAC		2260	297_{25}	255_{79}	223_{100}	208_{40}	194_{36}	602	5193
Mefenamic acid TMS		1980	313_{48}	223_{100}	208_{57}	180_{37}	73_{37}	681	5495
Mefenamic acid 2ME		2065	269_{100}	238_{25}	222_{45}	208_{24}	194_{44}	465	5191
Mefenamic acid-M (HO-) ME	UME	2400	271_{63}	224_{20}	209_{100}	194_{15}	180_{25}	474	6300
Mefenamic acid-M (HO-) 2ME	UME	2345	285_{100}	252_{54}	238_{79}	224_{72}	210_{44}	539	6301
Mefenorex	U UHY	1575	196_1	120_{100}	91_{20}	84_3	58_6	246	1719
Mefenorex AC		1935	253_1	162_{40}	120_{100}	91_{22}	84_{11}	394	1083
Mefenorex HFB		1735	407_1	316_{100}	240_{61}	118_{55}	91_{84}	1033	5063
Mefenorex PFP		1710	266_{100}	204_6	190_{57}	118_{45}	91_{51}	872	5064
Mefenorex TFA		1715	216_{100}	154_5	140_{51}	118_{41}	91_{39}	647	5065
Mefenorex -HCl	U UHY	1190	174_1	160_1	91_{21}	84_{100}	56_{71}	162	1082
Mefenorex-M (AM)		1160	134_1	120_1	91_6	77_1	65_4	115	54
Mefenorex-M (AM)	U	1160	134_1	120_1	91_6	65_4	44_{100}	115	5514
Mefenorex-M (AM) AC	U+UHYAC	1505	177_1	118_{19}	91_{11}	86_{31}	65_5	165	55
Mefenorex-M (AM) AC	UAAC U+UHYAC	1505	177_1	118_{19}	91_{11}	86_{31}	44_{100}	165	5515
Mefenorex-M (AM) HFB		1355	331_1	240_{79}	169_{21}	118_{100}	91_{53}	762	5047

Mefenorex-M (AM) PFP   Table 1-8-1: Compounds in order of names

Name	Detected	RI	Typical ions and intensities					Page	Entry
Mefenorex-M (AM) PFP		1330	281 $_1$	190 $_{73}$	118 $_{100}$	91 $_{36}$	65 $_{12}$	521	4379
Mefenorex-M (AM) TFA		1095	231 $_1$	140 $_{100}$	118 $_{92}$	91 $_{45}$	69 $_{19}$	309	4000
Mefenorex-M (AM) TMS		1190	192 $_6$	116 $_{100}$	100 $_{10}$	91 $_{11}$	73 $_{87}$	235	5581
Mefenorex-M (AM) formyl artifact		1100	147 $_2$	146 $_6$	125 $_5$	91 $_{12}$	56 $_{100}$	125	3261
Mefenorex-M (HO-) -HCl	UHY	1590	190 $_1$	133 $_1$	107 $_9$	84 $_{100}$	56 $_{31}$	196	1725
Mefenorex-M (HO-) -HCl AC	U+UHYAC	1630	232 $_1$	218 $_2$	176 $_5$	107 $_9$	84 $_{100}$	316	1729
Mefenorex-M (HO-) iso-1 2AC	U+UHYAC	2115	311 $_1$	162 $_{39}$	120 $_{100}$	107 $_{24}$	58 $_{47}$	671	1731
Mefenorex-M (HO-methoxy-)	UHY U+UHYAC	2145	256 $_1$	220 $_1$	137 $_{10}$	120 $_{100}$	84 $_{59}$	410	1728
Mefenorex-M (HO-methoxy-) AC	U+UHYAC	2360	298 $_1$	257 $_1$	162 $_{33}$	137 $_9$	120 $_{100}$	611	1727
Mefenorex-M (HO-methoxy-) -HCl	UHY	1775	220 $_1$	137 $_5$	120 $_6$	84 $_{100}$	56 $_{21}$	278	1726
Mefenorex-M-D5 AC	UAAC U+UHYAC	1480	182 $_3$	122 $_{46}$	92 $_{30}$	90 $_{100}$	66 $_{16}$	181	5690
Mefenorex-M-D5 HFB		1330	244 $_{100}$	169 $_{14}$	122 $_{46}$	92 $_{41}$	69 $_{40}$	787	6316
Mefenorex-M-D5 PFP		1320	194 $_{100}$	123 $_{42}$	119 $_{32}$	92 $_{46}$	69 $_{11}$	543	5566
Mefenorex-M-D5 TFA		1085	144 $_{100}$	123 $_{53}$	122 $_{56}$	92 $_{51}$	69 $_{28}$	330	5570
Mefenorex-M-D5 TMS		1180	212 $_1$	197 $_8$	120 $_{40}$	92 $_{11}$	73 $_{57}$	251	5582
Mefenorex-M-D11 PFP		1610	292 $_1$	194 $_{100}$	128 $_{82}$	98 $_{43}$	70 $_{14}$	575	7284
Mefenorex-M-D11 TFA		1615	242 $_1$	144 $_{100}$	128 $_{82}$	98 $_{43}$	70 $_{14}$	353	7283
Mefexamide		2185	280 $_2$	263 $_2$	155 $_3$	99 $_4$	86 $_{100}$	519	1480
Mefloquine		2280	359 $_1$	246 $_2$	196 $_1$	84 $_{100}$	56 $_{19}$	946	3205
Mefloquine -H2O		2220	224 $_1$	196 $_1$	97 $_{100}$	69 $_{28}$		884	3206
Mefloquine -H2O AC		2420	402 $_{28}$	360 $_7$	318 $_9$	126 $_{50}$	84 $_{100}$	1023	3207
Mefruside ME		2880	353 $_1$	311 $_1$	204 $_1$	110 $_3$	85 $_{100}$	1004	3057
Mefruside 2ME	UME	2860	367 $_1$	325 $_2$	218 $_1$	110 $_3$	85 $_{100}$	1040	3056
Mefruside -SO2NH	UME	2150	260 $_1$	218 $_2$	175 $_3$	111 $_8$	85 $_{100}$	626	3058
Melatonin		2450	232 $_{17}$	173 $_{86}$	160 $_{100}$	145 $_{25}$	117 $_{25}$	312	5913
Melatonin HFB		2295	428 $_9$	369 $_{100}$	356 $_{12}$	159 $_{15}$	144 $_8$	1075	5921
Melatonin PFP		2240	378 $_8$	319 $_{100}$	172 $_{13}$	159 $_{35}$	144 $_{23}$	945	5916
Melatonin TFA		2260	328 $_{11}$	269 $_{100}$	256 $_{13}$	159 $_{29}$	144 $_{19}$	748	5914
Melatonin TMS		2610	304 $_{14}$	245 $_{48}$	232 $_{64}$	202 $_4$	73 $_{100}$	635	6033
Melatonin 2TFA		2070	424 $_{13}$	269 $_{100}$	256 $_{44}$	159 $_{53}$	144 $_{29}$	1067	5915
Melatonin 2TMS		2640	376 $_{11}$	361 $_4$	245 $_{30}$	232 $_{98}$	73 $_{80}$	940	6032
Melatonin artifact (deacetyl-) 2HFB		2295	582 $_{17}$	369 $_{100}$	356 $_{72}$	159 $_{74}$	69 $_{45}$	1192	5922
Melatonin artifact (deacetyl-) 2PFP		2030	482 $_{40}$	360 $_{12}$	319 $_{100}$	306 $_{76}$	159 $_{35}$	1149	5923
Melatonin artifact (deacetyl-) 2TFA		2020	382 $_{34}$	269 $_{100}$	256 $_{86}$	159 $_{87}$	144 $_{38}$	958	5924
Melatonin artifact-1 HFB		2065	410 $_{100}$	213 $_{31}$	186 $_{22}$	169 $_{16}$	69 $_{30}$	1041	5926
Melatonin artifact-1 PFP		2010	360 $_{100}$	213 $_{38}$	186 $_{34}$	119 $_{37}$	69 $_{29}$	884	5920
Melatonin artifact-1 TFA		1990	310 $_{100}$	213 $_{38}$	186 $_{31}$	170 $_{25}$	69 $_{47}$	666	5918
Melatonin artifact-2 HFB		2590	410 $_{28}$	241 $_{100}$	226 $_7$	198 $_6$	170 $_5$	1040	5925
Melatonin artifact-2 PFP		2540	360 $_{40}$	241 $_{100}$	198 $_8$	184 $_7$	121 $_{14}$	884	5917
Melatonin artifact-2 TFA		2525	310 $_{58}$	241 $_{100}$	198 $_6$	184 $_7$	169 $_7$	665	5919
Melitracene	G U UHY U+UHYAC	2285	291 $_1$	217 $_1$	202 $_1$	58 $_{100}$		572	356
Melitracene-M (nor-) AC	U+UHYAC	2760	319 $_7$	246 $_{100}$	231 $_{37}$	86 $_{25}$		711	1179
Melitracene-M (nor-HO-dihydro-) 2AC	U+UHYAC	3030	379 $_{39}$	265 $_{44}$	223 $_{100}$	114 $_{26}$		950	1180
Melitracene-M (ring)	U+UHYAC	1900*	208 $_{11}$	193 $_{100}$	178 $_{61}$			238	1178
Meloxicam artifact-1 AC		1745	211 $_3$	169 $_{23}$	152 $_{57}$	104 $_{92}$	76 $_{100}$	246	6077
Meloxicam artifact-2 AC		2065	253 $_{17}$	211 $_{100}$	152 $_{43}$	118 $_{46}$	77 $_{53}$	393	6076
Melperone	G P-I U+UHYAC	1890	263 $_2$	125 $_{58}$	112 $_{100}$			436	174
Melperone-M	U+UHYAC	1490*	180 $_{25}$	125 $_{49}$	123 $_{35}$	95 $_{17}$	56 $_{100}$	174	85
Melperone-M (dihydro-) AC	U+UHYAC	2050	307 $_2$	264 $_9$	246 $_4$	123 $_9$	112 $_{100}$	652	176
Melperone-M (dihydro-) -H2O	UHY U+UHYAC	1835	247 $_1$	228 $_1$	133 $_3$	112 $_{100}$		369	175
Melperone-M (dihydro-oxo-) -H2O	UHY U+UHYAC	2220	261 $_{15}$	148 $_{100}$	137 $_{23}$	126 $_{85}$	98 $_{81}$	427	6511
Melperone-M (HO-) -H2O	UHY U+UHYAC	1900	261 $_6$	125 $_{25}$	112 $_{100}$			427	552
Memantine	G U UHY	1250	179 $_{12}$	164 $_7$	122 $_{86}$	108 $_{100}$		172	1557
Memantine AC	U+UHYAC	1600	221 $_{100}$	164 $_{72}$	150 $_{69}$	122 $_{12}$	107 $_{19}$	280	1482
Memantine-M (4-HO-)	U UHY	1550	195 $_{18}$	138 $_{36}$	108 $_{100}$			209	1560
Memantine-M (7-HO-)	UHY	1540	195 $_{12}$	180 $_2$	122 $_{35}$	108 $_{100}$		209	1559
Memantine-M (deamino-HO-)	U UHY	1525*	180 $_{20}$	165 $_{18}$	123 $_{85}$	109 $_{90}$	71 $_{100}$	176	1558
Memantine-M (HO-) AC	U+UHYAC	1860	237 $_{100}$	204 $_{29}$	164 $_{68}$	150 $_{91}$	107 $_{59}$	335	1554
Memantine-M (HO-) 2AC	U+UHYAC	1995	279 $_{81}$	237 $_{30}$	219 $_{61}$	164 $_{80}$	150 $_{100}$	514	1555
Memantine-M (HO-methyl-)	U UHY	1570	195 $_{10}$	164 $_{55}$	138 $_{30}$	120 $_{20}$	108 $_{100}$	209	1561
Memantine-M (HO-methyl-) 2AC	U+UHYAC	2090	279 $_{100}$	206 $_{87}$	164 $_{27}$	150 $_{68}$		515	1556
Menthol	G	1225*	138 $_{21}$	123 $_{30}$	95 $_{74}$	81 $_{86}$	71 $_{100}$	137	1826
MeOPP		1880	192 $_{34}$	150 $_{100}$	135 $_{20}$	120 $_{28}$	92 $_9$	198	6622
MeOPP AC	U+UHYAC	2185	234 $_{81}$	162 $_{100}$	149 $_{35}$	134 $_{36}$	120 $_{35}$	321	6609
MeOPP HFB	U+UHYHFB	1965	388 $_{100}$	373 $_{34}$	191 $_{75}$	135 $_{74}$	120 $_{66}$	979	6617
MeOPP ME		1840	206 $_{100}$	191 $_{17}$	162 $_{14}$	135 $_{77}$	120 $_{63}$	232	6623
MeOPP TFA	U+UHYTFA	1940	288 $_{100}$	273 $_{36}$	191 $_{20}$	135 $_{22}$	120 $_{28}$	552	6612
MeOPP TMS		2070	264 $_{100}$	249 $_{29}$	101 $_{45}$	86 $_{46}$	73 $_{81}$	441	6884
MeOPP-M (4-aminophenol N-acetyl-) HFB	UHYHFB PHFB	1735	347 $_{24}$	305 $_{39}$	169 $_{13}$	108 $_{100}$	69 $_{31}$	833	5099
MeOPP-M (4-aminophenol N-acetyl-) TFA		1630	247 $_{11}$	205 $_{30}$	108 $_{100}$	80 $_{19}$	69 $_{34}$	365	5092
MeOPP-M (4-aminophenol)	UHY	1240	109 $_{99}$	80 $_{41}$	52 $_{90}$			102	826
MeOPP-M (4-aminophenol) 2AC	U+UHYAC PAC	1765	193 $_{10}$	151 $_{53}$	109 $_{100}$	80 $_{24}$		199	188
MeOPP-M (4-methoxyaniline) HFB	U+UHYHFB	1400	319 $_{100}$	304 $_8$	300 $_6$	150 $_6$	122 $_{78}$	706	6620
MeOPP-M (4-methoxyaniline) TFA	U+UHYTFA	1335	219 $_{100}$	204 $_{16}$	149 $_{11}$	122 $_{76}$	109 $_{19}$	269	6615

Table 1-8-1: Compounds in order of names   Mescaline TMS

Name	Detected	RI	Typical ions and intensities					Page	Entry
MeOPP-M (aminophenol) 2HFB	U+UHYHFB	1405	501_{25}	482_6	304_{81}	169_{100}	109_{83}	1164	6621
MeOPP-M (aminophenol) 2TFA	U+UHYTFA	1280	301_{44}	204_{42}	176_{14}	109_{44}	69_{90}	617	6616
MeOPP-M (deethylene-) 2AC	U+UHYAC	2120	250_7	191_{51}	165_4	149_{33}	136_{100}	381	6611
MeOPP-M (deethylene-) 2HFB	U+UHYHFB	1765	558_7	345_{13}	332_{28}	240_{19}	135_{100}	1187	6619
MeOPP-M (deethylene-) 2TFA	U+UHYTFA	1765	358_{20}	245_{19}	232_{48}	135_{100}	120_{33}	876	6614
MeOPP-M (methoxyaniline) AC	PME UME	1630	165_{59}	123_{74}	108_{100}	95_{10}	80_{20}	147	5046
MeOPP-M (O-demethyl-) 2AC	U+UHYAC	2350	262_{39}	220_{60}	177_{21}	148_{100}	135_{59}	430	6610
MeOPP-M (O-demethyl-) 2HFB	U+UHYHFB	1990	570_{83}	551_{16}	373_{100}	344_{46}	317_{19}	1189	6618
MeOPP-M (O-demethyl-) 2TFA	U+UHYTFA	1915	370_{77}	273_{85}	244_{51}	217_{29}	120_{64}	918	6613
Mephenesin	P-I G	1660*	182_{13}	133_6	108_{100}	91_{23}	77_{19}	180	2804
Mephenesin 2AC	U+UHYAC	1805*	266_3	159_{100}	108_{46}	91_{38}	57_{29}	451	2805
Mephenesin 2TMS		1755*	326_{12}	205_{16}	147_{42}	133_{48}	73_{100}	741	4563
Mephentermine		1235	148_2	133_1	91_7	72_{100}	56_5	144	3721
Mephentermine AC		1505	148_1	132_4	114_{40}	91_{10}	72_{100}	229	3722
Mephentermine TFA		1335	244_1	168_{100}	110_{86}	91_{21}	56_{27}	416	3727
Mephenytoin	P G U+UHYAC	1780	218_6	189_{93}	104_{100}			264	1084
Mephenytoin-M (HO-)	U UHY	2400	234_2	205_{100}	152_{15}	120_{92}	109_{19}	320	2926
Mephenytoin-M (HO-) iso-1 AC	U+UHYAC	2390	276_1	247_{15}	205_{100}	120_{46}	91_7	496	2924
Mephenytoin-M (HO-) iso-2 AC	U+UHYAC	2540	276_1	247_{84}	205_{100}	134_{26}	107_{15}	496	4191
Mephenytoin-M (HO-methoxy-)	U UHY	2380	264_{10}	235_{100}	150_{83}	135_{29}		439	2927
Mephenytoin-M (HO-methoxy-) 2AC	U+UHYAC	2630	348_1	306_2	264_6	191_{100}	120_{53}	838	2925
Mephenytoin-M (nor-)	U UHY U+UHYAC	1950	204_5	175_{74}	132_6	104_{100}	77_{31}	226	2928
Mephenytoin-M (nor-) AC	U+UHYAC	1900	246_{25}	175_{100}	144_{33}	104_{53}	77_{20}	362	2929
Mephenytoin-M (nor-HO-) 2AC	U+UHYAC	2495	304_4	262_{52}	191_{100}	160_7	120_{19}	635	4173
Mepindolol		2390	262_{18}	147_{100}	114_{18}	100_{18}	72_{40}	432	1358
Mepindolol TMS		2320	334_2	188_3	147_{100}	118_{11}	72_{36}	781	6171
Mepindolol TMSTFA		2455	430_{12}	284_{79}	146_{54}	129_{85}	73_{80}	1080	6169
Mepindolol 2AC		2750	346_2	286_{50}	184_{100}	140_{85}	98_{93}	832	1359
Mepindolol 2TMSTFA		2565	502_2	284_{64}	218_{14}	129_{73}	73_{100}	1165	6170
Mepindolol formyl artifact	U	2410	274_{77}	186_{17}	147_{100}	86_{36}	72_{16}	486	1722
Mepindolol -H2O AC		2680	286_{38}	184_{100}	140_{95}	98_{81}		544	1705
Mepivacaine	P G U+UHYAC	2075	246_1	176_1	120_1	98_{100}	70_{10}	364	1085
Mepivacaine TMS		1980	318_2	261_{10}	248_4	98_{100}	73_{21}	705	4564
Mepivacaine-M (HO-)	UHY	2410	262_2	98_{100}	96_8	70_{16}		432	1086
Mepivacaine-M (HO-) AC	U+UHYAC	2450	304_1	98_{100}	96_9	70_{20}		636	1087
Mepivacaine-M (HO-piperidyl-) AC	U+UHYAC	2590	304_8	156_5	129_{15}	114_{100}	86_{11}	636	2970
Mepivacaine-M (nor-) AC	U+UHYAC	2170	274_2	154_{58}	126_{69}	84_{100}		487	2968
Mepivacaine-M (oxo-)	U UHY U+UHYAC	2400	260_7	218_3	112_{100}			422	2969
Mepivacaine-M (oxo-HO-piperidyl-) AC	U+UHYAC	2630	318_{17}	258_1	170_{64}	128_{100}	111_{12}	704	3049
Meprobamate	P G U+UHYAC	1785	144_{19}	114_{27}	96_{33}	83_{88}	55_{99}	265	1088
Meprobamate artifact-1	P G U	1535*	84_{100}	56_{81}				96	1089
Meprobamate artifact-2	P U UHY U+UHYAC	1720*	173_2	101_9	84_{99}	56_{90}		160	580
Meptazinol		1920	233_7	107_7	98_{19}	84_{83}	58_{100}	317	3546
Meptazinol AC		1945	275_3	107_5	98_{33}	84_{89}	58_{100}	492	3549
Meptazinol HFB		1810	429_2	303_1	98_{34}	84_{85}	58_{100}	1078	6136
Meptazinol PFP		1655	379_{13}	253_6	98_{45}	84_{99}	58_{130}	948	6127
Meptazinol TFA		1795	329_{17}	203_6	98_{41}	84_{100}	58_{86}	754	6206
Meptazinol TMS		2005	305_{11}	98_{32}	84_{100}	73_{32}	58_{100}	642	6207
Meptazinol-M (nor-)		1995	219_{13}	159_{11}	107_{15}	84_{23}	70_{100}	271	3547
Meptazinol-M (nor-) 2AC		2395	303_{41}	159_{27}	126_{38}	87_{100}	70_{71}	630	3551
Meptazinol-M (oxo-)		2410	247_{89}	204_{27}	148_{100}	87_{61}	55_{62}	368	3548
Meptazinol-M (oxo-) AC		2350	289_{49}	204_{100}	176_{46}	148_{66}	87_{58}	559	3550
Mepyramine	G U	2220	285_4	215_8	121_{50}	78_{20}	58_{50}	540	1656
Mepyramine HY	UHY U+UHYAC	1690	208_1	163_1	137_4	71_{50}	58_{100}	239	1660
Mepyramine-M (N-dealkyl-)	U	2120	214_{61}	136_{11}	121_{100}	78_{32}		256	1657
Mepyramine-M (N-dealkyl-) AC	U+UHYAC	2150	256_{42}	214_{62}	163_{38}	107_{100}	78_{40}	407	1659
Mepyramine-M (N-demethoxybenzyl-)	U	1580	165_3	119_8	107_{12}	78_{26}	58_{100}	148	1658
Mequitazine	G U UHY U+UHYAC	2765	322_{32}	212_6	198_8	180_5	124_{99}	723	1483
Mequitazine-M (HO-sulfoxide) AC	U+UHYAC	3230	396_2	354_5	180_3	124_{100}	70_{11}	1006	1672
Mequitazine-M (ring)	P G U UHY U+UHYAC	2010	199_{100}	167_{45}				216	10
Mequitazine-M (sulfone)	U UHY U+UHYAC	3250	354_7	244_6	180_3	124_{100}	70_9	861	1671
Mequitazine-M (sulfoxide)	U UHY U+UHYAC	3120	338_8	321_5	198_{19}	124_{100}	70_{11}	798	1670
Mequitazine-M AC	U+UHYAC	2550	257_{23}	215_{100}	183_7			409	12
Mequitazine-M 2AC	U+UHYAC	2865	315_{54}	273_{34}	231_{100}	202_{11}		688	2618
Mercaptodimethur		1915	225_{12}	184_7	168_{100}	153_{54}	109_{20}	291	3450
Mercaptodimethur-M/artifact (decarbamoyl-)		1535*	168_{100}	153_{88}	109_{87}	91_{32}	77_{16}	154	3451
Mesalazine ME2AC	U+UHYAC	1890	251_{13}	209_{100}	177_{91}	135_{91}	107_{20}	383	4485
Mesalazine MEAC	U+UHYAC	1945	209_{45}	177_{46}	135_{100}	107_{24}		240	4486
Mescaline		1690	211_{24}	182_{100}	167_{55}	151_{15}	148_{13}	247	1090
Mescaline AC		2070	253_{16}	194_{100}	179_{53}	151_9	77_5	394	1484
Mescaline HFB		1865	407_{35}	226_4	194_{36}	181_{100}	69_{11}	1033	5066
Mescaline PFP		1835	357_{33}	194_{37}	181_{100}	151_{10}	119_{19}	872	5067
Mescaline TFA		1830	307_{23}	194_{46}	181_{100}	148_{18}	126_{10}	648	5068
Mescaline TMS		1895	283_1	268_9	181_9	102_{100}	73_{65}	530	4959

Name	Detected	RI	Typical ions and intensities					Page	Entry
Mescaline 2AC		2125	295 $_{15}$	194 $_{100}$	181 $_{55}$	179 $_{54}$	151 $_{10}$	592	6943
Mescaline 2TMS		2080	355 $_{1}$	340 $_{13}$	174 $_{100}$	100 $_{6}$	73 $_{28}$	867	5683
Mescaline formyl artifact		1700	223 $_{25}$	181 $_{100}$	167 $_{6}$	148 $_{10}$	77 $_{4}$	285	3244
Mescaline precursor (trimethoxyphenylacetonitrile)		1610	207 $_{85}$	192 $_{100}$	164 $_{39}$	124 $_{24}$	78 $_{45}$	233	3273
Mescaline-D9		1685	220 $_{26}$	191 $_{100}$	190 $_{57}$	173 $_{50}$	152 $_{11}$	275	6907
Mescaline-D9 AC		2065	262 $_{18}$	203 $_{100}$	190 $_{51}$	185 $_{51}$	157 $_{10}$	432	6944
Mescaline-D9 HFB		1855	416 $_{28}$	203 $_{40}$	190 $_{100}$	185 $_{23}$	157 $_{7}$	1054	6939
Mescaline-D9 PFP		1820	366 $_{30}$	203 $_{40}$	190 $_{100}$	185 $_{25}$	119 $_{11}$	905	6934
Mescaline-D9 TFA		1825	316 $_{29}$	203 $_{30}$	190 $_{100}$	185 $_{20}$	157 $_{6}$	694	6929
Mescaline-D9 TMS		1885	292 $_{1}$	277 $_{5}$	190 $_{9}$	102 $_{100}$	73 $_{39}$	577	6946
Mescaline-D9 2AC		2120	304 $_{9}$	203 $_{100}$	190 $_{44}$	185 $_{46}$	157 $_{8}$	636	6945
Mescaline-D9 2TMS		2070	364 $_{1}$	174 $_{100}$	73 $_{30}$	349 $_{4}$	190 $_{12}$	898	6947
Mescaline-D9 formyl artifact		1690	232 $_{34}$	203 $_{2}$	190 $_{100}$	152 $_{7}$	140 $_{4}$	314	6911
Mescaline-M (deamino-COOH) ME	UGLUCSPEMEAC	1840*	240 $_{80}$	225 $_{42}$	181 $_{100}$	148 $_{17}$	137 $_{13}$	344	7135
Mesoridazine	P-I G U+UHYAC	3330	386 $_{12}$	370 $_{4}$	126 $_{7}$	98 $_{100}$	70 $_{11}$	973	4484
Mesoridazine	G P U+UHYAC	3330	386 $_{12}$	126 $_{9}$	98 $_{100}$	70 $_{15}$	42 $_{42}$	973	2200
Mesoridazine-M (side chain sulfone)	G P-I U+UHYAC	3415	402 $_{13}$	290 $_{5}$	197 $_{6}$	98 $_{100}$	70 $_{10}$	1023	394
Mesterolone		2545*	304 $_{100}$	218 $_{90}$	200 $_{42}$			637	1091
Mesterolone enol 2TMS		2530*	448 $_{3}$	433 $_{8}$	157 $_{42}$	141 $_{100}$	73 $_{71}$	1111	3982
Mestranol		2630*	310 $_{35}$	242 $_{13}$	227 $_{100}$	174 $_{46}$	115 $_{48}$	668	2806
Mestranol AC		2690*	352 $_{62}$	242 $_{9}$	227 $_{100}$	173 $_{29}$	147 $_{22}$	856	2807
Mesulphen	U UHY U+UHYAC	2250*	244 $_{100}$	227 $_{6}$	211 $_{75}$	184 $_{11}$	121 $_{10}$	355	5377
Mesulphen-M (di-HO-) 2AC	UGLUCAC	2830*	360 $_{100}$	277 $_{25}$	258 $_{35}$	227 $_{33}$	184 $_{23}$	883	5387
Mesulphen-M (di-HOOC-) 2ME	UME	2805*	332 $_{74}$	304 $_{100}$	273 $_{22}$	214 $_{14}$	184 $_{14}$	767	5392
Mesulphen-M (di-HOOC-) -CO2 ME	UME	2380*	274 $_{100}$	243 $_{53}$	215 $_{61}$	171 $_{43}$	121 $_{27}$	483	5388
Mesulphen-M (HO-)	U UHY	2430*	260 $_{100}$	197 $_{17}$	184 $_{17}$	242 $_{12}$	227 $_{11}$	419	5378
Mesulphen-M (HO-) AC	UGLUCAC	2535*	302 $_{100}$	227 $_{35}$	198 $_{24}$	259 $_{13}$	242 $_{26}$	622	5379
Mesulphen-M (HO-aryl-sulfoxide)	U UHY	2585*	276 $_{100}$	243 $_{60}$	227 $_{7}$	211 $_{20}$	165 $_{32}$	494	5385
Mesulphen-M (HO-aryl-sulfoxide) ME	UGLUCME	2625*	290 $_{31}$	242 $_{97}$	211 $_{100}$	183 $_{48}$	139 $_{33}$	562	5386
Mesulphen-M (HO-di-sulfoxide)	UHY	2785*	292 $_{37}$	275 $_{100}$	258 $_{41}$	197 $_{48}$	184 $_{67}$	574	5383
Mesulphen-M (HO-di-sulfoxide) AC	UGLUCAC	2895*	334 $_{69}$	291 $_{100}$	275 $_{25}$	209 $_{24}$	184 $_{64}$	779	5384
Mesulphen-M (HO-HOOC-) MEAC	UMEAC	2825*	346 $_{100}$	303 $_{16}$	271 $_{21}$	227 $_{54}$	184 $_{23}$	830	5393
Mesulphen-M (HO-HOOC-di-sulfoxide) MEAC	UMEAC	3025*	378 $_{12}$	314 $_{35}$	303 $_{100}$	272 $_{61}$	255 $_{45}$	945	5395
Mesulphen-M (HO-HOOC-sulfoxide) MEAC	UMEAC	2995*	362 $_{23}$	314 $_{91}$	272 $_{100}$	255 $_{64}$	196 $_{30}$	890	5394
Mesulphen-M (HOOC-) ME	UME	2545*	288 $_{100}$	185 $_{17}$	214 $_{13}$	257 $_{40}$	229 $_{32}$	551	5389
Mesulphen-M (HOOC-) -CO2	U UHY U+UHYAC	2235*	230 $_{100}$	197 $_{82}$	171 $_{13}$	152 $_{15}$		306	5396
Mesulphen-M (HOOC-di-sulfoxide) ME	UME	2895*	320 $_{100}$	304 $_{17}$	289 $_{42}$	272 $_{18}$	241 $_{6}$	714	5391
Mesulphen-M (HOOC-sulfoxide) ME	UME	2665*	304 $_{13}$	256 $_{100}$	225 $_{58}$	197 $_{35}$	152 $_{10}$	633	5390
Mesulphen-M (HO-sulfoxide)	UGLUC	2705*	276 $_{39}$	260 $_{13}$	228 $_{100}$	199 $_{50}$	184 $_{30}$	495	5381
Mesulphen-M (HO-sulfoxide) AC	UGLUCAC	2725*	318 $_{30}$	270 $_{85}$	228 $_{80}$	211 $_{100}$	184 $_{31}$	701	5382
Mesulphen-M (sulfoxide)	U	2400*	260 $_{35}$	244 $_{27}$	212 $_{100}$			419	5380
Mesuximide	P G U+UHYAC	1705	203 $_{44}$	118 $_{100}$	103 $_{23}$	91 $_{12}$	77 $_{20}$	224	1827
Mesuximide-M (di-HO-) 2AC	U+UHYAC	2260	319 $_{4}$	277 $_{23}$	235 $_{100}$	185 $_{21}$	150 $_{78}$	707	2920
Mesuximide-M (HO-)	U UHY	2220	219 $_{38}$	134 $_{100}$	119 $_{26}$	91 $_{11}$	65 $_{12}$	269	2915
Mesuximide-M (HO-) iso-1 AC	U+UHYAC	1960	261 $_{3}$	219 $_{38}$	134 $_{100}$	105 $_{5}$	77 $_{8}$	425	2916
Mesuximide-M (HO-) iso-2 AC	U+UHYAC	1995	261 $_{3}$	219 $_{47}$	134 $_{100}$	119 $_{18}$	77 $_{12}$	426	2917
Mesuximide-M (nor-)	P U+UHYAC	1750	189 $_{18}$	118 $_{100}$	103 $_{25}$	77 $_{22}$	58 $_{3}$	192	2914
Mesuximide-M (nor-) TMS		1730	261 $_{1}$	246 $_{13}$	146 $_{6}$	118 $_{100}$	103 $_{23}$	426	7423
Mesuximide-M (nor-HO-)	U UHY	2300	205 $_{37}$	134 $_{100}$	119 $_{35}$	91 $_{5}$	65 $_{11}$	228	2921
Mesuximide-M (nor-HO-) iso-1 AC	U+UHYAC	2120	247 $_{3}$	205 $_{45}$	134 $_{100}$	105 $_{5}$	94 $_{10}$	366	2918
Mesuximide-M (nor-HO-) iso-2 AC	U+UHYAC	2200	247 $_{3}$	205 $_{43}$	134 $_{100}$	119 $_{14}$	77 $_{8}$	366	2919
Metaclazepam	U+UHYAC	2640	392 $_{6}$	349 $_{100}$	347 $_{76}$	319 $_{16}$	163 $_{4}$	991	2144
Metaclazepam-M (amino-Br-Cl-benzophenone)	UHY	2270	311 $_{100}$	309 $_{71}$	276 $_{88}$	274 $_{90}$	195 $_{41}$	658	2151
Metaclazepam-M (amino-Br-Cl-benzophenone) AC	U+UHYAC	2500	353 $_{44}$	351 $_{32}$	311 $_{85}$	276 $_{100}$	274 $_{100}$	846	2149
Metaclazepam-M (amino-Br-Cl-HO-benzophenone) AC	U+UHYAC	2570	369 $_{27}$	367 $_{26}$	327 $_{83}$	325 $_{86}$	290 $_{29}$	907	7415
Metaclazepam-M (amino-Br-Cl-HO-benzophenone) 2AC	U+UHYAC	2685	409 $_{28}$	367 $_{50}$	325 $_{45}$	292 $_{96}$	290 $_{100}$	1037	2150
Metaclazepam-M (nor-)	U UHY U+UHYAC	2690	378 $_{22}$	349 $_{39}$	335 $_{100}$	333 $_{78}$	305 $_{14}$	945	2145
Metaclazepam-M (O-demethyl-)		2730	378 $_{26}$	347 $_{18}$	321 $_{100}$	319 $_{65}$	227 $_{13}$	945	2146
Metaclazepam-M (O-demethyl-) AC	U+UHYAC	2820	420 $_{11}$	337 $_{33}$	335 $_{100}$	333 $_{79}$	163 $_{13}$	1062	2147
Metaclazepam-M/artifact-1	U UHY U+UHYAC	2230	320 $_{100}$	318 $_{79}$	283 $_{47}$	239 $_{66}$	75 $_{34}$	701	2152
Metaclazepam-M/artifact-2	U UHY U+UHYAC	2250	334 $_{98}$	332 $_{79}$	297 $_{50}$	253 $_{100}$	75 $_{86}$	766	2153
Metaclazepam-M/artifact-3	U UHY U+UHYAC	2590	334 $_{35}$	299 $_{63}$	227 $_{100}$	163 $_{18}$	75 $_{22}$	778	2154
Metalaxyl		1890	279 $_{17}$	249 $_{61}$	206 $_{100}$	160 $_{68}$	130 $_{45}$	512	3452
Metaldehyde		1020*	131 $_{3}$	117 $_{8}$	89 $_{100}$	87 $_{19}$		164	1092
Metamfepramone	G U+UHYAC	1355	177 $_{1}$	105 $_{3}$	77 $_{12}$	72 $_{100}$	56 $_{6}$	166	1398
Metamfepramone iso-1 TMS		1470	249 $_{41}$	219 $_{11}$	176 $_{10}$	158 $_{22}$	73 $_{100}$	377	4565
Metamfepramone iso-2 TMS		1490	249 $_{43}$	219 $_{16}$	176 $_{11}$	158 $_{24}$	73 $_{100}$	377	4566
Metamfepramone-M (dihydro-)	G P U UHY	1430	161 $_{1}$	115 $_{1}$	105 $_{2}$	77 $_{6}$	72 $_{100}$	172	1113
Metamfepramone-M (dihydro-) AC	U+UHYAC	1495	162 $_{1}$	91 $_{2}$	72 $_{100}$	117 $_{2}$	105 $_{2}$	279	1114
Metamfepramone-M (dihydro-) TFA		1185	260 $_{1}$	162 $_{2}$	134 $_{4}$	91 $_{5}$	72 $_{100}$	489	4003
Metamfepramone-M (dihydro-) TMS		1485	251 $_{1}$	236 $_{1}$	163 $_{5}$	149 $_{5}$	72 $_{100}$	387	4568
Metamfepramone-M (HO-norephedrine) 3AC	U+UHYAC	2135	234 $_{6}$	165 $_{8}$	123 $_{11}$	86 $_{100}$	58 $_{45}$	580	4961
Metamfepramone-M (nor-)		1130	163 $_{1}$	148 $_{1}$	105 $_{3}$	77 $_{9}$	58 $_{60}$	143	5935
Metamfepramone-M (nor-) AC	U+UHYAC	1650	205 $_{1}$	105 $_{4}$	100 $_{39}$	77 $_{13}$	58 $_{80}$	228	5932

Table 1-8-1: Compounds in order of names   Metamfetamine-M (nor-3-HO-) 2HFB

Name	Detected	RI	Typical ions and intensities					Page	Entry
Metamfepramone-M (nor-) HFB		1440	359 $_1$	254 $_{65}$	210 $_{24}$	105 $_{100}$	77 $_{45}$	879	5936
Metamfepramone-M (nor-) PFP		1390	204 $_3$	105 $_5$	77 $_{10}$	58 $_{80}$		659	5934
Metamfepramone-M (nor-) TFA		1370	259 $_1$	154 $_{68}$	110 $_{31}$	105 $_{100}$	77 $_{49}$	416	5933
Metamfepramone-M (nor-) TMS		1570	220 $_5$	205 $_5$	130 $_{100}$	105 $_6$	73 $_{90}$	325	5937
Metamfepramone-M (nor-dihydro-)	G UHY	1375	146 $_1$	131 $_1$	105 $_3$	77 $_{12}$	58 $_{100}$	148	748
Metamfepramone-M (nor-dihydro-) TMSTFA		1620	318 $_1$	227 $_9$	179 $_{100}$	110 $_8$	73 $_{79}$	773	6038
Metamfepramone-M (nor-dihydro-) 2AC	PAC U+UHYAC	1795	249 $_1$	100 $_{57}$	58 $_{100}$	148 $_2$	117 $_2$	375	749
Metamfepramone-M (nor-dihydro-) 2HFB	UHYHFB	1500	344 $_{10}$	254 $_{100}$	210 $_{27}$	169 $_{18}$	69 $_{36}$	1186	5097
Metamfepramone-M (nor-dihydro-) 2PFP		1370	338 $_1$	294 $_1$	204 $_{100}$	160 $_{25}$	119 $_{14}$	1123	2577
Metamfepramone-M (nor-dihydro-) 2TFA		1345	338 $_1$	244 $_4$	154 $_{100}$	110 $_{72}$	69 $_{47}$	872	3997
Metamfepramone-M (nor-dihydro-) 2TMS		1620	294 $_3$	163 $_4$	147 $_8$	130 $_{100}$	73 $_{84}$	663	4543
Metamfepramone-M (nor-dihydro-) formyl artifact	G U	1430	177 $_1$	121 $_{10}$	107 $_6$	71 $_{100}$	56 $_{76}$	165	4500
Metamfepramone-M (nor-dihydro-) -H2O AC	U+UHYAC	1560	189 $_1$	148 $_3$	121 $_8$	100 $_{49}$	58 $_{100}$	192	5646
Metamfepramone-M (norephedrine) TMSTFA		1890	240 $_8$	198 $_3$	179 $_{100}$	117 $_5$	73 $_{88}$	708	6146
Metamfepramone-M (norephedrine) 2AC	U+UHYAC	1805	235 $_1$	107 $_{13}$	86 $_{100}$	176 $_5$	134 $_7$	325	2476
Metamfepramone-M (norephedrine) 2HFB	UHYHFB	1455	543 $_1$	330 $_{14}$	240 $_{100}$	169 $_{44}$	69 $_{57}$	1183	5098
Metamfepramone-M (norephedrine) 2PFP	UHYPFP	1380	443 $_1$	280 $_9$	190 $_{100}$	119 $_{59}$	105 $_{26}$	1104	5094
Metamfepramone-M (norephedrine) 2TFA	UTFA	1355	343 $_1$	230 $_6$	203 $_5$	140 $_{100}$	69 $_{29}$	819	5091
Metamfepramone-M (nor-HO-) 2AC	U+UHYAC	1885	263 $_6$	100 $_{71}$	58 $_{150}$	250 $_4$	150 $_9$	434	4960
Metamfetamine	U	1195	148 $_1$	134 $_2$	115 $_1$	91 $_9$	58 $_{100}$	127	1093
Metamfetamine AC	U+UHYAC	1575	191 $_1$	117 $_2$	100 $_{42}$	91 $_6$	58 $_{100}$	196	1094
Metamfetamine HFB		1460	254 $_{100}$	210 $_{44}$	169 $_{15}$	118 $_{41}$	91 $_{38}$	827	5069
Metamfetamine PFP		1415	204 $_{100}$	160 $_{46}$	118 $_{35}$	91 $_{25}$	69 $_4$	589	5070
Metamfetamine TFA		1300	245 $_1$	154 $_{100}$	118 $_{48}$	110 $_{55}$	91 $_{23}$	358	3998
Metamfetamine TMS		1325	206 $_4$	130 $_{100}$	91 $_{17}$	73 $_{83}$	59 $_{13}$	279	6214
Metamfetamine R-(-)-enantiomer HFBP		2000	351 $_{50}$	294 $_4$	266 $_{100}$	169 $_{11}$	121 $_7$	1103	6516
Metamfetamine S-(+)-enantiomer HFBP		2120	351 $_{31}$	91 $_9$	118 $_6$	266 $_{100}$	294 $_4$	1103	6517
Metamfetamine-D5		1190	154 $_1$	139 $_2$	119 $_1$	92 $_9$	62 $_{100}$	134	7291
Metamfetamine-D5 HFB		1440	258 $_{100}$	213 $_{28}$	169 $_{10}$	120 $_{14}$	92 $_{13}$	845	6771
Metamfetamine-D5 TFA		1295	250 $_1$	158 $_{100}$	120 $_{22}$	113 $_{35}$	92 $_{16}$	381	7292
Metamfetamine-D5 TMS		1320	211 $_6$	134 $_{100}$	118 $_1$	92 $_6$	73 $_{33}$	296	7293
Metamfetamine-D5 R-(-)-enantiomer HFBP		2105	355 $_{56}$	294 $_4$	266 $_{100}$	169 $_8$	92 $_8$	1110	6520
Metamfetamine-D5 S-(+)-enantiomer HFBP		2105	355 $_{48}$	294 $_5$	266 $_{100}$	191 $_7$		1110	6521
Metamfetamine-M (4-HO-) ME		1475	178 $_1$	121 $_4$	91 $_2$	77 $_3$	58 $_{100}$	172	6719
Metamfetamine-M (4-HO-) MEAC		1820	221 $_1$	148 $_{24}$	121 $_7$	100 $_{22}$	58 $_{100}$	279	6720
Metamfetamine-M (4-HO-) MEHFB		1665	375 $_1$	254 $_{55}$	210 $_{32}$	148 $_{67}$	121 $_{100}$	935	6722
Metamfetamine-M (4-HO-) MEPFP		1510	325 $_3$	204 $_{60}$	160 $_{29}$	148 $_{48}$	121 $_{100}$	734	7601
Metamfetamine-M (4-HO-) METFA		1645	275 $_2$	154 $_{59}$	148 $_{51}$	121 $_{100}$	110 $_{31}$	489	6721
Metamfetamine-M (deamino-oxo-di-HO-) 2AC	U+UHYAC	1735*	250 $_3$	208 $_{15}$	166 $_{45}$	123 $_{100}$		379	4210
Metamfetamine-M (deamino-oxo-HO-methoxy-)	UHY	1510*	180 $_{19}$	137 $_{100}$	122 $_{19}$	107 $_2$	94 $_{16}$	175	4247
Metamfetamine-M (deamino-oxo-HO-methoxy-) AC	U+UHY	1600*	222 $_2$	180 $_{22}$	137 $_{100}$			280	4211
Metamfetamine-M (deamino-oxo-HO-methoxy-) ME	UHYME	1540*	194 $_{25}$	151 $_{100}$	135 $_4$	107 $_{18}$	65 $_4$	204	4353
Metamfetamine-M (di-HO-) 3AC	U+UHYAC	2190	307 $_1$	234 $_7$	150 $_{12}$	100 $_{59}$	58 $_{100}$	649	4244
Metamfetamine-M (HO-)		1885	150 $_1$	135 $_1$	107 $_5$	77 $_5$	58 $_{100}$	148	1766
Metamfetamine-M (HO-) TFA		1770	261 $_1$	154 $_{100}$	134 $_{68}$	110 $_{42}$	107 $_{41}$	425	6180
Metamfetamine-M (HO-) TMSTFA		1690	333 $_3$	206 $_{72}$	179 $_{100}$	154 $_{53}$	73 $_{50}$	773	6228
Metamfetamine-M (HO-) 2AC	U+UHYAC	1995	249 $_1$	176 $_6$	134 $_7$	100 $_{43}$	58 $_{100}$	376	1767
Metamfetamine-M (HO-) 2HFB		1670	538 $_1$	330 $_{17}$	254 $_{100}$	210 $_{32}$	169 $_{22}$	1186	5076
Metamfetamine-M (HO-) 2PFP		1605	295 $_1$	280 $_{18}$	204 $_{100}$	160 $_{47}$	119 $_{39}$	1123	5077
Metamfetamine-M (HO-) 2TFA		1585	357 $_1$	230 $_{22}$	154 $_{100}$	110 $_{42}$	69 $_{29}$	872	5078
Metamfetamine-M (HO-) 2TMS		1620	309 $_{10}$	206 $_{70}$	179 $_{100}$	154 $_{32}$	73 $_{40}$	663	6190
Metamfetamine-M (HO-methoxy-)	UHY	1810	195 $_1$	137 $_4$	122 $_2$	94 $_3$	58 $_{100}$	208	4246
Metamfetamine-M (HO-methoxy-) 2HFB	UHFB	1760	360 $_{28}$	333 $_6$	254 $_{100}$	210 $_{23}$	169 $_{18}$	1193	6492
Metamfetamine-M (HO-methoxy-) iso-1 2AC	U+UHYAC	2095	279 $_1$	206 $_{15}$	164 $_{23}$	100 $_{39}$	58 $_{90}$	512	6757
Metamfetamine-M (HO-methoxy-) iso-2 2AC	U+UHYAC	2115	279 $_1$	206 $_{19}$	164 $_{26}$	100 $_{48}$	58 $_{100}$	512	4243
Metamfetamine-M (nor-)		1160	134 $_1$	120 $_1$	91 $_6$	77 $_1$	65 $_4$	115	54
Metamfetamine-M (nor-)	U	1160	134 $_1$	120 $_1$	91 $_6$	65 $_4$	44 $_{100}$	115	5514
Metamfetamine-M (nor-) AC	U+UHYAC	1505	177 $_1$	118 $_{19}$	91 $_{11}$	86 $_{31}$	65 $_5$	165	55
Metamfetamine-M (nor-) AC	UAAC U+UHYAC	1505	177 $_1$	118 $_{19}$	91 $_{11}$	86 $_{31}$	44 $_{100}$	165	5515
Metamfetamine-M (nor-) HFB		1355	331 $_1$	240 $_{79}$	169 $_{21}$	118 $_{100}$	91 $_{53}$	762	5047
Metamfetamine-M (nor-) PFP		1330	281 $_1$	190 $_{73}$	118 $_{100}$	91 $_{36}$	65 $_{12}$	521	4379
Metamfetamine-M (nor-) TFA		1095	231 $_1$	140 $_{100}$	118 $_{92}$	91 $_{45}$	69 $_{19}$	309	4000
Metamfetamine-M (nor-) TMS		1190	192 $_6$	116 $_{100}$	100 $_{10}$	91 $_{11}$	73 $_{87}$	235	5581
Metamfetamine-M (nor-) formyl artifact		1100	147 $_2$	146 $_6$	125 $_5$	91 $_{12}$	56 $_{100}$	125	3261
Metamfetamine-M (nor-)-D5 AC	UAAC U+UHYAC	1480	182 $_3$	122 $_{46}$	92 $_{30}$	90 $_{100}$	66 $_{16}$	181	5690
Metamfetamine-M (nor-)-D5 HFB		1330	244 $_{100}$	169 $_{14}$	122 $_{46}$	92 $_{41}$	69 $_{40}$	787	6316
Metamfetamine-M (nor-)-D5 PFP		1320	194 $_{100}$	123 $_{42}$	119 $_{32}$	92 $_{46}$	69 $_{11}$	543	5566
Metamfetamine-M (nor-)-D5 TFA		1085	144 $_{100}$	123 $_{53}$	122 $_{56}$	92 $_{51}$	69 $_{28}$	330	5570
Metamfetamine-M (nor-)-D5 TMS		1180	212 $_1$	197 $_8$	120 $_{100}$	92 $_{11}$	73 $_{57}$	251	5582
Metamfetamine-M (nor-)-D11 PFP		1610	292 $_1$	194 $_{100}$	128 $_{82}$	98 $_{43}$	70 $_{14}$	575	7284
Metamfetamine-M (nor-)-D11 TFA		1615	242 $_1$	144 $_{100}$	128 $_{82}$	98 $_{43}$	70 $_{14}$	353	7283
Metamfetamine-M (nor-3-HO-) TMSTFA		1630	319 $_8$	206 $_{86}$	191 $_{32}$	140 $_{100}$	73 $_{58}$	708	6141
Metamfetamine-M (nor-3-HO-) 2AC	U+UHYAC	1930	235 $_2$	176 $_{48}$	134 $_{52}$	107 $_{21}$	86 $_{100}$	324	4387
Metamfetamine-M (nor-3-HO-) 2HFB		1620	330 $_{30}$	303 $_{11}$	240 $_{100}$	169 $_{15}$	69 $_{29}$	1182	5737

Table 1-8-1: Compounds in order of names

Name	Detected	RI	Typical ions and intensities					Page	Entry
Metamfetamine-M (nor-3-HO-) 2PFP		1520	280 $_{36}$	253 $_9$	190 $_{100}$	119 $_{19}$	69 $_8$	1104	5738
Metamfetamine-M (nor-3-HO-) 2TFA		<1000	230 $_4$	203 $_1$	140 $_{14}$	115 $_1$		819	6224
Metamfetamine-M (nor-3-HO-) 2TMS	UHYTMS	1850	280 $_{11}$	179 $_3$	116 $_{100}$	100 $_{12}$	73 $_{72}$	594	5693
Metamfetamine-M (nor-3-HO-) formyl artifact ME		1290	177 $_2$	162 $_4$	121 $_4$	77 $_5$	56 $_{100}$	165	5129
Metamfetamine-M (nor-4-HO-)		1480	151 $_{10}$	107 $_{69}$	91 $_{10}$	77 $_{42}$	56 $_{100}$	129	1802
Metamfetamine-M (nor-4-HO-) AC	U+UHYAC	1890	193 $_1$	134 $_{100}$	107 $_{26}$	86 $_{25}$	77 $_{16}$	201	1803
Metamfetamine-M (nor-4-HO-) ME		1225	165 $_1$	122 $_{16}$	107 $_2$	91 $_3$	77 $_7$	148	3249
Metamfetamine-M (nor-4-HO-) ME		1225	165 $_1$	122 $_{16}$	91 $_3$	77 $_7$	44 $_{100}$	148	5517
Metamfetamine-M (nor-4-HO-) TFA		1670	247 $_4$	140 $_{15}$	134 $_{54}$	107 $_{100}$	77 $_{15}$	366	6335
Metamfetamine-M (nor-4-HO-) 2AC	U+UHYAC	1900	235 $_1$	176 $_{72}$	134 $_{100}$	107 $_{47}$	86 $_{71}$	324	1804
Metamfetamine-M (nor-4-HO-) 2HFB		<1000	330 $_{48}$	303 $_{15}$	240 $_{100}$	169 $_{44}$	69 $_{42}$	1182	6326
Metamfetamine-M (nor-4-HO-) 2PFP		<1000	280 $_{77}$	253 $_{16}$	190 $_{100}$	119 $_{56}$	69 $_{16}$	1104	6325
Metamfetamine-M (nor-4-HO-) 2TFA		<1000	230 $_{94}$	203 $_{14}$	140 $_{130}$	92 $_{15}$	69 $_{76}$	819	6324
Metamfetamine-M (nor-4-HO-) 2TMS		<1000	280 $_7$	179 $_9$	149 $_8$	116 $_{100}$	73 $_{78}$	594	6327
Metamfetamine-M (nor-4-HO-) formyl art.		1220	163 $_3$	148 $_4$	107 $_{30}$	77 $_{13}$	56 $_{100}$	142	6323
Metamfetamine-M (nor-4-HO-) formyl artifact ME		1255	177 $_6$	162 $_4$	121 $_{60}$	77 $_{12}$	56 $_{100}$	166	3250
Metamfetamine-M (nor-HO-methoxy-)	UHY	1465	181 $_9$	138 $_{100}$	122 $_{18}$	94 $_{24}$	77 $_{16}$	178	4351
Metamfetamine-M (nor-HO-methoxy-) ME	UHYME	1550	195 $_1$	152 $_{100}$	137 $_{17}$	107 $_{16}$	77 $_{14}$	208	4352
Metamfetamine-M 2AC	U+UHYAC	2065	265 $_3$	206 $_{27}$	164 $_{100}$	137 $_{23}$	86 $_{33}$	445	3498
Metamfetamine-M 2AC	U+UHYAC	1820*	266 $_3$	206 $_9$	164 $_{100}$	150 $_{10}$	137 $_{30}$	450	6409
Metamfetamine-M 2HFB	UHFB	1690	360 $_{82}$	333 $_{15}$	240 $_{100}$	169 $_{42}$	69 $_{39}$	1189	6512
Metamitron		2195	202 $_{85}$	174 $_{40}$	133 $_{10}$	104 $_{100}$	77 $_{37}$	222	3860
Metamizol	G P U	1995	215 $_{34}$	123 $_{99}$	91 $_{18}$	56 $_{40}$		670	197
Metamizol-M (bis-dealkyl-)	P U UHY	1955	203 $_{23}$	93 $_{14}$	84 $_{59}$	56 $_{100}$		224	219
Metamizol-M (bis-dealkyl-) AC	P U U+UHYAC	2270	245 $_{30}$	203 $_{13}$	84 $_{50}$	56 $_{100}$		359	183
Metamizol-M (bis-dealkyl-) 2AC	U+UHYAC	2280	287 $_8$	245 $_{31}$	203 $_{15}$	84 $_{56}$	56 $_{100}$	547	3333
Metamizol-M (bis-dealkyl-) artifact	U UHY	1945	180 $_{13}$	119 $_{99}$	91 $_{45}$			173	424
Metamizol-M (dealkyl-)	P U UHY	1980	217 $_{17}$	123 $_{14}$	98 $_7$	83 $_{17}$	56 $_{100}$	261	220
Metamizol-M (dealkyl-) AC	P U+UHYAC	2395	259 $_{20}$	217 $_7$	123 $_8$	56 $_{99}$		416	184
Metamizol-M (dealkyl-) ME artifact	P G U-I	1895	231 $_{36}$	123 $_7$	111 $_{17}$	97 $_{56}$	56 $_{100}$	310	189
Metamizol-M/artifact (ME)	G	1320	151 $_{100}$	119 $_{56}$	106 $_{76}$	92 $_{36}$	65 $_{66}$	129	3909
Metandienone	U+UHYAC	2690*	300 $_1$	282 $_2$	242 $_3$	161 $_{16}$	122 $_{100}$	617	2813
Metandienone enol 2TMS		2670*	444 $_{33}$	339 $_9$	206 $_{39}$	143 $_{11}$	73 $_{100}$	1106	3985
Metaraminol		1670	167 $_4$	121 $_{25}$	95 $_{56}$	77 $_{100}$	65 $_{53}$	153	4655
Metaraminol 3AC		2065	293 $_1$	233 $_7$	191 $_5$	86 $_{22}$	69 $_{100}$	579	1486
Metaraminol formyl artifact		1840	179 $_{20}$	160 $_9$	135 $_{100}$	107 $_{56}$	77 $_{52}$	171	4651
Metaraminol -H2O 2AC		1745	233 $_{15}$	93 $_{10}$	69 $_{99}$			315	1479
Metazachlor		2260	277 $_{13}$	228 $_8$	209 $_{100}$	133 $_{93}$	81 $_{84}$	500	3878
Metenolone		2800*	302 $_{22}$	287 $_{16}$	136 $_{72}$	123 $_{100}$	82 $_{25}$	625	2825
Metenolone TMS		2580*	374 $_{13}$	359 $_{14}$	331 $_{10}$	136 $_{29}$	73 $_{100}$	934	3987
Metenolone acetate	U+UHYAC	2825*	344 $_5$	302 $_5$	161 $_{22}$	136 $_{78}$	123 $_{100}$	826	2815
Metenolone enantate		2835*	344 $_5$	302 $_8$	161 $_{24}$	136 $_{78}$	123 $_{100}$	1051	2814
Metenolone enol 2TMS		2530*	446 $_{73}$	208 $_{17}$	193 $_{12}$	129 $_7$	73 $_{100}$	1109	3986
Metformine HFB		1350	307 $_{100}$	292 $_{67}$	278 $_{38}$	95 $_{18}$	69 $_{57}$	733	5740
Metformine PFP		1300	257 $_{100}$	242 $_{49}$	228 $_{32}$	175 $_{15}$	69 $_{36}$	489	5741
Metformine TFA		1285	207 $_{100}$	192 $_{51}$	178 $_{29}$	125 $_{13}$	69 $_{44}$	291	5724
Metformine 2PFP		1250	403 $_{100}$	388 $_{61}$	284 $_{18}$	175 $_{27}$	69 $_{43}$	1063	5742
Metformine 2TFA		1220	303 $_{100}$	288 $_{44}$	234 $_{10}$	125 $_{16}$	69 $_{32}$	717	5723
Metformine artifact-1	U+UHYAC	1380	153 $_{100}$	138 $_{54}$	124 $_{45}$	110 $_{20}$	69 $_{71}$	133	6311
Metformine artifact-1 AC	U+UHYAC	1660	195 $_{80}$	153 $_{20}$	138 $_{100}$	124 $_{32}$	110 $_{34}$	207	6510
Metformine artifact-2		1650	154 $_{100}$	139 $_{39}$	125 $_{30}$	111 $_{37}$	69 $_{43}$	134	6312
Metformine artifact-3		1675	182 $_{100}$	167 $_{78}$	153 $_{41}$	138 $_{29}$	124 $_{33}$	181	6313
Metformine artifact-4		1485	167 $_{100}$	152 $_{51}$	138 $_{38}$	124 $_{25}$	69 $_{70}$	153	6638
Metformine artifact-4 propionylated		1840	223 $_{87}$	167 $_{73}$	152 $_{100}$	138 $_{30}$	96 $_{31}$	285	6639
Methabenzthiazuron ME		1985	235 $_8$	162 $_1$	136 $_4$	109 $_4$	72 $_{100}$	323	3941
Methacrylic acid methylester		<1000*	100 $_{44}$	69 $_{100}$				99	4283
Methadol		2185	296 $_1$	253 $_3$	165 $_3$	115 $_2$	72 $_{100}$	673	5617
Methadol AC	G P U+UHYAC	2230	353 $_1$	338 $_1$	225 $_4$	91 $_6$	72 $_{100}$	859	5616
Methadone	P G U UHY U+UHYAC	2160	309 $_1$	294 $_3$	223 $_2$	165 $_2$	72 $_{100}$	663	241
Methadone TMS	U UHY U+UHYAC	2260	381 $_2$	296 $_{15}$	165 $_4$	73 $_{37}$	72 $_{100}$	958	4567
Methadone intermediate-1		1750	193 $_{100}$	165 $_{67}$	115 $_{29}$	105 $_{29}$	77 $_{32}$	200	2835
Methadone intermediate-2		2095	278 $_3$	190 $_5$	165 $_{12}$	115 $_8$	58 $_{100}$	509	2838
Methadone intermediate-3		2130	278 $_2$	263 $_3$	192 $_{14}$	165 $_{31}$	72 $_{100}$	509	2836
Methadone intermediate-3 artifact		1920	253 $_2$	167 $_6$	165 $_4$	91 $_2$	72 $_{100}$	396	2837
Methadone-D9		2150	318 $_1$	303 $_7$	178 $_{11}$	165 $_{15}$	78 $_{200}$	705	7820
Methadone-M (bis-nor-) -H2O	U+UHYAC	1940	263 $_1$	208 $_{100}$	193 $_{78}$	179 $_{38}$	130 $_{50}$	436	5295
Methadone-M (bis-nor-) -H2O AC	U+UHYAC	2220	305 $_{100}$	290 $_{74}$	262 $_{25}$	236 $_{29}$	220 $_{25}$	640	5292
Methadone-M (bis-nor-HO-) -H2O AC	U+UHYAC	2380	321 $_1$	266 $_{42}$	224 $_{100}$	209 $_{35}$	207 $_{32}$	719	5298
Methadone-M (bis-nor-HO-) -H2O 2AC	U+UHYAC	2645	363 $_{100}$	348 $_{43}$	320 $_{63}$	278 $_{22}$	149 $_{15}$	895	5299
Methadone-M (EDDP)	U UHY U+UHYAC	2040	277 $_{105}$	276 $_{100}$	262 $_{43}$	220 $_{23}$	165 $_{10}$	503	242
Methadone-M (HO-) AC	U+UHYAC	2540	367 $_1$	352 $_4$	239 $_{12}$	222 $_{12}$	72 $_{100}$	911	6026
Methadone-M (HO-EDDP) AC	U+UHYAC	2350	335 $_{100}$	304 $_{27}$	292 $_{26}$	276 $_{16}$	234 $_{18}$	785	5297
Methadone-M (nor-) -H2O	U UHY U+UHYAC	2040	277 $_{105}$	276 $_{100}$	262 $_{43}$	220 $_{23}$	165 $_{10}$	503	242
Methadone-M (nor-EDDP) AC	U+UHYAC	2220	305 $_{100}$	290 $_{74}$	262 $_{25}$	236 $_{29}$	220 $_{25}$	640	5292

Table 1-8-1: Compounds in order of names | | | | | | | | p-Methoxyamfetamine

Name	Detected	RI	Typical ions and intensities					Page	Entry
Methadone-M (nor-HO-) -H2O AC	U+UHYAC	2350	335 $_{100}$	304 $_{27}$	292 $_{26}$	276 $_{16}$	234 $_{18}$	785	5297
Methadone-M (nor-HO-EDDP) 2AC	U+UHYAC	2645	363 $_{100}$	348 $_{43}$	320 $_{63}$	278 $_{22}$	149 $_{15}$	895	5299
Methadone-M (N-oxide) artifact	U+UHYAC	1900*	208 $_{100}$	193 $_{68}$	178 $_{31}$	165 $_{33}$	130 $_{49}$	238	5294
Methadone-M/artifact	U+UHYAC	1960	277 $_9$	235 $_{53}$	208 $_{86}$	193 $_{100}$	130 $_{71}$	498	5296
Methadone-M/artifact		2120	265 $_{100}$	193 $_{67}$	179 $_{36}$	130 $_{46}$	115 $_{50}$	445	5715
Methadone-M/artifact AC	U+UHYAC	2260	319 $_{100}$	304 $_{39}$	276 $_{42}$	234 $_{18}$	115 $_{13}$	711	5293
Methamidophos		1195	141 $_{30}$	110 $_6$	94 $_{100}$	79 $_{11}$	64 $_{25}$	120	4088
Methanol		<1000*	32 $_{75}$	31 $_{100}$	30 $_{75}$	28 $_{50}$		89	1628
Methaqualone	P G U+UHYAC UME	2155	250 $_{38}$	235 $_{100}$	132 $_9$	91 $_{25}$	65 $_{14}$	380	1095
Methaqualone HFB		2360	446 $_{38}$	427 $_{10}$	399 $_7$	277 $_{100}$	235 $_{88}$	1108	5071
Methaqualone PFP		2345	396 $_{16}$	277 $_{69}$	235 $_{100}$	130 $_{19}$	91 $_{26}$	1004	5072
Methaqualone TFA		2360	346 $_{23}$	277 $_{46}$	235 $_{100}$	160 $_5$	130 $_{10}$	831	5073
Methaqualone-M (2-carboxy-)	U	2400	280 $_{32}$	235 $_{100}$	146 $_{73}$	132 $_{17}$		516	1099
Methaqualone-M (2-carboxy-) -CO2	U	2165	236 $_{68}$	219 $_{98}$	132 $_{17}$	91 $_{38}$	65 $_{35}$	328	1096
Methaqualone-M (2-formyl-)	U UHY U+UHYAC	2240	264 $_{15}$	235 $_{100}$	132 $_{21}$	91 $_{12}$		439	1097
Methaqualone-M (2'-HO-methyl-)	U	2410	266 $_{65}$	251 $_{33}$	235 $_{49}$	160 $_{100}$	132 $_{19}$	449	1100
Methaqualone-M (2-HO-methyl-)	U UHY	2360	266 $_{53}$	235 $_{100}$	132 $_{20}$	91 $_{38}$		449	1098
Methaqualone-M (2'-HO-methyl-) AC	U+UHYAC	2505	308 $_{41}$	265 $_{100}$	247 $_{30}$	132 $_6$	77 $_{12}$	655	3755
Methaqualone-M (2-HO-methyl-) AC	U+UHYAC	2475	308 $_{31}$	265 $_{73}$	235 $_{100}$			655	1104
Methaqualone-M (3'-HO-)	UHY	2490	266 $_{56}$	251 $_{100}$	249 $_{35}$	148 $_{17}$		449	1101
Methaqualone-M (3'-HO-) AC	U+UHYAC	2555	308 $_{48}$	266 $_{83}$	251 $_{100}$	143 $_{11}$	77 $_9$	655	3757
Methaqualone-M (4'-HO-)	U UHY	2500	266 $_{60}$	251 $_{100}$	249 $_{41}$	143 $_{24}$		449	1102
Methaqualone-M (4'-HO-) AC	U+UHYAC	2570	308 $_{40}$	266 $_{46}$	251 $_{100}$	143 $_{39}$	77 $_{36}$	655	1105
Methaqualone-M (4'-HO-5'-methoxy-)		2560	296 $_{100}$	281 $_{91}$	279 $_{33}$	249 $_{24}$	143 $_{43}$	597	1106
Methaqualone-M (5'-HO-) AC	U+UHYAC	2540	308 $_{39}$	266 $_{54}$	251 $_{100}$	143 $_7$	77 $_{12}$	655	3756
Methaqualone-M (6-HO-)		2525	266 $_{74}$	251 $_{98}$	249 $_{46}$	132 $_{24}$	91 $_{46}$	449	1103
Methaqualone-M (HO-methoxy-) AC	U+UHYAC	2640	338 $_{21}$	296 $_{100}$	281 $_{78}$	191 $_{37}$	143 $_{27}$	797	3758
Metharbital	P G U UHY U+UHYAC	1455	170 $_{100}$	155 $_{97}$	126 $_{12}$	112 $_{34}$		215	73
Metharbital ME		1420	184 $_{96}$	169 $_{100}$	126 $_{38}$	112 $_{25}$		249	74
Metharbital-M (HO-)	U UHY	1800	186 $_{10}$	170 $_{60}$	155 $_{100}$	128 $_{30}$	113 $_{24}$	255	2961
Metharbital-M (HO-) AC	U+UHYAC	1870	228 $_3$	196 $_{11}$	170 $_{100}$	155 $_{89}$	112 $_{15}$	406	2962
Metharbital-M (nor-)	G P U UHY U+UHYAC	1500	156 $_{100}$	141 $_{97}$	112 $_{20}$	98 $_{22}$	83 $_{12}$	184	72
Methcathinone		1130	163 $_1$	148 $_1$	105 $_3$	77 $_9$	58 $_{60}$	143	5935
Methcathinone AC	U+UHYAC	1650	205 $_1$	105 $_4$	100 $_{39}$	77 $_{13}$	58 $_{80}$	228	5932
Methcathinone HFB		1440	359 $_1$	254 $_{65}$	210 $_{24}$	105 $_{100}$	77 $_{45}$	879	5936
Methcathinone PFP		1390	204 $_3$	105 $_5$	77 $_{10}$	58 $_{80}$		659	5934
Methcathinone TFA		1370	259 $_1$	154 $_{68}$	110 $_{31}$	105 $_{100}$	77 $_{49}$	416	5933
Methcathinone TMS		1570	220 $_5$	205 $_5$	130 $_{100}$	105 $_6$	73 $_{90}$	325	5937
Methcathinone-M (HO-) 2AC	U+UHYAC	1885	263 $_6$	100 $_{71}$	58 $_{150}$	250 $_4$	150 $_9$	434	4960
Methenamine		1210	140 $_{99}$	112 $_{20}$				119	1107
Methidathion		2120	302 $_3$	145 $_{100}$	125 $_{14}$	93 $_{16}$	85 $_{67}$	621	3856
Methiomeprazine	U+UHYAC	2725	344 $_9$	298 $_1$	245 $_1$	100 $_2$	58 $_{100}$	824	1828
Methitural		2240	288 $_{17}$	214 $_{28}$	171 $_{64}$	155 $_{27}$	74 $_{100}$	552	1487
Methocarbamol	G	2050	241 $_8$	198 $_9$	124 $_{78}$	118 $_{100}$	109 $_{57}$	347	1982
Methocarbamol AC	U+UHYAC	2145	283 $_2$	240 $_1$	160 $_{100}$	124 $_{25}$	57 $_{68}$	528	1991
Methocarbamol -CHNO AC		2000*	240 $_{10}$	124 $_{43}$	117 $_{100}$	109 $_{42}$	77 $_{15}$	344	1992
Methocarbamol-M (guaifenesin)	P G	1610*	198 $_{15}$	124 $_{100}$	109 $_{75}$			215	796
Methocarbamol-M (guaifenesin) 2AC	U+UHYAC	1865*	282 $_3$	159 $_{99}$	124 $_{13}$	99 $_{18}$		526	799
Methocarbamol-M (guaifenesin) 2TMS		1850*	342 $_{16}$	196 $_{33}$	149 $_{32}$	103 $_{47}$	73 $_{100}$	818	4551
Methocarbamol-M (HO-) 2AC	U+UHYAC	2560	341 $_1$	160 $_{80}$	140 $_{14}$	99 $_{13}$	57 $_{100}$	810	4504
Methocarbamol-M (HO-guaifensin) 3AC	U+UHYAC	2235*	340 $_2$	298 $_1$	159 $_{100}$	140 $_{10}$	99 $_8$	807	797
Methocarbamol-M (HO-methoxy-) 2AC	U+UHYAC	2620	371 $_5$	170 $_{16}$	160 $_{100}$	69 $_{19}$	57 $_{87}$	922	4502
Methocarbamol-M (HO-methoxy-guaifensin) 3AC	U+UHYAC	2265*	370 $_2$	230 $_7$	212 $_6$	170 $_{26}$	159 $_{100}$	919	798
Methocarbamol-M (O-demethyl-) 2AC		2430	311 $_2$	269 $_3$	160 $_{96}$	121 $_7$	57 $_{100}$	670	4503
Methocarbamol-M (O-demethyl-guaifensin) 3AC	U+UHYAC	1920*	310 $_1$	268 $_{13}$	159 $_{100}$	117 $_{12}$	110 $_{19}$	666	800
Methohexital	P U	1780	261 $_{10}$	247 $_{29}$	221 $_{63}$	178 $_{36}$	79 $_{100}$	431	1108
Methohexital ME		1735	276 $_9$	235 $_{85}$	178 $_{69}$	79 $_{98}$	53 $_{100}$	497	1109
Methohexital-D5		1775	266 $_{10}$	252 $_{37}$	238 $_{28}$	221 $_{100}$	178 $_{48}$	456	6881
Methohexital-M (HO-)	UHY	1880	278 $_7$	245 $_{19}$	219 $_{80}$	79 $_{100}$	53 $_{86}$	506	2959
Methomyl		1515	162 $_1$	115 $_2$	105 $_{64}$	88 $_{27}$	58 $_{100}$	140	3903
Methoprotryne		2235	271 $_{17}$	256 $_{100}$	226 $_{29}$	212 $_{34}$	171 $_{28}$	475	3857
Methorphan	G P-I U+UHYAC	2145	271 $_{31}$	214 $_{16}$	171 $_{14}$	150 $_{29}$	59 $_{100}$	476	227
Methorphan-M (bis-demethyl-) 2AC	U+UHYAC	2710	327 $_{11}$	240 $_8$	199 $_{12}$	87 $_{100}$	72 $_{62}$	745	228
Methorphan-M (nor-) AC	U+UHYAC	2590	299 $_{13}$	213 $_{42}$	171 $_{22}$	87 $_{100}$	72 $_{52}$	612	4477
Methorphan-M (O-demethyl-)	UHY	2255	257 $_{38}$	200 $_{17}$	150 $_{28}$	59 $_{100}$		411	475
Methorphan-M (O-demethyl-) AC	U+UHYAC	2280	299 $_{100}$	231 $_{42}$	200 $_{20}$	150 $_{48}$	59 $_{15}$	612	230
Methorphan-M (O-demethyl-) HFB		2100	453 $_{100}$	396 $_{15}$	385 $_{91}$	169 $_{10}$	150 $_{27}$	1118	6151
Methorphan-M (O-demethyl-) PFP	UHYPFP	2060	403 $_{93}$	335 $_{78}$	303 $_{14}$	150 $_{100}$	119 $_{58}$	1027	4305
Methorphan-M (O-demethyl-) TFA		2015	353 $_{69}$	285 $_{80}$	150 $_{100}$	115 $_{26}$	69 $_{72}$	858	4006
Methorphan-M (O-demethyl-) TMS	UHYTMS	2230	329 $_{31}$	272 $_{20}$	150 $_{39}$	73 $_{26}$	59 $_{100}$	756	4304
Methorphan-M (O-demethyl-HO-) 2AC	U+UHYAC	2580	357 $_{68}$	247 $_{22}$	215 $_{17}$	150 $_{100}$	59 $_{30}$	874	1187
Methorphan-M (O-demethyl-methoxy-) AC	U+UHYAC	2520	329 $_{48}$	261 $_{23}$	229 $_{23}$	150 $_{100}$	59 $_{28}$	756	4476
Methorphan-M (O-demethyl-oxo-) AC	U+UHYAC	2695	313 $_{16}$	240 $_{11}$	199 $_{98}$	157 $_{12}$	73 $_{100}$	681	4475
p-Methoxyamfetamine		1225	165 $_1$	122 $_{16}$	107 $_2$	91 $_3$	77 $_7$	148	3249

Table 1-8-1: Compounds in order of names

Name	Detected	RI	Typical ions and intensities					Page	Entry
p-Methoxyamfetamine		1225	165_1	122_{16}	91_3	77_7	44_{100}	148	5517
p-Methoxyamfetamine HFB		1560	361_2	240_5	169_5	148_{53}	121_{100}	886	6769
p-Methoxyamfetamine PFP		1460	311_7	190_4	148_{43}	121_{100}	91_7	670	6775
p-Methoxyamfetamine TFA		1460	261_6	148_{33}	140_4	121_{100}	91_8	425	6774
Methoxyaniline AC	PME UME	1630	165_{59}	123_{74}	108_{100}	95_{10}	80_{20}	147	5046
4-Methoxyaniline HFB	U+UHYHFB	1400	319_{100}	304_8	300_6	150_6	122_{78}	706	6620
4-Methoxyaniline TFA	U+UHYTFA	1335	219_{100}	204_{16}	149_{11}	122_{76}	109_{19}	269	6615
3-Methoxybenzoic acid methylester		1490*	166_{58}	135_{97}	107_{43}	92_{23}	77_{40}	149	1110
4-Methoxybenzoic acid ET		1415*	180_{21}	77_{21}	107_{12}	152_{19}	135_{100}	174	6447
4-Methoxybenzoic acid ME		1270*	166_{34}	135_{100}	107_{15}	92_{16}	77_{26}	150	6446
Methoxychlor		2450*	344_2	274_3	227_{100}	212_5	152_5	823	1488
Methoxychlor -HCl		2340*	308_{100}	273_{13}	238_{80}	223_{28}	152_{24}	653	3858
p-Methoxyetilamfetamine		1660	192_1	149_1	121_5	91_2	72_{100}	202	5831
p-Methoxyetilamfetamine AC	U+UHYAC	1855	235_1	148_{32}	121_7	114_{26}	72_{100}	326	5322
p-Methoxyetilamfetamine HFB	UHFB	1785	389_2	268_{100}	240_{46}	148_{62}	121_{43}	983	5834
p-Methoxyetilamfetamine ME		1780	206_1	121_5	86_{100}	72_3	58_{20}	235	5835
p-Methoxyetilamfetamine PFP	UPFP	1765	339_1	218_{100}	190_{44}	148_{59}	121_{49}	802	5833
p-Methoxyetilamfetamine TFA	UTFA	1775	289_1	168_{100}	148_{63}	140_{41}	121_{62}	557	5832
p-Methoxyetilamfetamine TMS		2065	264_1	250_{14}	144_{100}	121_{17}	73_{84}	447	5836
Methoxyhydroxyphenylglycol (MHPG) 3AC		2030*	310_9	268_{14}	208_{52}	166_{100}	153_{91}	666	1111
p-Methoxymetamfetamine		1475	178_1	121_4	91_2	77_3	58_{100}	172	6719
p-Methoxymetamfetamine AC		1820	221_1	148_{24}	121_7	100_{22}	58_{100}	279	6720
p-Methoxymetamfetamine ET		1780	206_1	121_5	86_{100}	72_3	58_{20}	235	5835
p-Methoxymetamfetamine HFB		1665	375_1	254_{55}	210_{32}	148_{67}	121_{100}	935	6722
p-Methoxymetamfetamine PFP		1510	325_3	204_{60}	160_{29}	148_{48}	121_{100}	734	7601
p-Methoxymetamfetamine TFA		1645	275_2	154_{59}	148_{51}	121_{100}	110_{31}	489	6721
2-Methoxyphenylpiperazine-M (O-demethyl-) 2AC	U+UHYAC	2140	262_{10}	220_{18}	148_{100}	120_{51}	86_{12}	431	170
4-Methoxyphenylpiperazine		1880	192_{34}	150_{100}	135_{20}	120_{28}	92_9	198	6622
4-Methoxyphenylpiperazine HFB	U+UHYHFB	1965	388_{100}	373_{34}	191_{75}	135_{74}	120_{66}	979	6617
4-Methoxyphenylpiperazine ME		1840	206_{100}	191_{17}	162_{14}	135_{77}	120_{63}	232	6623
4-Methoxyphenylpiperazine TFA	U+UHYTFA	1940	288_{100}	273_{36}	191_{20}	135_{22}	120_{28}	552	6612
4-Methoxyphenylpiperazine TMS		2070	264_{100}	249_{29}	101_{45}	86_{46}	73_{81}	441	6884
4-Methoxyphenylpiperazine 2AC	U+UHYAC	2185	234_{81}	162_{100}	149_{35}	134_{36}	120_{35}	321	6609
4-Methoxyphenylpiperazine-M (aminophenol) 2HFB	U+UHYHFB	1405	501_{25}	482_6	304_{81}	169_{100}	109_{83}	1164	6621
4-Methoxyphenylpiperazine-M (aminophenol) 2TFA	U+UHYTFA	1280	301_{44}	204_{42}	176_{14}	109_{90}	69_{90}	617	6616
4-Methoxyphenylpiperazine-M (deethylene-) 2AC	U+UHYAC	2120	250_7	191_{51}	165_4	149_{33}	136_{100}	381	6611
4-Methoxyphenylpiperazine-M (deethylene-) 2HFB	U+UHYHFB	1765	558_7	345_{13}	332_{28}	240_{19}	135_{100}	1187	6619
4-Methoxyphenylpiperazine-M (deethylene-) 2TFA	U+UHYTFA	1765	358_{20}	245_{19}	232_{48}	135_{100}	120_{33}	876	6614
4-Methoxyphenylpiperazine-M (methoxyaniline) HFB	U+UHYHFB	1400	319_{100}	304_8	300_6	150_6	122_{78}	706	6620
4-Methoxyphenylpiperazine-M (methoxyaniline) TFA	U+UHYTFA	1335	219_{100}	204_{16}	149_{11}	122_{76}	109_{19}	269	6615
4-Methoxyphenylpiperazine-M (O-demethyl-) 2AC	U+UHYAC	2350	262_{39}	220_{60}	177_{21}	148_{100}	135_{59}	430	6610
4-Methoxyphenylpiperazine-M (O-demethyl-) 2HFB	U+UHYHFB	1990	570_{83}	551_{16}	373_{100}	344_{46}	317_{19}	1189	6618
4-Methoxyphenylpiperazine-M (O-demethyl-) 2TFA	U+UHYTFA	1915	370_{77}	273_{85}	244_{51}	217_{29}	120_{64}	918	6613
3-Methoxytyramine 2AC	U+UHYAC	2070	251_2	209_7	150_{100}	137_{15}		384	1273
4-Methycatechol 2TMS		1325*	268_{63}	180_{27}	165_{18}	149_{21}	73_{100}	461	6022
Methylacetate		<1000*	74_{23}	59_{10}	43_{100}	29_{23}		94	3777
Methylamine		<1000	31_{40}	30_{82}	28_{100}			89	3619
17-Methylandrostane-ol-3-one		2555*	304_{22}	289_{46}	247_{41}	231_{45}	55_{100}	636	3895
17-Methylandrostane-ol-3-one TMS		2610*	376_1	361_6	306_6	143_{100}	73_{40}	941	3925
17-Methylandrostane-ol-3-one enol TMS		2565*	376_{74}	347_{23}	143_{100}	127_{67}	73_{53}	941	3924
17-Methylandrostane-ol-3-one enol 2TMS		2580*	448_{13}	358_4	216_{10}	143_{81}	73_{100}	1111	3978
4-Methylbenzoic acid ET		1350*	164_{16}	136_{22}	119_{100}	91_{44}	65_{15}	145	6473
4-Methylbenzoic acid ME		1210*	150_{31}	119_{100}	91_{53}	89_8	65_{22}	127	6472
N-Methyl-Brolamfetamine		1885	230_2	77_5	143_2	199_2	58_{100}	545	6429
N-Methyl-Brolamfetamine AC	U+UHYAC	2225	329_1	256_{15}	199_3	100_{57}	58_{100}	753	6430
N-Methyl-Brolamfetamine-M (HO-) 2AC	U+UHYAC	2350	387_4	314_{26}	242_{57}	100_{59}	58_{100}	975	7059
N-Methyl-Brolamfetamine-M (N,O-bisdemethyl-) iso-1 AC	U+UHYAC	2120	301_{18}	242_{100}	215_{11}	185_{13}	86_{20}	617	7070
N-Methyl-Brolamfetamine-M (N,O-bisdemethyl-) iso-1 2AC	U+UHYAC	2235	343_{12}	301_{23}	284_{57}	242_{100}	86_{81}	819	7065
N-Methyl-Brolamfetamine-M (N,O-bisdemethyl-) iso-2 AC	U+UHYAC	2180	301_{29}	242_{100}	215_{14}	86_{36}		618	7071
N-Methyl-Brolamfetamine-M (N,O-bisdemethyl-) iso-2 2AC	U+UHYAC	2275	343_2	284_{56}	242_{100}	215_{13}	86_{23}	819	7066
N-Methyl-Brolamfetamine-M (N,O-bisdemethyl-deamino-oxo-) AC	U+UHYAC	1930*	300_8	258_{94}	215_{100}			612	7063
N-Methyl-Brolamfetamine-M (N,O-bisdemethyl-deamino-oxo-) iso-1	U+UHYAC	1870*	260_{66}	258_{72}	217_{99}	215_{100}		411	7068
N-Methyl-Brolamfetamine-M (N,O-bisdemethyl-deamino-oxo-) iso-2	U+UHYAC	1885*	260_{99}	258_{100}	217_{97}	215_{93}		411	7069
N-Methyl-Brolamfetamine-M (N,O-bisdemethyl-HO-) 3AC	U+UHYAC	2385	401_1	359_5	317_7	258_{10}	86_7	1019	7067
N-Methyl-Brolamfetamine-M (N,O-bisdemethyl-HO-) -H2O 2AC	U+UHYAC	2280	341_{32}	299_{62}	257_{100}	242_{72}		809	7072
N-Methyl-Brolamfetamine-M (N,O-bisdemethyl-HO-deamino-oxo-) 3AC	U+UHYAC	2145*	402_2	360_9	315_{13}	300_8	231_{24}	1022	7064
N-Methyl-Brolamfetamine-M (N-demethyl-)		1800	273_1	232_6	230_6	199_1	77_7	480	2548
N-Methyl-Brolamfetamine-M (N-demethyl-)		1800	273_1	230_6	105_1	77_7	44_{100}	480	5527
N-Methyl-Brolamfetamine-M (N-demethyl-) AC		2150	315_3	256_{20}	229_1	162_4	86_{15}	688	2549
N-Methyl-Brolamfetamine-M (N-demethyl-) AC		2150	315_3	256_{20}	162_4	86_{22}	44_{100}	688	5528
N-Methyl-Brolamfetamine-M (N-demethyl-) HFB		1945	469_{24}	256_{88}	240_{55}	229_{100}	199_{29}	1137	6008
N-Methyl-Brolamfetamine-M (N-demethyl-) PFP		1905	419_{21}	256_{69}	229_{94}	190_{55}	119_{87}	1059	6007
N-Methyl-Brolamfetamine-M (N-demethyl-) TFA		1935	369_{28}	256_{81}	229_{100}	199_{40}	69_{88}	914	6006
Methyl-Brolamfetamine-M (N-demethyl-) TMS		1920	345_1	272_2	229_1	116_{80}	73_{63}	827	6009

Table 1-8-1: Compounds in order of names — N-Methyl-DOB-M (N-demethyl-) TFA

Name	Detected	RI	Typical ions and intensities					Page	Entry
N-Methyl-Brolamfetamine-M (N-demethyl-) formyl artifact		1790	285_3	254_{15}	229_5	199_3	56_{100}	537	3242
N-Methyl-Brolamfetamine-M (N-demethyl-deamino-HO-) AC	U+UHYAC	1950*	316_7	274_{22}	214_{96}	186_{17}		692	7061
N-Methyl-Brolamfetamine-M (N-demethyl-deamino-oxo-)	U+UHYAC	1835*	272_7	229_{11}				477	7062
N-Methyl-Brolamfetamine-M (N-demethyl-HO-) 2AC	U+UHYAC	2270	373_3	86_{100}	313_{14}	271_{37}		929	7081
N-Methyl-Brolamfetamine-M (N-demethyl-HO-) -H2O	U+UHYAC	1960*	273_{31}	271_{29}	258_{43}	256_{42}		473	7073
N-Methyl-Brolamfetamine-M (N-demethyl-HO-) -H2O AC	U+UHYAC	2130*	313_{24}	271_{79}	256_{100}			678	7074
N-Methyl-Brolamfetamine-M (O,O-bisdemethyl-) 3AC	U+UHYAC	2330	385_4	270_8	228_{15}	100_{100}	58_{70}	969	7058
N-Methyl-Brolamfetamine-M (O-demethyl-) iso-1 2AC	U+UHYAC	2285	357_6	284_{24}	242_{17}	100_{83}	58_{100}	871	7056
N-Methyl-Brolamfetamine-M (O-demethyl-) iso-2 2AC	U+UHYAC	2295	357_3	284_{10}	242_{14}	100_{72}	58_{80}	871	7057
N-Methyl-Brolamfetamine-M (O-demethyl-HO-) 3AC	U+UHYAC	2430	415_1	373_9	100_{74}	58_{100}	258_{13}	1052	7060
N-Methyl-Brolamfetamine-M (tridemethyl-) 3AC	U+UHYAC	2325	371_2	329_{35}	287_{54}	228_{100}	86_{44}	921	7075
N-Methyl-Brolamfetamine-M (tridemethyl-) artifact 2AC	U+UHYAC	2225	311_{23}	269_{98}	227_{59}	212_{39}	133_{18}	670	7184
2-Methylbutane		<1000*	72_7	57_{57}	43_{100}	41_{95}	29_{44}	93	3811
2-Methyl-2-butene		<1000*	70_{35}	55_{100}	42_{33}	39_{42}	29_{28}	92	3814
3-Methyl-1-butene		<1000*	70_{23}	55_{100}	42_{30}	39_{50}	27_{49}	92	3810
4-Methylcatechol		1155*	124_{100}	106_{15}	95_{11}	78_{84}	51_{31}	109	5762
4-Methylcatechol HFB		1035*	320_{50}	169_{28}	151_{25}	123_{100}	95_{50}	714	5990
4-Methylcatechol PFP		1035*	270_{64}	151_{27}	123_{100}	95_{68}	77_{39}	467	5988
4-Methylcatechol TFA		<1000*	220_{53}	151_{13}	123_{100}	95_{38}	69_{56}	272	5987
4-Methylcatechol 2AC	U+UHYAC	1450*	208_2	166_{13}	124_{100}	106_6	78_{13}	236	2451
4-Methylcatechol 2HFB		1165*	516_{52}	319_{35}	263_{49}	169_{61}	69_{100}	1173	5991
4-Methylcatechol 2PFP		<1000*	416_{62}	269_{42}	253_{36}	213_{47}	119_{100}	1053	5989
N-Methylcytisine		1995	204_{16}	160_5	146_{10}	117_7	58_{100}	227	5597
4-Methyldibenzofuran		1620*	182_{100}	181_{91}	152_{14}	127_3	91_6	180	2561
4-Methyl-2,5-dimethoxyphenethylamine		1605	195_{20}	166_{100}	151_{60}	135_{27}	91_{19}	208	6904
4-Methyl-2,5-dimethoxyphenethylamine AC		1940	237_{14}	178_{100}	165_{35}	163_{40}	135_{51}	334	6912
4-Methyl-2,5-dimethoxyphenethylamine HFB		1710	391_{49}	226_6	178_{85}	165_{100}	135_{46}	987	6937
4-Methyl-2,5-dimethoxyphenethylamine PFP		1680	341_{53}	178_{79}	165_{100}	135_{54}	91_{21}	810	6932
4-Methyl-2,5-dimethoxyphenethylamine TFA		1685	291_{43}	178_{73}	165_{100}	135_{57}	91_{22}	568	6927
4-Methyl-2,5-dimethoxyphenethylamine TMS		1735	267_7	237_8	166_{20}	102_{100}	73_{91}	456	6914
4-Methyl-2,5-dimethoxyphenethylamine 2AC		2010	279_{11}	135_{21}	72_9	178_{100}	163_{34}	511	6913
4-Methyl-2,5-dimethoxyphenethylamine 2TMS		2020	339_2	324_{13}	174_{100}	100_{23}	86_{36}	805	6915
4-Methyl-2,5-dimethoxyphenethylamine formyl artifact		1530	207_{25}	176_{100}	165_{69}	135_{39}	91_{16}	234	6909
4-Methyl-2,5-dimethoxyphenethylamine-M (deamino-COOH) ME		1755*	224_{100}	209_{19}	177_{12}	165_{35}	135_8	288	7229
4-Methyl-2,5-dimethoxyphenethylamine-M (deamino-HO-) AC	U+UHYAC	1740*	238_{27}	178_{100}	163_{57}	135_{33}	79_{27}	336	7216
4-Methyl-2,5-dimethoxyphenethylamine-M (deamino-oxo-)		1730*	194_{54}	165_{100}	151_{25}	135_{85}	91_{51}	205	7232
4-Methyl-2,5-dimethoxyphenethylamine-M (HO-) 2AC	U+UHYAC	2390	295_{33}	236_{100}	223_6	193_{35}	163_{12}	591	7219
4-Methyl-2,5-dimethoxyphenethylamine-M (HO-) 2TFA		1950	403_{42}	290_{100}	277_{32}	177_{57}	163_{25}	1025	7228
4-Methyl-2,5-dimethoxyphenethylamine-M (HO-) 3AC	U+UHYAC	2400	337_{27}	244_{23}	236_{100}	193_{46}	125_{30}	793	7220
4-Methyl-2,5-dimethoxyphenethylamine-M (O-demethyl- N-acetyl-) 2AC	U+UHYAC	2250	307_5	265_7	206_{25}	164_{100}	149_{13}	648	7223
4-Methyl-2,5-dimethoxyphenethylamine-M (O-demethyl- N-acetyl-) iso.-1 AC	U+UHYAC	2130	265_6	223_{36}	164_{100}	151_{14}	91_4	444	7221
4-Methyl-2,5-dimethoxyphenethylamine-M (O-demethyl- N-acetyl-) iso.-1 TFA		1990	319_{12}	260_{100}	247_4	191_{18}	163_{26}	707	7224
4-Methyl-2,5-dimethoxyphenethylamine-M (O-demethyl- N-acetyl-) iso.-2 AC	U+UHYAC	2200	265_7	223_{25}	164_{100}	151_{14}	91_6	444	7222
4-Methyl-2,5-dimethoxyphenethylamine-M (O-demethyl- N-acetyl-) iso.-2 TFA		2050	319_{10}	260_{100}	247_{13}	245_{22}	163_{39}	707	7225
4-Methyl-2,5-dimethoxyphenethylamine-M (O-demethyl-) 3AC	U+UHYAC	2250	307_5	265_7	206_{25}	164_{100}	149_{13}	648	7223
4-Methyl-2,5-dimethoxyphenethylamine-M (O-demethyl-) iso-1 2AC	U+UHYAC	2130	265_6	223_{36}	164_{100}	151_{14}	91_4	444	7221
4-Methyl-2,5-dimethoxyphenethylamine-M (O-demethyl-) iso-1 2TFA		1780	373_{28}	260_{100}	247_{24}	191_{30}	163_{49}	929	7226
4-Methyl-2,5-dimethoxyphenethylamine-M (O-demethyl-) iso-2 2AC	U+UHYAC	2200	265_7	223_{25}	164_{100}	151_{14}	91_6	444	7222
4-Methyl-2,5-dimethoxyphenethylamine-M (O-demethyl-) iso-2 2TFA		1850	373_{18}	260_{100}	247_{48}	217_{15}	163_{39}	929	7227
4-Methyl-2,5-dimethoxyphenethylamine-M (O-demethyl-deamino-COOH) iso.-1 MEAC		1860*	252_{28}	210_{100}	178_{40}	150_{100}	122_{12}	389	7230
4-Methyl-2,5-dimethoxyphenethylamine-M (O-demethyl-deamino-COOH) iso.-2 MEAC		1900*	252_{21}	210_{100}	193_{10}	163_7	151_{55}	389	7231
4-Methyl-2,5-dimethoxyphenethylamine-M (O-demethyl-deamino-HO-) iso.-1 2AC	U+UHYAC	1875*	266_5	224_{13}	164_{100}	154_{46}	114_{15}	450	7217
4-Methyl-2,5-dimethoxyphenethylamine-M (O-demethyl-deamino-HO-) iso.-2 2AC	U+UHYAC	1890*	266_8	224_{12}	164_{100}	121_{10}	206_6	450	7218
N-Methyl-DOB		1885	230_2	77_5	143_2	199_2	58_{100}	545	6429
N-Methyl-DOB AC	U+UHYAC	2225	329_1	256_{15}	199_3	100_{57}	58_{100}	753	6430
N-Methyl-DOB-M (HO-) 2AC	U+UHYAC	2350	387_4	314_{26}	242_{57}	100_{59}	58_{100}	975	7059
N-Methyl-DOB-M (N,O-bisdemethyl-) iso-1 AC	U+UHYAC	2120	301_{18}	242_{100}	215_{11}	185_{13}	86_{20}	617	7070
N-Methyl-DOB-M (N,O-bisdemethyl-) iso-1 2AC	U+UHYAC	2235	343_{12}	301_{23}	284_{57}	242_{100}	86_{81}	819	7065
N-Methyl-DOB-M (N,O-bisdemethyl-) iso-2 AC	U+UHYAC	2180	301_{29}	242_{100}	215_{14}	86_{36}		618	7071
N-Methyl-DOB-M (N,O-bisdemethyl-) iso-2 2AC	U+UHYAC	2275	343_2	284_{56}	242_{100}	215_{13}	86_{23}	819	7066
N-Methyl-DOB-M (N,O-bisdemethyl-deamino-oxo-) AC	U+UHYAC	1930*	300_8	258_{94}	215_{100}			612	7063
N-Methyl-DOB-M (N,O-bisdemethyl-deamino-oxo-) iso-1	U+UHYAC	1870*	260_{66}	258_{72}	217_{99}	215_{100}		411	7068
N-Methyl-DOB-M (N,O-bisdemethyl-deamino-oxo-) iso-2	U+UHYAC	1885*	260_{99}	258_{100}	217_{97}	215_{93}		411	7069
N-Methyl-DOB-M (N,O-bisdemethyl-HO-) 3AC	U+UHYAC	2385	401_1	359_5	317_7	258_{10}	86_7	1019	7067
N-Methyl-DOB-M (N,O-bisdemethyl-HO-) -H2O 2AC	U+UHYAC	2280	341_{32}	299_{62}	257_{100}	242_{72}		809	7072
N-Methyl-DOB-M (N,O-bisdemethyl-HO-deamino-oxo-) 3AC	U+UHYAC	2145*	402_2	360_9	315_{13}	300_8	231_{24}	1022	7064
N-Methyl-DOB-M (N-demethyl-)		1800	273_1	232_6	230_6	199_1	77_7	480	2548
N-Methyl-DOB-M (N-demethyl-)		1790	285_3	254_{15}	229_5	199_3	56_{100}	537	3242
N-Methyl-DOB-M (N-demethyl-)		1800	273_1	230_6	105_3	77_7	44_{100}	480	5527
N-Methyl-DOB-M (N-demethyl-) AC		2150	315_3	256_{20}	229_1	162_4	86_{15}	688	2549
N-Methyl-DOB-M (N-demethyl-) AC		2150	315_3	256_{20}	162_4	86_{22}	44_{100}	688	5528
N-Methyl-DOB-M (N-demethyl-) HFB		1945	469_{24}	256_{88}	240_{55}	229_{100}	199_{29}	1137	6008
N-Methyl-DOB-M (N-demethyl-) PFP		1905	419_{21}	256_{69}	229_{94}	190_{55}	119_{87}	1059	6007
N-Methyl-DOB-M (N-demethyl-) TFA		1935	369_{28}	256_{81}	229_{100}	199_{40}	69_{88}	914	6006

N-Methyl-DOB-M (N-demethyl-) TMS

Table 1-8-1: Compounds in order of names

Name	Detected	RI	Typical ions and intensities					Page	Entry
N-Methyl-DOB-M (N-demethyl-) TMS		1920	345 $_1$	272 $_2$	229 $_1$	116 $_{80}$	73 $_{63}$	827	6009
N-Methyl-DOB-M (N-demethyl-deamino-HO-) AC	U+UHYAC	1950*	316 $_7$	274 $_{22}$	214 $_{96}$	186 $_{17}$		692	7061
N-Methyl-DOB-M (N-demethyl-deamino-oxo-)	U+UHYAC	1835*	272 $_7$	229 $_{11}$				477	7062
N-Methyl-DOB-M (N-demethyl-HO-) 2AC	U+UHYAC	2270	373 $_3$	86 $_{100}$	313 $_{14}$	271 $_{37}$		929	7081
N-Methyl-DOB-M (N-demethyl-HO-) -H2O	U+UHYAC	1960*	273 $_{31}$	271 $_{29}$	258 $_{43}$	256 $_{42}$		473	7073
N-Methyl-DOB-M (N-demethyl-HO-) -H2O AC	U+UHYAC	2130*	313 $_{24}$	271 $_{79}$	256 $_{100}$			678	7074
N-Methyl-DOB-M (O,O-bisdemethyl-) 3AC	U+UHYAC	2330	385 $_4$	270 $_8$	228 $_{15}$	100 $_{100}$	58 $_{70}$	969	7058
N-Methyl-DOB-M (O-demethyl-) iso-1 2AC	U+UHYAC	2285	357 $_6$	284 $_{24}$	242 $_{14}$	100 $_{63}$	58 $_{100}$	871	7056
N-Methyl-DOB-M (O-demethyl-) iso-2 2AC	U+UHYAC	2295	357 $_3$	284 $_{10}$	242 $_{14}$	100 $_{72}$	58 $_{80}$	871	7057
N-Methyl-DOB-M (O-demethyl-HO-) 3AC	U+UHYAC	2430	415 $_1$	373 $_9$	100 $_{74}$	58 $_{100}$	258 $_{13}$	1052	7060
N-Methyl-DOB-M (tridemethyl-) 3AC	U+UHYAC	2325	371 $_2$	329 $_{35}$	287 $_{54}$	228 $_{100}$	86 $_{44}$	921	7075
N-Methyl-DOB-M (tridemethyl-) artifact 2AC	U+UHYAC	2225	311 $_{23}$	269 $_{98}$	227 $_{59}$	212 $_{39}$	133 $_{18}$	670	7184
Methyldopa ME3AC		2330	351 $_1$	292 $_9$	250 $_{16}$	208 $_{35}$	102 $_{100}$	847	5120
Methyldopa ME4AC		2400	393 $_1$	320 $_7$	186 $_{51}$	144 $_{100}$	123 $_{79}$	994	5121
Methyldopa 2ME		1870	239 $_1$	180 $_{10}$	138 $_{23}$	137 $_{19}$	102 $_{100}$	341	5114
Methyldopa 3ME		1940	253 $_1$	194 $_{12}$	137 $_{11}$	116 $_{100}$	56 $_{87}$	394	5116
Methyldopa 3ME		1900	253 $_1$	194 $_{16}$	152 $_{50}$	151 $_{52}$	102 $_{100}$	394	5115
Methyldopa 4ME		1960	267 $_1$	208 $_{18}$	151 $_{16}$	116 $_{100}$	56 $_{96}$	455	5117
Methyldopa 4ME		2010	267 $_1$	208 $_{11}$	130 $_{100}$	70 $_{74}$	56 $_{48}$	455	5118
Methyldopa 5ME		2030	281 $_1$	222 $_{12}$	130 $_{100}$	70 $_{78}$	56 $_{43}$	523	5119
Methyldopa artifact (acetic acid adduct -2H2O) AC		2050	277 $_1$	235 $_4$	165 $_9$	123 $_{100}$	77 $_{10}$	500	5123
Methyldopa artifact (acetic acid adduct -2H2O) 2AC		2075	319 $_1$	277 $_4$	235 $_7$	165 $_{34}$	123 $_{100}$	707	5122
Methyldopa impurity 2AC	U+UHYAC	1735*	250 $_3$	208 $_{15}$	166 $_{45}$	123 $_{100}$		379	4210
Methyldopa-M	UHY	1465	181 $_9$	138 $_{100}$	122 $_{18}$	94 $_{24}$	77 $_{16}$	178	4351
Methyldopa-M (decarboxy-) 2AC	U+UHYAC	2065	265 $_3$	206 $_{27}$	164 $_{100}$	137 $_{23}$	86 $_{33}$	445	3498
Methyldopa-M (decarboxy-deamino-oxo-)	UHY	1510*	180 $_{19}$	137 $_{100}$	122 $_{19}$	107 $_2$	94 $_{16}$	175	4247
Methyldopa-M AC	U+UHYAC	1600*	222 $_2$	180 $_{22}$	137 $_{100}$			280	4211
Methyldopa-M ME	UHYME	1540*	194 $_{25}$	151 $_{100}$	135 $_4$	107 $_{18}$	65 $_4$	204	4353
Methyldopa-M ME	UHYME	1550	195 $_1$	152 $_{100}$	137 $_{17}$	107 $_{16}$	77 $_{14}$	208	4352
Methyldopa-M 2HFB	UHFB	1690	360 $_{82}$	333 $_{15}$	240 $_{100}$	169 $_{42}$	69 $_{39}$	1189	6512
5-Methyl-1-hexene		<1000*	98 $_9$	83 $_{12}$	70 $_{32}$	56 $_{97}$	41 $_{100}$	99	3822
2-Methyl-1-pentene		<1000*	84 $_{24}$	69 $_{28}$	56 $_{94}$	41 $_{100}$	27 $_{76}$	96	3817
N-Methyl-1-phenylethylamine		1460	134 $_3$	120 $_{100}$	105 $_9$	77 $_{13}$	58 $_{25}$	115	6221
N-Methyl-1-phenylethylamine AC		1430	177 $_{22}$	162 $_4$	120 $_{44}$	105 $_{20}$	77 $_{13}$	166	6229
2-Methyl-1-propanol (isobutanol)		<1000*	74 $_{13}$	55 $_4$	43 $_{100}$			94	1042
2-Methyl-2-propanol		<1000*	59 $_{100}$	57 $_{10}$	43 $_{20}$			94	2446
Methylene blue artifact		2680	285 $_{100}$	270 $_{38}$	254 $_6$	225 $_5$	142 $_{20}$	539	3387
2,2'-Methylene-bis-(4-methyl-6-tert.-butylphenol)	P	2340*	340 $_{17}$	177 $_{100}$	161 $_{79}$	149 $_{49}$	121 $_{29}$	809	5337
3,4-Methylenedioxybenzoic acid ET		1560*	194 $_{48}$	166 $_{27}$	149 $_{100}$	121 $_{23}$	65 $_{16}$	203	6471
3,4-Methylenedioxybenzoic acid ME		1445*	180 $_{52}$	149 $_{100}$	121 $_{27}$	65 $_{18}$	63 $_{19}$	173	6470
3,4-Methylenedioxybenzylalcohol		1420*	152 $_{10}$	135 $_{47}$	122 $_{37}$	93 $_{70}$	65 $_{50}$	131	7616
3,4-Methylenedioxybenzylalcohol HFB		1400*	348 $_{28}$	271 $_2$	135 $_{100}$	105 $_6$	77 $_{19}$	837	7620
3,4-Methylenedioxybenzylalcohol PFP		1325*	298 $_{30}$	149 $_8$	135 $_{100}$	105 $_7$	77 $_{28}$	604	7619
3,4-Methylenedioxybenzylalcohol TFA		1295*	248 $_{40}$	149 $_4$	135 $_{100}$	105 $_8$	77 $_{24}$	370	7618
3,4-Methylenedioxybenzylalcohol TMS		1560*	224 $_{45}$	209 $_{27}$	179 $_{17}$	135 $_{100}$	73 $_{24}$	287	7617
Methylenedioxybenzylchloride		1295*	170 $_{25}$	135 $_{100}$	105 $_8$	77 $_{41}$		156	6635
Methylenedioxybenzylpiperazine		1890	220 $_{21}$	178 $_{14}$	164 $_{13}$	135 $_{100}$	85 $_{36}$	274	6624
Methylenedioxybenzylpiperazine (demethylene-methyl-) Cl-artifact		1625*	172 $_{41}$	137 $_{100}$	122 $_6$	108 $_{13}$	77 $_7$	159	6636
Methylenedioxybenzylpiperazine AC	U+UHYAC	2350	262 $_{16}$	190 $_{11}$	176 $_{21}$	135 $_{100}$	85 $_{45}$	431	6625
Methylenedioxybenzylpiperazine HFB	U+UHYHFB	2190	416 $_8$	281 $_{15}$	148 $_9$	135 $_{100}$	105 $_{11}$	1053	6631
Methylenedioxybenzylpiperazine TFA	U+UHYTFA	2350	316 $_9$	181 $_{14}$	148 $_6$	135 $_{100}$	105 $_6$	693	6628
Methylenedioxybenzylpiperazine TMS		2080	292 $_{87}$	157 $_{53}$	135 $_{100}$	102 $_{57}$	73 $_{85}$	576	6887
Methylenedioxybenzylpiperazine artifact (piperonylacetate)		1530*	194 $_{63}$	152 $_{66}$	135 $_{100}$	122 $_{17}$	104 $_{18}$	204	6637
Methylenedioxybenzylpiperazine Cl-artifact		1295*	170 $_{25}$	135 $_{100}$	105 $_8$	77 $_{41}$		156	6635
Methylenedioxybenzylpiperazine-M (deethylene-) 2AC	U+UHYAC	2320	278 $_4$	235 $_{64}$	177 $_{11}$	150 $_{25}$	135 $_{100}$	506	6626
Methylenedioxybenzylpiperazine-M (deethylene-) 2HFB	U+UHYHFB	2080	586 $_3$	389 $_{24}$	346 $_7$	240 $_7$	135 $_{100}$	1193	6632
Methylenedioxybenzylpiperazine-M (deethylene-) 2TFA	U+UHYTFA	2230	386 $_8$	289 $_{19}$	246 $_{10}$	150 $_4$	135 $_{100}$	972	6629
Methylenedioxybenzylpiperazine-M (piperonylamine) AC	U+UHYAC	2015	193 $_{100}$	150 $_{80}$	135 $_{54}$	121 $_{37}$	93 $_{27}$	199	6627
Methylenedioxybenzylpiperazine-M (piperonylamine) HFB	U+UHYHFB	1640	347 $_{100}$	317 $_6$	289 $_{10}$	178 $_6$	135 $_{75}$	834	6633
Methylenedioxybenzylpiperazine-M (piperonylamine) PFP		1755	297 $_{100}$	267 $_8$	239 $_{13}$	178 $_7$	135 $_{51}$	601	7632
Methylenedioxybenzylpiperazine-M (piperonylamine) TFA	U+UHYTFA	1775	247 $_{100}$	217 $_9$	189 $_{16}$	148 $_{12}$	135 $_{80}$	365	6630
Methylenedioxybenzylpiperazine-M (piperonylamine) 2AC		2230	235 $_{23}$	192 $_{35}$	150 $_{24}$	135 $_{17}$	93 $_9$	323	7631
Methylenedioxybenzylpiperazine-M (piperonylamine) 2TMS		2130	295 $_{27}$	280 $_{100}$	206 $_{15}$	179 $_{50}$	73 $_{51}$	592	7630
Methylenedioxybenzylpiperazine-M (piperonylamine) formyl artifact		1560	163 $_{22}$	135 $_{100}$	121 $_3$	105 $_{10}$	77 $_{37}$	142	7629
3,4-Methylenedioxymethylnitrostyrene		2025	207 $_{35}$	160 $_{50}$	131 $_{11}$	103 $_{100}$	77 $_{69}$	232	2842
Methylenedioxypyrrolidinopropiophenone		1995	178 $_1$	149 $_2$	121 $_2$	98 $_{100}$	56 $_{13}$	367	5422
Methylephedrine	G P U UHY	1430	161 $_1$	115 $_1$	105 $_2$	77 $_6$	72 $_{100}$	172	1113
Methylephedrine AC	U+UHYAC	1495	162 $_1$	91 $_2$	72 $_{100}$	117 $_2$	105 $_2$	279	1114
Methylephedrine TFA		1185	260 $_1$	162 $_2$	134 $_4$	91 $_5$	72 $_{100}$	489	4003
Methylephedrine TMS		1485	251 $_1$	236 $_1$	163 $_5$	149 $_2$	72 $_{100}$	387	4568
Methylephedrine-M (nor-)	G UHY	1375	146 $_1$	131 $_1$	105 $_3$	77 $_{12}$	58 $_{100}$	148	748
Methylephedrine-M (nor-) TMSTFA		1620	318 $_1$	227 $_9$	179 $_{100}$	110 $_8$	73 $_{79}$	773	6038
Methylephedrine-M (nor-) 2AC	PAC U+UHYAC	1795	249 $_1$	100 $_{57}$	58 $_{100}$	148 $_2$	117 $_2$	375	749
Methylephedrine-M (nor-) 2HFB	UHYHFB	1500	344 $_{10}$	254 $_{100}$	210 $_{27}$	169 $_{18}$	69 $_{36}$	1186	5097

Table 1-8-1: Compounds in order of names 17-Methyltestosterone AC

Name	Detected	RI	Typical ions and intensities					Page	Entry
Methylephedrine-M (nor-) 2PFP		1370	338_1	294_3	204_{100}	160_{25}	119_{14}	1123	2577
Methylephedrine-M (nor-) 2TFA		1345	338_1	244_4	154_{100}	110_{72}	69_{47}	872	3997
Methylephedrine-M (nor-) 2TMS		1620	294_3	163_4	147_8	130_{100}	73_{84}	663	4543
Methylephedrine-M (nor-) formyl artifact	G U	1430	177_1	121_{10}	107_6	71_{100}	56_{76}	165	4500
Methylephedrine-M (nor-) -H2O AC	U+UHYAC	1560	189_1	148_3	121_8	100_{49}	58_{100}	192	5646
1-Methylethenylcyclopropane		<1000*	82_{24}	67_{100}	53_{14}	39_{44}	27_{36}	95	3818
2-Methylhexane		<1000*	100_4	85_{31}	57_{31}	43_{100}	27_{43}	100	3819
3-Methylhexane		<1000*	100_4	70_{39}	57_{41}	43_{100}	29_{54}	100	3820
1-Methylnaphthalene	G	1230*	142_{100}	141_{87}	115_{25}	89_4	71_6	121	2555
2-Methylnaphthalene		1250*	142_{100}	141_{88}	115_{23}	89_3	71_7	121	2556
Methylnitrostyrene		1560	163_{16}	146_{12}	115_{100}	105_{59}	91_{81}	142	2839
Methylparaben		1510*	152_{30}	121_{100}	93_{30}	65_{40}		131	1115
Methylparaben AC	U+UHYAC	1500*	194_{13}	152_{79}	121_{100}	93_{15}	65_{16}	204	1829
Methylparaben ME		1495*	166_{38}	135_{100}	107_{22}	92_{32}	77_{57}	149	1116
Methylparaben-M (4-hydroxyhippuric acid) ME	U	1820	209_{100}	177_{32}	149_{34}	121_{87}		240	817
Methylparaben-M (HO-) AC	U+UHYAC	1570*	210_{10}	168_{63}	136_{100}	108_{10}		243	2974
Methylparaben-M (methoxy-)	UHY U+UHYAC	1480*	182_{76}	150_{98}	122_{100}	120_{43}	107_{28}	180	2975
3-Methylpentane		<1000*	86_{13}	71_{13}	57_{100}	56_{75}	41_{83}	97	2552
2-Methylpentane		<1000*	86_3	71_{28}	57_{12}	43_{100}	27_{29}	97	3816
Methylpentynol		<1000*	97_2	83_{28}	69_{100}	55_{12}		99	1117
1-Methylphenanthrene		1880*	192_{100}	189_{26}	165_7	95_9	83_6	198	2564
Methylphenidate		1740	172_1	115_2	91_5	84_{100}	56_{10}	316	1118
Methylphenidate AC	U+UHYAC	2085	244_1	174_1	126_{58}	91_4	84_{100}	490	1119
Methylphenidate TFA		1730	329_1	180_{100}	150_{22}	91_{15}	67_{42}	753	4005
Methylphenobarbital	P G U UHY U+UHYAC	1895	246_{10}	218_{100}	146_{23}	117_{39}		363	1120
Methylphenobarbital ET		1900	274_3	246_{100}	218_{15}	146_{29}	117_{28}	486	2449
Methylphenobarbital ME	PME UME	1860	260_2	232_{100}	175_{20}	146_{24}	117_{34}	421	1121
Methylphenobarbital-M (HO-)	U UHY	2370	262_{58}	233_{76}	162_{48}	134_{100}	77_{53}	430	1122
Methylphenobarbital-M (HO-) AC	U+UHYAC	2330	304_6	262_{100}	233_{65}	162_{29}	134_{28}	635	2930
Methylphenobarbital-M (HO-) 2ME	UME	2200	290_{95}	261_{100}	233_{78}	176_{27}	148_{93}	564	856
Methylphenobarbital-M (HO-methoxy-)	U UHY U+UHYAC	2310	292_{90}	263_{70}	231_{100}	188_{35}	164_{63}	575	2931
Methylphenobarbital-M (HO-methoxy-) 2ME	UME	2300	320_{100}	291_{91}	263_{49}	206_{10}	178_{49}	715	6407
Methylphenobarbital-M (nor-)	P G U+UHYAC	1965	232_{14}	204_{100}	161_{18}	146_{12}	117_{37}	312	854
Methylphenobarbital-M (nor-) 2TMS		2015	376_2	361_{34}	261_{15}	146_{100}	73_{46}	939	4582
Methylphenobarbital-M (nor-HO-)	U UHY	2295	248_{70}	220_{61}	219_{100}	148_{55}		370	855
Methylphenobarbital-M (nor-HO-) AC	U+UHYAC	2360	290_8	248_{100}	219_{54}	148_8	120_5	563	2507
Methylphenobarbital-M (nor-HO-) 3ME	UME	2200	290_{95}	261_{100}	233_{78}	176_{27}	148_{93}	564	856
Methylphenobarbital-M (nor-HO-methoxy-) 3ME	UME	2300	320_{100}	291_{91}	263_{49}	206_{10}	178_{49}	715	6407
2-Methylphenoxyacetic acid		1440*	166_{70}	121_{46}	107_{100}	91_{96}	77_{87}	150	2269
1-Methylpiperazine		<1000	100_{31}	70_4	58_{100}	42_{38}		100	3614
Methylprednisolone		3100*	374_1	342_1	239_{12}	136_{100}	91_{35}	934	5247
Methylprednisolone 2AC	U+UHYAC	3200*	386_1	344_1	195_7	136_{100}	91_{23}	1124	5249
Methylprednisolone -C2H4O2	P	2780*	314_1	136_{100}	121_{28}	91_{30}	77_{14}	686	5248
2-Methylpropane		<1000*	58_5	43_{100}	41_{50}	27_{28}		91	3809
Methylpseudoephedrine		1385	117_1	105_5	91_3	77_{23}	72_{100}	172	7416
Methylpseudoephedrine AC		1450	162_1	146_2	117_4	105_5	72_{100}	279	7417
Methylpseudoephedrine TFA		1215	162_2	147_1	134_4	117_3	72_{100}	489	7420
Methylpseudoephedrine TMS		1465	163_5	149_6	102_6	91_6	72_{180}	387	7419
Methylpseudoephedrine-M (nor-)	G P U	1385	146_1	105_4	91_6	77_9	58_{100}	148	2473
Methylpsychotrine		4030	478_{20}	286_{25}	272_{57}	244_{100}	206_{31}	1146	5613
1-Methylpyrene		2250*	216_{100}	215_{47}	177_6	107_{14}		260	2569
Methylpyrrolidinobutyrophenone		1790	231_1	202_1	119_{10}	112_{100}	91_{21}	310	6990
Methylpyrrolidinobutyrophenone impurity-1		1760	227_{23}	119_{28}	108_{100}	91_{20}	80_{25}	298	6991
Methylpyrrolidinobutyrophenone impurity-2		1820	229_{16}	145_{10}	110_{100}	91_{26}	70_{64}	304	6992
Methylpyrrolidinobutyrophenone-M (carboxy-) ET		2210	260_1	177_2	149_6	112_{100}	70_{11}	559	6994
Methylpyrrolidinobutyrophenone-M (carboxy-) ME		2080	163_4	135_3	112_{100}	104_6	70_9	490	7001
Methylpyrrolidinobutyrophenone-M (carboxy-) TMS		2220	318_1	221_1	178_3	112_{100}	104_4	774	7005
Methylpyrrolidinobutyrophenone-M (carboxy-deamino-oxo-) ET		1720*	234_4	189_7	177_{100}	149_{33}	104_{13}	319	6995
Methylpyrrolidinobutyrophenone-M (carboxy-deamino-oxo-) ME		1650*	220_6	163_{100}	135_{19}	120_6	104_{12}	273	7002
Methylpyrrolidinobutyrophenone-M (carboxy-dihydro-) 2TMS		2140	392_1	280_1	178_3	163_3	112_{50}	1035	7006
Methylpyrrolidinobutyrophenone-M (carboxy-oxo-) ET		2390	303_1	258_2	149_6	126_{100}	104_{10}	629	6996
Methylpyrrolidinobutyrophenone-M (carboxy-oxo-) ME		2280	289_3	254_6	163_5	126_{100}	104_7	558	6998
Methylpyrrolidinobutyrophenone-M (carboxy-oxo-) TMS		2400	347_1	332_7	221_3	178_8	126_{100}	835	7003
Methylpyrrolidinobutyrophenone-M (carboxy-oxo-dihydro-) ET		2470	305_1	260_2	142_{11}	126_{100}	98_6	640	6997
Methylpyrrolidinobutyrophenone-M (carboxy-oxo-dihydro-) ETAC		2545	268_2	226_4	179_2	126_{100}	98_5	835	7054
Methylpyrrolidinobutyrophenone-M (carboxy-oxo-dihydro-) ME		2350	260_1	165_1	126_{80}	98_7	69_{11}	569	6999
Methylpyrrolidinobutyrophenone-M (carboxy-oxo-dihydro-) 2TMS		2430	406_{10}	332_3	280_3	178_5	126_{100}	1064	7004
Methylpyrrolidinobutyrophenone-M (HO-) AC		2170	238_1	177_3	112_{100}	89_3	70_4	559	7024
Methylpyrrolidinobutyrophenone-M (HO-) TMS		2145	319_1	304_1	178_2	112_{100}	104_2	711	7055
Methylpyrrolidinobutyrophenone-M (oxo-)		2010	245_2	162_8	126_{100}	119_7	91_{11}	359	6993
Methylsalicylate	P U+UHYAC	1200*	152_{39}	120_{94}	92_{100}	65_{53}		130	954
Methylstearate	G P	2130*	298_{18}	255_{13}	143_{19}	87_{56}	74_{100}	608	970
17-Methyltestosterone		2645*	302_{100}	229_{37}	161_{36}	124_{80}	91_{40}	625	3894
17-Methyltestosterone AC		2770*	344_{23}	302_{100}	284_{78}	269_{63}	91_{64}	826	3920

17-Methyltestosterone TMS   Table 1-8-1: Compounds in order of names

Name	Detected	RI	Typical ions and intensities					Page	Entry
17-Methyltestosterone TMS		2590*	374_{73}	302_{100}	229_{30}	124_{85}	79_{80}	934	3927
17-Methyltestosterone enol 2TMS		2665*	446_{44}	356_{5}	301_{33}	143_{9}	73_{100}	1109	3979
Methylthalidomide		2470	272_{43}	229_{57}	130_{40}	104_{74}	76_{100}	478	2114
Methylthalidomide ME		2330	286_{91}	255_{38}	213_{45}	130_{100}	102_{76}	543	2082
4-Methylthio-amfetamine		1300	181_{1}	138_{36}	122_{15}	91_{13}	44_{100}	177	5942
4-Methylthio-amfetamine		1300	181_{1}	138_{36}	122_{15}	91_{13}	78_{13}	177	5941
4-Methylthio-amfetamine AC		1700	223_{5}	164_{100}	137_{22}	122_{13}	86_{26}	284	5717
4-Methylthio-amfetamine HFB		1775	377_{23}	240_{17}	164_{83}	137_{120}	69_{29}	941	5743
4-Methylthio-amfetamine PFP		1760	327_{5}	190_{8}	164_{44}	137_{100}	122_{17}	743	5744
4-Methylthio-amfetamine TFA		1750	277_{7}	164_{40}	137_{100}	122_{18}	69_{17}	499	5720
4-Methylthio-amfetamine TMS		1750	238_{13}	137_{14}	116_{100}	100_{13}	73_{73}	394	5721
4-Methylthio-amfetamine 2AC		1760	265_{2}	164_{100}	137_{28}	122_{14}	86_{43}	443	5940
4-Methylthio-amfetamine derivative ME		1940	239_{11}	164_{12}	138_{20}	102_{100}	58_{24}	340	5719
4-Methylthio-amfetamine formyl artifact		1560	193_{14}	137_{34}	122_{8}	78_{4}	56_{100}	200	5718
4-Methylthio-amfetamine-M (deamino-HO-) AC		1460*	224_{7}	164_{100}	137_{36}	122_{15}	117_{21}	288	6898
4-Methylthio-amfetamine-M (deamino-HO-) PFP	U+UHYPFP	1560*	328_{45}	191_{4}	164_{100}	137_{94}	119_{26}	747	6952
4-Methylthio-amfetamine-M (deamino-oxo-)		1335*	180_{28}	137_{100}	122_{21}			174	6899
4-Methylthio-amfetamine-M (HO-) formyl artifact 2AC ???	U+UHYAC	2240	251_{53}	195_{8}	152_{65}	137_{100}	122_{16}	383	6902
4-Methylthio-amfetamine-M (HO-) iso-1 2PFP	U+UHYPFP	1780	475_{5}	326_{4}	285_{100}	256_{5}	190_{5}	1156	6949
4-Methylthio-amfetamine-M (HO-) iso-2 2AC	U+UHYAC	2260	281_{2}	222_{100}	150_{50}	123_{17}	86_{37}	521	6896
4-Methylthio-amfetamine-M (HO-) iso-2 2PFP	U+UHYPFP	1790	475_{4}	326_{3}	285_{100}	190_{6}	152_{6}	1157	6950
4-Methylthio-amfetamine-M (methylthiobenzoic acid)		1995*	168_{100}	151_{48}	135_{7}	125_{13}	108_{15}	154	7313
4-Methylthio-amfetamine-M (methylthiobenzoic acid) ME		1610*	182_{36}	151_{100}	123_{21}	108_{19}	79_{16}	179	6900
4-Methylthio-amfetamine-M (methylthiobenzoic acid) TMS		1770*	240_{39}	225_{80}	181_{87}	151_{100}	108_{35}	342	6901
4-Methylthio-amfetamine-M (ring-HO-) 2AC	U+UHYAC	2240	281_{7}	222_{100}	180_{17}	153_{92}	86_{58}	521	6895
4-Methylthio-amfetamine-M (ring-HO-) 2PFP	U+UHYPFP	1860	475_{18}	326_{5}	312_{100}	285_{40}	190_{16}	1157	6951
4-Methylthio-amfetamine-M/artifact (Sulfoxide) AC		2360	239_{1}	222_{69}	165_{52}	137_{100}	86_{17}	340	6897
4-Methylthio-amfetamine-M/artifcat (Sulfone) AC		2455	255_{3}	196_{8}	180_{10}	107_{23}	86_{100}	401	6903
4-Methylthiobenzoic acid		1995*	168_{100}	151_{48}	135_{7}	125_{13}	108_{15}	154	7313
Methylthionium chloride artifact		2680	285_{100}	270_{38}	254_{6}	225_{5}	142_{20}	539	3387
N-Methyl-trimethylsilyl-trifluoroacetamide		<1000	199_{2}	184_{14}	134_{39}	77_{100}	73_{88}	217	5694
Methyprylone	P G U	1525	183_{3}	155_{92}	140_{98}	98_{62}	83_{100}	182	1123
Methyprylone enol AC	U+UHYAC	1610	225_{11}	183_{100}	155_{74}	127_{29}	83_{13}	291	112
Methyprylone-M (HO-) AC	U+UHYAC	1720	241_{1}	213_{6}	153_{68}	98_{100}	83_{13}	348	115
Methyprylone-M (HO-) -H2O	U UHY	1540	181_{4}	166_{47}	153_{57}	98_{29}	83_{100}	178	1124
Methyprylone-M (HO-) -H2O enol AC	U+UHYAC	1470	223_{2}	195_{15}	166_{87}	153_{100}	83_{80}	285	123
Methyprylone-M (oxo-)	U UHY U+UHYAC	1870	197_{2}	182_{7}	168_{26}	98_{53}	83_{100}	212	113
Metipranolol		2220	309_{1}	294_{1}	265_{2}	152_{18}	72_{100}	662	4257
Metipranolol AC	U+UHYAC	2260	351_{13}	336_{21}	152_{50}	98_{50}	72_{100}	850	1600
Metipranolol TMS		2260	366_{3}	308_{2}	265_{5}	152_{18}	72_{100}	958	6176
Metipranolol TMSTFA		2395	477_{1}	284_{100}	242_{11}	129_{50}	73_{58}	1146	6175
Metipranolol 2AC		2670	393_{3}	333_{13}	200_{100}	140_{80}	98_{67}	996	1361
Metipranolol formyl artifact		2240	321_{4}	306_{16}	127_{100}	112_{87}	86_{69}	720	1360
Metipranolol -H2O AC		2660	333_{23}	248_{21}	152_{31}	140_{99}	98_{59}	775	1388
Metipranolol-M (deamino-HO-) 2AC	U+UHYAC	2240*	352_{6}	310_{3}	159_{100}	152_{10}	99_{11}	854	1599
Metipranolol-M/artifact (deacetyl-)		2190	267_{1}	223_{1}	152_{32}	116_{15}	72_{100}	457	4258
Metipranolol-M/artifact (phenol) AC	U+UHYAC	1610*	236_{15}	194_{19}	152_{100}			329	1598
Metixene	G U+UHYAC-I	2500	309_{23}	197_{40}	165_{6}	99_{99}		661	553
Metixene-M (nor-) AC	U+UHYAC	2960	337_{8}	197_{100}	165_{10}	152_{3}	112_{2}	793	554
Metobromuron		2040	258_{2}	197_{2}	170_{4}	91_{11}	61_{100}	412	3887
Metobromuron ME		1735	272_{11}	212_{100}	184_{70}	105_{41}	76_{36}	477	3975
Metobromuron-M/artifact (HOOC-) ME		1800	229_{100}	197_{69}	170_{30}	91_{93}	63_{71}	302	3888
Metoclopramide	P-I G UHY	2610	299_{1}	227_{2}	184_{12}	99_{25}	86_{100}	611	1125
Metoclopramide AC	PAC U+UHYAC	2735	341_{1}	269_{2}	226_{4}	184_{5}	86_{100}	812	1126
Metoclopramide TMS		2655	371_{1}	273_{10}	256_{15}	99_{20}	86_{100}	924	4615
Metoclopramide 2TMS		2400	443_{1}	428_{3}	414_{5}	256_{29}	86_{100}	1105	4569
Metoclopramide-M (deethyl-)	UHY	2095	71_{43}	58_{100}				473	1127
Metoclopramide-M (deethyl-) 2AC	U+UHYAC	2900	355_{1}	312_{2}	226_{28}	184_{20}	58_{100}	865	1897
Metofenazate-M/artifact (deacyl-)	UHY-I	3360	403_{27}	246_{72}	171_{31}	143_{84}	70_{100}	1026	4252
Metofenazate-M/artifact (deacyl-)	UHY-I	3360	403_{50}	246_{80}	143_{25}	70_{75}	42_{100}	1026	592
Metofenazate-M/artifact (deacyl-) AC	U+UHYAC-I	3470	445_{33}	246_{41}	185_{51}	125_{60}	70_{100}	1108	373
Metofenazate-M/artifact (trimethoxybenzoic acid)		1780*	212_{100}	197_{57}	169_{13}	141_{27}		248	1949
Metofenazate-M/artifact (trimethoxybenzoic acid) ET		1770*	240_{100}	225_{44}	212_{17}	195_{45}	141_{24}	344	5219
Metofenazate-M/artifact (trimethoxybenzoic acid) ME		1740*	226_{100}	211_{49}	195_{23}	155_{21}		293	1950
Metolazone 2ME		3910	393_{15}	378_{100}	287_{7}	179_{8}	91_{16}	993	3108
Metolazone 3ME		3780	407_{3}	392_{100}	284_{15}	249_{4}	118_{9}	1033	6891
Metolazone artifact ME		3310	377_{45}	362_{100}	282_{19}	267_{21}	91_{25}	941	3109
Metolazone artifact 2ME		3245	391_{64}	376_{100}	283_{24}	268_{12}	91_{24}	987	3110
Metonitazene		3350	382_{3}	380_{5}	352_{3}	121_{31}	86_{100}	960	1128
Metoprolol	P-I G U UHY	2080	267_{1}	252_{3}	223_{10}	107_{15}	72_{100}	457	1129
Metoprolol TMS		2115	339_{1}	324_{6}	223_{11}	101_{11}	72_{100}	805	4570
Metoprolol TMSTFA		2255	435_{1}	420_{2}	284_{100}	235_{19}	73_{34}	1090	6150
Metoprolol 2AC	U+UHYAC	2480	351_{1}	291_{6}	200_{100}	98_{27}	72_{52}	850	1133
Metoprolol 2TMS		2330	396_{2}	224_{6}	144_{100}	101_{8}	73_{52}	1046	4571

Table 1-8-1: Compounds in order of names — Mirtazapine-M (oxo-)

Name	Detected	RI	Typical ions and intensities					Page	Entry
Metoprolol formyl artifact	P G U UHY	2120	279_{15}	264_{43}	127_{90}	112_{74}	56_{100}	515	1130
Metoprolol -H2O AC	U+UHYAC	2330	291_{13}	206_{56}	189_{60}	140_{78}	98_{100}	572	1134
Metoprolol-M	U UHY	2200	295_{1}	280_{2}	251_{4}	107_{7}	72_{100}	588	1132
Metoprolol-M (HO-) 3AC	U+UHYAC	2730	409_{1}	349_{3}	200_{99}	140_{67}	72_{75}	1039	1136
Metoprolol-M (HO-) artifact	U	2240	295_{36}	280_{50}	250_{52}	128_{100}	56_{61}	594	1131
Metoprolol-M (O-demethyl-) 3AC	U+UHYAC	2620	319_{12}	200_{100}	140_{55}	98_{50}	72_{60}	949	1585
Metoxuron ME		1855	242_{22}	170_{7}	155_{4}	85_{12}	72_{100}	351	4156
Metoxuron artifact (HOOC-)		1810	183_{97}	168_{100}	140_{54}	112_{22}	76_{38}	220	2515
Metoxuron artifact (HOOC-) ME		1920	215_{100}	200_{52}	183_{54}	156_{46}	59_{83}	256	2516
Metribuzin		1870	214_{4}	198_{100}	144_{13}	103_{13}	57_{21}	255	3859
Metronidazole	G P U	1725	171_{35}	124_{72}	81_{100}	54_{97}		158	1137
Metronidazole AC	U+UHYAC	1695	213_{21}	171_{37}	87_{100}			252	1138
Metronidazole TMS		1665	243_{10}	228_{47}	182_{50}	167_{62}	73_{100}	354	4572
Metronidazole-M (HO-methyl-)		2010	187_{3}	170_{100}	140_{13}	126_{23}	97_{23}	189	1830
Metronidazole-M (HO-methyl-) AC	U+UHYAC	1875	229_{1}	212_{15}	170_{14}	123_{7}	87_{100}	303	1831
Metronidazole-M (HO-methyl-) 2AC	U+UHYAC	1870	271_{9}	229_{23}	212_{11}	170_{9}	87_{100}	473	1832
Metronidazole-M (HOOC-) ME		1515	199_{38}	153_{83}	125_{15}	109_{19}	53_{100}	217	1833
Metyrapone		1930	226_{19}	120_{96}	106_{100}	92_{55}	78_{52}	295	5235
Mevinphos		1415*	224_{1}	192_{25}	164_{8}	127_{100}	109_{24}	286	4054
Mexazolam		2600	319_{79}	263_{73}	262_{100}	191_{44}	163_{38}	890	4023
Mexazolam artifact AC		2550	361_{12}	261_{11}	191_{6}	163_{5}	101_{100}	886	4024
Mexazolam HY	UHY	2180	265_{62}	230_{100}	139_{43}	111_{50}		441	543
Mexazolam HYAC	U+UHYAC	2300	307_{42}	265_{58}	230_{100}	139_{16}	111_{14}	647	290
Mexiletine		1425	179_{7}	122_{8}	105_{7}	91_{10}	58_{99}	172	1490
Mexiletine AC	U+UHYAC	1780	221_{1}	122_{5}	100_{99}	77_{8}	58_{62}	279	1491
Mexiletine-M (deamino-di-HO-) iso-1 2AC	U+UHYAC	1910*	280_{1}	238_{1}	138_{9}	101_{100}	91_{6}	517	2899
Mexiletine-M (deamino-di-HO-) iso-2 2AC	U+UHYAC	1930*	280_{1}	238_{1}	138_{9}	101_{100}	91_{6}	517	2900
Mexiletine-M (deamino-di-HO-) iso-3 2AC	U+UHYAC	1940*	280_{2}	238_{1}	138_{22}	101_{100}	91_{6}	517	3042
Mexiletine-M (deamino-HO-) AC	U+UHYAC	1530*	222_{1}	122_{19}	101_{100}	91_{9}	77_{13}	282	3041
Mexiletine-M (deamino-oxo-)	U UHY U+UHYAC	1350*	178_{75}	135_{40}	121_{28}	105_{91}	91_{40}	168	3040
Mexiletine-M (deamino-oxo-HO-) iso-1 AC	U+UHYAC	1700*	236_{7}	194_{50}	176_{26}	136_{100}	121_{96}	329	2898
Mexiletine-M (deamino-oxo-HO-) iso-2 AC	U+UHYAC	1735*	236_{21}	194_{90}	151_{49}	136_{100}	121_{39}	329	3044
Mexiletine-M (deamino-oxo-HO-) iso-3 AC	U+UHYAC	1760*	236_{10}	194_{36}	137_{100}	121_{12}	91_{3}	329	3045
Mexiletine-M (HO-) iso-1 2AC	U+UHYAC	2100	279_{1}	160_{6}	120_{9}	100_{100}	58_{50}	512	2901
Mexiletine-M (HO-) iso-2 2AC	U+UHYAC	2180	279_{1}	178_{2}	138_{14}	100_{100}	58_{44}	513	3043
Mexiletine-M (HO-) iso-3 2AC	U+UHYAC	2420	279_{1}	178_{2}	138_{11}	100_{100}	58_{56}	513	2902
Mezlocilline-M/artifact		1560	164_{20}	108_{51}	85_{100}	79_{40}	56_{58}	144	7649
Mezlocilline-M/artifact AC		1590	206_{64}	164_{85}	108_{47}	99_{70}	85_{100}	230	7659
Mezlocilline-M/artifact ME2AC		1930	314_{35}	230_{38}	198_{100}	156_{43}	97_{77}	684	7652
Mezlocilline-M/artifact ME2TFA		1755	422_{28}	326_{100}	267_{36}	196_{54}	165_{32}	1064	7656
Mezlocilline-M/artifact MEAC		1980	272_{40}	230_{91}	215_{38}	100_{53}	97_{100}	478	7651
Mezlocilline-M/artifact MEPFP		1750	376_{100}	317_{65}	246_{70}	243_{62}	215_{53}	939	7657
Mezlocilline-M/artifact TFA		1420	260_{26}	191_{64}	79_{100}	69_{75}	56_{61}	418	7658
Mezlocilline-M/artifact TMS		1535	236_{14}	221_{100}	157_{23}	100_{83}	73_{47}	327	7650
Mianserin	P-I G U+UHYAC	2210	264_{42}	193_{100}	178_{26}	165_{29}	72_{42}	440	357
Mianserin-D3		2205	267_{62}	220_{23}	193_{100}	178_{30}	165_{33}	457	7800
Mianserin-M (HO-)	U UHY	2485	280_{100}	236_{16}	209_{89}	152_{9}	72_{36}	518	1139
Mianserin-M (HO-) AC	U+UHYAC	2580	322_{76}	278_{12}	209_{100}	197_{22}	72_{25}	723	358
Mianserin-M (HO-methoxy-)	U UHY	2530	310_{100}	266_{25}	239_{66}	224_{21}	72_{26}	668	2246
Mianserin-M (HO-methoxy-) AC	U+UHYAC	2560	352_{3}	310_{31}	280_{15}	208_{100}	178_{36}	855	2260
Mianserin-M (nor-)	U UHY	2230	250_{100}	208_{83}	193_{98}	178_{26}	165_{20}	381	2245
Mianserin-M (nor-) AC	U+UHYAC	2595	292_{31}	249_{17}	207_{45}	193_{37}	100_{100}	576	359
Mianserin-M (nor-HO-) 2AC	U+UHYAC	3005	350_{49}	265_{26}	209_{44}	100_{99}		845	360
Miconazole	U+UHYAC	2955	414_{5}	333_{5}	159_{97}	121_{33}	81_{49}	1049	1492
Midazolam	P G U+UHYAC	2580	325_{26}	310_{100}	297_{10}	222_{8}	283_{6}	734	294
Midazolam-M (di-HO-) 2AC		3020	441_{15}	399_{68}	340_{50}	326_{77}	310_{100}	1099	297
Midazolam-M (HO-)	P-I UHY	2830	341_{11}	310_{100}	283_{10}	249_{6}	75_{10}	810	295
Midazolam-M (HO-) AC	U+UHYAC	2820	383_{27}	340_{56}	310_{98}			962	296
Midazolam-M/artifact	UHY	2030	249_{97}	154_{33}	123_{45}	95_{41}		373	512
Midazolam-M/artifcat AC	U+UHYAC	2195	291_{52}	249_{100}	123_{57}	95_{61}		567	286
Midodrine TMSTFA		2220	422_{3}	332_{7}	239_{100}	75_{40}	73_{54}	1065	6193
Midodrine 2AC		2610	338_{11}	278_{21}	222_{66}	167_{100}	100_{43}	798	6192
Midodrine 3TMS		2430	470_{1}	455_{5}	309_{23}	239_{31}	174_{100}	1138	6194
Minaprine		2820	298_{1}	213_{6}	186_{48}	113_{57}	100_{100}	608	4623
Minaprine AC		2870	340_{1}	228_{6}	212_{5}	113_{70}	100_{100}	808	4624
Mirex		2600*	540_{1}	508_{6}	402_{11}	272_{100}	237_{48}	1181	3454
Mirtazapine	P G U+UHYAC	2250	265_{5}	208_{16}	195_{100}	180_{9}	167_{9}	446	4487
Mirtazapine-M (HO-)	UHY	2655	281_{19}	237_{10}	224_{18}	211_{100}	195_{31}	523	4498
Mirtazapine-M (HO-) AC	U+UHYAC	2650	323_{10}	266_{15}	253_{100}	211_{59}	71_{34}	728	4490
Mirtazapine-M (nor-)	U UHY	2325	251_{54}	250_{100}	207_{22}	193_{34}	180_{25}	385	4497
Mirtazapine-M (nor-) AC	U+UHYAC	2700	293_{18}	250_{21}	209_{68}	195_{100}	100_{65}	581	4488
Mirtazapine-M (nor-HO-) 2AC	U+UHYAC	2980	351_{33}	266_{52}	210_{66}	100_{100}	56_{37}	849	4489
Mirtazapine-M (nor-HO-methoxy-) 2AC	U+UHYAC	3195	381_{67}	297_{100}	296_{93}	241_{44}	100_{120}	956	4706
Mirtazapine-M (oxo-)	U+UHYAC	2655	279_{93}	250_{100}	208_{38}	195_{100}	180_{56}	511	5261

Table 1-8-1: Compounds in order of names

Name	Detected	RI	Typical ions and intensities					Page	Entry
Mitotane	G P U	2230*	318_6	235_{100}	199_{12}	165_{25}		700	1783
Mitotane -HCl	P U	1800*	282_{49}	247_{18}	212_{100}	176_{34}		525	1888
Mitotane-M (dichlorophenylmethane)	P U	1900*	236_{54}	201_{100}	165_{82}	82_{31}		327	1743
Mitotane-M (HO-) -2HCl	P U	1790*	264_8	235_{100}	199_{19}	165_{46}		438	1884
Mitotane-M (HO-HOOC-)	P U	2040*	296_1	251_{100}	139_{88}	111_{28}		595	1893
Mitotane-M (HOOC-) ME	P U	2530*	294_{16}	259_{15}	235_{100}	199_{26}	165_{66}	584	1889
Mitotane-M/artifact (dehydro-)	G P U	2100*	316_{46}	281_7	246_{100}	210_7	176_{11}	691	1784
Mizolastine ME		3600	446_{23}	306_8	198_{48}	179_{81}	109_{100}	1109	7752
Mizolastine TMS		3720	504_{15}	489_{11}	237_{100}	109_{85}		1167	7753
2,3-MMBDB		1660	192_3	135_2	96_4	86_{100}	71_8	278	5418
2,3-MMBDB-M (demethylenyl-methyl-) AC		1890	264_1	222_1	180_1	123_2	86_{70}	446	5753
MMDA		1700	209_1	166_{27}	120_1	77_3	65_2	241	3272
MMDA		1700	209_1	166_{27}	77_3	65_2	44_{100}	241	5520
MMDA AC		2050	251_7	192_{59}	165_{15}	86_{10}	77_7	384	3264
MMDA AC		2050	251_7	192_{59}	165_{15}	86_{10}	44_{100}	384	5521
MMDA formyl artifact		1685	221_{17}	165_{49}	120_3	77_7	56_{100}	277	3258
Moclobemide	G P U+UHYAC	2210	268_1	139_8	113_{12}	100_{100}	70_7	460	4629
Moclobemide TMS		2160	339_4	325_8	254_{11}	139_{43}	100_{100}	807	7682
Moclobemide-M/artifact (chlorobenzoic acid)	G UHY U+UHYAC	1400*	156_{61}	139_{100}	111_{54}	85_4	75_{39}	136	2726
Moclobemide-M/artifact (N-oxide) -C4H9NO		1615	181_{40}	139_{100}	111_{48}	75_{26}		176	5262
Moexipril ME	UME	3575	512_1	439_7	305_{38}	234_{100}	190_{23}	1171	4742
Moexipril TMS		3345	570_1	497_3	363_{16}	234_{100}	91_{27}	1189	4980
Moexipril 2ME		3590	526_1	453_4	305_{15}	248_{100}	190_{10}	1177	4743
Moexipril -H2O	G	3805	480_{73}	463_{35}	330_{64}	190_{100}	91_{61}	1148	4746
Moexiprilate 2ME	UME	3510	498_1	439_3	305_{34}	220_{100}	190_{23}	1163	4744
Moexiprilate 3ME	UME	3580	512_1	453_2	305_{12}	234_{100}	190_{11}	1171	4745
Moexiprilate -H2O ME	UME	3775	466_{86}	449_{59}	330_{63}	190_{120}	91_{83}	1133	4747
Moexiprilate -H2O TMS	UTMS	3630	524_{49}	509_{35}	190_{85}	91_{100}	73_{48}	1176	4981
Moexiprilate-M/artifact (HOOC-) 2ET	UET	2025	307_2	234_{100}	160_{12}	117_{17}	91_{32}	651	4741
Moexiprilate-M/artifact (HOOC-) 2ME	UME	1870	279_2	220_{100}	160_{10}	117_{28}	91_{57}	512	4734
Moexiprilate-M/artifact (HOOC-) 3ME	UME	1935	293_2	234_{100}	174_8	130_{16}	91_{50}	582	4735
Moexipril-M/artifact (deethyl-) 2ME	UME	3510	498_1	439_3	305_{34}	220_{100}	190_{23}	1163	4744
Moexipril-M/artifact (deethyl-) 3ME	UME	3580	512_1	453_2	305_{12}	234_{100}	190_{11}	1171	4745
Moexipril-M/artifact (deethyl-) -H2O ME	UME	3775	466_{86}	449_{59}	330_{63}	190_{120}	91_{83}	1133	4747
Moexipril-M/artifact (deethyl-) -H2O TMS	UTMS	3630	524_{49}	509_{35}	190_{85}	91_{100}	73_{48}	1176	4981
Moexipril-M/artifact (deethyl-HOOC-) 2ME	UME	1870	279_2	220_{100}	160_{10}	117_{28}	91_{57}	512	4734
Moexipril-M/artifact (deethyl-HOOC-) 3ME	UME	1935	293_2	234_{100}	174_8	130_{16}	91_{50}	582	4735
Moexipril-M/artifact (HOOC-) ET	UET	2025	307_2	234_{100}	160_{12}	117_{17}	91_{32}	651	4740
Moexipril-M/artifact (HOOC-) ME	UME	1930	293_3	234_{36}	220_{100}	160_{11}	91_{23}	582	4736
Moexipril-M/artifact (HOOC-) 2ME	UME	1985	307_2	248_{34}	234_{100}	174_5	91_6	651	4737
Mofebutazone		2240	232_{96}	189_{49}	176_{49}	108_{100}	77_{98}	312	2015
Mofebutazone AC	U+UHYAC	2060	274_5	232_{100}	189_{24}	176_{37}	108_{49}	486	2020
Mofebutazone 2AC	U+UHYAC	2220	316_{11}	274_{63}	232_{100}	189_{95}	108_{55}	693	2021
Mofebutazone 2ME		1960	260_{45}	204_{58}	121_{100}	83_{33}	77_{74}	422	6403
Mofebutazone-M (4-HO-) AC	U+UHYAC	2210	290_{92}	220_{92}	125_{46}	108_{100}	57_{61}	564	2016
Mofebutazone-M (4-HO-) ME		2065	262_{27}	234_{10}	122_{98}	121_{100}	77_{47}	431	2036
Mofebutazone-M (4-HO-) 2AC	U+UHYAC	2110	332_8	290_{100}	220_{83}	125_{84}	108_{80}	768	2017
Mofebutazone-M (4-HO-) 2ME		2075	276_{62}	220_{22}	121_{67}	77_{100}	71_{64}	497	6404
Mofebutazone-M (HOOC-)		1930	206_{41}	108_{100}	92_{20}	77_{29}	65_{19}	381	2019
Mofebutazone-M (HOOC-) ME		2070	264_{62}	232_{26}	204_{32}	134_{100}	108_{99}	440	2022
Mofebutazone-M (HOOC-) MEAC		2250	306_1	264_{100}	232_{29}	134_{69}	108_{66}	645	2024
Mofebutazone-M (HOOC-) 2ME		2100	278_{26}	264_{16}	232_{13}	121_{100}	105_{51}	508	2023
Mofebutazone-M (HOOC-) -CO2	U+UHYAC	1600	206_4	120_{15}	99_{100}	77_{78}	71_{77}	232	2018
Monalazone artifact 2ME		1920	229_{50}	198_{48}	135_{100}	103_{40}	76_{55}	303	2479
Monalazone artifact 3ME	UME	1850	243_{54}	199_{18}	135_{100}	104_{46}	76_{51}	354	2480
Monalide		1995	239_{15}	197_{38}	168_5	127_{55}	85_{100}	340	2723
Monocrotophos		1665	223_3	192_{13}	127_{100}	97_{25}	67_{49}	283	4132
Monocrotophos TFA		1540	319_1	236_3	193_{27}	127_{100}	67_{73}	706	4133
Monoisooctyladipate		2280*	259_7	241_{15}	147_{27}	129_{100}	57_{68}	415	2360
Monolinuron		1910	214_9	153_4	126_{11}	99_{10}	61_{100}	254	3889
Monolinuron ME		1675	228_{12}	168_{100}	140_{69}	111_{21}	77_{21}	299	3976
Monolinuron-M/artifact (HOOC-) ME		1690	185_{100}	153_{90}	140_{63}	126_{38}	99_{53}	186	3890
Monuron ME		1610	212_6	140_4	111_4	72_{100}		248	3942
Moperone	UHY U+UHYAC	2800	355_3	337_3	217_{89}	204_{100}	123_{28}	867	177
Moperone -H2O	UHY U+UHYAC	2710	337_{21}	199_{15}	186_{34}	172_{100}	123_{29}	794	178
Moperone-M	U+UHYAC	1490*	180_{25}	125_{49}	123_{35}	95_{17}	56_{100}	174	85
Moperone-M	U UHY U+UHYAC	3110	329_{42}	234_{30}	185_{44}	123_{100}		751	556
Moperone-M (N-dealkyl-) -H2O AC	U+UHYAC	2105	215_{99}	173_{35}				257	559
Moperone-M (N-dealkyl-oxo-) -2H2O	U UHY U+UHYAC	1600	169_{99}	91_{100}				156	163
Moperone-M (N-dealkyl-oxo-HO-) -2H2O	UHY	1875	185_{99}	156_{76}				186	555
Moperone-M (N-dealykl-oxo-HO-) -2H2O AC	U+UHYAC	2055	227_{75}	185_{99}				298	558
MOPPP		1705	233_1	135_7	98_{100}	92_{12}	77_{13}	316	6547
MOPPP-M (deamino-oxo-)		1440*	178_3	135_{100}	107_{12}	92_{12}	77_{22}	167	6540
MOPPP-M (demethyl-)		2010	219_1	121_3	98_{100}	93_2	69_3	271	6545

Table 1-8-1: Compounds in order of names   Moxaverine-M (O-demethyl-HO-phenyl-) iso-2 2AC

Name	Detected	RI	Typical ions and intensities					Page	Entry
MOPPP-M (demethyl-) ET		1955	247_1	149_2	121_5	98_{100}	69_4	368	6543
MOPPP-M (demethyl-) HFB		1805	317_4	169_7	98_{100}	69_{14}		1052	6544
MOPPP-M (demethyl-) TMS		2005	276_2	193_2	135_3	98_{100}	73_4	571	6776
MOPPP-M (demethyl-3-HO-) 2ET		2165	290_1	193_1	165_2	137_3	98_{100}	571	6525
MOPPP-M (demethyl-3-methoxy-) ET		2135	277_1	208_1	179_2	151_7	98_{100}	502	6524
MOPPP-M (demethyl-3-methoxy-) HFB		1960	445_1	347_4	98_{100}	69_{19}		1107	6532
MOPPP-M (demethyl-3-methoxy-) ME		2070	165_1	98_{100}	79_3	69_4	56_{11}	436	6538
MOPPP-M (demethyl-3-methoxy-) TMS		1960	321_1	306_3	223_3	165_1	98_{100}	719	6533
MOPPP-M (demethyl-3-methoxy-deamino-oxo-) ET		1680*	222_4	179_{100}	151_{81}	123_{19}		281	6523
MOPPP-M (demethyl-deamino-oxo-) ET		1530*	192_3	149_{100}	121_{95}	93_{25}	65_{21}	197	6539
MOPPP-M (dihydro-)		1935	234_1	135_3	98_{100}	77_9	56_{21}	326	6697
MOPPP-M (dihydro-) AC		1970	218_3	135_9	98_{100}	77_7	56_{18}	502	6702
MOPPP-M (dihydro-) TMS		1880	292_2	218_1	135_2	98_{100}	73_{17}	652	6707
MOPPP-M (oxo-)		2120	164_{15}	135_{15}	121_{14}	112_{100}		367	6542
MOPPP-M (parahydroxybenzoic acid) ET		1585*	166_{53}	151_4	138_{65}	121_{100}	93_{10}	151	6541
MOPPP-M (parahydroxybenzoic acid) 2ET		1520*	194_{17}	166_7	149_{100}	121_{76}	93_{17}	205	6646
Morazone		----	377_3	201_{100}	176_{36}	56_{64}		943	1226
Morazone-M (carboxy-phenazone) -CO2	P G U UHY U+UHYAC	1845	188_{100}	96_{81}	77_{51}			190	199
Morazone-M/artifact (HO-methoxy-phenmetrazine)	UHY	1900	223_6	151_3	107_5	71_{100}	56_5	285	3518
Morazone-M/artifact (HO-methoxy-phenmetrazine) 2AC	U+UHYAC	2320	307_6	265_{22}	113_{86}	86_{24}	71_{100}	649	1887
Morazone-M/artifact (HO-phenmetrazine) iso-1	UHY	1830	193_{11}	121_7	107_6	71_{100}	56_{62}	201	562
Morazone-M/artifact (HO-phenmetrazine) iso-1 2AC	U+UHYAC	2150	277_9	234_{17}	113_{55}	85_{26}	71_{100}	501	849
Morazone-M/artifact (HO-phenmetrazine) iso-2	UHY	1865	193_8	163_6	121_6	71_{100}	56_{43}	202	3517
Morazone-M/artifact (HO-phenmetrazine) iso-2 2AC	U+UHYAC	2200	277_3	234_8	113_{83}	85_{39}	71_{100}	501	848
Morazone-M/artifact (phenmetrazine)	U UHY	1440	177_8	105_5	77_{10}	71_{100}	56_{54}	166	851
Morazone-M/artifact (phenmetrazine) AC	U+UHYAC	1810	219_5	176_{10}	113_{77}	86_{50}	71_{100}	271	198
Morazone-M/artifact (phenmetrazine) TFA		1530	273_4	167_{85}	105_{36}	98_{47}	70_{100}	481	4002
Morazone-M/artifact (phenmetrazine) TMS		1620	249_6	143_{19}	115_{36}	100_{100}	73_{64}	377	5446
Morazone-M/artifact-1	UHY U+UHYAC	1670	204_{72}	176_{17}	92_{100}	77_{17}	65_{18}	225	560
Morazone-M/artifact-2	UHY U+UHYAC	1680	188_{100}	159_3	91_{30}	77_{65}	55_{47}	190	561
Morazone-M/artifact-2 AC	U+UHYAC	1690	230_7	188_{100}	159_6	91_{11}	77_{55}	306	3520
Morazone-M/artifact-3	UHY	1920	202_{64}	110_{72}	82_{23}	77_{52}	56_{100}	222	3519
Morphine	G UHY	2455	285_{100}	268_{15}	162_{58}	124_{20}		539	474
Morphine ME	P G U UHY	2375	299_{100}	229_{26}	162_{46}	124_{23}		611	473
Morphine TFA		2285	381_{55}	268_{100}	146_{14}	115_{13}	69_{23}	955	5569
Morphine 2AC	G PHYAC U+UHYAC	2620	369_{59}	327_{100}	310_{36}	268_{47}	162_{11}	915	225
Morphine 2HFB		2375	677_{10}	480_{11}	464_{100}	407_9	169_8	1201	6120
Morphine 2PFP		2360	577_{51}	558_7	430_8	414_{100}	119_{22}	1191	2251
Morphine 2TFA		2250	477_{71}	364_{100}	307_6	115_8	69_{31}	1145	4008
Morphine 2TMS	UHYTMS	2560	429_{19}	236_{21}	196_{15}	146_{21}	73_{100}	1079	2463
Morphine Cl-artifact 2AC	U+UHYAC	2680	403_{59}	361_{100}	344_{63}	302_{90}	204_{55}	1026	2992
Morphine-D3 ME		2495	344_{100}	285_{91}	232_{52}	193_{36}	156_{37}	825	7300
Morphine-D3 ME		2370	302_{100}	232_{26}	165_{47}	127_{24}		625	7295
Morphine-D3 TFA		2275	384_{39}	271_{100}	211_8	165_6	152_7	967	5572
Morphine-D3 2AC		2510	372_{87}	330_{100}	313_{57}	271_{87}	218_{56}	928	7294
Morphine-D3 2HFB		2375	680_4	483_9	467_{100}	169_{23}	414_7	1202	6126
Morphine-D3 2PFP		2350	580_{16}	433_7	417_{100}	269_5	119_8	1192	5567
Morphine-D3 2TFA		2240	480_{32}	383_6	367_{100}	314_6	307_6	1148	5571
Morphine-D3 2TMS		2550	432_{100}	290_{30}	239_{66}	199_{44}	73_{110}	1085	5578
Morphine-M (nor-) 2PFP		2440	563_{100}	400_{10}	355_{38}	327_7	209_{15}	1188	3534
Morphine-M (nor-) 3AC	U+UHYAC	2955	397_8	355_9	209_{41}	87_{100}	72_{33}	1010	1194
Morphine-M (nor-) 3PFP	UHYPFP	2405	709_{80}	533_{28}	388_{29}	367_{51}	355_{100}	1203	3533
Morphine-M (nor-) 3TMS	UHYTMS	2605	487_9	416_9	222_{18}	131_{10}	73_{50}	1155	3525
Morpholine		<1000	87_{12}	57_{23}	42_6			97	3612
Moxaverine	P U+UHYAC	2530	307_{61}	292_{99}	248_{11}	91_{10}		650	1493
Moxaverine-M (O-demethyl-) iso-1	UHY	2560	293_{29}	278_{100}	250_{43}	232_8	139_8	580	3215
Moxaverine-M (O-demethyl-) iso-1 AC	U+UHYAC	2610	335_{26}	320_{16}	292_{30}	278_{100}	250_{39}	784	3219
Moxaverine-M (O-demethyl-) iso-2	UHY	2645	293_{40}	292_{100}	276_{11}	248_5	204_2	580	3216
Moxaverine-M (O-demethyl-) iso-2 AC	U+UHYAC	2630	335_6	292_{100}	276_{29}	248_5	204_3	784	3220
Moxaverine-M (O-demethyl-di-HO-) iso-1 3AC	U+UHYAC	2910	451_{33}	393_{55}	336_{100}	306_{62}	290_{47}	1114	3231
Moxaverine-M (O-demethyl-di-HO-) iso-2 3AC	U+UHYAC	3075	451_{20}	408_{58}	392_{98}	349_{84}	306_{100}	1114	3233
Moxaverine-M (O-demethyl-di-HO-methoxy-) 3AC	U+UHYAC	3530	481_{18}	438_{43}	422_{52}	379_{100}	364_{42}	1149	3235
Moxaverine-M (O-demethyl-HO-ethyl-) -H2O iso-1	UHY	2625	291_{49}	290_{38}	276_{100}	248_{63}	230_{17}	568	3217
Moxaverine-M (O-demethyl-HO-ethyl-) -H2O iso-1 AC	U+UHYAC	2660	333_{53}	318_{19}	290_{31}	276_{100}	248_{44}	773	3221
Moxaverine-M (O-demethyl-HO-ethyl-) -H2O iso-2	UHY	2710	291_{45}	290_{100}	274_{14}	246_7	230_6	568	3218
Moxaverine-M (O-demethyl-HO-ethyl-) -H2O iso-2 AC	U+UHYAC	2680	333_{10}	290_{100}	274_{24}	246_8	230_8	773	3222
Moxaverine-M (O-demethyl-HO-ethyl-) iso-1 AC	U+UHYAC	2760	351_{32}	336_{54}	308_{100}	276_{52}	248_{38}	848	3223
Moxaverine-M (O-demethyl-HO-ethyl-) iso-1 2AC	U+UHYAC	2815	393_{52}	378_{29}	350_{73}	308_{100}	276_{58}	994	3225
Moxaverine-M (O-demethyl-HO-ethyl-) iso-2 AC	U+UHYAC	2795	351_{54}	336_{100}	308_{94}	276_{75}	91_{16}	849	3226
Moxaverine-M (O-demethyl-HO-ethyl-) iso-2 2AC	U+UHYAC	2830	393_{11}	350_{44}	308_{77}	290_{100}	274_{36}	994	3228
Moxaverine-M (O-demethyl-HO-methoxy-phenyl-) iso-1 2AC	U+UHYAC	2860	423_{17}	381_{33}	350_{56}	338_{100}	306_{48}	1066	3229
Moxaverine-M (O-demethyl-HO-methyl-) iso-2 2AC	U+UHYAC	3120	423_{34}	408_{28}	380_{100}	348_{38}	321_{25}	1066	3234
Moxaverine-M (O-demethyl-HO-phenyl-) iso-1 2AC	U+UHYAC	2895	393_{11}	350_{48}	334_{36}	290_{100}	274_{20}	994	3230
Moxaverine-M (O-demethyl-HO-phenyl-) iso-2 2AC	U+UHYAC	2930	393_{12}	350_{100}	334_{21}	308_{48}	292_7	995	3232

Table 1-8-1: Compounds in order of names

Name	Detected	RI	Typical ions and intensities				Page	Entry
Moxaverine-M (O-demethyl-oxo-ethyl-) iso-1 AC	U+UHYAC	2775	349_{11}	306_{100}	290_{33}	264_{17} 91_7	842	3224
Moxaverine-M (O-demethyl-oxo-ethyl-) iso-2 AC	U+UHYAC	2785	349_{55}	306_{99}	292_{100}	264_{42} 91_7	842	3225
Moxonidine AC	U+UHYAC	2380	283_{62}	248_{77}	206_{79}	176_{42} 86_{100}	528	6806
Moxonidine 2AC	U+UHYAC	2455	325_{23}	290_{24}	248_{33}	128_{100} 86_{47}	734	1277
MPBP		1790	231_1	202_1	119_{10}	112_{100} 91_{21}	310	6990
MPBP impurity-1		1760	227_{23}	119_{28}	108_{100}	91_{20} 80_{25}	298	6991
MPBP impurity-2		1820	229_{16}	145_{10}	110_{100}	91_{26} 70_{64}	304	6992
MPBP-M (carboxy-) ET		2210	260_1	177_2	149_6	112_{100} 70_{11}	559	6998
MPBP-M (carboxy-) ME		2080	163_4	135_3	112_{100}	104_6 70_9	490	7001
MPBP-M (carboxy-) TMS		2220	318_1	221_1	178_3	112_{100} 104_4	774	7005
MPBP-M (carboxy-deamino-oxo-) ET		1720*	234_4	189_7	177_{100}	149_{33} 104_{13}	319	6995
MPBP-M (carboxy-deamino-oxo-) ME		1650*	220_6	163_{100}	135_{19}	120_6 104_{12}	273	7002
MPBP-M (carboxy-dihydro-) 2TMS		2140	392_1	280_1	178_3	163_3 112_{50}	1035	7006
MPBP-M (carboxy-oxo-) ET		2390	303_1	258_2	149_6	126_{100} 104_{10}	629	6996
MPBP-M (carboxy-oxo-) ME		2280	289_3	258_4	163_5	126_{100} 104_7	558	6998
MPBP-M (carboxy-oxo-) TMS		2400	347_1	332_7	221_3	178_8 126_{100}	835	7003
MPBP-M (carboxy-oxo-dihydro-) ET		2470	305_1	260_2	142_{11}	126_{100} 98_6	640	6997
MPBP-M (carboxy-oxo-dihydro-) ETAC		2545	268_2	226_4	179_2	126_{100} 98_5	835	7054
MPBP-M (carboxy-oxo-dihydro-) ME		2350	260_1	165_1	126_{80}	98_7 69_{11}	569	6999
MPBP-M (carboxy-oxo-dihydro-) 2TMS		2430	406_{10}	332_3	280_3	178_5 126_{100}	1064	7004
MPBP-M (HO-) AC		2170	238_1	177_3	112_{100}	89_3 70_4	559	7024
MPBP-M (HO-) TMS		2145	319_1	304_1	178_2	112_{100} 104_2	711	7055
MPBP-M (oxo-)		2010	245_2	162_8	126_{100}	119_7 91_{11}	359	6993
MPCP		2150	273_{29}	272_{30}	230_{100}	121_{35} 84_{22}	482	3594
MPHP		1965	140_{100}	119_4	91_8	84_5 65_5	418	6647
MPHP-M (carboxy-)		2305	289_1	202_1	149_5	140_{100} 121_2	560	6651
MPHP-M (carboxy-) ET		2335	260_1	177_3	149_4	140_{100} 104_8	699	6666
MPHP-M (carboxy-) ME		2260	246_1	163_4	140_{100}	104_8 84_5	630	6662
MPHP-M (carboxy-) TMS		2390	304_1	221_2	178_7	140_{100} 104_6	888	6655
MPHP-M (carboxy-HO-alkyl-) ET		2545	177_3	156_{100}	149_4	138_4 104_8	775	6667
MPHP-M (carboxy-HO-alkyl-) ME		2460	163_6	156_{100}	149_9	138_5 104_7	709	6663
MPHP-M (carboxy-HO-alkyl-) MEAC		2715	198_{100}	163_4	138_{49}	104_4	888	6672
MPHP-M (carboxy-HO-alkyl-) iso-1 2TMS		2625	434_1	228_{100}	221_4	178_5 138_{23}	1112	6657
MPHP-M (carboxy-HO-alkyl-) iso-2 2TMS		2635	434_2	228_{100}	221_4	178_{10} 138_{62}	1113	6759
MPHP-M (di-HO-) 2AC		2600	198_{100}	178_5	138_{47}		937	6649
MPHP-M (di-HO-) 2TMS		2525	420_4	228_{100}	207_5	138_{60} 73_{25}	1091	6654
MPHP-M (dihydro-)		1965	260_1	77_9	119_2	140_{100} 91_{13}	428	6699
MPHP-M (dihydro-) AC		1990	244_2	140_{100}	119_5	91_7 77_4	631	6704
MPHP-M (dihydro-) TMS		1900	318_2	244_1	140_{100}	98_8 73_{14}	778	6709
MPHP-M (HO-alkyl-) iso-1 AC		2250	316_1	198_{100}	138_{60}	119_{12} 91_{17}	699	6693
MPHP-M (HO-alkyl-) iso-2 AC		2445	198_{100}	138_{65}	119_{10}	91_{19}	699	6694
MPHP-M (HO-tolyl-)		2250	218_1	140_{100}	135_4	77_6	492	6673
MPHP-M (HO-tolyl-) AC		2315	260_1	177_4	140_{100}	89_{10}	699	6675
MPHP-M (HO-tolyl-) TFA		2085	314_2	231_6	140_{100}	89_{11}	924	6674
MPHP-M (oxo-)		2165	190_3	154_{100}	119_9	98_{34} 91_{18}	482	6652
MPHP-M (oxo-carboxy-) ET		2525	286_2	177_3	154_{100}	98_{31} 86_{11}	765	6665
MPHP-M (oxo-carboxy-) ME		2445	286_1	221_6	163_3	154_{100} 98_{32}	698	6659
MPHP-M (oxo-carboxy-) TMS		2160	360_1	221_1	178_5	154_{100} 104_6	937	6656
MPHP-M (oxo-carboxy-dihydro-) ET		2620	288_1	179_1	154_{100}	112_{10} 98_{30}	776	6664
MPHP-M (oxo-carboxy-dihydro-) ME		2555	165_3	154_{100}	98_{28}	86_{14}	710	6660
MPHP-M (oxo-carboxy-dihydro-) MEAC		2725	330_1	154_{100}	98_{21}	86_7	888	6671
MPHP-M (oxo-carboxy-dihydro-) 2ME		2430	179_8	154_{100}	148_4	120_{10} 98_{37}	775	6668
MPHP-M (oxo-carboxy-HO-alkyl-) ET		2640	177_4	170_{100}	149_4	142_{22} 104_{10}	835	6653
MPHP-M (oxo-carboxy-HO-alkyl-) ME		2575	170_{100}	142_{25}	163_{22}	104_{22} 98_{22}	774	6661
MPHP-M (oxo-carboxy-HO-alkyl-) MEAC		2890	344_3	261_2	212_{53}	170_{22} 152_{100}	936	6670
MPHP-M (oxo-carboxy-HO-alkyl-) 2TMS		2695	448_{13}	242_{100}	221_{14}	214_{33} 98_{51}	1131	6658
MPHP-M (oxo-HO-alkyl-) AC		2425	331_1	212_{69}	170_{24}	152_{100} 98_{11}	765	6648
MPHP-M (oxo-HO-tolyl-) AC		2515	248_1	177_4	154_{100}	98_{32}	765	6650
MPHP-M (oxo-HO-tolyl-) HFB		2305	331_3	154_{100}	98_{31}		1152	6669
MPPP		1725	216_1	119_4	98_{100}	91_{17} 56_{53}	262	5736
MPPP-M (carboxy-)		2200	247_1	149_2	121_7	98_{100} 56_{13}	367	6500
MPPP-M (carboxy-) ET		2320	275_2	230_6	177_{27}	149_{13} 98_{100}	490	6498
MPPP-M (carboxy-) ME		2030	163_1	135_1	104_3	98_{80} 56_{12}	426	6502
MPPP-M (carboxy-) TMS		2195	304_1	221_1	178_4	104_6 98_{100}	709	6793
MPPP-M (carboxy-deamino-oxo-) ET		1620*	220_1	177_{100}	149_{58}	121_{17} 104_{15}	273	6494
MPPP-M (carboxy-deamino-oxo-) ME		1635*	206_7	177_{100}	149_{31}	121_8 104_9	230	6496
MPPP-M (carboxy-oxo-) ET		2335	289_1	244_3	149_7	112_{100} 84_4	558	6499
MPPP-M (dihydro-)		1765	218_1	105_1	98_{100}	77_{10} 56_{21}	272	6696
MPPP-M (dihydro-) AC		1815	202_2	119_5	98_{100}	91_7 56_{15}	427	6701
MPPP-M (dihydro-) TMS		1730	276_3	202_1	115_3	98_{100} 73_{17}	573	6706
MPPP-M (HO-)		2020	233_1	135_4	98_{100}	77_{19} 56_{35}	316	6503
MPPP-M (HO-) AC		2115	177_3	56_{34}	98_{100}	89_{17}	491	6504
MPPP-M (HO-) TMS		2095	290_8	135_4	98_{100}		641	6794
MPPP-M (oxo-)		1920	231_1	119_7	112_{100}	84_4 69_9	309	6501

Table 1-8-1: Compounds in order of names — Nalbuphine 3TMS

Name	Detected	RI	Typical ions and intensities					Page	Entry
MPPP-M (p-dicarboxy-) ET		1715*	194 $_{19}$	166 $_{40}$	149 $_{100}$	121 $_{17}$	65 $_{15}$	204	6497
MPPP-M (p-dicarboxy-) ETME		1560*	208 $_{17}$	193 $_{11}$	180 $_{37}$	163 $_{100}$	149 $_{43}$	237	6493
MPPP-M (p-dicarboxy-) 2ET		1645*	222 $_{12}$	194 $_{18}$	177 $_{100}$	166 $_{22}$	149 $_{60}$	281	6495
MSTFA		<1000	199 $_{2}$	184 $_{14}$	134 $_{39}$	77 $_{100}$	73 $_{88}$	217	5694
4-MTA		1300	181 $_{1}$	138 $_{36}$	122 $_{15}$	91 $_{13}$	44 $_{100}$	177	5942
4-MTA		1300	181 $_{1}$	138 $_{36}$	122 $_{15}$	91 $_{13}$	78 $_{13}$	177	5941
4-MTA AC		1700	223 $_{5}$	164 $_{100}$	137 $_{22}$	122 $_{13}$	86 $_{26}$	284	5717
4-MTA HFB		1775	377 $_{23}$	240 $_{17}$	164 $_{83}$	137 $_{120}$	69 $_{29}$	941	5743
4-MTA PFP		1760	327 $_{5}$	190 $_{8}$	164 $_{44}$	137 $_{100}$	122 $_{17}$	743	5744
4-MTA TFA		1750	277 $_{7}$	164 $_{40}$	137 $_{100}$	122 $_{18}$	69 $_{17}$	499	5720
4-MTA TMS		1750	238 $_{13}$	137 $_{14}$	116 $_{100}$	100 $_{13}$	73 $_{73}$	394	5721
4-MTA 2AC		1760	265 $_{2}$	164 $_{100}$	137 $_{28}$	122 $_{14}$	86 $_{43}$	443	5940
4-MTA formyl artifact		1560	193 $_{14}$	137 $_{34}$	122 $_{8}$	78 $_{4}$	56 $_{100}$	200	5718
4-MTA-M (deamino-HO-) AC		1460*	224 $_{7}$	164 $_{100}$	137 $_{36}$	122 $_{15}$	117 $_{21}$	288	6898
4-MTA-M (deamino-HO-) PFP	U+UHYPFP	1560*	328 $_{45}$	191 $_{4}$	164 $_{100}$	137 $_{94}$	119 $_{26}$	747	6952
4-MTA-M (deamino-oxo-)		1335*	180 $_{28}$	137 $_{100}$	122 $_{21}$			174	6899
4-MTA-M (HO-) iso-1 2PFP	U+UHYPFP	1780	475 $_{4}$	326 $_{4}$	285 $_{100}$	256 $_{5}$	190 $_{5}$	1156	6949
4-MTA-M (HO-) iso-2 2AC	U+UHYAC	2260	281 $_{2}$	222 $_{100}$	150 $_{50}$	123 $_{17}$	86 $_{37}$	521	6896
4-MTA-M (HO-) iso-2 2PFP	U+UHYPFP	1790	475 $_{4}$	326 $_{3}$	285 $_{100}$	190 $_{6}$	152 $_{6}$	1157	6950
4-MTA-M (methylthiobenzoic acid)		1995*	168 $_{100}$	151 $_{48}$	135 $_{7}$	125 $_{13}$	108 $_{15}$	154	7313
4-MTA-M (methylthiobenzoic acid) ME		1610*	182 $_{36}$	151 $_{100}$	123 $_{21}$	108 $_{19}$	79 $_{16}$	179	6900
4-MTA-M (methylthiobenzoic acid) TMS		1770*	240 $_{39}$	225 $_{80}$	181 $_{87}$	151 $_{100}$	108 $_{35}$	342	6901
4-MTA-M (ring-HO-) 2AC	U+UHYAC	2240	281 $_{7}$	222 $_{100}$	180 $_{17}$	153 $_{92}$	86 $_{58}$	521	6895
4-MTA-M (ring-HO-) 2PFP	U+UHYPFP	1860	475 $_{18}$	326 $_{5}$	312 $_{100}$	285 $_{40}$	190 $_{16}$	1157	6951
4-MTA-M/artifact (Sulfone) AC		2455	255 $_{3}$	196 $_{8}$	180 $_{10}$	107 $_{23}$	86 $_{100}$	401	6903
4-MTA-M/artifact (Sulfoxide) AC		2360	239 $_{1}$	222 $_{69}$	165 $_{52}$	137 $_{100}$	86 $_{17}$	340	6897
Muzolimine		2445	271 $_{17}$	256 $_{5}$	173 $_{16}$	137 $_{10}$	99 $_{100}$	473	4175
Muzolimine ME		2170	285 $_{31}$	173 $_{14}$	137 $_{9}$	113 $_{100}$	84 $_{13}$	537	4178
Muzolimine MEAC		2520	327 $_{14}$	312 $_{5}$	173 $_{23}$	155 $_{18}$	113 $_{100}$	743	4231
Muzolimine METFA		2290	381 $_{8}$	209 $_{46}$	173 $_{100}$	137 $_{43}$	102 $_{58}$	954	4230
Muzolimine TMS		2210	343 $_{76}$	328 $_{18}$	171 $_{100}$	156 $_{28}$	73 $_{77}$	819	4181
Muzolimine 2AC		2625	355 $_{2}$	313 $_{20}$	173 $_{33}$	141 $_{52}$	99 $_{100}$	864	4176
Muzolimine 2ME		2190	299 $_{40}$	173 $_{11}$	127 $_{100}$	98 $_{15}$	55 $_{19}$	609	4179
Muzolimine 2TFA		2020	463 $_{32}$	448 $_{8}$	173 $_{100}$	102 $_{28}$	69 $_{40}$	1130	4177
Muzolimine 2TMS		2265	415 $_{26}$	400 $_{7}$	242 $_{18}$	214 $_{21}$	73 $_{100}$	1052	4182
Muzolimine 3ME		2235	313 $_{57}$	298 $_{22}$	173 $_{8}$	141 $_{100}$	84 $_{35}$	679	4180
Mycophenolic acid	U+UHYAC	3000*	320 $_{79}$	302 $_{34}$	247 $_{69}$	207 $_{100}$	159 $_{31}$	715	6421
Mycophenolic acid ME	P	2260*	334 $_{45}$	316 $_{53}$	247 $_{100}$	229 $_{39}$	207 $_{73}$	780	6420
Mycophenolic acid 2ME		2270*	348 $_{54}$	316 $_{19}$	275 $_{26}$	243 $_{48}$	221 $_{100}$	839	6795
Myristic acid	P	1760*	228 $_{14}$	185 $_{18}$	129 $_{34}$	73 $_{100}$	60 $_{98}$	302	1140
Myristic acid ET		1720*	256 $_{6}$	213 $_{16}$	157 $_{24}$	101 $_{53}$	88 $_{100}$	408	5401
Myristic acid ME	PME	1710*	242 $_{3}$	199 $_{8}$	143 $_{11}$	87 $_{65}$	74 $_{100}$	353	1141
Myristic acid TMS		2280*	300 $_{5}$	285 $_{63}$	149 $_{38}$	117 $_{71}$	73 $_{100}$	617	4644
Myristic acid glycerol ester	G	2260*	285 $_{6}$	271 $_{10}$	211 $_{66}$	98 $_{83}$	55 $_{100}$	625	5587
Myristic acid isopropyl ester	G	1830*	270 $_{2}$	228 $_{55}$	211 $_{31}$	102 $_{80}$	60 $_{100}$	472	6469
Myristicin	G	1400*	192 $_{100}$	165 $_{23}$	147 $_{17}$	119 $_{29}$	91 $_{31}$	197	4374
Myristicin-M (1-HO-) AC	U+UHYAC	2020*	250 $_{100}$	208 $_{30}$	154 $_{47}$	149 $_{49}$	133 $_{79}$	379	7150
Myristicin-M (demethyl-) AC	U+UHYAC	1655*	220 $_{13}$	178 $_{100}$	147 $_{30}$	119 $_{20}$	91 $_{29}$	273	7145
Myristicin-M (demethylenyl-) 2AC	U+UHYAC	1880*	264 $_{1}$	222 $_{30}$	180 $_{100}$	147 $_{15}$	91 $_{34}$	439	7148
Myristicin-M (demethylenyl-methyl-) AC	U+UHYAC	1755*	236 $_{35}$	194 $_{100}$	179 $_{43}$	119 $_{15}$	91 $_{12}$	328	7140
Myristicin-M (di-HO-) 2AC	U+UHYAC	2210*	310 $_{7}$	250 $_{70}$	208 $_{41}$	165 $_{100}$	77 $_{33}$	666	7149
Nabumetone		1875*	228 $_{60}$	185 $_{21}$	171 $_{100}$	141 $_{25}$	128 $_{38}$	301	7534
Nabumetone-M/artifact (O-demethyl-)		1925*	214 $_{54}$	171 $_{28}$	157 $_{100}$	128 $_{16}$	115 $_{14}$	256	7536
Nabumetone-M/artifact (O-demethyl-) AC		1990*	256 $_{25}$	214 $_{85}$	171 $_{41}$	157 $_{100}$	128 $_{21}$	407	7535
Nadolol		2540	309 $_{1}$	294 $_{10}$	265 $_{3}$	86 $_{100}$	57 $_{11}$	663	2612
Nadolol 3AC	U+UHYAC	2650	435 $_{1}$	420 $_{35}$	183 $_{7}$	112 $_{17}$	86 $_{100}$	1091	1363
Nadolol 3TMS		2250	525 $_{1}$	510 $_{3}$	147 $_{8}$	86 $_{100}$	73 $_{30}$	1177	5488
Nadolol formyl artifact		2560	321 $_{8}$	306 $_{100}$	201 $_{14}$	141 $_{24}$	70 $_{34}$	720	1362
Nadolol-M/artifact (deisobutyl-) -2H2O 2AC		2540	301 $_{56}$	259 $_{10}$	241 $_{36}$	98 $_{100}$	57 $_{51}$	619	1706
Naftidrofuryl	G P U+UHYAC	2840	383 $_{1}$	368 $_{1}$	141 $_{10}$	99 $_{29}$	86 $_{100}$	965	2826
Naftidrofuryl-M (deethyl-)	U UHY	2780	355 $_{3}$	296 $_{16}$	198 $_{47}$	141 $_{100}$	58 $_{85}$	867	2827
Naftidrofuryl-M (di-oxo-HOOC-) ME	UHYME	2810*	326 $_{42}$	198 $_{54}$	153 $_{67}$	141 $_{100}$	71 $_{20}$	739	2830
Naftidrofuryl-M (HO-HOOC-) MEAC	U+UHYAC	2740*	356 $_{1}$	283 $_{35}$	153 $_{63}$	141 $_{100}$	73 $_{75}$	869	2831
Naftidrofuryl-M (HO-oxo-HOOC-) MEAC	U+UHYAC	2920*	297 $_{36}$	198 $_{47}$	153 $_{62}$	141 $_{100}$	115 $_{30}$	919	2832
Naftidrofuryl-M (oxo-HOOC-) ME	UME U+UHYAC	2760*	312 $_{44}$	198 $_{50}$	153 $_{74}$	141 $_{100}$	115 $_{30}$	676	2829
Naftidrofuryl-M/artifact (HOOC-) ME	UHYME U+UHYAC	2390*	298 $_{25}$	153 $_{52}$	141 $_{100}$	84 $_{63}$	71 $_{94}$	607	2828
Nalbuphine	G	2960	357 $_{15}$	302 $_{100}$	284 $_{4}$			874	3061
Nalbuphine AC	U+UHYAC	3030	399 $_{10}$	344 $_{100}$	326 $_{3}$	302 $_{5}$		1016	3063
Nalbuphine 2AC	U+UHYAC	3110	441 $_{9}$	386 $_{100}$	344 $_{8}$	296 $_{3}$		1101	3064
Nalbuphine 2HFB		2560	680 $_{10}$	662 $_{23}$	405 $_{33}$	263 $_{67}$	169 $_{100}$	1204	6135
Nalbuphine 2PFP		2700	649 $_{11}$	594 $_{100}$	486 $_{11}$	400 $_{10}$	119 $_{26}$	1199	6124
Nalbuphine 3AC	U+UHYAC	3080	483 $_{14}$	440 $_{8}$	428 $_{100}$	368 $_{22}$	326 $_{10}$	1151	3065
Nalbuphine 3PFP		2510	795 $_{4}$	740 $_{87}$	576 $_{97}$	412 $_{100}$	357 $_{89}$	1205	6125
Nalbuphine 3TMS		2860	573 $_{14}$	518 $_{9}$	429 $_{5}$	101 $_{4}$	73 $_{70}$	1189	6205

Nalbuphine-M (N-dealkyl-) Table 1-8-1: Compounds in order of names

Name	Detected	RI	Typical ions and intensities					Page	Entry
Nalbuphine-M (N-dealkyl-)		2930	289 $_{100}$	272 $_{36}$	242 $_{12}$	202 $_{20}$	115 $_{13}$	558	3062
Nalbuphine-M (N-dealkyl-) 2AC		2970	373 $_{39}$	331 $_{50}$	313 $_{19}$	227 $_{41}$	87 $_{100}$	931	3066
Nalbuphine-M (N-dealkyl-) 3AC		3020	415 $_{34}$	373 $_{66}$	296 $_{25}$	227 $_{56}$	87 $_{100}$	1052	3067
Naled		1640*	301 $_{17}$	299 $_{10}$	189 $_{12}$	145 $_{83}$	109 $_{100}$	944	3430
Nalorphine	UHY	2620	311 $_{100}$	294 $_{11}$	282 $_{6}$	241 $_{12}$	188 $_{14}$	672	1736
Nalorphine AC		2800	353 $_{100}$	294 $_{55}$	241 $_{16}$	230 $_{9}$		859	1738
Nalorphine 2AC	U+UHYAC	2820	395 $_{86}$	353 $_{100}$	336 $_{50}$	294 $_{30}$	230 $_{18}$	1002	1737
Nalorphine 2TMS		2400	455 $_{34}$	440 $_{14}$	414 $_{15}$	324 $_{10}$	73 $_{40}$	1122	5497
Naloxone	G P-I UHY	2715	327 $_{100}$	286 $_{15}$	242 $_{25}$			745	563
Naloxone AC	U+UHYAC	2840	369 $_{54}$	327 $_{100}$	286 $_{14}$	242 $_{18}$		916	361
Naloxone ME		2825	341 $_{100}$	300 $_{14}$	256 $_{22}$			813	565
Naloxone MEAC		2890	383 $_{98}$	340 $_{29}$	324 $_{22}$	242 $_{15}$		964	567
Naloxone PFP		2530	473 $_{76}$	388 $_{36}$	119 $_{75}$	96 $_{85}$	70 $_{100}$	1141	4329
Naloxone TMS		2660	399 $_{100}$	358 $_{13}$	316 $_{9}$	166 $_{16}$	73 $_{54}$	1015	4307
Naloxone 2AC	U+UHYAC	2750	411 $_{29}$	369 $_{100}$	352 $_{21}$	310 $_{23}$	285 $_{30}$	1045	2982
Naloxone 2ET		2830	383 $_{99}$	270 $_{100}$				965	564
Naloxone 2ME		2885	355 $_{98}$	256 $_{28}$	82 $_{84}$			866	566
Naloxone 2PFP		2470	619 $_{8}$	472 $_{8}$	284 $_{3}$	119 $_{69}$	82 $_{100}$	1196	4327
Naloxone 2TMS	UHYTMS	2680	471 $_{29}$	456 $_{17}$	355 $_{11}$	96 $_{10}$	73 $_{100}$	1139	4308
Naloxone enol 2AC	U+UHYAC	2810	411 $_{93}$	369 $_{83}$	330 $_{76}$	270 $_{71}$	82 $_{100}$	1045	2984
Naloxone enol 2PFP		2360	619 $_{100}$	472 $_{55}$	456 $_{50}$	371 $_{19}$	119 $_{61}$	1196	4328
Naloxone enol 2TMS	UHYTMS	2700	471 $_{100}$	456 $_{25}$	366 $_{13}$	82 $_{16}$	73 $_{71}$	1139	4309
Naloxone enol 3AC	U+UHYAC	2770	453 $_{5}$	411 $_{14}$	369 $_{33}$	327 $_{100}$	242 $_{11}$	1118	2983
Naloxone enol 3PFP		2270	765 $_{83}$	618 $_{48}$	602 $_{65}$	454 $_{22}$	119 $_{100}$	1204	4328
Naloxone enol 3TMS	UHYTMS	2645	543 $_{50}$	528 $_{59}$	438 $_{57}$	355 $_{29}$	73 $_{100}$	1183	4306
Naloxone-M (dihydro-) 2AC	U+UHYAC	2820	413 $_{15}$	371 $_{6}$	242 $_{4}$	82 $_{100}$		1048	1188
Naloxone-M (dihydro-) 3AC	U+UHYAC	2855	455 $_{11}$	413 $_{41}$	327 $_{16}$	254 $_{17}$	82 $_{100}$	1121	3720
beta-Naltrexol 3TMS		2720	559 $_{84}$	73 $_{100}$	55 $_{33}$	544 $_{11}$	372 $_{43}$	1187	6491
Naltrexone	UHY	2880	341 $_{88}$	300 $_{24}$	256 $_{19}$	243 $_{18}$	55 $_{100}$	813	4310
Naltrexone AC		2980	383 $_{44}$	341 $_{89}$	300 $_{12}$	243 $_{19}$	55 $_{100}$	964	4313
Naltrexone 2AC	U+UHYAC	2870	425 $_{36}$	383 $_{43}$	341 $_{25}$	324 $_{13}$	55 $_{100}$	1070	4311
Naltrexone 2TMS		2760	485 $_{72}$	470 $_{13}$	388 $_{9}$	73 $_{67}$	55 $_{53}$	1152	6276
Naltrexone enol 2AC	U+UHYAC	3060	425 $_{58}$	383 $_{43}$	342 $_{12}$	110 $_{12}$	55 $_{100}$	1070	4314
Naltrexone enol 3AC	U+UHYAC	2960	467 $_{60}$	425 $_{64}$	408 $_{14}$	324 $_{10}$	55 $_{100}$	1135	4312
Naltrexone enol 3TMS		2700	557 $_{100}$	542 $_{68}$	484 $_{9}$	73 $_{90}$	55 $_{29}$	1186	6275
Naltrexone-M (dihydro-) 3AC	U+UHYAC	2990	469 $_{26}$	427 $_{36}$	413 $_{39}$	228 $_{13}$	55 $_{100}$	1138	4331
Naltrexone-M (dihydro-) 3TMS		2720	559 $_{84}$	73 $_{100}$	55 $_{33}$	544 $_{11}$	372 $_{43}$	1187	6491
Naltrexone-M (dihydro-methoxy-) 3AC	U+UHYAC	3200	499 $_{83}$	457 $_{57}$	440 $_{24}$	303 $_{17}$	55 $_{100}$	1163	4332
Naltrexone-M (methoxy-)	UHY	2920	371 $_{92}$	330 $_{19}$	286 $_{23}$	274 $_{22}$	55 $_{100}$	924	4330
Naltrexone-M (methoxy-) AC	U+UHYAC	3150	413 $_{83}$	372 $_{22}$	328 $_{15}$	274 $_{31}$	55 $_{100}$	1048	4316
Naltrexone-M (methoxy-) 2AC	U+UHYAC	3130	455 $_{89}$	412 $_{27}$	396 $_{20}$	273 $_{9}$	55 $_{100}$	1121	4315
Naltrexone-M (methoxy-) enol 2AC	U+UHYAC	3300	455 $_{69}$	414 $_{30}$	384 $_{8}$	110 $_{11}$	55 $_{100}$	1121	4318
Naltrexone-M (methoxy-) enol 3AC	U+UHYAC	3180	497 $_{78}$	454 $_{27}$	396 $_{9}$	256 $_{13}$	55 $_{100}$	1162	4317
Nandrolone		2395*	274 $_{93}$	256 $_{17}$	110 $_{100}$	91 $_{70}$	79 $_{62}$	487	3748
Nandrolone TMS		2760*	346 $_{5}$	255 $_{14}$	237 $_{9}$	108 $_{95}$	91 $_{100}$	833	3004
Naphazoline	G	2100	210 $_{46}$	209 $_{100}$	141 $_{24}$			245	1142
Naphthalene		1190*	128 $_{100}$	102 $_{7}$	77 $_{3}$	64 $_{6}$	51 $_{7}$	111	2554
1-Naphthaleneacetic acid		1805*	186 $_{30}$	141 $_{100}$	115 $_{27}$	63 $_{6}$		188	3647
1-Naphthaleneacetic acid ME		1720*	200 $_{26}$	141 $_{100}$	115 $_{19}$	70 $_{4}$		219	3648
alpha-Naphthoflavone		2810	272 $_{64}$	244 $_{8}$	170 $_{100}$	122 $_{12}$	114 $_{51}$	478	6460
1-Naphthol		1500*	144 $_{75}$	115 $_{100}$	89 $_{17}$	74 $_{7}$	63 $_{21}$	123	928
1-Naphthol AC	U+UHYAC	1555*	186 $_{13}$	144 $_{100}$	115 $_{47}$	89 $_{8}$	63 $_{7}$	188	932
1-Naphthol HFB		1310*	340 $_{46}$	169 $_{25}$	143 $_{28}$	115 $_{100}$	89 $_{11}$	806	7476
1-Naphthol PFP		1510*	290 $_{45}$	171 $_{100}$	143 $_{20}$	115 $_{49}$	89 $_{8}$	562	7468
1-Naphthol TMS		1525*	216 $_{100}$	201 $_{95}$	185 $_{51}$	115 $_{39}$	73 $_{21}$	260	7460
Naphthoxyacetic acid methylester		1765*	216 $_{100}$	157 $_{29}$	127 $_{68}$	115 $_{71}$	63 $_{11}$	259	4046
N-(1-Naphthyl-)phthalimide		2545	273 $_{100}$	228 $_{43}$	202 $_{10}$	140 $_{13}$	76 $_{49}$	481	3646
Naproxen	G P U+UHYAC	1780*	230 $_{42}$	185 $_{100}$	170 $_{18}$	141 $_{18}$	115 $_{19}$	307	1733
Naproxen ET		1830*	258 $_{45}$	185 $_{100}$	153 $_{7}$	141 $_{8}$	115 $_{7}$	413	4356
Naproxen ME	PME UME U+UHYAC	1800*	244 $_{41}$	185 $_{100}$	170 $_{15}$	141 $_{16}$	115 $_{12}$	356	1734
Naproxen TMS		1735*	302 $_{34}$	287 $_{20}$	243 $_{55}$	185 $_{100}$	73 $_{75}$	624	5218
Naproxen -CO2	G P U+UHYAC	1660*	184 $_{100}$	169 $_{24}$	141 $_{68}$	115 $_{32}$		185	1735
Naproxen-M (HO-) 2ME	UME	2120*	274 $_{100}$	259 $_{43}$	215 $_{89}$	184 $_{31}$	171 $_{36}$	485	6295
Naproxen-M (O-demethyl-) MEAC	U+UHYAC	2085*	272 $_{12}$	230 $_{9}$	171 $_{100}$	141 $_{5}$	115 $_{5}$	479	4358
Naproxen-M (O-demethyl-) 2ME	PME UME U+UHYAC	1800*	244 $_{41}$	185 $_{100}$	170 $_{15}$	141 $_{16}$	115 $_{12}$	356	1734
Naproxen-M (O-demethyl-) -CH2O2 AC	U+UHYAC	1810*	212 $_{14}$	170 $_{100}$	153 $_{3}$	141 $_{11}$	115 $_{13}$	249	4357
Naptalam -H2O		2545	273 $_{100}$	228 $_{43}$	202 $_{10}$	140 $_{13}$	76 $_{49}$	481	3646
Naratriptan		3210	335 $_{28}$	320 $_{15}$	170 $_{15}$	97 $_{53}$	70 $_{100}$	784	7505
Naratriptan HFB		2970	531 $_{30}$	516 $_{4}$	438 $_{1}$	97 $_{8}$	70 $_{100}$	1178	7507
Naratriptan TFA		2995	431 $_{16}$	416 $_{2}$	168 $_{7}$	96 $_{7}$	70 $_{100}$	1083	7506
Naratriptan TMS		3220	407 $_{24}$	392 $_{25}$	242 $_{40}$	96 $_{100}$	70 $_{74}$	1034	7503
Naratriptan 2TMS		3360	479 $_{18}$	464 $_{22}$	242 $_{36}$	97 $_{90}$	70 $_{71}$	1147	7504
Narceine ME		2960	459 $_{1}$	234 $_{1}$	178 $_{1}$	58 $_{100}$		1126	5151
Narceine artifact 2ME		1870*	254 $_{37}$	223 $_{100}$	207 $_{16}$	191 $_{62}$	77 $_{30}$	398	5152

Table 1-8-1: Compounds in order of names   Nifedipine-M (dehydro-demethyl-HO-) -H2O

Name	Detected	RI	Typical ions and intensities					Page	Entry
Narceine -H2O		3260	427 $_6$	234 $_9$	58 $_{100}$			1074	5153
Narcobarbital	P U UHY	1805	223 $_{100}$	181 $_{86}$	138 $_{10}$	124 $_{16}$		622	1143
Narconumal	P G U	1560	209 $_5$	181 $_{100}$	167 $_{11}$	124 $_{24}$	97 $_{19}$	289	1144
Narconumal ME		1520	238 $_2$	220 $_7$	195 $_{100}$	138 $_{50}$	111 $_{26}$	337	1145
Nealbarbital	P G U	1720	223 $_7$	181 $_{18}$	167 $_{45}$	141 $_{56}$	57 $_{100}$	337	1146
Nealbarbital 2ME		1620	250 $_{15}$	209 $_{52}$	195 $_{78}$	169 $_{100}$	57 $_{62}$	452	1147
Nebivolol 2TMSTFA		2900	645 $_1$	494 $_{11}$	404 $_{18}$	177 $_{54}$	73 $_{70}$	1198	6204
Nebivolol 3AC		3540	531 $_5$	471 $_{27}$	428 $_{100}$	412 $_{13}$	231 $_{15}$	1178	6107
Neburon ME		2070	288 $_{10}$	202 $_5$	174 $_8$	114 $_{63}$	57 $_{100}$	551	4158
NECA 2AC		2735	392 $_5$	333 $_{13}$	262 $_{66}$	136 $_{75}$	85 $_{100}$	992	3092
NECA -2H2O		2930	272 $_{27}$	228 $_{100}$	172 $_{30}$	136 $_{28}$	66 $_{30}$	478	3093
NECA 3AC		3265	434 $_5$	375 $_{15}$	363 $_{16}$	304 $_{50}$	85 $_{100}$	1088	3091
Nefazodone	U+UHYAC	4510	469 $_4$	454 $_{20}$	303 $_{100}$	274 $_{42}$	260 $_{36}$	1138	5305
Nefazodone-M (deamino-HO-)		2340	291 $_{32}$	198 $_{57}$	171 $_{23}$	120 $_{100}$	91 $_{18}$	570	5301
Nefazodone-M (deamino-HO-) AC	U+UHYAC	2500	333 $_{32}$	240 $_{45}$	120 $_{100}$	91 $_{32}$	77 $_{22}$	774	5302
Nefazodone-M (HO-ethyl-deamino-HO-) 2AC	U+UHYAC	2650	391 $_{11}$	298 $_{23}$	238 $_9$	120 $_{100}$	91 $_{13}$	989	5303
Nefazodone-M (HO-phenyl-) AC	U+UHYAC	4890	527 $_1$	512 $_4$	361 $_{100}$	332 $_{19}$	209 $_{50}$	1177	5306
Nefazodone-M (HO-phenyl-deamino-HO-) 2AC	U+UHYAC	2830	391 $_8$	349 $_{14}$	240 $_{100}$	178 $_{19}$	136 $_{55}$	989	5304
Nefazodone-M (N-dealkyl-)		1910	196 $_{24}$	154 $_{100}$	138 $_{12}$	111 $_9$	75 $_{12}$	211	6885
Nefazodone-M (N-dealkyl-) AC	U+UHYAC	2265	238 $_{32}$	195 $_{15}$	166 $_{100}$	154 $_{31}$	111 $_{27}$	336	405
Nefazodone-M (N-dealkyl-) HFB	U+UHYHFB	1960	392 $_{100}$	195 $_{38}$	166 $_{41}$	139 $_{36}$	111 $_{25}$	991	6604
Nefazodone-M (N-dealkyl-) ME		1820	210 $_{100}$	166 $_{19}$	139 $_{81}$	111 $_{33}$	70 $_{99}$	244	6886
Nefazodone-M (N-dealkyl-) TFA	U+UHYTFA	1920	292 $_{100}$	250 $_{12}$	195 $_{77}$	166 $_{79}$	139 $_{66}$	574	6597
Nefazodone-M (N-dealkyl-) TMS		2035	268 $_{61}$	253 $_{29}$	128 $_{63}$	101 $_{68}$	86 $_{71}$	461	6888
Nefazodone-M (N-dealkyl-HO-) TFA	U+UHYTFA	2035	308 $_{100}$	272 $_{19}$	211 $_{46}$	182 $_{36}$	155 $_{46}$	653	6599
Nefazodone-M (N-dealkyl-HO-) iso-1 AC	U+UHYAC	2335	254 $_{65}$	211 $_{23}$	182 $_{100}$	166 $_{71}$	154 $_{33}$	398	5308
Nefazodone-M (N-dealkyl-HO-) iso-1 2AC	U+UHYAC	2515	296 $_{30}$	254 $_{66}$	182 $_{100}$	154 $_{36}$		597	406
Nefazodone-M (N-dealkyl-HO-) iso-1 2TFA	U+UHYTFA	2040	404 $_{28}$	307 $_{74}$	278 $_{34}$	265 $_{13}$	154 $_{40}$	1028	6600
Nefazodone-M (N-dealkyl-HO-) iso-2 AC	U+UHYAC	2345	254 $_{79}$	211 $_{15}$	182 $_{100}$	169 $_{40}$	154 $_{40}$	398	5307
Nefazodone-M (N-dealkyl-HO-) iso-2 2AC	U+UHYAC	2525	296 $_{20}$	254 $_{74}$	182 $_{100}$	169 $_{72}$	154 $_{24}$	597	32
Nefazodone-M (N-dealkyl-HO-) iso-2 2HFB	U+UHYHFB	2145	604 $_{43}$	585 $_9$	407 $_{100}$	378 $_{22}$	154 $_{22}$	1195	6605
Nefazodone-M (N-dealkyl-HO-) iso-2 2TFA	U+UHYTFA	2045	404 $_{13}$	307 $_{33}$	278 $_{10}$	265 $_4$	154 $_{22}$	1028	6598
Nefopam	G P	2035	253 $_2$	225 $_8$	179 $_{21}$	165 $_9$	58 $_{100}$	395	243
Nefopam-M (HO-) iso-1 AC	U+UHYAC	2250	311 $_{11}$	238 $_{61}$	195 $_{100}$	165 $_{55}$	87 $_{25}$	672	1164
Nefopam-M (HO-) iso-2 AC	U+UHYAC	2285	311 $_{17}$	268 $_{23}$	208 $_{64}$	195 $_{67}$	178 $_{100}$	672	245
Nefopam-M (nor-) AC	U+UHYAC	2080	281 $_8$	208 $_{99}$	194 $_{29}$	179 $_{19}$	87 $_{61}$	523	244
Nefopam-M (nor-di-HO-) -H2O iso-1 2AC	U+UHYAC	2610	337 $_{26}$	295 $_{14}$	266 $_{66}$	87 $_{100}$		792	1166
Nefopam-M (nor-di-HO-) -H2O iso-2 2AC	U+UHYAC	2640	337 $_{53}$	295 $_{27}$	195 $_{35}$	87 $_{100}$		792	1167
Nevirapine		2520	266 $_{57}$	265 $_{100}$	251 $_{87}$	237 $_{38}$	133 $_{28}$	451	7436
Nevirapine AC		2465	308 $_{15}$	265 $_{53}$	251 $_{100}$	133 $_{74}$	78 $_{55}$	656	7437
Nevirapine TMS		2435	338 $_{44}$	337 $_{49}$	323 $_{68}$	249 $_{32}$	73 $_{100}$	798	7438
Nicardipine		3900	479 $_1$	462 $_2$	147 $_{51}$	134 $_{71}$	91 $_{100}$	1147	1724
Nicardipine ME		3800	493 $_1$	476 $_6$	148 $_{42}$	134 $_{45}$	91 $_{100}$	1160	4878
Nicardipine-M	UME	2495	332 $_{60}$	315 $_{100}$	301 $_{22}$	285 $_{68}$	212 $_9$	768	4883
Nicardipine-M	UME	2250	316 $_{53}$	299 $_{100}$	269 $_{86}$	241 $_{13}$	127 $_{27}$	693	4882
Nicardipine-M (deamino-HOOC-) ME	UME	2950	404 $_2$	373 $_2$	315 $_3$	282 $_{100}$	192 $_6$	1028	4879
Nicardipine-M (deamino-HOOC-) 2ME	UME	2970	418 $_{19}$	359 $_{62}$	301 $_{100}$	296 $_{96}$	224 $_{16}$	1058	4880
Nicardipine-M (dehydro-deamino-HO-)	UME	2665	374 $_7$	313 $_{79}$	312 $_{100}$	299 $_{48}$	252 $_{23}$	933	4894
Nicardipine-M (dehydro-deamino-HOOC-) ME	UME	2645	402 $_1$	371 $_6$	312 $_{100}$	281 $_{13}$	139 $_7$	1022	4881
Nicardipine-M -H2O	UHY U+UHYAC UME	2650	328 $_{26}$	311 $_{100}$	281 $_{42}$	222 $_6$	139 $_5$	748	3658
Nicardipine-M -H2O TMS	UTMS	2615	386 $_{32}$	371 $_{11}$	192 $_{100}$	177 $_{30}$	151 $_{10}$	972	5004
Nicardipine-M/artifact (debenzylmethylaminoethyl-) ME	UME	2690	346 $_8$	331 $_6$	315 $_7$	287 $_8$	224 $_{100}$	832	4871
Nicardipine-M/artifact (debenzylmethylaminoethyl-) 2ME	UME	2695	360 $_4$	329 $_9$	301 $_{100}$	238 $_{63}$	224 $_9$	884	4884
Nicardipine-M/artifact (dehydro-debenzylmethylaminoethyl-) TMS	UTMS	2455	402 $_{10}$	387 $_{100}$	313 $_{35}$	281 $_8$	152 $_7$	1023	5001
Nicardipine-M/artifact ME	UME	2300	344 $_{31}$	327 $_{100}$	313 $_{59}$	297 $_{78}$	252 $_{13}$	824	4873
Nicardipine-M/artifact 2TMS	UTMS	2375	460 $_6$	445 $_4$	267 $_{100}$	193 $_{14}$	144 $_{13}$	1128	5003
Nicardipine-M/artifact -CO2	U UHY U+UHYAC UME	2175	286 $_{30}$	269 $_{100}$	239 $_{48}$	180 $_8$	139 $_{13}$	542	3656
Nicergoline-M/artifact (alcohol)		2515	300 $_{25}$	270 $_{24}$	198 $_{36}$	181 $_{100}$	168 $_{45}$	616	5251
Nicergoline-M/artifact (alcohol) AC		2610	342 $_{25}$	312 $_{26}$	198 $_{41}$	181 $_{100}$	168 $_{48}$	818	5253
Nicergoline-M/artifact (HOOC-)		1020	201 $_{84}$	183 $_{49}$	156 $_{30}$	76 $_{68}$	51 $_{100}$	220	5252
Nicergoline-M/artifact (HOOC-) ME	UME	1095	215 $_{42}$	184 $_{79}$	156 $_{70}$	136 $_{30}$	76 $_{100}$	256	5250
Nicethamide	U	1535	178 $_{58}$	177 $_{71}$	149 $_{24}$	106 $_{100}$	78 $_{74}$	168	1148
Niclosamide ME		2920	340 $_8$	305 $_5$	169 $_{100}$	126 $_{15}$	111 $_{13}$	806	4155
Nicomorphine		4060	495 $_{47}$	389 $_{21}$	373 $_{41}$	106 $_{100}$	78 $_{91}$	1161	5501
Nicomorphine HY	G UHY	2455	285 $_{100}$	268 $_{15}$	162 $_{58}$	124 $_{20}$		539	474
Nicomorphine HY2AC	G PHYAC U+UHYAC	2620	369 $_{59}$	327 $_{100}$	310 $_{36}$	268 $_{47}$	162 $_{11}$	915	225
Nicotinamide	G	1605	122 $_{100}$	106 $_{56}$	78 $_{94}$			107	1149
Nicotine	G P U U+UHYAC	1380	162 $_{16}$	133 $_{26}$	84 $_{100}$			142	1150
Nicotine-M (cotinine)	P U+UHYAC	1715	176 $_{36}$	118 $_{12}$	98 $_{100}$			163	692
Nicotinic acid ME		1390	137 $_{70}$	106 $_{97}$	78 $_{83}$			117	1151
Nifedipine	G P UME	2575	346 $_9$	329 $_{100}$	284 $_{84}$	268 $_{37}$	224 $_{49}$	832	2485
Nifedipine ME		2550	360 $_6$	343 $_{100}$	298 $_{74}$	282 $_{31}$	238 $_{39}$	885	4876
Nifedipine-M (dehydro-2-HOOC-) ME	UME	2695	388 $_1$	357 $_6$	342 $_{100}$	195 $_6$	139 $_{10}$	979	4877
Nifedipine-M (dehydro-demethyl-HO-) -H2O	P U+UHYAC UME	2485	328 $_1$	297 $_3$	282 $_{100}$	267 $_{12}$	250 $_{12}$	748	2489

139

Nifedipine-M (dehydro-HO-) Table 1-8-1: Compounds in order of names

Name	Detected	RI	Typical ions and intensities					Page	Entry
Nifedipine-M (dehydro-HO-)	U UHY U+UHYAC	2600	296 100	265 93	237 27	167 36	126 26	884	2492
Nifedipine-M (dehydro-HO-HOOC-)	U UHY	2910	298 4	282 100	251 67	223 16	126 10	830	2490
Nifedipine-M (dehydro-HO-HOOC-) AC	U+UHYAC	2890	388 100	346 64	274 30	233 31	177 63	979	2493
Nifedipine-M (dehydro-HO-HOOC-) -H2O -C2H2O2	U UHY U+UHYAC	2390	224 100	196 95	154 33	127 32	63 49	468	2491
Nifedipine-M (dehydro-HOOC-) TMS	UTMS	2630	446 2	431 100	400 53	387 21	296 14	1108	5012
Nifedipine-M/artifact (dehydro-)	P G U+UHYAC UME	2255	344 1	313 5	298 100	267 6	252 7	824	2486
Nifedipine-M/artifact (dehydro-demethyl-)	P U UHY U+UHYAC	2290	298 8	283 100	252 88	224 33	126 20	758	2488
Nifedipine-M/artifact (dehydro-demethyl-) TMS	UTMS	2410	402 1	387 33	356 100	252 34	152 18	1023	5011
Nifedipine-M/artifact (dehydro-demethyl-) -CO2	U UHY U+UHYAC UME	2080	286 1	255 5	240 100	225 15	209 22	543	2487
Nifenalol		1870	224 1	209 1	191 10	77 8	72 100	289	4344
Nifenalol TMSTFA		2050	392 1	335 6	224 100	126 25	73 63	992	6172
Nifenalol 2AC		2305	308 1	248 55	206 64	114 69	72 100	656	1365
Nifenalol formyl artifact		1900	236 1	221 9	191 32	118 6	85 100	330	1364
Nifenalol -H2O AC	U+UHYAC	2265	248 11	206 10	191 6	114 31	72 100	371	1707
Nifenazone	G U UHY U+UHYAC	3080	308 30	202 8	56 100			656	200
Nifenazone-M (deacyl-)	P U UHY	1955	203 23	93 14	84 59	56 100		224	219
Nifenazone-M (deacyl-) AC	P U U+UHYAC	2270	245 30	203 13	84 50	56 100		359	183
Nifenazone-M (deacyl-) artifact	U UHY	1945	180 13	119 99	91 45			173	424
Nifenazone-M (dealkyl-) 2AC	U+UHYAC	2280	287 8	245 31	203 15	84 56	56 100	547	3333
Niflumic acid		2085	282 82	263 48	237 99	168 8	145 16	526	1404
Niflumic acid ME		1960	296 70	295 95	263 57	236 100	145 20	596	1497
Niflumic acid MEAC		1995	338 9	296 94	295 100	263 52	236 64	796	5101
Niflumic acid TMS		1840	354 59	353 62	263 44	236 100	168 14	861	5045
Niflumic acid -CO2		2055	238 55	237 99	217 9	168 5	145 4	335	1422
Niflumic acid-M (di-HO-) 3ME	UME	2330	356 100	324 33	309 51	281 32	226 9	868	6382
Niflumic acid-M (HO-) iso-1 2ME	UME	2140	326 100	325 75	293 29	251 58	196 24	738	6380
Niflumic acid-M (HO-) iso-2 2ME	UME	2170	326 100	294 44	279 60	251 29	121 14	738	6381
Nilvadipine	U+UHYAC	2800	385 4	342 29	298 17	263 46	221 100	970	4630
Nilvadipine ME		2780	399 5	356 12	340 16	312 100	235 59	1014	4886
Nilvadipine-M (dehydro-deisopropyl-HO-) ME	UME	2705	371 11	340 10	312 100	249 65	217 10	922	4889
Nilvadipine-M/artifact (dehydro-)	U+UHYAC UME	2565	383 19	341 39	324 100	310 27	164 18	962	4887
Nilvadipine-M/artifact (dehydro-deisopropyl-) ME	UME	2520	355 43	340 100	324 51	308 18	164 13	864	4888
Nilvadipine-M/artifact (dehydro-deisopropyl-) TMS	UTMS	2645	413 6	398 100	324 8	261 3	164 7	1047	5008
Nimesulide		2550	308 60	229 100	183 36	154 76	77 35	653	7556
Nimesulide AC		2595	350 2	308 100	229 76	154 35	77 38	844	7558
Nimesulide ME		2535	322 46	243 100	197 39	168 36	91 58	721	7557
Nimesulide TMS		2580	380 75	365 57	228 84	137 78	73 100	952	7552
Nimesulide artifact (-SO2CH2) AC		2430	272 27	230 100	200 17	179 13	154 10	478	7559
Nimetazepam		2485	295 84	294 100	248 72	91 100		589	569
Nimetazepam HY		2520	256 95	255 100	193 25	105 35	77 60	406	3071
Nimetazepam-M (nor-)	G P-I U+UHYAC-I	2760	281 100	253 86	234 62	222 47	206 64	520	568
Nimetazepam-M (nor-) TMS		2315	353 62	352 98	338 13	306 28	73 100	857	5500
Nimetazepam-M (nor-) HY	UHY-I U+UHYAC-I	2365	242 100	241 84	195 28	105 49	77 63	350	298
Nimetazepam-M (nor-) HYAC	U+UHYAC-I	2400	284 16	242 70	241 100	179 25	77 50	532	2904
Nimodipine		2845	418 13	359 17	296 100	254 43	196 26	1059	2582
Nimodipine ME		2990	432 11	345 55	287 100	268 43	210 71	1084	4890
Nimodipine-M (dehydro-deisopropyl-O-demethyl-) ME	UME	2665	374 7	313 79	312 100	299 48	252 21	933	4894
Nimodipine-M (dehydro-deisopropyl-O-demethyl-HOOC-) 2ME	UME	2645	402 1	371 2	312 100	281 13	139 7	1022	4881
Nimodipine-M (dehydro-demethoxyethyl-HO-) -H2O	U+UHYAC UME	2740	356 22	314 58	297 100	266 61	223 36	868	4895
Nimodipine-M (dehydro-O-demethyl-HOOC-) ME	UME	2740	430 2	371 14	340 62	298 100	281 56	1080	4891
Nimodipine-M -H2O ME	UHY U+UHYAC UME	2650	328 26	311 100	281 42	222 6	139 5	748	3658
Nimodipine-M -H2O TMS	UTMS	2615	386 32	371 11	192 100	177 30	151 10	972	5004
Nimodipine-M/artifact (dehydro-)	UME	2655	416 5	357 37	340 63	298 100	281 59	1054	5043
Nimodipine-M/artifact (dehydro-deisopropyl-) ME	UME	2550	388 3	357 12	314 85	313 97	281 39	980	4892
Nimodipine-M/artifact (dehydro-demethoxyethyl-) ME	UME	2390	372 18	330 53	313 100	298 61	252 57	927	4893
Nimodipine-M/artifact (deisopropyl-demethoxyethyl-) 2ME	UME	2690	346 8	331 6	315 7	287 8	224 100	832	4871
Nimodipine-M/artifact (deisopropyl-demethoxyethyl-) 3ME	UME	2695	360 9	329 9	301 100	238 63	224 9	884	4884
Nimodipine-M/artifact 2ME	UME	2300	344 31	327 100	313 59	297 78	252 13	824	4873
Nimodipine-M/artifact 2TMS	UTMS	2375	460 6	445 4	267 100	193 14	144 13	1128	5003
Nimodipine-M/artifact -CO2 ME	U UHY U+UHYAC UME	2175	286 30	269 100	239 48	180 8	139 13	542	3656
Nisoldipine		2730	388 6	371 100	284 22	270 28	210 28	980	4284
Nisoldipine ME		2770	402 8	385 100	340 13	298 14	284 16	1024	4896
Nisoldipine-M (dehydro-deisobutyl-2-HOOC-) 2ME	UME	2695	388 1	357 9	342 100	195 6	139 10	979	4877
Nisoldipine-M (dehydro-deisobutyl-HO-) -H2O	P U+UHYAC UME	2485	328 1	297 3	282 100	267 12	250 12	748	2489
Nisoldipine-M (dehydro-demethyl- di-HO-) -H2O	U+UHYAC UME	2785	340 37	297 17	268 100	251 28	59 47	972	4898
Nisoldipine-M (dehydro-HO-)	U+UHYAC UME	2615	402 1	356 63	313 30	284 100	59 61	1023	4287
Nisoldipine-M (dehydro-HO-demethyl-) -H2O	UME	2665	370 5	324 27	268 100	251 13	222 19	919	5090
Nisoldipine-M (dehydro-HOOC-) ME	UME	2715	430 1	399 6	384 100	101 47	59 65	1080	4897
Nisoldipine-M (HO-)		2785	404 6	387 100	284 22	270 30	210 36	1029	4285
Nisoldipine-M/artifact (dehydro-)	UME	2450	386 1	340 42	284 100	236 10	57 37	973	4286
Nisoldipine-M/artifact (dehydro-deisobutyl-) ME	P G U+UHYAC UME	2255	344 1	313 5	298 100	267 6	252 7	824	2486
Nisoldipine-M/artifact (dehydro-deisobutyl-) TMS	UTMS	2410	402 1	387 33	356 100	252 34	152 18	1023	5011
Nisoldipine-M/artifact (dehydro-deisobutyl-) -CO2	U UHY U+UHYAC UME	2080	286 1	255 5	240 100	225 15	209 22	543	2487
Nisoldipine-M/artifact (deisobutyl-) ME	G P UME	2575	346 9	329 100	284 84	268 37	224 49	832	2485

Table 1-8-1: Compounds in order of names   **Nordazepam-M (HO-) iso-2 HY2AC**

Name	Detected	RI	Typical ions and intensities					Page	Entry
Nitrazepam	G P-I U+UHYAC-I	2760	281 $_{100}$	253 $_{86}$	234 $_{62}$	222 $_{47}$	206 $_{64}$	520	568
Nitrazepam TMS		2315	353 $_{62}$	352 $_{98}$	338 $_{13}$	306 $_{28}$	73 $_{100}$	857	5500
Nitrazepam HY	UHY-I U+UHYAC-I	2365	242 $_{100}$	241 $_{84}$	195 $_{28}$	105 $_{49}$	77 $_{63}$	350	298
Nitrazepam HYAC	U+UHYAC-I	2400	284 $_{16}$	242 $_{70}$	241 $_{100}$	179 $_{25}$	77 $_{50}$	532	2904
Nitrazepam iso-1 ME		2485	295 $_{84}$	294 $_{100}$	248 $_{72}$	91 $_{100}$		589	569
Nitrazepam iso-2 ME		2690	276 $_{82}$	275 $_{86}$	249 $_{98}$	231 $_{56}$		494	570
Nitrazepam-M (amino-)	UGLUC	2785	251 $_{100}$	223 $_{66}$	222 $_{64}$			384	571
Nitrazepam-M (amino-) AC	UGLUCAC	3150	293 $_{100}$	265 $_{75}$	223 $_{25}$	222 $_{28}$		579	572
Nitrazepam-M (amino-) HY	UHY-I	2225	212 $_{100}$	211 $_{93}$	195 $_{12}$	107 $_{19}$	77 $_{14}$	249	573
Nitrazepam-M (amino-) HY2AC	U+UHYAC	2985	296 $_{98}$	254 $_{75}$	212 $_{87}$	211 $_{69}$		598	299
Nitrendipine	G P U+UHYAC UME	2700	360 $_{17}$	331 $_{11}$	238 $_{100}$	210 $_{44}$	150 $_{10}$	885	2583
Nitrendipine ET		2765	388 $_{13}$	329 $_{31}$	315 $_{100}$	266 $_{43}$	238 $_{26}$	980	4874
Nitrendipine ME		2740	374 $_{11}$	315 $_{32}$	301 $_{100}$	252 $_{64}$	224 $_{55}$	933	4870
Nitrendipine-M (dehydro-deethyl-HO-) -H2O	UHY U+UHYAC UME	2650	328 $_{26}$	311 $_{100}$	281 $_{42}$	222 $_{6}$	139 $_{5}$	748	3658
Nitrendipine-M (dehydro-demethyl-) -CO2	U	2330	300 $_{9}$	283 $_{14}$	253 $_{58}$	251 $_{100}$	139 $_{77}$	615	3657
Nitrendipine-M (dehydro-demethyl-deethyl-HO-) -H2O TMS	UTMS	2615	386 $_{32}$	371 $_{11}$	192 $_{100}$	177 $_{30}$	151 $_{10}$	972	5004
Nitrendipine-M (dehydro-demethyl-HO-) -H2O	UHY U+UHYAC UME	2690	342 $_{48}$	325 $_{100}$	297 $_{83}$	266 $_{40}$	139 $_{8}$	816	3659
Nitrendipine-M/artifact (deethyl-) ME	UME	2690	346 $_{8}$	331 $_{6}$	315 $_{7}$	287 $_{8}$	224 $_{100}$	832	4871
Nitrendipine-M/artifact (deethyl-) 2ME	UME	2695	360 $_{9}$	329 $_{9}$	301 $_{100}$	238 $_{63}$	224 $_{9}$	884	4884
Nitrendipine-M/artifact (dehydro-)	UME	2370	358 $_{54}$	341 $_{100}$	313 $_{65}$	281 $_{18}$	252 $_{13}$	877	4872
Nitrendipine-M/artifact (dehydro-deethyl-) ME	UME	2300	344 $_{31}$	327 $_{100}$	313 $_{59}$	297 $_{78}$	252 $_{13}$	824	4873
Nitrendipine-M/artifact (dehydro-deethyl-) TMS	UTMS	2455	402 $_{10}$	387 $_{100}$	313 $_{35}$	281 $_{8}$	152 $_{7}$	1023	5001
Nitrendipine-M/artifact (dehydro-deethyl-) -CO2	U UHY U+UHYAC UME	2175	286 $_{30}$	269 $_{100}$	239 $_{48}$	180 $_{8}$	139 $_{13}$	542	3656
Nitrendipine-M/artifact (dehydro-deethyl-demethyl-) 2TMS	UTMS	2375	460 $_{6}$	445 $_{4}$	267 $_{100}$	193 $_{14}$	144 $_{13}$	1128	5003
Nitrendipine-M/artifact (dehydro-demethyl-) ET	UET	2470	372 $_{39}$	355 $_{64}$	327 $_{100}$	299 $_{28}$	281 $_{43}$	927	4875
Nitrendipine-M/artifact (dehydro-demethyl-) TMS	UTMS	2530	416 $_{16}$	401 $_{100}$	371 $_{10}$	327 $_{19}$	178 $_{12}$	1054	5002
Nitrofen		2205	283 $_{100}$	253 $_{16}$	202 $_{51}$	139 $_{25}$	75 $_{12}$	527	3861
Nitrofurantoin ME		2250	252 $_{30}$	206 $_{11}$	167 $_{16}$	140 $_{32}$	114 $_{100}$	388	5226
4-Nitrophenol	P-I UHY	1530	139 $_{100}$	109 $_{23}$	93 $_{18}$	65 $_{57}$		119	829
4-Nitrophenol AC	U+UHYAC	1500	181 $_{67}$	139 $_{100}$	109 $_{78}$	65 $_{76}$	63 $_{78}$	176	830
4-Nitrophenol ME		1455	153 $_{100}$	123 $_{37}$	92 $_{55}$	77 $_{72}$		132	831
Nitrothal-isopropyl		2005	295 $_{8}$	254 $_{59}$	236 $_{100}$	212 $_{66}$	194 $_{79}$	590	3455
Nitroxoline	G P U+UHYAC	1750	190 $_{100}$	160 $_{83}$	116 $_{87}$	89 $_{94}$	63 $_{48}$	193	1918
NMPEA		1460	134 $_{3}$	120 $_{100}$	105 $_{9}$	77 $_{13}$	58 $_{25}$	115	6221
NMPEA AC		1430	177 $_{22}$	162 $_{4}$	120 $_{44}$	105 $_{20}$	77 $_{13}$	166	6229
Nomifensine	UHY	2150	238 $_{25}$	194 $_{100}$	178 $_{23}$	165 $_{18}$		338	574
Nomifensine AC	U+UHYAC	2470	280 $_{37}$	222 $_{100}$	194 $_{30}$	178 $_{35}$		518	362
Nomifensine TMS		2065	310 $_{22}$	266 $_{57}$	237 $_{88}$	193 $_{51}$	73 $_{90}$	668	5478
Nomifensine-M (HO-)	UHY	2450	254 $_{31}$	210 $_{63}$	194 $_{29}$	86 $_{101}$		400	575
Nomifensine-M (HO-) iso-1 2AC	U+UHYAC	2850	338 $_{55}$	310 $_{48}$	280 $_{100}$	268 $_{67}$	226 $_{56}$	798	363
Nomifensine-M (HO-) iso-2 2AC	U+UHYAC	2880	338 $_{34}$	308 $_{56}$	280 $_{100}$	268 $_{72}$	194 $_{90}$	798	364
Nomifensine-M (HO-methoxy-) 2AC	U+UHYAC	2970	368 $_{31}$	310 $_{93}$	268 $_{99}$	224 $_{100}$		913	365
Nomifensine-M (HO-methoxy-) iso-1	UHY	2505	284 $_{84}$	241 $_{80}$	210 $_{100}$	86 $_{74}$		535	576
Nomifensine-M (HO-methoxy-) iso-2	UHY	2590	284 $_{100}$	241 $_{79}$	210 $_{73}$	86 $_{90}$		535	577
Nonadecane		1900*	268 $_{3}$	99 $_{12}$	85 $_{35}$	71 $_{58}$	57 $_{100}$	462	2363
Nonadecanoic acid ME		2200*	312 $_{35}$	269 $_{16}$	143 $_{18}$	87 $_{66}$	74 $_{100}$	678	3038
Nonane		900*	128 $_{3}$	85 $_{19}$	57 $_{69}$	43 $_{100}$	29 $_{57}$	111	3784
Nonivamide		2530	293 $_{23}$	195 $_{13}$	151 $_{16}$	137 $_{100}$	122 $_{9}$	583	5896
Nonivamide AC		2585	335 $_{3}$	293 $_{39}$	195 $_{25}$	151 $_{19}$	137 $_{100}$	785	5897
Nonivamide HFB		2385	489 $_{25}$	404 $_{17}$	391 $_{35}$	347 $_{100}$	333 $_{86}$	1157	5900
Nonivamide PFP		2320	439 $_{8}$	354 $_{22}$	341 $_{41}$	297 $_{100}$	283 $_{85}$	1096	5899
Nonivamide TFA		2305	389 $_{6}$	304 $_{20}$	291 $_{33}$	247 $_{100}$	233 $_{98}$	983	5898
Nonivamide TMS		2880	365 $_{47}$	350 $_{9}$	209 $_{100}$	179 $_{39}$	73 $_{97}$	902	6028
Nonivamide 2TMS		2640	437 $_{11}$	422 $_{8}$	339 $_{56}$	209 $_{100}$	73 $_{53}$	1093	6027
Norcinnamolaurine	U	2955	283 $_{1}$	176 $_{20}$	149 $_{2}$	118 $_{4}$	91 $_{6}$	529	5660
Norcinnamolaurine 2AC	UAC	2930	367 $_{4}$	324 $_{9}$	218 $_{100}$	176 $_{94}$	118 $_{4}$	909	5662
Norcodeine 2AC	U+UHYAC	2945	369 $_{14}$	327 $_{3}$	223 $_{37}$	87 $_{100}$	72 $_{36}$	916	226
Norcodeine-M (O-demethyl-) 2PFP		2440	563 $_{100}$	400 $_{10}$	355 $_{38}$	327 $_{7}$	209 $_{15}$	1188	3534
Norcodeine-M (O-demethyl-) 3AC	U+UHYAC	2955	397 $_{8}$	355 $_{9}$	209 $_{41}$	87 $_{100}$	72 $_{33}$	1010	1194
Norcodeine-M (O-demethyl-) 3PFP	UHYPFP	2405	709 $_{80}$	533 $_{28}$	388 $_{29}$	367 $_{51}$	355 $_{100}$	1203	3533
Norcodeine-M (O-demethyl-) 3TMS	UHYTMS	2605	487 $_{9}$	416 $_{9}$	222 $_{18}$	131 $_{10}$	73 $_{50}$	1155	3525
Nordazepam	P G U	2520	270 $_{86}$	269 $_{97}$	242 $_{100}$	241 $_{82}$	77 $_{17}$	468	463
Nordazepam TMS		2300	342 $_{62}$	341 $_{100}$	327 $_{19}$	269 $_{4}$	73 $_{30}$	817	4573
Nordazepam enol AC		2545	312 $_{55}$	270 $_{34}$	241 $_{100}$	227 $_{8}$	205 $_{9}$	674	6102
Nordazepam enol ME		2225	284 $_{78}$	283 $_{100}$	91 $_{61}$			532	464
Nordazepam HY	UHY	2050	231 $_{80}$	230 $_{95}$	154 $_{23}$	105 $_{38}$	77 $_{100}$	308	419
Nordazepam HYAC	U+UHYAC PHYAC	2245	273 $_{31}$	230 $_{100}$	154 $_{13}$	105 $_{23}$	77 $_{50}$	480	273
Nordazepam-D5		2515	275 $_{83}$	273 $_{72}$	247 $_{100}$	218 $_{11}$	212 $_{11}$	489	6851
Nordazepam-M (HO-)	UGLUC	2750	286 $_{82}$	258 $_{100}$	230 $_{11}$	166 $_{7}$	139 $_{8}$	541	2113
Nordazepam-M (HO-) AC	U+UHYAC	3000	328 $_{22}$	286 $_{90}$	258 $_{100}$	166 $_{8}$	139 $_{7}$	747	2111
Nordazepam-M (HO-) HY	UHY	2400	247 $_{72}$	246 $_{100}$	230 $_{11}$	121 $_{26}$	65 $_{22}$	365	2112
Nordazepam-M (HO-) HYAC	U+UHYAC	2270	289 $_{18}$	247 $_{86}$	246 $_{100}$	105 $_{7}$	77 $_{35}$	556	3143
Nordazepam-M (HO-) iso-1 HY2AC	U+UHYAC	2560	331 $_{48}$	289 $_{64}$	247 $_{100}$	230 $_{41}$	154 $_{13}$	762	2125
Nordazepam-M (HO-) iso-2 HY2AC	U+UHYAC	2610	331 $_{46}$	289 $_{54}$	246 $_{100}$	154 $_{11}$	121 $_{11}$	762	1751

Table 1-8-1: Compounds in order of names

Name	Detected	RI	Typical ions and intensities					Page	Entry
Nordazepam-M (HO-methoxy-) HY2AC	U+UHYAC	2700	361_{7}	319_{27}	276_{38}	260_{5}	246_{4}	886	1752
Norephedrine	P U	1370	132_{2}	118_{8}	107_{11}	91_{10}	77_{100}	130	2475
Norephedrine TMSTFA		1890	240_{8}	198_{3}	179_{100}	117_{5}	73_{88}	708	6146
Norephedrine 2AC	U+UHYAC	1805	235_{1}	107_{13}	86_{100}	176_{5}	134_{7}	325	2476
Norephedrine 2HFB	UHYHFB	1455	543_{1}	330_{14}	240_{100}	169_{44}	69_{57}	1183	5098
Norephedrine 2PFP	UHYPFP	1380	443_{1}	280_{9}	190_{100}	119_{59}	105_{26}	1104	5094
Norephedrine 2TFA	UTFA	1355	343_{1}	230_{6}	203_{5}	140_{100}	69_{29}	819	5091
Norephedrine 2TMS		1555	280_{5}	163_{4}	147_{10}	116_{100}	73_{83}	594	4574
Norephedrine formyl artifact		1240	117_{2}	105_{4}	91_{4}	77_{10}	57_{100}	143	4650
Norephedrine-M (HO-) 3AC	U+UHYAC	2135	234_{6}	165_{8}	123_{11}	86_{100}	58_{45}	580	4961
Norepinephrine artifact (3,4-dihydroxybenzoic acid) ME2AC		1750*	252_{1}	210_{19}	168_{100}	137_{60}	109_{8}	388	5254
Norethisterone AC		2720*	340_{65}	298_{92}	283_{69}	91_{91}	56_{99}	808	1498
Norethisterone acetate		2720*	340_{65}	298_{92}	283_{69}	91_{91}	56_{99}	808	1498
Norethisterone -H2O		2480*	280_{24}	265_{77}	149_{75}	91_{100}	77_{94}	519	4260
Norfenefrine	G	1670	153_{7}	124_{100}	95_{60}	77_{66}	65_{22}	132	4662
Norfenefrine 3AC	U+UHYAC	2085	279_{1}	236_{11}	220_{27}	165_{27}	73_{100}	511	1152
Norfenefrine 3TMS		1785	369_{1}	354_{10}	267_{13}	102_{100}	73_{92}	917	4575
Norfenefrine formyl artifact	G	2040	165_{22}	146_{22}	136_{100}	107_{39}	77_{51}	147	4664
Norfenefrine-M (deamino-HO-) 3AC	U+UHYAC	1790*	280_{3}	220_{19}	178_{50}	136_{97}	123_{78}	516	1153
Norgestrel		2780*	312_{80}	245_{78}	229_{56}	135_{52}	91_{100}	677	4631
Norgestrel AC		2820*	354_{8}	325_{50}	245_{20}	91_{100}	77_{80}	863	5234
Norgestrel -H2O		2760*	294_{100}	265_{17}	185_{83}	159_{24}	131_{24}	587	4632
Normethadone	U+UHYAC	2105	295_{1}	224_{2}	165_{2}	72_{5}	58_{100}	594	246
Normethadone-M (HO-) AC	U+UHYAC	2505	353_{1}	294_{6}	72_{36}	58_{100}		859	1198
Normethadone-M (nor-) enol 2AC	U+UHYAC	2665	365_{3}	323_{4}	267_{52}	193_{41}	86_{100}	902	1199
Normethadone-M (nor-) -H2O	U UHY U+UHYAC	2030	263_{98}	220_{15}				436	1197
Normethadone-M (nor-dihydro-) -H2O AC	U+UHYAC	2850	307_{6}	266_{33}	193_{34}	86_{101}		651	1200
d-Norpseudoephedrine	U UHY	1360	132_{4}	117_{9}	105_{22}	79_{54}	77_{100}	129	1154
d-Norpseudoephedrine TMSTFA		1630	213_{7}	191_{7}	179_{100}	149_{5}	73_{80}	707	6260
d-Norpseudoephedrine 2AC	U+UHYAC	1740	235_{2}	176_{4}	129_{8}	107_{9}	86_{100}	324	1155
d-Norpseudoephedrine 2HFB		1335	330_{16}	303_{6}	240_{100}	169_{19}	119_{12}	1182	7418
d-Norpseudoephedrine formyl artifact		1280	117_{2}	105_{2}	91_{4}	77_{6}	57_{100}	143	4649
Nortriptyline	P-I G U UHY	2255	263_{27}	220_{67}	202_{100}	189_{39}	91_{30}	436	38
Nortriptyline AC	PAC U+UHYAC	2660	305_{10}	232_{100}	219_{11}	202_{8}	86_{19}	640	41
Nortriptyline HFB		2420	459_{5}	240_{100}	232_{49}	217_{36}	202_{36}	1126	7685
Nortriptyline PFP		2405	409_{3}	232_{100}	217_{71}	203_{69}	190_{69}	1038	7684
Nortriptyline TFA		2410	359_{3}	232_{76}	217_{54}	202_{70}	140_{100}	880	7683
Nortriptyline TMS		2340	335_{1}	320_{1}	203_{5}	116_{100}	73_{52}	785	5440
Nortriptyline-D3		2250	266_{6}	220_{41}	215_{51}	202_{100}	189_{45}	453	7794
Nortriptyline-D3 AC		2655	308_{11}	232_{100}	217_{46}	202_{47}	89_{23}	657	7795
Nortriptyline-D3 HFB		2415	462_{2}	243_{58}	232_{100}	217_{40}	203_{33}	1130	7798
Nortriptyline-D3 PFP		2400	412_{2}	232_{100}	217_{53}	203_{47}	193_{46}	1047	7797
Nortriptyline-D3 TFA		2405	362_{2}	232_{100}	217_{53}	202_{48}	143_{47}	891	7796
Nortriptyline-D3 TMS		2335	338_{1}	323_{10}	202_{33}	119_{100}	73_{73}	800	7799
Nortriptyline-M (HO-)	U-I UGLUC	2390	279_{8}	261_{6}	218_{100}	203_{39}	91_{10}	513	39
Nortriptyline-M (HO-) -H2O	UHY	2600	261_{14}	218_{99}	215_{100}	202_{66}	189_{23}	427	2270
Nortriptyline-M (HO-) -H2O AC	U+UHYAC	2670	303_{20}	230_{100}	215_{74}	202_{35}	86_{18}	629	42
Nortriptyline-M (nor-HO-) -H2O AC	U+UHYAC	2710	289_{15}	230_{100}	215_{70}	202_{31}	189_{5}	558	1873
Noscapine	G U+UHYAC	3130	412_{1}	220_{100}	205_{14}	147_{2}	77_{2}	1048	2525
Noscapine artifact	U+UHYAC	1780*	194_{92}	176_{52}	165_{100}	147_{69}	77_{29}	204	2326
Noxiptyline		2270	224_{2}	208_{11}	178_{3}	71_{17}	58_{100}	587	366
Noxiptyline-M (HO-dibenzocycloheptanone) -H2O	U+UHYAC	2000*	206_{37}	178_{100}	152_{15}			231	1172
Noxiptyline-M (nor-di-HO-) -H2O 2AC	U+UHYAC	3020	378_{1}	336_{4}	220_{52}	178_{100}	100_{91}	946	1174
Noxiptyline-M (nor-HO-) -H2O AC	U+UHYAC	2750	320_{1}	221_{9}	205_{56}	178_{99}	100_{31}	715	1173
Noxiptyline-M/artifact (dibenzocycloheptanone)	G U+UHYAC	1850*	208_{100}	180_{86}	165_{45}	152_{24}	193_{10}	237	1171
Nuarimol		2390	314_{34}	235_{51}	203_{51}	139_{72}	107_{100}	683	3649
Octacosane		2800*	394_{1}	99_{11}	85_{38}	71_{57}	57_{100}	1000	3797
Octadecane		1800*	254_{4}	141_{5}	85_{42}	71_{66}	57_{100}	400	2351
2-Octadecyloxyethanol		2085*	283_{5}	224_{4}	111_{19}	97_{36}	57_{100}	687	2357
Octamethyldiphenylbicyclohexasiloxane	U	2110*	539_{100}	389_{54}	327_{29}	197_{28}	135_{57}	1185	6457
Octamylamine AC		1570	186_{3}	128_{8}	100_{45}	58_{100}		349	5144
Octane		800*	114_{4}	85_{27}	57_{30}	43_{100}	29_{64}	105	3782
Octanoic acid hexadecylester		2500*	368_{3}	224_{4}	145_{100}	88_{26}	57_{65}	914	6565
Octodrine AC		1140	171_{2}	156_{2}	128_{5}	86_{100}	60_{43}	159	5255
Octopamine		1720	153_{2}	123_{100}	107_{20}	95_{52}	77_{64}	133	4665
Octopamine 3AC		2245	236_{2}	220_{20}	165_{28}	123_{91}	73_{100}	511	2808
Ofloxacin ME	UME	3750	375_{82}	305_{19}	290_{12}	246_{28}	71_{100}	936	4692
Ofloxacin -CO2	U+UHYAC	3285	317_{33}	247_{16}	231_{12}	121_{7}	71_{100}	697	4691
Olanzapine	P-I G U+UHYAC	2765	312_{24}	242_{100}	229_{85}	213_{55}	198_{21}	676	4675
Olanzapine AC	U+UHYAC	2780	354_{18}	284_{100}	242_{87}	83_{79}	70_{50}	862	4676
Olanzapine-M (nor-) 2AC	U+UHYAC	3200	382_{100}	339_{59}	284_{84}	254_{59}	213_{46}	959	4677
Oleamide	P U UHY U+UHYAC	2385	114_{5}	281_{3}	128_{7}	72_{59}	59_{100}	525	5345
Oleic acid ET		2095*	310_{5}	264_{33}	101_{82}	88_{100}	55_{72}	669	5405
Oleic acid ME		2085*	296_{3}	264_{15}	222_{9}	97_{30}	55_{100}	600	2667

Table 1-8-1: Compounds in order of names Oxazepam artifact-1

Name	Detected	RI	Typical ions and intensities					Page	Entry
Oleic acid TMS		2620*	354 $_4$	339 $_{38}$	129 $_{50}$	117 $_{62}$	73 $_{110}$	863	4522
Oleic acid glycerol ester 2AC	U+UHYAC	2790*	380 $_{34}$	264 $_{71}$	159 $_{100}$	81 $_{49}$	69 $_{85}$	1099	5602
Omethoate	G P-I	1585	213 $_7$	156 $_{90}$	110 $_{99}$	79 $_{52}$	58 $_{70}$	251	1501
Omoconazole		2925	422 $_{13}$	387 $_{21}$	267 $_{44}$	111 $_{75}$	69 $_{100}$	1064	5560
Omoconazole HY		2110	268 $_{15}$	233 $_{12}$	173 $_{100}$	145 $_{18}$	95 $_{95}$	458	6079
Omoconazole HYAC		2185	310 $_{14}$	268 $_{52}$	233 $_{24}$	95 $_{53}$	69 $_{100}$	665	6078
Opipramol	G P UHY	3055	363 $_{100}$	232 $_{45}$	218 $_{67}$	206 $_{85}$	70 $_{56}$	895	578
Opipramol AC	U+UHYAC	3170	405 $_{36}$	232 $_{25}$	218 $_{42}$	206 $_{80}$	70 $_{29}$	1032	367
Opipramol TMS		3150	435 $_{47}$	232 $_{14}$	206 $_{100}$	113 $_{32}$	73 $_{43}$	1091	4576
Opipramol-M (HO-) 2AC	U+UHYAC	3330	463 $_{100}$	403 $_{24}$	264 $_{20}$	185 $_{35}$	70 $_{37}$	1132	2675
Opipramol-M (HO-methoxy-ring)	U UHY	2340	239 $_{100}$	224 $_{47}$	209 $_{42}$	180 $_{74}$		339	423
Opipramol-M (HO-methoxy-ring) AC	U+UHYAC	2420	281 $_{42}$	239 $_{100}$	224 $_{28}$	196 $_{29}$	162 $_{16}$	521	2506
Opipramol-M (HO-ring)	UHY	2240	209 $_{100}$	180 $_{16}$	152 $_7$			241	2511
Opipramol-M (HO-ring) AC	U+UHYAC	2450	251 $_{33}$	209 $_{100}$	180 $_{74}$	152 $_{11}$		383	425
Opipramol-M (HO-ring) 2AC	U+UHYAC	2490	293 $_{21}$	251 $_{25}$	209 $_{79}$	208 $_{100}$	178 $_{17}$	579	2672
Opipramol-M (N-dealkyl-) AC	U+UHYAC PHYAC-I	3190	361 $_{100}$	232 $_{45}$	193 $_{84}$	141 $_{44}$	99 $_{43}$	889	427
Opipramol-M (N-dealkyl-) ME		2685	333 $_{40}$	232 $_{23}$	218 $_{24}$	113 $_{57}$	70 $_{100}$	778	3193
Opipramol-M (N-dealkyl-di-HO-oxo-) 2AC	U+UHYAC	3300	449 $_{100}$	264 $_{27}$	222 $_{34}$	171 $_{56}$	98 $_{31}$	1112	2674
Opipramol-M (N-dealkyl-HO-oxo-) AC	U+UHYAC	3050	391 $_{100}$	232 $_{37}$	206 $_{49}$	171 $_{41}$	98 $_{35}$	989	2673
Opipramol-M (ring)	P U UHY U+UHYAC	1985	193 $_{100}$	165 $_{19}$	139 $_5$	113 $_3$	96 $_9$	200	309
Opipramol-M (ring) AC	U+UHYAC	2040	235 $_{27}$	193 $_{100}$	192 $_{68}$	165 $_{17}$		323	2671
Opipramol-M/artifact (acridine)	U UHY U+UHYAC	1800	179 $_{100}$	151 $_{14}$				170	421
Orciprenaline TMSTFA		2180	379 $_1$	322 $_3$	241 $_{12}$	211 $_{100}$	73 $_{79}$	948	6168
Orciprenaline 2TMSTFA		2150	451 $_1$	436 $_1$	283 $_{97}$	126 $_9$	73 $_{90}$	1115	6167
Orciprenaline 3TMS		1740	412 $_3$	356 $_{62}$	322 $_2$	147 $_1$	72 $_{70}$	1075	5484
Orciprenaline 3TMSTFA		2100	523 $_1$	355 $_{98}$	126 $_{13}$	73 $_{90}$		1176	6166
Orciprenaline 4AC		2370	379 $_1$	319 $_{47}$	277 $_{100}$	235 $_{38}$	72 $_{78}$	948	1342
Orciprenaline 4TMS		1975	484 $_2$	144 $_{60}$	102 $_3$	73 $_{50}$		1163	5485
Orlistat-M/artifact (alcohol) -H2CO3		2820*	292 $_{40}$	160 $_{24}$	142 $_{25}$	114 $_{100}$	69 $_{60}$	577	5862
Orlistat-M/artifact (alcohol) -H2O		2540*	336 $_{27}$	181 $_{50}$	155 $_{100}$	109 $_{39}$	55 $_{44}$	790	5861
Ornidazole		1825	219 $_{28}$	172 $_{37}$	112 $_{56}$	81 $_{85}$	53 $_{100}$	268	1834
Ornidazole AC	U+UHYAC	1815	261 $_{36}$	219 $_{69}$	173 $_{35}$	135 $_{83}$	53 $_{100}$	424	1836
Ornidazole -HCl		1730	183 $_{66}$	166 $_{16}$	152 $_{24}$	108 $_{28}$	54 $_{100}$	181	1835
Orphenadrine	P-I G U	1935	181 $_4$	165 $_4$	73 $_{22}$	58 $_{100}$		466	1156
Orphenadrine HY	UHY	1760*	198 $_{45}$	180 $_{58}$	165 $_{28}$	119 $_{87}$	77 $_{100}$	215	1159
Orphenadrine HYAC	U+UHYAC	1750*	240 $_2$	180 $_{99}$	165 $_{37}$			345	1161
Orphenadrine-M	UHY	1560*	182 $_{82}$	167 $_{100}$	108 $_{43}$	107 $_{45}$		180	1157
Orphenadrine-M (methyl-benzophenone)	UHY U+UHYAC	1700*	196 $_{80}$	195 $_{100}$	165 $_{13}$	91 $_{67}$	77 $_{78}$	211	1158
Orphenadrine-M (nor-)	UHY	1900	255 $_{47}$	180 $_{100}$	165 $_{41}$	86 $_{72}$		404	1160
Orphenadrine-M HYAC	U+UHYAC	2005	239 $_{34}$	180 $_{100}$	165 $_{33}$			339	1162
Oryzalin		2680	346 $_8$	317 $_{100}$	275 $_{50}$	258 $_9$	75 $_9$	831	4055
Oryzalin -SO2NH		2025	267 $_{12}$	238 $_{100}$	222 $_{11}$	196 $_{64}$	138 $_8$	454	4056
Oseltamivir AC		2590	354 $_1$	212 $_{18}$	142 $_{100}$	100 $_{46}$	96 $_{47}$	862	7429
Oseltamivir HFB		2375	421 $_6$	333 $_{12}$	212 $_{26}$	142 $_{72}$	96 $_{100}$	1169	7432
Oseltamivir PFP		2385	412 $_9$	371 $_{16}$	212 $_{34}$	142 $_{78}$	96 $_{100}$	1124	7431
Oseltamivir TFA		2410	362 $_5$	321 $_{11}$	212 $_{26}$	142 $_{67}$	96 $_{100}$	1036	7430
Oseltamivir 2TMS		2330	441 $_5$	325 $_{44}$	312 $_{28}$	254 $_{34}$	73 $_{100}$	1123	7435
Oseltamivir formyl artifact		2350	324 $_{12}$	253 $_{54}$	142 $_{60}$	112 $_{86}$	96 $_{100}$	733	7433
Oseltamivir formyl artifact ME		2465	338 $_8$	250 $_{61}$	225 $_{60}$	126 $_{100}$	83 $_{75}$	800	7434
Oxabolone		2640*	290 $_{68}$	147 $_{39}$	126 $_{88}$	91 $_{77}$	55 $_{100}$	565	3947
Oxabolone AC		2820*	332 $_{100}$	290 $_{63}$	272 $_{92}$	147 $_{48}$	79 $_{42}$	770	3948
Oxabolone 2TMS		2695*	434 $_1$	419 $_{100}$	329 $_2$	303 $_2$	73 $_{45}$	1089	3950
Oxabolone cipionate		3660*	414 $_{30}$	290 $_{71}$	147 $_{34}$	125 $_{86}$	55 $_{100}$	1051	3946
Oxabolone cipionate TMS		3580*	486 $_1$	471 $_{100}$	329 $_2$	181 $_4$	73 $_{40}$	1154	3949
Oxaceprol ME		1635	187 $_7$	128 $_{54}$	86 $_{100}$	68 $_{28}$		189	2283
Oxaceprol MEAC		1690	229 $_1$	198 $_1$	169 $_9$	110 $_{37}$	68 $_{100}$	303	2709
Oxadiazon		2125	344 $_{32}$	302 $_{37}$	258 $_{58}$	175 $_{100}$	57 $_{54}$	823	4036
Oxadixyl		2280	278 $_7$	233 $_{12}$	163 $_{75}$	132 $_{54}$	105 $_{100}$	507	2517
Oxamyl -C2H3NO		1630	162 $_{13}$	145 $_{13}$	115 $_{13}$	99 $_9$	72 $_{100}$	140	3904
Oxapadol		2625	278 $_{100}$	248 $_{89}$	219 $_{83}$	105 $_{80}$	77 $_{83}$	506	1502
Oxatomide		3200	426 $_7$	219 $_{27}$	204 $_{66}$	167 $_{93}$	125 $_{100}$	1072	1673
Oxatomide-M (carbinol)	UHY	1645*	184 $_{45}$	165 $_{14}$	152 $_7$	105 $_{100}$	77 $_{63}$	184	1333
Oxatomide-M (carbinol) AC	U+UHYAC	1700*	226 $_{20}$	184 $_{20}$	165 $_{100}$	105 $_{14}$	77 $_{35}$	294	1241
Oxatomide-M (carbinol) ME	UHY	1655*	198 $_{70}$	167 $_{94}$	121 $_{100}$	105 $_{56}$	77 $_{71}$	215	6779
Oxatomide-M (HO-BPH) iso-1	UHY	2065*	198 $_{93}$	121 $_{72}$	105 $_{100}$	93 $_{22}$	77 $_{66}$	214	1627
Oxatomide-M (HO-BPH) iso-1 AC	U+UHYAC	2010*	240 $_{27}$	198 $_{100}$	121 $_{47}$	105 $_{85}$	77 $_{80}$	343	2196
Oxatomide-M (HO-BPH) iso-2	P-I U UHY	2080*	198 $_{50}$	121 $_{100}$	105 $_{17}$	93 $_{14}$	77 $_{28}$	214	732
Oxatomide-M (HO-BPH) iso-2 AC	U+UHYAC	2050*	240 $_{20}$	198 $_{100}$	121 $_{94}$	105 $_{41}$	77 $_{51}$	343	2197
Oxatomide-M (N-dealkyl-)	U UHY	2120	252 $_{12}$	207 $_{58}$	167 $_{100}$	152 $_{33}$	85 $_{49}$	392	1602
Oxatomide-M (norcyclizine) AC	U+UHYAC	2525	294 $_{16}$	208 $_{56}$	167 $_{100}$	152 $_{30}$	85 $_{78}$	586	1601
Oxazepam	P G UGLUC	2320	268 $_{98}$	239 $_{57}$	233 $_{52}$	205 $_{66}$	77 $_{100}$	542	579
Oxazepam TMS		2635	356 $_9$	341 $_{100}$	312 $_{56}$	239 $_{12}$	135 $_{21}$	876	4577
Oxazepam 2TMS		2200	430 $_{36}$	429 $_{63}$	340 $_{10}$	313 $_{14}$	73 $_{70}$	1080	5499
Oxazepam artifact-1	P-I UHY U+UHYAC	2060	240 $_{59}$	239 $_{100}$	205 $_{81}$	177 $_{16}$	151 $_9$	342	300

Table 1-8-1: Compounds in order of names

Name	Detected	RI	Typical ions and intensities					Page	Entry
Oxazepam artifact-2	UHY U+UHYAC	2070	254_{77}	253_{100}	219_{98}			397	301
Oxazepam artifact-3	G P U	2500	298_{78}	240_{100}	203_{65}			604	1257
Oxazepam HY	UHY	2050	231_{80}	230_{95}	154_{23}	105_{38}	77_{100}	308	419
Oxazepam HYAC	U+UHYAC PHYAC	2245	273_{31}	230_{100}	154_{13}	105_{23}	77_{50}	480	273
Oxazepam iso-1 2ME		2425	314_{26}	271_{101}	239_{45}	205_{27}		684	581
Oxazepam iso-2 -H2O 2ME		2575	296_{80}	295_{100}	267_{97}	239_{58}	205_{61}	596	582
Oxazepam-M	P G U	2520	270_{86}	269_{97}	242_{100}	241_{82}	77_{17}	468	463
Oxazepam-M (HO-) artifact AC	U+UHYAC	2515	312_{30}	270_{100}	253_{46}	235_{77}	206_{9}	674	1747
Oxazepam-M (HO-) HYAC	U+UHYAC	2270	289_{18}	247_{86}	246_{100}	105_{7}	77_{35}	556	3143
Oxazepam-M TMS		2300	342_{62}	341_{100}	327_{19}	269_{4}	73_{30}	817	4573
Oxazolam		2540	328_{4}	283_{11}	251_{99}	70_{43}		748	1168
Oxazolam HYAC	U+UHYAC PHYAC	2245	273_{31}	230_{100}	154_{13}	105_{23}	77_{50}	480	273
Oxazolam-M	P G UGLUC	2320	268_{98}	239_{57}	233_{52}	205_{66}	77_{100}	542	579
Oxazolam-M TMS		2635	356_{9}	341_{100}	312_{56}	239_{12}	135_{21}	876	4577
Oxazolam-M 2TMS		2200	430_{36}	429_{63}	340_{10}	313_{14}	73_{70}	1080	5499
Oxazolam-M HY	UHY	2050	231_{80}	230_{95}	154_{23}	105_{38}	77_{100}	308	419
Oxcarbazepine		2375	252_{36}	209_{65}	180_{100}	152_{28}	89_{19}	388	6065
Oxcarbazepine artifact (acridinecarboxylic acid) (ME)		2165	237_{100}	206_{58}	178_{81}	151_{27}	75_{9}	333	6066
Oxcarbazepine enol AC		2575	294_{10}	252_{92}	209_{110}	180_{95}	152_{27}	585	6067
Oxcarbazepine-M/artifact (ring) AC	U+UHYAC	2450	251_{33}	209_{100}	180_{74}	152_{11}		383	425
Oxedrine 3AC		2175	293_{6}	233_{44}	191_{67}	149_{100}	86_{93}	580	1530
Oxeladin	P U+UHYAC	2180	335_{1}	320_{1}	219_{1}	144_{8}	86_{100}	786	1163
Oxetacaine AC		2550	318_{3}	287_{2}	188_{69}	91_{20}	87_{100}	1169	6070
Oxiconazole		3290	427_{2}	392_{18}	240_{8}	159_{100}	81_{55}	1073	2824
Oxilofrine (erythro-)		1875	148_{1}	95_{1}	77_{4}	71_{6}	58_{100}	178	1971
Oxilofrine (erythro-) ME2AC		2000	279_{1}	247_{1}	206_{3}	100_{60}	58_{100}	513	2348
Oxilofrine (erythro-) 3AC	U+UHYAC	2145	307_{1}	247_{1}	205_{1}	100_{72}	58_{100}	649	750
Oxilofrine (erythro-) formyl artifact		1790	133_{2}	121_{10}	107_{6}	71_{100}	56_{76}	201	4499
Oxilofrine (erythro-) -H2O 2AC	U+UHYAC	1990	247_{9}	205_{1}	163_{24}	107_{8}	58_{100}	367	1972
Oxomemazine	G U UHY U+UHYAC	2830	330_{11}	271_{3}	180_{2}	152_{2}	58_{100}	759	1768
Oxomemazine-M (bis-nor-)	UHY	2785	272_{100}	244_{22}	231_{8}	180_{9}	152_{7}	623	1770
Oxomemazine-M (bis-nor-) AC	U+UHYAC	3035	344_{42}	272_{66}	244_{100}	231_{68}	114_{25}	824	1772
Oxomemazine-M (nor-)	UHY	2720	316_{38}	271_{63}	231_{100}	180_{10}	152_{8}	693	1769
Oxomemazine-M (nor-) AC	U+UHYAC	3125	358_{18}	272_{17}	244_{41}	128_{100}	86_{18}	877	1771
Oxprenolol	P-I G	1970	265_{1}	250_{1}	221_{5}	150_{3}	72_{100}	447	4256
Oxprenolol TMS		1850	337_{1}	322_{1}	221_{4}	150_{5}	72_{60}	795	5475
Oxprenolol TMSTFA		2135	433_{1}	418_{1}	284_{71}	129_{55}	73_{60}	1086	6163
Oxprenolol 2AC	PAC-I	2390	349_{1}	289_{14}	200_{94}	98_{26}	72_{100}	843	1336
Oxprenolol 2TMS		2070	394_{1}	222_{10}	144_{60}	101_{5}	73_{47}	1040	5476
Oxprenolol formyl artifact	P-I G	1985	277_{16}	262_{76}	248_{43}	148_{47}	56_{99}	502	1339
Oxprenolol -H2O AC	PAC-I U+UHYAC	2260	289_{12}	188_{16}	140_{16}	98_{16}	72_{100}	560	1335
Oxprenolol-M (deamino-HO-) 2AC	U+UHYAC	1900*	308_{2}	249_{8}	159_{100}	99_{16}		655	1334
Oxprenolol-M (deamino-HO-dealkyl-)	UHY	1700*	184_{26}	135_{4}	110_{100}	81_{8}	64_{5}	184	2683
Oxprenolol-M (deamino-HO-dealkyl-) 3AC	U+UHYAC	1920*	310_{1}	268_{13}	159_{100}	117_{12}	110_{19}	666	800
Oxprenolol-M (HO-) -H2O iso-1 2AC	U+UHYAC	2520	347_{5}	305_{7}	200_{44}	72_{100}		835	1337
Oxprenolol-M (HO-) -H2O iso-2 2AC	U+UHYAC	2570	347_{13}	305_{5}	204_{19}	72_{100}		836	1338
Oxprenolol-M (HO-) iso-1 3AC	U+UHYAC	3050	407_{5}	347_{50}	305_{16}	204_{69}	72_{99}	1034	1340
Oxprenolol-M (HO-) iso-2 3AC	U+UHYAC	3100	407_{10}	347_{27}	200_{72}	140_{40}	72_{100}	1034	1341
Oxybenzone	UHY	2135*	228_{64}	227_{90}	151_{100}	105_{11}	77_{26}	300	3662
Oxybenzone AC	U+UHYAC	2225*	270_{5}	227_{100}	151_{56}	105_{12}	77_{29}	469	3663
Oxybenzone-M (O-demethyl-)	UHY	2280*	214_{61}	213_{83}	137_{100}	105_{21}	77_{33}	255	3660
Oxybenzone-M (O-demethyl-) 2AC	U+UHYAC	2315*	298_{3}	256_{45}	213_{100}	137_{21}	77_{18}	605	3661
Oxyberberine		2995	351_{100}	336_{70}	322_{49}	308_{28}	292_{26}	847	5661
Oxybuprocaine		2425	236_{8}	192_{18}	136_{8}	99_{42}	86_{100}	658	1943
Oxybuprocaine AC		2640	335_{1}	278_{7}	234_{11}	99_{48}	86_{100}	846	1944
Oxybuprocaine-M (HOOC-) AC		2060	251_{51}	220_{10}	195_{22}	167_{100}	136_{83}	385	1946
Oxybuprocaine-M (HOOC-) MEAC		2100	265_{41}	234_{6}	223_{14}	167_{100}	136_{29}	445	1945
Oxybutynine		2505	357_{5}	342_{88}	189_{41}	107_{55}	55_{100}	874	3724
Oxycodone	G	2540	315_{100}	258_{19}	230_{35}	140_{17}	70_{37}	689	583
Oxycodone AC	U+UHYAC	2555	357_{99}	314_{41}	298_{15}	240_{14}		873	247
Oxycodone HFB		2330	511_{100}	314_{70}	240_{62}	115_{52}	69_{77}	1170	6152
Oxycodone PFP		2350	461_{100}	314_{56}	240_{45}	212_{31}	119_{31}	1129	6119
Oxycodone TFA		2290	411_{100}	314_{33}	240_{28}	115_{34}	54_{67}	1043	4013
Oxycodone TMS		2555	387_{100}	372_{23}	330_{6}	229_{11}	73_{34}	977	4322
Oxycodone enol 2AC		2560	399_{100}	357_{33}	314_{15}	240_{21}		1015	248
Oxycodone enol 2TMS		2510	459_{100}	444_{22}	368_{12}	312_{16}	73_{74}	1127	4321
Oxycodone-D6		2535	321_{100}	236_{44}	204_{33}	143_{24}	115_{26}	720	7296
Oxycodone-D6 TMS		2555	393_{97}	379_{17}	276_{12}	236_{30}	73_{100}	996	7297
Oxycodone-M (dihydro-) 2AC	U+UHYAC	2570	401_{60}	359_{100}	242_{64}	70_{60}		1021	1189
Oxycodone-M (nor-) enol 3AC	U+UHYAC	2680	427_{63}	385_{100}	343_{23}	87_{26}		1074	1190
Oxycodone-M (nor-dihydro-) 2AC	U+UHYAC	2900	387_{14}	343_{100}	258_{59}	87_{24}	72_{21}	976	1191
Oxycodone-M (nor-dihydro-) 3AC	U+UHYAC	2935	429_{3}	387_{36}	242_{99}	87_{19}	72_{13}	1078	1192
Oxycodone-M (O-demethyl-)		2555	301_{100}	244_{9}	216_{55}	203_{22}	70_{50}	619	7166
Oxycodone-M (O-demethyl-) AC	U+UHYAC	2650	343_{16}	301_{100}	216_{43}	203_{30}	70_{42}	821	7167

Table 1-8-1: Compounds in order of names Paracetamol-M (methoxy-) AC

Name	Detected	RI	Typical ions and intensities					Page	Entry
Oxycodone-M (O-demethyl-) TMS		2560	373 78	288 42	259 34	73 84	70 76	931	7169
Oxycodone-M (O-demethyl-) 2AC	U+UHYAC	2620	385 28	343 100	300 49	284 29	70 24	970	7168
Oxycodone-M (O-demethyl-) 2TMS		2570	445 18	288 9	229 10	216 7	73 100	1108	7170
Oxycodone-M (O-demethyl-) 3TMS		2525	517 7	502 8	412 3	355 10	73 100	1174	7171
Oxydemeton-S-Methyl	G P-I	1860*	218 1	169 60	125 47	109 100	79 26	361	1500
Oxyfedrine-M (N-dealkyl-)	U UHY	1360	132 4	117 9	105 22	79 54	77 100	129	1154
Oxyfedrine-M (N-dealkyl-) TMSTFA		1630	213 7	191 7	179 100	149 5	73 80	707	6260
Oxyfedrine-M (N-dealkyl-) 2AC	U+UHYAC	1740	235 2	176 4	129 8	107 9	86 100	324	1155
Oxyfedrine-M (N-dealkyl-) 2HFB		1335	330 16	303 6	240 100	169 19	119 12	1182	7418
Oxyfedrine-M (N-dealkyl-) formyl artifact		1280	117 2	105 2	91 4	77 6	57 100	143	4649
Oxymetazoline	U+UHYAC	2195	260 99	245 93	217 44	81 51		424	1503
Oxymetazoline 2AC		2760	344 60	302 100	287 51	230 40	203 40	826	1504
Oxymetholone		3005*	332 19	275 21	174 58	91 84	55 100	771	2823
Oxymetholone enol 3TMS		2870*	548 100	490 23	405 7	281 10	73 75	1184	3983
Oxymorphone		2555	301 100	244 9	216 55	203 22	70 50	619	7166
Oxymorphone AC	U+UHYAC	2650	343 16	301 100	216 43	203 30	70 42	821	7167
Oxymorphone TMS		2560	373 78	288 42	259 34	73 84	70 76	931	7169
Oxymorphone 2AC	U+UHYAC	2620	385 28	343 100	300 49	284 29	70 24	970	7168
Oxymorphone 2TMS		2570	445 18	288 9	229 10	216 7	73 100	1108	7170
Oxymorphone 3TMS		2525	517 7	502 8	412 3	355 10	73 100	1174	7171
Oxypertine		3445	379 11	217 11	175 101	132 17	70 30	950	368
Oxypertine-M (HO-phenylpiperazine) 2AC	U+UHYAC	2350	262 39	220 60	177 21	148 100	135 59	430	6610
Oxypertine-M (phenylpiperazine) AC	U+UHYAC	1920	204 48	161 21	132 99	56 77		227	1276
Oxyphenbutazone		9999	324 30	199 90	135 25	93 100	77 70	732	1513
Oxyphenbutazone AC	U+UHYAC	2700	366 33	324 49	199 100	93 80	77 72	905	1506
Oxyphenbutazone artifact (phenyldiazophenol)		2070	198 46	121 41	93 100	77 72	65 54	214	1027
Oxyphenbutazone artifact (phenyldiazophenol) ME		2020	212 41	135 35	107 91	77 100	64 26	249	4205
Oxyphenbutazone iso-1 2ME	UME	2545	352 100	213 66	148 18	107 36	77 65	855	1505
Oxyphenbutazone iso-2 2ME		2720	352 100	309 26	190 41	160 46	77 63	855	1507
Oxyphencyclimine -H2O		2405	326 90	243 100	171 16	127 31	105 41	741	6308
Oxyphencyclimine-M/artifact (HOOC-) ME		1755	248 1	189 81	166 23	105 100	77 69	371	6309
Palmitamide	P U UHY U+UHYAC	2130	255 1	212 2	128 5	72 37	59 100	404	5344
Palmitic acid	G P U UHY U+UHYAC	1965*	256 29	213 20	185 15	129 38	73 100	408	822
Palmitic acid ET		1950*	284 4	241 6	157 18	101 56	88 100	537	5403
Palmitic acid ME	G P U UHY U+UHYAC	1940*	270 10	227 7	143 13	87 58	74 100	472	1801
Palmitic acid TMS		2470*	328 6	313 62	132 46	117 93	73 100	751	4668
Palmitic acid glycerol ester	G	2420*	313 3	299 11	239 49	98 100	57 99	760	5588
Palmitic acid glycerol ester 2AC	U+UHYAC	2645*	354 4	239 34	159 100	98 29	84 18	1051	5412
Palmitic acid glycerol ester 2TMS	G	2620*	460 3	372 77	205 25	147 62	73 170	1143	7449
Palmitoleic acid TMS		2450*	326 3	311 33	129 52	117 69	73 100	742	4669
Pangamic acid-M/artifact (gluconic acid) ME5AC		1975*	361 3	289 14	173 58	145 72	115 100	1062	5227
Panthenol		1920	205 1	175 5	157 14	133 100	102 24	229	1522
Panthenol 3AC	U+UHYAC	2045	331 1	217 92	175 77	145 51	115 100	765	1509
Panthenol artifact		1920	189 27	159 58	145 17	71 100		191	823
Papaverine	G P U+UHYAC	2820	339 75	338 90	324 76	308 18	293 9	803	824
Papaverine-M (bis-demethyl-) iso-1 2AC	U+UHYAC	2970	395 31	353 56	310 100	294 15	179 10	1002	3689
Papaverine-M (bis-demethyl-) iso-2 2AC	U+UHYAC	2995	395 29	353 50	310 100	294 25	196 8	1002	3690
Papaverine-M (bis-demethyl-) iso-3 2AC	U+UHYAC	3050	395 28	353 50	310 100	294 21	179 4	1002	3691
Papaverine-M (bis-demethyl-) iso-4 2AC	U+UHYAC	3065	395 29	353 61	310 100	294 22	179 10	1002	3692
Papaverine-M (O-demethyl-)	UHY	2805	325 77	324 100	310 80	266 13	153 9	735	3684
Papaverine-M (O-demethyl-) iso-1 AC	U+UHYAC	2860	367 72	324 100	310 50	278 7	153 13	909	3685
Papaverine-M (O-demethyl-) iso-2 AC	U+UHYAC	2895	367 73	324 100	310 60	296 45	254 53	909	3686
Papaverine-M (O-demethyl-) iso-3 AC	U+UHYAC	2910	367 74	324 100	308 26	254 11	153 10	909	3687
Papaverine-M (O-demethyl-) iso-4 AC	U+UHYAC	2940	367 47	324 100	310 77	294 14	137 5	909	3688
Paracetamol	G P U	1780	151 34	109 100	81 16	80 23		129	825
Paracetamol AC	U+UHYAC PAC	1765	193 10	151 53	109 100	80 24		199	188
Paracetamol HFB	UHYHFB PHFB	1735	347 24	305 39	169 13	108 100	69 31	833	5099
Paracetamol ME	PME UME	1630	165 59	123 74	108 100	95 10	80 20	147	5046
Paracetamol PFP		1675	297 19	255 31	119 38	108 100	80 28	601	5095
Paracetamol TFA		1630	247 11	205 30	108 100	80 19	69 34	365	5092
Paracetamol 2AC	U+UHYAC	2085	235 10	193 11	151 30	109 100		323	827
Paracetamol 2TMS		1780	295 50	280 68	206 83	116 15	73 100	592	4578
Paracetamol Cl-artifact AC	U+UHYAC	2030	227 6	185 74	143 100	114 4	79 12	296	2993
Paracetamol HY	UHY	1240	109 99	80 41	52 90			102	826
Paracetamol HYME	UHYME	1100	123 27	109 100	94 7	80 96	53 47	108	3766
Paracetamol-D4 AC		1760	197 4	155 30	113 100	84 8		212	6550
Paracetamol-D4 HFB		1730	351 21	309 26	169 8	112 100	69 25	847	6552
Paracetamol-D4 ME		1625	169 30	127 47	112 100	99 9	84 13	156	6554
Paracetamol-D4 PFP		1675	301 13	259 16	119 9	112 100	84 19	601	6553
Paracetamol-D4 TFA		1625	251 30	209 44	112 100	84 13	69 33	382	6559
Paracetamol-D4 2TMS		1775	299 35	284 47	210 57	116 15	73 100	612	6551
Paracetamol-M (HO-) 3AC	U+UHYAC	2150	251 6	209 23	167 87	125 100		383	2384
Paracetamol-M (HO-methoxy-) AC	U+UHYAC	2170	239 12	197 86	155 100	140 42	110 9	339	2383
Paracetamol-M (methoxy-) AC	U+UHYAC	1940	223 12	181 79	139 100			283	201

Table 1-8-1: Compounds in order of names

Name	Detected	RI	Typical ions and intensities					Page	Entry
Paracetamol-M (methoxy-) Cl-artifact AC	U+UHYAC	2060	257_6	215_{77}	173_{100}	158_{21}	130_5	409	2994
Paracetamol-M 2AC	U+UHYAC	2270	262_{20}	220_{35}	188_{17}	160_{74}	146_{100}	428	2387
Paracetamol-M 3AC	U+UHYAC	2340	304_{15}	261_{31}	219_{46}	160_{100}	146_{72}	632	2388
Paracetamol-M conjugate 2AC	U+UHYAC	3050	396_{20}	354_7	246_{100}	204_{73}	162_{71}	1004	2389
Paracetamol-M conjugate 3AC	U+UHYAC	3030	438_{35}	353_{40}	246_{72}	204_{97}	162_{100}	1094	1387
Paracetamol-M iso-1 3AC	U+UHYAC	2200	305_{26}	263_{57}	221_{14}	160_{69}	146_{100}	637	2385
Paracetamol-M iso-2 3AC	U+UHYAC	2220	305_{34}	263_{100}	221_{82}	162_{54}	146_{99}	637	2386
Parafluorofentanyl		2560	354_1	263_{100}	220_{11}	207_{25}	164_{45}	862	6029
Parahydroxybenzoic acid ET		1585*	166_{53}	151_4	138_{65}	121_{100}	93_{10}	151	6541
Parahydroxybenzoic acid 2ET		1520*	194_{17}	166_7	149_{100}	121_{76}	93_{17}	205	6646
Paraldehyde		<1000*	131_3	117_9	89_{15}	87_8		113	1915
Paramethadione		1110	157_1	129_{59}	72_{17}	57_{100}		138	274
Paraoxon	P-I	1890	275_{52}	149_{53}	109_{100}	99_{45}	81_{58}	488	1464
Parathion-ethyl	P-I G U	1970	291_{49}	186_{21}	139_{53}	109_{100}	97_{96}	567	828
Parathion-ethyl-M (4-nitrophenol)	P-I UHY	1530	139_{100}	109_{23}	93_{18}	65_{57}		119	829
Parathion-ethyl-M (4-nitrophenol) AC	U+UHYAC	1500	181_{67}	139_{100}	109_{78}	65_{76}	63_{78}	176	830
Parathion-ethyl-M (4-nitrophenol) ME		1455	153_{100}	123_{37}	92_{55}	77_{72}		132	831
Parathion-ethyl-M (amino-)	P U	1900	261_{84}	125_{100}	109_{81}	80_{41}		425	1325
Parathion-ethyl-M (paraoxon)	P-I	1890	275_{52}	149_{53}	109_{100}	99_{45}	81_{58}	488	1464
Parathion-methyl		1855	263_{95}	233_{10}	125_{88}	109_{100}	79_{36}	433	1510
Parathion-methyl-M (4-nitrophenol)	P-I UHY	1530	139_{100}	109_{23}	93_{18}	65_{57}		119	829
Parathion-methyl-M (4-nitrophenol) AC	U+UHYAC	1500	181_{67}	139_{100}	109_{78}	65_{76}	63_{78}	176	830
Parathion-methyl-M (4-nitrophenol) ME		1455	153_{100}	123_{37}	92_{55}	77_{72}		132	831
Paroxetine	G	2850	329_{34}	192_{43}	138_{34}	109_{36}	70_{100}	754	5264
Paroxetine AC	U+UHYAC	2980	371_{32}	234_{100}	138_{53}	86_{46}	70_{58}	923	5265
Paroxetine HFB		2685	525_{33}	266_{17}	138_{100}	135_{79}	109_{81}	1177	7686
Paroxetine ME		2600	343_6	206_{15}	191_{12}	84_{15}	58_{100}	821	5275
Paroxetine PFP		2680	475_7	338_5	175_{19}	138_{100}	109_{77}	1144	6320
Paroxetine TFA		2700	425_{10}	288_8	166_{30}	138_{100}	109_{81}	1069	6319
Paroxetine TMS		2710	401_{12}	264_{30}	249_{80}	116_{100}	73_{98}	1021	4579
Paroxetine-M (demethylenyl-3-methyl-) 2AC	U+UHYAC	3030	415_{14}	373_{42}	234_{100}	192_{20}	86_{38}	1053	5263
Paroxetine-M (demethylenyl-4-methyl-) 2AC	U+UHYAC	3020	415_8	373_{22}	234_{100}	192_{21}	86_{24}	1053	5343
Paroxetine-M/artifact (dephenyl-) 2AC		2230	293_{19}	123_{53}	87_{45}	233_{69}	220_{100}	581	5309
PCC		1525	192_4	191_7	164_8	149_{100}	122_{13}	198	3581
PCC -HCN		1190	165_{85}	164_{80}	150_{100}	136_{64}	122_{39}	149	3582
PCDI		1570	203_{25}	160_{100}	146_{26}	91_{73}	77_{17}	225	3599
PCDI intermediate (DMCC)		<1000	152_7	151_{13}	137_{18}	109_{100}	84_{12}	132	3580
PCDI precursor (dimethylamine)		<1000	45_{50}	44_{100}	28_{70}			90	3618
PCE AC		1920	245_8	188_{42}	158_{78}	117_{17}	91_{100}	360	3622
PCE artifact (phenylcyclohexene)	U+UHYAC	1270*	158_{100}	143_{56}	129_{93}	115_{60}	91_{30}	138	3606
PCEEA		1755	247_{34}	204_{79}	188_{38}	159_{100}	91_{93}	369	7076
PCEEA AC		2110	289_1	246_7	232_8	159_{100}	91_{65}	561	7367
PCEEA-M (carboxy-) TMS		1975	305_{21}	262_{100}	188_{19}	159_{58}	91_{58}	641	7376
PCEEA-M (carboxy-3'-HO-) 2TMS		2200	393_1	378_3	350_{38}	246_{60}	157_{100}	996	7379
PCEEA-M (carboxy-4'-cis-HO-) 2TMS		2250	393_3	262_{100}	246_{22}	157_{42}	91_{23}	996	7378
PCEEA-M (carboxy-4'-trans-HO-) 2TMS		2285	393_1	276_4	262_{100}	247_4	157_{55}	996	7377
PCEEA-M (N-dealkyl-) AC		1850	217_6	174_{19}	158_{100}	132_{78}	104_{27}	262	7016
PCEEA-M (N-dealkyl-) TFA		1630	271_{10}	228_{31}	202_{22}	158_{100}	115_{61}	474	7039
PCEEA-M (N-dealkyl-3'-HO-) iso-1 2AC		2055	275_2	216_{25}	190_{16}	174_{32}	156_{100}	491	7012
PCEEA-M (N-dealkyl-3'-HO-) iso-1 2TFA		1690	270_{37}	383_6	240_{13}	172_{14}	156_{78}	962	7041
PCEEA-M (N-dealkyl-3'-HO-) iso-2 2AC		2065	275_2	233_3	216_{14}	190_{15}	156_{100}	491	7013
PCEEA-M (N-dealkyl-3'-HO-) iso-2 2TFA		1730	383_4	270_{35}	240_{17}	172_{13}	156_{100}	962	7040
PCEEA-M (N-dealkyl-4'-HO-) -H2O AC		1680	215_{15}	172_{100}	156_9	119_{15}	103_{19}	257	7021
PCEEA-M (N-dealkyl-4'-HO-) iso-1 2AC		2090	275_1	215_{10}	172_{34}	156_{100}	132_{48}	491	7014
PCEEA-M (N-dealkyl-4'-HO-) iso-1 2TFA		1700	383_2	269_7	240_{25}	172_{17}	156_{100}	962	7042
PCEEA-M (N-dealkyl-4'-HO-) iso-2 2AC		2100	275_3	215_{10}	172_{25}	156_{100}	132_{36}	491	7015
PCEEA-M (N-dealkyl-4'-HO-) iso-2 2TFA		1735	383_2	269_7	240_{19}	172_{16}	156_{100}	962	7043
PCEEA-M (O-deethyl-) AC		1905	261_{17}	218_{100}	159_{15}	91_{32}	87_{29}	428	7077
PCEEA-M (O-deethyl-) TFA		1690	315_{64}	286_{10}	272_{100}	238_{16}	91_{54}	689	7387
PCEEA-M (O-deethyl-) TMS		1860	291_{24}	248_{60}	188_{32}	159_{100}	91_{58}	573	7380
PCEEA-M (O-deethyl-3'-HO-) 2AC		2225	260_{100}	319_{12}	276_{15}	157_{67}	87_{38}	710	7078
PCEEA-M (O-deethyl-3'-HO-) 2TFA		1775	427_{12}	314_{100}	272_{17}	157_3	91_1	1073	7389
PCEEA-M (O-deethyl-3'-HO-) 2TMS		2110	336_{16}	276_{24}	247_{19}	157_{100}	129_{12}	950	7381
PCEEA-M (O-deethyl-3'-HO-HO-phenyl-) 3AC		2470	377_1	318_{14}	276_{56}	234_{75}	155_{100}	942	7375
PCEEA-M (O-deethyl-4'-cis-HO-) 2TMS		2160	379_3	364_2	248_{100}	246_{16}	157_{75}	950	7382
PCEEA-M (O-deethyl-4'-HO-) 2TFA		1825	427_4	314_{12}	272_{100}	157_{28}	91_{40}	1073	7388
PCEEA-M (O-deethyl-4'-HO-) -H2O AC		1860	259_{38}	230_{45}	200_{94}	186_{100}		417	7386
PCEEA-M (O-deethyl-4'-HO-) -H2O TFA		1650	313_{73}	284_{100}	200_{91}	170_{34}	141_{20}	680	7390
PCEEA-M (O-deethyl-4'-HO-) iso-1 2AC		2270	319_5	259_{99}	218_{110}	157_{54}	87_{77}	710	7079
PCEEA-M (O-deethyl-4'-HO-) iso-2 2AC		2280	319_3	259_{86}	218_{100}	157_{54}	87_{58}	710	7080
PCEEA-M (O-deethyl-4'-HO-HO-phenyl-) 3AC		2650	377_3	317_{77}	276_{100}	234_{54}	173_{77}	943	7374
PCEEA-M (O-deethyl-4'-trans-HO-) 2TMS		2180	379_2	276_{13}	248_{90}	157_{100}	91_{25}	951	7383
PCEEA-M (O-deethyl-HO-phenyl-) 2AC		2340	319_{17}	276_{100}	234_{39}	175_{30}	107_{30}	710	7373
PCEEA-M (O-deethyl-HO-phenyl-) 2TMS		2225	379_{14}	336_{35}	247_{100}	207_{52}	179_{40}	951	7384

Table 1-8-1: Compounds in order of names   PCPIP

Name	Detected	RI	Typical ions and intensities					Page	Entry
PCEPA		1915	261_{13}	232_{12}	218_{100}	117_{29}	91_{53}	428	5877
PCEPA AC		2210	303_{2}	260_{11}	246_{14}	158_{48}	91_{100}	631	5878
PCEPA TFA		2040	357_{1}	260_{4}	159_{100}	91_{94}	81_{22}	874	5879
PCEPA-M (3'-HO-) AC		2080	319_{16}	276_{16}	260_{100}	234_{12}	218_{24}	712	7007
PCEPA-M (3'-HO-) TFA		1980	373_{8}	260_{100}	218_{19}	186_{14}	157_{14}	931	7052
PCEPA-M (4'-HO-) TFA		2010	373_{10}	260_{8}	218_{100}	186_{4}	157_{16}	932	7053
PCEPA-M (4'-HO-) -H2O		1870	259_{35}	230_{100}	216_{20}	203_{25}	186_{32}	418	7009
PCEPA-M (4'-HO-) iso-1 AC		2140	319_{1}	259_{3}	244_{1}	218_{7}	91_{5}	712	7010
PCEPA-M (4'-HO-) iso-2 AC		2145	319_{4}	259_{70}	244_{13}	218_{100}	87_{15}	712	7011
PCEPA-M (carboxy-) TMS		2045	319_{25}	276_{100}	188_{8}	159_{18}	144_{26}	712	7027
PCEPA-M (carboxy-) -H2O		1930	229_{100}	200_{9}	186_{90}	158_{57}	144_{52}	304	7018
PCEPA-M (carboxy-2"-HO-) 2TMS		2210	407_{12}	364_{7}	317_{9}	290_{5}	276_{100}	1035	7032
PCEPA-M (carboxy-2"-HO-) -H2O AC		1975	287_{26}	244_{100}	172_{17}	159_{15}	117_{17}	547	7026
PCEPA-M (carboxy-2"-HO-) -H2O TFA		1905	341_{12}	200_{5}	186_{65}	159_{95}	91_{100}	811	7048
PCEPA-M (carboxy-3'-HO-) 2TMS		2275	407_{3}	364_{100}	276_{32}	246_{85}	157_{36}	1035	7028
PCEPA-M (carboxy-3'-HO-) iso-1 -H2O AC		2080	287_{34}	244_{34}	228_{26}	202_{21}	157_{100}	548	7022
PCEPA-M (carboxy-3'-HO-) iso-1 -H2O TFA		1960	341_{7}	271_{3}	228_{29}	186_{31}	157_{100}	811	7044
PCEPA-M (carboxy-3'-HO-) iso-2 -H2O AC		2105	287_{32}	244_{43}	228_{41}	202_{26}	157_{100}	548	7023
PCEPA-M (carboxy-3'-HO-) iso-2 -H2O TFA		1985	341_{7}	271_{5}	228_{29}	186_{38}	157_{100}	811	7045
PCEPA-M (carboxy-4'-cis-HO-) 2TMS		2310	407_{4}	276_{100}	246_{13}	157_{14}	144_{12}	1035	7029
PCEPA-M (carboxy-4'-HO-) 2TMS		2370	407_{20}	364_{170}	275_{25}	247_{26}	179_{22}	1035	7031
PCEPA-M (carboxy-4'-HO-) iso-1 -H2O AC		2160	287_{6}	227_{100}	198_{44}	184_{10}	156_{71}	548	7020
PCEPA-M (carboxy-4'-HO-) iso-1 -H2O TFA		1970	341_{4}	271_{5}	227_{48}	157_{74}	91_{100}	811	7046
PCEPA-M (carboxy-4'-HO-) iso-2 -H2O AC		2175	287_{2}	157_{83}	144_{34}	227_{100}	198_{33}	548	7019
PCEPA-M (carboxy-4'-HO-) iso-2 -H2O TFA		2010	341_{2}	271_{5}	227_{28}	157_{64}	91_{100}	811	7047
PCEPA-M (carboxy-4'-trans-HO-) 2TMS		2335	407_{2}	276_{100}	157_{14}	144_{11}	91_{14}	1035	7030
PCEPA-M (carboxy-HO-phenyl-) 2TMS		2470	495_{2}	364_{100}	335_{5}	245_{12}	179_{5}	1161	7131
PCEPA-M (HO-phenyl-) AC		2150	319_{29}	276_{100}	234_{33}	175_{18}	107_{26}	713	7000
PCEPA-M (N-dealkyl-) AC		1850	217_{6}	174_{19}	158_{100}	132_{78}	104_{27}	262	7016
PCEPA-M (N-dealkyl-) TFA		1630	271_{10}	228_{31}	202_{22}	158_{100}	115_{61}	474	7039
PCEPA-M (N-dealkyl-3'-HO-) iso-1 2AC		2055	275_{2}	216_{25}	190_{16}	174_{32}	156_{100}	491	7012
PCEPA-M (N-dealkyl-3'-HO-) iso-1 2TFA		1690	270_{37}	383_{6}	240_{13}	172_{14}	156_{78}	962	7041
PCEPA-M (N-dealkyl-3'-HO-) iso-2 2AC		2065	275_{2}	233_{3}	216_{14}	190_{15}	156_{100}	491	7013
PCEPA-M (N-dealkyl-3'-HO-) iso-2 2TFA		1730	383_{4}	270_{35}	240_{17}	172_{13}	156_{100}	962	7040
PCEPA-M (N-dealkyl-4'-HO-) -H2O AC		1680	215_{15}	172_{100}	156_{9}	119_{15}	103_{19}	257	7021
PCEPA-M (N-dealkyl-4'-HO-) iso-1 2AC		2090	275_{1}	215_{10}	172_{34}	156_{100}	132_{48}	491	7014
PCEPA-M (N-dealkyl-4'-HO-) iso-1 2TFA		1700	383_{2}	269_{7}	240_{25}	172_{17}	156_{100}	962	7042
PCEPA-M (N-dealkyl-4'-HO-) iso-2 2AC		2100	275_{3}	215_{10}	172_{29}	156_{100}	132_{36}	491	7015
PCEPA-M (N-dealkyl-4'-HO-) iso-2 2TFA		1735	383_{2}	269_{7}	240_{19}	172_{16}	156_{100}	962	7043
PCEPA-M (O-deethyl-) AC		1980	275_{23}	232_{100}	172_{23}	101_{16}	91_{46}	492	6985
PCEPA-M (O-deethyl-) TFA		1830	329_{19}	286_{100}	216_{4}	172_{14}	159_{11}	754	7038
PCEPA-M (O-deethyl-) TMS		1955	305_{22}	262_{100}	189_{19}	172_{8}	159_{13}	642	7033
PCEPA-M (O-deethyl-) 2AC		2590	317_{5}	274_{14}	260_{28}	158_{69}	91_{100}	699	7835
PCEPA-M (O-deethyl-3'-HO-) 2AC		2165	334_{3}	274_{100}	157_{51}	118_{10}		776	6988
PCEPA-M (O-deethyl-3'-HO-) 2TFA		1900	441_{16}	328_{100}	286_{26}	172_{10}		1100	7051
PCEPA-M (O-deethyl-3'-HO-) 2TMS		2195	393_{6}	350_{100}	262_{23}	246_{13}	157_{28}	997	7034
PCEPA-M (O-deethyl-3'-HO-HO-phenyl-) 3AC		2495	391_{18}	332_{40}	290_{92}	248_{100}	101_{36}	989	7025
PCEPA-M (O-deethyl-4'-cis-HO-) 2TMS		2240	393_{2}	262_{100}	246_{11}	157_{11}	132_{12}	997	7035
PCEPA-M (O-deethyl-4'-HO-) 2TFA		1940	441_{7}	328_{11}	286_{100}	172_{11}	157_{16}	1100	7050
PCEPA-M (O-deethyl-4'-HO-) -H2O AC		1955	273_{39}	244_{56}	200_{39}	186_{46}	158_{100}	482	7017
PCEPA-M (O-deethyl-4'-HO-) -H2O TFA		1795	327_{30}	298_{41}	214_{32}	186_{35}	158_{100}	744	7049
PCEPA-M (O-deethyl-4'-HO-) iso-1 2AC		2200	333_{4}	273_{73}	232_{100}	172_{41}	91_{53}	776	6986
PCEPA-M (O-deethyl-4'-HO-) iso-2 2AC		2210	333_{3}	273_{76}	232_{100}	172_{43}	91_{46}	776	6987
PCEPA-M (O-deethyl-4'-HO-HO-phenyl-) 3AC		2730	391_{7}	331_{86}	290_{100}	248_{48}	173_{37}	990	7008
PCEPA-M (O-deethyl-4'-trans-HO-) 2TMS		2255	393_{2}	262_{100}	157_{11}	132_{10}	117_{6}	997	7036
PCEPA-M (O-deethyl-HO-phenyl-) 2AC		2230	333_{17}	290_{100}	248_{39}	107_{46}	101_{16}	776	6989
PCEPA-M (O-deethyl-HO-phenyl-) 2TMS		2300	393_{22}	350_{100}	322_{26}	247_{31}	179_{36}	997	7037
PCM		1960	245_{29}	202_{100}	168_{14}	117_{23}	91_{86}	360	3592
PCM intermediate (MCC)		1560	194_{9}	164_{9}	151_{100}	124_{41}	56_{21}	206	3578
PCM intermediate (MCC) -HCN		1260	167_{81}	152_{52}	108_{100}	94_{36}	81_{77}	153	3579
PCM precursor (morpholine)		<1000	87_{12}	57_{23}	42_{6}			97	3612
PCME		1480	189_{17}	146_{100}	132_{20}	117_{14}	91_{16}	193	3595
PCME AC		1870	231_{4}	174_{42}	158_{97}	91_{100}	74_{50}	310	3620
PCME artifact (phenylcyclohexene)	U+UHYAC	1270*	158_{100}	143_{56}	129_{93}	115_{60}	91_{30}	138	3606
PCME intermediate (MECC)		<1000	138_{3}	137_{10}	123_{13}	95_{100}	82_{15}	118	3593
PCME intermediate (MECC) -HCN		<1000	111_{25}	91_{13}	82_{27}	68_{100}	55_{67}	103	3597
PCME precursor (methylamine)		<1000	31_{40}	30_{82}	28_{100}			89	3619
PCMEA		1790	233_{20}	190_{85}	159_{60}	117_{23}	91_{100}	317	5871
PCMEA AC		2120	275_{1}	232_{3}	159_{56}	118_{20}	91_{100}	492	5872
PCMEA TFA		1915	329_{1}	159_{88}	117_{13}	91_{100}	81_{22}	754	5873
PCMPA		1895	247_{13}	204_{100}	132_{17}	117_{22}	91_{36}	369	5874
PCMPA AC		2200	289_{2}	246_{9}	232_{15}	158_{48}	91_{100}	561	5875
PCMPA TFA		1960	343_{1}	246_{2}	159_{82}	91_{100}	81_{22}	822	5876
PCPIP		2020	258_{20}	215_{45}	99_{100}	70_{57}	56_{68}	415	3605

PCPIP artifact (phenylcyclohexene)   Table 1-8-1: Compounds in order of names

Name	Detected	RI	Typical ions and intensities					Page	Entry
PCPIP artifact (phenylcyclohexene)	U+UHYAC	1270*	158 $_{100}$	143 $_{56}$	129 $_{93}$	115 $_{60}$	91 $_{30}$	138	3606
PCPIP intermediate (PICC)		1680	207 $_9$	180 $_9$	123 $_8$	99 $_{100}$	70 $_{42}$	235	3587
PCPIP intermediate (PICC) -HCN		1380	180 $_{26}$	165 $_8$	123 $_{13}$	110 $_{17}$	70 $_{100}$	176	3588
PCPIP precursor (1-methylpiperazine)		<1000	100 $_{31}$	70 $_4$	58 $_{100}$	42 $_{38}$		100	3614
PCPR		1625	217 $_{15}$	174 $_{100}$	104 $_{23}$	91 $_{48}$	58 $_{20}$	262	3604
PCPR AC		1965	259 $_9$	202 $_{44}$	158 $_{83}$	102 $_{42}$	91 $_{100}$	418	3621
PCPR artifact (phenylcyclohexene)	U+UHYAC	1270*	158 $_{100}$	143 $_{56}$	129 $_{93}$	115 $_{60}$	91 $_{30}$	138	3606
PCPR intermediate (PRCC)		<1000	139 $_{21}$	110 $_{65}$	96 $_{40}$	69 $_{19}$	54 $_{89}$	119	3600
PCPR intermediate (PRCC) -HCN		<1000	139 $_{21}$	110 $_{65}$	96 $_{40}$	54 $_{89}$	41 $_{100}$	119	5539
PCPR precursor (propylamine)		<1000	59 $_7$	42 $_{27}$	30 $_{100}$			91	3616
PCPR-M (2''-HO-) AC		1965	275 $_{25}$	232 $_{100}$	188 $_{19}$	159 $_{55}$	91 $_{35}$	492	7391
PCPR-M (2''-HO-3'-HO-) 2AC		2250	333 $_7$	290 $_9$	274 $_{59}$	216 $_{17}$	157 $_{100}$	776	7402
PCPR-M (2''-HO-4'-HO-) iso-1 2AC		2290	333 $_1$	273 $_{29}$	232 $_{43}$	157 $_{100}$	91 $_{61}$	776	7403
PCPR-M (2''-HO-4'-HO-) iso-2 2AC		2300	333 $_4$	273 $_{68}$	232 $_{85}$	174 $_{66}$	157 $_{100}$	777	7404
PCPR-M (2''-HO-4'-HO-HO-phenyl-) 3AC		2610	391 $_1$	331 $_{16}$	290 $_{34}$	215 $_{54}$	173 $_{100}$	990	7401
PCPR-M (3'-HO-) iso-1 AC		1975	275 $_{15}$	232 $_{13}$	216 $_{100}$	174 $_{25}$	157 $_{24}$	493	7392
PCPR-M (3'-HO-) iso-2 AC		1985	275 $_{15}$	232 $_{15}$	216 $_{100}$	174 $_{34}$	157 $_{30}$	493	7393
PCPR-M (3'-HO-HO-phenyl-) iso-1 2AC		2345	333 $_{12}$	290 $_{14}$	274 $_{100}$	232 $_{92}$	173 $_{39}$	777	7397
PCPR-M (3'-HO-HO-phenyl-) iso-2 2AC		2360	333 $_{13}$	290 $_{18}$	274 $_{100}$	232 $_{76}$	173 $_{46}$	777	7398
PCPR-M (4'-HO-) iso-1 AC		2020	275 $_6$	215 $_{79}$	174 $_{100}$	157 $_{30}$	91 $_{31}$	493	7394
PCPR-M (4'-HO-) iso-2 AC		2030	275 $_6$	215 $_{70}$	174 $_{100}$	157 $_{34}$	91 $_{32}$	493	7395
PCPR-M (4'-HO-HO-phenyl-) iso-1 2AC		2385	333 $_6$	273 $_{60}$	232 $_{100}$	190 $_{40}$	173 $_{29}$	777	7399
PCPR-M (4'-HO-HO-phenyl-) iso-2 2AC		2400	333 $_4$	273 $_{67}$	232 $_{100}$	190 $_{90}$	173 $_{64}$	777	7400
PCPR-M (HO-phenyl-) AC		2070	275 $_{17}$	232 $_{100}$	190 $_{56}$	175 $_{20}$	107 $_{44}$	493	7396
PCPR-M (N-dealkyl-) AC		1850	217 $_6$	174 $_{19}$	158 $_{100}$	132 $_{78}$	104 $_{27}$	262	7016
PCPR-M (N-dealkyl-) TFA		1630	271 $_{10}$	228 $_{31}$	202 $_{22}$	158 $_{100}$	115 $_{61}$	474	7039
PCPR-M (N-dealkyl-3'-HO-) iso-1 2AC		2055	275 $_2$	216 $_{25}$	190 $_{14}$	174 $_{32}$	156 $_{100}$	491	7012
PCPR-M (N-dealkyl-3'-HO-) iso-1 2TFA		1690	270 $_{37}$	383 $_6$	240 $_{13}$	172 $_{14}$	156 $_{78}$	962	7041
PCPR-M (N-dealkyl-3'-HO-) iso-2 2AC		2065	275 $_2$	233 $_3$	216 $_{14}$	190 $_{15}$	156 $_{100}$	491	7013
PCPR-M (N-dealkyl-3'-HO-) iso-2 2TFA		1730	383 $_4$	270 $_{35}$	240 $_{17}$	172 $_{13}$	156 $_{100}$	962	7040
PCPR-M (N-dealkyl-4'-cis-HO-) 2TMS		1985	335 $_6$	320 $_4$	246 $_{11}$	204 $_{100}$	73 $_{40}$	786	7405
PCPR-M (N-dealkyl-4'-HO-) -H2O AC		1680	215 $_{15}$	172 $_{100}$	156 $_9$	119 $_{15}$	103 $_{19}$	257	7021
PCPR-M (N-dealkyl-4'-HO-) iso-1 2AC		2090	275 $_1$	215 $_{10}$	172 $_{34}$	156 $_{100}$	132 $_{48}$	491	7014
PCPR-M (N-dealkyl-4'-HO-) iso-1 2TFA		1700	383 $_2$	269 $_7$	240 $_{25}$	172 $_{17}$	156 $_{100}$	962	7042
PCPR-M (N-dealkyl-4'-HO-) iso-2 2AC		2100	275 $_3$	215 $_{10}$	172 $_{29}$	156 $_{100}$	132 $_{36}$	491	7015
PCPR-M (N-dealkyl-4'-HO-) iso-2 2TFA		1735	383 $_2$	269 $_7$	240 $_{19}$	172 $_{16}$	156 $_{100}$	962	7043
PCPR-M (N-dealkyl-4'-trans-HO-) 2TMS		2000	335 $_5$	320 $_1$	246 $_1$	204 $_{100}$	73 $_{32}$	786	7406
Pecazine	G U UHY U+UHYAC	2545	310 $_{86}$	199 $_{67}$	112 $_{68}$	96 $_{66}$	58 $_{100}$	667	369
Pecazine-M (HO-) AC	U+UHYAC	2750	368 $_{57}$	326 $_4$	215 $_{40}$	112 $_{92}$	58 $_{100}$	913	1278
Pecazine-M (nor-) AC	U+UHYAC	2985	338 $_{61}$	212 $_{101}$	198 $_{59}$	98 $_{29}$		798	1279
Pecazine-M (nor-HO-) 2AC	U+UHYAC	3415	396 $_{50}$	354 $_{21}$	228 $_{100}$	214 $_{45}$	98 $_{16}$	1006	1280
Pecazine-M (ring)	P G U UHY U+UHYAC	2010	199 $_{100}$	167 $_{45}$				216	10
Pecazine-M AC	U+UHYAC	2550	257 $_{23}$	215 $_{100}$	183 $_7$			409	12
Pecazine-M 2AC	U+UHYAC	2865	315 $_{54}$	273 $_{34}$	231 $_{100}$	202 $_{11}$		688	2618
Pemoline 2ME		1590	204 $_{46}$	190 $_{20}$	118 $_{100}$	90 $_{44}$		226	832
Pemoline-M (mandelic acid)		1890*	152 $_{10}$	107 $_{100}$	79 $_{75}$	77 $_{55}$	51 $_{63}$	131	5759
Pemoline-M (mandelic acid) ME		1485*	166 $_7$	107 $_{100}$	79 $_{59}$	77 $_{40}$		149	1071
Penbutolol	G	2130	291 $_3$	276 $_6$	161 $_4$	86 $_{70}$	57 $_{18}$	573	2596
Penbutolol TMS		2100	363 $_3$	348 $_{16}$	247 $_5$	101 $_{16}$	86 $_{100}$	896	5491
Penbutolol 2AC		2205	375 $_2$	315 $_6$	158 $_{74}$	98 $_{59}$	56 $_{99}$	938	1367
Penbutolol formyl artifact		2150	303 $_{10}$	288 $_{100}$	159 $_3$	141 $_9$	91 $_9$	631	1366
Penbutolol-M (deisobutyl-HO-) -H2O 2AC	U+UHYAC	2240	317 $_8$	275 $_{34}$	216 $_{34}$	178 $_{100}$	98 $_{33}$	698	1708
Penbutolol-M (di-HO-) 3AC	U+UHYAC	2890	449 $_2$	434 $_{35}$	374 $_{13}$	332 $_{11}$	86 $_{100}$	1112	1709
Penbutolol-M (HO-) 2AC		2520	391 $_2$	376 $_{10}$	158 $_{61}$	86 $_{52}$	56 $_{100}$	990	1382
Penbutolol-M (HO-) artifact		2425	319 $_4$	304 $_{15}$	178 $_7$	86 $_{100}$	57 $_{28}$	713	1381
Pencycuron ME		2575	342 $_4$	273 $_4$	125 $_{100}$	106 $_{19}$	77 $_{22}$	818	3971
Penfluridol		3350	523 $_{24}$	292 $_{100}$	201 $_{26}$	109 $_{27}$		1176	584
Penfluridol-M	U	----	291 $_{18}$	274 $_7$	154 $_{35}$	84 $_{41}$	56 $_{100}$	567	585
Penfluridol-M (deamino-carboxy-)	P-I UHY U+UHYAC	2230*	276 $_7$	216 $_{17}$	203 $_{100}$	183 $_{22}$		496	169
Penfluridol-M (deamino-carboxy-) ME	P-I UHYME U+UHYAC	2125*	290 $_{12}$	258 $_{13}$	216 $_{30}$	203 $_{100}$	183 $_{22}$	563	3372
Penfluridol-M (deamino-HO-)	UHY-I	2120*	262 $_{16}$	203 $_{100}$	201 $_{16}$	183 $_{12}$		430	515
Penfluridol-M (deamino-HO-) AC	U+UHYAC	2150*	304 $_7$	244 $_{22}$	216 $_{41}$	203 $_{100}$	183 $_{17}$	635	307
Penfluridol-M (N-dealkyl-)	UHY	2210	279 $_7$	261 $_{14}$	82 $_6$	56 $_{100}$		509	586
Penfluridol-M (N-dealkyl-) AC	U+UHYAC	2240	321 $_{20}$	303 $_8$	278 $_{23}$	99 $_{25}$	57 $_{100}$	718	165
Penfluridol-M (N-dealkyl-oxo-) -2H2O	U UHY U+UHYAC	1920	257 $_{98}$	222 $_{11}$	167 $_{11}$			409	164
Penoxalin		2020	281 $_{15}$	252 $_{100}$				522	1221
Pentachloroaniline		1845	265 $_{100}$	263 $_{58}$	230 $_6$	192 $_8$	132 $_8$	432	3470
Pentachlorobenzene		1515*	250 $_{100}$	248 $_{72}$	213 $_{16}$	178 $_8$	108 $_{21}$	369	3471
2,2',4,5,5'-Pentachlorobiphenyl		2155*	326 $_{100}$	324 $_{60}$	289 $_{11}$	254 $_{37}$	184 $_{11}$	729	882
1,2,3,7,8-Pentachlorodibenzo-p-dioxin (PCDD)		----*	356 $_{100}$	354 $_{60}$	291 $_{38}$	228 $_{32}$	178 $_{21}$	860	3494
Pentachlorophenol		1760*	266 $_{100}$	264 $_{58}$	228 $_8$	200 $_8$	165 $_{13}$	437	833
Pentachlorophenol ME	UME	1815*	280 $_{100}$	278 $_{62}$	265 $_{100}$	263 $_{64}$	235 $_{54}$	504	834
Pentadecane	P	1500*	212 $_{10}$	169 $_6$	85 $_{43}$	71 $_{61}$	57 $_{100}$	251	2766
Pentadecanoic acid ET		1840*	270 $_6$	227 $_9$	157 $_{20}$	101 $_{54}$	88 $_{100}$	472	5402

Table 1-8-1: Compounds in order of names — Perhexiline-M (di-HO-) 3AC

Name	Detected	RI	Typical ions and intensities					Page	Entry
Pentadecanoic acid ME		1830*	256 $_6$	213 $_6$	143 $_{12}$	87 $_{58}$	74 $_{100}$	408	3036
Pentafluoropropionic acid		<1000*	147 $_9$	119 $_{49}$	100 $_{87}$	97 $_{31}$	69 $_{100}$	144	5543
Pentafluoropropionic acid		<1000*	147 $_9$	119 $_{49}$	100 $_{87}$	69 $_{100}$	45 $_{77}$	144	5549
Pentamidine		3010	306 $_{14}$	188 $_{17}$	132 $_{25}$	102 $_{26}$	69 $_{100}$	808	1948
Pentane		500*	72 $_{10}$	57 $_{13}$	43 $_{100}$	41 $_{66}$	29 $_{34}$	93	3812
Pentanochlor		1935	239 $_{15}$	197 $_{12}$	141 $_{100}$	106 $_{13}$	71 $_{60}$	340	4037
Pentazocine	G P-I UHY	2280	285 $_{33}$	217 $_{84}$	110 $_{73}$	70 $_{100}$		540	587
Pentazocine AC	U+UHYAC	2330	327 $_{37}$	312 $_{26}$	259 $_{100}$	110 $_{49}$	70 $_{31}$	746	249
Pentazocine PFP		2120	431 $_{18}$	363 $_{100}$	348 $_{55}$	110 $_{51}$	69 $_{59}$	1083	4320
Pentazocine TFA		2075	381 $_{10}$	366 $_{16}$	313 $_{62}$	110 $_{46}$	69 $_{100}$	957	4007
Pentazocine TMS		2320	357 $_{32}$	342 $_{37}$	289 $_{100}$	244 $_{53}$	73 $_{110}$	875	4319
Pentazocine artifact (+H2O)	UHY	2375	303 $_{14}$	288 $_7$	230 $_{100}$	58 $_{11}$		631	588
Pentazocine artifact (+H2O) AC	U+UHYAC	2435	345 $_7$	330 $_7$	272 $_{100}$	229 $_3$	173 $_4$	829	252
Pentazocine-M (dealkyl-) 2AC	U+UHYAC	2380	301 $_9$	87 $_{100}$	72 $_{29}$			620	251
Pentazocine-M (HO-)	U	2545	301 $_{21}$	268 $_{23}$	217 $_{99}$	110 $_{45}$	70 $_{31}$	621	589
Pentazocine-M AC	U+UHYAC	2350	323 $_6$	109 $_{100}$	94 $_{26}$			729	250
Pentetrazole		1540	138 $_{23}$	109 $_{16}$	82 $_{33}$	55 $_{100}$		118	835
Pentifylline	G U	2240	264 $_{45}$	193 $_{24}$	180 $_{100}$	137 $_{19}$	109 $_{33}$	440	836
Pentifylline-M (di-HO-) -H2O	UHY U+UHYAC	2285	278 $_{50}$	261 $_{24}$	207 $_{25}$	194 $_{100}$	123 $_{23}$	507	1930
Pentifylline-M (di-HO-) iso-1 2AC	U+UHYAC	2680	380 $_{14}$	251 $_{19}$	181 $_{100}$	180 $_{71}$		953	1215
Pentifylline-M (di-HO-) iso-2 2AC	U+UHYAC	2820	380 $_{20}$	278 $_7$	265 $_{22}$	193 $_{27}$	180 $_{100}$	953	1928
Pentifylline-M (HO-)	G P UHY	2505	280 $_{25}$	236 $_{20}$	193 $_{32}$	180 $_{100}$	109 $_{31}$	518	1213
Pentifylline-M (HO-) AC	U+UHYAC	2560	322 $_{32}$	262 $_{15}$	193 $_{35}$	180 $_{100}$		723	1214
Pentobarbital	P G U+UHYAC	1740	197 $_4$	156 $_{100}$	141 $_{84}$	98 $_{10}$	69 $_{12}$	295	837
Pentobarbital (ME)	P G	1700	211 $_4$	170 $_{99}$	155 $_{100}$	141 $_{11}$	112 $_{10}$	345	2584
Pentobarbital 2ME	PME UME	1630	225 $_6$	184 $_{100}$	169 $_{85}$			400	839
Pentobarbital 2TMS		1850	370 $_1$	355 $_{40}$	300 $_{51}$	285 $_{100}$	73 $_{39}$	921	4580
Pentobarbital-D5		1735	197 $_7$	161 $_{100}$	143 $_{58}$	100 $_7$		311	6882
Pentobarbital-D5 2TMS		1845	375 $_2$	360 $_{43}$	305 $_{60}$	290 $_{100}$	100 $_{62}$	938	7299
Pentobarbital-M (HO-)	U	1955	227 $_3$	197 $_{26}$	195 $_{21}$	156 $_{100}$	141 $_{70}$	352	838
Pentobarbital-M (HO-) (ME)	P U	1865	209 $_{19}$	170 $_{100}$	155 $_{100}$	112 $_{10}$	69 $_{59}$	407	3341
Pentobarbital-M (HO-) 2ME	PME UME	1820	223 $_{25}$	184 $_{95}$	169 $_{100}$	112 $_{13}$	69 $_{40}$	471	3340
Pentobarbital-M (HO-) -H2O	U+UHYAC	1890	224 $_2$	195 $_{31}$	156 $_{30}$	141 $_{30}$	69 $_{100}$	289	840
Pentobarbital-M (HO-) -H2O (ME)	U+UHYAC	1870	209 $_{23}$	170 $_{24}$	155 $_{16}$	69 $_{100}$		337	3825
Pentorex	U	1250	148 $_2$	105 $_4$	91 $_4$	58 $_{100}$		144	841
Pentorex AC	UAAC	1580	148 $_3$	131 $_3$	105 $_6$	100 $_{52}$	58 $_{100}$	229	842
Pentoxifylline	P G U	2435	278 $_{56}$	221 $_{100}$	193 $_{62}$	180 $_{68}$	109 $_{37}$	507	843
Pentoxifylline TMS		2505	350 $_{12}$	253 $_{60}$	237 $_{35}$	143 $_{56}$	73 $_{100}$	846	4581
Pentoxifylline-M (dihydro-)	G P UHY	2505	280 $_{25}$	236 $_{20}$	193 $_{32}$	180 $_{100}$	109 $_{31}$	518	1213
Pentoxifylline-M (dihydro-) AC	U+UHYAC	2560	322 $_{32}$	262 $_{15}$	193 $_{35}$	180 $_{100}$		723	1214
Pentoxifylline-M (dihydro-) -H2O	U+UHYAC	2300	262 $_{25}$	193 $_{35}$	181 $_{100}$	137 $_{24}$	109 $_{45}$	431	1732
Pentoxifylline-M (dihydro-HO-) 2AC	U+UHYAC	2680	380 $_{14}$	251 $_{19}$	181 $_{100}$	180 $_{71}$		953	1215
Pentoxyverine	G U+UHYAC	2390	318 $_1$	144 $_8$	115 $_5$	91 $_{25}$	86 $_{100}$	778	6480
Pentoxyverine-M (deethyl-) AC	G U+UHYAC	2600	347 $_2$	217 $_{12}$	145 $_{43}$	100 $_{60}$	58 $_{100}$	836	6485
Pentoxyverine-M (deethyl-di-HO-) 3AC	G U+UHYAC	3120	463 $_5$	289 $_{21}$	141 $_{70}$	100 $_{100}$	58 $_{116}$	1131	6487
Pentoxyverine-M (deethyl-HO-) 2AC	G U+UHYAC	2860	405 $_2$	231 $_5$	143 $_{48}$	100 $_{50}$	58 $_{100}$	1031	6486
Pentoxyverine-M (HO-) AC	G U+UHYAC	2575	376 $_3$	143 $_{13}$	128 $_7$	86 $_{100}$		990	6484
Pentoxyverine-M/artifact (alcohol) AC	G U+UHYAC	1115	203 $_1$	188 $_1$	144 $_2$	100 $_4$	86 $_{100}$	225	6481
Pentoxyverine-M/artifact (HOOC-)	G U+UHYAC	1765*	190 $_{12}$	145 $_{100}$	115 $_{22}$	103 $_{17}$	91 $_{73}$	193	6482
Pentoxyverine-M/artifact (HOOC-) ME	G U+UHYAC	1485*	204 $_{11}$	145 $_{100}$	128 $_8$	115 $_9$	91 $_{44}$	226	6483
Perazine	P G U+UHYAC	2790	339 $_{44}$	238 $_{22}$	141 $_{33}$	113 $_{80}$	70 $_{100}$	804	370
Perazine-M (aminoethyl-aminopropyl-) 2AC	U+UHYAC	3310	383 $_{54}$	198 $_{71}$	185 $_{60}$	100 $_{97}$	86 $_{100}$	963	2678
Perazine-M (aminopropyl-) AC	U+UHYAC	2720	298 $_{56}$	212 $_{39}$	198 $_{100}$	180 $_{31}$	100 $_{87}$	605	2076
Perazine-M (aminopropyl-HO-) 2AC	U+UHYAC	3100	356 $_{88}$	215 $_{65}$	214 $_{96}$	100 $_{100}$	72 $_{19}$	869	2677
Perazine-M (di-HO-) 2AC	U+UHYAC	3600	455 $_{83}$	230 $_{52}$	141 $_4$	113 $_{71}$	70 $_{100}$	1121	2679
Perazine-M (HO-)	UHY	3175	355 $_{42}$	215 $_{43}$	155 $_{48}$	113 $_{59}$	70 $_{100}$	866	590
Perazine-M (HO-) AC	U+UHYAC	3190	397 $_{31}$	214 $_{18}$	141 $_{43}$	113 $_{91}$	70 $_{100}$	1010	371
Perazine-M (HO-methoxy-) AC	U+UHYAC	3230	427 $_{84}$	258 $_{29}$	244 $_{45}$	113 $_{100}$	70 $_{80}$	1074	2684
Perazine-M (N-deethyl-) 2AC	U+UHYAC	3400	397 $_{39}$	238 $_8$	212 $_{12}$	198 $_{26}$	100 $_{100}$	1011	1323
Perazine-M (nor-) AC	U+UHYAC	3210	367 $_{62}$	238 $_{47}$	199 $_{59}$	141 $_{100}$	99 $_{79}$	910	1316
Perazine-M (nor-HO-) 2AC	U+UHYAC	3700	425 $_{100}$	214 $_{24}$	141 $_{50}$	99 $_{30}$	56 $_{20}$	1070	2685
Perazine-M (ring)	P G U UHY U+UHYAC	2010	199 $_{100}$	167 $_{45}$				216	10
Perazine-M AC	U+UHYAC	2550	257 $_{23}$	215 $_{100}$	183 $_7$			409	12
Perazine-M 2AC	U+UHYAC	2865	315 $_{54}$	273 $_{34}$	231 $_{100}$	202 $_{11}$		688	2618
Perfluorotributylamine (PFTBA)		----	614 $_2$	502 $_6$	219 $_{28}$	131 $_{23}$	69 $_{100}$	1201	2134
Pergolide		2820	314 $_{100}$	285 $_{41}$	267 $_{12}$	194 $_{12}$	154 $_{51}$	686	5627
Pergolide HFB		2835	510 $_{86}$	482 $_{100}$	350 $_{17}$	232 $_{24}$	87 $_{65}$	1169	5856
Pergolide PFP		2830	460 $_{92}$	431 $_{100}$	300 $_{14}$	232 $_{11}$	87 $_{27}$	1128	5855
Pergolide TFA		2835	410 $_{73}$	381 $_{100}$	250 $_{17}$	154 $_{22}$	87 $_{45}$	1041	5854
Pergolide TMS		3205	386 $_{100}$	357 $_{23}$	226 $_{13}$	87 $_{16}$	73 $_{63}$	974	5857
Perhexiline		2245	277 $_1$	194 $_7$	98 $_1$	84 $_{100}$	55 $_{11}$	504	3303
Perhexiline AC		2540	319 $_1$	236 $_2$	126 $_{100}$	84 $_{71}$	55 $_{20}$	713	3304
Perhexiline-M (di-HO-)	U UHY	2660	309 $_1$	210 $_9$	98 $_6$	84 $_{100}$	56 $_9$	664	3398
Perhexiline-M (di-HO-) 3AC	U+UHYAC	3285	435 $_1$	294 $_3$	126 $_{100}$	84 $_{78}$		1091	3401

Perhexiline-M (di-HO-) -H2O   Table 1-8-1: Compounds in order of names

Name	Detected	RI	Typical ions and intensities					Page	Entry
Perhexiline-M (di-HO-) -H2O	U UHY	2510	291_2	208_6	192_{11}	84_{100}	56_{10}	573	3397
Perhexiline-M (di-HO-) -H2O 2AC	U+UHYAC	2820	375_9	315_6	234_6	126_{100}	84_{82}	938	3400
Perhexiline-M (HO-)	U UHY	2485	293_1	210_6	97_5	84_{100}	56_{10}	584	3396
Perhexiline-M (HO-) 2AC	U+UHYAC	2790	377_1	294_1	236_3	126_{100}	84_{72}	944	3399
Periciazine	G UHY	3265	365_{101}	264_{39}	223_{32}	142_{43}	114_{85}	900	591
Periciazine AC	U+UHYAC	3390	407_{28}	263_{12}	184_{38}	156_{100}	114_{35}	1034	372
Periciazine TMS		3250	437_{22}	263_{19}	223_{32}	186_{100}	73_{64}	1093	5436
Periciazine-M/artifact (-COOH) METMS		3285	470_{26}	296_{25}	214_{49}	186_{100}	73_{72}	1138	5439
Periciazine-M/artifact (ring)	U UHY U+UHYAC	2555	224_{100}	192_{32}				286	1281
Periciazine-M/artifact (ring) TMS		2310	296_{34}	281_3	223_6	73_{100}		596	5437
Periciazine-M/artifact (ring-COOH) METMS		2430	329_{19}	314_5	197_{39}	73_{100}		753	5438
Perindopril ET	UET	2415	396_1	323_{16}	172_{100}	124_7	98_{38}	1008	4754
Perindopril ME	UME	2450	382_1	309_{12}	172_{100}	124_5	98_{35}	961	4748
Perindopril METMS		2620	454_1	439_9	367_{28}	186_{100}	98_{58}	1120	4986
Perindopril TMS		2480	440_1	425_7	367_{23}	172_{100}	98_{59}	1099	4985
Perindopril 2ET	UET	2440	424_1	351_5	200_{100}	172_{17}	126_{16}	1068	4755
Perindopril 2ME	UME	2495	396_1	323_{11}	186_{100}	158_4	112_{11}	1007	4749
Perindopril 2TMS		2595	512_1	497_{16}	439_{16}	244_{100}	240_{26}	1171	4987
Perindoprilate 2ET	UET	2415	396_1	323_{16}	172_{100}	124_7	98_{38}	1008	4754
Perindoprilate 2ME	UME	2435	368_1	309_9	158_{100}	124_5	98_{40}	913	4750
Perindoprilate 2TMS	UTMS	2590	484_1	469_{14}	367_{32}	216_{100}	98_{29}	1151	4988
Perindoprilate 3ET	UET	2440	424_1	351_5	200_{100}	172_{17}	126_{16}	1068	4755
Perindoprilate 3ME	UME	2470	382_1	323_7	172_{100}	112_{11}	86_9	961	4751
Perindoprilate -H2O isopropylate		2440	364_{78}	277_{79}	249_{100}	222_{68}	98_{67}	898	4756
Perindoprilate-M/artifact -H2O ME	UME	2560	336_{53}	277_{64}	249_{76}	222_{100}	98_{22}	790	4753
Perindoprilate-M/artifact -H2O TMS	UTMS	2645	394_{56}	379_{42}	277_{93}	249_{100}	98_{53}	1000	4989
Perindopril-M/artifact (deethyl-) 2ET	UET	2415	396_1	323_{16}	172_{100}	124_7	98_{38}	1008	4754
Perindopril-M/artifact (deethyl-) 2ME	UME	2435	368_1	309_9	158_{100}	124_5	98_{40}	913	4750
Perindopril-M/artifact (deethyl-) 2TMS	UTMS	2590	484_1	469_{14}	367_{32}	216_{100}	98_{29}	1151	4988
Perindopril-M/artifact (deethyl-) 3ET	UET	2440	424_1	351_5	200_{100}	172_{17}	126_{16}	1068	4755
Perindopril-M/artifact (deethyl-) 3ME	UME	2470	382_1	323_7	172_{100}	112_{11}	86_9	961	4751
Perindopril-M/artifact (deethyl-) -H2O ME	UME	2560	336_{53}	277_{64}	249_{76}	222_{100}	98_{22}	790	4753
Perindopril-M/artifact (deethyl-) -H2O TMS	UTMS	2645	394_{56}	379_{42}	277_{93}	249_{100}	98_{53}	1000	4989
Perindopril-M/artifact (deethyl-) -H2O isopropylate		2440	364_{78}	277_{79}	249_{100}	222_{68}	98_{67}	898	4756
Perindopril-M/artifact -H2O	G UME	2590	350_{66}	277_{92}	249_{100}	222_{90}	98_{56}	846	4752
Permethrin iso-1		2640*	390_4	183_{100}	163_7	127_6	77_8	984	3000
Permethrin iso-2		2670*	390_4	183_{100}	163_{30}	127_6		984	3001
Perphenazine	UHY-I	3360	403_{27}	246_{72}	171_{31}	143_{84}	70_{100}	1026	4252
Perphenazine	UHY-I	3360	403_{50}	246_{80}	143_{25}	70_{75}	42_{100}	1026	592
Perphenazine AC	U+UHYAC-I	3470	445_{33}	246_{41}	185_{51}	125_{60}	70_{100}	1108	373
Perphenazine TMS		3340	475_{10}	372_{22}	246_{100}	232_{22}	73_{41}	1144	5444
Perphenazine-M (amino-) AC	U+UHYAC	2990	332_8	233_{19}	100_{100}			767	1255
Perphenazine-M (dealkyl-) AC	U+UHYAC	3500	401_{59}	233_{61}	141_{99}	99_{68}		1020	1282
Perphenazine-M (ring)	U-I UHY-I U+UHYAC-I	2100	233_{100}	198_{55}				314	311
Perthane		2225*	306_5	223_{100}	193_{12}	178_{15}	165_{18}	644	3473
Perthane -HCl		2095*	270_8	223_{100}	193_6	179_8	165_{10}	469	3474
Pethidine	P G U UHY U+UHYAC	1760	247_{36}	218_{20}	172_{46}	71_{100}		368	253
Pethidine-M (deethyl-) (ME)	U	1800	233_{14}	218_5	158_{65}	71_{100}		317	593
Pethidine-M (HO-) AC	U+UHYAC	2205	305_9	230_6	188_{10}	71_{100}		640	1195
Pethidine-M (nor-)	U UHY	1885	233_{21}	158_{28}	91_{37}	77_{38}	57_{100}	317	594
Pethidine-M (nor-) AC	U+UHYAC	2240	275_{34}	232_{36}	202_{30}	187_{100}	158_{37}	491	254
Pethidine-M (nor-) HFB		1690	429_{33}	410_6	356_{42}	341_{100}	143_{16}	1078	7823
Pethidine-M (nor-) PFP		1660	379_{37}	360_7	306_{51}	291_{100}	143_{54}	947	7822
Pethidine-M (nor-) TFA		1680	329_{42}	256_{47}	241_{100}	143_{68}	103_{36}	754	7821
Pethidine-M (nor-) TMS		1650	305_{57}	290_{36}	276_{52}	232_{34}	73_{100}	641	7824
Pethidine-M (nor-HO-) 2AC	U+UHYAC	2600	333_{48}	290_{24}	245_{73}	203_{100}	57_{78}	774	1196
Phenacetin	G U+UHYAC	1680	179_{66}	137_{51}	109_{97}	108_{100}	80_{18}	171	186
Phenacetin TMS		1535	251_{34}	236_{51}	222_{16}	162_{54}	73_{100}	385	5451
Phenacetin-M	UHY	1240	109_{99}	80_{41}	52_{90}			102	826
Phenacetin-M (deethyl-)	G P U	1780	151_{34}	109_{100}	81_{16}	80_{23}		129	825
Phenacetin-M (deethyl-) ME	PME UME	1630	165_{59}	123_{74}	108_{100}	95_{10}	80_{20}	147	5046
Phenacetin-M (deethyl-) 2TMS		1780	295_{50}	280_{68}	206_{83}	116_{15}	73_{100}	592	4578
Phenacetin-M (deethyl-) Cl-artifact AC	U+UHYAC	2030	227_6	185_{74}	143_{100}	114_4	79_{12}	296	2993
Phenacetin-M (deethyl-) HYME	UHYME	1100	123_{27}	109_{100}	94_7	80_{96}	53_{47}	108	3766
Phenacetin-M (HO-) AC	U+UHYAC	1755	237_{15}	195_{31}	153_{100}	124_{55}		333	187
Phenacetin-M (hydroquinone)	UHY	<1000*	110_{100}	81_{27}				103	814
Phenacetin-M (hydroquinone) 2AC	U+UHYAC	1395*	194_8	152_{26}	110_{100}			203	815
Phenacetin-M (p-phenetidine)	UHY	1280	137_{68}	108_{99}	80_{38}	65_9		117	844
Phenacetin-M AC	U+UHYAC PAC	1765	193_{10}	151_{53}	109_{100}	80_{24}		199	188
Phenacetin-M HFB	UHYHFB PHFB	1735	347_{24}	305_{39}	169_{13}	108_{100}	69_{31}	833	5099
Phenacetin-M PFP		1675	297_{19}	255_{31}	119_{38}	108_{100}	80_{28}	601	5095
Phenacetin-M TFA		1630	247_{11}	205_{30}	108_{100}	80_{19}	69_{34}	365	5092
Phenalenone		1790*	180_{100}	152_{60}	126_7	76_{31}	63_{16}	174	3186
Phenallymal	P G U UHY U+UHYAC	2045	244_{22}	215_{100}	141_5	104_{57}		355	845

Table 1-8-1: Compounds in order of names   Phenolphthalein-M (methoxy-) 2AC

Name	Detected	RI	Typical ions and intensities					Page	Entry
Phenanthrene		1780*	178_{100}	176_{17}	152_6	89_6	76_7	167	2563
Phenazone	P G U UHY U+UHYAC	1845	188_{100}	96_{81}	77_{51}			190	199
Phenazone artifact	P	3390	388_{24}	269_{19}	177_{46}	77_{50}	56_{100}	981	4713
Phenazone-M (HO-)	U UHY	1855	204_{35}	120_{18}	85_{100}	56_{50}		226	218
Phenazone-M (HO-) iso-1 AC	U+UHYAC	2095	246_2	204_{19}	119_1	91_3	56_{100}	362	190
Phenazone-M (HO-) iso-2 AC	U+UHYAC	2190	246_{88}	204_{33}	159_{37}	112_{68}	77_{100}	363	3214
Phenazopyridine	G	2480	213_{100}	184_8	136_{27}	108_{84}	81_{76}	252	846
Phenazopyridine AC		2700	255_{100}	213_{12}	150_{14}	108_{69}	77_{31}	402	1837
Phencyclidine	P	1910	243_{21}	242_{23}	200_{100}	91_{62}	84_{36}	354	255
Phencyclidine artifact (phenylcyclohexene)	U+UHYAC	1270*	158_{100}	143_{56}	129_{93}	115_{60}	91_{30}	138	3606
Phencyclidine intermediate (PCC)		1525	192_4	191_7	164_8	149_{100}	122_{13}	198	3581
Phencyclidine intermediate (PCC) -HCN		1190	165_{85}	164_{80}	150_{100}	136_{64}	122_{39}	149	3582
Phencyclidine precursor (piperidine)		<1000	85_{11}	84_{26}	70_5	56_{18}		96	3615
Phencyclidine-M (3'HO-4''HO-) 2AC	UGLUCSPEAC	2550	359_{13}	316_{12}	300_{100}	157_{23}	91_{38}	881	7132
Phencyclidine-M (4'HO-4''HO-) iso-1 2AC	UGLUCSPEAC	2600	359_9	299_{72}	258_{100}	157_{16}	91_{49}	881	7133
Phencyclidine-M (4'HO-4''HO-) iso-2 2AC	UGLUCSPEAC	2610	359_7	299_{98}	258_{100}	157_{30}	91_{81}	881	7134
Phendimetrazine	G U UHY U+UHYAC	1480	191_{10}	85_{55}	57_{100}			196	847
Phendimetrazine-M (nor-)	U UHY	1440	177_8	105_5	77_{10}	71_{100}	56_{54}	166	851
Phendimetrazine-M (nor-) AC	U+UHYAC	1810	219_5	176_{10}	113_{77}	86_{50}	71_{100}	271	198
Phendimetrazine-M (nor-) TFA		1530	273_4	167_{85}	105_{36}	98_{47}	70_{100}	481	4002
Phendimetrazine-M (nor-) TMS		1620	249_6	143_{19}	115_{36}	100_{100}	73_{64}	377	5446
Phendimetrazine-M (nor-HO-) iso-1	UHY	1830	193_{11}	121_7	107_6	71_{100}	56_{62}	201	562
Phendimetrazine-M (nor-HO-) iso-1 2AC	U+UHYAC	2150	277_9	234_{17}	113_{55}	85_{26}	71_{100}	501	849
Phendimetrazine-M (nor-HO-) iso-2	UHY	1865	193_8	163_6	121_6	71_{100}	56_{43}	202	3517
Phendimetrazine-M (nor-HO-) iso-2 2AC	U+UHYAC	2200	277_3	234_8	113_{83}	85_{39}	71_{100}	501	848
Phendimetrazine-M (nor-HO-methoxy-)	UHY	1900	223_6	151_3	107_5	71_{100}	56_5	285	3518
Phendimetrazine-M (nor-HO-methoxy-) 2AC	U+UHYAC	2320	307_6	265_{22}	113_{86}	86_{24}	71_{100}	649	1887
Phendipham-M/artifact (phenol) TFA		1460	263_{100}	231_{22}	218_{49}	69_{53}	59_{88}	433	4128
p-Phenetidine		1280	137_{68}	108_{99}	80_{38}	65_9		117	844
p-Phenetidine AC	G U+UHYAC	1680	179_{66}	137_{51}	109_{97}	108_{100}	80_{18}	171	186
Phenglutarimide	U UHY U+UHYAC	2235	288_3	216_6	98_{11}	86_{100}		555	595
Phenglutarimide-M (deethyl-)	UHY	2370	260_{82}	189_{100}				422	1283
Phenglutarimide-M (deethyl-) AC	U+UHYAC	2530	302_{13}	260_{24}	189_{100}			624	1284
Phenindamine	U UHY U+UHYAC	2180	261_{51}	260_{100}	218_8	202_{20}	182_{13}	427	1674
Phenindamine-M (HO-)	UHY	2300	277_{67}	276_{100}	233_{11}	200_{15}	189_{10}	502	1678
Phenindamine-M (HO-) AC	U+UHYAC	2580	319_{86}	318_{100}	276_{61}	234_{24}	57_{36}	709	1675
Phenindamine-M (nor-)	UHY	2210	247_{80}	246_{100}	217_{43}	202_{68}	168_{46}	367	1679
Phenindamine-M (nor-) AC	U+UHYAC	2640	289_{100}	259_{44}	246_{48}	218_{42}	202_{64}	559	1676
Phenindamine-M (nor-HO-)	UHY	2590	263_{60}	262_{100}	233_{22}	191_{20}	184_{23}	435	1681
Phenindamine-M (nor-HO-) 2AC	U+UHYAC	3000	347_{100}	305_{63}	262_{44}	234_{51}	189_{26}	835	1677
Phenindamine-M (N-oxide)	UHY	2230	277_{38}	260_{100}	215_{13}	202_{14}	189_{16}	502	1680
Pheniramine	P G U+UHYAC	1805	240_1	196_2	169_{100}	72_{21}	58_{75}	346	852
Pheniramine-M (nor-)	U UHY	2080	226_7	182_{10}	169_{100}	168_{79}		295	2148
Pheniramine-M (nor-) AC	U+UHYAC	2250	268_9	225_3	182_{41}	169_{99}		462	853
Phenkapton		2535*	376_{23}	341_{14}	153_{54}	121_{82}	97_{100}	938	3475
Phenmedipham-M/artifact (HOOC-) ME		1370	165_{100}	133_{75}	120_{74}	106_{42}	77_{61}	147	3905
Phenmedipham-M/artifact (phenol)		1625	167_{100}	135_{78}	122_{57}	108_{20}	81_{40}	152	3906
Phenmedipham-M/artifact (phenol) 2ME		1560	195_{100}	164_8	136_{47}	108_{34}	72_{57}	207	4093
Phenmedipham-M/artifact (tolylcarbamic acid) 2ME		1340	179_{100}	134_{24}	120_{52}	91_{51}	72_{40}	171	4094
Phenmetrazine	U UHY	1440	177_8	105_5	77_{10}	71_{100}	56_{54}	166	851
Phenmetrazine AC	U+UHYAC	1810	219_5	176_{10}	113_{77}	86_{50}	71_{100}	271	198
Phenmetrazine TFA		1530	273_4	167_{85}	105_{36}	98_{47}	70_{100}	481	4002
Phenmetrazine TMS		1620	249_6	143_{19}	115_{36}	100_{100}	73_{64}	377	5446
Phenmetrazine-M (HO-) iso-1	UHY	1830	193_{11}	121_7	107_6	71_{100}	56_{62}	201	562
Phenmetrazine-M (HO-) iso-1 2AC	U+UHYAC	2150	277_9	234_{17}	113_{55}	85_{26}	71_{100}	501	849
Phenmetrazine-M (HO-) iso-2	UHY	1865	193_8	163_6	121_6	71_{100}	56_{43}	202	3517
Phenmetrazine-M (HO-) iso-2 2AC	U+UHYAC	2200	277_3	234_8	113_{83}	85_{39}	71_{100}	501	848
Phenmetrazine-M (HO-methoxy-)	UHY	1900	223_6	151_3	107_5	71_{100}	56_5	285	3518
Phenmetrazine-M (HO-methoxy-) 2AC	U+UHYAC	2320	307_6	265_{22}	113_{86}	86_{24}	71_{100}	649	1887
Phenobarbital	P G U+UHYAC	1965	232_{14}	204_{100}	161_{18}	146_{12}	117_{37}	312	854
Phenobarbital ME	P G U UHY U+UHYAC	1895	246_{10}	218_{100}	146_{23}	117_{39}		363	1120
Phenobarbital 2ET		1920	288_2	260_{100}	232_{17}	146_{35}	117_{27}	553	2450
Phenobarbital 2ME	PME UME	1860	260_2	232_{100}	175_{20}	146_{24}	117_{34}	421	1121
Phenobarbital 2TMS		2015	376_2	361_{34}	261_{15}	146_{100}	73_{46}	939	4582
Phenobarbital-D5		1960	237_{24}	209_{100}	179_{11}	166_{20}	122_{45}	333	6883
Phenobarbital-D5 2TMS		2015	382_1	366_6	266_5	151_{100}	122_{17}	958	7298
Phenobarbital-M (HO-)	U UHY	2295	248_{70}	220_{61}	219_{100}	148_{55}		370	855
Phenobarbital-M (HO-) AC	U+UHYAC	2360	290_8	248_{100}	219_{54}	148_8	120_5	563	2507
Phenobarbital-M (HO-) 3ME	UME	2200	290_{95}	261_{100}	233_{78}	176_{27}	148_{93}	564	856
Phenobarbital-M (HO-methoxy-) 3ME	UME	2300	320_{100}	291_{91}	263_{49}	206_{10}	178_{49}	715	6407
Phenol	UHY	<1000*	94_{100}	66_{41}				98	4219
Phenolphthalein 2AC	U+UHYAC	3375*	402_{10}	360_{100}	318_{98}	274_{84}	225_{24}	1022	3077
Phenolphthalein 2ME	UME	3060*	346_{75}	302_{87}	271_{100}	239_{23}	135_8	832	3078
Phenolphthalein-M (methoxy-) 2AC	U+UHYAC	3395*	432_5	390_{100}	348_{30}	304_{22}	273_{35}	1084	3402

Phenopyrazone 2AC   Table 1-8-1: Compounds in order of names

Name	Detected	RI	Typical ions and intensities					Page	Entry
Phenopyrazone 2AC		2475	336_3	294_{36}	252_{100}	145_{47}	77_{85}	787	5130
Phenothiazine	P G U UHY U+UHYAC	2010	199_{100}	167_{45}				216	10
Phenothiazine-M (di-HO-) 2AC	U+UHYAC	2865	315_{54}	273_{34}	231_{100}	202_{11}		688	2618
Phenothiazine-M AC	U+UHYAC	2550	257_{23}	215_{100}	183_7			409	12
Phenothrin		2835*	350_7	250_2	183_{75}	123_{100}	81_{25}	846	3882
Phenoxyacetic acid methylester	U	1495*	166_{44}	107_{100}	77_{41}			151	858
Phenoxybenzamine		2240	303_1	268_1	254_1	196_{83}	91_{100}	628	2037
Phenoxybenzamine artifact-1		2225	268_1	254_{15}	192_{55}	182_{11}	91_{100}	458	2038
Phenoxybenzamine artifact-2		2270	268_1	254_3	220_{51}	91_{100}	77_{11}	458	2039
1-Phenoxy-2-propanol	G	1280*	152_{19}	108_{14}	94_{100}	77_{27}	66_{14}	131	6450
Phenprocoumon	G P U	2440*	280_{43}	251_{100}	189_{22}	121_{26}	91_{34}	516	859
Phenprocoumon AC	U+UHYAC	2475*	322_{11}	280_{69}	251_{100}	189_{36}	121_{28}	722	860
Phenprocoumon TMS		2585*	352_{35}	323_{100}	261_{15}	193_{53}	73_{73}	854	4583
Phenprocoumon HY		1980*	254_{22}	225_{75}	136_{33}	121_{100}	91_{54}	400	4822
Phenprocoumon HYAC	U+UHYAC	2095*	296_1	278_{24}	225_{66}	121_{100}	91_{49}	599	4824
Phenprocoumon HYME		2025*	268_{13}	239_{50}	150_{18}	135_{100}	91_{27}	462	4823
Phenprocoumon iso-1 ME	PME UME UHYME	2375*	294_{62}	279_{96}	265_{78}	203_{100}	121_{19}	585	4417
Phenprocoumon iso-2 ME	PME UME UHYME	2395*	294_{64}	265_{88}	203_{52}	121_{13}	91_{100}	586	861
Phenprocoumon-M (di-HO-) 3ET	UET	2730*	396_{27}	352_{62}	323_{100}	295_{35}	201_{33}	1007	4821
Phenprocoumon-M (di-HO-) 3ME	UME UGLUCME	2770*	354_{32}	325_{100}	279_8	201_2	151_{13}	862	4421
Phenprocoumon-M (di-HO-) 3TMS	UTMS	2730*	528_{23}	499_{100}	484_{21}	412_{10}	73_{24}	1178	5034
Phenprocoumon-M (HO-) iso-1 2ET	UET	2745*	352_{47}	323_{100}	295_{21}	201_{15}	121_9	854	4818
Phenprocoumon-M (HO-) iso-1 2ME	UME UHYME	2655*	324_{80}	309_{33}	295_{100}	233_{62}	91_{55}	731	4418
Phenprocoumon-M (HO-) iso-1 2TMS	UTMS	2650*	440_{14}	425_4	411_{100}	193_{14}	73_{21}	1098	5033
Phenprocoumon-M (HO-) iso-2 2ET	UET	2760*	352_{70}	337_{29}	323_{100}	295_{32}	165_{22}	855	4819
Phenprocoumon-M (HO-) iso-2 2ME	UME UGLUCME	2675*	324_{26}	295_{100}	279_8	201_3	121_{23}	731	4420
Phenprocoumon-M (HO-) iso-2 2TMS	UTMS	2675*	440_{23}	425_{12}	411_{100}	281_{22}	73_{21}	1098	5032
Phenprocoumon-M (HO-) iso-3 2ET	UET	2770*	352_{63}	323_{100}	295_{75}	165_{14}	137_{21}	855	4820
Phenprocoumon-M (HO-) iso-3 2ME	UME UHYME	2705*	324_{38}	295_{100}	233_9	151_8	91_{19}	731	4419
Phenprocoumon-M (HO-methoxy-) 2ME	UME UGLUCME	2770*	354_{32}	325_{100}	279_8	201_2	151_{13}	862	4421
Phentermine		1170	134_2	91_{42}	65_{12}	58_{99}		127	1511
Phentermine AC		1510	191_1	134_6	117_8	100_{59}	58_{100}	196	1512
Phentermine HFB		1365	330_1	254_{100}	214_7	132_{14}	91_{30}	828	5074
Phentermine PFP		1335	280_1	204_{100}	164_9	132_{14}	91_{21}	589	5075
Phentermine TFA		1100	230_1	154_{100}	132_{16}	114_{10}	59_{43}	358	3999
Phentermine TMS		1195	221_1	206_7	130_{100}	114_{17}	73_{46}	279	5102
Phentolamine ME		2475	295_{90}	136_{65}	120_{100}	91_{63}	65_{51}	593	5204
Phentolamine 2ME		2500	309_{100}	202_{18}	189_{13}	146_{12}	85_{27}	662	5205
Phentolamine artifact AC		2310	313_{43}	254_{100}	212_{68}	167_{12}	91_{46}	682	5201
Phentolamine-M/artifact (N-alkyl-)		2080	199_{100}	183_{13}	154_9	91_{25}	77_{19}	218	5203
Phentolamine-M/artifact (N-alkyl-) AC		2140	241_{29}	199_{100}	183_{10}	154_{15}	91_{20}	347	5199
Phentolamine-M/artifact (N-alkyl-) ME		1985	213_{100}	182_{17}	154_{31}	91_{36}	77_{30}	252	5202
Phentolamine-M/artifact (N-alkyl-) 2AC	U+UHYAC	2280	283_7	241_{22}	199_{100}	183_{10}	154_{10}	529	5200
Phenylacetaldehyde	U	1200*	120_{67}	91_{100}	65_{15}			106	4221
Phenylacetamide	U	1390	135_{23}	91_{100}	65_{24}			115	4223
Phenylacetic acid	U UHY U+UHYAC	1280*	136_{31}	91_{100}	65_{20}			116	4222
Phenylacetic acid ET	UET	1200*	164_{17}	91_{100}	65_{14}			146	4227
Phenylacetic acid ME	UME	1120*	150_{46}	91_{100}	65_{41}			128	4226
Phenylacetone		<1000*	134_{13}	91_{54}	65_{22}			114	3240
Phenylacetone		<1000*	134_{13}	91_{54}	65_{22}	43_{100}		114	5516
Phenylalanine MEAC		1870	162_{93}	120_{52}	91_{56}	88_{100}	65_{17}	277	2581
N-Phenylalphanaphthylamine		2180	219_{100}	109_{23}				270	868
N-Phenylbetanaphthylamine		2190	219_{100}	191_3	115_{12}	109_{12}	77_7	270	2579
N-Phenylbetanaphthylamine AC		2270	261_{46}	219_{100}	217_{56}	127_8	115_{15}	426	2580
Phenylbutazone	G P U	2375	308_{47}	252_{14}	183_{100}	77_{89}		657	862
Phenylbutazone ME	P UME	2290	322_{73}	266_{22}	183_{100}	118_{19}	77_{91}	724	863
Phenylbutazone TMS		2575	380_8	337_8	246_{29}	77_{39}	73_{100}	953	5442
Phenylbutazone artifact	P	2435	324_{26}	183_{88}	119_{38}	77_{100}		732	864
Phenylbutazone artifact AC		2435	366_{67}	184_{69}	183_{100}	105_{33}	77_{94}	903	5188
Phenylbutazone artifact TMS		2330	396_{15}	325_9	183_{11}	143_{19}	73_{100}	1006	5443
Phenylbutazone-M (HO-)		9999	324_{30}	199_{90}	135_{25}	93_{100}	77_{70}	732	1513
Phenylbutazone-M (HO-) AC	U+UHYAC	2700	366_{33}	324_{49}	199_{100}	93_{80}	77_{72}	905	1506
Phenylbutazone-M (HO-) artifact (phenyldiazophenol)		2070	198_{46}	121_{41}	93_{100}	77_{72}	65_{54}	214	1027
Phenylbutazone-M (HO-) artifact (phenyldiazophenol) ME		2020	212_{41}	135_{35}	107_{91}	77_{100}	64_{26}	249	4205
Phenylbutazone-M (HO-) iso-1 2ME	UME	2545	352_{100}	213_{66}	148_{18}	107_{36}	77_{65}	855	1505
Phenylbutazone-M (HO-) iso-2 2ME		2720	352_{100}	309_{26}	190_{41}	160_{46}	77_{63}	855	1507
Phenylbutazone-M (HO-alkyl-) ME	P UME	2500	338_{58}	266_{15}	183_{100}	162_{47}	77_{49}	799	6383
Phenylbutazone-M (HOOC-) 2ME	P UME	2590	366_{100}	266_{49}	183_{74}	105_{34}	77_{130}	905	6385
Phenylbutazone-M (oxo-) ME	P UME	2480	336_{100}	266_{43}	183_{64}	105_{28}	77_{120}	789	6384
1-(1-Phenylcyclohexyl)-2-ethoxyethylamine		1755	247_{34}	204_{79}	188_{38}	159_{100}	91_{93}	369	7076
1-(1-Phenylcyclohexyl)-2-ethoxyethylamine AC		2110	289_1	246_7	232_8	159_{100}	91_{65}	561	7367
1-(1-Phenylcyclohexyl)-2-ethoxyethylamine-M (carboxy-) TMS		1975	305_{21}	262_{100}	188_{19}	159_{58}	91_{58}	641	7376
1-(1-Phenylcyclohexyl)-2-ethoxyethylamine-M (carboxy-3'-HO-) 2TMS		2200	393_1	378_3	350_{38}	246_{60}	157_{100}	996	7379
1-(1-Phenylcyclohexyl)-2-ethoxyethylamine-M (carboxy-4'-cis-HO-) 2TMS		2250	393_3	262_{100}	246_{22}	157_{42}	91_{23}	996	7378

Table 1-8-1: Compounds in order of names   1-(1-Phenylcyclohexyl)-propanamine-M (2"-HO-4'-HO-) iso-1 2AC

Name	Detected	RI	Typical ions and intensities					Page	Entry
1-(1-Phenylcyclohexyl)-2-ethoxyethylamine-M (carboxy-4'-trans-HO-) 2TMS		2285	393 $_1$	276 $_4$	262 $_{100}$	247 $_4$	157 $_{55}$	996	7377
1-(1-Phenylcyclohexyl)-2-ethoxyethylamine-M (O-deethyl-) AC		1905	261 $_{17}$	218 $_{100}$	159 $_{15}$	91 $_{32}$	87 $_{29}$	428	7077
1-(1-Phenylcyclohexyl)-2-ethoxyethylamine-M (O-deethyl-) TFA		1690	315 $_{64}$	286 $_{10}$	272 $_{100}$	238 $_{16}$	91 $_{54}$	689	7387
1-(1-Phenylcyclohexyl)-2-ethoxyethylamine-M (O-deethyl-) TMS		1860	291 $_{24}$	248 $_{60}$	188 $_{32}$	159 $_{100}$	91 $_{58}$	573	7380
1-(1-Phenylcyclohexyl)-2-ethoxyethylamine-M (O-deethyl-3'-HO-) 2AC		2225	260 $_{100}$	319 $_{12}$	276 $_{15}$	157 $_{67}$	87 $_{38}$	710	7078
1-(1-Phenylcyclohexyl)-2-ethoxyethylamine-M (O-deethyl-3'-HO-) 2TFA		1775	427 $_{12}$	314 $_{100}$	272 $_{17}$	157 $_3$	91 $_1$	1073	7389
1-(1-Phenylcyclohexyl)-2-ethoxyethylamine-M (O-deethyl-3'-HO-) 2TMS		2110	336 $_{16}$	276 $_{24}$	247 $_{19}$	157 $_{100}$	129 $_{12}$	950	7381
1-(1-Phenylcyclohexyl)-2-ethoxyethylamine-M (O-deethyl-3'-HO-HO-phenyl-) 3AC		2470	377 $_1$	318 $_{14}$	276 $_{56}$	234 $_{75}$	155 $_{100}$	942	7375
1-(1-Phenylcyclohexyl)-2-ethoxyethylamine-M (O-deethyl-4'-cis-HO-) 2TMS		2160	379 $_3$	364 $_2$	248 $_{100}$	246 $_{16}$	157 $_{75}$	950	7382
1-(1-Phenylcyclohexyl)-2-ethoxyethylamine-M (O-deethyl-4'-HO-) 2TFA		1825	427 $_4$	314 $_{12}$	272 $_{100}$	157 $_{28}$	91 $_{40}$	1073	7388
1-(1-Phenylcyclohexyl)-2-ethoxyethylamine-M (O-deethyl-4'-HO-) -H2O AC		1860	259 $_{38}$	230 $_{45}$	200 $_{94}$	186 $_{100}$		417	7386
1-(1-Phenylcyclohexyl)-2-ethoxyethylamine-M (O-deethyl-4'-HO-) -H2O TFA		1650	313 $_{73}$	284 $_{100}$	200 $_{91}$	170 $_{34}$	141 $_{20}$	680	7390
1-(1-Phenylcyclohexyl)-2-ethoxyethylamine-M (O-deethyl-4'-HO-) iso-1 2AC		2270	319 $_5$	259 $_{99}$	218 $_{110}$	157 $_{54}$	87 $_{77}$	710	7079
1-(1-Phenylcyclohexyl)-2-ethoxyethylamine-M (O-deethyl-4'-HO-) iso-2 2AC		2280	319 $_3$	259 $_{86}$	218 $_{100}$	157 $_{54}$	87 $_{58}$	710	7080
1-(1-Phenylcyclohexyl)-2-ethoxyethylamine-M (O-deethyl-4'-HO-HO-phenyl-) 3AC		2650	377 $_3$	317 $_{77}$	276 $_{100}$	234 $_{54}$	173 $_{77}$	943	7374
1-(1-Phenylcyclohexyl)-2-ethoxyethylamine-M (O-deethyl-4'-trans-HO-) 2TMS		2180	379 $_2$	276 $_{13}$	248 $_{90}$	157 $_{100}$	91 $_{25}$	951	7383
1-(1-Phenylcyclohexyl)-2-ethoxyethylamine-M (O-deethyl-HO-phenyl-) 2AC		2340	319 $_{17}$	276 $_{100}$	234 $_{39}$	175 $_{30}$	107 $_{30}$	710	7373
1-(1-Phenylcyclohexyl)-2-ethoxyethylamine-M (O-deethyl-HO-phenyl-) 2TMS		2225	379 $_{14}$	336 $_{35}$	247 $_{100}$	207 $_{52}$	179 $_{40}$	951	7384
1-(1-Phenylcyclohexyl)-2-ethoxypropylamine		1915	261 $_{13}$	232 $_{12}$	218 $_{100}$	117 $_{29}$	91 $_{53}$	428	5877
1-(1-Phenylcyclohexyl)-2-ethoxypropylamine AC		2210	303 $_2$	260 $_{11}$	246 $_{14}$	158 $_{48}$	91 $_{100}$	631	5878
1-(1-Phenylcyclohexyl)-2-ethoxypropylamine TFA		2040	357 $_1$	260 $_4$	159 $_{100}$	91 $_{94}$	81 $_{22}$	874	5879
1-(1-Phenylcyclohexyl)-2-ethoxypropylamine-M (3'-HO-) AC		2080	319 $_{16}$	276 $_{16}$	260 $_{100}$	234 $_{12}$	218 $_{24}$	712	7007
1-(1-Phenylcyclohexyl)-2-ethoxypropylamine-M (3'-HO-) TFA		1980	373 $_8$	260 $_{100}$	218 $_{19}$	186 $_{14}$	157 $_{14}$	931	7052
1-(1-Phenylcyclohexyl)-2-ethoxypropylamine-M (4'-HO-) TFA		2010	373 $_{10}$	260 $_8$	218 $_{100}$	186 $_4$	157 $_{16}$	932	7053
1-(1-Phenylcyclohexyl)-2-ethoxypropylamine-M (4'-HO-) -H2O		1870	259 $_{35}$	230 $_{100}$	216 $_{20}$	203 $_{25}$	186 $_{32}$	418	7009
1-(1-Phenylcyclohexyl)-2-ethoxypropylamine-M (4'-HO-) iso-1 AC		2140	319 $_1$	259 $_3$	244 $_1$	218 $_7$	91 $_5$	712	7010
1-(1-Phenylcyclohexyl)-2-ethoxypropylamine-M (4'-HO-) iso-2 AC		2145	319 $_4$	259 $_{70}$	244 $_{13}$	218 $_{100}$	87 $_{15}$	712	7011
1-(1-Phenylcyclohexyl)-2-ethoxypropylamine-M (carboxy-) TMS		2045	319 $_{25}$	276 $_{100}$	188 $_8$	159 $_{18}$	144 $_{26}$	712	7027
1-(1-Phenylcyclohexyl)-2-ethoxypropylamine-M (carboxy-) -H2O		1930	229 $_{100}$	200 $_9$	186 $_{90}$	158 $_{57}$	144 $_{52}$	304	7018
1-(1-Phenylcyclohexyl)-2-ethoxypropylamine-M (carboxy-2"-HO-) 2TMS		2210	407 $_2$	364 $_7$	317 $_2$	290 $_5$	276 $_{100}$	1035	7032
1-(1-Phenylcyclohexyl)-2-ethoxypropylamine-M (carboxy-2"-HO-) -H2O AC		1975	287 $_{26}$	244 $_{100}$	172 $_{17}$	159 $_{15}$	117 $_{17}$	547	7026
1-(1-Phenylcyclohexyl)-2-ethoxypropylamine-M (carboxy-2"-HO-) -H2O TFA		1905	341 $_{12}$	200 $_5$	186 $_{65}$	159 $_{95}$	91 $_{100}$	811	7048
1-(1-Phenylcyclohexyl)-2-ethoxypropylamine-M (carboxy-3'-HO-) 2TMS		2275	407 $_3$	364 $_{100}$	276 $_{32}$	246 $_{85}$	157 $_{36}$	1035	7028
1-(1-Phenylcyclohexyl)-2-ethoxypropylamine-M (carboxy-3'-HO-) iso-1 -H2O AC		2080	287 $_{34}$	244 $_{34}$	228 $_{26}$	202 $_{21}$	157 $_{100}$	548	7022
1-(1-Phenylcyclohexyl)-2-ethoxypropylamine-M (carboxy-3'-HO-) iso-1 -H2O TFA		1960	341 $_7$	271 $_3$	228 $_{29}$	186 $_{31}$	157 $_{100}$	811	7044
1-(1-Phenylcyclohexyl)-2-ethoxypropylamine-M (carboxy-3'-HO-) iso-2 -H2O AC		2105	287 $_{32}$	244 $_{43}$	228 $_{41}$	202 $_{26}$	157 $_{100}$	548	7023
1-(1-Phenylcyclohexyl)-2-ethoxypropylamine-M (carboxy-3'-HO-) iso-2 -H2O TFA		1985	341 $_7$	271 $_5$	228 $_{29}$	186 $_{38}$	157 $_{100}$	811	7045
1-(1-Phenylcyclohexyl)-2-ethoxypropylamine-M (carboxy-4'-cis-HO-) 2TMS		2310	407 $_4$	276 $_{100}$	246 $_{13}$	157 $_{14}$	144 $_{12}$	1035	7029
1-(1-Phenylcyclohexyl)-2-ethoxypropylamine-M (carboxy-4'-HO-) 2TMS		2370	407 $_{20}$	364 $_{170}$	275 $_{25}$	247 $_{26}$	179 $_{22}$	1035	7031
1-(1-Phenylcyclohexyl)-2-ethoxypropylamine-M (carboxy-4'-HO-) iso-1 -H2O AC		2160	287 $_6$	227 $_{100}$	198 $_{44}$	184 $_{10}$	156 $_{71}$	548	7020
1-(1-Phenylcyclohexyl)-2-ethoxypropylamine-M (carboxy-4'-HO-) iso-1 -H2O TFA		1970	341 $_4$	271 $_5$	227 $_{48}$	157 $_{74}$	91 $_{100}$	811	7046
1-(1-Phenylcyclohexyl)-2-ethoxypropylamine-M (carboxy-4'-HO-) iso-2 -H2O AC		2175	287 $_2$	157 $_{83}$	144 $_{34}$	227 $_{100}$	198 $_{33}$	548	7019
1-(1-Phenylcyclohexyl)-2-ethoxypropylamine-M (carboxy-4'-HO-) iso-2 -H2O TFA		2010	341 $_2$	271 $_5$	227 $_{28}$	157 $_{64}$	91 $_{100}$	811	7047
1-(1-Phenylcyclohexyl)-2-ethoxypropylamine-M (carboxy-4'-trans-HO-) 2TMS		2335	407 $_2$	276 $_{100}$	157 $_{14}$	144 $_{11}$	91 $_{14}$	1035	7030
1-(1-Phenylcyclohexyl)-2-ethoxypropylamine-M (carboxy-HO-phenyl-) 2TMS		2470	495 $_2$	364 $_{100}$	335 $_5$	245 $_{12}$	179 $_5$	1161	7131
1-(1-Phenylcyclohexyl)-2-ethoxypropylamine-M (HO-phenyl-) AC		2150	319 $_{29}$	276 $_{100}$	234 $_{33}$	175 $_{18}$	107 $_{26}$	713	7000
1-(1-Phenylcyclohexyl)-2-ethoxypropylamine-M (O-deethyl-) AC		1980	275 $_{23}$	232 $_{100}$	172 $_{23}$	101 $_{16}$	91 $_{46}$	492	6985
1-(1-Phenylcyclohexyl)-2-ethoxypropylamine-M (O-deethyl-) TFA		1830	329 $_{19}$	286 $_{100}$	216 $_4$	172 $_{14}$	159 $_{11}$	754	7038
1-(1-Phenylcyclohexyl)-2-ethoxypropylamine-M (O-deethyl-) TMS		1955	305 $_{22}$	262 $_{100}$	189 $_{19}$	172 $_8$	159 $_{13}$	642	7033
1-(1-Phenylcyclohexyl)-2-ethoxypropylamine-M (O-deethyl-) 2AC		2590	317 $_5$	274 $_{14}$	260 $_{28}$	158 $_{69}$	91 $_{100}$	699	7835
1-(1-Phenylcyclohexyl)-2-ethoxypropylamine-M (O-deethyl-3'-HO-) 2AC		2165	334 $_3$	274 $_{100}$	157 $_{51}$	118 $_{10}$		776	6988
1-(1-Phenylcyclohexyl)-2-ethoxypropylamine-M (O-deethyl-3'-HO-) 2TFA		1900	441 $_{16}$	328 $_{100}$	286 $_{26}$	172 $_{10}$		1100	7051
1-(1-Phenylcyclohexyl)-2-ethoxypropylamine-M (O-deethyl-3'-HO-) 2TMS		2195	393 $_6$	350 $_{100}$	262 $_{23}$	246 $_{13}$	157 $_{28}$	997	7034
1-(1-Phenylcyclohexyl)-2-ethoxypropylamine-M (O-deethyl-3'-HO-HO-phenyl-) 3AC		2495	391 $_{18}$	332 $_{40}$	290 $_{92}$	248 $_{100}$	101 $_{36}$	989	7025
1-(1-Phenylcyclohexyl)-2-ethoxypropylamine-M (O-deethyl-4'-HO-) 2TFA		1940	441 $_7$	328 $_{11}$	286 $_{100}$	172 $_{11}$	157 $_{16}$	1100	7050
1-(1-Phenylcyclohexyl)-2-ethoxypropylamine-M (O-deethyl-4'-HO-) -H2O AC		1955	273 $_{39}$	244 $_{56}$	200 $_{39}$	186 $_{46}$	158 $_{100}$	482	7017
1-(1-Phenylcyclohexyl)-2-ethoxypropylamine-M (O-deethyl-4'-HO-) -H2O TFA		1795	327 $_{30}$	298 $_{41}$	214 $_{32}$	186 $_{35}$	158 $_{100}$	744	7049
1-(1-Phenylcyclohexyl)-2-ethoxypropylamine-M (O-deethyl-4'-HO-) iso-1 2AC		2200	333 $_4$	273 $_{73}$	232 $_{100}$	172 $_{41}$	91 $_{53}$	776	6986
1-(1-Phenylcyclohexyl)-2-ethoxypropylamine-M (O-deethyl-4'-HO-) iso-2 2AC		2210	333 $_3$	273 $_{76}$	232 $_{100}$	172 $_{43}$	91 $_{46}$	776	6987
1-(1-Phenylcyclohexyl)-2-ethoxypropylamine-M (O-deethyl-4'-HO-HO-phenyl-) 3AC		2730	391 $_7$	331 $_{86}$	290 $_{100}$	248 $_{48}$	173 $_{37}$	990	7008
1-(1-Phenylcyclohexyl)-2-ethoxypropylamine-M (O-deethyl-HO-phenyl-) 2AC		2230	333 $_{17}$	290 $_{100}$	248 $_{39}$	107 $_{46}$	101 $_{16}$	776	6989
1-(1-Phenylcyclohexyl)-2-methoxyethylamine		1790	233 $_{20}$	190 $_{85}$	159 $_{60}$	117 $_{23}$	91 $_{100}$	317	5871
1-(1-Phenylcyclohexyl)-2-methoxyethylamine AC		2120	275 $_1$	232 $_3$	159 $_{56}$	118 $_{20}$	91 $_{100}$	492	5872
1-(1-Phenylcyclohexyl)-2-methoxyethylamine TFA		1915	329 $_1$	159 $_{88}$	117 $_{13}$	91 $_{100}$	81 $_{22}$	754	5873
1-(1-Phenylcyclohexyl)-2-methoxypropylamine		1895	247 $_{13}$	204 $_{100}$	132 $_{17}$	117 $_{22}$	91 $_{36}$	369	5874
1-(1-Phenylcyclohexyl)-2-methoxypropylamine AC		2200	289 $_2$	246 $_9$	232 $_{15}$	158 $_{48}$	91 $_{100}$	561	5875
1-(1-Phenylcyclohexyl)-2-methoxypropylamine TFA		1960	343 $_1$	246 $_2$	159 $_{82}$	91 $_{100}$	81 $_{22}$	822	5876
1-(1-Phenylcyclohexyl)-3-ethoxypropylamine-M (O-deethyl-4'-cis-HO-) 2TMS		2240	393 $_2$	262 $_{100}$	246 $_{11}$	157 $_{11}$	132 $_{12}$	997	7035
1-(1-Phenylcyclohexyl)-3-ethoxypropylamine-M (O-deethyl-4'-trans-HO-) 2TMS		2255	393 $_2$	262 $_{100}$	157 $_{11}$	132 $_{10}$	117 $_6$	997	7036
1-(1-Phenylcyclohexyl)-3-ethoxypropylamine-M (O-deethyl-HO-phenyl-) 2TMS		2300	393 $_{22}$	350 $_{100}$	322 $_{26}$	247 $_{31}$	179 $_{36}$	997	7037
1-(1-Phenylcyclohexyl)-propanamine		1625	217 $_{15}$	174 $_{100}$	104 $_{23}$	91 $_{48}$	58 $_{20}$	262	3604
1-(1-Phenylcyclohexyl)-propanamine AC		1965	259 $_9$	202 $_{44}$	158 $_{83}$	102 $_{42}$	91 $_{100}$	418	3621
1-(1-Phenylcyclohexyl)-propanamine-M (2"-HO-3'-HO-) 2AC		2250	333 $_7$	290 $_9$	274 $_{59}$	216 $_{17}$	157 $_{100}$	776	7402
1-(1-Phenylcyclohexyl)-propanamine-M (2"-HO-4'-HO-) iso-1 2AC		2290	333 $_1$	273 $_{29}$	232 $_{43}$	157 $_{100}$	91 $_{61}$	776	7403

1-(1-Phenylcyclohexyl)-propanamine-M (2"-HO-4'-HO-) iso-2 2AC   Table 1-8-1: Compounds in order of names

Name	Detected	RI	Typical ions and intensities					Page	Entry
1-(1-Phenylcyclohexyl)-propanamine-M (2"-HO-4'-HO-) iso-2 2AC		2300	333 $_4$	273 $_{68}$	232 $_{85}$	174 $_{66}$	157 $_{100}$	777	7404
1-(1-Phenylcyclohexyl)-propanamine-M (2"-HO-4'-HO-HO-phenyl-) 3AC		2610	391 $_1$	331 $_{16}$	290 $_{34}$	215 $_{54}$	173 $_{100}$	990	7401
1-(1-Phenylcyclohexyl)-propanamine-M (3'-HO-) iso-1 AC		1975	275 $_{15}$	232 $_{15}$	216 $_{100}$	174 $_{25}$	157 $_{24}$	493	7392
1-(1-Phenylcyclohexyl)-propanamine-M (3'-HO-) iso-2 AC		1985	275 $_{15}$	232 $_{15}$	216 $_{100}$	174 $_{34}$	157 $_{30}$	493	7393
1-(1-Phenylcyclohexyl)-propanamine-M (3'-HO-HO-phenyl-) iso-1 2AC		2345	333 $_{12}$	290 $_{14}$	274 $_{100}$	232 $_{92}$	173 $_{39}$	777	7397
1-(1-Phenylcyclohexyl)-propanamine-M (3'-HO-HO-phenyl-) iso-2 2AC		2360	333 $_{13}$	290 $_{18}$	274 $_{100}$	232 $_{76}$	173 $_{46}$	777	7398
1-(1-Phenylcyclohexyl)-propanamine-M (4'-HO-) iso-1 AC		2020	275 $_6$	215 $_{79}$	174 $_{100}$	157 $_{30}$	91 $_{31}$	493	7394
1-(1-Phenylcyclohexyl)-propanamine-M (4'-HO-) iso-2 AC		2030	275 $_6$	215 $_{70}$	174 $_{100}$	157 $_{34}$	91 $_{32}$	493	7395
1-(1-Phenylcyclohexyl)-propanamine-M (4'-HO-HO-phenyl-) iso-1 2AC		2385	333 $_6$	273 $_{60}$	232 $_{100}$	190 $_{40}$	173 $_{29}$	777	7399
1-(1-Phenylcyclohexyl)-propanamine-M (4'-HO-HO-phenyl-) iso-2 2AC		2400	333 $_4$	273 $_{67}$	232 $_{100}$	190 $_{90}$	173 $_{64}$	777	7400
1-(1-Phenylcyclohexyl)-propanamine-M (carboxy-2"-HO-) -H2O AC		1965	275 $_{25}$	232 $_{100}$	188 $_{19}$	159 $_{55}$	91 $_{35}$	492	7391
1-(1-Phenylcyclohexyl)-propanamine-M (HO-phenyl-) AC		2070	275 $_{17}$	232 $_{100}$	190 $_{56}$	175 $_{20}$	107 $_{44}$	493	7396
p-Phenylenediamine	G	1280	108 $_{100}$	91 $_3$	80 $_{35}$	53 $_{13}$		102	5330
p-Phenylenediamine ME		1000	122 $_{38}$	108 $_{100}$	93 $_6$	80 $_{40}$		108	5333
p-Phenylenediamine 2AC	U+UHYAC	2690	192 $_{39}$	150 $_{15}$	108 $_{100}$	80 $_{59}$	52 $_{45}$	198	5331
p-Phenylenediamine 2HFB		1775	500 $_{57}$	481 $_{16}$	331 $_{12}$	303 $_{100}$	108 $_{51}$	1164	5332
p-Phenylenediamine 2ME		1060	136 $_{16}$	122 $_{10}$	108 $_{100}$	93 $_6$	80 $_{36}$	116	5334
p-Phenylenediamine 2PFP		1600	400 $_{65}$	281 $_{17}$	253 $_{82}$	119 $_{28}$	108 $_{100}$	1017	5858
p-Phenylenediamine 2TFA		1800	300 $_{100}$	203 $_{59}$	133 $_{16}$	108 $_{54}$	69 $_{30}$	613	5397
Phenylephrine		1810	167 $_{17}$	121 $_{28}$	95 $_{47}$	77 $_{100}$	65 $_{48}$	153	4666
Phenylephrine TFA		1755	263 $_1$	141 $_{61}$	123 $_{51}$	95 $_{100}$	77 $_{46}$	434	6158
Phenylephrine 2TFA		1755	359 $_4$	232 $_{30}$	140 $_{100}$	121 $_{19}$	69 $_{41}$	879	6157
Phenylephrine 2TMSTFA		1835	407 $_1$	392 $_1$	267 $_{100}$	140 $_7$	73 $_{77}$	1034	6156
Phenylephrine 3AC	U+UHYAC	2110	293 $_1$	250 $_3$	220 $_1$	165 $_2$	86 $_{100}$	580	1514
Phenylephrine 3TMS		2110	383 $_1$	368 $_9$	267 $_5$	116 $_{100}$	73 $_{55}$	965	4584
Phenylephrine formyl artifact		1810	179 $_{32}$	160 $_{23}$	135 $_{100}$	107 $_{44}$	77 $_{42}$	171	4652
Phenylethanol	G UHY	<1000*	122 $_{21}$	91 $_{100}$	77 $_5$	65 $_{20}$	51 $_9$	108	4216
Phenylethanol AC	U+UHYAC	1060*	104 $_{100}$	91 $_{45}$	77 $_{17}$	65 $_{24}$	51 $_{21}$	146	4217
Phenylethanol-M (acid)	U UHY U+UHYAC	1280*	136 $_{31}$	91 $_{100}$	65 $_{20}$			116	4222
Phenylethanol-M (acid) ET	UET	1200*	164 $_{17}$	91 $_{100}$	65 $_{41}$			146	4227
Phenylethanol-M (acid) ME	UME	1120*	150 $_{46}$	91 $_{100}$	65 $_{41}$			128	4226
Phenylethanol-M (aldehyde)	U	1200*	120 $_{67}$	91 $_{100}$	65 $_{15}$			106	4221
Phenylethanol-M (homovanillic acid)	U	1610*	182 $_{35}$	137 $_{100}$	122 $_{12}$	94 $_7$	65 $_9$	179	3368
Phenylethanol-M (homovanillic acid) HFB		1770*	378 $_{79}$	333 $_{39}$	181 $_{60}$	107 $_{100}$	69 $_{72}$	945	5975
Phenylethanol-M (homovanillic acid) ME	UME	1750*	196 $_{25}$	137 $_{100}$	122 $_{16}$	107 $_2$	94 $_{12}$	210	812
Phenylethanol-M (homovanillic acid) MEAC	U+UHYAC	1700*	238 $_2$	196 $_{47}$	137 $_{100}$	122 $_6$	107 $_6$	336	2973
Phenylethanol-M (homovanillic acid) MEHFB		1570*	392 $_{100}$	333 $_{84}$	169 $_{24}$	107 $_{78}$	69 $_{60}$	991	5974
Phenylethanol-M (homovanillic acid) MEPFP		1570*	342 $_{96}$	283 $_{100}$	195 $_{35}$	119 $_{37}$	107 $_{51}$	815	5972
Phenylethanol-M (homovanillic acid) METMS		1670*	268 $_{41}$	238 $_{69}$	209 $_{52}$	179 $_{100}$	73 $_{95}$	461	6016
Phenylethanol-M (homovanillic acid) PFP		1685*	328 $_{85}$	283 $_{45}$	181 $_{58}$	119 $_{33}$	107 $_{100}$	747	5973
Phenylethanol-M (homovanillic acid) 2ME	UME	1720*	210 $_{31}$	151 $_{100}$	123 $_4$	107 $_6$		244	5959
Phenylethanol-M (homovanillic acid) 2TMS		1760*	326 $_{32}$	267 $_{20}$	209 $_{46}$	179 $_{34}$	73 $_{100}$	740	6015
Phenylethanol-M (HO-phenylacetic acid)	U	1565*	152 $_{26}$	107 $_{100}$	77 $_{18}$			130	818
Phenylethanol-M (HO-phenylacetic acid) AC		1565*	194 $_4$	152 $_{46}$	107 $_{100}$	77 $_{18}$		203	5819
Phenylethanol-M (HO-phenylacetic acid) HFB		1495*	348 $_{40}$	303 $_{100}$	275 $_{35}$	169 $_{30}$	69 $_{70}$	837	5956
Phenylethanol-M (HO-phenylacetic acid) ME	UME	1570*	166 $_{18}$	121 $_3$	107 $_{100}$	77 $_{19}$	59 $_3$	150	4224
Phenylethanol-M (HO-phenylacetic acid) MEAC		1550*	149 $_3$	208 $_3$	166 $_{37}$	107 $_{100}$	77 $_{12}$	236	5820
Phenylethanol-M (HO-phenylacetic acid) MEHFB		1405*	362 $_{36}$	303 $_{100}$	275 $_{33}$	78 $_{49}$	59 $_{56}$	890	5957
Phenylethanol-M (HO-phenylacetic acid) MEPFP		1220*	312 $_{34}$	253 $_{100}$	225 $_{28}$	119 $_{39}$	78 $_{43}$	674	5955
Phenylethanol-M (HO-phenylacetic acid) METFA		1120*	262 $_{33}$	203 $_{100}$	175 $_{21}$	69 $_{56}$	59 $_{35}$	429	5750
Phenylethanol-M (HO-phenylacetic acid) METMS		1485*	238 $_{48}$	223 $_{17}$	179 $_{100}$	163 $_{68}$	73 $_{78}$	336	6018
Phenylethanol-M (HO-phenylacetic acid) TFA		1450*	248 $_{36}$	203 $_{100}$	175 $_{29}$	77 $_{37}$	69 $_{57}$	369	5954
Phenylethanol-M (HO-phenylacetic acid) 2ME	UME	1420*	180 $_{17}$	121 $_{100}$	91 $_9$	77 $_{10}$		175	4228
Phenylethanol-M (HO-phenylacetic acid) 2PFP		1340*	430 $_{22}$	253 $_{100}$	225 $_{13}$	119 $_{45}$	69 $_{52}$	1079	5675
Phenylethanol-M (HO-phenylacetic acid) 2TMS		1675*	296 $_{16}$	281 $_{18}$	252 $_{18}$	179 $_{35}$	73 $_{100}$	599	5821
Phenylethanol-M (phenylacetamide)	U	1390	135 $_{23}$	91 $_{100}$	65 $_{24}$			115	4223
N-Phenylisopropyl-adenosine 3AC		3730	511 $_1$	420 $_{84}$	259 $_{18}$	162 $_{100}$	139 $_{46}$	1170	3090
Phenylmercuric acetate		9999*	327 $_{10}$	238 $_7$	93 $_{100}$	63 $_{50}$		796	865
Phenylmethylbarbital	P G U UHY U+UHYAC	1880	218 $_{51}$	132 $_{100}$	104 $_{89}$	78 $_{27}$		264	866
Phenylmethylbarbital 2ME		1790	246 $_{39}$	132 $_{100}$	104 $_{82}$			363	867
2-Phenyl-2-oxazoline	U	1065	147 $_{62}$	117 $_{100}$	105 $_{14}$	77 $_{37}$	51 $_{24}$	124	4371
Phenylpropanolamine	P U	1370	132 $_2$	118 $_8$	107 $_{11}$	91 $_{10}$	77 $_{100}$	130	2475
Phenylpropanolamine (HO-) 3AC	U+UHYAC	2135	234 $_6$	165 $_2$	123 $_{11}$	86 $_{100}$	58 $_{45}$	580	4961
Phenylpropanolamine TMSTFA		1890	240 $_8$	198 $_3$	179 $_{100}$	117 $_5$	73 $_{88}$	708	6146
Phenylpropanolamine 2AC	U+UHYAC	1805	235 $_1$	107 $_{13}$	86 $_{100}$	176 $_5$	134 $_7$	325	2476
Phenylpropanolamine 2HFB	UHYHFB	1455	543 $_1$	330 $_{14}$	240 $_{100}$	169 $_{44}$	69 $_{57}$	1183	5098
Phenylpropanolamine 2PFP	UHYPFP	1380	443 $_1$	280 $_9$	190 $_{100}$	119 $_{59}$	105 $_{26}$	1104	5094
Phenylpropanolamine 2TFA	UTFA	1355	343 $_1$	230 $_6$	203 $_5$	140 $_{100}$	69 $_{29}$	819	5091
Phenylpropanolamine 2TMS		1555	280 $_5$	163 $_4$	147 $_{10}$	116 $_{100}$	73 $_{83}$	594	4574
Phenylpropanolamine formyl artifact		1240	117 $_2$	105 $_4$	91 $_4$	77 $_{10}$	57 $_{100}$	143	4650
Phenyltoloxamine	G U+UHYAC	1950	255 $_5$	210 $_2$	152 $_2$	72 $_3$	58 $_{70}$	404	1682
Phenyltoloxamine-M (deamino-HO-)	UHY	1830*	228 $_{100}$	183 $_{76}$	165 $_{67}$	91 $_{33}$	77 $_{29}$	301	1694
Phenyltoloxamine-M (deamino-HO-) AC	U+UHYAC	2080*	270 $_5$	181 $_9$	165 $_9$	128 $_6$	87 $_{100}$	471	1685
Phenyltoloxamine-M (HO-) iso-1	UHY	2280	271 $_6$	226 $_1$	152 $_3$	107 $_4$	58 $_{100}$	476	1695

Table 1-8-1: Compounds in order of names — Pholcodine-M/artifact (O-dealkyl-) 2TMS

Name	Detected	RI	Typical ions and intensities					Page	Entry
Phenyltoloxamine-M (HO-) iso-1 AC	U+UHYAC	2260	313_4	268_2	91_3	72_{14}	58_{100}	681	1686
Phenyltoloxamine-M (HO-) iso-2	UHY	2300	271_{10}	226_3	152_6	91_7	58_{100}	476	1696
Phenyltoloxamine-M (HO-) iso-2 AC	U+UHYAC	2280	313_5	268_1	152_2	107_3	58_{100}	682	1687
Phenyltoloxamine-M (HO-methoxy-)	UHY	2320	301_5	271_4	152_4	72_7	58_{100}	620	1698
Phenyltoloxamine-M (HO-methoxy-) AC	U+UHYAC	2380	343_6	298_2	256_4	137_5	58_{100}	822	1689
Phenyltoloxamine-M (nor-)	UHY	2140	241_{31}	210_{71}	165_{41}	91_{54}	58_{100}	348	1697
Phenyltoloxamine-M (nor-) AC	U+UHYAC	2350	283_1	195_4	165_{12}	100_{100}	58_{17}	530	1688
Phenyltoloxamine-M (nor-HO-) iso-1	UHY	2320	257_{13}	226_{12}	197_{10}	152_{11}	58_{100}	410	1700
Phenyltoloxamine-M (nor-HO-) iso-1 2AC	U+UHYAC	2580	341_1	226_1	115_3	100_{100}	58_{24}	813	1690
Phenyltoloxamine-M (nor-HO-) iso-2	UHY	2340	257_{20}	226_{18}	197_{10}	91_{12}	58_{100}	410	1699
Phenyltoloxamine-M (nor-HO-) iso-2 2AC	U+UHYAC	2610	341_2	226_2	107_5	100_{100}	58_{26}	813	1691
Phenyltoloxamine-M (nor-HO-methoxy-) 2AC	U+UHYAC	2770	371_7	329_3	256_2	100_{100}	58_{12}	924	2413
Phenyltoloxamine-M (N-oxide) -(CH3)2NOH	UHY U+UHYAC	1500*	210_{95}	195_{39}	181_{53}	165_{100}	91_{36}	245	2201
Phenyltoloxamine-M (O-dealkyl-)	UHY	1680*	184_{100}	165_{52}	152_{15}	106_{37}	78_{59}	185	1692
Phenyltoloxamine-M (O-dealkyl-) AC	U+UHYAC	1740*	226_{27}	184_{100}	165_{43}	152_{23}	106_{29}	295	1683
Phenyltoloxamine-M (O-dealkyl-HO-) iso-1 2AC	U+UHYAC	2105*	284_7	242_{24}	200_{100}	151_8	122_{16}	533	2821
Phenyltoloxamine-M (O-dealkyl-HO-) iso-2	UHY	2220*	200_{100}	152_{21}	122_{47}	107_{86}	94_{66}	219	1693
Phenyltoloxamine-M (O-dealkyl-HO-) iso-2 2AC	U+UHYAC	2130*	284_{15}	242_{35}	200_{98}	115_{24}	107_{100}	533	1684
Phenytoin	P G U UHY	2350	252_{55}	223_{50}	180_{100}	104_{51}	77_{52}	389	869
Phenytoin AC	U+UHYAC	2300	294_{33}	223_{20}	208_{100}	180_{22}	77_{46}	585	871
Phenytoin ME	PME UME	2245	266_{71}	237_{38}	180_{100}	104_{66}	77_{51}	450	874
Phenytoin 2ME (2,3)		2225	280_{38}	251_{76}	134_{60}	77_{71}	72_{100}	517	4512
Phenytoin 2ME (N,N)	PME UME	2275	280_{28}	203_{67}	194_{72}	118_{100}	77_{56}	517	4513
Phenytoin 2TMS		2350	396_4	381_{15}	281_{62}	165_{79}	73_{100}	1006	4585
Phenytoin-M (3'-HO-) 3ME	UME UHYME	2445	310_{50}	281_{100}	238_{23}	233_{24}	134_{39}	666	4511
Phenytoin-M (4'-HO-) 3ME	UME UHYME	2490	310_{55}	233_{100}	224_{44}	148_{54}	118_{24}	667	4510
Phenytoin-M (HO-)	P-I U UHY	2795	268_{100}	239_{76}	196_{77}	120_{36}	104_{17}	460	870
Phenytoin-M (HO-) (ME)2AC	U+UHYAC	2690	366_{19}	324_{100}	224_{70}			904	872
Phenytoin-M (HO-) AC	U+UHYAC	2785	310_{17}	268_{100}	239_{39}	196_{29}	120_{10}	666	3047
Phenytoin-M (HO-) 2AC	U+UHYAC	2775	352_{10}	310_{91}	268_{13}	224_{100}	196_{12}	852	873
Phenytoin-M (HO-) 2ME	UME UHYME	2720	296_{100}	267_{90}	219_{74}	210_{66}	180_{45}	598	2833
Phenytoin-M (HO-methoxy-)	UHY	2770	298_{100}	269_{51}	226_{20}	150_{13}	104_{12}	605	3422
Phenytoin-M (HO-methoxy-) (ME)2AC	U+UHYAC	2640	396_7	354_{100}	300_{27}	254_{38}	151_{28}	1005	3423
Phenytoin-M (HO-methoxy-) 2AC	U+UHYAC	2800	382_8	340_{100}	268_{30}	254_{50}	196_{19}	959	3424
Phenytoin-M (HO-methoxy-) 2ME	UHYME	2740	326_{100}	297_{30}	282_{21}	249_{31}	196_{24}	739	2834
Phenytoin-M (HO-methoxy-) 3ME (2,3)	UHYME	2540	340_{31}	311_{44}	263_{18}	83_{42}	72_{83}	807	4515
Phenytoin-M (HO-methoxy-) 3ME (N,N)	UHYME	2585	340_{66}	263_{100}	254_{32}	178_{39}	118_{41}	807	4514
Phloroglucinol ME2AC	U+UHYAC	1705*	224_7	182_{14}	140_{100}	111_{34}	69_{19}	287	5633
Phloroglucinol 2MEAC	U+UHYAC	1485*	196_{28}	154_{100}	125_{95}	94_{28}	69_{44}	211	5632
Phloroglucinol 3AC	U+UHYAC	1850*	252_3	210_{11}	168_{25}	126_{100}	69_{15}	388	5634
Phloroglucinol 3ME		1230*	168_{100}	139_{83}	125_{22}	109_{23}	95_{13}	155	5628
Pholcodine	P G U UHY	3070	398_5	114_{100}	100_{70}	70_5		1012	1976
Pholcodine AC	U+UHYAC	3260	440_2	114_{100}	100_{70}	70_5		1098	1977
Pholcodine HFB		2830	594_1	354_{13}	277_{14}	114_{100}	100_{74}	1194	6164
Pholcodine PFP		2980	544_1	354_{14}	277_{14}	114_{100}	100_{48}	1183	3523
Pholcodine TFA		2800	494_1	380_3	277_5	114_{100}	100_{63}	1160	4015
Pholcodine TMS		3140	470_8	356_1	114_{100}	100_{66}	73_{20}	1139	3524
Pholcodine-M (demorpholino-HO-) 2AC	U+UHYAC	2860	413_{12}	354_{17}	327_{20}	215_{12}	87_{100}	1048	2127
Pholcodine-M (demorpholino-HO-) 2PFP	UHYPFP	2515	621_{26}	458_{100}	294_8	191_8	119_{13}	1196	3535
Pholcodine-M (demorpholino-HO-) 2TMS	UHYTMS	2755	473_{37}	442_{17}	280_9	180_{35}	73_{100}	1142	3527
Pholcodine-M (demorpholino-HO-) -H2O AC	U+UHYAC	2575	353_{100}	310_8	294_{63}	241_{28}	204_{27}	859	3712
Pholcodine-M (HO-) -H2O AC	U+UHYAC	3290	438_{40}	351_3	277_5	112_{29}	98_{100}	1095	3503
Pholcodine-M (nor-) AC		3620	426_2	340_1	114_{62}	100_{100}	70_5	1072	3500
Pholcodine-M (nor-) PFP	UHYPFP	3270	530_5	502_2	114_{48}	100_{100}	56_8	1178	3538
Pholcodine-M (nor-) 2AC	U+UHYAC	3650	468_3	382_1	114_{64}	100_{100}	56_8	1137	2124
Pholcodine-M (nor-) 2PFP	UHYPFP	3010	676_2	513_1	380_{13}	114_{71}	100_{100}	1201	3537
Pholcodine-M (nor-) 2TMS	UHYTMS	3260	528_2	468_2	114_{100}	100_{29}	73_{15}	1178	3528
Pholcodine-M (nor-demorpholino-HO-) 3AC	U+UHYAC	3275	441_3	357_1	296_1	209_4	87_{100}	1100	3499
Pholcodine-M (nor-demorpholino-HO-) 3PFP	UHYPFP	2560	753_{21}	590_8	355_{23}	191_{100}	119_{39}	1204	3536
Pholcodine-M (nor-demorpholino-HO-) 3TMS	UHYTMS	2735	531_5	399_3	280_4	131_{13}	73_{100}	1179	3526
Pholcodine-M (nor-HO-) -H2O 2AC	U+UHYAC	3665	466_{42}	296_4	207_6	112_{17}	98_{100}	1134	3502
Pholcodine-M (nor-oxo-) 2AC	U+UHYAC	3380	482_1	382_7	128_{100}	70_6		1149	3522
Pholcodine-M (nor-oxo-) 2TMS	UHYTMS	3400	542_3	486_{13}	396_{14}	357_{15}	128_{100}	1182	3529
Pholcodine-M (O-dealkyl-) TFA		2285	381_{55}	268_{100}	146_{14}	115_{13}	69_{23}	955	5569
Pholcodine-M (O-dealkyl-) 2AC	G PHYAC U+UHYAC	2620	369_{59}	327_{100}	310_{36}	268_{47}	162_{11}	915	225
Pholcodine-M (O-dealkyl-) 2TFA		2250	477_{71}	364_{100}	307_6	115_8	69_{31}	1145	4008
Pholcodine-M (O-dealkyl-)-D3 TFA		2275	384_{39}	271_{100}	211_8	165_6	152_7	967	5572
Pholcodine-M (O-dealkyl-)-D3 2TFA		2240	480_{32}	383_6	367_{100}	314_6	307_6	1148	5571
Pholcodine-M (oxo-) AC	U+UHYAC	3350	454_2	268_1	180_1	128_{100}	70_6	1120	3501
Pholcodine-M (oxo-) TMS	UHYTMS	3615	484_2	469_1	356_1	128_{100}	70_6	1151	3530
Pholcodine-M/artifact (O-dealkyl-)	G UHY	2455	285_{100}	268_{15}	162_{58}	124_{20}		539	474
Pholcodine-M/artifact (O-dealkyl-) 2HFB		2375	677_{10}	480_{11}	464_{100}	407_9	169_8	1201	6120
Pholcodine-M/artifact (O-dealkyl-) 2PFP		2360	577_{51}	558_7	430_8	414_{100}	119_{22}	1191	2251
Pholcodine-M/artifact (O-dealkyl-) 2TMS	UHYTMS	2560	429_{19}	236_{21}	196_{15}	146_{21}	73_{100}	1079	2463

Pholcodine-M/artifact (O-dealkyl-) Cl-artifact 2AC Table 1-8-1: Compounds in order of names

Name	Detected	RI	Typical ions and intensities					Page	Entry
Pholcodine-M/artifact (O-dealkyl-) Cl-artifact 2AC	U+UHYAC	2680	403 $_{59}$	361 $_{100}$	344 $_{63}$	302 $_{90}$	204 $_{55}$	1026	2992
Pholcodine-M/artifact (O-dealkyl-)-D3 2HFB		2375	680 $_4$	483 $_9$	467 $_{100}$	169 $_{23}$	414 $_7$	1202	6126
Pholcodine-M/artifact (O-dealkyl-)-D3 2PFP		2350	580 $_{16}$	433 $_7$	417 $_{100}$	269 $_5$	119 $_8$	1192	5567
Pholcodine-M/artifact (O-dealkyl-)-D3 2TMS		2550	432 $_{100}$	290 $_{30}$	239 $_{66}$	199 $_{44}$	73 $_{110}$	1085	5578
Pholcodine-M/artifact 2PFP		2440	563 $_{100}$	400 $_{10}$	355 $_{38}$	327 $_7$	209 $_{15}$	1188	3534
Pholcodine-M/artifact 3AC	U+UHYAC	2955	397 $_8$	355 $_9$	209 $_{41}$	87 $_{100}$	72 $_{33}$	1010	1194
Pholcodine-M/artifact 3PFP	UHYPFP	2405	709 $_{80}$	533 $_{28}$	388 $_{29}$	367 $_{51}$	355 $_{100}$	1203	3533
Pholcodine-M/artifact 3TMS	UHYTMS	2605	487 $_9$	416 $_9$	222 $_{18}$	131 $_{10}$	73 $_{50}$	1155	3525
Pholedrine		1885	150 $_1$	135 $_1$	107 $_5$	77 $_5$	58 $_{100}$	148	1766
Pholedrine TFA		1770	261 $_1$	154 $_{100}$	134 $_{68}$	110 $_{42}$	107 $_{41}$	425	6180
Pholedrine TMSTFA		1690	333 $_3$	206 $_{72}$	179 $_{100}$	154 $_{53}$	73 $_{50}$	773	6228
Pholedrine 2AC	U+UHYAC	1995	249 $_1$	176 $_6$	134 $_7$	100 $_{43}$	58 $_{100}$	376	1767
Pholedrine 2HFB		1670	538 $_1$	330 $_{17}$	254 $_{100}$	210 $_{32}$	169 $_{22}$	1186	5076
Pholedrine 2PFP		1605	295 $_1$	280 $_{18}$	204 $_{100}$	160 $_{47}$	119 $_{39}$	1123	5077
Pholedrine 2TFA		1585	357 $_1$	230 $_{22}$	154 $_{100}$	110 $_{42}$	69 $_{29}$	872	5078
Pholedrine 2TMS		1620	309 $_{10}$	206 $_{70}$	179 $_{100}$	154 $_{32}$	73 $_{40}$	663	6190
Phorate		1675*	260 $_{26}$	231 $_{11}$	121 $_{38}$	97 $_{26}$	75 $_{100}$	419	3476
Phosalone	G P-I	2535	367 $_{59}$	182 $_{100}$	154 $_{24}$	121 $_{54}$	97 $_{42}$	907	2722
Phosalone impurity	G	1385*	234 $_{28}$	154 $_{46}$	121 $_{100}$	97 $_{96}$	65 $_{52}$	317	3299
Phosalone impurity	G	2235*	384 $_5$	231 $_{67}$	153 $_{55}$	125 $_{51}$	97 $_{100}$	965	3837
Phosalone impurity	G	1050	214 $_{28}$	186 $_{100}$	121 $_{69}$	97 $_{79}$	93 $_{50}$	253	6361
Phosalone impurity (dichloro-)	G	2645	401 $_{45}$	216 $_{100}$	154 $_{54}$	121 $_{79}$	97 $_{36}$	1019	6365
Phosalone-M (thiol) AC	U+UHYAC	2135	257 $_{41}$	214 $_{12}$	182 $_{43}$	169 $_{100}$	111 $_{17}$	408	6366
Phosalone-M/artifact	U	1800	169 $_{100}$	113 $_{26}$	78 $_{25}$	63 $_7$		155	4372
Phosalone-M/artifact AC	U+UHYAC	1595	211 $_{10}$	169 $_{100}$	113 $_{13}$	76 $_{11}$	125 $_7$	245	6362
Phosalone-M/artifact ME		1750	183 $_{100}$	154 $_{45}$	92 $_{65}$	76 $_{20}$	63 $_{16}$	181	2440
Phosalone-M/artifact HY2AC	U+UHYAC	1850	227 $_{17}$	185 $_{15}$	167 $_{39}$	143 $_{100}$	114 $_{17}$	296	6364
Phosalone-M/artifact HY3AC	U+UHYAC	2160	269 $_6$	227 $_{20}$	185 $_{100}$	129 $_{10}$	86 $_{15}$	463	6363
Phosdrin		1415*	224 $_1$	192 $_{25}$	164 $_8$	127 $_{100}$	109 $_{24}$	286	4054
Phosmet		2380	317 $_{21}$	160 $_{100}$	133 $_{16}$	104 $_{15}$	77 $_{30}$	695	3477
Phosphamidon iso-1	G P U	1820	264 $_{63}$	227 $_8$	193 $_{12}$	127 $_{100}$	72 $_{42}$	609	2533
Phosphamidon iso-2	G	1900	264 $_{67}$	138 $_{38}$	127 $_{100}$	109 $_{55}$	72 $_{76}$	609	2534
Phosphine		<1000*	34 $_{100}$	31 $_{33}$				89	4194
Phosphoric acid 3TMS		1060*	314 $_{17}$	299 $_{100}$	211 $_{15}$	133 $_{15}$	73 $_{61}$	684	4678
Phoxim	G	2005	298 $_{16}$	168 $_{16}$	135 $_{47}$	109 $_{100}$	81 $_{57}$	604	4077
Phoxim artifact-1	P-I U	1400*	198 $_{26}$	170 $_{38}$	138 $_{95}$	111 $_{100}$	81 $_{80}$	213	1442
Phoxim artifact-2	G	1670	306 $_5$	278 $_{24}$	222 $_{64}$	194 $_{100}$	99 $_{70}$	643	4087
Phoxim-M/artifact	U UHY U+UHYAC	1480	146 $_{63}$	116 $_{97}$	89 $_{100}$	51 $_{30}$		124	4370
Phoxim-M/artifact	P U UHY U+UHYAC	1350	196 $_9$	171 $_{96}$	143 $_{100}$	111 $_{42}$	97 $_{50}$	210	4369
Phthalic acid butyl-2-ethylhexyl ester		1950*	223 $_7$	205 $_4$	149 $_{100}$	104 $_4$	57 $_7$	781	713
Phthalic acid butyl-2-methylpropyl ester		1970*	278 $_1$	223 $_4$	205 $_4$	149 $_{100}$	76 $_4$	507	2995
Phthalic acid butyloctyl ester		1950*	223 $_6$	205 $_4$	149 $_{100}$	122 $_4$	104 $_{11}$	781	2361
Phthalic acid decyldodecyl ester		2990*	474 $_1$	335 $_6$	307 $_8$	149 $_{100}$	57 $_{11}$	1143	3542
Phthalic acid decylhexyl ester		2665*	390 $_1$	307 $_5$	251 $_{10}$	233 $_2$	149 $_{100}$	986	6402
Phthalic acid decyloctyl ester		2675*	418 $_1$	307 $_9$	279 $_{12}$	149 $_{100}$	57 $_{28}$	1059	3544
Phthalic acid decyltetradecyl ester		3250*	502 $_1$	363 $_4$	307 $_7$	149 $_{100}$	57 $_{14}$	1166	3543
Phthalic acid diisodecyl ester		2800*	446 $_1$	307 $_{24}$	167 $_2$	149 $_{100}$	57 $_7$	1109	3541
Phthalic acid diisohexyl ester		2380*	334 $_1$	251 $_2$	233 $_2$	149 $_{100}$	104 $_3$	781	6397
Phthalic acid diisononyl ester		2700*	418 $_1$	293 $_{35}$	167 $_{19}$	149 $_{100}$	71 $_{54}$	1059	1232
Phthalic acid diisooctyl ester		2520*	390 $_1$	279 $_{15}$	167 $_{41}$	149 $_{100}$	57 $_{29}$	986	723
Phthalic acid dimethyl ester		1450*	194 $_7$	163 $_{100}$	133 $_9$	104 $_7$	77 $_{15}$	203	4948
Phthalic acid dioctyl ester		2655*	390 $_1$	279 $_{12}$	261 $_2$	167 $_2$	149 $_{100}$	986	6401
Phthalic acid ethyl methyl ester		1520*	208 $_2$	176 $_{11}$	163 $_{100}$	149 $_{58}$	77 $_{25}$	237	4940
Phthalic acid ethylhexyl methyl ester		2010*	181 $_{24}$	163 $_{100}$	149 $_{48}$	83 $_{11}$	70 $_{34}$	576	5319
Phthalic acid hexyloctyl ester		2500*	362 $_1$	279 $_4$	251 $_5$	149 $_{100}$	85 $_5$	893	6053
Physcion		2660*	284 $_{100}$	255 $_{12}$	241 $_9$	213 $_6$	128 $_{13}$	532	3556
Physcion ME		2775*	298 $_{100}$	280 $_{48}$	252 $_{55}$	135 $_{22}$	115 $_{12}$	605	3567
Physcion 2AC		2920*	368 $_1$	326 $_{24}$	284 $_{100}$	255 $_6$	128 $_3$	912	3569
Physcion 2ME		2845*	312 $_{65}$	297 $_{100}$	295 $_{27}$	267 $_7$	142 $_{13}$	675	3568
Physostigmine	G U	2240	275 $_{33}$	218 $_{100}$	174 $_{95}$	160 $_{76}$	132 $_{13}$	492	875
Physostigmine-M/artifact	G UHY	1835	218 $_{84}$	188 $_{13}$	174 $_{97}$	160 $_{100}$	146 $_{13}$	266	876
Physostigmine-M/artifact AC	U+UHYAC	2010	260 $_{89}$	218 $_{100}$	174 $_{83}$	160 $_{76}$	132 $_{10}$	423	2616
Phytanic acid		2035*	312 $_5$	250 $_{20}$	157 $_{44}$	87 $_{100}$	71 $_{87}$	678	6063
Phytanic acid ME		2015*	326 $_8$	171 $_{39}$	143 $_{21}$	101 $_{100}$	74 $_{89}$	742	6062
PIA 3AC		3730	511 $_1$	420 $_{84}$	259 $_{18}$	162 $_{100}$	139 $_{46}$	1170	3090
PICC		1680	207 $_9$	180 $_9$	123 $_8$	99 $_{100}$	70 $_{42}$	235	3587
PICC -HCN		1380	180 $_{26}$	165 $_8$	123 $_{13}$	110 $_{17}$	70 $_{100}$	176	3588
Picloram ME		1875	254 $_{23}$	223 $_{19}$	196 $_{100}$	168 $_{17}$	63 $_{40}$	396	3651
Picloram -CO2		1440	196 $_{100}$	161 $_{33}$	134 $_{22}$	98 $_{13}$	86 $_{11}$	209	3650
Picosulfate-M (bis-methoxy-bis-phenol)	UHY	2820	337 $_{100}$	322 $_{69}$	307 $_8$	259 $_{14}$		792	2458
Picosulfate-M (bis-methoxy-bis-phenol) 2AC	U+UHYAC	2950	421 $_{83}$	379 $_{100}$	364 $_{54}$	337 $_{25}$	322 $_{46}$	1063	2456
Picosulfate-M (bis-methoxy-bis-phenol) 2ME	UGLUCEXME	2760	365 $_{100}$	350 $_{61}$	287 $_{41}$	249 $_{13}$	220 $_{11}$	901	6813
Picosulfate-M (bis-phenol)	UHY	2655	277 $_{99}$	199 $_{51}$				500	107
Picosulfate-M (bis-phenol) 2AC	G U+UHYAC PAC-I	2835	361 $_{99}$	319 $_{63}$	277 $_{74}$	199 $_{45}$		887	106

Table 1-8-1: Compounds in order of names   Piperonylpiperazine TMS

Name	Detected	RI	Typical ions and intensities					Page	Entry
Picosulfate-M (bis-phenol) 2ME	UGLUCEXME	2595	305_{100}	290_{27}	227_{49}	182_6	169_6	639	6811
Picosulfate-M (methoxy-bis-phenol)	UHY	2680	307_{100}	306_{49}	292_{19}	229_{35}	69_{22}	648	109
Picosulfate-M (methoxy-bis-phenol) 2AC	U+UHYAC	2870	391_{46}	349_{100}	307_{48}	292_{12}	229_{23}	988	1750
Picosulfate-M (methoxy-bis-phenol) 2ME	UGLUCEXME	2695	335_{100}	320_{40}	257_{57}	220_7	139_{13}	783	6812
Pilocarpine	G U+UHYAC	2160	208_9	121_4	109_9	95_{100}		238	2233
Pilocarpine-M (1-HO-ethyl-) AC	U+UHYAC	2390	266_7	206_{38}	177_{31}	95_{100}	82_{19}	451	4360
Pilocarpine-M (2-HO-ethyl-) AC	U+UHYAC	2200	266_{14}	206_{16}	124_{14}	95_{100}	87_{30}	451	4359
Pimozide		3870	461_{42}	230_{101}	187_{38}	133_{35}	82_{38}	1129	596
Pimozide TMS		4155	533_{53}	302_{91}	259_{36}	203_{53}	73_{100}	1179	4586
Pimozide-M (benzimidazolone)	UHY-I	1950	134_{100}	106_{36}	79_{62}	67_{20}		114	491
Pimozide-M (benzimidazolone) 2AC	U+UHYAC-I	1730	218_{11}	176_{19}	134_{98}	106_{11}		263	171
Pimozide-M (deamino-carboxy-)	P-I UHY U+UHYAC	2230*	276_7	216_{17}	203_{100}	183_{22}		496	169
Pimozide-M (deamino-carboxy-) ME	P-I UHYME U+UHYAC	2125*	290_{12}	258_{13}	216_{30}	203_{100}	183_{22}	563	3372
Pimozide-M (deamino-HO-)	UHY-I	2120*	262_{16}	203_{100}	201_{16}	183_{12}		430	515
Pimozide-M (deamino-HO-) AC	U+UHYAC	2150*	304_7	244_{22}	216_{41}	203_{100}	183_{17}	635	307
Pimozide-M (N-dealkyl-)	UHY	2415	217_9	134_{99}	106_{42}	79_{87}		261	87
Pimozide-M (N-dealkyl-) AC	U+UHYAC	2770	259_{60}	216_{15}	134_{64}	125_{42}	82_{100}	417	89
Pimozide-M (N-dealkyl-) ME	UHY	2290	231_{12}	134_{99}	106_{37}	79_{81}		310	86
Pimozide-M (N-dealkyl-) 2AC	U+UHYAC	2750	301_{28}	259_{43}	134_{28}	125_{28}	82_{100}	620	88
Pinaverium bromide artifact-1		2450	281_6	212_3	114_5	100_{100}	70_7	524	6441
Pinaverium bromide artifact-2		2110	310_7	231_{100}	229_{99}	185_{12}	107_{15}	664	6442
Pinaverium bromide artifact-3		1975	266_{24}	231_{100}	229_{98}	185_{11}	107_{13}	447	6443
Pinaverium bromide artifact-4		1915	260_{43}	231_{91}	229_{100}	181_7	107_{23}	418	6444
Pinaverium bromide artifact-5		1695	230_{100}	215_{21}	187_6	108_{49}		305	6445
Pinazepam		2585	308_{93}	307_{96}	280_{100}	217_{28}	91_{76}	653	3072
Pinazepam HY		2330	269_{19}	268_{22}	227_{82}	190_{23}	77_{100}	463	3073
Pinazepam HYAC		2400	311_{11}	268_{41}	227_{70}	190_{24}	77_{100}	670	3076
Pinazepam-M	P G U	2520	270_{86}	269_{97}	242_{100}	241_{82}	77_{17}	468	463
Pinazepam-M TMS		2300	342_{62}	341_{100}	327_{19}	269_4	73_{30}	817	4573
Pinazepam-M HY	UHY	2050	231_{80}	230_{95}	154_{23}	105_{38}	77_{100}	308	419
Pindolol	G	2240	248_{27}	204_8	133_{100}	116_{15}	72_{73}	372	1227
Pindolol TMSTFA		2415	416_{10}	284_{110}	246_{13}	129_{99}	73_{87}	1055	6160
Pindolol 2AC		2750	332_7	200_{97}	186_{58}	140_{57}	98_{49}	769	878
Pindolol 2TMSTFA		2485	488_5	318_5	284_{69}	129_{57}	73_{80}	1156	6161
Pindolol formyl artifact		2260	260_{34}	133_{84}	127_{98}	86_{55}		423	877
Pindone		1825*	230_{29}	173_{100}	146_{45}	105_{43}	89_{49}	307	3652
Pipamperone	P-I G U UHY U+UHYAC	3040	331_{36}	194_{30}	165_{100}	138_{75}	123_{79}	937	179
Pipamperone 2TMS		3100	519_1	403_{36}	296_{56}	211_{76}	73_{100}	1174	4587
Pipamperone artifact	U+UHYAC	1350*	164_{43}	133_{38}	123_{100}	95_{41}		145	1914
Pipamperone-M	U+UHYAC	1490*	180_{25}	125_{49}	123_{35}	95_{17}	56_{100}	174	85
Pipamperone-M (dihydro-) -H2O	UHY U+UHYAC	3000	315_{33}	224_{59}	139_{100}	98_{75}	70_{28}	881	5586
Pipamperone-M (HO-)	UHY	3250	347_{49}	292_{15}	165_{100}	154_{75}	123_{40}	990	597
Pipamperone-M (HO-) AC	U+UHYAC	3290	389_{63}	292_{27}	194_{37}	165_{100}	123_{70}	1086	599
Pipamperone-M (N-dealkyl-) AC	PAC-I U+UHYAC	2500	209_{100}	150_{67}	124_{64}	84_{24}	82_{46}	396	598
Pipazetate-M (alcohol)	U UHY	1830	156_2	112_5	98_{100}	96_2	70_4	161	2274
Pipazetate-M (alcohol) AC	U+UHYAC	1710	215_1	156_3	142_1	98_{100}		258	2276
Pipazetate-M (HO-alcohol) AC	U+UHYAC	1800	231_1	156_2	142_1	112_5	98_{100}	310	2277
Pipazetate-M (HO-ring)	U UHY	2800	216_{100}	187_{68}	168_{50}	140_9		259	2272
Pipazetate-M (HO-ring) AC	U+UHYAC	2575	258_{30}	216_{100}	187_{13}	183_{13}	155_{10}	412	2271
Pipazetate-M (ring-sulfone)	U UHY U+UHYAC	2750	232_{100}	200_{42}	184_{20}	168_{22}		311	2273
Pipazetate-M/artifact (ring)	U+UHYAC	2045	200_{100}	168_{36}	156_{13}			219	386
Piperacilline TMS		2900	589_{11}	574_8	486_{15}	446_{12}	73_{100}	1193	4617
Piperacilline 2TMS		2780	661_{15}	646_2	369_7	147_{11}	73_{100}	1200	4616
Piperacilline-M/artifact AC	U+UHYAC	2660	289_5	246_{25}	132_3	100_{16}	58_{100}	557	4288
Piperacilline-M/artifact 2AC	U+UHYAC	2530	331_4	288_{22}	113_{20}	100_{28}	58_{100}	763	4289
Piperazine 2AC		1750	170_{15}	85_{33}	69_{25}	56_{100}		157	879
Piperazine 2HFB		1290	478_3	459_9	309_{100}	281_{22}	252_{41}	1146	6634
Piperazine 2TFA		1005	278_{10}	209_{59}	152_{25}	69_{56}	56_{100}	505	4129
Piperidine		<1000	85_{11}	84_{26}	70_5	56_{18}		96	3615
Piperonol		1420*	152_{100}	135_{47}	122_{37}	93_{70}	65_{56}	131	7616
Piperonol AC		1530*	194_{63}	152_{66}	135_{100}	122_{17}	104_{18}	204	6637
Piperonol HFB		1400*	348_{28}	271_2	135_{100}	105_6	77_{19}	837	7620
Piperonol PFP		1325*	298_{30}	149_8	135_{100}	105_7	77_{28}	604	7619
Piperonol TFA		1295*	248_{40}	149_4	135_{100}	105_8	77_{24}	370	7618
Piperonol TMS		1560*	224_{45}	209_{27}	179_{17}	135_{100}	73_{24}	287	7617
Piperonyl butoxide	G P	2375*	338_5	193_{11}	176_{100}	149_{35}	57_{49}	799	3478
Piperonylacetate		1530*	194_{63}	152_{66}	135_{100}	122_{17}	104_{18}	204	6637
Piperonylic acid ET		1560*	194_{48}	166_{27}	149_{100}	121_{23}	65_{16}	203	6471
Piperonylic acid ME		1445*	180_{52}	149_{100}	121_{27}	63_{19}	63_7	173	6470
Piperonylpiperazine		1890	220_{21}	178_9	164_{13}	135_{100}	85_{36}	274	6624
Piperonylpiperazine AC	U+UHYAC	2350	262_{16}	190_{11}	176_{21}	135_{100}	85_{45}	431	6625
Piperonylpiperazine HFB	U+UHYHFB	2190	416_8	281_{15}	148_9	135_{100}	105_{11}	1053	6631
Piperonylpiperazine TFA	U+UHYTFA	2350	316_9	181_{14}	148_6	135_{100}	105_6	693	6628
Piperonylpiperazine TMS		2080	292_{87}	157_{53}	135_{100}	102_{57}	73_{85}	576	6887

Table 1-8-1: Compounds in order of names

Name	Detected	RI	Typical ions and intensities					Page	Entry
Piperonylpiperazine Cl-artifact		1295*	170_{25}	135_{100}	105_8	77_{41}		156	6635
Piperonylpiperazine-M (deethylene-) 2AC	U+UHYAC	2320	278_4	235_{64}	177_{11}	150_{25}	135_{100}	506	6626
Piperonylpiperazine-M (deethylene-) 2HFB	U+UHYHFB	2080	586_3	389_{24}	346_7	240_7	135_{100}	1193	6632
Piperonylpiperazine-M (deethylene-) 2TFA	U+UHYTFA	2230	386_8	289_{10}	246_{10}	150_4	135_{100}	972	6629
Piperonylpiperazine-M (piperonylamine) AC	U+UHYAC	2015	193_{100}	150_{80}	135_{54}	121_{37}	93_{27}	199	6627
Piperonylpiperazine-M (piperonylamine) HFB	U+UHYHFB	1640	347_{100}	317_6	289_{10}	178_6	135_{75}	834	6633
Piperonylpiperazine-M (piperonylamine) PFP		1755	297_{100}	267_8	239_{13}	178_7	135_{51}	601	7632
Piperonylpiperazine-M (piperonylamine) TFA	U+UHYTFA	1775	247_{100}	217_9	189_{16}	148_{12}	135_{80}	365	6630
Piperonylpiperazine-M (piperonylamine) TMS		2130	295_{27}	280_{15}	206_{15}	179_{50}	73_{51}	592	7630
Piperonylpiperazine-M (piperonylamine) 2AC		2230	235_{23}	192_{35}	150_{100}	135_{17}	93_9	323	7631
Piperonylpiperazine-M (piperonylamine) formyl artifact		1560	163_{22}	135_{100}	121_3	105_{10}	77_{37}	142	7629
Pipradrol-M (BPH)	U+UHYAC	1610*	182_{31}	152_3	105_{100}	77_{70}	51_{39}	180	1624
Pipradrol-M (HO-BPH) iso-1	UHY	2065*	198_{93}	121_{72}	105_{100}	93_{22}	77_{66}	214	1627
Pipradrol-M (HO-BPH) iso-1 AC	U+UHYAC	2010*	240_{27}	198_{100}	121_{47}	105_{85}	77_{80}	343	2196
Pipradrol-M (HO-BPH) iso-2	P-I U UHY	2080*	198_{50}	121_{100}	105_{17}	93_{14}	77_{28}	214	732
Pipradrol-M (HO-BPH) iso-2 AC	U+UHYAC	2050*	240_{20}	198_{100}	121_{94}	105_{41}	77_{51}	343	2197
Pipradrol		2400	248_1	182_2	165_4	105_{18}	84_{100}	456	7337
Pipradrol TMS		2365	324_2	239_4	165_7	84_{100}	73_{24}	805	7343
Pipradrol 2AC		2630	249_1	183_2	165_2	126_{49}	84_{100}	850	7339
Pipradrol -H2O AC	U+UHYAC	2520	291_{26}	249_{100}	206_{11}	191_{16}	165_{24}	570	7338
Pipradrol -H2O HFB		2330	445_{100}	368_{75}	248_{37}	206_{20}	276_5	1107	7342
Pipradrol -H2O PFP		2320	395_{100}	318_{56}	248_{57}	206_{50}	165_{39}	1001	7341
Pipradrol -H2O TFA		2350	345_{100}	276_{22}	268_{51}	248_{50}	206_{44}	829	7340
Pipradrol-M (HO-) -H2O 2AC	U+UHYAC	2700	349_{37}	307_{83}	265_{100}	222_9	152_{10}	843	7815
Piracetam	G P-I U+UHYAC	1520	142_{20}	125_5	98_{100}	84_{71}	70_{78}	121	374
Piracetam 2TMS		1670	286_1	271_5	188_{19}	171_{31}	73_{100}	544	4588
Pirbuterol 2TMS		1915	369_5	309_{39}	299_{79}	209_{100}	75_{33}	968	6189
Pirbuterol 3AC		2130	351_{27}	281_{82}	238_{100}	189_{80}	86_{80}	905	6061
Pirbuterol 3TMS		2010	441_6	371_{100}	280_{14}	266_{30}	266_{30}	1122	6188
Pirbuterol artifact 2AC		2250	250_{49}	190_{83}	175_{20}	148_{69}	55_{100}	380	6054
Pirenzepin		3005	351_{13}	211_{12}	113_{100}	70_{45}		850	375
Piretanide (ME)AC	U+UHYAC	3110	418_{32}	376_{14}	295_{71}	266_{100}	218_{17}	1058	6412
Piretanide 2ME	UME	3010	390_{90}	295_{100}	266_{61}	219_{33}	77_{46}	985	3100
Piretanide 2MEAC	U+UHYAC	3070	432_{18}	313_{13}	295_{91}	266_{100}	236_{22}	1084	6413
Piretanide 3ME	UME	2965	404_{41}	295_{100}	266_{25}	219_{19}	77_{25}	1028	3101
Piretanide -SO2NH ME	UME	2485	297_{88}	296_{100}	220_{42}	180_{32}	77_{39}	602	3102
Pirimicarb		1850	238_{36}	166_{100}	138_{11}	123_8	72_{49}	338	3480
Pirimiphos-methyl		1960	305_{78}	290_{100}	276_{85}	233_{40}	125_{80}	638	3479
Piritramide	P-I U+UHYAC	3560	386_{98}	345_{10}	301_{15}	138_{44}		1082	256
Pirlindole		2300	226_{47}	198_{100}	183_{10}	167_8	99_{12}	296	6099
Pirlindole AC		2645	268_{78}	240_{100}	223_{25}	197_{85}	115_{10}	462	6101
Pirlindole ME		2290	240_{41}	212_{100}	114_{18}			346	6100
Pirlindole TMS		2335	298_{29}	270_{100}	226_6	198_{16}	73_{31}	608	6200
Piroxycam ME		2760	345_9	330_{42}	280_{45}	250_{79}	121_{100}	827	5154
Piroxycam 2ME		2790	359_2	330_{38}	250_{12}	162_{52}	121_{100}	879	5155
Piroxycam artifact	PME UME	1600	197_{56}	133_{62}	132_{65}	104_{89}	76_{100}	212	2863
Pirprofen		2175	251_{100}	210_{59}	206_{44}	169_{12}	138_6	383	1838
Pirprofen ET		2110	279_{88}	249_{79}	238_{36}	206_{100}	204_{80}	510	1853
Pirprofen ME		2055	265_{75}	224_{59}	206_{100}	169_{13}	103_{27}	442	2234
Pirprofen artifact		1870	237_{72}	222_{100}	206_{72}	196_{25}	164_{18}	742	1840
Pirprofen artifact ME		1670	213_{30}	154_{100}	119_{42}			251	1846
Pirprofen artifact 2ME		1750	227_{33}	168_{100}	133_{32}			297	1847
Pirprofen -CO2		1760	207_{100}	190_9	166_{99}	103_{16}		232	1839
Pirprofen-M (diol) ET		2500	313_{100}	254_{97}	226_{26}	211_{44}	166_{13}	679	1856
Pirprofen-M (diol) ME		2550	299_{68}	240_{100}	211_{28}	166_{17}	103_9	610	1850
Pirprofen-M (diol) ME2AC		2545	383_{21}	324_{17}	264_{100}	204_{14}	166_9	963	1852
Pirprofen-M (diol) MEAC		2530	341_{23}	281_{32}	264_{100}	222_{10}	166_{13}	810	1851
Pirprofen-M (epoxide) ET		2280	295_{28}	281_8	222_{100}	166_9	103_7	589	1855
Pirprofen-M (epoxide) ME		2260	281_{47}	222_{100}	166_{29}	103_{23}		520	1849
Pirprofen-M (HO-) -H2O ME		1945	263_{100}	204_{100}	169_{47}	141_{12}	115_8	434	1848
Pirprofen-M (pyrrole) artifact		1770	235_{92}	220_{100}	205_{89}	169_{21}	115_{15}	322	1843
Pirprofen-M/artifact (pyrrole)		2040	249_{100}	204_{90}	169_{49}	141_{11}	115_9	373	1841
Pirprofen-M/artifact (pyrrole) ET		1990	277_{85}	204_{100}	169_{32}	141_8	115_6	499	1854
Pirprofen-M/artifact (pyrrole) ME		1945	263_{100}	204_{100}	169_{47}	141_{12}	115_8	434	1848
Pirprofen-M/artifact (pyrrole) -CH2O2		1680	203_{100}	168_{23}	141_{36}	115_{17}		223	1842
Pirprofen-M/artifact (pyrrole) -CO2		1800	205_{100}	169_{19}	164_{85}	141_8	102_{23}	228	1844
Pitofenone		3120	367_1	152_1	112_3	98_{100}	55_7	910	3994
Pivalic acid anhydride		<1000*	146_1	85_{44}	57_{100}			188	2758
Pizotifen		2340	295_{99}	223_{10}	197_{20}	96_{62}	58_{42}	591	1515
PMA		1225	165_1	122_{16}	107_2	91_3	77_7	148	3249
PMA		1225	165_1	122_{16}	91_3	77_7	44_{100}	148	5517
PMA AC	U+UHYAC	1720	207_1	148_{41}	121_{16}	91_5	86_{10}	235	3265
PMA AC	U+UHYAC	1720	207_1	148_{41}	121_{16}	86_{10}	44_{100}	234	5537
PMA HFB		1560	361_2	240_5	169_5	148_{53}	121_{100}	886	6769

Table 1-8-1: Compounds in order of names　　　　　　　　　　　　　　　　　　　　　　　　　　　　**Polythiazide 3ME**

Name	Detected	RI	Typical ions and intensities					Page	Entry
PMA PFP		1460	311_7	190_4	148_{43}	121_{100}	91_7	670	6775
PMA TFA		1460	261_6	148_{33}	140_4	121_{100}	91_8	425	6774
PMA formyl artifact		1255	177_6	162_4	121_{60}	77_{12}	56_{100}	166	3250
PMA precursor (4-methoxyphenylacetone)		1205*	164_{10}	121_{100}	91_8	77_{16}		146	3277
PMA-M (O-demethyl-)		1480	151_{10}	107_{69}	91_{10}	77_{42}	56_{100}	129	1802
PMA-M (O-demethyl-) AC	U+UHYAC	1890	193_1	134_{100}	107_{26}	86_{25}	77_{16}	201	1803
PMA-M (O-demethyl-) 2AC	U+UHYAC	1900	235_1	176_{72}	134_{100}	107_{47}	86_{71}	324	1804
PMA-M (O-demethyl-) 2HFB		<1000	330_{48}	303_{15}	240_{100}	169_{44}	69_{42}	1182	6326
PMA-M (O-demethyl-) 2PFP		<1000	280_{77}	253_{16}	190_{100}	119_{56}	69_{16}	1104	6325
PMA-M (O-demethyl-) 2TFA		<1000	230_{94}	203_{14}	140_{130}	92_{15}	69_{76}	819	6324
PMA-M (O-demethyl-) 2TMS		<1000	280_7	179_9	149_8	116_{100}	73_{78}	594	6327
PMA-M (O-demethyl-methoxy-)	UHY	1465	181_9	138_{100}	122_{18}	94_{24}	77_{16}	178	4351
PMA-M (O-demethyl-methoxy-) ME	UHYME	1550	195_1	152_{100}	137_{17}	107_{16}	77_{14}	208	4352
PMA-M 2AC	U+UHYAC	2065	265_3	206_{27}	164_{100}	137_{23}	86_{33}	445	3498
PMA-M 2HFB	UHFB	1690	360_{82}	333_{15}	240_{100}	169_{42}	69_{39}	1189	6512
PMEA		1660	192_1	149_1	121_5	91_2	72_{100}	202	5831
PMEA AC	U+UHYAC	1855	235_1	148_{32}	121_7	114_{26}	72_{100}	326	5322
PMEA HFB	UHFB	1785	389_2	268_{100}	240_{46}	148_{62}	121_{43}	983	5834
PMEA ME		1780	206_1	121_5	86_{100}	72_3	58_{20}	235	5835
PMEA PFP	UPFP	1765	339_1	218_{100}	190_{44}	148_{59}	121_{49}	802	5833
PMEA TFA	UTFA	1775	289_1	168_{100}	148_{63}	140_{41}	121_{62}	557	5832
PMEA TMS		2065	264_1	250_{14}	144_{100}	121_{17}	73_{84}	447	5836
PMMA		1475	178_1	121_4	91_2	77_3	58_{100}	172	6719
PMMA AC		1820	221_1	148_{24}	121_7	100_{22}	58_{100}	279	6720
PMMA ET		1780	206_1	121_5	86_{100}	72_3	58_{20}	235	5835
PMMA HFB		1665	375_1	254_{55}	210_{32}	148_{67}	121_{100}	935	6722
PMMA PFP		1510	325_3	204_{60}	160_{29}	148_{48}	121_{100}	734	7601
PMMA TFA		1645	275_2	154_{59}	148_{51}	121_{100}	110_{31}	489	6721
PMMA-M (bis-demethyl-)		1480	151_{10}	107_{69}	91_{10}	77_{42}	56_{100}	129	1802
PMMA-M (bis-demethyl-) AC	U+UHYAC	1890	193_1	134_{100}	107_{26}	86_{25}	77_{16}	201	1803
PMMA-M (bis-demethyl-) ME		1225	165_1	122_{16}	107_2	93_3	77_7	148	3249
PMMA-M (bis-demethyl-) ME		1225	165_1	122_{16}	91_3	77_7	44_{100}	148	5517
PMMA-M (bis-demethyl-) TFA		1670	247_4	140_{15}	134_{54}	107_{100}	77_{15}	366	6335
PMMA-M (bis-demethyl-) 2AC	U+UHYAC	1900	235_1	176_{72}	134_{100}	107_{47}	86_{71}	324	1804
PMMA-M (bis-demethyl-) 2HFB		<1000	330_{48}	303_{15}	240_{100}	169_{44}	69_{42}	1182	6326
PMMA-M (bis-demethyl-) 2PFP		<1000	280_{77}	253_{16}	190_{100}	119_{56}	69_{16}	1104	6325
PMMA-M (bis-demethyl-) 2TFA		<1000	230_{94}	203_{14}	140_{130}	92_{15}	69_{76}	819	6324
PMMA-M (bis-demethyl-) 2TMS		<1000	280_7	179_9	149_8	116_{100}	73_{78}	594	6327
PMMA-M (bis-demethyl-) formyl art.		1220	163_3	148_4	107_{30}	77_{13}	56_{100}	142	6323
PMMA-M (bis-demethyl-methoxy-)	UHY	1465	181_9	138_{100}	122_{18}	94_{24}	77_{16}	178	4351
PMMA-M (bis-demethyl-methoxy-) ME	UHYME	1550	195_1	152_{100}	137_{17}	107_{16}	77_{14}	208	4352
PMMA-M (nor-) AC	U+UHYAC	1720	207_1	148_{41}	121_{16}	91_5	86_{10}	235	3265
PMMA-M (nor-) AC	U+UHYAC	1720	207_1	148_{41}	121_5	86_{10}	44_{100}	234	5537
PMMA-M (O-demethyl-)		1885	150_1	135_1	107_5	77_5	58_{100}	148	1766
PMMA-M (O-demethyl-) TFA		1770	261_1	154_{100}	134_{68}	110_{42}	107_{41}	425	6180
PMMA-M (O-demethyl-) TMSTFA		1690	333_3	206_{72}	179_{100}	154_{53}	73_{50}	773	6228
PMMA-M (O-demethyl-) 2AC	U+UHYAC	1995	249_1	176_6	134_7	100_{43}	58_{100}	376	1767
PMMA-M (O-demethyl-) 2HFB		1670	538_1	330_{17}	254_{100}	210_{32}	169_{22}	1186	5076
PMMA-M (O-demethyl-) 2PFP		1605	295_1	280_{18}	204_{100}	160_{47}	119_{39}	1123	5077
PMMA-M (O-demethyl-) 2TFA		1585	357_1	230_{22}	154_{100}	110_{42}	69_{29}	872	5078
PMMA-M (O-demethyl-) 2TMS		1620	309_{10}	206_{70}	179_{100}	154_{32}	73_{40}	663	6190
PMMA-M (O-demethyl-HO-alkyl-)		1875	148_1	95_1	77_4	71_6	58_{100}	178	1971
PMMA-M (O-demethyl-HO-alkyl-) (erythro) 3AC	U+UHYAC	2145	307_1	247_1	205_1	100_{72}	58_{100}	649	750
PMMA-M (O-demethyl-HO-alkyl-) (threo-) 3AC	U+UHYAC	2160	307_1	123_2	100_{39}	77_2	58_{100}	649	6758
PMMA-M (O-demethyl-HO-alkyl-) ME2AC		2000	279_1	247_1	206_3	100_{60}	58_{100}	513	2348
PMMA-M (O-demethyl-HO-alkyl-) formyl artifact		1790	133_2	121_{10}	107_6	71_{100}	56_{76}	201	4499
PMMA-M (O-demethyl-HO-alkyl-) -H2O 2AC	U+UHYAC	1990	247_9	205_8	163_{24}	107_8	56_{100}	367	1972
PMMA-M (O-demethyl-HO-aryl-) 3AC	U+UHYAC	2190	307_1	234_7	150_{12}	100_{59}	58_{100}	649	4244
PMMA-M (O-demethyl-methoxy-)	UHY	1810	195_1	137_4	122_2	94_3	58_{100}	208	4246
PMMA-M (O-demethyl-methoxy-) 2HFB	UHFB	1760	360_{28}	333_6	254_{100}	210_{23}	169_{18}	1193	6492
PMMA-M (O-demethyl-methoxy-) iso-1 2AC	U+UHYAC	2095	279_1	206_{15}	164_{23}	100_{39}	58_{90}	512	6757
PMMA-M (O-demethyl-methoxy-) iso-2 2AC	U+UHYAC	2115	279_1	206_{19}	164_{26}	100_{48}	58_{100}	512	4243
PMMA-M 2AC	U+UHYAC	2065	265_3	206_{27}	164_{100}	137_{23}	86_{33}	445	3498
PMMA-M 2HFB	UHFB	1690	360_{82}	333_{15}	240_{100}	169_{42}	69_{39}	1189	6512
Polychlorinated biphenyl (3Cl)		1860*	258_{97}	256_{100}	186_{91}	150_{39}	75_{38}	405	2615
Polychlorinated biphenyl (4Cl)		1945*	292_{100}	290_{75}	255_{27}	220_{65}	184_{10}	561	881
Polychlorinated biphenyl (5Cl)		2155*	326_{100}	324_{60}	289_{11}	254_{37}	184_{11}	729	882
Polychlorinated biphenyl (6Cl)		2290*	362_{80}	360_{100}	358_{52}	288_{34}	218_{12}	875	884
Polychlorinated biphenyl (6Cl)		2330*	362_{79}	360_{100}	358_{45}	288_{27}	218_9	875	2633
Polychlorinated biphenyl (7Cl)		2460*	394_{100}	392_{44}	324_{48}	252_{14}	162_{15}	990	885
Polychlorocamphene		2245*	410_1	376_3	341_{22}	195_{44}	89_{99}	1040	880
Polyethylene glycol (PEG 300)	G P U	1300*	283_1	239_3	163_4	133_{19}	89_{100}	527	29
Polyethylene glycol (PEG 300) AC		1300*	263_2	219_9	175_7	131_{11}	87_{100}	433	4639
Polythiazide 3ME	UEXME	3205	481_1	352_{100}	244_{13}	145_4	129_5	1149	3119

Table 1-8-1: Compounds in order of names

Name	Detected	RI	Typical ions and intensities					Page	Entry
Potasan (E838) HY	G UHY	2015*	176_{74}	148_{100}	147_{71}	120_{12}	91_{27}	162	2571
Potasan (E838) HYAC	U+UHYAC	2005*	218_{11}	176_{92}	148_{100}	120_{9}	91_{29}	263	2572
PPP		1595	202_{1}	188_{1}	133_{1}	98_{50}	56_{17}	224	5943
PPP-M (4-HO-)		2010	219_{1}	121_{3}	98_{100}	93_{2}	69_{3}	271	6545
PPP-M (4-HO-) ET		1955	247_{1}	149_{2}	121_{5}	98_{100}	69_{4}	368	6543
PPP-M (4-HO-) HFB		1805	317_{4}	169_{7}	98_{100}	69_{14}		1052	6544
PPP-M (4-HO-) TMS		2005	276_{2}	193_{2}	135_{3}	98_{100}	73_{4}	571	6776
PPP-M (cathinone) AC		1610	191_{2}	134_{2}	105_{35}	86_{100}	77_{48}	195	5901
PPP-M (cathinone) HFB		1395	345_{1}	240_{6}	169_{4}	105_{100}	77_{36}	827	5904
PPP-M (cathinone) PFP		1335	190_{6}	119_{7}	105_{100}	77_{40}	69_{5}	588	5903
PPP-M (cathinone) TFA		1350	245_{1}	140_{7}	105_{100}	77_{48}	69_{10}	358	5902
PPP-M (cathinone) TMS		1590	206_{14}	191_{15}	116_{100}	77_{27}	73_{74}	278	5905
PPP-M (dihydro-)		1680	204_{1}	105_{4}	98_{100}	77_{15}	56_{24}	230	6695
PPP-M (dihydro-) AC		1720	188_{2}	115_{5}	105_{7}	98_{100}	77_{9}	368	6700
PPP-M (dihydro-) TMS		1665	262_{3}	188_{1}	115_{3}	98_{100}	73_{13}	503	6705
PPP-M (norephedrine)	P U	1370	132_{2}	118_{3}	107_{1}	91_{10}	77_{100}	130	2475
PPP-M (norephedrine) 2TMS		1555	280_{5}	163_{4}	147_{10}	116_{100}	73_{83}	594	4574
PPP-M (oxo-)		1820	217_{1}	112_{100}	105_{4}	84_{7}	69_{24}	261	6546
PPP-M TMSTFA		1890	240_{8}	198_{3}	179_{100}	117_{5}	73_{88}	708	6146
PPP-M 2AC	U+UHYAC	1805	235_{1}	107_{13}	86_{100}	176_{5}	134_{7}	325	2476
PPP-M 2HFB	UHYHFB	1455	543_{1}	330_{14}	240_{100}	169_{44}	69_{57}	1183	5098
PPP-M 2PFP	UHYPFP	1380	443_{1}	280_{9}	190_{100}	119_{59}	105_{26}	1104	5094
PPP-M 2TFA	UTFA	1355	343_{1}	230_{6}	140_{100}	69_{29}		819	5091
Prajmaline artifact	G P-I U UHY	2925	368_{1}	340_{4}	224_{100}	196_{26}	126_{28}	914	2711
Prajmaline artifact AC	U+UHYAC	2950	410_{2}	382_{16}	266_{100}	238_{30}	144_{35}	1042	2715
Prajmaline artifact HFB		2545	564_{1}	420_{100}	393_{8}	280_{17}	194_{43}	1188	7580
Prajmaline artifact PFP		2370	514_{2}	486_{13}	370_{100}	342_{49}	279_{73}	1172	7579
Prajmaline artifact TFA		2390	464_{3}	436_{24}	350_{9}	320_{100}	279_{38}	1133	7578
Prajmaline artifact TMS		2690	440_{2}	425_{6}	296_{100}	268_{18}	73_{35}	1099	7577
Prajmaline artifact 2AC	U+UHYAC	3050	452_{6}	409_{4}	393_{13}	308_{100}	126_{24}	1117	7575
Prajmaline artifact 2TMS		2680	512_{2}	368_{66}	296_{48}	198_{100}	73_{84}	1171	7576
Prajmaline-M (HO-) artifact	UHY	3130	384_{12}	313_{7}	224_{100}	196_{15}	126_{15}	968	2713
Prajmaline-M (HO-) artifact 2AC	U+UHYAC	3060	468_{2}	440_{8}	266_{100}	238_{26}	126_{17}	1137	2717
Prajmaline-M (HO-methoxy-) artifact	UHY	3200	414_{13}	343_{7}	224_{100}	196_{18}	126_{12}	1051	2714
Prajmaline-M (HO-methoxy-) artifact 2AC	U+UHYAC	3300	498_{13}	470_{8}	303_{28}	266_{100}	126_{11}	1163	2718
Prajmaline-M (methoxy-) artifact	P-I U UHY	2895	398_{11}	370_{15}	297_{4}	254_{100}	126_{13}	1013	2712
Prajmaline-M (methoxy-) artifact AC	U+UHYAC	2920	440_{7}	398_{4}	340_{19}	296_{100}	126_{22}	1098	2716
Pramipexole		1920	211_{42}	151_{42}	127_{33}	70_{40}	56_{100}	247	7495
Pramipexole 2AC		2550	194_{56}	152_{100}	126_{17}	110_{4}	99_{3}	591	7496
Pramipexole 2HFB		2300	348_{200}	300_{10}	179_{18}	135_{21}		1195	7499
Pramipexole 2PFP		2270	298_{100}	272_{12}	179_{10}	153_{10}	135_{11}	1166	7498
Pramipexole 2TFA		2220	248_{27}	222_{5}	179_{9}	135_{14}	69_{100}	1025	7497
Pramipexole 2TMS		2230	355_{3}	270_{9}	198_{25}	115_{100}	73_{97}	867	7500
Pramiverine		2270	292_{3}	278_{11}	215_{59}	98_{100}	70_{48}	584	2653
Pramiverine AC	U+UHYAC	2705	335_{24}	292_{12}	234_{18}	180_{100}	98_{51}	786	2658
Pratol		2610*	268_{100}	253_{21}	225_{9}	132_{78}	117_{17}	459	5598
Pratol AC		2610*	310_{64}	268_{100}	253_{17}	132_{40}	117_{9}	665	5599
Pratol ME		2600*	282_{100}	267_{15}	150_{15}	132_{64}	89_{14}	526	5600
Prazepam	G P-I U+UHYAC	2650	324_{26}	295_{21}	269_{29}	91_{75}	55_{100}	731	600
Prazepam HY	UHY U+UHYAC	2410	285_{64}	270_{45}	105_{67}	77_{83}	56_{100}	538	302
Prazepam HYAC	U+UHYAC PHYAC	2245	273_{31}	230_{100}	154_{13}	105_{23}	77_{50}	480	273
Prazepam-M	P G U	2520	270_{86}	269_{97}	242_{100}	241_{82}	77_{17}	468	463
Prazepam-M (dealkyl-HO-)	UGLUC	2750	286_{82}	258_{100}	230_{11}	166_{7}	139_{8}	541	2113
Prazepam-M (dealkyl-HO-) AC	U+UHYAC	3000	328_{22}	286_{90}	258_{100}	166_{8}	139_{7}	747	2111
Prazepam-M (dealkyl-HO-) HY	UHY	2400	247_{72}	246_{100}	230_{11}	121_{26}	65_{22}	365	2112
Prazepam-M (dealkyl-HO-) HYAC	U+UHYAC	2270	289_{18}	247_{86}	246_{100}	105_{7}	77_{35}	556	3143
Prazepam-M (dealkyl-HO-) iso-1 HY2AC	U+UHYAC	2560	331_{48}	289_{64}	247_{100}	230_{41}	154_{13}	762	2125
Prazepam-M (dealkyl-HO-) iso-2 HY2AC	U+UHYAC	2610	331_{46}	289_{54}	246_{100}	154_{11}	121_{11}	762	1751
Prazepam-M (dealkyl-HO-methoxy-) HY2AC	U+UHYAC	2700	361_{7}	319_{27}	276_{38}	260_{5}	246_{4}	886	1752
Prazepam-M (HO-) AC	UGLUCAC	2920	382_{18}	340_{40}	311_{99}	257_{100}	55_{42}	959	2512
Prazepam-M (HO-) HYAC	U+UHYAC	2595	343_{8}	283_{11}	257_{100}	241_{28}	228_{11}	820	2513
Prazepam-M TMS		2300	342_{62}	341_{100}	327_{19}	269_{4}	73_{30}	817	4573
Prazepam-M HY	UHY	2050	231_{80}	230_{55}	154_{23}	105_{38}	77_{100}	308	419
PRCC -HCN		<1000	139_{21}	110_{65}	96_{40}	69_{19}	54_{89}	119	3600
PRCC -HCN		<1000	139_{21}	110_{65}	96_{40}	54_{89}	41_{100}	119	5539
Prednisolone	G P U	2800*	300_{18}	122_{99}	91_{13}			885	886
Prednisolone 3AC		3400*	372_{5}	314_{8}	147_{12}	122_{100}		1154	704
Prednisolone acetate		3560*	402_{11}	342_{8}	147_{31}	122_{94}	121_{100}	1024	3296
Prednisone		2610*	358_{22}	256_{10}	160_{30}	121_{100}	91_{90}	878	5256
Prednisone -C2H4O2		2610*	298_{37}	254_{20}	160_{60}	121_{72}	91_{100}	607	5257
Prednylidene		3330*	342_{38}	309_{11}	147_{25}	122_{100}	121_{71}	928	2809
Prednylidene artifact		3100*	312_{7}	159_{10}	122_{100}	91_{34}	77_{17}	673	2810
Pregabaline 2TMS		1995	303_{1}	288_{4}	147_{13}	102_{100}	73_{69}	630	7280
Pregabaline -H2O		1440	141_{24}	111_{14}	98_{11}	84_{43}	56_{100}	120	7276

Table 1-8-1: Compounds in order of names   Procaine

Name	Detected	RI	Typical ions and intensities					Page	Entry
Pregabaline -H2O AC		1500	183_3	142_{28}	126_{76}	124_{89}	84_{100}	182	7277
Pregabaline -H2O PFP		1450	287_{15}	246_{18}	202_{22}	176_{55}	55_{100}	546	7279
Pregabaline -H2O TFA		1520	237_2	196_{36}	126_{51}	83_{79}	69_{100}	333	7278
Pregabaline -H2O TMS		1445	213_7	198_{100}	156_{21}	140_7	102_6	253	7281
Pregnandiol -H2O AC	U+UHYAC	2910*	344_{24}	284_{46}	107_{68}	93_{100}	67_{92}	826	5585
Prenalterol		1990	225_3	210_2	181_7	110_6	72_{100}	292	1857
Prenalterol 3AC		2430	351_1	291_{12}	200_{100}	140_{20}	72_{46}	850	1860
Prenalterol formyl artifact		2040	237_{43}	222_{76}	86_{50}	72_{46}	56_{100}	335	1858
Prenalterol -H2O 2AC		2410	291_{70}	207_{65}	150_{35}	140_{46}	98_{100}	569	1859
Prenylamine	U UHY	2560	329_1	238_{64}	165_{14}	91_{40}	58_{100}	756	1518
Prenylamine AC		2925	371_1	280_{32}	238_{29}	100_{42}	58_{100}	925	1519
Prenylamine-M (AM)		1160	134_1	120_1	91_6	77_1	65_4	115	54
Prenylamine-M (AM)	U	1160	134_1	120_1	91_6	65_4	44_{100}	115	5514
Prenylamine-M (AM) AC	U+UHYAC	1505	177_1	118_{19}	91_{11}	86_{31}	65_5	165	55
Prenylamine-M (AM) AC	UAAC U+UHYAC	1505	177_1	118_{19}	91_{11}	86_{31}	44_{100}	165	5515
Prenylamine-M (AM) HFB		1355	331_1	240_{79}	169_{21}	118_{100}	91_{53}	762	5047
Prenylamine-M (AM) PFP		1330	281_1	190_{73}	118_{100}	91_{36}	65_{12}	521	4379
Prenylamine-M (AM) TFA		1095	231_1	140_{100}	118_{92}	91_{45}	69_{19}	309	4000
Prenylamine-M (AM) TMS		1190	192_6	116_{100}	100_{10}	91_{11}	73_{87}	235	5581
Prenylamine-M (AM) formyl artifact		1100	147_2	146_6	125_5	91_{12}	56_{100}	125	3261
Prenylamine-M (AM)-D5 AC	UAAC U+UHYAC	1480	182_3	122_{46}	92_{30}	90_{100}	66_{16}	181	5690
Prenylamine-M (AM)-D5 HFB		1330	244_{100}	169_{14}	122_{46}	92_{41}	69_{40}	787	6316
Prenylamine-M (AM)-D5 PFP		1320	194_{100}	123_{42}	119_{32}	92_{46}	69_{11}	543	5566
Prenylamine-M (AM)-D5 TFA		1085	144_{100}	123_{53}	122_{56}	92_{51}	69_{28}	330	5570
Prenylamine-M (AM)-D5 TMS		1180	212_1	197_8	120_{100}	92_{11}	73_{57}	251	5582
Prenylamine-M (AM)-D11 TFA		1610	292_1	194_{100}	128_{82}	98_{43}	70_{14}	575	7284
Prenylamine-M (AM)-D11 TFA		1615	242_1	144_{100}	128_{82}	98_{43}	70_{14}	353	7283
Prenylamine-M (deamino-HO-) -H2O	UHY U+UHYAC	1940*	194_{49}	167_{100}	165_{67}	152_{34}	116_{17}	205	3388
Prenylamine-M (HO-) 2AC	U+UHYAC	3200	429_2	338_{46}	296_{77}	100_{14}	58_{100}	1079	3403
Prenylamine-M (HO-methoxy-) 2AC	U+UHYAC	3310	459_4	368_{35}	326_{78}	270_7	58_{100}	1127	3404
Prenylamine-M (N-dealkyl-) AC	U+UHYAC	2320	253_{15}	193_{12}	165_{19}	152_{10}	73_{100}	395	3391
Prenylamine-M (N-dealkyl-HO-) 2AC	U+UHYAC	2635	311_{54}	269_{14}	239_{21}	183_{63}	73_{100}	672	3392
Prenylamine-M (N-dealkyl-HO-methoxy-) 2AC	U+UHYAC	2700	341_{10}	299_{54}	213_{74}	152_{14}	73_{100}	812	3393
Pridinol		2290	295_5	180_7	113_{13}	98_{101}		594	601
Pridinol -H2O	UHY U+UHYAC	2220	277_{24}	163_{31}	110_{100}			503	1285
Pridinol-M (amino-) -H2O AC	U+UHYAC	2250	251_{14}	208_{11}	192_{100}	84_8		385	1286
Pridinol-M (amino-HO-) -H2O 2AC	U+UHYAC	2645	309_{15}	208_{100}				660	1288
Pridinol-M (di-HO-) -H2O 2AC	U+UHYAC	2980	393_{10}	309_{10}	208_{100}			995	1289
Pridinol-M (HO-) -H2O AC	U+UHYAC	2615	335_{71}	292_{18}	209_{52}	110_{100}		785	1287
Prilocaine	G P UHY	1850	220_1	107_4	86_{100}	65_2		274	1216
Prilocaine AC	U+UHYAC	2060	262_5	156_{66}	128_{33}	107_{14}	86_{99}	432	1520
Prilocaine TMS		1850	292_1	235_{18}	207_{24}	86_{100}	73_{55}	577	4589
Prilocaine 2TMS		1910	349_1	235_8	206_{11}	158_{88}	73_{100}	898	4618
Prilocaine artifact	G P U+UHYAC	1840	232_8	217_9	118_{41}	84_{100}	56_{63}	313	4259
Prilocaine-M (deacyl-) AC	U+UHYAC	1350	149_{30}	127_{33}	107_{100}	106_{92}	77_{33}	126	3929
Prilocaine-M (HO-)	UHY	2155	236_1	123_8	86_{100}			331	3934
Prilocaine-M (HO-) 2AC	U+UHYAC	2435	320_1	156_{66}	128_{52}	86_{100}	56_9	716	3932
Prilocaine-M (HO-deacyl-)	UHY	1160	123_{100}	106_{16}	94_{29}	78_{51}		108	3933
Prilocaine-M (HO-deacyl-) 2AC	U+UHYAC	1810	207_4	165_{41}	123_{100}	94_7	77_5	233	3931
Prilocaine-M (HO-deacyl-) 3AC	U+UHYAC	1770	249_2	207_{12}	165_{44}	151_1	123_{100}	374	3930
Primidone	P G U+UHYAC	2260	218_{11}	190_{100}	161_{28}	146_{93}	117_{50}	265	887
Primidone AC	U+UHYAC	2115	260_7	232_{30}	189_{12}	146_{100}	117_{42}	421	889
Primidone 2ME	UME PME	2060	246_7	218_{100}	146_{91}	117_{53}	103_{19}	363	6405
Primidone-M (diamide)	P U+UHYAC	1935	163_{99}	148_{10}	103_{27}	91_{36}		231	888
Primidone-M (HO-methoxy-phenobarbital) 3ME	UME	2300	320_{100}	291_{91}	263_{49}	206_{10}	178_{49}	715	6407
Primidone-M (HO-phenobarbital)	U UHY	2295	248_{70}	220_{61}	219_{100}	148_{55}		370	855
Primidone-M (HO-phenobarbital) AC	U+UHYAC	2360	290_8	248_{100}	219_{54}	148_8	120_5	563	2507
Primidone-M (HO-phenobarbital) 3ME	UME	2200	290_{95}	261_{100}	233_{78}	176_{27}	148_{93}	564	856
Primidone-M (phenobarbital)	P G U+UHYAC	1965	232_{14}	204_{100}	161_{18}	146_{12}	117_{37}	312	854
Primidone-M (phenobarbital) 2ET		1920	288_2	260_{100}	232_{17}	146_{35}	117_{27}	553	2450
Primidone-M (phenobarbital) 2ME	PME UME	1860	260_2	232_{100}	175_{20}	146_{24}	117_{34}	421	1121
Primidone-M (phenobarbital) 2TMS		2015	376_2	361_{34}	261_{15}	146_{100}	73_{46}	939	4582
Probarbital	P G U	1555	169_{12}	156_{93}	141_{100}	98_{22}		215	890
Probarbital 2ME		1485	197_{18}	184_{92}	169_{100}	112_{23}		295	891
Probenecide ET		2220	313_3	284_{100}	213_{29}	149_{31}	103_8	680	3080
Probenecide ME		2205	299_4	270_{100}	199_{51}	135_{69}	76_{17}	611	3079
Probucol		3195*	279_{86}	263_{11}	223_{21}	73_{12}	57_{100}	1173	7531
Probucol artifact AC		2680*	442_1	410_9	238_{37}	223_{67}	57_{100}	1102	7532
Probucol artifact-1		1850*	278_{89}	263_{68}	219_8	207_{15}	57_{100}	504	7530
Probucol artifact-2		2680*	410_{23}	395_5	190_8	162_{10}	57_{100}	1040	7529
Probucol artifact-3		2800*	442_{54}	427_5	237_{11}	178_{18}	57_{100}	1102	7528
Procainamide	P U+UHYAC	2270	235_2	120_{20}	99_{48}	86_{100}		326	893
Procainamide AC	U+UHYAC	2550	277_1	275_1	120_4	86_{100}	58_{15}	503	2896
Procaine	U+UHYAC	2025	221_1	164_5	120_{24}	99_{39}	86_{100}	331	892

Procaine AC Table 1-8-1: Compounds in order of names

Name	Detected	RI	Typical ions and intensities					Page	Entry
Procaine AC	U+UHYAC	2350	278_1	206_2	120_9	99_{23}	86_{100}	508	3297
Procaine-M (PABA) AC	U+UHYAC	2145	179_{31}	137_{100}	120_{92}	92_{16}	65_{24}	169	3298
Procaine-M (PABA) ME		1550	151_{55}	120_{100}	92_{28}	65_{26}		128	23
Procaine-M (PABA) MEAC		1985	193_{33}	151_{60}	120_{100}	92_{16}	65_{19}	199	24
Procaine-M (PABA) 2TMS		1645	281_{50}	236_7	148_{13}	73_{55}		522	5487
Procarterol 2TMS		2295	419_2	335_{100}	100_{62}	73_{30}	58_{28}	1089	6230
Procarterol 3TMS		2390	491_2	407_{76}	100_{100}	73_{90}	58_{83}	1168	6217
Procarterol -H2O AC		2610	314_{19}	272_{18}	247_{26}	100_{100}	58_{26}	686	1861
Prochloraz		2405	310_3	235_{50}	143_{100}	130_9	87_{33}	934	3886
Prochlorperazine	G U UHY U+UHYAC	2970	373_{15}	141_{37}	113_{69}	70_{100}		930	376
Prochlorperazine-M (amino-) AC	U+UHYAC	2990	332_8	233_{19}	100_{100}			767	1255
Prochlorperazine-M (nor-) AC	U+UHYAC	3500	401_{59}	233_{61}	141_{99}	99_{68}		1020	1282
Prochlorperazine-M (ring)	U-I UHY-I U+UHYAC-I	2100	233_{100}	198_{55}				314	311
Procyclidine	P-I	2320	287_1	269_2	204_{13}	84_{100}	55_8	549	602
Procyclidine TMS		2305	344_1	269_1	186_2	84_{70}	73_{10}	882	5453
Procyclidine artifact (dehydro-)		2290	285_3	202_{100}	105_{28}	82_{61}	55_{31}	541	4238
Procyclidine artifact (dehydro-) TMS		2420	357_{24}	272_{40}	182_{100}	115_{42}	73_{69}	875	5454
Procyclidine -H2O		2160	269_{13}	268_{40}	186_{100}	96_{56}	84_{50}	466	4237
Procyclidine-M (amino-HO-) iso-1 -H2O 2AC	U+UHYAC	2560	315_3	255_{20}	196_{100}	168_{35}	155_{22}	690	1290
Procyclidine-M (amino-HO-) iso-2 -H2O 2AC	U+UHYAC	2625	315_8	255_{60}	196_{100}	132_{66}	115_{60}	691	4242
Procyclidine-M (HO-) -H2O	UHY	2360	285_{23}	284_{31}	186_{100}	96_{42}	84_{46}	541	4239
Procyclidine-M (HO-) iso-1 -H2O AC	U+UHYAC	2450	327_{34}	326_{42}	186_{100}	96_{60}	84_{59}	746	1291
Procyclidine-M (HO-) iso-2 -H2O AC	U+UHYAC	2500	327_{27}	326_{35}	186_{100}	96_{46}	84_{63}	746	4241
Procyclidine-M (oxo-) -H2O	UHY U+UHYAC	2490	283_8	200_{100}	130_{17}	115_{61}	86_{55}	530	4240
Procymidone		2065	283_{53}	255_5	124_5	96_{100}	67_{43}	527	3481
Procymidone artifact (deschloro-)		1935	249_{54}	220_3	111_5	96_{100}	67_{41}	374	3482
Profenamine	G P-I U+UHYAC	2335	312_1	213_5	199_{11}	100_{100}		676	1317
Profenamine-M (bis-deethyl-) AC	U+UHYAC	2450	298_{43}	212_{100}	180_{42}	100_7	58_5	605	1319
Profenamine-M (bis-deethyl-HO-) 2AC	U+UHYAC	2900	356_{65}	270_{100}	228_{83}	196_{26}	100_{13}	869	2619
Profenamine-M (deethyl-) AC	U+UHYAC	2515	326_{23}	212_{100}	128_6	72_{93}		741	1318
Profenamine-M (deethyl-HO-) 2AC	U+UHYAC	2880	384_{11}	270_{27}	128_{75}	72_{100}		967	1320
Profenofos		2155*	372_{27}	337_{62}	206_{47}	139_{83}	97_{100}	926	3483
Profluralin		1830	347_{16}	330_{46}	318_{100}	264_{20}	55_{69}	834	3880
Progesterone		2780*	314_{69}	272_{44}	124_{100}			687	894
Proglumetacin artifact ME	PME UME	2130	233_{38}	174_{100}				315	1230
Proglumetacin artifact 2ME	PME UME	2090	247_{38}	188_{100}	173_{11}	145_8		367	6294
Proglumetacin-M/artifact (HO-indometacin) 2ME	UME	2880	401_{27}	262_4	139_{100}	111_{23}		1020	6293
Proglumetacin-M/artifact (HOOC-) ME		2445	348_1	247_2	220_{13}	105_{100}	98_{54}	840	5258
Proglumetacin-M/artifact (indometacin)	G P-I	2550	313_{30}	139_{101}	111_{24}			871	1038
Proglumetacin-M/artifact (indometacin) ET		2820	385_{40}	312_{19}	158_6	139_{100}	111_{20}	969	3168
Proglumetacin-M/artifact (indometacin) ME	P(ME) G(ME) U(ME)	2770	371_9	312_6	139_{100}	111_{17}		922	1039
Proglumetacin-M/artifact (indometacin) TMS		2650	429_{23}	370_4	312_{17}	139_{100}	73_{22}	1078	5462
Proglumetacin-M/artifact -H2O iso-1 AC		1765	212_7	139_{83}	98_{40}	70_{100}	56_{98}	250	5260
Proglumetacin-M/artifact -H2O iso-2 AC		1900	212_{20}	139_{85}	125_{64}	98_{62}	70_{100}	250	5259
Proline MEAC		1465	171_8	128_1	112_{32}	70_{100}	68_7	158	2708
Proline-M (HO-) ME2AC		1690	229_1	198_1	169_9	110_{37}	68_{100}	303	2709
Proline-M (HO-) MEAC		1635	187_7	128_{54}	86_{100}	68_{28}		189	2283
Prolintane	G U UHY U+UHYAC	1720	216_1	174_{100}	126_{100}	91_{30}	65_{12}	262	2729
Prolintane-M (di-HO-phenyl) 2AC	U+UHYAC	2295	332_1	290_2	248_1	126_{100}	123_2	777	4112
Prolintane-M (HO-methoxy-phenyl) AC	U+UHYAC	2215	304_1	262_2	137_2	126_{100}	55_3	641	4109
Prolintane-M (HO-phenyl-)	UHY	2135	232_1	190_3	126_{100}	107_3	96_3	317	4103
Prolintane-M (HO-phenyl-) AC	U+UHYAC	2110	274_1	232_4	190_2	126_{100}	107_6	493	4108
Prolintane-M (oxo-)	U UHY U+UHYAC	1895	231_1	188_2	140_{100}	98_{37}	91_{18}	310	4102
Prolinterol-M (oxo-di-HO-) 2AC	U+UHYAC	2485	347_1	279_7	198_{100}	156_{27}	128_{27}	836	4115
Prolintane-M (oxo-di-HO-methoxy-) 2AC	U+UHYAC	2560	377_5	234_{32}	198_{100}	192_{41}	156_{27}	943	4116
Prolintane-M (oxo-di-HO-phenyl-)		2475	263_2	178_{14}	140_{100}	98_{48}	86_{24}	436	4107
Prolintane-M (oxo-di-HO-phenyl-) 2AC	U+UHYAC	2460	347_1	220_5	178_6	140_{100}	98_{17}	836	4114
Prolintane-M (oxo-di-HO-phenyl-) 2ME	UHYME	2260	291_3	206_{35}	140_{100}	98_{32}	86_{12}	572	4106
Prolintane-M (oxo-HO-alkyl-)		2200	188_2	156_{32}	91_{17}	86_{100}	71_{34}	368	4104
Prolintane-M (oxo-HO-alkyl-) AC	U+UHYAC	2255	198_{67}	156_{58}	138_{100}	91_{32}	86_{81}	560	4110
Prolintane-M (oxo-HO-methoxy-phenyl-)	UHY	2240	277_2	192_{28}	140_{100}	98_{35}	86_{13}	502	4105
Prolintane-M (oxo-HO-methoxy-phenyl-) AC	U+UHYAC	2360	319_2	234_4	192_{18}	140_{100}	86_{10}	710	4113
Prolintane-M (oxo-HO-phenyl-) AC	U+UHYAC	2275	289_1	204_3	140_{100}	98_{24}	86_{10}	560	4111
Prolintane-M (oxo-tri-HO-) 3AC	U+UHYAC	2630	405_2	198_{100}	156_{23}	128_{21}	84_8	1031	4117
Promazine	P G U UHY U+UHYAC	2315	284_{15}	199_{19}	86_{23}	58_{100}		535	377
Promazine-M (bis-nor-) AC	U+UHYAC	2720	298_{56}	212_{39}	198_{100}	180_{31}	100_{87}	605	2076
Promazine-M (bis-nor-HO-) 2AC	U+UHYAC	3100	356_{88}	215_{65}	214_{96}	100_{100}	72_{19}	869	2677
Promazine-M (HO-)	UHY	2685	300_9	254_3	215_7	86_{16}	58_{100}	615	605
Promazine-M (HO-) AC	U+UHYAC	2710	342_{38}	257_{25}	215_{48}	86_{47}	58_{100}	817	378
Promazine-M (nor-)	UHY	2405	270_{65}	238_{26}	213_{35}	199_{99}	44_{100}	470	604
Promazine-M (nor-) AC	U+UHYAC	2805	312_{32}	198_{19}	180_{12}	114_{100}		675	379
Promazine-M (nor-HO-) 2AC	U+UHYAC	3195	370_{21}	328_4	214_{29}	114_{100}	86_{18}	919	380
Promazine-M (ring)	P G U UHY U+UHYAC	2010	199_{100}	167_{45}				216	10
Promazine-M (sulfoxide)	U	2705	300_{10}	284_9	212_{53}	58_{100}		615	603

Table 1-8-1: Compounds in order of names   Propofol AC

Name	Detected	RI	Typical ions and intensities					Page	Entry
Promazine-M AC	U+UHYAC	2550	257_{23}	215_{100}	183_7			409	12
Promazine-M 2AC	U+UHYAC	2865	315_{54}	273_{34}	231_{100}	202_{11}		688	2618
Promecarb		1665	207_1	150_{69}	135_{100}	91_{16}	58_{10}	235	3484
Promecarb-M/artifact (decarbamoyl-)		1290*	150_{67}	135_{100}	107_{25}	91_{42}	77_{18}	128	3485
Promethazine	P G U+UHYAC	2270	284_2	213_4	198_5	180_4	72_{100}	535	381
Promethazine-M (bis-nor-) AC	U+UHYAC	2450	298_{43}	212_{100}	180_{42}	100_7	58_5	605	1319
Promethazine-M (bis-nor-HO-) 2AC	U+UHYAC	2900	356_{65}	270_{100}	228_{83}	196_{28}	100_{13}	869	2619
Promethazine-M (di-HO-) 2AC	U+UHYAC	3075	400_7	329_{10}	244_3	230_3	72_{100}	1018	2621
Promethazine-M (HO-)	UHY	2590	300_5	229_4	214_3	196_3	72_{100}	616	609
Promethazine-M (HO-) AC	U+UHYAC	2690	342_{13}	271_{12}	214_5	196_6	72_{100}	817	383
Promethazine-M (HO-methoxy-) AC	U+UHYAC	2800	372_6	301_5	253_6	226_2	72_{100}	927	2617
Promethazine-M (nor-)	P UHY	2250	270_3	213_{100}	198_{21}	180_{16}	58_{52}	470	607
Promethazine-M (nor-) AC	U+UHYAC	2540	312_{52}	212_{100}	180_{73}	114_{71}	58_{48}	675	382
Promethazine-M (nor-di-HO-) 3AC	U+UHYAC	3360	428_{24}	328_{41}	244_{19}	114_{100}	58_{69}	1076	3334
Promethazine-M (nor-HO-)	UHY	2580	286_7	229_{30}	212_{100}	180_{34}	58_{46}	543	608
Promethazine-M (nor-HO-) AC	U+UHYAC	2960	328_{56}	228_{100}	196_{42}	114_{55}	58_{82}	749	2620
Promethazine-M (nor-HO-) 2AC	U+UHYAC	3015	370_{44}	270_{100}	228_{79}	114_{63}		919	384
Promethazine-M (nor-sulfoxide) AC	U+UHYAC	2810	328_8	312_{10}	212_{100}	100_{23}	58_{85}	749	610
Promethazine-M (ring)	P G U UHY U+UHYAC	2010	199_{100}	167_{45}				216	10
Promethazine-M AC	U+UHYAC	2550	257_{23}	215_{100}	183_7			409	12
Promethazine-M 2AC	U+UHYAC	2865	315_{54}	273_{34}	231_{100}	202_{11}		688	2618
Promethazine-M/artifact (sulfoxide)	G P U+UHYAC	2710	300_2	284_4	213_{18}	72_{100}		616	606
Prometryn		1930	241_{100}	226_{58}	184_{97}	106_{31}	58_{70}	348	3862
Propachlor		1600	211_6	196_7	176_{29}	120_{100}	77_{44}	246	3486
Propafenone	P-I G	2740	312_2	297_4	121_4	91_9	72_{100}	814	2391
Propafenone 2AC	U+UHYAC	2980	425_3	322_{63}	200_{71}	140_{100}	72_{92}	1071	2259
Propafenone 2TMS		2860	485_1	283_2	144_{100}	91_{12}	73_{51}	1152	4590
Propafenone 3TMS		2840	557_1	370_9	206_3	144_{100}	73_{66}	1186	4591
Propafenone artifact	P-I G	2760	353_{10}	324_{16}	128_{100}	98_{69}	91_{77}	860	895
Propafenone -H2O	G UHY	2300	323_{14}	294_{22}	230_{20}	98_{92}	91_{100}	729	897
Propafenone -H2O AC	U+UHYAC	2930	365_7	322_{18}	140_{82}	98_{100}	91_{87}	902	902
Propafenone-M (deamino-di-HO-) 3AC	U+UHYAC	2950*	442_6	224_3	159_{100}	137_{13}	91_{26}	1103	903
Propafenone-M (deamino-HO-) 2AC	U+UHYAC	2715*	384_{26}	159_{100}	121_{78}	91_{56}		968	901
Propafenone-M (HO-) -H2O	UHY	2720	339_{21}	310_{14}	98_{100}	91_{72}		804	898
Propafenone-M (HO-) -H2O 2AC	U+UHYAC	3050	423_{10}	282_{61}	140_{57}	98_{100}	72_{83}	1066	904
Propafenone-M (O-dealkyl-)	G P-I U+UHYAC	1830*	226_{35}	207_{14}	121_{100}	91_{28}	65_{32}	294	896
Propafenone-M (O-dealkyl-) AC	U+UHYAC	2130*	268_6	225_{38}	208_{32}	121_{100}	91_{15}	460	3726
Propafenone-M (O-dealkyl-HO-) iso-1	UHY	2345*	242_{35}	223_{22}	121_{100}	107_{67}	65_{26}	351	3344
Propafenone-M (O-dealkyl-HO-) iso-1 AC	U+UHYAC	2215*	284_{30}	242_{96}	137_{100}	91_{79}		533	899
Propafenone-M (O-dealkyl-HO-) iso-2	UHY	2355*	242_{43}	223_{27}	121_{100}	107_{78}	65_{30}	351	3345
Propafenone-M (O-dealkyl-HO-) iso-2 AC	U+UHYAC	2370*	284_{17}	242_{26}	224_{38}	121_{100}	65_{17}	533	3350
Propafenone-M (O-dealkyl-HO-methoxy-)	UHY	2400*	272_{36}	151_9	137_{100}	121_{52}	65_{31}	479	3346
Propafenone-M (O-dealkyl-HO-methoxy-) AC	U+UHYAC	2580*	314_{19}	272_{89}	167_{99}	137_{16}	91_{52}	685	900
Propallylonal	P G U UHY U+UHYAC	1875	209_{84}	167_{100}	124_{26}			550	921
Propallylonal 2ME	PME	1745	237_{99}	195_{100}	138_{43}			692	923
Propallylonal-M (desbromo-)	P G U UHY U+UHYAC	1610	210_6	195_{18}	167_{100}	124_{43}		245	63
Propallylonal-M (desbromo-) 2TMS		1620	354_3	339_{65}	297_{32}	100_{37}	73_{80}	862	5458
Propallylonal-M (desbromo-dihydro-HO-) 2ME		----	241_5	214_{21}	198_{53}	183_{91}	169_{100}	408	924
Propallylonal-M (desbromo-HO-)	U UHY U+UHYAC	1770	226_5	184_{99}	169_{99}	141_{63}		294	922
Propallylonal-M (desbromo-oxo-) 2ME		1720	239_2	212_{29}	197_{58}	169_{100}	112_{39}	399	925
Propamocarb		1875	188_2	143_2	129_3	72_3	58_{100}	191	2730
Propamocarb TFA		1290	284_1	225_1	126_1	69_9	58_{100}	535	4135
1,2-Propane diol		<1000*	76_1	61_2	45_{29}	40_6	32_{100}	95	6454
1,2-Propane diol dibenzoate		2240*	284_1	227_{10}	162_{20}	105_{100}	77_{90}	532	1760
1,2-Propane diol dipivalate	PPIV	1350*	143_8	127_2	103_{10}	85_{27}	57_{100}	357	6423
1,2-Propane diol phenylboronate		1240*	162_{67}	147_{10}	118_{22}	104_{47}	91_{48}	141	1898
1,3-Propane diol dibenzoate		2300*	284_1	227_4	162_4	105_{100}	77_{35}	532	1761
1,3-Propane diol dipivalate		1420*	143_{15}	103_{22}	85_{39}	57_{100}		357	1905
1,3-Propane diol phenylboronate		1370*	162_{93}	132_5	104_{100}	91_{27}	77_{22}	141	1899
1-Propanol		<1000*	60_8	59_{15}	42_{15}	31_{100}		91	6456
Propazine		1740	229_{61}	214_{100}	187_{33}	172_{73}	58_{36}	304	2398
Propetamphos		1780	281_2	236_{24}	194_{37}	138_{100}	110_{35}	521	2518
Propetamphos-M/artifact (HOOC-) ME		1675	253_1	208_{24}	138_{100}	122_{17}	110_{80}	393	7539
Propham		1430	179_{50}	137_{46}	120_{34}	93_{100}	65_{26}	171	3487
Propiconazole		2330	340_1	259_{60}	191_{23}	173_{83}	69_{100}	810	3488
Propiconazole artifact (dichlorophenylethanone)	U+UHYAC	1280*	188_{16}	173_{100}	145_{19}	109_{11}	75_{14}	190	3489
Propionic acid anhydride		<1000*	79_2	57_{100}	44_6			112	2757
Propiophenone		<1000*	134_{19}	105_{100}	77_{60}	74_3	51_{24}	114	7282
Propivan		1840	277_1	205_2	99_{13}	86_{40}	58_{14}	504	1523
Propiverine		2460	367_1	309_{31}	225_{85}	183_{100}	105_{65}	911	6080
Propiverine-M/artifact (carbinol)		2430	325_{18}	183_{100}	105_{89}	98_{79}	77_{71}	736	6081
Propiverine-M/artifact (carbinol) AC		2455	367_3	183_{21}	165_8	98_{58}	96_{100}	911	6082
Propofol	G P U	1320*	178_{24}	163_{100}	121_{15}	117_{17}	91_{13}	169	3305
Propofol AC		1510*	220_9	178_{49}	163_{100}	135_9	91_{17}	274	3306

Table 1-8-1: Compounds in order of names

Name	Detected	RI	Typical ions and intensities					Page	Entry
Propofol ME		1290*	192_{38}	177_{100}	149_{22}	119_{36}	91_{31}	198	3521
Propofol TMS		1305*	250_{33}	235_{100}	219_{6}	161_{6}	73_{86}	382	6874
Propoxur	G P U	1585	209_{1}	152_{17}	110_{100}	81_{7}		241	926
Propoxur TFA		1530	305_{8}	263_{13}	206_{62}	109_{47}	69_{100}	638	4130
Propoxur HYAC	U+UHYAC	1390*	194_{2}	152_{21}	110_{100}	81_{5}	52_{10}	205	1223
Propoxur HYME	UHYME	1380*	166_{5}	151_{10}	110_{100}	81_{15}	64_{43}	151	2536
Propoxur impurity-M (HO-)	UHY	1440*	186_{10}	146_{32}	144_{100}	79_{30}		187	2537
Propoxur impurity-M (HO-) AC	U+UHYAC	1520*	228_{3}	186_{22}	146_{47}	144_{100}	79_{10}	299	1225
Propoxur impurity-M (HO-) ME	UHYME	1530*	200_{16}	185_{20}	144_{100}	98_{24}	63_{35}	219	2540
Propoxur impurity-M (O-dealkyl-HO-)	UHY	1490*	146_{34}	144_{100}	115_{8}	98_{27}	63_{43}	122	2539
Propoxur-M (HO-) HY	UHY	1470*	168_{16}	126_{100}	108_{3}	97_{11}		155	2538
Propoxur-M (HO-) HY2AC	U+UHYAC	1680*	252_{8}	210_{17}	168_{42}	126_{100}	97_{11}	390	1224
Propoxur-M (O-dealkyl-) HY	UHY	<1000*	110_{100}	92_{11}	81_{15}	64_{42}	53_{15}	103	2535
Propoxur-M/artifact (isopropoxyphenol)	P G U	1070*	152_{13}	137_{1}	110_{100}	81_{7}	64_{7}	132	2632
Propoxyphene	G P	2205	250_{2}	193_{3}	178_{2}	91_{15}	58_{100}	805	476
Propoxyphene artifact		1755*	208_{56}	193_{41}	130_{38}	115_{100}	91_{42}	238	477
Propoxyphene-M (HY)	UHY	2395	281_{9}	190_{76}	119_{96}	105_{100}	56_{96}	520	480
Propoxyphene-M (nor-) -H2O	UHY	2240	251_{30}	217_{95}	119_{99}			387	479
Propoxyphene-M (nor-) -H2O AC	U+UHYAC	2365	293_{18}	220_{99}	205_{38}			583	232
Propoxyphene-M (nor-) -H2O N-prop.	U UHY U+UHYAC	2555	307_{8}	234_{75}	105_{100}	100_{74}	91_{67}	651	231
Propoxyphene-M (nor-) N-prop.	P U	2400	307_{16}	220_{68}	100_{100}	57_{83}		736	478
Propranolol	P-I G U UHY	2160	259_{2}	215_{4}	144_{11}	115_{12}	72_{100}	417	927
Propranolol TMSTFA		2320	427_{2}	284_{60}	242_{12}	129_{44}	73_{24}	1074	6154
Propranolol 2AC	U+UHYAC	2605	343_{5}	283_{14}	200_{100}	140_{80}	98_{81}	822	931
Propranolol formyl artifact	P G U	2205	271_{65}	256_{41}	183_{24}	127_{100}	112_{44}	474	3413
Propranolol -H2O	UHY	2220	241_{5}	98_{45}	56_{100}			349	930
Propranolol -H2O AC	U+UHYAC	2330	283_{24}	198_{52}	140_{100}	127_{81}	98_{87}	530	935
Propranolol-M (1-naphthol)		1500*	144_{75}	115_{100}	89_{17}	74_{7}	63_{21}	123	928
Propranolol-M (1-naphthol) AC	U+UHYAC	1555*	186_{13}	144_{100}	115_{47}	89_{8}	63_{7}	188	932
Propranolol-M (1-naphthol) HFB		1310*	340_{46}	169_{25}	143_{28}	115_{100}	89_{11}	806	7476
Propranolol-M (1-naphthol) PFP		1510*	290_{45}	171_{100}	143_{20}	115_{49}	89_{8}	562	7468
Propranolol-M (1-naphthol) TMS		1525*	216_{100}	201_{95}	185_{51}	115_{39}	73_{21}	260	7460
Propranolol-M (4-HO-1-naphthol) 2AC	U+UHYAC	1900*	244_{14}	202_{19}	160_{100}	131_{21}	103_{8}	355	933
Propranolol-M (deamino-di-HO-) 3AC	U+UHYAC	2565*	360_{3}	318_{1}	159_{100}			884	936
Propranolol-M (deamino-HO-)	UHY	2065*	218_{21}	144_{100}	115_{37}			264	929
Propranolol-M (deamino-HO-) 2AC	U+UHYAC	2195*	302_{11}	159_{100}	144_{48}	115_{44}		623	934
Propranolol-M (HO-) 3AC	U+UHYAC	2940	401_{1}	341_{12}	186_{66}	140_{100}	98_{75}	1021	939
Propranolol-M (HO-) -H2O iso-1 2AC	U+UHYAC	2750	341_{9}	197_{20}	140_{100}	98_{55}		813	937
Propranolol-M (HO-) -H2O iso-2 2AC	U+UHYAC	2900	341_{12}	197_{27}	140_{100}	98_{75}		813	938
Propylamine		<1000	59_{7}	42_{27}	30_{100}			91	3616
Propylbenzene		<1000*	120_{20}	105_{4}	91_{100}	65_{17}		107	3786
4-Propyl-2,5-dimethoxyphenethylamine		1720	223_{21}	194_{100}	179_{13}	165_{51}	135_{15}	286	6906
4-Propyl-2,5-dimethoxyphenethylamine AC		2090	265_{14}	206_{100}	193_{28}	177_{60}	135_{13}	446	6920
4-Propyl-2,5-dimethoxyphenethylamine HFB		1895	419_{49}	206_{76}	193_{100}	177_{42}	163_{13}	1061	6940
4-Propyl-2,5-dimethoxyphenethylamine PFP		1865	369_{54}	206_{77}	193_{100}	177_{42}	119_{15}	915	6935
4-Propyl-2,5-dimethoxyphenethylamine TFA		1870	319_{36}	206_{69}	193_{100}	177_{39}	149_{38}	708	6930
4-Propyl-2,5-dimethoxyphenethylamine TMS		1860	295_{5}	265_{5}	194_{20}	102_{100}	73_{64}	595	6922
4-Propyl-2,5-dimethoxyphenethylamine 2AC		2160	307_{20}	206_{100}	193_{24}	177_{42}	135_{13}	650	6921
4-Propyl-2,5-dimethoxyphenethylamine 2TMS		2130	367_{1}	352_{5}	174_{100}	100_{11}	86_{28}	912	6923
4-Propyl-2,5-dimethoxyphenethylamine formyl artifact		1755	235_{23}	204_{100}	193_{62}	163_{9}	135_{11}	325	6908
Propylene glycol dipivalate	PPIV	1350*	143_{8}	127_{2}	103_{10}	85_{27}	57_{100}	357	6423
Propylhexedrine	U UHY	1170	155_{1}	140_{3}	58_{100}			135	940
Propylhexedrine AC	U+UHYAC	1570	197_{12}	182_{52}	140_{16}	100_{100}	58_{67}	213	942
Propylhexedrine HFB	UHFB UHYHFB	1440	351_{1}	254_{100}	210_{25}	182_{53}	69_{45}	848	5100
Propylhexedrine PFP	UPFP UHYPFP	1385	301_{1}	204_{100}	182_{31}	160_{24}	119_{18}	620	5095
Propylhexedrine TFA	UTFA	1385	251_{1}	182_{27}	154_{100}	110_{22}	69_{16}	386	5093
Propylhexedrine-M (HO-)	U UHY	1475	171_{3}	156_{5}	58_{100}			159	941
Propylhexedrine-M (HO-) 2AC	U+UHYAC	1915	255_{2}	240_{3}	195_{3}	100_{64}	58_{100}	404	943
Propylparaben	U UHY	1630*	180_{11}	138_{67}	121_{100}	93_{12}	65_{14}	175	2971
Propylparaben AC	U+UHYAC	1610*	222_{8}	180_{32}	138_{100}	121_{84}	93_{9}	281	2972
4-Propylthio-2,5-dimethoxyphenethylamine		2470	255_{26}	226_{100}	183_{63}	169_{31}	153_{34}	403	6855
4-Propylthio-2,5-dimethoxyphenethylamine AC		2410	297_{22}	238_{100}	225_{30}	183_{15}	153_{14}	603	6858
4-Propylthio-2,5-dimethoxyphenethylamine HFB		2175	451_{14}	238_{24}	225_{100}	181_{23}	153_{21}	1114	6861
4-Propylthio-2,5-dimethoxyphenethylamine PFP		2160	401_{23}	238_{17}	225_{100}	181_{17}	153_{19}	1020	6862
4-Propylthio-2,5-dimethoxyphenethylamine TFA		2170	351_{26}	238_{17}	225_{100}	181_{24}	153_{23}	847	6863
4-Propylthio-2,5-dimethoxyphenethylamine 2AC		2470	339_{14}	238_{100}	225_{30}	181_{22}	153_{17}	803	6859
4-Propylthio-2,5-dimethoxyphenethylamine 2TMS		2395	399_{1}	384_{4}	369_{4}	225_{7}	174_{100}	1016	6860
4-Propylthio-2,5-dimethoxyphenethylamine deuteroformyl artifact		2060	269_{22}	238_{25}	225_{100}	183_{13}	153_{24}	466	6857
4-Propylthio-2,5-dimethoxyphenethylamine formyl artifact		2050	267_{35}	236_{22}	225_{100}	183_{26}	153_{46}	455	6856
4-Propylthio-2,5-dimethoxyphenethylamine-M (deamino-HO-)	UGLUC	2000*	256_{69}	225_{100}	183_{23}	150_{56}	135_{23}	407	6864
4-Propylthio-2,5-dimethoxyphenethylamine-M (deamino-HO-) AC	U+UHYAC	2080*	298_{73}	238_{100}	181_{45}	147_{22}		606	6869
4-Propylthio-2,5-dimethoxyphenethylamine-M (deamino-HOOC-)		2110*	270_{100}	225_{55}	213_{46}	181_{34}	153_{21}	469	6872
4-Propylthio-2,5-dimethoxyphenethylamine-M (deamino-HOOC-) ME		1950*	284_{100}	227_{50}	225_{74}	183_{24}	153_{25}	534	6873
4-Propylthio-2,5-dimethoxyphenethylamine-M (deamino-oxo-)		2190*	254_{42}	225_{100}	183_{14}	153_{24}	137_{8}	399	7235

Table 1-8-1: Compounds in order of names

Name	Detected	RI	Typical ions and intensities					Page	Entry
4-Propylthio-2,5-dimethoxyphenethylamine-M (HO- N-acetyl-)	UGLUC	2525	313_{40}	254_{100}	242_{44}	210_{38}	183_{21}	680	6866
4-Propylthio-2,5-dimethoxyphenethylamine-M (HO- N-acetyl-) TFA		2345	409_{27}	350_{100}	337_{9}	236_{5}	181_{13}	1038	6871
4-Propylthio-2,5-dimethoxyphenethylamine-M (HO- sulfone N-acetyl-)	UGLUC	2740	286_{73}	345_{31}	164_{100}	151_{27}	120_{18}	828	6865
4-Propylthio-2,5-dimethoxyphenethylamine-M (HO- sulfone) 2AC	U+UHYAC	2760	387_{31}	340_{42}	328_{100}	268_{36}	108_{33}	976	6868
4-Propylthio-2,5-dimethoxyphenethylamine-M (HO-) 2AC	U+UHYAC	2585	355_{51}	296_{72}	283_{10}	236_{92}	101_{100}	865	6867
4-Propylthio-2,5-dimethoxyphenethylamine-M (HO-) 2TFA		2110	463_{80}	434_{43}	350_{60}	337_{100}	231_{67}	1130	6870
4-Propylthio-2,5-dimethoxyphenethylamine-M (HO-) 3AC	U+UHYAC	2630	397_{69}	296_{99}	283_{12}	236_{100}	101_{64}	1010	6875
4-Propylthio-2,5-dimethoxyphenethylamine-M (S-depropyl-) AC	U+UHYAC	2170	255_{18}	196_{100}	183_{41}	181_{34}	153_{21}	401	6831
4-Propylthio-2,5-dimethoxyphenethylamine-M (S-depropyl-) iso-1 2AC		2240	297_{29}	210_{14}	196_{100}	183_{35}	181_{29}	602	6823
4-Propylthio-2,5-dimethoxyphenethylamine-M (S-depropyl-) iso-2 2AC	U+UHYAC	2360	297_{16}	255_{11}	238_{20}	196_{100}	183_{37}	602	6826
4-Propylthio-2,5-dimethoxyphenethylamine-M (S-depropyl-methyl- N-acetyl-)	U+UHYAC	2230	269_{19}	210_{100}	197_{35}	195_{21}	167_{27}	465	6832
4-Propylthio-2,5-dimethoxyphenethylamine-M (S-depropyl-methyl- sulfone) AC	U+UHYAC	2580	301_{7}	242_{100}	230_{4}	196_{7}	124_{7}	618	6829
4-Propylthio-2,5-dimethoxyphenethylamine-M (S-depropyl-methyl- sulfoxide) AC	U+UHYAC	2460	285_{16}	268_{23}	226_{33}	211_{100}	197_{31}	538	6830
Propyphenazone	G P U+UHYAC	1910	230_{41}	215_{100}	56_{65}			308	202
Propyphenazone-M (di-HO-) 2AC	U+UHYAC	2680	346_{26}	303_{28}	273_{100}	231_{44}	56_{62}	832	2594
Propyphenazone-M (HO-methyl-)	UHY	2410	246_{51}	231_{100}	215_{9}	77_{16}		364	912
Propyphenazone-M (HO-methyl-) AC	U+UHYAC	2240	288_{62}	273_{82}	245_{99}	232_{94}	190_{39}	553	206
Propyphenazone-M (HOOC-) ME	UME	2160	274_{31}	215_{100}	56_{50}			486	917
Propyphenazone-M (HO-phenyl-)	UHY	2300	246_{56}	231_{100}	96_{39}	56_{100}		364	911
Propyphenazone-M (HO-phenyl-) AC	U+UHYAC	2530	288_{45}	273_{43}	246_{46}	231_{100}	56_{84}	554	208
Propyphenazone-M (HO-phenyl-) ME	UME	2310	260_{36}	245_{64}	122_{26}	96_{22}	56_{100}	423	915
Propyphenazone-M (HO-propyl-)	P U UHY	2210	246_{24}	231_{12}	215_{100}	124_{28}	56_{75}	364	910
Propyphenazone-M (HO-propyl-) AC	U+UHYAC	2305	288_{23}	245_{19}	228_{24}	215_{100}	56_{51}	554	207
Propyphenazone-M (isopropanolyl-)	UGLUC	2020	246_{13}	231_{100}	213_{13}			364	913
Propyphenazone-M (isopropenyl-)	P U UHY	1970	228_{38}	136_{98}	95_{50}			302	907
Propyphenazone-M (nor-)	P G U UHY	1765	216_{52}	174_{100}	77_{68}			260	905
Propyphenazone-M (nor-) AC	U+UHYAC	1820	258_{6}	216_{35}	201_{100}	185_{6}	77_{17}	414	203
Propyphenazone-M (nor-) ME	UME	1735	230_{30}	215_{100}	200_{20}	185_{17}	75_{55}	308	914
Propyphenazone-M (nor-) TMS		1860	288_{18}	273_{100}	185_{10}	77_{14}	73_{17}	554	4620
Propyphenazone-M (nor-di-HO-)	UHY	2090	248_{100}	206_{19}	136_{19}	109_{48}		371	909
Propyphenazone-M (nor-di-HO-) AC	U+UHYAC	2250	290_{31}	248_{100}	206_{21}	136_{9}	109_{33}	564	1882
Propyphenazone-M (nor-di-HO-) 2AC	U+UHYAC	2400	332_{26}	290_{100}	274_{10}	232_{16}	206_{26}	769	2593
Propyphenazone-M (nor-di-HO-) 3ME	UHYME	2240	290_{54}	275_{100}	260_{40}			565	3768
Propyphenazone-M (nor-HO-)	UHY	1780	232_{93}	190_{48}	121_{35}	93_{99}	77_{63}	312	906
Propyphenazone-M (nor-HO-) AC	U+UHYAC	1895	274_{97}	232_{4}	214_{25}	190_{100}		486	204
Propyphenazone-M (nor-HO-) AC	U+UHYAC	2190	274_{19}	232_{100}	190_{69}	121_{10}	93_{14}	486	2595
Propyphenazone-M (nor-HO-phenyl-)	UHY	2080	232_{81}	190_{100}	121_{51}	93_{68}	65_{60}	313	908
Propyphenazone-M (nor-HO-phenyl-) 2AC	U+UHYAC	2165	316_{11}	274_{48}	259_{12}	232_{82}	217_{100}	693	205
Propyphenazone-M (nor-HO-phenyl-) iso-1 2ME	UME	2030	260_{45}	245_{100}	230_{50}	215_{28}		423	916
Propyphenazone-M (nor-HO-phenyl-) iso-2 2ME		2060	260_{44}	245_{100}	230_{48}	215_{13}	77_{35}	423	3767
Propyphenazone-M (nor-HO-propyl-) 2AC	U+UHYAC	2120	316_{11}	274_{63}	214_{100}	201_{97}	77_{60}	693	1933
Propyzamide		1790	255_{26}	173_{100}	145_{34}	109_{19}	84_{16}	401	3490
Propyzamide artifact (deschloro-)		1645	221_{24}	206_{7}	139_{100}	111_{29}	75_{16}	275	3491
Proquazone		2670	278_{89}	235_{100}	221_{57}	77_{40}		507	944
Prothiofos		2190*	344_{1}	309_{58}	267_{63}	113_{100}	63_{92}	822	3492
Prothipendyl	P G U+UHYAC	2350	285_{20}	227_{13}	200_{21}	86_{20}	58_{100}	539	385
Prothipendyl-M (bis-nor-) AC	U U+UHYAC	2830	299_{45}	227_{20}	213_{25}	200_{100}	100_{39}	610	387
Prothipendyl-M (bis-nor-HO-) 2AC	U+UHYAC	3030	357_{58}	315_{22}	258_{22}	216_{100}	100_{40}	872	1883
Prothipendyl-M (HO-)	U UHY	2720	301_{16}	230_{7}	216_{11}	86_{18}	58_{100}	619	612
Prothipendyl-M (HO-) AC	U+UHYAC	2780	343_{24}	230_{13}	216_{8}	86_{17}	58_{100}	821	388
Prothipendyl-M (HO-ring)	U UHY	2800	216_{100}	187_{68}	168_{50}	140_{9}		259	2272
Prothipendyl-M (HO-ring) AC	U+UHYAC	2575	258_{30}	216_{100}	187_{13}	183_{13}	155_{10}	412	2275
Prothipendyl-M (nor-) AC	U+UHYAC	2880	313_{34}	227_{28}	200_{29}	181_{19}	114_{100}	679	389
Prothipendyl-M (nor-HO-) 2AC	U+UHYAC	3070	371_{14}	258_{9}	216_{26}	114_{100}	86_{20}	923	390
Prothipendyl-M (ring)	U+UHYAC	2045	200_{100}	168_{36}	156_{13}			219	386
Prothipendyl-M (sulfoxide)	P U	2750	301_{2}	285_{3}	216_{9}	86_{13}	58_{100}	619	611
Protocatechuic acid ME2AC		1750*	252_{1}	210_{19}	168_{100}	137_{60}	109_{8}	388	5254
Protopine		2730	353_{1}	190_{7}	163_{20}	148_{100}	89_{27}	858	5776
Protopine-M (demethylene-methyl-) iso-1		2990	355_{4}	190_{5}	165_{14}	148_{100}	136_{5}	865	6738
Protopine-M (demethylene-methyl-) iso-1 AC		3050	397_{2}	312_{4}	190_{10}	165_{13}	148_{100}	1010	6740
Protopine-M (demethylene-methyl-) iso-2		3010	355_{8}	190_{8}	165_{21}	148_{100}	136_{36}	865	6739
Protopine-M (demethylene-methyl-) iso-2 AC		3070	397_{2}	190_{9}	165_{10}	148_{100}		1010	6741
Protriptyline	G UHY	2250	263_{2}	191_{100}	84_{6}	70_{60}		436	613
Protriptyline AC	U+UHYAC	2690	305_{11}	191_{100}	114_{15}			641	391
Protriptyline TMS	G UHY	2350	335_{1}	320_{2}	191_{23}	142_{42}	116_{100}	785	5455
Protriptyline-M (HO-) 2AC	U+UHYAC	2895	363_{11}	321_{6}	249_{39}	207_{100}	114_{15}	895	393
Protriptyline-M (nor-) AC	U+UHYAC	2780	291_{48}	218_{45}	191_{100}	100_{25}	86_{20}	570	392
Proxyphylline		2080	238_{62}	194_{99}	180_{90}	137_{23}	109_{67}	336	945
Proxyphylline AC	U+UHYAC	2180	280_{100}	237_{37}	220_{69}	193_{46}	180_{65}	516	946
Proxyphylline TMS		2080	310_{13}	295_{30}	180_{29}	117_{55}	73_{100}	667	4592
Proxyphylline-M (HO-) 2AC	U+UHYAC	2455	338_{52}	236_{14}	194_{31}	180_{100}	159_{26}	797	1433
Pseudoephedrine	G P U	1385	146_{1}	105_{4}	91_{6}	77_{9}	58_{100}	148	2473
Pseudoephedrine TMSTFA		1460	213_{1}	191_{7}	179_{100}	140_{4}	73_{80}	773	6155
Pseudoephedrine 2AC	U+UHYAC	1820	189_{1}	148_{1}	117_{2}	100_{45}	58_{100}	376	2474

Pseudoephedrine 2PFP   Table 1-8-1: Compounds in order of names

Name	Detected	RI	Typical ions and intensities					Page	Entry
Pseudoephedrine 2PFP		1430	438 $_1$	338 $_1$	294 $_4$	204 $_{100}$	160 $_{27}$	1123	2578
Pseudoephedrine 2TFA		1440	338 $_1$	244 $_3$	154 $_{100}$	110 $_{61}$	69 $_{48}$	872	4016
Pseudoephedrine 2TMS		1605	294 $_2$	163 $_5$	149 $_6$	130 $_{100}$	73 $_{46}$	663	4593
Pseudoephedrine formyl artifact	G	1300	162 $_1$	117 $_3$	91 $_7$	71 $_{100}$	56 $_{30}$	166	4653
Pseudotropine AC		1230	183 $_{15}$	140 $_5$	124 $_{100}$	94 $_{59}$	82 $_{92}$	183	5435
Pseudotropine benzoate		2040	245 $_{13}$	124 $_{100}$	94 $_{44}$	82 $_{79}$	77 $_{42}$	360	5124
Psilocine		1995	204 $_{14}$	160 $_5$	146 $_5$	130 $_2$	58 $_{100}$	227	2470
Psilocine AC	U+UHYAC	2270	246 $_6$	202 $_1$	146 $_3$	130 $_1$	58 $_{100}$	364	2471
Psilocine HFB		2110	400 $_1$	342 $_2$	145 $_3$	117 $_5$	58 $_{60}$	1017	6317
Psilocine PFP		2095	350 $_2$	292 $_1$	145 $_3$	58 $_{60}$	186 $_1$	845	6350
Psilocine TFA		2080	300 $_2$	242 $_1$	117 $_2$	69 $_3$	58 $_{80}$	615	6349
Psilocine 2AC	U+UHYAC	2340	288 $_7$	246 $_1$	202 $_1$	122 $_3$	58 $_{100}$	554	2472
Psilocine 2TMS		2250	348 $_6$	333 $_1$	290 $_{29}$	73 $_{29}$	58 $_{100}$	840	6348
Psilocine-M (4-hydroxyindoleacetic acid) MEAC		2315	247 $_{32}$	205 $_{60}$	173 $_{40}$	145 $_{100}$	117 $_{30}$	366	6346
Psilocine-M (4-hydroxytryptophol) 2AC		2370	261 $_8$	201 $_{19}$	159 $_{100}$	146 $_{45}$	117 $_{30}$	426	6347
Psilocybin artifact		1995	204 $_{14}$	160 $_5$	146 $_5$	130 $_2$	58 $_{100}$	227	2470
Psilocybin artifact AC	U+UHYAC	2270	246 $_6$	202 $_1$	146 $_3$	130 $_1$	58 $_{100}$	364	2471
Psilocybin artifact HFB		2110	400 $_1$	342 $_2$	145 $_3$	117 $_5$	58 $_{60}$	1017	6317
Psilocybin artifact PFP		2095	350 $_2$	292 $_1$	145 $_3$	58 $_{60}$	186 $_1$	845	6350
Psilocybin artifact TFA		2080	300 $_2$	242 $_1$	117 $_2$	69 $_3$	58 $_{80}$	615	6349
Psilocybin artifact 2AC	U+UHYAC	2340	288 $_7$	246 $_1$	202 $_1$	122 $_3$	58 $_{100}$	554	2472
Psilocybin artifact 2TMS		2250	348 $_6$	333 $_1$	290 $_{29}$	73 $_{29}$	58 $_{100}$	840	6348
Psilocybin-M (4-hydroxyindoleacetic acid) MEAC		2315	247 $_{32}$	205 $_{60}$	173 $_{40}$	145 $_{100}$	117 $_{30}$	366	6346
Psilocybin-M (4-hydroxytryptophol) 2AC		2370	261 $_8$	201 $_{19}$	159 $_{100}$	146 $_{45}$	117 $_{30}$	426	6347
PVP		2185	231 $_1$	188 $_1$	126 $_{100}$	97 $_6$	77 $_{14}$	310	7441
PVP-M (carboxy-oxo-) AC	UGLUCSPEAC	2215	319 $_1$	246 $_{13}$	214 $_{73}$	172 $_{100}$	101 $_{32}$	709	7833
PVP-M (carboxy-oxo-) HFB	UGLUCSPEHFB	1980	473 $_1$	442 $_4$	368 $_{48}$	336 $_{100}$	101 $_{66}$	1141	7828
PVP-M (carboxy-oxo-) ME	UGLUCSPEME	1980	260 $_3$	186 $_{100}$	105 $_9$	101 $_{48}$		569	7834
PVP-M (carboxy-oxo-) TMS	UGLUCSPETMS	2025	349 $_1$	334 $_8$	318 $_7$	244 $_{100}$	173 $_6$	843	7827
PVP-M (carboxy-oxo-) 2TFA	UGLUCSPETFA	2010	373 $_1$	342 $_6$	268 $_{58}$	236 $_{100}$	101 $_{34}$	930	7832
PVP-M (di-HO-) 2AC		2440	269 $_5$	251 $_3$	184 $_{100}$	124 $_{63}$	121 $_{11}$	836	7766
PVP-M (di-HO-) iso-1 2TMS		2345	392 $_2$	214 $_{100}$	193 $_3$	124 $_6$	73 $_{17}$	1035	7771
PVP-M (di-HO-) iso-2 2TMS		2350	392 $_2$	214 $_{100}$	193 $_4$	124 $_6$	73 $_5$	1036	7825
PVP-M (HO-alkyl-) AC		2025	227 $_2$	184 $_{100}$	124 $_{51}$	105 $_{12}$	95 $_5$	560	7760
PVP-M (HO-alkyl-) TMS		1950	304 $_3$	214 $_{100}$	124 $_9$	105 $_4$	73 $_8$	712	7773
PVP-M (HO-alkyl-oxo-) AC		2170	303 $_1$	198 $_{100}$	138 $_{87}$	110 $_{19}$	96 $_{24}$	629	7764
PVP-M (HO-alkyl-oxo-) TMS		2260	318 $_8$	228 $_{100}$	214 $_8$	138 $_{92}$	105 $_{10}$	774	7772
PVP-M (HO-phenyl-) AC		2110	126 $_{100}$	121 $_4$	96 $_4$	84 $_4$		560	7763
PVP-M (HO-phenyl-) ME		1990	261 $_1$	135 $_9$	126 $_{100}$	110 $_{12}$	96 $_4$	428	7759
PVP-M (HO-phenyl-) TMS		2095	304 $_2$	73 $_7$	150 $_3$	193 $_2$	126 $_{80}$	712	7770
PVP-M (HO-phenyl-carboxy-oxo-) MEAC	UGLUCSPEMEAC	2550	349 $_1$	276 $_8$	214 $_{67}$	172 $_{100}$	135 $_{11}$	842	7830
PVP-M (HO-phenyl-carboxy-oxo-) 2AC	UGLUCSPEAC	2635	377 $_1$	304 $_9$	214 $_{56}$	172 $_{100}$	101 $_{31}$	942	7831
PVP-M (HO-phenyl-carboxy-oxo-) 2ME	UGLUCSPEME	2360	290 $_2$	186 $_{100}$	135 $_8$	101 $_{31}$	59 $_{17}$	719	7829
PVP-M (HO-phenyl-N,N-bisdealkyl-) MEAC		1970	249 $_8$	186 $_{24}$	135 $_{100}$	114 $_{29}$	72 $_{81}$	376	7757
PVP-M (HO-phenyl-N,N-bisdealkyl-) 2AC		2080	277 $_3$	163 $_3$	121 $_{32}$	114 $_{60}$	72 $_{100}$	501	7762
PVP-M (HO-phenyl-N,N-bisdealkyl-) 2TMS		1860	337 $_1$	322 $_3$	144 $_{100}$	98 $_{61}$	86 $_{29}$	795	7768
PVP-M (HO-phenyl-oxo-) AC		2320	303 $_1$	220 $_4$	140 $_{100}$	121 $_{11}$	98 $_{32}$	629	7765
PVP-M (HO-phenyl-oxo-) ME		2225	275 $_2$	192 $_{14}$	140 $_{100}$	135 $_{13}$	98 $_{23}$	491	7758
PVP-M (HO-phenyl-oxo-) TMS		2320	318 $_2$	250 $_{11}$	193 $_{19}$	140 $_{100}$	98 $_{19}$	775	7769
PVP-M (N,N-bisdealkyl-) AC		1590	219 $_3$	134 $_6$	114 $_{64}$	105 $_{21}$	72 $_{100}$	271	7761
PVP-M (N,N-bisdealkyl-) TMS		1375	234 $_5$	191 $_4$	156 $_8$	144 $_{100}$	113 $_9$	378	7767
PVP-M (oxo-)		1875	245 $_3$	140 $_{100}$	105 $_8$	98 $_{38}$	86 $_{11}$	359	7756
PYCC		1255	178 $_9$	150 $_{14}$	135 $_{100}$	121 $_{13}$	70 $_{25}$	169	3583
PYCC -HCN		1180	151 $_{93}$	150 $_{100}$	136 $_{76}$	122 $_{71}$	95 $_{70}$	130	3585
Pyranocoumarin		2670*	322 $_{100}$	265 $_{70}$	249 $_{39}$	148 $_{17}$	72 $_{87}$	722	4047
Pyranocoumarin-M (demethyl-HO-dihydro-) iso-1 -H2O 2ME	UME	2780*	336 $_7$	293 $_{100}$	165 $_6$	150 $_{41}$	115 $_{13}$	788	4827
Pyranocoumarin-M (demethyl-HO-dihydro-) iso-2 -H2O 2ME	UME	2805*	336 $_4$	293 $_{100}$	173 $_{10}$	145 $_{12}$	121 $_{13}$	788	4828
Pyranocoumarin-M (demethyl-HO-dihydro-) iso-3 -H2O 2ME	UME	2830*	336 $_3$	293 $_{100}$	217 $_3$	151 $_{17}$	115 $_{10}$	789	4829
Pyranocoumarin-M (di-HO-) 2ET	UET	2990*	410 $_{100}$	367 $_{78}$	339 $_{45}$	179 $_{36}$	151 $_{19}$	1041	4838
Pyranocoumarin-M (O-demethyl-) artifact ME	UME	2580*	322 $_{18}$	279 $_{100}$	189 $_7$	121 $_{11}$	91 $_{15}$	723	1030
Pyranocoumarin-M (O-demethyl-) artifact TMS		2675*	380 $_{53}$	337 $_{100}$	261 $_{26}$	193 $_{28}$	73 $_{62}$	952	4970
Pyranocoumarin-M (O-demethyl-) artifact enol 2TMS		2790*	452 $_{12}$	437 $_{10}$	437 $_{100}$	247 $_{13}$	73 $_{10}$	1116	4971
Pyranocoumarin-M (O-demethyl-di-HO-) artifact 3ME	UME	3150*	382 $_{33}$	339 $_{100}$	325 $_{49}$	231 $_{44}$	151 $_{14}$	959	4830
Pyranocoumarin-M (O-demethyl-dihydro-) artifact ME	UME	2660*	324 $_{18}$	291 $_{62}$	215 $_{47}$	177 $_{62}$	91 $_{100}$	731	1032
Pyranocoumarin-M (O-demethyl-dihydro-) artifact 2TMS		2785*	454 $_{12}$	439 $_{40}$	364 $_{100}$	335 $_{74}$	73 $_{79}$	1119	4972
Pyranocoumarin-M (O-demethyl-dihydro-) -H2O	UHY U+UHYAC UHYME	2550*	292 $_{100}$	263 $_{72}$	249 $_{43}$	198 $_{17}$	121 $_{32}$	575	1031
Pyranocoumarin-M (O-demethyl-HO-) artifact 2TMS	UTMS	3015*	468 $_{48}$	425 $_{100}$	337 $_{47}$	115 $_{14}$	73 $_{34}$	1136	4967
Pyranocoumarin-M (O-demethyl-HO-) artifact enol 3TMS	UTMS	3105*	540 $_{26}$	497 $_{100}$	395 $_{27}$	335 $_{20}$	73 $_{92}$	1181	4969
Pyranocoumarin-M (O-demethyl-HO-) iso-1 artifact 2ET	UET	2810*	380 $_{44}$	337 $_{100}$	309 $_{61}$	233 $_{12}$	165 $_{13}$	952	4832
Pyranocoumarin-M (O-demethyl-HO-) iso-1 artifact 2ME	UME	2810*	352 $_{21}$	309 $_{100}$	277 $_5$	151 $_5$	91 $_{14}$	853	1033
Pyranocoumarin-M (O-demethyl-HO-) iso-1 artifact 2TMS	UTMS	2795*	468 $_{33}$	425 $_{100}$	268 $_{23}$	193 $_{49}$	73 $_{28}$	1136	4968
Pyranocoumarin-M (O-demethyl-HO-) iso-2 artifact 2ET	UET	2870*	380 $_{73}$	337 $_{100}$	309 $_{66}$	187 $_{41}$	121 $_{59}$	952	4833
Pyranocoumarin-M (O-demethyl-HO-) iso-2 artifact 2ME	UME	2830*	352 $_{32}$	309 $_{100}$	295 $_{31}$	201 $_{43}$	121 $_{56}$	853	4825
Pyranocoumarin-M (O-demethyl-HO-) iso-3 artifact 2ET	UET	2870*	380 $_{35}$	337 $_{100}$	309 $_{54}$	165 $_{18}$	137 $_{28}$	953	4834

Table 1-8-1: Compounds in order of names — Quetiapine-M (-COOH) ME

Name	Detected	RI	Typical ions and intensities					Page	Entry
Pyranocoumarin-M (O-demethyl-HO-) iso-3 artifact 2ME	UME	2870*	352 $_{21}$	309 $_{100}$	295 $_{12}$	206 $_{8}$	91 $_{21}$	853	4826
Pyrazinamide	P-I	1460	123 $_{92}$	80 $_{100}$	53 $_{81}$			108	947
Pyrazophos		2590	373 $_{10}$	265 $_{10}$	232 $_{34}$	221 $_{100}$	97 $_{11}$	929	3863
Pyrene		1990*	202 $_{100}$	200 $_{21}$	174 $_{2}$	150 $_{1}$	101 $_{10}$	222	2567
Pyridate		2985	378 $_{2}$	350 $_{6}$	283 $_{19}$	205 $_{100}$	57 $_{91}$	946	3864
Pyridine		<1000	79 $_{100}$	52 $_{66}$				95	1549
Pyridostigmine bromide -CH3Br		1320	166 $_{9}$	95 $_{1}$	78 $_{1}$	72 $_{100}$	56 $_{3}$	151	4348
Pyridoxic acid lactone		1700	165 $_{100}$	147 $_{25}$	136 $_{66}$	119 $_{21}$	108 $_{26}$	146	5645
Pyridoxine	PAC U+UHYAC	1945	295 $_{1}$	253 $_{23}$	193 $_{54}$	151 $_{100}$	123 $_{49}$	590	5089
Pyrilamine	G U	2220	285 $_{4}$	215 $_{8}$	121 $_{100}$	78 $_{20}$	58 $_{50}$	540	1656
Pyrilamine HY	UHY U+UHYAC	1690	208 $_{1}$	163 $_{1}$	137 $_{4}$	71 $_{50}$	58 $_{100}$	239	1660
Pyrilamine-M (N-dealkyl-)	U	2120	214 $_{61}$	136 $_{11}$	121 $_{100}$	78 $_{32}$		256	1657
Pyrilamine-M (N-dealkyl-) AC	U+UHYAC	2150	256 $_{42}$	214 $_{62}$	163 $_{38}$	107 $_{100}$	78 $_{40}$	407	1659
Pyrilamine-M (N-demethoxybenzyl-)	U	1580	165 $_{3}$	119 $_{8}$	107 $_{12}$	78 $_{26}$	58 $_{100}$	148	1658
Pyrimethamine		2185	248 $_{86}$	247 $_{100}$	219 $_{20}$	212 $_{17}$		370	2025
Pyrimethamine AC	U+UHYAC	2580	290 $_{56}$	289 $_{80}$	247 $_{100}$	219 $_{8}$	212 $_{8}$	563	2026
Pyrithyldione	P G U UHY U+UHYAC	1520	167 $_{2}$	152 $_{23}$	139 $_{86}$	98 $_{77}$	83 $_{100}$	153	948
Pyritinol		9999	368 $_{6}$	199 $_{24}$	166 $_{100}$	151 $_{52}$	106 $_{57}$	912	950
Pyritinol 3ME		9999	410 $_{8}$	165 $_{91}$	136 $_{100}$			1041	951
Pyritinol-M		9999	199 $_{6}$	151 $_{100}$	122 $_{17}$	106 $_{22}$		217	952
Pyritinol-M	U UHY U+UHYAC	1800	207 $_{100}$					232	949
Pyrocatechol	UHY	<1000*	110 $_{100}$	92 $_{11}$	81 $_{15}$	64 $_{42}$	53 $_{15}$	103	2535
Pyrrobutamine	U UHY U+UHYAC	2370	311 $_{2}$	240 $_{43}$	205 $_{100}$	125 $_{25}$	91 $_{30}$	671	2204
Pyrrobutamine-M (oxo-)	U UHY U+UHYAC	2920	325 $_{9}$	240 $_{65}$	205 $_{100}$	115 $_{88}$	98 $_{71}$	734	2205
Pyrrocaine	G U UHY U+UHYAC	1830	232 $_{14}$	218 $_{37}$	132 $_{19}$	85 $_{33}$	71 $_{101}$	313	1040
Pyrrolidine		<1000	71 $_{19}$	70 $_{26}$	43 $_{79}$			92	3608
Pyrrolidine AC		1320	113 $_{61}$	98 $_{3}$	85 $_{19}$	70 $_{100}$	60 $_{26}$	104	6459
Pyrrolidinopropiophenone		1595	202 $_{1}$	188 $_{1}$	133 $_{1}$	98 $_{50}$	56 $_{17}$	224	5943
Pyrrolidinopropiophenone-M (oxo-)		1820	217 $_{1}$	112 $_{100}$	105 $_{4}$	84 $_{7}$	69 $_{24}$	261	6546
Pyrrolidinovalerophenone		2185	231 $_{1}$	188 $_{1}$	126 $_{100}$	97 $_{6}$	77 $_{14}$	310	7441
Pyrrolidinovalerophenone-M (carboxy-oxo-) AC	UGLUCSPEAC	2215	319 $_{1}$	246 $_{13}$	214 $_{73}$	172 $_{100}$	101 $_{32}$	709	7833
Pyrrolidinovalerophenone-M (carboxy-oxo-) HFB	UGLUCSPEHFB	1980	473 $_{1}$	442 $_{4}$	368 $_{48}$	336 $_{100}$	101 $_{66}$	1141	7828
Pyrrolidinovalerophenone-M (carboxy-oxo-) ME	UGLUCSPEME	1980	260 $_{3}$	186 $_{100}$	105 $_{9}$	101 $_{48}$		569	7834
Pyrrolidinovalerophenone-M (carboxy-oxo-) TMS	UGLUCSPETMS	2025	349 $_{1}$	334 $_{8}$	318 $_{7}$	244 $_{100}$	173 $_{6}$	843	7827
Pyrrolidinovalerophenone-M (carboxy-oxo-) 2TFA	UGLUCSPETFA	2010	373 $_{1}$	342 $_{6}$	268 $_{58}$	236 $_{100}$	101 $_{34}$	930	7832
Pyrrolidinovalerophenone-M (di-HO-) 2AC		2440	269 $_{5}$	251 $_{3}$	184 $_{100}$	124 $_{63}$	121 $_{1}$	836	7766
Pyrrolidinovalerophenone-M (di-HO-) iso-1 2TMS		2345	392 $_{2}$	214 $_{100}$	193 $_{3}$	124 $_{6}$	73 $_{17}$	1035	7771
Pyrrolidinovalerophenone-M (di-HO-) iso-2 2TMS		2350	392 $_{2}$	214 $_{100}$	193 $_{4}$	124 $_{6}$	73 $_{5}$	1036	7825
Pyrrolidinovalerophenone-M (HO-alkyl-) AC		2025	227 $_{2}$	184 $_{100}$	124 $_{51}$	105 $_{12}$	95 $_{5}$	560	7760
Pyrrolidinovalerophenone-M (HO-alkyl-) TMS		1950	304 $_{3}$	214 $_{100}$	124 $_{9}$	105 $_{4}$	73 $_{8}$	712	7773
Pyrrolidinovalerophenone-M (HO-alkyl-oxo-) AC		2170	303 $_{1}$	198 $_{100}$	138 $_{87}$	110 $_{5}$	96 $_{24}$	629	7764
Pyrrolidinovalerophenone-M (HO-alkyl-oxo-) TMS		2260	318 $_{8}$	228 $_{100}$	214 $_{8}$	138 $_{92}$	105 $_{10}$	774	7772
Pyrrolidinovalerophenone-M (HO-phenyl-) AC		2110	126 $_{100}$	121 $_{4}$	96 $_{4}$	84 $_{4}$		560	7763
Pyrrolidinovalerophenone-M (HO-phenyl-) ME		1990	261 $_{1}$	135 $_{9}$	126 $_{100}$	110 $_{12}$	96 $_{4}$	428	7759
Pyrrolidinovalerophenone-M (HO-phenyl-) TMS		2095	304 $_{2}$	73 $_{7}$	150 $_{3}$	193 $_{2}$	126 $_{80}$	712	7770
Pyrrolidinovalerophenone-M (HO-phenyl-carboxy-oxo-) MEAC	UGLUCSPEMEAC	2550	349 $_{1}$	276 $_{8}$	214 $_{67}$	172 $_{100}$	135 $_{11}$	842	7830
Pyrrolidinovalerophenone-M (HO-phenyl-carboxy-oxo-) 2AC	UGLUCSPEAC	2635	377 $_{1}$	304 $_{9}$	214 $_{56}$	172 $_{100}$	101 $_{31}$	942	7831
Pyrrolidinovalerophenone-M (HO-phenyl-carboxy-oxo-) 2ME	UGLUCSPEME	2360	290 $_{2}$	186 $_{100}$	135 $_{8}$	101 $_{31}$	59 $_{17}$	719	7829
Pyrrolidinovalerophenone-M (HO-phenyl-N,N-bisdealkyl-) MEAC		1970	249 $_{8}$	186 $_{24}$	135 $_{100}$	114 $_{9}$	72 $_{81}$	376	7757
Pyrrolidinovalerophenone-M (HO-phenyl-N,N-bisdealkyl-) 2AC		2080	277 $_{3}$	163 $_{3}$	121 $_{32}$	114 $_{60}$	72 $_{100}$	501	7762
Pyrrolidinovalerophenone-M (HO-phenyl-N,N-bisdealkyl-) 2TMS		1860	337 $_{1}$	322 $_{3}$	144 $_{100}$	98 $_{61}$	86 $_{29}$	795	7768
Pyrrolidinovalerophenone-M (HO-phenyl-oxo-) AC		2320	303 $_{1}$	220 $_{4}$	140 $_{100}$	121 $_{11}$	98 $_{32}$	629	7765
Pyrrolidinovalerophenone-M (HO-phenyl-oxo-) ME		2225	275 $_{2}$	192 $_{14}$	140 $_{100}$	135 $_{13}$	98 $_{23}$	491	7758
Pyrrolidinovalerophenone-M (HO-phenyl-oxo-) TMS		2320	318 $_{2}$	250 $_{11}$	193 $_{19}$	140 $_{100}$	98 $_{19}$	775	7769
Pyrrolidinovalerophenone-M (N,N-bisdealkyl-) AC		1590	219 $_{3}$	134 $_{6}$	114 $_{64}$	105 $_{21}$	72 $_{100}$	271	7761
Pyrrolidinovalerophenone-M (N,N-bisdealkyl-) TMS		1375	234 $_{5}$	191 $_{4}$	156 $_{8}$	144 $_{100}$	113 $_{9}$	378	7767
Pyrrolidinovalerophenone-M (oxo-)		1875	245 $_{3}$	140 $_{100}$	105 $_{8}$	98 $_{38}$	86 $_{11}$	359	7756
Quazepam	U U+UHYAC	2440	386 $_{100}$	359 $_{64}$	323 $_{49}$	303 $_{36}$	245 $_{40}$	971	2130
Quazepam HY	UHY U+UHYAC	1985	331 $_{100}$	312 $_{10}$	262 $_{34}$	166 $_{16}$	123 $_{16}$	761	2131
Quazepam-M (dealkyl-oxo-)	G P-I UGLUC	2470	288 $_{100}$	287 $_{64}$	260 $_{93}$	259 $_{75}$		551	508
Quazepam-M (dealkyl-oxo-) TMS		2470	360 $_{44}$	359 $_{48}$	341 $_{30}$	197 $_{12}$	73 $_{70}$	883	4621
Quazepam-M (dealkyl-oxo-) HY	UHY	2030	249 $_{97}$	154 $_{33}$	123 $_{45}$	95 $_{41}$		373	512
Quazepam-M (dealkyl-oxo-) HYAC	U+UHYAC	2195	291 $_{52}$	249 $_{100}$	123 $_{57}$	95 $_{61}$		567	286
Quazepam-M (HO-) HYAC	U+UHYAC	2250	389 $_{81}$	347 $_{100}$	278 $_{53}$	166 $_{62}$	125 $_{61}$	981	2133
Quazepam-M (oxo-)	U UHY U+UHYAC	2255	370 $_{58}$	342 $_{100}$	307 $_{10}$	259 $_{29}$	109 $_{7}$	918	2132
Quazepam-M (oxo-) HY	UHY U+UHYAC	1985	331 $_{100}$	312 $_{10}$	262 $_{34}$	166 $_{16}$	123 $_{16}$	761	2131
Quazepam-M/artifact	U	2480	400 $_{44}$	323 $_{21}$	244 $_{100}$	209 $_{22}$		1017	2140
Quercetin 4AC		3510*	470 $_{1}$	428 $_{24}$	386 $_{48}$	344 $_{57}$	302 $_{100}$	1138	4671
Quercetin 4ME		3510*	358 $_{100}$	343 $_{5}$	329 $_{5}$			877	4672
Quercetin 5TMS		3090*	662 $_{1}$	647 $_{100}$	575 $_{27}$	559 $_{15}$	487 $_{6}$	1201	2514
Quetiapine	G	3280	383 $_{1}$	321 $_{24}$	239 $_{57}$	210 $_{100}$	144 $_{48}$	963	6448
Quetiapine AC	U+UHYAC	3320	425 $_{2}$	321 $_{25}$	239 $_{54}$	210 $_{100}$	186 $_{56}$	1070	6431
Quetiapine artifact (desulfo-) AC	U+UHYAC	3345	393 $_{1}$	289 $_{10}$	219 $_{12}$	207 $_{100}$	178 $_{15}$	995	6437
Quetiapine-M (-COOH) ME	U+UHYAC	3240	411 $_{2}$	321 $_{33}$	239 $_{71}$	210 $_{100}$	172 $_{87}$	1044	6432

Quetiapine-M (N-CH2-COOH) ME　　　　　　　　　　　　　Table 1-8-1: Compounds in order of names

Name	Detected	RI	Typical ions and intensities					Page	Entry
Quetiapine-M (N-CH2-COOH) ME	U+UHYAC P	2900	367_{17}	308_{10}	239_{52}	227_{65}	210_{100}	908	6433
Quetiapine-M (N-dealkyl-)	U+UHY	2670	295_{9}	265_{11}	239_{25}	227_{100}	210_{58}	590	6438
Quetiapine-M (N-dealkyl-) AC	U+UHYAC	2970	337_{64}	294_{8}	251_{60}	239_{64}	210_{100}	792	6434
Quetiapine-M (N-dealkyl-) artifact (desulfo-)	U+UHYAC	2640	263_{6}	207_{93}	195_{100}	178_{43}	151_{23}	435	6439
Quetiapine-M (N-dealkyl-) artifact (desulfo-) AC	U+UHYAC	2970	305_{10}	219_{33}	207_{100}	194_{21}	178_{24}	639	6436
Quetiapine-M (N-dealkyl-HO-) 2AC	U+UHYAC	3960	395_{82}	352_{36}	267_{78}	242_{74}	226_{100}	1001	6435
Quinalphos		2070	298_{18}	157_{57}	146_{100}	118_{48}	90_{41}	604	3453
Quinalphos HY		2020	146_{100}	118_{91}	91_{42}	76_{3}	64_{23}	124	7413
Quinalphos HYME		1750	160_{100}	131_{71}	104_{24}	90_{11}	77_{17}	138	7414
Quinapril ET	UET	3105	466_{1}	393_{2}	234_{100}	160_{17}	91_{33}	1134	4763
Quinapril ME	UME	3110	452_{1}	379_{2}	234_{100}	190_{11}	91_{51}	1117	4757
Quinapril TMS		3125	510_{1}	495_{1}	437_{5}	234_{80}	91_{12}	1170	4992
Quinapril 2ET	UET	3140	494_{1}	421_{1}	262_{100}	130_{23}	91_{53}	1160	4764
Quinapril 2ME		3120	466_{1}	393_{2}	248_{100}	174_{13}	91_{24}	1134	4758
Quinapril -H2O	G	3380	420_{3}	316_{37}	270_{35}	130_{64}	91_{100}	1062	4761
Quinaprilate 2ET	UET	3105	466_{1}	393_{2}	234_{100}	160_{17}	91_{33}	1134	4763
Quinaprilate 2ME	UME	3030	438_{1}	379_{2}	220_{100}	160_{11}	91_{52}	1095	4759
Quinaprilate 2TMS	UTMS	3160	554_{1}	539_{7}	437_{31}	278_{100}	91_{29}	1185	4990
Quinaprilate 3ET	UET	3140	494_{1}	421_{1}	262_{100}	130_{23}	91_{53}	1160	4764
Quinaprilate 3ME	UME	3080	452_{1}	393_{2}	234_{100}	174_{10}	91_{23}	1117	4760
Quinaprilate -H2O ME	UME	3310	406_{2}	302_{35}	270_{34}	130_{70}	91_{100}	1033	4762
Quinaprilate -H2O TMS	UME	3255	464_{15}	449_{40}	360_{100}	270_{54}	91_{37}	1132	4991
Quinaprilate-M/artifact (HOOC-) 2ET	UET	2025	307_{2}	234_{100}	160_{12}	117_{17}	91_{32}	651	4740
Quinaprilate-M/artifact (HOOC-) 2ME	UME	1870	279_{2}	220_{100}	160_{10}	117_{28}	91_{57}	512	4734
Quinaprilate-M/artifact (-HOOC-) 3ME	UME	1935	293_{2}	234_{100}	174_{8}	130_{16}	91_{50}	582	4735
Quinapril-M/artifact (deethyl-) 2ET	UET	3105	466_{1}	393_{2}	234_{100}	160_{17}	91_{33}	1134	4763
Quinapril-M/artifact (deethyl-) 2ME	UME	3030	438_{1}	379_{2}	220_{100}	160_{11}	91_{52}	1095	4759
Quinapril-M/artifact (deethyl-) 2TMS	UTMS	3160	554_{1}	539_{7}	437_{31}	278_{100}	91_{29}	1185	4990
Quinapril-M/artifact (deethyl-) 3ET	UET	3140	494_{1}	421_{1}	262_{100}	130_{23}	91_{53}	1160	4764
Quinapril-M/artifact (deethyl-) 3ME	UME	3080	452_{1}	393_{2}	234_{100}	174_{10}	91_{23}	1117	4760
Quinapril-M/artifact (deethyl-) -H2O ME	UME	3310	406_{2}	302_{35}	270_{34}	130_{70}	91_{100}	1033	4762
Quinapril-M/artifact (deethyl-) -H2O TMS	UME	3255	464_{15}	449_{40}	360_{100}	270_{54}	91_{37}	1132	4991
Quinapril-M/artifact (deethyl-HOOC-) 2ME	UME	1870	279_{2}	220_{100}	160_{10}	117_{28}	91_{57}	512	4734
Quinapril-M/artifact (deethyl-HOOC-) 3ME	UME	1935	293_{2}	234_{100}	174_{8}	130_{16}	91_{50}	582	4735
Quinapril-M/artifact (HOOC-) ET	UET	2025	307_{2}	234_{100}	160_{12}	117_{17}	91_{32}	651	4740
Quinapril-M/artifact (HOOC-) ME	UME	1930	293_{3}	234_{36}	220_{100}	160_{11}	91_{23}	582	4736
Quinapril-M/artifact (HOOC-) 2ME	UME	1985	307_{2}	248_{34}	234_{100}	174_{5}	91_{6}	651	4737
Quinestrol		3025*	364_{22}	338_{33}	296_{15}	270_{100}	213_{38}	898	1524
Quinethazone 4ME	UEXME	2980	345_{1}	316_{100}	273_{4}	208_{34}	173_{13}	827	6854
Quinidine	G U P	2790	324_{11}	189_{6}	173_{12}	136_{40}		732	661
Quinidine AC	U+UHYAC	2750	366_{24}	307_{14}	189_{14}	136_{100}		906	664
Quinidine TMS		2790	396_{47}	381_{17}	261_{55}	136_{100}	73_{81}	1007	4594
Quinidine-M	U UHY	2940	338_{19}	323_{8}	152_{60}	124_{17}	122_{14}	796	662
Quinidine-M (di-HO-dihydro-) 3AC	U+UHYAC	3350	484_{85}	425_{100}	365_{31}	254_{23}	194_{32}	1151	667
Quinidine-M (HO-) 2AC	U+UHYAC	3185	424_{100}	365_{65}	305_{47}	194_{25}		1068	666
Quinidine-M (N-oxide)	U UHY	2950	340_{22}	324_{12}	189_{28}	152_{100}	136_{38}	808	663
Quinidine-M (N-oxide) AC	U+UHYAC	2935	382_{24}	189_{11}	152_{100}			960	665
Quinine	G P-I U	2800	324_{1}	189_{3}	136_{100}	117_{3}	81_{6}	732	668
Quinine AC	U+UHYAC	2760	366_{3}	309_{4}	189_{4}	136_{100}		906	669
Quinine TMS		2690	396_{1}	381_{3}	261_{8}	136_{100}	73_{28}	1007	4595
Quinine-M (di-HO-dihydro-) 3AC	U+UHYAC	3360	484_{40}	425_{100}	365_{25}	254_{8}	188_{10}	1151	3747
Quinine-M (HO-) 2AC	U+UHYAC	3195	424_{100}	409_{40}	365_{85}	305_{39}	194_{22}	1068	3746
Quinine-M (N-oxide) AC	U+UHYAC	2945	382_{1}	231_{49}	189_{33}	152_{100}	55_{22}	960	3745
Quinomethionate		2080	234_{70}	206_{90}	174_{27}	148_{46}	116_{100}	318	3323
Quintozene		1790	293_{47}	249_{70}	237_{100}	212_{47}	142_{35}	577	3865
Ramifenazone	G	2045	245_{28}	230_{12}	137_{31}	83_{25}	56_{100}	360	530
Ramifenazone AC		2400	287_{35}	244_{37}	137_{31}	56_{99}		548	194
Ramifenazone-M (nor-) 2AC	U+UHYAC	2365	315_{5}	273_{45}	231_{48}	123_{100}	70_{94}	689	195
Ramifenazone-M (nor-HO-) -H2O 2AC	U+UHYAC	2160	313_{3}	271_{9}	229_{28}	77_{16}	56_{100}	680	531
Ramipril ET	UET	2920	444_{1}	371_{11}	234_{100}	160_{9}	91_{22}	1106	4771
Ramipril ME	UME	2880	430_{1}	357_{9}	234_{100}	160_{15}	91_{39}	1081	4765
Ramipril METMS		3020	502_{1}	487_{2}	415_{7}	248_{100}	91_{16}	1165	4994
Ramipril TMS		2935	488_{1}	473_{9}	415_{28}	234_{100}	91_{56}	1156	4993
Ramipril 2ET	UET	2990	472_{1}	399_{9}	262_{100}	248_{17}	188_{8}	1141	4772
Ramipril 2ME	UME	2910	444_{1}	371_{14}	248_{100}	174_{11}	91_{12}	1106	4766
Ramiprilate 2ET	UET	2920	444_{1}	371_{11}	234_{100}	160_{9}	91_{22}	1106	4771
Ramiprilate 2ME	UME	2830	416_{1}	357_{5}	220_{100}	160_{7}	91_{24}	1055	4767
Ramiprilate 2TMS	UTMS	2975	532_{1}	517_{13}	415_{47}	278_{100}	91_{25}	1179	4995
Ramiprilate 3ET	UET	2990	472_{1}	399_{9}	262_{100}	248_{17}	188_{8}	1141	4772
Ramiprilate 3ME	UME	2865	430_{1}	371_{6}	234_{100}	174_{9}	91_{20}	1081	4768
Ramiprilate-M/artifact -H2O ME	UME	2925	384_{16}	280_{100}	248_{82}	193_{20}	91_{68}	968	4770
Ramiprilate-M/artifact -H2O TMS	UTMS	3025	442_{21}	427_{36}	338_{100}	248_{78}	91_{51}	1103	4996
Ramipril-M (deethyl-) artifact 2TMS	UTMS	2975	532_{1}	517_{13}	415_{47}	278_{100}	91_{25}	1179	4995
Ramipril-M/artifact (deethyl-) 2ET	UET	2920	444_{1}	371_{11}	234_{100}	160_{9}	91_{22}	1106	4771

Table 1-8-1: Compounds in order of names   **Rosiglitazone artifact 3TMS**

Name	Detected	RI	Typical ions and intensities					Page	Entry
Ramipril-M/artifact (deethyl-) 2ME	UME	2830	416 $_1$	357 $_5$	220 $_{100}$	160 $_7$	91 $_{22}$	1055	4767
Ramipril-M/artifact (deethyl-) 3ET	UET	2990	472 $_1$	399 $_9$	262 $_{100}$	248 $_{17}$	188 $_8$	1141	4772
Ramipril-M/artifact (deethyl-) 3ME	UME	2865	430 $_1$	371 $_6$	234 $_{100}$	174 $_9$	91 $_{20}$	1081	4768
Ramipril-M/artifact (deethyl-) -H2O ME	UME	2925	384 $_{16}$	280 $_{100}$	248 $_{82}$	193 $_{20}$	91 $_{68}$	968	4770
Ramipril-M/artifact (deethyl-) -H2O TMS	UTMS	3025	442 $_{21}$	427 $_{36}$	338 $_{100}$	248 $_{78}$	91 $_{51}$	1103	4996
Ramipril-M/artifact (HOOC-) ME	UME	1930	293 $_3$	234 $_{36}$	220 $_{100}$	160 $_{11}$	91 $_{23}$	582	4736
Ramipril-M/artifact (HOOC-) 2ME	UME	1985	307 $_2$	248 $_{34}$	234 $_{100}$	174 $_5$	91 $_6$	651	4737
Ramipril-M/artifact -H2O	G	2980	398 $_{13}$	294 $_{82}$	248 $_{100}$	209 $_{29}$	91 $_{87}$	1012	4769
Ranitidine	G	2985	297 $_{25}$	169 $_{43}$	137 $_{100}$	110 $_{68}$	94 $_{100}$	685	5411
Reboxetine		2375	313 $_3$	176 $_{76}$	175 $_{100}$	91 $_{55}$	56 $_{95}$	682	6368
Reboxetine AC		2650	355 $_1$	236 $_8$	218 $_{68}$	176 $_{100}$	91 $_{23}$	866	6370
Reboxetine HFB		2505	509 $_1$	371 $_{30}$	240 $_5$	138 $_{15}$	91 $_{100}$	1169	6373
Reboxetine ME		2315	327 $_1$	190 $_{100}$	99 $_{30}$	91 $_{19}$	71 $_{10}$	745	6369
Reboxetine PFP		2480	459 $_1$	321 $_{29}$	190 $_6$	138 $_{14}$	91 $_{100}$	1126	6372
Reboxetine TFA		2465	409 $_1$	271 $_{30}$	138 $_{14}$	110 $_7$	91 $_{100}$	1038	6371
Reboxetine TMS		2525	385 $_1$	248 $_{86}$	158 $_{30}$	73 $_{97}$	56 $_{100}$	971	6374
Reframidine		2735	323 $_{33}$	322 $_{42}$	280 $_{40}$	188 $_{100}$		726	6737
Remifentanil		2600	376 $_2$	319 $_9$	227 $_{68}$	212 $_{54}$	168 $_{100}$	940	6567
Remoxipride		2520	370 $_1$	243 $_2$	228 $_2$	98 $_{100}$	70 $_8$	918	4693
Repaglinide		3160	466 $_{28}$	423 $_{100}$	245 $_{66}$	186 $_{75}$	172 $_{75}$	1134	5863
Repaglinide TMS		3390	524 $_{10}$	482 $_{37}$	245 $_{85}$	186 $_{100}$	172 $_{94}$	1177	5864
Repaglinide 2TMS		3285	596 $_{36}$	581 $_{18}$	229 $_{100}$	186 $_{30}$	172 $_{79}$	1194	5865
Reserpine		9999	608 $_{25}$	397 $_{31}$	365 $_{28}$	212 $_{42}$	195 $_{100}$	1195	1516
Reserpine-M (trimethoxybenzoic acid)		1780*	212 $_{100}$	197 $_{57}$	169 $_{13}$	141 $_{27}$		248	1949
Reserpine-M (trimethoxybenzoic acid) ET		1770*	240 $_{100}$	225 $_{44}$	212 $_{17}$	195 $_{45}$	141 $_{24}$	344	5219
Reserpine-M (trimethoxybenzoic acid) ME		1740*	226 $_{100}$	211 $_{49}$	195 $_{23}$	155 $_{21}$		293	1950
Reserpine-M (trimethoxyhippuric acid)		2085	251 $_{27}$	223 $_{10}$	195 $_{100}$	152 $_9$		464	1951
Reserpine-M (trimethoxyhippuric acid) ME		2350	283 $_{62}$	268 $_4$	195 $_{100}$	152 $_8$		528	1952
Resmethrin		2300*	338 $_7$	171 $_{50}$	143 $_{34}$	128 $_{44}$	123 $_{100}$	799	4035
Rhein		2675*	284 $_{100}$	255 $_{10}$	241 $_9$	139 $_{10}$	128 $_{12}$	531	3557
Rhein ME		2660*	298 $_{100}$	267 $_{62}$	239 $_{46}$	183 $_{11}$	155 $_{19}$	604	3558
Rhein MEAC		2945*	340 $_{24}$	298 $_{100}$	267 $_{27}$	239 $_{13}$	81 $_{18}$	806	3570
Rhein 2ME		2740*	312 $_{100}$	294 $_{48}$	266 $_{56}$	251 $_{41}$	126 $_{25}$	674	3571
Rhein 3ME		2855*	326 $_{26}$	311 $_{100}$	235 $_{12}$	151 $_{13}$	75 $_{10}$	738	3572
Ribavarine 3AC		2490	328 $_2$	298 $_6$	266 $_6$	139 $_{38}$	113 $_{100}$	748	7331
Ribavarine 4TMS		2240	532 $_2$	517 $_8$	401 $_{100}$	241 $_{28}$	217 $_{25}$	1179	7330
Ribavarine -H2O 3TMS		2015	442 $_{15}$	333 $_6$	281 $_{14}$	217 $_{37}$	73 $_{100}$	1103	7329
Ricinoleic acid ME		2260*	198 $_5$	166 $_{30}$	124 $_{33}$	74 $_{44}$	55 $_{100}$	677	5183
Ricinoleic acid -H2O	G	2140*	280 $_3$	95 $_{33}$	81 $_{46}$	67 $_{64}$	55 $_{70}$	519	2551
Ricinoleic acid -H2O ET		2150*	308 $_{10}$	263 $_6$	95 $_{51}$	81 $_{85}$	67 $_{100}$	658	5642
Ricinoleic acid -H2O ME		2110*	294 $_{10}$	263 $_6$	95 $_{43}$	81 $_{67}$	67 $_{100}$	587	1068
Ritodrine 3TMS		2620	488 $_2$	267 $_{12}$	236 $_{100}$	193 $_{72}$	73 $_{99}$	1167	6219
Ritodrine 3TMSTFA		2620	584 $_2$	267 $_{100}$	193 $_5$	179 $_5$	73 $_{37}$	1195	6186
Ritodrine -H2O 2HFB		2215	661 $_{21}$	344 $_{72}$	316 $_{100}$	303 $_{54}$	169 $_{21}$	1200	6185
Ritodrine -H2O 2PFP		2170	561 $_6$	294 $_{65}$	266 $_{100}$	253 $_{61}$	119 $_{69}$	1188	6132
Ritodrine -H2O 3AC	U+UHYAC	2930	395 $_5$	233 $_{47}$	204 $_{82}$	162 $_{79}$	121 $_{100}$	1003	5618
Ritodrine -H2O 3TFA		2230	558 $_1$	356 $_{77}$	217 $_{100}$	103 $_{54}$	69 $_{98}$	1186	6220
Ritodrine-M/artifact (N-dealkyl-) 2AC	U+UHYAC	1950	221 $_2$	162 $_{26}$	120 $_{100}$	107 $_{30}$	77 $_9$	277	1015
Rizatriptan		2525	269 $_3$	211 $_3$	156 $_4$	142 $_6$	58 $_{100}$	466	5841
Rizatriptan TFA		2475	365 $_1$	307 $_2$	156 $_2$	143 $_4$	58 $_{80}$	900	5842
Rizatriptan TMS		2840	341 $_2$	283 $_3$	142 $_{10}$	73 $_{26}$	58 $_{100}$	815	5843
Rizatriptan-M (deamino-HO-) PFP		2455	388 $_{38}$	224 $_{45}$	211 $_{100}$	156 $_{33}$	143 $_{15}$	979	5849
Rizatriptan-M (deamino-HO-) 2PFP		2330	534 $_{18}$	370 $_{100}$	357 $_{50}$	302 $_{70}$	142 $_{74}$	1180	5848
Rizatriptan-M (deamino-HO-) 2TFA		2390	434 $_{36}$	320 $_{30}$	307 $_{100}$	156 $_{18}$	143 $_{16}$	1087	5847
Rizatriptan-M (deamino-HO-) 2TMS		2860	386 $_{23}$	283 $_{100}$	215 $_5$	142 $_{42}$	73 $_{59}$	973	5846
Rizatriptan-M (deamino-HOOC-) ME		2525	270 $_{44}$	211 $_{100}$	202 $_5$	143 $_{14}$	115 $_6$	469	5844
Rizatriptan-M (deamino-HOOC-) 2TMS		2910	400 $_{42}$	283 $_{100}$	215 $_{23}$	142 $_{51}$	73 $_{77}$	1018	5845
RO 15-4513		3140	326 $_8$	300 $_{50}$	254 $_{57}$	226 $_{100}$	198 $_{22}$	739	3682
RO 15-4513 artifact		3160	300 $_{50}$	254 $_{57}$	226 $_{100}$	198 $_{22}$	145 $_9$	615	3677
Rofecoxib		2760*	314 $_{100}$	285 $_{15}$	257 $_{90}$	178 $_{66}$	131 $_{50}$	683	7489
Rofecoxib -SO2CH2		2470*	236 $_{52}$	234 $_{76}$	205 $_{100}$	177 $_{84}$	151 $_{20}$	327	7490
Rolicyclidine		1830	229 $_{17}$	186 $_{100}$	152 $_{22}$	91 $_{37}$	70 $_{31}$	305	3596
Rolicyclidine intermediate		1255	178 $_9$	150 $_{14}$	135 $_{100}$	121 $_{13}$	70 $_{25}$	169	3583
Rolicyclidine intermediate (PYCC) -HCN		1180	151 $_{93}$	150 $_{100}$	136 $_{76}$	122 $_{71}$	95 $_{70}$	130	3585
Rolicyclidine precursor (pyrrolidine)		<1000	71 $_{19}$	70 $_{26}$	43 $_{79}$			92	3608
Ropinirole		2000	231 $_6$	160 $_{10}$	130 $_6$	114 $_{130}$	86 $_{11}$	424	7517
Ropinirole AC		2020	273 $_1$	160 $_4$	130 $_3$	114 $_{100}$	86 $_{10}$	625	7518
Ropivacaine		2250	274 $_1$	148 $_1$	126 $_{100}$	120 $_8$	84 $_{16}$	487	5407
Rosiglitazone		3080	357 $_3$	135 $_{15}$	121 $_{100}$	107 $_{24}$	78 $_{24}$	873	7726
Rosiglitazone ME		3045	371 $_{12}$	135 $_{55}$	121 $_{100}$	107 $_{55}$	78 $_{69}$	923	7725
Rosiglitazone 2TMS		3110	501 $_{24}$	429 $_{15}$	393 $_{21}$	135 $_{79}$	121 $_{100}$	1165	7728
Rosiglitazone artifact		2185	223 $_5$	151 $_1$	107 $_{100}$	91 $_4$	91 $_4$	283	7730
Rosiglitazone artifact ME		2160	237 $_3$	176 $_2$	151 $_1$	107 $_{100}$	77 $_{15}$	332	7729
Rosiglitazone artifact 3TMS		2235	439 $_{45}$	424 $_{10}$	274 $_5$	223 $_{11}$	73 $_{60}$	1096	7727

Table 1-8-1: Compounds in order of names

Name	Detected	RI	Typical ions and intensities					Page	Entry
Rotenone		3195*	394_{100}	379_{20}	351_5	203_{10}	192_{43}	998	4082
Roxatidine	P	2655	306_{19}	190_{10}	116_{93}	98_{33}	84_{100}	646	4196
Roxatidine AC	U+UHYAC	2710	348_{22}	158_{69}	116_{35}	98_{41}	84_{100}	840	4197
Roxatidine PFP		2470	452_8	262_{100}	107_{26}	98_{32}	84_{99}	1116	4199
Roxatidine TFA		2485	402_{25}	290_{13}	212_{45}	100_{40}	84_{100}	1024	4200
Roxatidine acetate	U+UHYAC	2710	348_{22}	158_{69}	116_{35}	98_{41}	84_{100}	840	4197
Roxatidine artifact (phenol)	U+UHYAC	1810	191_{24}	190_{33}	107_{72}	98_{37}	84_{100}	196	4201
Roxatidine HY PFP		2245	394_{12}	393_{14}	204_{72}	98_{45}	84_{100}	999	4204
Roxatidine HY TFA		2280	344_{10}	343_{14}	154_{49}	98_{46}	84_{100}	825	4203
Roxatidine HY formyl artifact		2150	260_{12}	179_{29}	148_{20}	98_{45}	84_{100}	424	4202
Roxatidine HYAC	U+UHYAC	2485	290_{17}	190_{11}	100_{88}	98_{34}	84_{100}	566	4198
Rutin-M/artifact (quercetin) 4AC		3510*	470_1	428_{24}	386_{48}	344_{57}	302_{100}	1138	4671
Rutin-M/artifact (quercetin) 4ME		3510*	358_{100}	343_5	329_5			877	4672
Rutin-M/artifact (quercetin) 4TMS		3090*	662_1	647_{100}	575_{27}	559_{15}	487_6	1201	2514
Rutin-M/artifact (rutinose) 7AC		2780*	473_1	391_5	273_{58}	153_{100}	111_{73}	1196	5158
Rutinose 7AC		2780*	473_1	391_5	273_{58}	153_{100}	111_{73}	1196	5158
Saccharin ME	PME UME	1600	197_{56}	133_{62}	132_{65}	104_{89}	76_{100}	212	2863
Saccharose 8AC	U+UHYAC	2950*	331_{71}	271_6	211_{67}	169_{100}	109_{71}	1202	1961
Saccharose 8HFB		1950	751_2	537_{10}	519_{94}	323_{59}	169_{100}	1208	5790
Saccharose 8PFP		1860*	601_4	437_7	419_{87}	273_{53}	119_{100}	1208	5789
Saccharose 8TFA		2010*	547_4	337_{29}	319_{100}	223_{87}	69_{86}	1207	5788
Saccharose 8TMS		2680*	451_4	437_{14}	361_{51}	289_{48}	73_{100}	1206	4335
Safrole		1200*	162_{100}	135_{33}	131_{50}	104_{70}	77_{60}	141	3048
Safrole-M (1-HO-) AC	U+UHYAC	1880*	220_{85}	177_{35}	149_{45}	131_{71}		273	7147
Safrole-M (demethylenyl-) 2AC	U+UHYAC	1680*	234_6	192_{42}	150_{100}	131_{24}	91_{36}	319	7144
Safrole-M (demethylenyl-methyl-) AC	U+UHYAC	1530*	206_9	164_{100}	149_{37}	91_{53}	77_{37}	231	7146
Safrole-M (di-HO-) 2AC	U+UHYAC	2015*	280_4	220_{55}	177_{43}	135_{100}	77_{40}	516	7143
Safrole-M (HO-) AC	U+UHYAC	1655*	220_{13}	178_{100}	147_{30}	119_{20}	91_{29}	273	7145
Safrole-M (HO-demethylenyl-methyl-) 2AC	U+UHYAC	1880*	264_2	222_{30}	180_{100}	147_{15}	91_{34}	439	7148
Salacetamide	U+UHYAC	1670	179_4	161_{12}	137_{27}	120_{100}	92_{89}	170	3723
Salbutamol 2AC		2230	323_1	308_1	188_3	135_7	86_{100}	728	2029
Salbutamol 3AC	U+UHYAC	2250	365_2	290_7	188_{11}	135_{11}	86_{100}	901	2028
Salbutamol 3TMS		1750	455_1	440_2	369_{75}	86_{100}	73_{91}	1122	5222
Salbutamol -H2O		1850	221_6	193_{18}	149_{23}	86_{100}	57_{31}	279	2027
Salicylamide	P G UHY	1460	137_{90}	120_{90}	92_{80}	65_{56}		117	755
Salicylamide AC	U+UHYAC	1660	179_{39}	137_{65}	120_{100}	92_{39}	63_{20}	170	193
Salicylamide 2ME		1480	165_{11}	135_{100}	105_6	92_{15}	77_{30}	148	6395
Salicylamide 2TMS		1725	281_3	266_{100}	250_{80}	176_{40}	73_{88}	522	4596
Salicylamide glycolic acid ether ME		1915	209_4	150_{52}	133_{70}	121_{100}	105_{53}	240	5146
Salicylamide-M (HO-) 2AC	U+UHYAC	1860	237_6	195_{11}	153_{28}	136_{100}	108_{21}	332	209
Salicylic acid	G P UHY	1295*	138_{40}	120_{90}	92_{100}	64_{52}		118	953
Salicylic acid AC	G P-I U+UHYAC	1545*	180_6	138_{79}	120_{99}	92_{37}		173	1443
Salicylic acid ET		1350*	166_{34}	120_{99}	92_{40}	65_{19}		149	955
Salicylic acid ME	P U+UHYAC	1200*	152_{39}	120_{94}	92_{100}	65_{53}		130	954
Salicylic acid MEAC	P(ME)	1400*	194_{60}	179_{40}	135_{100}	91_{10}		203	2637
Salicylic acid 2ME	PME UME	1210*	166_{28}	135_{100}	133_{47}	92_{30}	77_{52}	150	6391
Salicylic acid 2TMS		1195*	267_{37}	193_4	135_{17}	91_{11}	73_{60}	526	4523
Salicylic acid artifact (trimer)	G U+UHYAC	3190*	360_{39}	240_{58}	152_{36}	120_{100}	92_{75}	883	4496
Salicylic acid glycine conjugate	U	1825	195_{32}	177_{17}	121_{98}	120_{100}	92_{43}	206	956
Salicylic acid glycine conjugate ME	U	1810	209_{20}	149_{12}	121_{100}	92_{17}	65_{22}	240	957
Salicylic acid glycine conjugate MEAC	U+UHYAC	1885	251_3	209_{69}	177_{23}	149_{29}	121_{100}	383	2976
Salicylic acid glycine conjugate 2ME	UME	1845	223_{12}	135_{100}	90_{58}	77_{44}		283	958
Salicylic acid-M (3-HO-) 3ME	UME	1385*	196_{73}	165_{81}	163_{100}	122_{26}	107_{20}	210	6393
Salicylic acid-M (5-HO-) 3ME	UME	1530*	196_{100}	181_{34}	165_{72}	163_{66}	107_{29}	210	6394
Salicylic acid-M (HO-) 2ME	PME UME	1210*	182_{36}	150_{100}	122_{12}	107_{30}	79_{18}	179	6392
Salsalate ME		1740*	272_{10}	240_3	152_4	121_{100}	93_{13}	478	7527
Sanguinarine artifact (dihydro-)		2945	333_{83}	332_{100}	318_9	260_4	138_6	772	5778
Sanguinarine artifact (N-demethyl-)		3130	317_{100}	259_6	201_{10}	174_6	158_{11}	696	5777
Scopolamine	G U	2315	303_{24}	154_{32}	138_{66}	108_{48}	94_{100}	629	959
Scopolamine AC	U+UHYAC	2450	345_{13}	154_{22}	138_{59}	108_{41}	94_{100}	829	1526
Scopolamine -H2O	U+UHYAC	2230	285_{18}	154_{22}	138_{38}	108_{43}	94_{100}	540	960
Scopolamine-M/artifact (deacyl-)		1210	155_{48}	126_{27}	96_{100}	94_{61}	81_{72}	135	3194
Scopolamine-M/artifact (deacyl-) AC		1410	197_9	154_5	138_{39}	94_{29}	81_{100}	213	3195
Scopolamine-M/artifact (HOOC-) -H2O ME		1510*	162_{100}	150_{38}	118_{48}	103_{38}	77_{18}	141	3196
Sebaic acid bisoctyl ester	U UHY U+UHYAC	2705*	426_1	315_2	297_4	185_{100}	112_{19}	1072	5408
Sebuthylazine		1855	229_{13}	214_{12}	200_{100}	173_8	132_{10}	304	3866
Secobarbital	P G U+UHYAC	1795	209_4	195_{25}	168_{99}	167_{79}	141_{11}	338	961
Secobarbital (ME)	P	1970	252_1	209_{26}	182_{100}	181_{91}	167_{35}	391	2289
Secobarbital 2ME		1690	248_3	196_{100}	181_{25}	138_{25}	111_{22}	452	964
Secobarbital 2TMS		1670	382_1	367_{40}	339_{34}	297_{67}	73_{100}	960	5470
Secobarbital-M (deallyl-)	U	1665	169_3	154_9	129_{100}			215	962
Secobarbital-M (HO-) -H2O	U+UHYAC	1970	236_7	168_{100}	167_{77}	69_{85}		330	963
Selegiline	G U+UHYAC	1450	172_1	115_1	96_{100}	91_{13}	56_{35}	190	2502
Selegiline-M (4-HO-amfetamine)		1480	151_{10}	107_{69}	91_{10}	77_{42}	56_{100}	129	1802

Table 1-8-1: Compounds in order of names — Sibutramine-M (nor-)

Name	Detected	RI	Typical ions and intensities					Page	Entry
Selegiline-M (4-HO-amfetamine) AC	U+UHYAC	1890	193_1	134_{100}	107_{26}	86_{25}	77_{16}	201	1803
Selegiline-M (4-HO-amfetamine) 2AC	U+UHYAC	1900	235_1	176_{72}	134_{100}	107_{47}	86_{71}	324	1804
Selegiline-M (4-HO-amfetamine) 2HFB		<1000	330_{48}	303_{15}	240_{100}	169_{44}	69_{42}	1182	6326
Selegiline-M (4-HO-amfetamine) 2PFP		<1000	280_{77}	253_{16}	190_{100}	119_{56}	69_{16}	1104	6325
Selegiline-M (4-HO-amfetamine) 2TFA		<1000	230_{94}	203_{14}	140_{130}	92_{15}	69_{76}	819	6324
Selegiline-M (4-HO-amfetamine) 2TMS		<1000	280_7	179_9	149_8	116_{100}	73_{78}	594	6327
Selegiline-M (bis-dealkyl-)		1160	134_1	120_1	91_6	77_1	65_4	115	54
Selegiline-M (bis-dealkyl-)	U	1160	134_1	120_1	91_6	65_4	44_{100}	115	5514
Selegiline-M (bis-dealkyl-) AC	U+UHYAC	1505	177_1	118_{19}	91_{11}	86_{31}	65_5	165	55
Selegiline-M (bis-dealkyl-) AC	UAAC U+UHYAC	1505	177_1	118_{19}	91_{11}	86_{31}	44_{100}	165	5515
Selegiline-M (bis-dealkyl-) HFB		1355	331_1	240_{79}	169_{21}	118_{100}	91_{53}	762	5047
Selegiline-M (bis-dealkyl-) PFP		1330	281_1	190_{73}	118_{100}	91_{36}	65_{12}	521	4379
Selegiline-M (bis-dealkyl-) TFA		1095	231_1	140_{100}	118_{92}	91_{45}	69_{19}	309	4000
Selegiline-M (bis-dealkyl-) TMS		1190	192_6	116_{100}	100_{10}	91_{11}	73_{87}	235	5581
Selegiline-M (bis-dealkyl-)-D5 AC	UAAC U+UHYAC	1480	182_3	122_{46}	92_{30}	90_{40}	66_{16}	181	5690
Selegiline-M (bis-dealkyl-)-D5 HFB		1330	244_{100}	169_{14}	122_{46}	92_{41}	69_{40}	787	6316
Selegiline-M (bis-dealkyl-)-D5 PFP		1320	194_{100}	123_{42}	119_{32}	92_{46}	69_{11}	543	5566
Selegiline-M (bis-dealkyl-)-D5 TFA		1085	144_{100}	123_{53}	122_{56}	92_{51}	69_{28}	330	5570
Selegiline-M (bis-dealkyl-)-D5 TMS		1180	212_1	197_8	120_{100}	92_{11}	73_{57}	251	5582
Selegiline-M (bis-dealkyl-)-D11 TFA		1610	292_1	194_{100}	128_{82}	98_{43}	70_{14}	575	7284
Selegiline-M (bis-dealkyl-)-D11 TFA		1615	242_1	144_{100}	128_{82}	98_{43}	70_{14}	353	7283
Selegiline-M (bis-dealkyl-4-HO-) TFA		1670	247_4	140_{15}	134_{54}	107_{100}	77_{15}	366	6335
Selegiline-M (bis-dealkyl-4-HO-) formyl art.		1220	163_3	148_4	107_{30}	77_{13}	56_{100}	142	6323
Selegiline-M (dealkyl-)	U	1195	148_1	134_2	115_1	91_9	58_{100}	127	1093
Selegiline-M (dealkyl-) AC	U+UHYAC	1575	191_1	117_2	100_{42}	91_6	58_{100}	196	1094
Selegiline-M (dealkyl-) HFB		1460	254_{100}	210_{44}	169_{15}	118_{41}	91_{38}	827	5069
Selegiline-M (dealkyl-) PFP		1415	204_{100}	160_{46}	118_{35}	91_{25}	69_4	589	5070
Selegiline-M (dealkyl-) TFA		1300	245_1	154_{100}	118_{48}	110_{55}	91_{23}	358	3998
Selegiline-M (dealkyl-) TMS		1325	206_4	130_{100}	91_{17}	73_{83}	59_{13}	279	6214
Selegiline-M (dealkyl-HO-)		1885	150_1	135_1	107_5	77_5	58_{100}	148	1766
Selegiline-M (dealkyl-HO-) TFA		1770	261_1	154_{100}	134_{68}	110_{42}	107_{41}	425	6180
Selegiline-M (dealkyl-HO-) TMSTFA		1690	333_3	206_{72}	179_{100}	154_{53}	73_{50}	773	6228
Selegiline-M (dealkyl-HO-) 2AC	U+UHYAC	1995	249_1	176_6	134_7	100_{43}	58_{100}	376	1767
Selegiline-M (dealkyl-HO-) 2HFB		1670	538_1	330_{17}	254_{100}	210_{32}	169_{22}	1186	5076
Selegiline-M (dealkyl-HO-) 2PFP		1605	295_1	280_{18}	204_{100}	160_{47}	119_{39}	1123	5077
Selegiline-M (dealkyl-HO-) 2TFA		1585	357_1	230_{22}	154_{100}	110_{42}	69_{29}	872	5078
Selegiline-M (dealkyl-HO-) 2TMS		1620	309_{10}	206_{70}	179_{100}	154_{32}	73_{40}	663	6190
Selegiline-M (HO-)	UHY	1580	107_3	96_{100}	56_{21}			224	2948
Selegiline-M (HO-) AC	U+UHYAC	1860	230_1	107_4	96_{100}	56_{17}		359	2950
Selegiline-M (nor-)	UHY	1350	128_1	115_1	91_9	82_{100}	65_6	161	2946
Selegiline-M (nor-) AC	U+UHYAC	1735	214_1	124_{31}	91_{10}	82_{100}	65_6	257	2949
Selegiline-M (nor-HO-)	UHY	1550	135_1	107_5	82_{100}			192	2947
Selegiline-M (nor-HO-) 2AC	U+UHYAC	2030	272_1	176_{10}	134_{19}	124_{33}	82_{100}	481	2951
Serotonin 3ME	G U UHY U+UHYAC	2040	218_{16}	160_{10}	145_7	117_{10}	58_{100}	266	4059
Sertraline	G P-I U	2260	304_{15}	274_{100}	262_{34}	159_{59}	115_{27}	638	4641
Sertraline AC	U+UHYAC	2760	347_{65}	290_{100}	274_{88}	159_{56}	74_{46}	834	4640
Sertraline HFB		2525	501_{95}	332_{26}	274_{100}	159_{57}	128_{49}	1164	7690
Sertraline PFP		2515	451_{61}	436_{19}	274_{100}	202_{40}	159_{66}	1114	7689
Sertraline TFA		2520	401_{99}	400_{100}	274_{90}	202_{44}	159_{56}	1019	7688
Sertraline TMS		2530	377_7	362_{14}	348_{36}	334_{22}	274_{100}	942	7691
Sertraline -CH5N	G P-I U+UHYAC	2275*	274_{100}	239_{43}	202_{57}	159_{57}	128_{85}	483	4682
Sertraline-M (di-HO-ketone) -H2O enol 2AC	U+UHYAC	2890*	388_8	346_{24}	304_{100}	275_8	176_{10}	978	4685
Sertraline-M (HO-) 2AC	U+UHYAC	3015	405_{36}	348_{24}	332_{67}	290_{100}	159_{12}	1030	4681
Sertraline-M (HO-ketone) AC	U+UHYAC	2660*	348_{22}	290_{75}	288_{100}	261_{57}	227_{35}	837	5311
Sertraline-M (HO-ketone) -H2O enol AC	U+UHYAC	2600*	330_{17}	290_{63}	288_{100}	218_{64}	189_{55}	757	4683
Sertraline-M (ketone)	U+UHYAC	2480*	290_{99}	248_{38}	227_{100}	199_{47}	163_{30}	562	5310
Sertraline-M (ketone) enol AC	U+UHYAC	2530*	332_5	290_{100}	247_5	212_{15}	189_6	767	4684
Sertraline-M (nor-)	UHY	2400	290_{14}	274_{26}	159_{45}	130_{67}	119_{100}	568	4643
Sertraline-M (nor-) AC	U+UHYAC	2700	333_6	274_{100}	239_{28}	159_{32}	115_{37}	772	4642
Sertraline-M (nor-) HFB		2325	487_7	274_{100}	203_{23}	159_{31}	128_{43}	1154	7194
Sertraline-M (nor-) PFP	UHYPFP	2350	437_{11}	274_{100}	203_{28}	159_{38}	128_{60}	1092	7189
Sertraline-M (nor-) TFA	UHYTFA	2300	387_{10}	274_{100}	202_{25}	159_{32}	128_{48}	975	7188
Sertraline-M (nor-) TMS		2350	362_{12}	348_{14}	274_{100}	217_{37}	73_{67}	894	7190
Sertraline-M/artifact	U+UHYAC	2320*	272_{87}	236_{30}	202_{100}	118_{11}	100_{36}	477	4686
Sethoxydim		2390	281_9	219_{31}	178_{100}	149_{62}	108_{24}	745	3653
Sibutramine		1870	137_2	128_2	114_{100}	72_{33}	58_{13}	514	5725
Sibutramine-M (bis-nor-)		1950	194_1	165_3	137_3	130_4	86_{100}	386	5729
Sibutramine-M (bis-nor-) AC	U+UHYAC	2155	293_1	165_2	137_5	128_{67}	86_{100}	581	5892
Sibutramine-M (bis-nor-) HFB		1940	363_{12}	240_{23}	165_{100}	137_{42}	69_{47}	1126	5747
Sibutramine-M (bis-nor-) PFP		1900	313_{10}	190_{16}	165_{100}	137_{36}	69_{28}	1009	5748
Sibutramine-M (bis-nor-) TFA		1875	263_{13}	165_{100}	137_{46}	102_{17}	69_{40}	835	5731
Sibutramine-M (bis-nor-) TMS		2450	308_2	266_2	158_{100}	102_{12}	73_{62}	729	5732
Sibutramine-M (bis-nor-) formyl artifcat		1920	263_1	221_7	179_4	165_{11}	98_{100}	435	5730
Sibutramine-M (nor-)		1840	128_2	115_2	100_{100}	58_{33}		446	5726

Table 1-8-1: Compounds in order of names

Name	Detected	RI	Typical ions and intensities					Page	Entry
Sibutramine-M (nor-) AC	U+UHYAC	2160	307 $_1$	165 $_1$	142 $_{75}$	100 $_{100}$	58 $_{20}$	650	5891
Sibutramine-M (nor-) HFB		1990	296 $_{100}$	254 $_{31}$	240 $_{50}$	210 $_{12}$	69 $_{51}$	1129	5745
Sibutramine-M (nor-) PFP		1975	246 $_{100}$	204 $_{35}$	190 $_{47}$	160 $_{11}$	69 $_{36}$	1044	5746
Sibutramine-M (nor-) TFA		1950	196 $_{100}$	154 $_{31}$	140 $_{34}$	128 $_{19}$	69 $_{40}$	887	5727
Sibutramine-M (nor-) TMS		2460	322 $_1$	172 $_{100}$	116 $_2$	73 $_{21}$		795	5728
Sigmodal	P G U	2055	237 $_{24}$	193 $_{11}$	167 $_{100}$	122 $_{19}$	78 $_{32}$	692	965
Sigmodal 2ME		1910	265 $_{28}$	195 $_{100}$	138 $_{18}$			823	966
Sildenafil		3400	474 $_1$	404 $_6$	99 $_{100}$	70 $_8$	56 $_{31}$	1143	5713
Sildenafil ME		3390	488 $_1$	418 $_{22}$	99 $_{100}$	70 $_8$	56 $_{31}$	1156	6522
Sildenafil TMS		4030	476 $_{37}$	454 $_4$	99 $_{100}$	73 $_{23}$	56 $_{33}$	1184	5714
Simazine	G P-I U	1690	201 $_{100}$	186 $_{67}$	173 $_{46}$	158 $_{28}$	68 $_{83}$	221	1326
Simazine-M (deethyl-)	U	1730	173 $_{100}$	158 $_{97}$	145 $_{77}$	130 $_{18}$	68 $_{78}$	160	4236
Skatole	U	1340	131 $_{55}$	130 $_{100}$	103 $_{10}$	77 $_{16}$	65 $_{11}$	113	4218
Skatole-M (HO-)	U	1370	147 $_{75}$	146 $_{100}$	117 $_{15}$			125	819
Sorbitol 6AC	U+UHYAC	2090*	361 $_{15}$	289 $_{39}$	187 $_{55}$	145 $_{68}$	115 $_{100}$	1087	1966
Sorbitol 6HFB		1540*	521 $_2$	478 $_5$	307 $_{58}$	240 $_{34}$	169 $_{100}$	1208	5808
Sorbitol 6PFP		1530*	378 $_4$	257 $_{23}$	219 $_{48}$	190 $_{29}$	119 $_{100}$	1207	5807
Sorbitol 6TFA		1435*	435 $_1$	321 $_3$	278 $_{14}$	265 $_{12}$	69 $_{100}$	1204	5806
Sotalol		9999	272 $_4$	239 $_1$	199 $_3$	122 $_4$	72 $_{100}$	479	1368
Sotalol TMSTFA		2410	425 $_1$	272 $_{100}$	193 $_4$	126 $_7$	73 $_{66}$	1098	6173
Sotalol -H2O AC	U+UHYAC	2675	296 $_{38}$	217 $_{29}$	175 $_{99}$	133 $_{46}$	84 $_{42}$	598	1369
Sotalol-M/artifact (amino-) -H2O 2AC	U+UHYAC	2500	260 $_{100}$	218 $_{71}$	203 $_{54}$	133 $_{37}$	84 $_{32}$	423	1710
Sparfloxacin		3455	392 $_{27}$	348 $_7$	322 $_{100}$	278 $_{40}$	70 $_6$	992	6104
Sparfloxacin -CO2		3190	348 $_{33}$	313 $_7$	278 $_{100}$	235 $_6$	208 $_4$	839	6105
Sparteine	G U	1785	234 $_{24}$	193 $_{27}$	137 $_{90}$	98 $_{95}$		321	967
Sparteine-M (oxo-)	U UHY U+UHYAC	2230	248 $_{42}$	149 $_{51}$	136 $_{100}$	98 $_{26}$	84 $_{18}$	373	2877
Sparteine-M (oxo-HO-)	U	2290	264 $_{74}$	165 $_{13}$	150 $_{45}$	136 $_{100}$	98 $_{36}$	441	2878
Sparteine-M (oxo-HO-) enol 2AC	U+UHYAC	2550	348 $_5$	306 $_{21}$	264 $_{62}$	134 $_{100}$	121 $_{50}$	840	2880
Sparteine-M (oxo-HO-) -H2O	U	2205	246 $_{13}$	148 $_4$	134 $_6$	98 $_{100}$	84 $_7$	365	2879
Spirapril ET		3440	421 $_{10}$	289 $_8$	234 $_{100}$	160 $_{21}$	91 $_{22}$	1160	7513
Spirapril ME		3390	407 $_5$	275 $_6$	234 $_{100}$	160 $_{16}$	91 $_{47}$	1148	7512
Spirapril -H2O		3595	448 $_{10}$	344 $_{78}$	298 $_{100}$	117 $_{49}$	91 $_{80}$	1110	7511
Spironolactone -CH3COSH	P UHY U+UHYAC	3250*	340 $_{100}$	325 $_{18}$	267 $_{80}$	227 $_{15}$		808	2344
Squalene	G P U UHY U+UHYAC	2800*	410 $_1$	341 $_1$	137 $_{10}$	81 $_{42}$	69 $_{100}$	1043	968
Stanozolol		3085	328 $_{22}$	175 $_6$	133 $_{21}$	96 $_{100}$	94 $_{95}$	751	2816
Stanozolol AC		2120	370 $_{18}$	257 $_{19}$	138 $_{97}$	96 $_{100}$	94 $_{95}$	921	2817
Stanozolol 2TMS		3025	472 $_{13}$	342 $_3$	168 $_7$	143 $_{100}$	75 $_8$	1141	3984
Stearamide	P U UHY U+UHYAC	2400	283 $_1$	240 $_2$	128 $_6$	72 $_{35}$	59 $_{100}$	531	5346
Stearic acid	P G U UHY U+UHYAC	2170*	284 $_{60}$	241 $_{25}$	185 $_{27}$	129 $_{40}$	73 $_{100}$	537	969
Stearic acid ET		2140*	312 $_9$	269 $_{10}$	157 $_{28}$	101 $_{62}$	88 $_{100}$	678	5406
Stearic acid ME	G P	2130*	298 $_{18}$	255 $_{13}$	143 $_{19}$	87 $_{56}$	74 $_{100}$	608	970
Stearic acid TMS		2640*	356 $_{19}$	341 $_{100}$	145 $_{33}$	117 $_{92}$	73 $_{120}$	871	4017
Stearic acid glycerol ester 2AC	U+UHYAC	2790*	382 $_6$	267 $_{26}$	159 $_{100}$	98 $_{30}$	84 $_{19}$	1104	5413
Stearic acid glycerol ester 2TMS		2780*	488 $_2$	400 $_{67}$	205 $_{23}$	147 $_{63}$	73 $_{150}$	1166	7450
Stearyl alcohol		2020*	270 $_1$	252 $_2$	224 $_4$	97 $_{65}$	55 $_{100}$	472	2356
Steviol		2600*	318 $_{41}$	300 $_{17}$	260 $_{17}$	121 $_{100}$	55 $_{45}$	705	3342
Steviol ME		2530*	332 $_{22}$	274 $_{10}$	254 $_{11}$	146 $_{20}$	121 $_{100}$	771	3343
Steviol MEAC		2580*	374 $_{25}$	332 $_{32}$	314 $_{40}$	146 $_{18}$	121 $_{100}$	934	4300
Stevioside artifact (isosteviol)		2620*	318 $_{25}$	300 $_{21}$	121 $_{66}$	109 $_{71}$	55 $_{100}$	705	3680
Stevioside artifact (isosteviol) ME		2520*	332 $_{73}$	300 $_{71}$	273 $_{100}$	121 $_{81}$	55 $_{71}$	771	3681
Stevioside-M (steviol)		2600*	318 $_{41}$	300 $_{17}$	260 $_{17}$	121 $_{100}$	55 $_{45}$	705	3342
Stevioside-M (steviol) ME		2530*	332 $_{22}$	274 $_{10}$	254 $_{11}$	146 $_{20}$	121 $_{100}$	771	3343
Stevioside-M (steviol) MEAC		2580*	374 $_{25}$	332 $_{32}$	314 $_{40}$	146 $_{18}$	121 $_{100}$	934	4300
Stigma-3,5-dien-7-one		3630*	410 $_{22}$	269 $_{11}$	187 $_{25}$	174 $_{100}$	161 $_{22}$	1042	5584
Stigmast-3,5-ene		3300*	396 $_{100}$	381 $_{25}$	147 $_{80}$	105 $_{64}$	81 $_{65}$	1008	5626
Stigmast-5-en-3-ol		3265*	414 $_{62}$	329 $_{39}$	303 $_{44}$	105 $_{88}$	55 $_{90}$	1052	5622
Stigmast-5-en-3-ol -H2O		3300*	396 $_{100}$	381 $_{25}$	147 $_{80}$	105 $_{64}$	81 $_{65}$	1008	5626
Stigmasterol		3210*	412 $_{23}$	271 $_{24}$	255 $_{28}$	69 $_{66}$	55 $_{70}$	1047	5621
Stigmasterol -H2O		3285*	394 $_{100}$	255 $_{49}$	145 $_{53}$	81 $_{75}$	55 $_{98}$	1000	5625
Strychnine		3120	334 $_{100}$	167 $_{32}$	130 $_{36}$	107 $_{28}$	79 $_{25}$	780	971
Sublimate		9999*	272 $_{74}$	202 $_{100}$				476	972
Sufentanil		2730	289 $_{100}$	158 $_7$	140 $_{25}$	110 $_{21}$	77 $_{13}$	974	6791
Sufentanil HY		2650	330 $_1$	233 $_{100}$	158 $_{23}$	140 $_{16}$	96 $_{18}$	760	6792
Sulazepam		2640	300 $_{100}$	273 $_{68}$	237 $_{71}$	227 $_{69}$	74 $_{71}$	613	4029
Sulazepam HY	UHY U+UHYAC	2100	245 $_{95}$	228 $_{38}$	193 $_{29}$	105 $_{38}$	77 $_{100}$	358	272
Sulazepam HYAC	U+UHYAC	2260	287 $_{11}$	244 $_{100}$	228 $_{39}$	182 $_{49}$	77 $_{70}$	546	2542
Sulfabenzamide AC		2720	318 $_3$	282 $_4$	118 $_{100}$	105 $_{54}$	77 $_{41}$	702	3164
Sulfabenzamide ME		2700	290 $_4$	226 $_8$	118 $_{100}$	105 $_{51}$	77 $_{39}$	563	3149
Sulfabenzamide MEAC		2750	332 $_{38}$	184 $_{14}$	118 $_{100}$	105 $_{54}$	77 $_{52}$	768	3165
Sulfabenzamide 2ME		2770	304 $_{16}$	240 $_4$	118 $_{100}$	105 $_{48}$	77 $_{39}$	634	3150
Sulfabenzamide 2MEAC		2650	346 $_{38}$	212 $_{12}$	118 $_{100}$	105 $_{30}$	77 $_{29}$	831	3166
Sulfabenzamide-M	G P UHY	2185	172 $_{57}$	156 $_{55}$	108 $_{50}$	92 $_{74}$	65 $_{100}$	159	973
Sulfabenzamide-M AC	U+UHYAC	2690	214 $_{31}$	172 $_{100}$	156 $_{54}$	108 $_{42}$	92 $_{46}$	254	974
Sulfabenzamide-M ME	UME	2135	186 $_{68}$	156 $_{61}$	108 $_{52}$	92 $_{100}$	65 $_{78}$	187	3136

Table 1-8-1: Compounds in order of names   Sulfurylamine

Name	Detected	RI	Typical ions and intensities					Page	Entry
Sulfabenzamide-M MEAC		2600	228_{59}	186_{100}	156_{63}	108_{30}	92_{33}	299	3148
Sulfabenzamide-M 4ME		2095	228_{44}	184_{30}	136_{100}	120_{70}	77_{29}	300	4098
Sulfadiazine ME		2625	199_{100}	184_2	108_{16}	92_{24}	65_{31}	438	3135
Sulfadiazine MEAC		3710	241_{100}	199_{14}	108_7	92_7		644	3158
Sulfadimethoxine 2TMS		3030	439_2	390_{35}	375_{100}	212_{26}	89_{25}	1119	5866
Sulfaethidole		2620	284_{69}	220_{21}	156_{33}	108_{38}	92_{100}	531	1862
Sulfaethidole AC		2490	326_{34}	283_{12}	213_{100}	136_{22}	108_{62}	737	1863
Sulfaethidole ME		3060	298_{100}	234_{11}	190_{16}	92_{27}	83_{27}	604	3151
Sulfaethidole 2ME		2840	234_{28}	161_{31}	106_{100}	92_{60}	65_{37}	675	3152
Sulfaethidole 2MEAC		3410	354_{27}	276_{57}	203_{71}	148_{100}	106_{86}	860	3159
Sulfaethidole-M	G P UHY	2185	172_{57}	156_{55}	108_{50}	92_{74}	65_{100}	159	973
Sulfaethidole-M AC	U+UHYAC	2690	214_{31}	172_{100}	156_{54}	108_{42}	92_{46}	254	974
Sulfaethidole-M ME	UME	2135	186_{68}	156_{61}	108_{52}	92_{100}	65_{78}	187	3136
Sulfaethidole-M MEAC		2600	228_{59}	186_{100}	156_{63}	108_{30}	92_{33}	299	3148
Sulfaethidole-M 4ME		2095	228_{44}	184_{30}	136_{100}	120_{70}	77_{29}	300	4098
Sulfaguanole ME		2905	323_{49}	249_{14}	203_{100}	178_8	57_{80}	725	3153
Sulfaguanole-M	G P UHY	2185	172_{57}	156_{55}	108_{50}	92_{74}	65_{100}	159	973
Sulfaguanole-M AC	U+UHYAC	2690	214_{31}	172_{100}	156_{54}	108_{42}	92_{46}	254	974
Sulfaguanole-M ME	UME	2135	186_{68}	156_{61}	108_{52}	92_{100}	65_{78}	187	3136
Sulfaguanole-M MEAC		2600	228_{59}	186_{100}	156_{63}	108_{30}	92_{33}	299	3148
Sulfaguanole-M 4ME		2095	228_{44}	184_{30}	136_{100}	120_{70}	77_{29}	300	4098
Sulfamerazine		2625	199_{100}	140_3	108_{28}	92_{55}	65_{56}	438	4267
Sulfamethizole ME	UME	2660	284_{85}	156_{49}	92_{100}	65_{86}		531	1322
Sulfamethizole-M	G P UHY	2185	172_{57}	156_{55}	108_{50}	92_{74}	65_{100}	159	973
Sulfamethizole-M AC	U+UHYAC	2690	214_{31}	172_{100}	156_{54}	108_{42}	92_{46}	254	974
Sulfamethizole-M ME	UME	2135	186_{68}	156_{61}	108_{52}	92_{100}	65_{78}	187	3136
Sulfamethizole-M MEAC		2600	228_{59}	186_{100}	156_{63}	108_{30}	92_{33}	299	3148
Sulfamethizole-M 4ME		2095	228_{44}	184_{30}	136_{100}	120_{70}	77_{29}	300	4098
Sulfamethoxazole ME	P	2500	267_2	203_{18}	162_{46}	108_{71}	92_{100}	453	3154
Sulfamethoxazole MEAC		3255	309_8	245_{48}	230_{67}	161_{100}	134_{86}	658	3160
Sulfamethoxazole 2ME	P	2460	281_2	203_{16}	162_{44}	108_{71}	92_{100}	521	3155
Sulfamethoxazole 2TMS		2515	397_1	382_7	228_{58}	178_{89}	73_{100}	1009	4597
Sulfamethoxazole impurity		1025	125_{100}	98_{12}	93_{23}	80_{28}	65_7	109	6351
Sulfamethoxazole-M	G P UHY	2185	172_{57}	156_{55}	108_{50}	92_{74}	65_{100}	159	973
Sulfamethoxazole-M AC	U+UHYAC	2690	214_{31}	172_{100}	156_{54}	108_{42}	92_{46}	254	974
Sulfamethoxazole-M ME	UME	2135	186_{68}	156_{61}	108_{52}	92_{100}	65_{78}	187	3136
Sulfamethoxazole-M MEAC		2600	228_{59}	186_{100}	156_{63}	108_{30}	92_{33}	299	3148
Sulfamethoxazole-M 4ME		2095	228_{44}	184_{30}	136_{100}	120_{70}	77_{29}	300	4098
Sulfametoxydiazine MEAC		3620	271_{100}	229_8	139_4	92_6	65_9	845	3161
Sulfametoxydiazine 3ME	PME	2925	322_1	229_{100}	138_6	92_{21}	65_{21}	721	3156
Sulfametoxydiazine-M	G P UHY	2185	172_{57}	156_{55}	108_{50}	92_{74}	65_{100}	159	973
Sulfametoxydiazine-M AC	U+UHYAC	2690	214_{31}	172_{100}	156_{54}	108_{42}	92_{46}	254	974
Sulfametoxydiazine-M ME	UME	2135	186_{68}	156_{61}	108_{52}	92_{100}	65_{78}	187	3136
Sulfametoxydiazine-M MEAC		2600	228_{59}	186_{100}	156_{63}	108_{30}	92_{33}	299	3148
Sulfametoxydiazine-M 4ME		2095	228_{44}	184_{30}	136_{100}	120_{70}	77_{29}	300	4098
Sulfanilamide	G P UHY	2185	172_{57}	156_{55}	108_{50}	92_{74}	65_{100}	159	973
Sulfanilamide AC	U+UHYAC	2690	214_{31}	172_{100}	156_{54}	108_{42}	92_{46}	254	974
Sulfanilamide ME	UME	2135	186_{68}	156_{61}	108_{52}	92_{100}	65_{78}	187	3136
Sulfanilamide MEAC		2600	228_{59}	186_{100}	156_{63}	108_{30}	92_{33}	299	3148
Sulfanilamide 4ME		2095	228_{44}	184_{30}	136_{100}	120_{70}	77_{29}	300	4098
Sulfaperin 2MEAC		3420	255_{100}	213_7	122_4	93_5	65_7	779	3162
Sulfaperin 3ME		2795	306_1	213_{100}	198_1	92_{18}	65_{21}	645	3157
Sulfaperin-M	G P UHY	2185	172_{57}	156_{55}	108_{50}	92_{74}	65_{100}	159	973
Sulfaperin-M AC	U+UHYAC	2690	214_{31}	172_{100}	156_{54}	108_{42}	92_{46}	254	974
Sulfaperin-M ME	UME	2135	186_{68}	156_{61}	108_{52}	92_{100}	65_{78}	187	3136
Sulfaperin-M MEAC		2600	228_{59}	186_{100}	156_{63}	108_{30}	92_{33}	299	3148
Sulfaperin-M 4ME		2095	228_{44}	184_{30}	136_{100}	120_{70}	77_{29}	300	4098
Sulfapyridine	P G U	2600	184_{100}	156_4	108_{24}	92_{51}	65_{60}	374	2864
Sulfaquinoxaline		3065	300_{11}	236_{90}	108_{63}	92_{100}	65_{76}	614	3917
Sulfaquinoxaline AC		3440	342_{15}	277_{100}	235_{40}	90_{36}	65_{41}	816	6058
Sulfathiourea-M	G P UHY	2185	172_{57}	156_{55}	108_{50}	92_{74}	65_{100}	159	973
Sulfathiourea-M AC	U+UHYAC	2690	214_{31}	172_{100}	156_{54}	108_{42}	92_{46}	254	974
Sulfathiourea-M ME	UME	2135	186_{68}	156_{61}	108_{52}	92_{100}	65_{78}	187	3136
Sulfathiourea-M MEAC		2600	228_{59}	186_{100}	156_{63}	108_{30}	92_{33}	299	3148
Sulfathiourea-M 4ME		2095	228_{44}	184_{30}	136_{100}	120_{70}	77_{29}	300	4098
Sulfinpyrazone		2285	278_{35}	105_{31}	77_{99}			1028	975
Sulfinpyrazone ME		2235	292_{36}	182_7	144_{14}	105_{27}	77_{100}	1058	3145
Sulforidazine	G P-I U+UHYAC	3415	402_{13}	290_5	197_6	98_{100}	70_{10}	1023	394
Sulforidazine-M (nor-) AC	U+UHYAC	3800	430_{17}	277_{10}	154_{101}	84_{37}		1080	1293
Sulforidazine-M (ring)	U+UHYAC	3180	277_{100}	198_{64}				499	1292
Sulfotep	G	1650*	322_{100}	266_{30}	238_{27}	202_{30}	97_{48}	721	2603
Sulfur mole	G	1885*	256_{20}	192_{47}	160_{38}	128_{28}	64_{100}	404	6455
Sulfuric acid 2TMS		<1000*	227_{23}	147_{100}	131_2	93_{20}	73_{35}	349	5695
Sulfurylamine		1625	96_{63}	82_7	80_{100}	64_{29}		98	6055

Sulfurylamine ME   **Table 1-8-1:** Compounds in order of names

Name	Detected	RI	Typical ions and intensities					Page	Entry
Sulfurylamine ME		1345	110_{82}	109_{69}	94_{100}	80_{42}	64_{54}	103	6057
Sulfurylamine 2ME		1140	124_{64}	94_{100}	78_{13}	60_{33}		109	6056
Sulindac		2890*	312_{45}	297_{100}	265_{7}	233_{43}	117_{10}	868	1527
Sulindac ME		3220*	370_{39}	354_{43}	295_{13}	248_{34}	233_{100}	919	1528
Sulpiride ME	PME-I UHYME U+UHYAC	3125	355_{1}	228_{1}	134_{1}	98_{100}	70_{4}	866	3211
Sulpiride 2ME	UHYME	2995	369_{1}	368_{1}	242_{1}	134_{4}	98_{100}	917	3144
Sulpiride -SO2NH	G U+UHYAC UHYME	2295	154_{1}	135_{3}	111_{2}	98_{100}	70_{6}	432	976
Sulprofos		2260*	322_{90}	280_{11}	156_{100}	139_{53}	113_{53}	721	3456
Sultiame	G P U UHY U+UHYAC	3000	290_{100}	225_{10}	184_{38}	168_{14}	104_{12}	562	3718
Sultiame ME	PME UME UHYME	2880	304_{100}	274_{29}	226_{10}	198_{12}	104_{15}	633	3729
Sultiame 2ME	UME UHYME	2815	318_{100}	274_{55}	226_{31}	210_{18}	104_{16}	702	3728
Sultiame -SO2NH	U UHY U+UHYAC	2035	211_{43}	146_{23}	119_{19}	105_{100}	77_{64}	246	3719
Sumatriptan		2745	295_{3}	237_{2}	156_{6}	143_{27}	58_{100}	591	7696
Sumatriptan AC		2855	337_{1}	237_{1}	156_{5}	143_{8}	58_{100}	793	7697
Sumatriptan HFB		2575	491_{1}	433_{2}	339_{3}	142_{24}	58_{100}	1158	7700
Sumatriptan ME		2700	309_{2}	156_{6}	143_{16}	115_{6}	58_{100}	661	7702
Sumatriptan PFP		2560	441_{1}	289_{2}	156_{3}	142_{13}	58_{100}	1100	7699
Sumatriptan TFA		2575	391_{1}	239_{3}	156_{5}	142_{22}	58_{100}	988	7698
Sumatriptan 2TMS		2745	439_{7}	381_{8}	273_{11}	215_{15}	58_{100}	1097	7701
Suxibuzone ME		3020	452_{29}	264_{29}	183_{28}	115_{100}	77_{69}	1116	2820
Suxibuzone artifact	G P U	2375	308_{47}	252_{14}	183_{100}	77_{89}		657	862
Suxibuzone artifact TMS		2575	380_{8}	337_{8}	246_{29}	77_{39}	73_{100}	953	5442
Suxibuzone-M/artifact (HO-alkyl-phenylbutazone) ME	P UME	2500	338_{58}	266_{15}	183_{100}	162_{47}	77_{49}	799	6383
Suxibuzone-M/artifact (HOOC-phenylbutazone) 2ME	P UME	2590	366_{100}	266_{49}	183_{74}	105_{34}	77_{130}	905	6385
Suxibuzone-M/artifact (oxo-phenylbutazone) ME	P UME	2480	336_{100}	266_{43}	183_{64}	105_{28}	77_{120}	789	6384
Suxibuzone-M/artifact (phenylbutazone) ME	P UME	2290	322_{73}	266_{22}	183_{100}	118_{19}	77_{91}	724	863
Swep	G P-I U UHY U+UHYAC	1850	219_{100}	187_{77}	174_{86}	160_{47}	133_{56}	267	850
Synephrine 3AC	U+UHYAC	2185	293_{1}	250_{7}	220_{24}	123_{61}	86_{130}	580	5176
Synephrine formyl artifact		1590	179_{1}	121_{3}	107_{4}	77_{4}	57_{100}	172	5432
Synephrine formyl artifact ME		1590	193_{1}	135_{12}	121_{12}	77_{29}	57_{250}	202	5434
Synephrine -H2O 2AC	U+UHYAC	2140	233_{26}	191_{54}	149_{100}	107_{21}	56_{36}	316	5433
Talbutal	P G U	1705	167_{100}	153_{14}	124_{32}	97_{32}		290	977
Talbutal 2ME		1600	234_{2}	195_{100}	181_{34}	138_{40}	111_{33}	392	978
Talinolol		2350	281_{5}	167_{9}	135_{5}	86_{100}	57_{21}	895	4268
Talinolol AC		2420	323_{14}	206_{15}	98_{63}	86_{76}	57_{100}	1032	4261
Talinolol TMS		1980	321_{5}	220_{2}	101_{7}	86_{100}	57_{25}	1091	6191
Talinolol formyl artifact		2425	293_{15}	135_{16}	86_{100}	70_{62}	57_{94}	938	4269
Tamoxifen		2610	371_{2}	253_{1}	91_{2}	72_{24}	58_{50}	925	5706
Tartaric acid 4TMS		1615*	423_{2}	292_{11}	219_{8}	147_{26}	73_{100}	1094	4301
TCDI		1535	209_{19}	165_{42}	123_{15}	97_{100}	81_{19}	242	3601
TCDI artifact/impurity		1310*	164_{100}	149_{36}	135_{75}	97_{22}	91_{21}	145	3590
TCDI intermediate (DMCC)		<1000	152_{7}	151_{13}	137_{18}	109_{100}	84_{12}	132	3580
TCDI precursor (bromothiophene)		<1000*	164_{99}	162_{100}	117_{4}	83_{80}	57_{31}	140	3609
TCDI precursor (dimethylamine)		<1000	45_{50}	44_{100}	28_{70}			90	3618
TCM		1975	251_{9}	208_{9}	165_{62}	123_{13}	97_{100}	385	3591
TCM artifact/impurity		1310*	164_{100}	149_{36}	135_{75}	97_{22}	91_{21}	145	3590
TCM intermediate (MCC)		1560	194_{9}	164_{9}	151_{100}	124_{41}	56_{21}	206	3578
TCM intermediate (MCC) -HCN		1260	167_{81}	152_{52}	108_{100}	94_{36}	81_{77}	153	3579
TCM precursor (bromothiophene)		<1000*	164_{99}	162_{100}	117_{4}	83_{80}	57_{31}	140	3609
TCM precursor (morpholine)		<1000	87_{12}	57_{23}	42_{6}			97	3612
TCPY		1810	235_{19}	192_{58}	165_{34}	97_{100}	70_{39}	325	3603
TCPY artifact/impurity		1310*	164_{100}	149_{36}	135_{75}	97_{22}	91_{21}	145	3590
TCPY intermediate		1255	178_{5}	150_{14}	135_{100}	121_{13}	70_{25}	169	3583
TCPY intermediate (PYCC) -HCN		1180	151_{93}	150_{100}	136_{76}	122_{71}	95_{70}	130	3585
TCPY precursor (bromothiophene)		<1000*	164_{99}	162_{100}	117_{4}	83_{80}	57_{31}	140	3609
TCPY precursor (pyrrolidine)		<1000	71_{19}	70_{26}	43_{79}			92	3608
Tebuthiuron ME		1900	242_{5}	208_{4}	171_{6}	126_{23}	72_{100}	352	4096
Tebuthiuron -C2H3NO ME		1500	185_{29}	170_{100}	156_{19}	102_{7}	88_{51}	186	4097
Tecnazene		1605	259_{41}	215_{55}	203_{100}	108_{64}	73_{42}	415	3461
Temazepam	P UGLUC	2625	300_{33}	271_{100}	256_{23}	228_{16}	77_{30}	614	417
Temazepam AC	UGLUCAC	2730	342_{6}	300_{40}	271_{100}	255_{16}	77_{17}	816	2099
Temazepam ME		2600	314_{60}	271_{100}	255_{46}			684	418
Temazepam TMS		2665	372_{23}	343_{100}	283_{26}	257_{38}	73_{54}	926	4598
Temazepam artifact-1	G	2475	256_{19}	241_{7}	179_{100}	163_{7}	77_{8}	405	5780
Temazepam artifact-2	G	2815	270_{64}	269_{100}	254_{12}	228_{26}	191_{5}	468	5779
Temazepam HY	UHY U+UHYAC	2100	245_{95}	228_{38}	193_{29}	105_{38}	77_{100}	358	272
Temazepam HYAC	U+UHYAC	2260	287_{11}	244_{100}	228_{39}	182_{49}	77_{70}	546	2542
Temazepam-M (HO-) HY	UHY	2580	261_{91}	260_{100}	244_{42}	209_{21}	121_{17}	424	2048
Temazepam-M (HO-) HYAC	U+UHYAC	2600	303_{77}	260_{100}	244_{47}	121_{11}		626	2060
Temazepam-M (nor-)	P G UGLUC	2320	268_{98}	239_{57}	233_{52}	205_{66}	77_{100}	542	579
Temazepam-M TMS		2635	356_{9}	341_{100}	312_{56}	239_{12}	135_{21}	876	4577
Temazepam-M 2TMS		2200	430_{36}	429_{63}	340_{10}	313_{14}	73_{70}	1080	5499
Temazepam-M artifact-1	P-I UHY U+UHYAC	2060	240_{59}	239_{100}	205_{81}	177_{16}	151_{9}	342	300
Temazepam-M HY	UHY	2050	231_{80}	230_{95}	154_{23}	105_{38}	77_{100}	308	419

Table 1-8-1: Compounds in order of names   1,2,3,5-Tetrachlorobenzene

Name	Detected	RI	Typical ions and intensities					Page	Entry
Temazepam-M HYAC	U+UHYAC PHYAC	2245	273 $_{31}$	230 $_{100}$	154 $_{13}$	105 $_{23}$	77 $_{50}$	480	273
Temephos		3205*	466 $_{100}$	357 $_{6}$	339 $_{5}$	203 $_{20}$	125 $_{22}$	1133	3459
Tenamfetamine	U UHY	1495	179 $_{1}$	136 $_{18}$	105 $_{7}$	77 $_{5}$		170	3241
Tenamfetamine	U UHY	1495	179 $_{1}$	136 $_{18}$	77 $_{5}$	51 $_{5}$	44 $_{100}$	171	5518
Tenamfetamine AC	U+UHYAC	1860	221 $_{3}$	162 $_{46}$	135 $_{12}$	86 $_{8}$	77 $_{12}$	277	3263
Tenamfetamine AC	U+UHYAC	1860	221 $_{3}$	162 $_{46}$	135 $_{12}$	86 $_{8}$	44 $_{100}$	277	5519
Tenamfetamine HFB	UHFB	1650	375 $_{4}$	240 $_{6}$	169 $_{10}$	162 $_{50}$	135 $_{100}$	935	5291
Tenamfetamine PFP	UPFP	1605	325 $_{5}$	190 $_{10}$	162 $_{46}$	135 $_{100}$	119 $_{18}$	733	5290
Tenamfetamine TFA	UTFA	1615	275 $_{4}$	162 $_{39}$	135 $_{100}$	105 $_{6}$	77 $_{12}$	488	5289
Tenamfetamine TMS		1735	251 $_{1}$	236 $_{4}$	135 $_{8}$	116 $_{100}$	73 $_{47}$	385	6334
Tenamfetamine formyl artifact	P-I	1520	191 $_{9}$	135 $_{27}$	105 $_{4}$	77 $_{17}$	56 $_{100}$	195	3252
Tenamfetamine R-(-)-enantiomer HFBP		2280	472 $_{2}$	294 $_{15}$	266 $_{100}$	162 $_{66}$	135 $_{8}$	1140	6640
Tenamfetamine S-(+)-enantiomer HFBP		2290	472 $_{2}$	294 $_{16}$	266 $_{90}$	162 $_{69}$	135 $_{9}$	1140	6641
Tenamfetamine-D5 AC		1840	226 $_{13}$	167 $_{100}$	166 $_{93}$	136 $_{48}$	78 $_{28}$	295	5688
Tenamfetamine-D5 HFB		1630	380 $_{10}$	244 $_{10}$	167 $_{31}$	166 $_{29}$	136 $_{100}$	952	6773
Tenamfetamine-D5 2AC		1910	268 $_{3}$	167 $_{100}$	166 $_{81}$	136 $_{47}$	90 $_{59}$	462	5689
Tenamfetamine-D5 R-(-)-enantiomer HFBP		2275	477 $_{18}$	341 $_{6}$	294 $_{27}$	266 $_{100}$	167 $_{59}$	1145	6798
Tenamfetamine-D5 S-(+)-enantiomer HFBP		2285	477 $_{9}$	341 $_{6}$	294 $_{26}$	266 $_{100}$	167 $_{51}$	1145	6799
Tenamfetamine-M (deamino-HO-) AC	U+UHYAC	1620*	222 $_{9}$	162 $_{100}$	135 $_{56}$	104 $_{10}$	77 $_{17}$	281	6410
Tenamfetamine-M 2 HFB	UHFB	1690	360 $_{82}$	333 $_{15}$	240 $_{100}$	169 $_{42}$	69 $_{39}$	1189	6512
Tenocyclidine		1910	249 $_{20}$	206 $_{30}$	165 $_{52}$	97 $_{100}$	84 $_{23}$	378	3589
Tenocyclidine artifact/impurity		1310*	164 $_{100}$	149 $_{36}$	135 $_{75}$	97 $_{22}$	91 $_{21}$	145	3590
Tenocyclidine intermediate (PCC)		1525	192 $_{4}$	191 $_{7}$	164 $_{8}$	149 $_{100}$	122 $_{13}$	198	3581
Tenocyclidine intermediate (PCC) -HCN		1190	165 $_{85}$	164 $_{80}$	150 $_{100}$	136 $_{64}$	122 $_{39}$	149	3582
Tenocyclidine precursor (bromothiophene)		<1000*	164 $_{99}$	162 $_{100}$	117 $_{4}$	83 $_{80}$	57 $_{31}$	140	3609
Tenocyclidine precursor (piperidine)		<1000	85 $_{11}$	84 $_{26}$	70 $_{5}$	56 $_{18}$		96	3615
Tenoxicam 2ME		2690	365 $_{7}$	350 $_{81}$	176 $_{32}$	135 $_{76}$	78 $_{100}$	899	4030
TEPP	G	1590*	290 $_{3}$	263 $_{89}$	235 $_{52}$	179 $_{5}$	161 $_{100}$	562	4086
Terbacil		1850	216 $_{4}$	201 $_{2}$	161 $_{100}$	160 $_{73}$	117 $_{37}$	259	3869
Terbinafine		2230	291 $_{6}$	276 $_{22}$	234 $_{11}$	141 $_{100}$	115 $_{36}$	572	7488
Terbufos		1795*	288 $_{2}$	231 $_{38}$	186 $_{9}$	97 $_{22}$	57 $_{100}$	551	3872
Terbumeton		1790	225 $_{39}$	210 $_{100}$	169 $_{78}$	154 $_{27}$	141 $_{19}$	292	3874
Terbutaline		2430	225 $_{1}$	192 $_{10}$	111 $_{11}$	86 $_{100}$	57 $_{34}$	292	2731
Terbutaline 2ME		2120	253 $_{1}$	220 $_{4}$	168 $_{11}$	139 $_{1}$	86 $_{100}$	395	2735
Terbutaline 2TMS		2050	354 $_{1}$	284 $_{23}$	264 $_{3}$	86 $_{70}$	73 $_{44}$	917	6184
Terbutaline 3AC	U+UHYAC	2375	351 $_{1}$	276 $_{9}$	192 $_{4}$	150 $_{10}$	86 $_{100}$	850	2732
Terbutaline 3TMS		2010	426 $_{4}$	356 $_{100}$	147 $_{5}$	86 $_{93}$	73 $_{48}$	1101	6183
Terbutaline artifact 2ME		2250	265 $_{5}$	250 $_{20}$	220 $_{53}$	164 $_{21}$	99 $_{100}$	447	2736
Terbutaline -H2O 2AC		2040	291 $_{10}$	249 $_{4}$	192 $_{14}$	150 $_{100}$	57 $_{12}$	570	2733
Terbutaline-M/artifact (N-dealkyl-) 3AC		2170	277 $_{6}$	235 $_{18}$	193 $_{35}$	150 $_{10}$	55 $_{100}$	500	2734
Terbutryn		1960	241 $_{72}$	226 $_{100}$	185 $_{74}$	170 $_{49}$	157 $_{51}$	348	3867
Terbutylazine		1805	229 $_{33}$	214 $_{100}$	173 $_{40}$	132 $_{15}$	68 $_{14}$	304	3875
Terephthalic acid diethyl ester		1645*	222 $_{12}$	194 $_{18}$	177 $_{100}$	166 $_{22}$	149 $_{60}$	281	6495
Terephthalic acid ethyl methyl ester		1560*	208 $_{17}$	193 $_{11}$	180 $_{37}$	163 $_{100}$	149 $_{43}$	237	6493
Terephthalic acid monoethyl ester		1715*	194 $_{19}$	166 $_{40}$	149 $_{100}$	121 $_{17}$	65 $_{15}$	204	6497
Terfenadine	G	3700	471 $_{20}$	280 $_{100}$	262 $_{10}$	183 $_{14}$	105 $_{48}$	1140	2237
Terfenadine AC	U+UHYAC	3600	452 $_{1}$	280 $_{100}$	262 $_{9}$	105 $_{22}$	57 $_{21}$	1172	2236
Terfenadine -2H2O	U+UHYAC	3460	435 $_{1}$	262 $_{100}$	115 $_{4}$	91 $_{12}$	57 $_{16}$	1091	2235
Terfenadine-M (benzophenone)	U+UHYAC	1610*	182 $_{31}$	152 $_{3}$	105 $_{100}$	77 $_{70}$	51 $_{39}$	180	1624
Terfenadine-M (N-dealkyl-) -H2O	UHY	2600	249 $_{53}$	248 $_{100}$	191 $_{24}$	165 $_{24}$	129 $_{64}$	377	2219
Terfenadine-M (N-dealkyl-) -H2O AC	U+UHYAC	2550	291 $_{100}$	205 $_{24}$	191 $_{26}$	91 $_{80}$	72 $_{36}$	571	2217
Terfenadine-M (N-dealkyl-oxo-) -2H2O	U+UHYAC	2190	245 $_{82}$	167 $_{100}$	152 $_{17}$	139 $_{21}$	115 $_{16}$	359	2218
Tertatolol		2310	295 $_{4}$	280 $_{17}$	251 $_{17}$	166 $_{53}$	86 $_{100}$	593	4362
Tertatolol AC		2350	337 $_{10}$	322 $_{47}$	166 $_{39}$	112 $_{20}$	86 $_{100}$	794	4361
Tertatolol TMSTFA		2510	463 $_{36}$	392 $_{13}$	242 $_{88}$	191 $_{24}$	166 $_{250}$	1131	6139
Tertatolol formyl artifact	P	2400	307 $_{12}$	292 $_{100}$	141 $_{13}$	96 $_{14}$	57 $_{23}$	650	4363
Testosterone		2620*	288 $_{88}$	246 $_{42}$	124 $_{100}$			555	979
Testosterone AC	U+UHYAC	2750*	330 $_{67}$	288 $_{35}$	228 $_{34}$	147 $_{64}$	124 $_{100}$	760	1864
Testosterone acetate	U+UHYAC	2750*	330 $_{67}$	288 $_{35}$	228 $_{34}$	147 $_{64}$	124 $_{100}$	760	1864
Testosterone dipropionate		3350*	400 $_{73}$	358 $_{37}$	288 $_{21}$	147 $_{6}$	124 $_{100}$	1018	1865
Testosterone enol 2TMS		2690*	432 $_{91}$	417 $_{9}$	209 $_{11}$	195 $_{3}$	73 $_{100}$	1085	3804
Testosterone propionate		2815*	344 $_{100}$	330 $_{23}$	288 $_{26}$	246 $_{14}$	124 $_{37}$	826	1866
Testosterone propionate enol AC		3020*	386 $_{37}$	344 $_{100}$	329 $_{31}$	302 $_{50}$	284 $_{49}$	974	1867
Tetrabenazine	G	2490	317 $_{25}$	274 $_{30}$	261 $_{100}$	191 $_{43}$		699	395
Tetrabenazine-M (O-bis-demethyl-) AC	U+UHYAC	2510	331 $_{13}$	296 $_{73}$	232 $_{36}$	191 $_{100}$	177 $_{49}$	765	396
Tetrabenazine-M (O-bis-demethyl-HO-) 2AC	U+UHYAC	2665	389 $_{77}$	330 $_{89}$	302 $_{99}$	288 $_{89}$	233 $_{67}$	983	398
Tetrabenazine-M (O-demethyl-HO-)	U UHY	2500	319 $_{61}$	318 $_{64}$	274 $_{56}$	205 $_{99}$	191 $_{89}$	710	615
Tetrabenazine-M (O-demethyl-HO-) AC	U+UHYAC	2585	361 $_{60}$	302 $_{100}$	274 $_{70}$	246 $_{48}$	205 $_{67}$	888	397
Tetrabromo-o-cresol	P	2190*	424 $_{100}$	420 $_{15}$	343 $_{40}$	263 $_{11}$	234 $_{8}$	1061	2738
Tetrabromo-o-cresol AC	U+UHYAC	2465*	466 $_{8}$	462 $_{1}$	424 $_{100}$	420 $_{17}$	343 $_{15}$	1129	2739
Tetrabromo-o-cresol ME	UME	2350*	438 $_{100}$	436 $_{68}$	423 $_{61}$	314 $_{32}$	74 $_{52}$	1087	2740
Tetracaine	G	2350	264 $_{1}$	221 $_{2}$	193 $_{6}$	71 $_{53}$	58 $_{100}$	441	1868
Tetracaine-M/artifact (HOOC-) ME		2015	207 $_{17}$	176 $_{7}$	164 $_{100}$	120 $_{3}$	105 $_{7}$	235	1869
1,2,3,5-Tetrachlorobenzene		1370*	216 $_{100}$	214 $_{76}$	179 $_{19}$	143 $_{11}$	108 $_{20}$	253	3472

Table 1-8-1: Compounds in order of names

Name	Detected	RI	Typical ions and intensities					Page	Entry
2,2',5,5'-Tetrachlorobiphenyl		1945*	292 100	290 75	255 27	220 65	184 10	561	881
2,3,7,8-Tetrachlorodibenzo-p-dioxin (TCDD)		----*	322 100	320 80	257 33	194 32	161 27	713	1465
2,3,7,8-Tetrachlorodibenzofuran (TCDF)		----*	306 100	304 85	241 25	171 36	152 21	632	3493
Tetrachloroethylene		<1000*	166 100	164 76	129 74	94 48	47 54	144	3783
Tetrachloromethane		<1000*	117 100	82 51	47 52			130	980
2,3,4,5-Tetrachlorophenol	U	1500*	232 100	230 79	194 15	166 20	131 26	305	3366
Tetrachlorvinphos		2120*	364 1	329 36	240 4	109 40	79 15	896	3190
Tetrachlorvinphos-M/artifact		1710*	256 5	207 100	179 26	143 10	109 15	405	3191
Tetradecane	P	1400*	198 10	99 14	85 33	71 56	57 100	216	2767
Tetradifon		2505*	354 19	227 59	159 100	111 68	75 49	860	3868
Tetraethylene glycol dipivalate	PPIV	1820*	175 1	129 100	113 5	85 18	57 75	892	6427
Tetrahexylammoniumhydrogensulfate artifact-1		1380	185 4	114 100	100 3	79 5	57 8	186	4947
Tetrahexylammoniumhydrogensulfate artifact-2		1725	269 2	198 100	128 42	98 14	58 37	467	4491
Tetrahydrocannabinol	G	2470*	314 85	299 100	271 41	243 29	231 45	687	981
Tetrahydrocannabinol AC	U+UHYAC-I	2450*	356 13	313 35	297 100	243 12	231 24	870	982
Tetrahydrocannabinol ET		2390*	342 90	327 100	313 39	271 27	259 30	818	2531
Tetrahydrocannabinol ME		2360*	328 82	313 100	285 28	257 24	245 27	751	2530
Tetrahydrocannabinol TMS		2405*	386 100	371 86	315 37	303 31	73 80	974	4599
Tetrahydrocannabinol iso-1 PFP		2150*	460 100	445 14	417 70	392 30	377 100	1128	5669
Tetrahydrocannabinol iso-2 PFP		2170*	460 100	445 65	417 75	389 87	297 80	1128	5668
Tetrahydrocannabinol-D3		2450*	317 96	302 100	274 43	258 26	234 55	700	5663
Tetrahydrocannabinol-D3 AC		2750*	359 9	316 18	300 100	274 9	234 10	882	7309
Tetrahydrocannabinol-D3 ME		2355*	331 81	316 100	288 34	257 29	248 39	766	6040
Tetrahydrocannabinol-D3 TMS		2385*	389 100	374 96	346 26	315 59	306 41	984	5670
Tetrahydrocannabinol-D3 iso-1 PFP		2130*	463 60	420 54	395 25	380 100	342 13	1131	5665
Tetrahydrocannabinol-D3 iso-1 TFA		2160*	413 51	370 31	345 14	330 100	232 10	1048	5667
Tetrahydrocannabinol-D3 iso-2 PFP		2150*	463 100	448 65	420 70	389 85	300 81	1132	5664
Tetrahydrocannabinol-D3 iso-2 TFA		2180*	413 100	398 74	370 71	339 62	300 73	1049	5666
Tetrahydrocannabinolic acid 2TMS		2635*	502 3	487 100	413 3	147 6	73 41	1166	4605
Tetrahydrocannabinol-M (11-HO-)		2775*	330 12	299 100	231 10	217 9	193 7	760	4661
Tetrahydrocannabinol-M (11-HO-) 2ME		2580*	358 13	313 100	257 3	231 3		879	4659
Tetrahydrocannabinol-M (11-HO-) 2PFP		2350*	622 15	607 5	551 9	458 100	415 24	1197	4658
Tetrahydrocannabinol-M (11-HO-) 2TFA		2450*	522 13	451 8	408 100	395 13	365 24	1175	4657
Tetrahydrocannabinol-M (11-HO-) 2TMS		2630*	474 5	459 4	403 2	371 100	73 14	1143	4656
Tetrahydrocannabinol-M (11-HO-) -H2O AC		2740*	354 48	312 100	297 19	269 31	91 21	863	4660
Tetrahydrocannabinol-M (HO-nor-delta-9-HOOC-) 2ME	UTHCME-I	2840*	388 42	373 65	329 100	201 24	189 28	981	3466
Tetrahydrocannabinol-M (nor-delta-9-HOOC-) 2ME	UTHCME UGLUCEXME	2620*	372 52	357 79	341 9	313 100	245 6	929	1439
Tetrahydrocannabinol-M (nor-delta-9-HOOC-) 2PFP		2440*	622 35	607 49	459 100	445 76	69 82	1197	4380
Tetrahydrocannabinol-M (nor-delta-9-HOOC-) 2TMS		2470*	488 40	473 51	398 12	371 100	73 66	1156	5671
Tetrahydrocannabinol-M (nor-delta-9-HOOC-)-D3 2ME		2590*	375 39	360 58	356 19	316 100	301 9	938	6187
Tetrahydrocannabinol-M (nor-delta-9-HOOC-)-D3 2PFP		2425*	625 20	610 57	462 100	448 62	432 43	1197	6039
Tetrahydrocannabinol-M (nor-delta-9-HOOC-)-D3 2TMS		2660*	491 44	476 55	374 100	300 15	73 28	1159	5672
Tetrahydrocannabinol-M (oxo-nor-delta-9-HOOC-) 2ME	UTHCME-I	2860*	386 55	371 73	327 100	314 22	189 11	974	3467
Tetrahydrofuran		<1000*	72 40	71 38	42 100	27 31		93	4185
Tetrahydrogestrinone		2660*	312 57	265 52	240 43	227 100	211 34	677	7573
Tetrahydrogestrinone TMS		2490*	384 75	299 98	281 39	270 51	73 90	968	7574
Tetramethrin		2735	331 1	164 100	123 31	107 9	81 13	765	3883
Tetramethylbenzene		1080*	134 25	119 100	105 12	91 23	77 16	115	3791
Tetramethylcitrate		1445*	189 11	157 100	133 4	125 38	59 16	370	5705
Tetrasul		2310*	324 55	322 44	252 100	217 7	108 25	721	3879
Tetrazepam	G P U	2400	288 52	259 17	253 100	225 13		552	616
Tetrazepam +H2O iso-1 ALHY		2350	267 76	196 14	179 35	168 100	140 11	454	2094
Tetrazepam +H2O iso-1 ALHYAC		2420	309 40	249 9	168 100	140 14		659	2095
Tetrazepam +H2O iso-2 ALHY		2370	267 36	168 100	140 14	77 15		454	2093
Tetrazepam +H2O iso-2 ALHYAC		2480	309 37	249 7	168 100	140 14	111 10	659	2096
Tetrazepam AC		2590	330 9	288 100	259 28	244 9	180 8	759	5699
Tetrazepam iso-1 HY	UHY U+UHYAC	2220	249 46	234 23	220 35	207 100		374	303
Tetrazepam iso-2 HY	G P U+UHYAC	2280	249 52	220 37	207 100	178 15	165 16	374	2059
Tetrazepam-M (di-HO-) -2H2O HY	UHY U+UHYAC	2100	245 95	228 38	193 29	105 38	77 100	358	272
Tetrazepam-M (di-HO-) -2H2O HYAC	U+UHYAC	2260	287 11	244 100	228 39	182 49	77 70	546	2542
Tetrazepam-M (di-HO-) iso-1 HY2AC	U+UHYAC	2600	365 37	264 61	246 100	206 53		899	2063
Tetrazepam-M (di-HO-) iso-2 HY2AC	U+UHYAC	2640	365 56	264 56	246 100	220 35	206 42	900	2061
Tetrazepam-M (HO-) -H2O	U+UHYAC	2430	286 100	228 32				542	2089
Tetrazepam-M (HO-) -H2O HY	U+UHYAC	2200	247 100	230 19	192 28	168 19	138 22	366	2062
Tetrazepam-M (HO-) iso-1	UGLUC	2570	304 43	275 52	261 19	235 100		634	618
Tetrazepam-M (HO-) iso-1 AC	U+UHYAC	2630	346 11	304 21	287 100	251 15		831	2056
Tetrazepam-M (HO-) iso-1 HY	UHY	2330	265 50	220 14	206 8	168 100	111 18	442	617
Tetrazepam-M (HO-) iso-1 HYAC	U+UHYAC	2380	307 100	248 73	234 49	220 40	194 24	647	304
Tetrazepam-M (HO-) iso-2	U+UHYAC	2580	304 43	275 52	261 19	235 100		634	2090
Tetrazepam-M (HO-) iso-2 AC	UGLUCAC	2640	346 12	304 21	287 100	251 16		831	620
Tetrazepam-M (HO-) iso-2 HY	UHY	2410	265 88	234 66	220 57	207 73	194 100	442	919
Tetrazepam-M (HO-) iso-2 HYAC	U+UHYAC	2470	307 100	264 42	248 80	207 59	194 50	647	305
Tetrazepam-M (HO-) iso-3 HY	UHY	2460	265 100	248 18	220 34	207 60	111 10	442	2085
Tetrazepam-M (HO-) iso-3 HYAC	U+UHYAC	2535	307 100	248 26	220 18	206 48		647	2087

Table 1-8-1: Compounds in order of names   Thiamazole TMS

Name	Detected	RI	Typical ions and intensities					Page	Entry
Tetrazepam-M (HO-) iso-4 HY	UHY	2475	265_{100}	234_{39}	220_{38}	207_{79}	194_{59}	442	2086
Tetrazepam-M (HO-) iso-4 HYAC	U+UHYAC	2560	307_{100}	264_{31}	248_{33}	218_{58}	206_{31}	648	2088
Tetrazepam-M (HO-oxo-) -H2O	P-I	2445	300_{80}	285_{6}	271_{17}	265_{100}	237_{13}	614	5781
Tetrazepam-M (nor-)	U+UHYAC	2530	274_{48}	273_{42}	245_{17}	239_{100}	211_{14}	485	2101
Tetrazepam-M (nor-) +H2O iso-1 ALHY2AC		2510	337_{12}	196_{100}	154_{30}	126_{11}		791	2097
Tetrazepam-M (nor-) +H2O iso-2 ALHY2AC		2540	337_{24}	260_{18}	196_{100}	154_{19}		791	2098
Tetrazepam-M (nor-) ALHY		2100	235_{60}	218_{37}	206_{100}	192_{37}	154_{17}	322	2092
Tetrazepam-M (nor-) HY	UHY	2130	235_{67}	220_{24}	193_{100}			323	2100
Tetrazepam-M (nor-HO-) HY2AC	U+UHYAC	2500	335_{42}	275_{60}	233_{100}	196_{63}	154_{68}	783	2064
Tetrazepam-M (oxo-)	U+UHYAC	2430	302_{18}	285_{100}	267_{36}	245_{27}		623	2057
Tetrazepam-M (oxo-) HY	U+UHYAC	2390	263_{100}	234_{25}	220_{24}	207_{33}	194_{19}	434	2058
Tetrazepam-M (tri-HO-) -2H2O	UGLUC	2670	300_{67}	272_{100}	237_{10}			614	619
Tetrazepam-M (tri-HO-) -2H2O AC	UGLUCAC	2790	342_{16}	300_{61}	272_{100}	237_{9}		816	621
Tetrazepam-M (tri-HO-) -2H2O HY	UHY	2580	261_{91}	260_{100}	244_{42}	209_{21}	121_{17}	424	2048
Tetrazepam-M (tri-HO-) -2H2O HYAC	U+UHYAC	2600	303_{77}	260_{100}	244_{47}	121_{11}		626	2060
Tetroxoprim		2840	334_{100}	276_{40}	245_{19}	123_{10}	59_{28}	780	1744
Tetryzoline	U UHY	1830	200_{100}	185_{86}	171_{35}			219	983
Tetryzoline AC	U+UHYAC	2110	242_{81}	200_{99}	185_{67}	86_{48}		353	986
Tetryzoline 2AC	U+UHYAC	2400	284_{48}	242_{45}	199_{100}			536	987
TFMPP		1620	230_{24}	188_{100}	172_{17}	145_{18}	56_{15}	307	5886
TFMPP AC		1890	272_{25}	200_{100}	188_{37}	172_{37}	56_{69}	479	5887
TFMPP HFB		1750	426_{57}	229_{36}	200_{54}	145_{29}	56_{80}	1071	6768
TFMPP PFP		1690	376_{79}	229_{48}	200_{67}	172_{47}	56_{100}	939	5889
TFMPP TFA		1690	326_{86}	229_{52}	200_{68}	172_{57}	56_{90}	738	5888
TFMPP TMS		1920	302_{100}	287_{32}	101_{76}	86_{70}	73_{98}	624	5890
TFMPP-M (deethylene-) 2AC	U+UHYAC	1865	288_{1}	229_{37}	187_{48}	174_{100}	145_{13}	552	6583
TFMPP-M (deethylene-) 2HFB	U+UHYHFB	1575	596_{1}	577_{5}	383_{82}	370_{100}	145_{22}	1194	6591
TFMPP-M (deethylene-) 2TFA	U+UHYTFA	1530	377_{20}	283_{95}	270_{100}	172_{45}	145_{51}	1004	6588
TFMPP-M (HO-) 2AC	U+UHYAC	2275	330_{11}	288_{45}	245_{11}	216_{100}	203_{46}	759	6578
TFMPP-M (HO-) 2HFB	U+UHYHFB	1985	638_{12}	441_{100}	412_{68}	399_{17}	188_{26}	1198	6589
TFMPP-M (HO-) 2TFA	U+UHYTFA	2005	438_{100}	419_{25}	396_{12}	341_{79}	312_{61}	1094	6585
TFMPP-M (HO-deethylene-) 3AC	U+UHYAC	2275	346_{5}	287_{31}	245_{58}	203_{79}	190_{100}	831	6584
TFMPP-M (HO-glucuronide) 4TMS		2920	375_{11}	318_{100}	276_{23}	245_{19}	73_{79}	1203	6767
TFMPP-M (HO-trifluoromethylaniline N-acetyl-) TFA	UTFA	1415	315_{12}	296_{7}	273_{47}	219_{5}	176_{100}	687	6807
TFMPP-M (HO-trifluoromethylaniline) 2TFA	U+UHYTFA	1395	369_{100}	350_{23}	300_{9}	272_{47}	177_{13}	914	6587
TFMPP-M (HO-trifluoromethylaniline) iso-1 2AC	U+UHYAC	1810	261_{23}	219_{74}	177_{100}	157_{50}	129_{6}	425	6581
TFMPP-M (HO-trifluoromethylaniline) iso-2 AC	U+UHYAC	1710	219_{41}	200_{8}	177_{39}	157_{100}	129_{20}	269	6582
TFMPP-M (HO-trifluoromethylaniline) iso-2 2AC	U+UHYAC	1840	261_{3}	219_{73}	177_{90}	157_{100}	129_{15}	425	6580
TFMPP-M (trifluoromethylaniline) AC	U+UHYAC	1400	203_{20}	184_{9}	161_{100}	142_{6}	114_{12}	223	6579
TFMPP-M (trifluoromethylaniline) HFB	U+UHYHFB	1130	357_{27}	338_{18}	188_{100}	160_{89}	145_{67}	871	6590
TFMPP-M (trifluoromethylaniline) TFA	U+UHYTFA	1230	257_{81}	238_{25}	188_{79}	160_{100}	145_{82}	409	6586
Thalidomide		2440	258_{32}	230_{19}	173_{73}	104_{67}	76_{100}	412	988
Thebacone		2500	341_{100}	298_{65}	242_{32}	162_{26}		813	258
Thebacone TMS		2475	371_{31}	356_{14}	313_{9}	234_{30}	73_{100}	925	6215
Thebacone Cl-artifact		2630	375_{100}	340_{47}	318_{28}	146_{13}	115_{10}	936	4401
Thebacone-M (dihydro-) AC	U+UHYAC	2435	343_{100}	300_{33}	284_{30}	226_{14}	70_{10}	822	233
Thebacone-M (dihydro-) 6-beta isomer TMS		2495	373_{100}	316_{27}	236_{6}	146_{7}	73_{200}	932	6762
Thebacone-M (N-demethyl-) 2TMS		2610	429_{88}	414_{26}	314_{37}	292_{30}	73_{150}	1079	6763
Thebacone-M (nor-dihydro-)	UHY	2440	287_{100}	244_{24}	242_{22}	150_{32}	115_{24}	547	4368
Thebacone-M (nor-dihydro-) AC	U+UHYAC	2700	329_{40}	243_{42}	183_{26}	87_{100}	72_{44}	755	3054
Thebacone-M (nor-dihydro-) 2AC	U+UHYAC	2750	371_{20}	285_{7}	243_{26}	87_{100}	72_{33}	924	235
Thebacone-M (O-demethyl-) AC	U+UHYAC	2625	369_{39}	327_{100}	284_{54}	228_{37}	162_{40}	916	1186
Thebacone-M (O-demethyl-) TMS		2520	429_{22}	414_{22}	357_{10}	234_{23}	73_{100}	1079	6208
Thebacone-M (O-demethyl-dihydro-) AC	U+UHYAC	2490	329_{100}	287_{56}	230_{10}	164_{20}	70_{21}	755	3055
Thebacone-M (O-demethyl-dihydro-) TFA		2250	383_{100}	286_{19}	270_{44}	213_{19}	69_{50}	963	6199
Thebacone-M (O-demethyl-dihydro-) 2AC	U+UHYAC	2545	371_{83}	329_{100}	286_{34}	212_{21}	70_{33}	924	234
Thebacone-M (O-demethyl-dihydro-) 2HFB		2260	679_{41}	482_{53}	466_{100}	360_{13}	169_{21}	1202	6197
Thebacone-M (O-demethyl-dihydro-) 2PFP		2330	579_{60}	432_{21}	416_{49}	310_{4}	119_{100}	1191	2460
Thebacone-M (O-demethyl-dihydro-) 2TFA		2190	479_{91}	382_{25}	366_{61}	260_{7}	69_{100}	1147	6198
Thebacone-M (O-demethyl-dihydro-) 6-alpha isomer 2TMS		2520	431_{33}	373_{7}	236_{23}	146_{15}	73_{100}	1083	2469
Thebacone-M (O-demethyl-dihydro-) 6-beta isomer 2TMS		2540	431_{87}	374_{28}	236_{7}	146_{8}	73_{200}	1083	6760
Thebaine		2545	311_{100}	296_{80}	242_{19}	211_{9}	165_{10}	672	257
Thebaol		2970*	254_{100}	239_{58}	211_{3}	152_{4}	139_{5}	399	2328
Thebaol AC		2950*	296_{36}	254_{100}	239_{43}	210_{2}	139_{2}	597	2327
Theobromine	P G U+UHYAC	1980	180_{100}	137_{12}	109_{41}	82_{38}		174	989
Theobromine TMS		2020	252_{20}	237_{100}	109_{15}	100_{12}	73_{25}	390	5452
Theophylline	P G U+UHYAC	2025	180_{100}	95_{85}	68_{69}			174	990
Theophylline TMS		1920	252_{61}	237_{100}	223_{14}	135_{7}	73_{37}	391	4600
THG		2660*	312_{57}	265_{52}	240_{43}	227_{100}	211_{34}	677	7573
THG TMS		2490*	384_{75}	299_{98}	281_{39}	270_{51}	73_{90}	968	7574
Thiamazole	G P-I	1615	114_{100}	99_{5}	81_{16}	72_{30}	69_{18}	104	4703
Thiamazole AC	GAC PAC-I U+UHYAC	1440	156_{40}	114_{100}	86_{10}	81_{18}	72_{20}	136	4704
Thiamazole ME	GME PME-I	1205	128_{100}	113_{8}	95_{22}	72_{18}	59_{7}	111	4687
Thiamazole TMS	GTMS PTMS-I	1400	186_{51}	171_{100}	116_{7}	113_{8}	73_{23}	188	4688

Thiamine artifact-1

Table 1-8-1: Compounds in order of names

Name	Detected	RI	Typical ions and intensities					Page	Entry
Thiamine artifact-1	UHY	1380	143_{31}	113_{42}	112_{100}	85_{30}	71_9	122	448
Thiamine artifact-2 2ME		1190	153_{64}	138_{38}	122_{100}	111_{11}	81_{66}	133	5142
Thiazafluron ME		1560	254_1	126_7	112_2	72_{100}		397	3944
Thiethylperazine	G P-I G U+UHYAC	3205	399_{60}	259_{20}	141_{38}	113_{59}	70_{100}	1015	1870
Thiethylperazine-M (nor-) AC	U+UHYAC	3650	427_{15}	291_{32}	259_{20}	141_{100}	99_{30}	1074	2231
Thiethylperazine-M (ring)	U UHY U+UHYAC	2750	259_{100}	230_{37}	198_{19}	186_{18}		415	1871
Thiethylperazine-M (sulfone)	UHY U+UHYAC	3400	431_{19}	305_{28}	127_{41}	113_{42}	70_{100}	1083	2232
Thiobutabarbital	P G U UHY U+UHYAC	1790	228_{22}	172_{100}	157_{85}	97_9	57_{16}	300	992
Thiobutabarbital-M (butabarbital)	P G U UHY U+UHYAC	1655	183_6	156_{100}	141_{84}			250	149
Thiocyclam		1495	181_1	135_{31}	103_2	71_{100}	56_{23}	176	4083
Thiocyclam -S		1040	149_{31}	103_{14}	84_9	70_{100}	56_{28}	126	4084
Thiofanox -C2H3NO		1085	161_{28}	115_{54}	83_{39}	61_{69}	55_{100}	140	3908
Thiometon		1695*	246_4	159_1	125_{37}	88_{100}	60_{46}	361	2519
Thionazine		1600	248_{42}	192_{37}	143_{59}	107_{87}	97_{100}	370	3877
Thiopental	P G U+UHYAC	1855	242_{31}	173_{69}	172_{100}	157_{87}	69_{21}	352	993
Thiopental (ME)	P	1820	256_{15}	186_{94}	171_{100}	143_{11}	112_{11}	407	4229
Thiopental 2ME	PME UME	1825	270_{21}	200_{100}	185_{55}	97_{19}	69_{28}	471	994
Thiopental iso-1 2TMS		1925	386_1	371_3	344_{19}	97_{24}	73_{100}	973	4611
Thiopental iso-2 2TMS		1995	386_2	371_7	301_{48}	97_{24}	73_{100}	973	4610
Thiopental-M (HO-)	P U	2050	258_{25}	173_{93}	172_{100}	157_{63}	69_{32}	413	4437
Thiopental-M (HO-) AC	U+UHYAC	2205	300_{24}	240_{25}	211_{39}	172_{62}	69_{100}	615	3142
Thiopental-M (HOOC-) 3ME	UME	----	314_{24}	283_5	241_7	200_{100}	185_{31}	685	995
Thiopental-M (HO-pentobarbital)	U	1955	227_3	197_{26}	195_{21}	156_{100}	141_{70}	352	838
Thiopental-M (HO-pentobarbital) (ME)	P U	1865	209_{19}	170_{100}	155_{100}	112_{10}	69_{59}	407	3341
Thiopental-M (HO-pentobarbital) 2ME	PME UME	1820	223_{25}	184_{95}	169_{100}	112_{13}	69_{40}	471	3340
Thiopental-M (HO-pentobarbital) -H2O	U+UHYAC	1890	224_2	195_{31}	156_{30}	141_{30}	69_{100}	289	840
Thiopental-M (pentobarbital)	P G U+UHYAC	1740	197_4	156_{100}	141_{84}	98_{10}	69_{12}	295	837
Thiopental-M (pentobarbital) (ME)	P G	1700	211_4	170_{99}	155_{100}	141_{11}	112_{10}	345	2584
Thiopental-M (pentobarbital) 2ME	PME UME	1630	225_6	184_{100}	169_{85}			400	839
Thiopental-M (pentobarbital) 2TMS		1850	370_1	355_{40}	300_{51}	285_{100}	73_{39}	921	4580
Thiophanate 4ME		2575	426_1	379_{10}	146_{10}	88_{100}		1072	3977
Thiophanate-methyl 4ME		2600	398_1	351_{13}	230_4	88_{100}	59_{16}	1011	3943
Thiophenecarboxylic acid		<1000*	128_5	127_{55}	111_{100}	83_7		110	4282
Thiophenylmethanol		<1000*	114_{100}	97_{62}	85_{78}	81_{24}		104	4280
Thiopropazate	U+UHYAC-I	3470	445_{33}	246_{41}	185_{51}	125_{60}	70_{100}	1108	373
Thiopropazate-M (amino-) AC	U+UHYAC	2990	332_8	233_{19}	100_{100}			767	1255
Thiopropazate-M (deacetyl-)	UHY-I	3360	403_{27}	246_{72}	171_{31}	143_{84}	70_{100}	1026	4252
Thiopropazate-M (deacetyl-)	UHY-I	3360	403_{50}	246_{80}	143_{25}	70_{75}	42_{100}	1026	592
Thiopropazate-M (deacetyl-) TMS		3340	475_{10}	372_{22}	246_{100}	232_{22}	73_{41}	1144	5444
Thiopropazate-M (dealkyl-) AC	U+UHYAC	3500	401_{59}	233_{61}	141_{99}	99_{68}		1020	1282
Thiopropazate-M (ring)	U-I UHY U+UHYAC-I	2100	233_{100}	198_{55}				314	311
Thioproperazine		3575	446_{30}	320_{25}	127_{36}	113_{50}	70_{100}	1109	399
Thioproperazine-M (ring)	U UHY U+UHYAC	3200	306_{101}	198_{77}				643	1294
Thioridazine	P G U+UHYAC	3125	370_8	126_9	98_{100}	70_{14}		920	400
Thioridazine-M	U+UHYAC	3360	404_{14}	292_2	126_4	98_{100}	70_5	1028	1993
Thioridazine-M (HO-) AC	U+UHYAC	3450	428_{16}	244_3	126_5	98_{100}	70_5	1076	1720
Thioridazine-M (HO-methoxy-piperidyl-) AC	U+UHYAC	3600	458_{43}	404_5	258_{11}	244_{12}	214_{100}	1124	1892
Thioridazine-M (HO-piperidyl-) AC	U+UHYAC	3460	428_{46}	258_8	244_6	156_{100}	96_{15}	1076	1890
Thioridazine-M (nor-) AC	U+UHYAC	3490	398_{20}	356_1	245_3	154_{100}	84_{17}	1012	1295
Thioridazine-M (nor-HO-piperidyl-) 2AC	U+UHYAC	3750	456_{32}	292_{37}	212_{52}	140_{17}	112_{100}	1122	1894
Thioridazine-M (oxo-)	U+UHYAC	3500	384_{100}	258_{32}	244_{82}	140_{10}	112_{80}	966	1321
Thioridazine-M (oxo-/side chain sulfone)	U+UHYAC	3800	416_{100}	290_{27}	277_{22}	140_{16}	112_{80}	1054	1895
Thioridazine-M (ring sulfone)	P U+UHYAC	3420	402_7	370_{15}	258_5	244_8	98_{100}	1023	1740
Thioridazine-M (ring)	G P U+UHYAC	2570	245_{100}	230_{33}	198_{40}	186_{27}	154_8	357	4388
Thioridazine-M (side chain sulfone)	G P-I U+UHYAC	3415	402_{13}	290_5	197_6	98_{100}	70_{10}	1023	394
Thioridazine-M/artifact (sulfoxide)	P-I G U+UHYAC	3330	386_{12}	370_4	126_7	98_{100}	70_{11}	973	4484
Thioridazine-M/artifact (sulfoxide)	G P U+UHYAC	3330	386_{12}	126_6	98_{100}	70_{15}	42_{100}	973	2200
Thiram		2260	240_{13}	208_2	121_{13}	88_{100}	76_{43}	341	3460
Tiabendazole		2090	201_{100}	174_{56}				220	1535
Tiapride	P G U+UHYAC	2820	328_1	311_1	213_7	134_3	86_{100}	749	1296
Tiapride-M (deethyl-) AC	U+UHYAC	3020	299_8	242_{14}	213_{39}	113_{58}	58_{100}	817	6414
Tiapride-M (O-demethyl-)	U+UHYAC	2580	314_1	242_2	199_2	86_{60}	58_{11}	685	1297
Tiapride-M (O-demethyl-N-oxide) -(C2H5)2NOH	U+UHYAC	2590	241_{100}	226_{23}	182_{13}	178_{42}	162_{30}	346	1298
Tiaprofenic acid ME	UME	2180*	274_{24}	215_{99}	153_{15}	105_{56}	77_{50}	484	1538
Tiaprofenic acid 2ME	UME	2320*	288_{12}	229_{100}	201_5	105_{18}	77_{21}	551	6396
Tiaprofenic acid artifact	U+UHYAC	1880*	230_{85}	215_{100}	153_{30}	105_{97}	77_{98}	306	2041
Tiaprofenic acid -CO2	G P U+UHYAC	1865*	216_{63}	201_{31}	139_{100}	105_{35}	77_{55}	259	1537
Tiaprofenic acid -CO2 HYAC	U+UHYAC	2050*	258_4	216_{92}	187_{19}	105_{65}	77_{100}	413	2043
Tiaprofenic acid HYAC	U+UHYAC	2150*	302_{18}	215_{100}	187_{11}	105_{31}	77_{35}	622	2042
Tiaprofenic acid-M (HO-) AC	U+UHYAC	2230*	318_6	231_{100}	153_{41}	105_{67}	77_{100}	701	2044
Tibolone		2550*	312_{28}	297_{12}	229_{17}	187_{20}	91_{29}	677	5827
Tibolone AC		2540*	354_{68}	339_{53}	229_{72}	105_{100}	91_{84}	863	6023
Tibolone TFA		2520*	408_{89}	306_{20}	294_{20}	229_{100}	187_{43}	1036	5828
Tibolone enol 2TMS		2700*	456_{42}	442_5	301_8	182_8	73_{60}	1123	5830

Table 1-8-1: Compounds in order of names TMA-2-M (O-bisdemethyl-) iso-1 3AC

Name	Detected	RI	Typical ions and intensities					Page	Entry
Tibolone -H2O		2395*	294_{100}	279_{31}	237_{34}	209_{33}	91_{54}	587	5829
Ticlopidine	U+UHYAC UHYME	2110	263_{12}	125_{18}	110_{100}			433	996
Ticlopidine-M (HO-) iso-1 AC	U+UHYAC	2380	321_9	286_2	141_9	138_9	110_{100}	717	6475
Ticlopidine-M (HO-) iso-2 AC	U+UHYAC	2400	321_6	286_2	141_9	138_7	110_{100}	717	6476
Ticlopidine-M (N-dealkyl-) AC	U+UHYAC	1690	181_{44}	139_6	128_{36}	110_{87}	85_{100}	177	6474
Tienilic acid ME	UEXME	2630*	344_{33}	309_{33}	261_{22}	243_6	111_{100}	823	6852
Tienylic acid ME		2570*	344_{11}	309_{16}	261_{12}	243_5	111_{100}	823	7421
Tienylic acid TMS		2605*	402_{11}	367_{24}	357_{14}	111_{68}	73_{100}	1022	7422
Tiletamine		1785	223_1	195_{40}	166_{100}	123_{25}	110_{38}	284	7452
Tiletamine AC		2160	265_{13}	237_8	194_{33}	166_{49}	151_{100}	443	7453
Tiletamine HFB		1965	419_1	375_1	362_2	151_{50}	97_{33}	1060	7456
Tiletamine ME		1890	237_1	209_{29}	166_{100}	123_{27}	110_{31}	334	7454
Tiletamine TFA		1955	319_4	275_5	262_7	151_{100}	97_{30}	706	7455
Tiletamine TMS		1820	295_2	267_{20}	250_{73}	166_{57}	73_{100}	592	7457
Tilidine	G U-I	1835	273_1	176_7	97_{100}	82_{59}		482	624
Tilidine-M (bis-nor-)	U UHY	1840	245_6	83_{29}	69_{100}			360	626
Tilidine-M (bis-nor-) AC	U+UHYAC	2100	287_5	244_6	111_{89}	69_{100}		548	259
Tilidine-M (bis-nor-HO-)	U	1950	261_{15}	244_6	103_{31}	85_{100}	69_{32}	427	627
Tilidine-M (bis-nor-oxime-)		1965	259_{49}	186_{100}	168_{70}			416	628
Tilidine-M (nitro-)		1990	275_{65}	258_{28}	184_{87}	103_{98}		490	629
Tilidine-M (nor-)	P U UHY	1820	259_7	83_{100}	68_{31}			417	625
Tilidine-M (nor-) AC	U+UHYAC	2165	301_3	258_4	125_{100}	83_{74}	69_{36}	621	260
Tilidine-M (phenylcyclohexenone)	U UHY U+UHYAC	1520*	172_{23}	104_{60}	68_{100}			160	630
Tilidine-M/artifact AC	U+UHYAC	1550*	212_{10}	170_{100}				249	1219
Tilidine-M/artifact 2AC	U+UHYAC	2280	271_{13}	211_{25}	169_{100}	103_{26}		474	1220
Timolol	G P	2265	316_1	301_7	130_{21}	114_{12}	86_{100}	694	2613
Timolol AC	U+UHYAC	2290	358_1	343_7	112_{31}	86_{45}	56_{100}	877	1371
Timolol TMS		2290	388_1	373_8	272_3	186_6	86_{100}	981	6162
Timolol formyl artifact		2275	328_2	313_{19}	271_7	86_{100}	57_{77}	750	1370
Timolol-M (deisobutyl-) 2AC	U+UHYAC	2620	344_2	284_4	158_{100}	98_{16}	56_{63}	824	1712
Timolol-M (deisobutyl-) -H2O AC	U+UHYAC	2205	284_{35}	254_{48}	157_{51}	98_{38}	56_{79}	532	1711
Tinidazole	U+UHYAC	2010	247_{15}	201_{100}	154_{15}	123_{38}	80_{46}	365	2737
Tinox iso-1		1395*	216_1	143_{10}	125_{10}	109_8	74_{100}	258	3463
Tinox iso-2		1500*	216_2	142_{12}	109_{20}	79_{23}	74_{100}	258	3464
Tioclomarole -H2O		3405*	428_{33}	303_{100}	297_{20}	255_{79}	92_{35}	1075	6090
Tioconazole		2800	386_2	351_1	305_4	177_{13}	131_{100}	971	2648
Tiotixene		3555	443_2	343_3	221_{19}	113_{100}	70_{43}	1105	401
Tiotixene artifact (ring)		2900	305_{100}	213_{14}	197_{78}	152_{19}		637	2244
Tiotropium-M/artifact (HOOC-) ME		2140*	254_2	195_{69}	177_2	111_{80}	83_{11}	396	7369
Tiotropium-M/artifact (HOOC-) MEAC		2240*	296_9	237_{36}	195_{100}	177_8	111_{61}	595	7370
Tiotropium-M/artifact (HOOC-) 2ME		2160*	268_1	237_3	209_{100}	195_{22}	111_{99}	458	7371
Tiropramide		3240	452_2	368_7	105_{22}	86_{100}	77_{12}	1135	5687
Tizanidine		2500	253_{100}	224_{19}	218_{62}	196_{79}	183_{12}	393	7250
Tizanidine AC		2545	295_{16}	260_{30}	218_{85}	196_{51}	86_{100}	588	7253
Tizanidine ME		2210	267_2	232_{100}	210_{11}	198_{13}	183_{17}	453	7251
Tizanidine TMS		2400	325_9	290_{28}	240_{54}	142_{100}	99_{72}	733	7260
Tizanidine 2AC		2575	337_{26}	260_{54}	218_{82}	128_{100}	86_{42}	790	7252
Tizanidine 2TMS		2375	397_{35}	362_{74}	240_{100}	214_{92}	73_{83}	1009	7259
Tizanidine artifact AC		1975	227_2	185_{100}	157_8	150_6	125_5	296	7255
Tizanidine artifact HFB		1705	381_{14}	346_{85}	212_{100}	184_{79}	157_{74}	954	7258
Tizanidine artifact PFP		1780	331_{28}	296_{100}	212_{54}	184_{25}	157_{22}	761	7257
Tizanidine artifact TFA		1665	281_{18}	246_{100}	212_{62}	184_{43}	157_{54}	520	7256
Tizanidine artifact 2AC		1950	269_1	227_{28}	185_{100}	157_6	150_3	463	7254
Tizanidine-M (dehydro-) AC		2175	293_5	251_{50}	216_{100}	179_{10}	134_4	578	7312
TMA		1680	225_1	182_{24}	167_8	151_2	107_1	292	3259
TMA		1680	225_1	182_{24}	167_8	151_2	44_{100}	292	5540
TMA AC		2020	267_9	208_{47}	193_{17}	181_{16}	86_{10}	455	3266
TMA AC		2020	267_9	208_{47}	193_{17}	86_{10}	44_{100}	456	5541
TMA formyl artifact		1680	237_{25}	181_{93}	148_9	77_6	56_{100}	335	3251
TMA intermediate (3,4,5-trimethoxyphenyl-nitropropene)		2050	253_{100}	206_{30}	191_{44}	161_{27}	77_{48}	393	2840
TMA intermediate (trimethoxyphenyl-2-nitroethene)		2145	239_{100}	192_{34}	177_{44}	149_{32}	92_{26}	339	2841
TMA precursor (3,4,5-trimethoxybenzaldehyde)		1550*	196_{100}	181_{51}	125_{52}	110_{51}	93_{37}	211	3279
TMA-2		1670	225_2	182_{100}	167_{33}	151_{12}	139_2	292	7348
TMA-2		1670	225_2	182_{99}	167_{37}	151_{13}	44_{100}	292	7366
TMA-2 AC	U+UHYAC	2140	267_{23}	208_{100}	181_{66}	151_{23}	86_6	456	7152
TMA-2 HFB		1780	421_7	208_{12}	181_{100}	151_{21}	136_{10}	1063	7347
TMA-2 PFP		1740	371_{22}	208_{11}	181_{200}	151_{23}	136_8	922	7346
TMA-2 TFA		1760	321_{21}	208_{10}	181_{200}	151_{48}	136_{10}	718	7345
TMA-2 TMS		1765	297_1	282_2	224_9	116_{100}	73_{46}	603	7349
TMA-2 2AC	U+UHYAC	2200	309_{10}	208_{100}	181_{65}	151_{21}	86_{11}	662	7161
TMA-2 formyl artifact		1650	237_{18}	206_{33}	181_{100}	151_{41}	56_{29}	335	7344
TMA-2-M (deamino-HO-) AC	U+UHYAC	1670*	268_8	208_{100}	193_{39}	181_{31}		461	7157
TMA-2-M (O-bisdemethyl-) artifact 2AC	U+UHYAC	2200	263_{28}	221_{100}	179_{65}	164_{44}	132_{10}	434	7183
TMA-2-M (O-bisdemethyl-) iso-1 3AC	U+UHYAC	2300	323_2	281_{27}	180_{100}	153_{50}	86_{22}	726	7162

Table 1-8-1: Compounds in order of names

Name	Detected	RI	Typical ions and intensities					Page	Entry
TMA-2-M (O-bisdemethyl-) iso-2 3AC	U+UHYAC	2305	323_2	281_6	180_{100}	153_{52}	86_{20}	727	7163
TMA-2-M (O-bisdemethyl-) iso-3 3AC	U+UHYAC	2330	323_4	281_{12}	180_{100}	153_{52}	86_{25}	727	7164
TMA-2-M (O-deamino-oxo-)	U+UHYAC	1540*	224_{25}	181_{100}	151_{37}	136_{28}		289	7165
TMA-2-M (O-demethyl-) iso-1 2AC	U+UHYAC	2215	295_9	253_{23}	194_{100}	167_{47}	86_{13}	592	7154
TMA-2-M (O-demethyl-) iso-2 2AC	U+UHYAC	2230	295_{17}	236_{32}	194_{100}	167_{54}	86_{14}	592	7153
TMA-2-M (O-demethyl-) iso-2 3AC	U+UHYAC	2280	337_5	236_{29}	194_{100}	167_{60}	86_{22}	793	7156
TMA-2-M (O-demethyl-) iso-3 2AC	U+UHYAC	2250	295_9	236_{52}	194_{100}	167_{40}	86_7	592	7155
TMA-2-M (O-demethyl-deamino-oxo-) iso-1 AC	U+UHYAC	1680*	252_8	210_{40}	167_{100}			390	7158
TMA-2-M (O-demethyl-deamino-oxo-) iso-2 AC	U+UHYAC	1705*	252_{11}	210_{32}	167_{100}	137_9		390	7159
TMA-2-M (O-demethyl-deamino-oxo-) iso-3 AC	U+UHYAC	1760*	252_6	210_{19}	167_{58}	137_6		390	7160
Tocainide		1730	192_{31}	176_{11}	147_7	121_{38}	57_{99}	198	1536
Tocainide AC	U+UHYAC	2040	234_{13}	147_{15}	121_{100}	106_{11}	87_{20}	321	1534
Tocainide-M (HO-) 2AC	U+UHYAC	2480	292_3	250_5	179_{19}	137_{100}	86_{34}	575	2897
alpha-Tocopherol	G P UHY	3030*	430_{100}	205_7	165_{58}	57_{16}		1082	2403
alpha-Tocopherol AC	G U+UHYAC	3070*	472_{10}	430_{100}	247_3	165_{58}	57_{16}	1141	2402
gamma-Tocopherol	P	2990*	416_{53}	203_6	191_{20}	151_{100}	57_6	1056	5816
Tofisopam		3020	382_{55}	341_{72}	326_{100}	269_6	77_{14}	960	4019
Tolazamide ME	UME	2540	325_5	268_{16}	155_{70}	113_{100}	85_{79}	735	4935
Tolazamide 2ME		2540	339_{32}	229_{26}	184_{31}	155_{34}	91_{100}	803	3139
Tolazamide artifact-1 ME	UME	1315	172_{12}	113_{76}	98_{80}	68_{70}	59_{100}	160	3140
Tolazamide artifact-1 2ME		1245	186_2	143_7	117_{100}	85_{57}	59_{26}	188	3141
Tolazamide artifact-2	UME	1620	197_{33}	155_{50}	91_{100}	65_{36}		212	4910
Tolazamide artifact-3	G P-I U+UHYAC UME	1700	171_{25}	155_{19}	107_6	91_{100}	65_{25}	158	2008
Tolazamide artifact-3 ME	UME	1740	185_{25}	155_{19}	121_6	91_{100}	65_{25}	186	3131
Tolazamide artifact-3 TMS	UTMS	1875	243_3	228_{100}	167_6	149_{12}	91_9	354	5022
Tolazamide artifact-3 2ME		1690	199_{34}	155_{15}	91_{100}	65_{23}		217	3130
Tolazamide artifact-4 ME		1845	212_1	179_{46}	122_{21}	91_{100}	72_{53}	247	3132
Tolazamide-M (HO-) artifact AC	U+UHYAC	2180	229_{16}	187_{80}	107_{80}	89_{99}	77_{44}	303	4915
Tolazamide-M (HO-) artifact ME	UME	2265	201_{40}	172_{45}	107_{100}	89_{81}	77_{91}	220	4913
Tolazamide-M (HO-) artifact 2ME	UME	2030	215_{60}	171_{29}	107_{100}	89_{80}	77_{72}	257	4914
Tolazamide-M (HO-) artifact 3TMS	UTMS	2000	403_{20}	388_{79}	272_{35}	258_{55}	73_{80}	1026	5018
Tolazamide-M (HOOC-) artifact 2ME		1920	229_{50}	198_{48}	135_{100}	103_{40}	76_{55}	303	2479
Tolazamide-M (HOOC-) artifact 3ME	UME	1850	243_{54}	199_{18}	135_{100}	104_{46}	76_{51}	354	2480
Tolazoline-M (HO-dihydro-) 2AC	U+UHYAC	2175	262_{10}	161_{36}	118_{100}	91_{49}		431	997
Tolbutamide ME		2320	284_{38}	155_{77}	129_{96}	91_{100}	72_{52}	534	3137
Tolbutamide TMS		2255	342_1	327_{12}	187_{17}	91_{29}	72_{100}	817	5017
Tolbutamide 2ME		2170	298_1	241_5	155_{79}	113_{77}	91_{100}	607	3138
Tolbutamide artifact-1	UME	1620	197_{33}	155_{50}	91_{100}	65_{36}		212	4910
Tolbutamide artifact-2	G P-I U+UHYAC UME	1700	171_{25}	155_{19}	107_6	91_{100}	65_{25}	158	2008
Tolbutamide artifact-2 ME	UME	1740	185_{25}	155_{19}	121_6	91_{100}	65_{25}	186	3131
Tolbutamide artifact-2 TMS	UTMS	1875	243_3	228_{100}	167_6	149_{12}	91_9	354	5022
Tolbutamide artifact-2 2ME		1690	199_{34}	155_{15}	91_{100}	65_{23}		217	3130
Tolbutamide artifact-3 ME		1845	212_1	179_{46}	122_{21}	91_{100}	72_{53}	247	3132
Tolbutamide-M (HO-) ME	UME	2645	300_{20}	171_{76}	129_{100}	107_{39}	72_{49}	615	4936
Tolbutamide-M (HO-) 2ME	UME	2740	314_7	215_{69}	197_{31}	134_{100}	89_{62}	685	4937
Tolbutamide-M (HO-) artifact AC	U+UHYAC	2180	229_{16}	187_{80}	107_{80}	89_{99}	77_{44}	303	4915
Tolbutamide-M (HO-) artifact ME	UME	2265	201_{40}	172_{45}	107_{100}	89_{81}	77_{91}	220	4913
Tolbutamide-M (HO-) artifact 2ME	UME	2030	215_{60}	171_{29}	107_{100}	89_{80}	77_{72}	257	4914
Tolbutamide-M (HO-) artifact 3TMS	UTMS	2000	403_{20}	388_{79}	272_{35}	258_{55}	73_{80}	1026	5018
Tolbutamide-M (HOOC-) 2ME	UME	2590	328_{14}	199_{56}	135_{57}	129_{100}	72_{47}	748	4938
Tolbutamide-M (HOOC-) artifact 2ME		1920	229_{50}	198_{48}	135_{100}	103_{40}	76_{55}	303	2479
Tolbutamide-M (HOOC-) artifact 3ME	UME	1850	243_{54}	199_{18}	135_{100}	104_{46}	76_{51}	354	2480
Tolclophos-methyl		1855*	300_1	265_{100}	250_{15}	125_{27}	79_{16}	612	3462
Tolfenamic acid ME		2255	275_{41}	243_{44}	208_{100}	180_{25}	89_{17}	488	6095
Tolfenamic acid MEAC		2285	317_{10}	275_{56}	243_{70}	208_{100}	180_{35}	696	6096
Toliprolol TMSTFA		1985	376_2	284_{44}	228_{11}	133_{80}	73_{70}	989	6174
Toliprolol 2AC		2155	307_1	247_{24}	190_{55}	140_{25}	72_{100}	651	1390
Toliprolol formyl artifact		1820	235_{21}	220_{27}	127_{29}	108_{58}	56_{100}	326	1389
Toliprolol -H2O AC	U+UHYAC	2230	247_{27}	200_{59}	140_{27}	98_{26}	72_{81}	368	1716
Toliprolol-M (deamino-di-HO-) 3AC	U+UHYAC	2200*	324_{15}	282_7	159_{100}	124_{11}	99_{10}	731	1714
Toliprolol-M (deamino-HO-) 2AC	U+UHYAC	1820*	266_7	196_1	159_{100}	108_{29}	99_{19}	451	1713
Toliprolol-M (HO-) 2AC	U+UHYAC	2210	323_2	308_{13}	124_8	98_{26}	72_{100}	729	1715
Toliprolol-M (HO-) 3AC	U+UHYAC	2550	365_1	305_{28}	200_{100}	98_{35}	72_{34}	902	1718
Toliprolol-M (HO-) -H2O 2AC	U+UHYAC	2290	305_{34}	262_{18}	220_{85}	140_{100}	98_{51}	640	1717
Tolmetin	U	1885	212_{99}	198_{37}	122_{37}	91_{23}		410	998
Tolmetin ET		2265	285_{31}	212_{100}	119_{20}	91_{23}		540	1000
Tolmetin ME	UME	2235	271_{45}	256_{19}	212_{100}	119_{14}	91_{18}	475	999
Tolmetin-M (HOOC-) 2ME	UME	2600	315_{34}	256_{100}	242_{26}	197_{12}	135_{13}	689	6298
Tolmetin-M (oxo-) ME	UME	2340	285_{25}	226_{100}	119_{24}			538	6297
Tolmetin-M (oxo-HOOC-) 2ME	UME	2640	329_{25}	270_{100}	211_4	163_{16}	135_7	753	6299
Tolperisone		1905	245_1	230_1	119_{37}	98_{150}	91_{25}	360	5643
Tolperisone artifact	G UHY U+UHYAC	1175*	160_{14}	132_{17}	119_{100}	91_{43}	65_{20}	139	5644
Tolperisone-M (dihydro-) AC	U+UHYAC	1970	289_5	246_3	210_6	137_9	98_{100}	561	7516
Tolperisone-M (dihydro-HO-) 2AC	U+UHYAC	2375	347_3	304_3	119_1	98_{100}	70_3	837	7514

Table 1-8-1: Compounds in order of namesTrandolaprilate 3ET

Name	Detected	RI	Typical ions and intensities					Page	Entry
Tolperisone-M (HO-) AC	U+UHYAC	2315	303_1	177_3	126_1	98_{80}	70_4	630	7515
Tolpropamine	U UHY U+UHYAC	1900	253_{22}	193_7	165_{11}	115_6	58_{100}	396	2206
Tolpropamine-M (bis-nor-) AC	U+UHYAC	2340	267_{13}	195_{11}	181_{17}	87_{100}	73_{38}	456	2211
Tolpropamine-M (bis-nor-HO-alkyl-) -H2O AC	U+UHYAC	2560	265_{11}	206_{100}	178_{18}	165_{13}	86_{19}	445	2210
Tolpropamine-M (HO-)	UHY	2150	269_{11}	178_3	165_5	91_5	58_{100}	466	2216
Tolpropamine-M (HO-alkyl-) AC	U+UHYAC	2250	311_{21}	206_3	178_8	115_5	58_{100}	673	2212
Tolpropamine-M (HO-phenyl-) AC	U+UHYAC	2230	311_4	178_3	165_4	115_3	58_{100}	673	2207
Tolpropamine-M (nor-)	UHY	2100	239_{88}	193_{100}	165_{51}	115_{23}	57_{65}	341	2214
Tolpropamine-M (nor-) AC	U+UHYAC	2360	281_{24}	193_{27}	165_{26}	115_{14}	87_{100}	524	2208
Tolpropamine-M (nor-HO-)	UHY	2200	255_{99}	193_{100}	165_{41}	115_{31}	91_{38}	404	2215
Tolpropamine-M (nor-HO-alkyl-) -H2O AC	U+UHYAC	2585	279_{29}	206_{100}	178_{16}	86_{39}	58_{18}	514	2209
Tolpropamine-M (N-oxide) -(CH3)2NOH	U UHY U+UHYAC	1750*	208_{88}	193_{88}	178_{45}	165_{45}	115_{100}	239	2213
Toluene		<1000*	92_{59}	91_{100}	65_{16}	51_{10}	39_{20}	98	1001
4-Toluenesulfonic acid ET		1750*	200_{37}	172_{18}	155_{47}	108_{17}	91_{100}	219	3147
4-Toluenesulfonic acid ethylester		1750*	200_{37}	172_{18}	155_{47}	108_{17}	91_{100}	219	3147
o-Toluidine AC		1300	149_{53}	106_{100}	91_{33}	79_{18}	77_{16}	126	5198
p-Toluidine	UHY	<1000	107_{87}	106_{100}	89_8	77_{25}	63_6	102	3405
p-Toluidine AC	U U+UHYAC	1410	149_{42}	107_{100}	91_3	77_{14}	65_4	126	3406
p-Toluidine-M (carbamoyl-)	UHY	<1000	133_{100}	106_{84}	78_{87}	52_{47}		128	3408
p-Toluidine-M (carbamoyl-) ME	UHYME	1100	147_{99}	132_{11}	106_{64}	78_{100}	52_{40}	146	3410
p-Toluidine-M (carbamoyl-HO-)	UHY	1300	149_{100}	121_{21}	104_{20}	93_{47}	66_{32}	151	3409
p-Toluidine-M (HO-)	UHY	1120	123_{100}	106_{20}	94_{54}	77_{33}		108	3407
p-Toluidine-M (HO-) 2AC	U+UHYAC	1960	207_8	165_{53}	123_{100}	94_{14}	77_{11}	233	3411
p-Toluidine-M (HO-) 3AC	U+UHYAC	1940	249_{11}	207_{38}	165_{63}	123_{100}	77_9	374	3412
Tolylfluanid		2045	346_{10}	238_{46}	181_{33}	137_{100}	92_{13}	830	3465
p-Tolylpiperazine		1660	176_{41}	134_{100}	119_{13}	91_{11}	65_6	164	7606
p-Tolylpiperazine AC		1985	218_{67}	175_{19}	146_{100}	133_{37}	119_{36}	266	7607
p-Tolylpiperazine HFB		1860	372_{25}	175_{54}	146_{36}	119_{36}	91_{30}	927	6770
p-Tolylpiperazine PFP		1825	322_{100}	175_{39}	146_{23}	119_{34}	91_{22}	722	7610
p-Tolylpiperazine TFA		1825	272_{100}	175_{30}	146_{20}	119_{25}	91_{21}	479	7609
p-Tolylpiperazine TMS		1805	248_{100}	233_{36}	206_{14}	134_{32}	73_{45}	372	7608
Topiramate		2240	324_{100}	266_9	189_{22}	110_{31}	59_{54}	801	5722
Topiramate ME		2140	353_1	338_{100}	220_{22}	171_{22}	127_{49}	857	5708
Topiramate TMS		2620	396_{15}	341_6	229_{18}	152_{21}	127_{29}	1044	5710
Topiramate 2TMS		2675	468_{100}	410_{21}	290_{14}	226_{13}	151_{27}	1150	5711
Topiramate artifact (-SO2NH)		1680*	245_{100}	229_{22}	171_{46}	127_{54}	69_{81}	421	5707
Topiramate artifact (-SO2NH) TMS		1900*	317_{31}	257_{13}	229_{65}	199_{20}	171_{100}	769	5709
Torasemide AC		2790	287_3	246_{40}	198_6	181_{100}	154_6	985	7335
Torasemide ME		2730	362_{44}	267_{10}	246_{12}	198_{14}	181_{100}	890	7332
Torasemide artifact ME		2430	277_{40}	246_{13}	198_9	181_{100}	154_7	499	7334
Torasemide artifact TMS		2520	335_{14}	263_{41}	246_{23}	228_{12}	181_{100}	783	7336
Torasemide artifact 2ME		2395	291_{54}	246_5	183_{81}	181_{100}	168_{22}	568	7333
Torasemide artifact 3ME	UEXME	2330	305_{36}	195_{100}	181_{100}	168_{15}	154_{11}	638	6853
Toxaphene (TM)		2245*	410_1	376_3	341_{22}	195_{44}	89_{99}	1040	880
Tramadol	P G U	1945	263_{12}	218_1	188_1	135_3	58_{100}	437	631
Tramadol AC	U+UHYAC	2100	305_1	188_{10}	135_1	116_2	58_{100}	641	4435
Tramadol TMS		2015	335_4	320_1	245_3	73_{17}	58_{100}	786	4601
Tramadol artifact	G P U+UHYAC	1630*	188_{100}	173_{23}	159_{41}	129_{23}	115_{19}	191	4436
Tramadol -H2O	G P UHY U+UHYAC	1905	245_1	200_2	141_2	128_3	58_{100}	361	262
Tramadol-M (bis-demethyl-) 2AC	U+UHYAC	2420	319_{43}	156_{46}	114_{91}	86_{74}	74_{100}	711	4441
Tramadol-M (bis-demethyl-) -H2O 2AC	U+UHYAC	2465	301_{29}	228_{17}	186_{35}	86_{100}		621	265
Tramadol-M (HO-)	U	2200	279_8	234_2	58_{100}			515	1754
Tramadol-M (HO-) 2 AC	U+UHYAC	2310	363_1	186_7	116_1	58_{100}		895	4439
Tramadol-M (HO-) -H2O	G P	2200	261_{18}	202_{29}	189_{64}	121_{68}	73_{100}	428	6756
Tramadol-M (N-demethyl-) AC	U+UHYAC	2370	291_{86}	200_{24}	135_{100}	114_{94}	86_{75}	572	4440
Tramadol-M (N-demethyl-) -H2O AC	U+UHYAC	2295	273_{30}	200_{100}	86_{98}	58_{60}		482	264
Tramadol-M (O-demethyl-)	U	1995	249_1	121_3	107_2	93_2	58_{100}	378	634
Tramadol-M (O-demethyl-) 2Ac	U+UHYAC	2200	333_1	216_1	174_1	121_1	58_{50}	777	4438
Tramadol-M (O-demethyl-) 2TMS		2010	393_1	303_2	84_{12}	73_{32}	58_{100}	997	7195
Tramadol-M (O-demethyl-) Ac		2080	291_1	248_1	163_1	121_1	58_{100}	572	2602
Tramadol-M (O-demethyl-) -H2O		1920	231_1	91_1	73_5	58_{100}		311	633
Tramadol-M (O-demethyl-) -H2O AC	U+UHYAC	2000	273_2	184_6	58_{100}			482	263
Tramazoline AC		2760	257_{71}	214_{100}	185_{59}	172_{53}	86_{90}	411	2811
Trandolapril ET	UET	2975	458_1	385_5	234_{100}	160_7	91_{12}	1125	4779
Trandolapril ME	UME	2970	444_1	371_6	234_{100}	160_{10}	91_{22}	1106	4773
Trandolapril METMS		3055	516_4	429_5	248_{100}	160_4	91_{19}	1173	5000
Trandolapril TMS		2970	502_6	487_{17}	429_{36}	234_{100}	91_{22}	1165	4999
Trandolapril 2ET	UET	3050	486_1	413_2	262_{100}	188_9	91_9	1154	4780
Trandolapril 2ME		2995	458_1	385_4	248_{100}	174_6	91_7	1125	4774
Trandolapril -H2O	G	3090	412_{17}	308_{86}	262_{100}	234_{54}	91_{36}	1047	4777
Trandolaprilate 2ET	UET	2975	458_1	385_5	234_{100}	160_7	91_{12}	1125	4779
Trandolaprilate 2ME	UME	2940	430_1	371_9	220_{100}	160_7	91_{27}	1081	4775
Trandolaprilate 2TMS	UTMS	3040	546_4	531_{28}	429_{39}	278_{60}	91_{41}	1184	4997
Trandolaprilate 3ET	UET	3050	486_1	413_2	262_{100}	188_9	91_9	1154	4780

Trandolaprilate 3ME **Table 1-8-1:** Compounds in order of names

Name	Detected	RI	Typical ions and intensities					Page	Entry
Trandolaprilate 3ME	UME	3005	444_1	385_{11}	234_{100}	174_6	91_{10}	1106	4776
Trandolaprilate-M/artifact 2ET	UET	2025	307_2	234_{100}	160_{12}	117_{17}	91_{32}	651	4740
Trandolaprilate-M/artifact 2ME	UME	1870	279_2	220_{100}	160_{10}	117_{28}	91_{57}	512	4734
Trandolaprilate-M/artifact 3ME	UME	1935	293_2	234_{100}	174_8	130_{16}	91_{50}	582	4735
Trandolaprilate-M/artifact -H2O ME	UME	3070	398_9	294_{100}	262_{50}	117_7	91_{10}	1013	4778
Trandolaprilate-M/artifact -H2O TMS	UTMS	3105	456_{19}	441_{54}	352_{100}	262_{43}	223_{26}	1122	4998
Trandolapril-M/artifact (deethyl-) 2ET	UET	2975	458_1	385_5	234_{100}	160_7	91_{12}	1125	4779
Trandolapril-M/artifact (deethyl-) 2ME	UME	2940	430_1	371_9	220_{100}	160_7	91_{21}	1081	4775
Trandolapril-M/artifact (deethyl-) 2TMS	UTMS	3040	546_4	531_{28}	429_{39}	278_{100}	91_{41}	1184	4997
Trandolapril-M/artifact (deethyl-) 3ET	UET	3050	486_1	413_2	262_{100}	188_9	91_9	1154	4780
Trandolapril-M/artifact (deethyl-) 3ME	UME	3005	444_1	385_{11}	234_{100}	174_6	91_{10}	1106	4776
Trandolapril-M/artifact (deethyl-) -H2O ME	UME	3070	398_9	294_{100}	262_{50}	117_7	91_{10}	1013	4778
Trandolapril-M/artifact (deethyl-) -H2O TMS	UTMS	3105	456_{19}	441_{54}	352_{100}	262_{43}	223_{26}	1122	4998
Trandolapril-M/artifact (deethyl-HOOC-) 2ME	UME	1870	279_2	220_{100}	160_{10}	117_{28}	91_{57}	512	4734
Trandolapril-M/artifact (deethyl-HOOC-) 3ME	UME	1935	293_2	234_{100}	174_8	130_{16}	91_{50}	582	4735
Trandolapril-M/artifact (HOOC-) ET	UET	2025	307_2	234_{100}	160_{12}	117_{17}	91_{32}	651	4740
Trandolapril-M/artifact (HOOC-) ME	UME	1930	293_3	234_{36}	220_{100}	160_{11}	91_{23}	582	4736
Trandolapril-M/artifact (HOOC-) 2ME	UME	1985	307_2	248_{34}	234_{100}	174_5	91_6	651	4737
Tranexamic acid ME		1280	171_{18}	140_{20}	112_{44}	81_{93}	67_{100}	159	5680
Tranexamic acid MEAC		1930	213_1	198_1	154_{16}	73_{100}	60_{100}	252	5681
Tranexamic acid 2TMS		1800	301_2	286_6	197_7	102_{100}	73_{34}	621	6218
Tranylcypromine	G P-I	1230	133_{89}	132_{96}	115_{60}	56_{100}		113	635
Tranylcypromine AC	U+UHYAC	1635	175_5	132_{100}	116_{99}	84_{34}	56_{83}	162	402
Tranylcypromine TMS		1220	205_{22}	190_9	128_9	100_{25}	73_{100}	229	5448
Tranylcypromine-M (HO-) 2AC	U+UHYAC	2080	233_9	191_{16}	148_{42}	132_{100}	84_{40}	316	3420
Trapidil		2250	205_{36}	176_{100}	162_{48}	109_{47}	72_{30}	229	6108
Trazodone	G P-I U+UHYAC	3345	371_3	356_5	278_{11}	205_{74}	70_{62}	923	403
Trazodone-M (4-amino-2-Cl-phenol) 2AC	U+UHYAC	2020	227_4	185_{57}	143_{100}	79_{14}	114_5	297	404
Trazodone-M (4-amino-2-Cl-phenol) 2TFA	U+UHYTFA	1440	335_{44}	266_3	238_{52}	210_{14}	143_{42}	781	6602
Trazodone-M (4-amino-2-Cl-phenol) 3AC	U+UHYAC	1900	269_4	227_{37}	185_{75}	143_{100}	79_7	463	6595
Trazodone-M (deamino-HO-) AC		1985	235_{100}	192_{18}	175_{30}	148_{48}	136_{66}	323	5312
Trazodone-M (HO-)		3350	387_{17}	372_{27}	294_{32}	205_{100}	154_{87}	976	5313
Trazodone-M (HO-) AC	U+UHYAC	3580	429_{18}	414_{18}	336_{25}	205_{99}		1078	407
Trazodone-M (N-dealkyl-)		1910	196_{24}	154_{100}	138_{12}	111_9	75_{12}	211	6885
Trazodone-M (N-dealkyl-) AC	U+UHYAC	2265	238_{32}	195_{15}	166_{100}	154_{31}	111_{27}	336	405
Trazodone-M (N-dealkyl-) HFB	U+UHYHFB	1960	392_{100}	195_{38}	166_{41}	139_{36}	111_{25}	991	6604
Trazodone-M (N-dealkyl-) ME		1820	210_{100}	166_{19}	139_{81}	111_{33}	70_{99}	244	6886
Trazodone-M (N-dealkyl-) TFA	U+UHYTFA	1920	292_{100}	250_{12}	195_{77}	166_{79}	139_{66}	574	6597
Trazodone-M (N-dealkyl-) TMS		2035	268_{61}	253_{29}	128_{63}	101_{68}	86_{71}	461	6888
Trazodone-M (N-dealkyl-HO-) TFA	U+UHYTFA	2035	308_{100}	272_{19}	211_{46}	182_{36}	155_{46}	653	6599
Trazodone-M (N-dealkyl-HO-) iso-1 AC	U+UHYAC	2335	254_{65}	211_{23}	182_{100}	166_{71}	154_{33}	398	5308
Trazodone-M (N-dealkyl-HO-) iso-1 2AC	U+UHYAC	2515	296_{30}	254_{56}	182_{100}	154_{36}		597	406
Trazodone-M (N-dealkyl-HO-) iso-1 2TFA	U+UHYTFA	2040	404_{28}	307_{74}	278_{34}	265_{13}	154_{40}	1028	6600
Trazodone-M (N-dealkyl-HO-) iso-2 AC	U+UHYAC	2345	254_{79}	211_{15}	182_{100}	169_{40}	154_{40}	398	5307
Trazodone-M (N-dealkyl-HO-) iso-2 2AC	U+UHYAC	2525	296_{20}	254_{74}	182_{100}	169_{72}	154_{24}	597	32
Trazodone-M (N-dealkyl-HO-) iso-2 2HFB	U+UHYHFB	2145	604_{43}	585_9	407_{100}	378_{22}	154_{22}	1195	6605
Trazodone-M (N-dealkyl-HO-) iso-2 2TFA	U+UHYTFA	2045	404_{13}	307_{33}	278_{10}	265_4	154_{22}	1028	6598
Tremulone		3630*	410_{22}	269_{11}	187_{25}	174_{100}	161_{22}	1042	5584
Triacontane		3000*	422_2	113_{10}	85_{38}	71_{62}	57_{100}	1065	2366
Triadimefon		1980	293_6	208_{31}	181_8	128_{25}	57_{100}	578	1531
Triadimenol		2045	295_1	238_2	168_{72}	112_{100}	57_{68}	590	3468
Triallate		1810	303_1	268_{41}	186_6	143_{18}	86_{100}	626	3870
Triamcinolone		3200*	326_5	91_{37}	270_5	122_{100}	121_{80}	1979	5679
Triamiphos		2200	294_{48}	251_7	207_4	160_{100}	135_{38}	586	623
Triamterene ME	UME	2875	267_{70}	266_{100}	251_{11}	193_9	133_8	454	3120
Triazolam	G	3080	342_{56}	313_{100}	279_{25}	238_{54}	203_{30}	815	636
Triazolam-M (HO-)		3000	358_8	328_{99}	293_{61}	265_{37}	239_{30}	876	1533
Triazolam-M (HO-) AC	U+UHYAC-I	3200	400_{44}	359_{71}	357_{99}	329_9	239_{10}	1017	1532
Triazolam-M (HO-) -CH2O		3000	328_{100}	293_{77}	265_{43}	239_{32}	75_{28}	746	2050
Triazolam-M HY		2865	331_{34}	296_{100}	139_{43}	111_{35}		761	306
Triazophos		2250	313_{19}	285_{38}	257_{55}	172_{62}	161_{100}	678	3871
Tribenzylamine		2160	287_9	210_{16}	196_{15}	91_{84}	65_9	549	4492
Tributoxyethylphosphate		2350*	355_2	299_{13}	199_{21}	125_{39}	57_{100}	1013	3051
Tributylamine		1250	185_4	142_{100}	100_{81}	58_{31}		187	4186
Tributylphosphate		1485*	211_3	155_{13}	111_2	99_{100}	57_{20}	452	5179
Trichlorfon		1450*	221_3	185_6	145_{29}	109_{96}	79_{100}	405	117
Trichlorfon ME		1395*	235_3	205_{36}	161_{23}	109_{80}	93_{100}	467	4148
Trichlormethiazide 4ME		2810	435_{100}	401_{33}	219_{18}	184_{24}	150_{19}	1090	3111
2,4,6-Trichloroaniline	U	1470	197_{98}	195_{100}	159_{12}	124_{28}	97_{15}	206	2642
2,4,4'-Trichlorobiphenyl		1860*	258_{97}	256_{100}	186_{91}	150_{39}	75_{38}	405	2615
Trichloroethane		<1000*	117_{14}	97_{100}	61_{59}	35_{25}	27_{90}	113	3780
Trichloroethanol	P UHY	<1000*	148_3	119_{20}	113_{60}	82_{46}	77_{100}	125	1413
Trichloroethylene		<1000*	130_{100}	95_{80}	60_{32}			112	1544
Trichlorofluoromethane		<1000*	136_1	117_2	101_{77}	82_5	66_{13}	116	3794

Table 1-8-1: Compounds in order of names

Name	Detected	RI	Typical ions and intensities					Page	Entry
Trichloroisobutyl salicylate		1820*	296 1	138 48	120 100	92 21	65 21	595	4270
Trichloroisobutyl salicylate ME	U+UHYAC	1890*	310 2	152 10	135 100	123 27	77 17	664	4271
Trichloromethoxypropionamide		1150	174 4	140 9	110 15	88 66	60 70	227	5684
Trichloronat		2005*	332 1	297 43	269 47	196 7	109 100	766	3469
2,4,5-Trichlorophenol	U	1440*	198 93	196 100	132 34	97 64	73 14	209	784
2,4,6-Trichlorophenol	U	1420*	198 98	196 100	160 11	132 37	97 31	209	3363
2,4,5-Trichlorophenoxyacetic acid (T)		1850*	254 60	209 17	196 100	167 30	109 22	396	2396
2,4,5-Trichlorophenoxyacetic acid (T) ME		1760*	268 47	233 100	209 33	179 18	145 22	457	1962
2,4,5-Trichlorophenoxyacetic acid (T) isobutylester		2280*	310 27	254 15	219 50	196 20	57 100	664	1956
2,4,5-Trichlorophenoxyacetic acid (T) octylester		2320*	366 59	254 79	209 20	71 100	57 93	903	1957
2,4,5-Trichlorophenoxyacetic acid (T)-M (trichlorophenol)	U	1440*	198 93	196 100	132 34	97 64	73 14	209	784
Triclopyr ME		1700	269 21	210 100	180 27	144 17	59 36	462	3654
Triclosan	U	2060*	288 100	252 25	218 90	146 64	114 69	550	691
Triclosan AC	U+UHYAC	2070*	330 8	290 99	288 100	252 9	218 25	757	1872
Tricosane		2300*	324 3	99 13	85 35	71 56	57 100	733	2364
Tridecane		1300*	184 5	99 7	85 30	71 55	57 100	185	2362
Tridemorph		1875	297 1	282 1	128 100	115 5	70 5	604	4085
Trietazine		1760	229 54	214 48	200 100	186 52	96 15	304	3876
Triethylamine		<1000	101 17	86 100	72 5	70 4	58 48	100	1907
Triethylcitrate AC		1880*	318 1	273 8	213 20	203 55	157 100	703	4478
Triethylene glycol dipivalate	PPIV	1685*	158 1	129 87	113 6	85 29	57 100	705	6426
Triflubazam		2275	334 62	289 56	215 24	77 100	51 78	779	4021
Triflubazam HY		1840	276 100	275 82	138 9	77 54	51 55	495	4020
Trifluoperazine	G U UHY U+UHYAC	2685	407 38	267 19	113 100	70 95		1034	408
Trifluoperazine-M (amino-) AC	U+UHYAC	2765	366 47	280 17	266 40	248 18	100 300	903	1267
Trifluoperazine-M (nor-) AC	U+UHYAC	3145	435 49	267 90	141 100	99 59		1090	1268
Trifluoperazine-M (ring)	U+UHYAC	2190	267 100	235 19				453	1266
Trifluoroacetaldehyde		<1000*	98 2	79 6	51 36	47 60		98	2997
Trifluoroacetic acid		<1000*	114 2	95 10	69 100	51 52	45 120	104	5547
Trifluoroacetic acid		<1000*	114 2	97 7	95 10	69 100	51 52	104	5546
3-Trifluoromethylaniline AC	U+UHYAC	1400	203 20	184 9	161 100	142 6	114 12	223	6579
3-Trifluoromethylaniline HFB	U+UHYHFB	1130	357 27	338 18	188 100	160 89	145 67	871	6590
3-Trifluoromethylaniline TFA	U+UHYTFA	1230	257 81	238 25	188 79	160 100	145 82	409	6586
3-Trifluoromethylaniline-M (HO- N-acetyl-) TFA	UTFA	1415	315 12	296 7	273 47	219 5	176 100	687	6807
3-Trifluoromethylaniline-M (HO-) 2TFA	U+UHYTFA	1395	369 100	350 23	300 9	272 42	177 13	914	6587
3-Trifluoromethylaniline-M (HO-) iso-1 2AC	U+UHYAC	1810	261 23	219 74	177 100	157 50	129 6	425	6581
3-Trifluoromethylaniline-M (HO-) iso-2 AC	U+UHYAC	1710	219 41	200 8	177 39	157 100	129 20	269	6582
3-Trifluoromethylaniline-M (HO-) iso-2 2AC	U+UHYAC	1840	261 3	219 73	177 90	157 100	129 15	425	6580
4-Trifluoromethylaniline AC	U+UHYAC	1420	203 22	184 4	161 100	142 20	111 12	223	7372
Trifluoromethylphenylpiperazine		1620	230 24	188 100	172 17	145 18	56 15	307	5886
Trifluoromethylphenylpiperazine AC		1890	272 25	200 100	188 37	172 37	56 69	479	5887
Trifluoromethylphenylpiperazine HFB		1750	426 57	229 36	200 54	145 29	56 80	1071	6768
Trifluoromethylphenylpiperazine PFP		1690	376 79	229 48	200 67	172 47	56 100	939	5889
Trifluoromethylphenylpiperazine TFA		1690	326 86	229 52	200 68	172 57	56 90	738	5888
Trifluoromethylphenylpiperazine TMS		1920	302 100	287 32	101 76	86 70	73 98	624	5890
Trifluoromethylphenylpiperazine-M (deethylene-) 2AC	U+UHYAC	1865	288 1	229 37	187 48	174 100	145 13	552	6583
Trifluoromethylphenylpiperazine-M (deethylene-) 2HFB	U+UHYHFB	1575	596 1	577 5	383 82	370 100	145 22	1194	6591
Trifluoromethylphenylpiperazine-M (deethylene-) 2TFA	U+UHYTFA	1530	377 20	283 95	270 42	172 45	145 51	1004	6588
Trifluoromethylphenylpiperazine-M (HO-) 2AC	U+UHYAC	2275	330 11	288 48	245 11	216 100	203 46	759	6578
Trifluoromethylphenylpiperazine-M (HO-) 2HFB	U+UHYHFB	1985	638 12	441 100	412 68	399 17	188 26	1198	6589
Trifluoromethylphenylpiperazine-M (HO-) 2TFA	U+UHYTFA	2005	438 100	419 25	396 12	341 79	312 61	1094	6585
Trifluoromethylphenylpiperazine-M (HO-deethylene-) 3AC	U+UHYAC	2275	346 5	287 31	245 58	203 79	190 100	831	6584
Trifluoromethylphenylpiperazine-M (HO-glucuronide) 4TMS		2920	375 11	318 100	276 23	245 19	73 79	1203	6767
Trifluoromethylphenylpiperazine-M (HO-trifluoromethylaniline N-acetyl-) TFA	UTFA	1415	315 12	296 7	273 47	219 5	176 100	687	6807
Trifluoromethylphenylpiperazine-M (HO-trifluoromethylaniline) 2TFA	U+UHYTFA	1395	369 100	350 23	300 9	272 42	177 13	914	6587
Trifluoromethylphenylpiperazine-M (HO-trifluoromethylaniline) iso-1 2AC	U+UHYAC	1810	261 23	219 74	177 100	157 50	129 6	425	6581
Trifluoromethylphenylpiperazine-M (HO-trifluoromethylaniline) iso-2 AC	U+UHYAC	1710	219 41	200 8	177 39	157 100	129 20	269	6582
Trifluoromethylphenylpiperazine-M (HO-trifluoromethylaniline) iso-2 2AC	U+UHYAC	1840	261 3	219 73	177 90	157 100	129 15	425	6580
Trifluoromethylphenylpiperazine-M (trifluoromethylaniline) AC	U+UHYAC	1400	203 20	184 9	161 100	142 6	114 12	223	6579
Trifluoromethylphenylpiperazine-M (trifluoromethylaniline) HFB	U+UHYHFB	1130	357 27	338 18	188 100	160 89	145 67	871	6590
Trifluoromethylphenylpiperazine-M (trifluoromethylaniline) TFA	U+UHYTFA	1230	257 81	238 25	188 79	160 100	145 82	409	6586
Trifluperidol	G	2700	409 3	271 100	258 74	123 54		1038	637
Trifluperidol TMS		2740	481 1	466 2	343 30	271 73	123 100	1149	5456
Trifluperidol 2TMS		2780	538 4	330 93	240 59	103 30	73 100	1185	5457
Trifluperidol-M	U+UHYAC	1490*	180 25	125 49	123 35	95 17	56 100	174	85
Trifluperidol-M	UHY	1950	257 7	223 12	173 12	145 20	56 100	409	638
Trifluperidol-M (N-dealkyl-)	UHY	1970	245 8	227 18	56 100			358	639
Trifluperidol-M (N-dealkyl-) AC	U+UHYAC	2035	287 32	269 18	244 27	99 26	57 100	546	167
Trifluperidol-M (N-dealkyl-oxo-) -2H2O	U UHY U+UHYAC	1570	223 100	154 17	127 6			283	168
Triflupromazine	P G U+UHYAC	2240	352 5	267 5	86 22	58 100		852	409
Triflupromazine-M (bis-nor) AC	U+UHYAC	2765	366 47	280 17	266 40	248 18	100 300	903	1267
Triflupromazine-M (bis-nor-HO-) 2AC	U+UHYAC	3070	424 21	342 17	282 31	100 100	72 26	1067	2639
Triflupromazine-M (bis-nor-HO-methoxy-) AC	U+UHYAC	3055	412 16	312 15	269 5	100 100	72 14	1046	5638
Triflupromazine-M (HO-)	UHY	2700	368 36	86 24	58 200	282 10	322 17	912	5635

Triflupromazine-M (HO-) AC Table 1-8-1: Compounds in order of names

Name	Detected	RI	Typical ions and intensities					Page	Entry
Triflupromazine-M (HO-) AC	U+UHYAC	2720	410_9	368_7	322_8	86_{20}	58_{100}	1041	1299
Triflupromazine-M (HO-methoxy-)	UHY	2730	398_{65}	352_7	312_{19}	86_{42}	58_{300}	1012	5636
Triflupromazine-M (HO-methoxy-) AC	U+UHYAC	2750	440_{30}	353_{11}	312_{19}	86_{42}	58_{300}	1098	5637
Triflupromazine-M (nor-) AC	U+UHYAC	2740	380_{25}	280_7	248_{10}	114_{100}	86_{19}	952	1300
Triflupromazine-M (nor-HO-) 2AC	U+UHYAC	3120	438_{22}	339_7	282_{21}	114_{100}	86_{13}	1094	1301
Triflupromazine-M (nor-HO-methoxy-) AC	U+UHYAC	3170	426_{50}	312_{39}	114_{300}	86_{48}		1071	5639
Triflupromazine-M (nor-HO-methoxy-) 2AC	U+UHYAC	3170	468_{16}	312_{15}	269_5	114_{100}	86_{14}	1136	2640
Triflupromazine-M (ring)	U+UHYAC	2190	267_{100}	235_{19}				453	1266
Trifluralin		1680	335_7	306_{100}	290_{14}	264_{85}	248_{11}	783	3873
Trihexylamine		1725	269_2	198_{100}	128_{42}	98_{14}	58_{37}	467	4491
Trihexyphenidyl	P-I G U	2250	301_1	218_5	98_{100}			621	92
Trihexyphenidyl-M (amino-HO-) iso-1 -H2O 2AC	U+UHYAC	2560	315_3	255_{20}	196_{100}	168_{35}	155_{22}	690	1290
Trihexyphenidyl-M (amino-HO-) iso-2 -H2O 2AC	U+UHYAC	2625	315_8	255_{60}	196_{100}	132_{66}	115_{60}	691	4242
Trihexyphenidyl-M (di-HO-) -H2O iso-1 2AC	U+UHYAC	2555	399_4	357_3	98_{100}			1016	1303
Trihexyphenidyl-M (di-HO-) -H2O iso-2 2AC	U+UHYAC	2665	399_5	338_3	194_{30}	98_{100}		1016	1304
Trihexyphenidyl-M (HO-)	U	2500	317_1	299_1	218_6	98_{100}		700	93
Trihexyphenidyl-M (HO-) AC	U+UHYAC	2635	359_1	316_5	218_{13}	98_{100}		881	1553
Trihexyphenidyl-M (HO-) -H2O AC	U+UHYAC	2505	341_{16}	298_2	200_{30}	98_{100}		815	1552
Trihexyphenidyl-M (tri-HO-) -H2O 3AC	U+UHYAC	2965	457_6	398_3	336_2	194_{25}	156_{100}	1124	1305
Trihexyphenidyl-M -2H2O -CO2 AC	U+UHYAC	2095*	242_{71}	200_{27}	182_{100}	167_{58}		353	1302
Trimebutine		2660	195_5	162_1	152_1	109_1	58_{100}	977	7634
Trimebutine-M (TMBA)		1780*	212_{100}	197_{57}	169_{13}	141_{27}		248	1949
Trimebutine-M (TMBA) ME		1740*	226_{100}	211_{49}	195_{23}	155_{21}		293	1950
Trimebutine-M/artifact (alcohol)		1070	175_{12}	160_{13}	115_7	91_{11}	58_{100}	202	7633
Trimethadion		1080	143_{73}	128_{26}	100_4	70_{14}	58_{100}	122	1003
Trimethadion-M (nor-)	U	1060	129_{15}	107_2	70_7	59_{100}		112	2923
Trimethoprim	P G U UHY	2590	290_{100}	259_{36}	123_{19}			564	1004
Trimethoprim 2AC	U+UHYAC	3000	374_{49}	359_{54}	332_{53}	317_{82}	275_{58}	934	1006
Trimethoprim 2TMS		2650	434_{50}	419_{100}	331_{10}	210_6	73_{42}	1088	4602
Trimethoprim 3TMS		2805	506_{36}	491_{100}	403_{11}	246_{10}	73_{90}	1168	4603
Trimethoprim iso-1 AC	PAC U+UHYAC	2700	332_{100}	317_{44}	289_{85}	275_{66}	259_{17}	769	1005
Trimethoprim iso-2 AC	U+UHYAC	2880	332_{100}	317_{21}	290_{28}	275_{32}	259_{34}	769	2576
2,3,5-Trimethoxyamfetamine		2040	225_2	182_{100}	167_{30}	151_7	107_5	291	2622
2,3,5-Trimethoxyamfetamine AC		2285	267_{15}	208_{100}	193_{30}	181_{37}	86_{15}	455	2625
2,3,5-Trimethoxyamfetamine 2ME		1990	253_{26}	208_{95}	181_{100}	167_{15}	72_{15}	395	2624
2,3,5-Trimethoxyamfetamine intermediate (propenyltrimethoxybenzene)		1620*	208_{100}	193_{65}	165_9	150_6	133_{12}	237	2626
2,4,5-Trimethoxyamfetamine AC	U+UHYAC	2140	267_{23}	208_{100}	181_{66}	151_{23}	86_6	456	7152
2,4,5-Trimethoxyamfetamine 2AC	U+UHYAC	2200	309_{10}	208_{100}	181_{65}	151_{21}	86_{11}	662	7161
2,4,5-Trimethoxyamfetamine-M (deamino-HO-) AC	U+UHYAC	1670*	268_8	208_{100}	193_{39}	181_{31}		461	7157
2,4,5-Trimethoxyamfetamine-M (O-bisdemethyl-) artifact 2AC	U+UHYAC	2200	263_{28}	221_{100}	179_{65}	164_{44}	132_{10}	434	7183
2,4,5-Trimethoxyamfetamine-M (O-bisdemethyl-) iso-1 3AC	U+UHYAC	2300	323_2	281_{27}	180_{100}	153_{50}	86_{22}	726	7162
2,4,5-Trimethoxyamfetamine-M (O-bisdemethyl-) iso-2 3AC	U+UHYAC	2305	323_2	281_6	180_{100}	153_{52}	86_{20}	727	7163
2,4,5-Trimethoxyamfetamine-M (O-bisdemethyl-) iso-3 3AC	U+UHYAC	2330	323_4	281_{12}	180_{100}	153_{52}	86_{25}	727	7164
2,4,5-Trimethoxyamfetamine-M (O-deamino-oxo-)	U+UHYAC	1540*	224_{25}	181_{100}	151_{37}	136_{28}		289	7165
2,4,5-Trimethoxyamfetamine-M (O-demethyl-) iso-1 2AC	U+UHYAC	2215	295_9	253_{23}	194_{100}	167_{47}	86_{13}	592	7154
2,4,5-Trimethoxyamfetamine-M (O-demethyl-) iso-2 2AC	U+UHYAC	2230	295_{17}	236_{32}	194_{100}	167_{54}	86_{14}	592	7153
2,4,5-Trimethoxyamfetamine-M (O-demethyl-) iso-2 3AC	U+UHYAC	2280	337_5	236_{29}	194_{100}	167_{60}	86_{22}	793	7156
2,4,5-Trimethoxyamfetamine-M (O-demethyl-) iso-3 2AC	U+UHYAC	2250	295_9	236_{52}	194_{100}	167_{40}	86_7	592	7155
2,4,5-Trimethoxyamfetamine-M (O-demethyl-deamino-oxo-) iso-1 AC	U+UHYAC	1680*	252_8	210_{40}	167_{100}			390	7158
2,4,5-Trimethoxyamfetamine-M (O-demethyl-deamino-oxo-) iso-2 AC	U+UHYAC	1705*	252_{11}	210_{32}	167_{100}	137_9		390	7159
2,4,5-Trimethoxyamfetamine-M (O-demethyl-deamino-oxo-) iso-3 AC	U+UHYAC	1760*	252_6	210_{19}	167_{58}	137_6		390	7160
3,4,5-Trimethoxyamfetamine		1680	225_1	182_{24}	167_8	151_2	107_1	292	3259
3,4,5-Trimethoxyamfetamine		1680	225_1	182_{24}	167_8	151_2	44_{100}	292	5540
3,4,5-Trimethoxyamfetamine AC		2020	267_9	208_{47}	193_{17}	181_{16}	86_{10}	455	3266
3,4,5-Trimethoxyamfetamine AC		2020	267_9	208_{47}	193_{17}	86_{10}	44_{100}	456	5541
3,4,5-Trimethoxyamfetamine formyl artifact		1680	237_{25}	181_{93}	148_9	77_6	56_{100}	335	3251
3,4,5-Trimethoxyamfetamine intermediate-1		2050	253_{100}	206_{30}	191_{44}	161_{27}	77_{48}	393	2840
3,4,5-Trimethoxyamfetamine intermediate-2		2145	239_{100}	192_{34}	177_{44}	149_{32}	92_{26}	339	2841
3,4,5-Trimethoxybenzaldehyd		1550*	196_{100}	181_{51}	125_{52}	110_{51}	93_{37}	211	3279
Trimethoxybenzoic acid		1780*	212_{100}	197_{57}	169_{13}	141_{27}		248	1949
Trimethoxybenzoic acid ET		1770*	240_{100}	225_{44}	212_{17}	195_{45}	141_{24}	344	5219
Trimethoxybenzoic acid ME		1740*	226_{100}	211_{49}	195_{23}	155_{21}		293	1950
Trimethoxybenzoic acid-M (glycine conjugate) ME		2350	283_{62}	268_4	195_{100}	152_8		528	1952
3,4,5-Trimethoxybenzyl alcohol		1650*	198_{100}	183_{22}	127_{51}	95_{39}	77_{22}	214	6059
3,4,5-Trimethoxybenzyl alcohol AC		1650*	240_{80}	198_{52}	181_{100}	169_{29}	123_{27}	343	6060
Trimethoxycocaine	UGLUCME	2550	393_{27}	212_{21}	182_{100}	94_{24}	82_{102}	995	5678
Trimethoxyhippuric acid ME		2350	283_{62}	268_4	195_{100}	152_8		528	1952
2,3,5-Trimethoxymetamfetamine AC		2310	281_9	224_{12}	208_{65}	100_{47}	58_{100}	523	2623
3,4,5-Trimethoxyphenyl-2-nitroethene		2145	239_{100}	192_{34}	177_{44}	149_{32}	92_{26}	339	2841
3,4,5-Trimethoxyphenyl-2-nitropropene		2050	253_{100}	206_{30}	191_{44}	161_{27}	77_{48}	393	2840
Trimethylamine		<1000	59_{38}	58_{100}	42_{59}	30_{34}		91	4187
1,2,3-Trimethylbenzene		<1000*	120_{42}	105_{100}	91_{11}	77_{16}	65_7	106	3788
1,2,4-Trimethylbenzene		<1000*	120_{45}	105_{100}	91_{11}	77_{15}	65_6	106	3826
Trimethylcitrate	UME	1410*	175_{14}	143_{100}	101_{86}	69_{18}	59_{44}	319	4451

Table 1-8-1: Compounds in order of names   Tryptophan-M (tryptamine) AC

Name	Detected	RI	Typical ions and intensities					Page	Entry
Bis-(Trimethylsilyl-)trifluoroacetamide		1100	257_3	192_{35}	188_{47}	100_{58}	73_{200}	409	5431
Trimetozine		2260	281_{45}	195_{100}				522	1529
Trimipramine	P G U+UHYAC	2225	294_7	249_{33}	193_{26}	99_{13}	58_{100}	587	410
Trimipramine artifact	G	2025	235_{100}	220_{60}	206_{43}	192_{19}	178_{27}	325	6561
Trimipramine-D3		2215	297_7	249_{29}	208_{19}	102_{16}	61_{100}	603	5426
Trimipramine-D3 artifact		2045*	249_{29}	234_7	208_{23}	194_{100}	167_9	377	6329
Trimipramine-M (bis-nor-) AC	U+UHYAC	2650	308_{13}	208_{100}	193_{32}	114_{30}	72_{20}	657	2865
Trimipramine-M (bis-nor-di-HO-) 3AC	U+UHYAC	3400	424_{14}	324_{100}	282_{30}	114_{30}	72_{20}	1068	2856
Trimipramine-M (bis-nor-HO-) 2AC	U+UHYAC	3050	366_{21}	266_{100}	224_{42}	209_{20}	114_5	906	2676
Trimipramine-M (bis-nor-HO-methoxy-) 2AC	U+UHYAC	3130	396_{13}	296_{100}	254_{49}	114_{30}	72_{20}	1007	2866
Trimipramine-M (di-HO-) 2AC	U+UHYAC	2900	410_{10}	365_{12}	323_{60}	99_6	58_{100}	1042	2293
Trimipramine-M (di-HO-ring)	UHY	2600	227_{100}	196_7				297	2296
Trimipramine-M (di-HO-ring) 2AC	U+UHYAC	2750	311_{28}	269_{23}	227_{100}	196_7		670	2292
Trimipramine-M (HO-)	P-I UHY	2575	310_{11}	265_{34}	250_{16}	224_{20}	58_{100}	669	640
Trimipramine-M (HO-) AC	PAC U+UHYAC	2660	352_{12}	307_{26}	265_{38}	99_7	58_{100}	856	411
Trimipramine-M (HO-methoxy-)	UHY	2590	340_{20}	295_{29}	254_{23}	99_{10}	58_{100}	809	2314
Trimipramine-M (HO-methoxy-) AC	U+UHYAC	2700	382_4	337_7	295_{23}	99_8	58_{100}	960	2291
Trimipramine-M (HO-methoxy-ring)	UHY	2390	241_{100}	226_{17}	210_{12}	180_{14}		347	2315
Trimipramine-M (HO-methoxy-ring) AC	U+UHYAC	2370	283_{10}	241_{100}	226_{17}	210_{12}	180_{14}	529	2867
Trimipramine-M (HO-ring)	UHY	2240	211_{100}	196_{15}	180_{10}	152_4		246	2295
Trimipramine-M (HO-ring) AC	U+UHYAC	2535	253_{26}	211_{100}	196_{19}	180_{11}	152_4	393	1218
Trimipramine-M (nor-)	U UHY	2245	280_{13}	249_{31}	234_{14}	208_{100}	193_{56}	519	6330
Trimipramine-M (nor-) AC	U+UHYAC	2680	322_{17}	208_{100}	193_{32}	128_{13}	86_6	724	2290
Trimipramine-M (nor-di-HO-) 3AC	U+UHYAC	3555	438_{15}	324_{100}	282_{33}	240_{16}	128_{32}	1095	413
Trimipramine-M (nor-HO-) 2AC	U+UHYAC	3155	380_{22}	266_{100}	224_{24}	128_{14}	86_7	953	412
Trimipramine-M (nor-HO-) -H2O AC	U+UHYAC	2670	320_{11}	206_{100}	128_6			716	991
Trimipramine-M (nor-HO-methoxy-) 2AC	U+UHYAC	3180	410_{37}	296_{100}	254_{49}	128_{14}	86_5	1042	2294
Trimipramine-M (N-oxide) -(CH3)2NOH		2045*	249_{29}	234_7	208_{23}	194_{100}	167_9	377	6329
Trimipramine-M (ring)	U U+UHYAC	1930	195_{100}	180_{40}	167_9	96_{33}	83_{22}	207	308
Trimipramine-M (ring) ME		1915	209_{70}	194_{100}	178_{13}	165_{11}		242	6352
Tripelenamine	U UHY U+UHYAC	1970	255_3	197_{18}	185_{11}	91_{79}	58_{100}	404	2030
Tripelenamine-M (benzylpyridylamine)	UHY U+UHYAC	1650	184_{100}	106_{90}	91_{57}	79_{47}	65_{27}	185	1603
Tripelenamine-M (HO-)	UHY	2400	271_{11}	213_{33}	91_{80}	72_{45}	58_{100}	476	1609
Tripelenamine-M (HO-) AC	U+UHYAC	2390	313_5	255_{17}	91_{80}	72_{19}	58_{100}	682	1606
Tripelenamine-M (nor-)	U UHY	2420	241_8	197_{28}	129_{81}	112_{18}	91_{100}	349	1610
Tripelenamine-M (nor-) AC	U+UHYAC	2420	283_8	197_{41}	183_7	91_{100}	78_8	530	1607
Tripelenamine-M (nor-HO-) 2AC	U+UHYAC	2860	341_7	255_{37}	213_{27}	177_{29}	91_{100}	814	1608
Tripelenamine-M/artifact-1	UHY U+UHYAC	1845	212_{30}	183_{100}	107_{26}	91_{39}	78_{47}	248	1604
Tripelenamine-M/artifact-2	UHY U+UHYAC	2220	239_{100}	210_9	148_{49}	134_{25}	91_{19}	339	1605
Triphenylphosphate		2340*	326_{100}	325_{81}	233_{15}	170_{14}	77_{35}	738	2871
Triphenylphosphine oxide	G	2460*	278_{38}	277_{100}	199_{19}	183_{16}	152_{11}	505	6676
Triprolidine		2315	278_{36}	208_{100}	193_{26}	96_{22}	84_{20}	509	6103
TRIS 4AC	U+UHYAC	1910	216_3	156_{36}	127_{26}	114_{66}	72_{100}	557	4635
Tris-(2-chloroethyl-)phosphate		1870*	284_1	249_{32}	205_{17}	143_{24}	63_{100}	531	4255
Trisalicyclide	G U+UHYAC	3190*	360_{39}	240_{58}	152_{36}	120_{100}	92_{75}	883	4496
Tritoqualine artifact-1		2130	281_{100}	252_{94}	224_{58}	196_{60}	168_{57}	522	5236
Tritoqualine artifact-1 AC		2325	323_{17}	281_{96}	252_{100}	224_{37}	196_{34}	727	5239
Tritoqualine artifact-1 2AC		2350	365_9	323_{36}	281_{100}	252_{71}	224_{29}	900	5240
Tritoqualine artifact-2		2170	311_{100}	282_{51}	254_{29}	222_{27}	166_{20}	669	5237
Tritoqualine artifact-2 AC		2335	353_{17}	311_{100}	282_{34}	254_{21}	222_{17}	669	5238
Trometamol 4AC	U+UHYAC	1910	216_3	156_{36}	127_{26}	114_{66}	72_{100}	557	4635
Tropacocaine		2040	245_{13}	124_{100}	94_{44}	82_{79}	77_{42}	360	5124
Tropicamide		2340	284_1	266_8	254_{43}	163_{15}	92_{100}	536	1983
Tropicamide AC	U+UHYAC	2410	326_8	266_{26}	163_{18}	104_{30}	92_{100}	741	2238
Tropicamide -CH2O		2230	254_{21}	163_{16}	121_4	92_{100}	65_{21}	400	1985
Tropicamide -H2O		2250	266_{86}	251_{16}	103_{100}	92_{29}	77_{27}	452	1984
Tropine AC		1240	183_{25}	140_{11}	124_{100}	94_{42}	82_{63}	183	5125
Tropisetrone		2720	284_{22}	144_{27}	124_{100}	94_{45}	82_{52}	536	4633
Tropisetrone AC		2800	326_{14}	144_{11}	124_{100}	94_{44}	82_{52}	741	4634
Trovafloxacine TMS		3400	473_7	431_{100}	207_{12}	165_{43}	57_{15}	1155	5712
Tryptamine		1730	160_{19}	130_{100}	103_{14}	77_{19}		139	1007
Tryptamine AC		2390	202_{16}	143_{100}	130_{95}	103_{10}	77_{13}	222	2905
Tryptamine 2AC		2440	244_{12}	143_{100}	130_{84}	103_7	77_{11}	356	2906
Tryptophan ME2AC		2170	302_8	243_{25}	201_{32}	130_{100}		624	1009
Tryptophan MEAC		2150	260_{13}	201_{33}	130_{100}			421	1008
Tryptophan-M (HO-skatole)	U	1370	147_{75}	146_{100}	117_{15}			125	819
Tryptophan-M (hydroxy indole acetic acid) ME		----	205_{32}	146_{100}				228	1010
Tryptophan-M (indole acetic acid) ME	UME	1900	189_{25}	130_{100}	103_5	77_8		192	1011
Tryptophan-M (indole formic acid) ME	UME	1940	175_{47}	144_{100}	116_{23}	89_{19}		161	1012
Tryptophan-M (indole formic acid) 2ME	UME	1760	189_{62}	158_{100}	130_{14}	103_{10}	77_{15}	192	4944
Tryptophan-M (indole lactic acid) ME		----	219_{16}	130_{100}				269	1013
Tryptophan-M (indole pyruvic acid) 2ME		----	231_{98}	216_{30}	188_{77}	129_{47}		309	1014
Tryptophan-M (tryptamine)		1730	160_{19}	130_{100}	103_{14}	77_{19}		139	1007
Tryptophan-M (tryptamine) AC		2390	202_{16}	143_{100}	130_{95}	103_{10}	77_{13}	222	2905

Tryptophan-M (tryptamine) 2AC Table 1-8-1: Compounds in order of names

Name	Detected	RI	Typical ions and intensities					Page	Entry
Tryptophan-M (tryptamine) 2AC		2440	244_{12}	143_{100}	130_{84}	103_7	77_{11}	356	2906
Tyramine		1745	137_{26}	108_{100}	77_{23}			118	1485
Tyramine 2AC	U+UHYAC	1950	221_2	162_{26}	120_{100}	107_{30}	77_9	277	1015
Umbelliferone	UHY	1780*	162_{94}	134_{100}	105_{23}	78_{28}	63_9	140	4366
Umbelliferone AC	U+UHYAC	1840*	204_{16}	162_{100}	134_{85}	105_9	77_{11}	225	4367
Umbelliferone HFB		1685*	358_{100}	330_{51}	169_{25}	133_{80}	105_{23}	875	7614
Umbelliferone ME		1750*	176_{100}	148_{76}	133_{86}	105_{12}	77_{16}	162	7611
Umbelliferone PFP		1550*	308_{100}	280_{50}	261_5	161_3	133_{49}	653	7613
Umbelliferone TFA		1540*	258_{100}	230_{50}	133_{49}	119_{14}	105_8	412	7615
Umbelliferone TMS		1925*	234_{87}	219_{100}	191_{20}	163_{58}	73_{25}	319	7612
Undecane		1100*	156_2	98_4	85_{19}	71_{32}	57_{82}	137	3792
Urapidil-M (N-dealkyl-) 2AC	U+UHYAC	2070	234_{36}	162_{100}	149_{29}	134_{38}	120_{33}	320	6808
Urea AC	U+UHYAC	1670	102_{25}	74_{14}	60_{30}	59_{100}		101	5335
Urea 2TMS	UTMS	1420	204_2	189_{55}	147_{100}	130_7	73_{37}	226	5673
Urea artifact	U UHY U+UHYAC	2880	129_{100}	86_{20}	70_{13}			112	4424
Valganciclovir 4TMS		3440	642_7	599_{10}	455_{18}	144_{100}	73_{91}	1198	7310
Valganciclovir 5TMS		3530	714_4	699_{12}	352_{13}	144_{100}	73_{59}	1203	7311
Valproic acid	P G U	1150*	144_1	115_3	102_{32}	73_{100}		123	1019
Valpromide	U+UHYAC	1205	144_1	143_1	114_9	101_{43}	72_{100}	122	4670
Valsartan 2ET		3745	491_2	406_{100}	334_{35}	278_{75}	192_{40}	1159	4840
Valsartan 2ME	P UME	3420	463_2	378_{66}	320_{31}	264_{100}	192_{48}	1132	4839
Vamidothion		2070	287_2	145_{55}	109_{36}	87_{100}	58_{69}	545	3457
Vanillic acid ME		1455*	182_{50}	151_{100}	123_{14}	108_{16}	77_{16}	180	5216
Vanillic acid MEAC		1640*	224_6	193_2	182_{96}	151_{100}	123_{11}	287	5328
Vanillin	G	1630*	152_{96}	151_{100}	123_{20}	109_{11}		131	1974
Vanillin AC		1650*	194_9	152_{100}	151_{86}	123_9	109_9	204	1973
Vanillin mandelic acid		1465*	198_{10}	167_{100}	152_{32}	151_{26}	109_7	213	5138
Vanillin mandelic acid ME		1690*	212_{18}	153_{95}	125_{33}	93_{100}	65_{64}	248	5139
Vanillin mandelic acid ME2AC		1930*	296_3	254_{20}	222_{50}	153_{100}	151_{43}	597	5141
Vanillin mandelic acid 2ME		1780*	226_{25}	167_{100}	139_{94}	124_{40}	108_{32}	293	1020
Vanillin mandelic acid 2MEAC		1830*	268_3	226_8	209_5	167_{100}	151_{10}	460	5140
Venlafaxine	P-I G U	2055	277_1	134_6	91_2	77_1	58_{50}	504	5266
Venlafaxine AC	U+UHYAC	2100	319_1	202_7	134_{17}	121_9	58_{100}	713	5267
Venlafaxine TMS		2075	334_1	178_7	171_{15}	134_{40}	58_{80}	844	7692
Venlafaxine -H2O	U+UHYAC	1950	259_1	121_1	115_2	91_1	58_{100}	418	5268
Venlafaxine-M (HO-) iso-1		2310	293_1	179_4	134_{21}	58_{100}		583	5278
Venlafaxine-M (HO-) iso-1 AC	U+UHYAC	2320	335_1	134_{10}	58_{100}			786	5270
Venlafaxine-M (HO-) iso-2		2350	293_1	179_3	134_{18}	91_6	58_{100}	583	5279
Venlafaxine-M (nor-)		2195	263_1	202_{32}	134_{100}	91_{63}	65_{43}	437	5276
Venlafaxine-M (nor-) AC	U+UHYAC	2510	305_1	219_9	207_{19}	134_{100}	58_7	642	5273
Venlafaxine-M (nor-) -H2O HFB		2425	441_3	214_{60}	201_{100}	159_{30}	121_{35}	1100	7695
Venlafaxine-M (nor-) -H2O PFP		2205	391_2	214_{20}	201_{85}	190_{91}	119_{100}	988	7694
Venlafaxine-M (nor-) -H2O TFA		2210	341_2	214_{28}	201_{100}	159_{47}	69_{90}	812	7693
Venlafaxine-M (nor-HO-) -H2O AC	U+UHYAC	2560	303_1	205_5	134_{100}	119_8	91_9	630	5274
Venlafaxine-M (O-demethyl-)		2210	263_1	165_3	120_{22}	107_5	58_{180}	437	5277
Venlafaxine-M (O-demethyl-) AC	U+UHYAC	2230	305_1	188_8	120_4	107_5	58_{100}	642	5269
Venlafaxine-M (O-demethyl-) 2TMS		2100	392_1	192_4	171_7	73_{33}	58_{100}	1036	7186
Venlafaxine-M (O-demethyl-) -H2O AC	U+UHYAC	2065	287_1	115_2	107_3	58_{80}	145_1	549	7185
Venlafaxine-M (O-demethyl-) -H2O HFB		1825	371_6	169_{38}	115_{22}	91_{10}	58_{100}	1126	7716
Venlafaxine-M (O-demethyl-) -H2O PFP		1845	162_2	146_2	119_{22}	81_{17}	58_{100}	1038	7715
Venlafaxine-M (O-demethyl-) -H2O TFA		1905	157_{10}	128_8	107_{19}	77_{15}	58_{100}	880	7714
Venlafaxine-M (O-demethyl-) -H2O TMS		1980	317_1	302_1	141_2	73_{11}	58_{100}	699	7187
Venlafaxine-M (O-demethyl-oxo-HO-) iso-1 2AC	U+UHYAC	2430	377_1	200_{11}	71_9	58_{100}		943	5271
Venlafaxine-M (O-demethyl-oxo-HO-) iso-2 2AC	U+UHYAC	2500	377_1	260_2	200_4	71_8	58_{100}	943	5272
Verapamil	P G U+UHYAC	3150	303_{100}	260_4	151_{10}	58_{15}		1120	1021
Verapamil-M (N-bis-dealkyl-) AC	U+UHYAC	2545	318_{48}	275_{100}	233_{40}			705	1923
Verapamil-M (N-dealkyl-)	U UHY	2100	290_{61}	247_{50}	164_{53}	70_{37}	57_{100}	566	1919
Verapamil-M (N-dealkyl-) AC	U+UHYAC	2460	332_{61}	289_{87}	247_{100}	216_{17}	114_{16}	770	1922
Verapamil-M (nor-)	U UHY	3180	289_{100}	260_6	151_{10}			1098	1920
Verapamil-M (nor-) AC	U+UHYAC	3570	482_1	289_{20}	260_2	164_{100}	151_{12}	1150	6400
Verapamil-M (nor-O-demethyl-) 2AC	U+UHYAC	3680	510_1	317_{15}	246_2	164_{100}	151_8	1170	6399
Verapamil-M (O-demethyl-) AC	U+UHYAC	3200	331_{100}	289_2	246_4	151_3	58_{11}	1150	1921
Veratric acid	P UHY	1730*	182_{100}	167_{29}	121_{20}	111_{29}	77_{38}	179	4407
Veratric acid ME	UHYME	1585*	196_{97}	181_{11}	165_{100}	137_{15}	137_{15}	211	4408
Vigabatrine		1510	129_1	111_{13}	84_{17}	67_9	56_{100}	112	7458
Viloxazine	G U UHY	1855	237_6	138_{25}	100_{59}	56_{100}		335	641
Viloxazine AC	U U+UHYAC	2220	279_{26}	142_{79}	100_{100}	86_{28}	56_{51}	513	414
Viloxazine HFB		1950	433_{92}	296_{53}	240_{100}	110_{47}	81_{55}	1086	7719
Viloxazine PFP		1930	383_{91}	246_{76}	190_{100}	110_{84}	56_{80}	963	7718
Viloxazine TFA		1940	333_{78}	196_{51}	140_{67}	81_{75}	69_{100}	773	7717
Viloxazine TMS		1800	309_3	172_{59}	157_{38}	73_{100}	56_{94}	662	5477
Viloxazine-M (di-oxo-)	U UHY	2325	265_{15}	138_{46}	110_{58}	100_{70}	56_{100}	443	642
Viloxazine-M (HO-) 2AC	U+UHYAC	2590	337_{10}	295_{48}	142_{95}	100_{100}	56_{50}	793	415
Viloxazine-M (O-deethyl-) 2AC	U+UHYAC	2360	293_1	251_{64}	142_{61}	100_{100}	56_{45}	580	3754

Table 1-8-1: Compounds in order of names Zolazepam

Name	Detected	RI	Typical ions and intensities					Page	Entry
Viminol		2760	344 $_3$	315 $_2$	142 $_{100}$	125 $_{16}$	86 $_{15}$	892	261
Viminol -H2O		2405	344 $_{26}$	315 $_{18}$	287 $_6$	170 $_{12}$	125 $_{100}$	825	4254
Viminol-M/artifact AC	U+UHYAC	2785	394 $_{12}$	335 $_{20}$	125 $_{100}$			998	1228
Vinbarbital	P G U UHY U+UHYAC	1765	195 $_{100}$	152 $_{13}$	141 $_{16}$	79 $_{11}$	67 $_{20}$	290	1022
Vinbarbital 2ME	PME UME	1670	223 $_{100}$	166 $_{13}$	138 $_7$			392	1023
Vinbarbital-M (HO-)	U	2070	211 $_{53}$	193 $_{28}$	167 $_{100}$	155 $_{26}$	85 $_{90}$	344	2964
Vinbarbital-M (HO-) -H2O	UHY U+UHYAC	2020	193 $_{100}$	169 $_7$	150 $_{25}$	85 $_{28}$		282	2963
Vinclozolin		1905	285 $_{100}$	212 $_{84}$	178 $_{67}$	124 $_5$	53 $_{84}$	537	3458
Vinylbital	P G U+UHYAC	1745	209 $_1$	195 $_1$	154 $_{100}$	83 $_{29}$	71 $_{15}$	290	1024
Vinylbital (ME)	P-I	1720	209 $_1$	195 $_1$	168 $_{100}$	97 $_6$	83 $_8$	338	1029
Vinylbital 2ME	PME UME	1655	223 $_2$	209 $_2$	182 $_{100}$	125 $_6$	97 $_{30}$	392	1025
Vinylbital-M (devinyl-)	U	1665	169 $_3$	154 $_9$	129 $_{100}$			215	962
Vinylbital-M (HO-)	U	1995	195 $_7$	154 $_{100}$	112 $_{28}$	83 $_{38}$	69 $_{46}$	344	1028
Vinylbital-M (HO-) -H2O	UHY U+UHYAC	1970	222 $_1$	196 $_7$	154 $_{31}$	129 $_{24}$	69 $_{100}$	282	4345
Vinyltoluene		<1000*	118 $_{29}$	117 $_{53}$	115 $_{17}$	91 $_4$	58 $_8$	105	3717
Viquidil AC		3305	366 $_{24}$	323 $_{23}$	214 $_{55}$	186 $_{91}$	160 $_{100}$	906	6091
Vitamin B1 artifact-2 2ME		1190	153 $_{64}$	138 $_{38}$	122 $_{100}$	111 $_{11}$	81 $_{66}$	133	5142
Vitamin B6	PAC U+UHYAC	1945	295 $_1$	253 $_{23}$	193 $_{54}$	151 $_{100}$	123 $_{49}$	590	5089
Warfarin	G	9999*	308 $_{32}$	265 $_{100}$	187 $_{12}$	121 $_{24}$	92 $_6$	654	3765
Warfarin AC		2670*	350 $_{10}$	308 $_{20}$	290 $_{18}$	265 $_{100}$	121 $_{17}$	845	4837
Warfarin ET	UET	2565*	336 $_{41}$	293 $_{100}$	265 $_{86}$	189 $_{21}$	121 $_{44}$	788	4831
Warfarin ME	UME	2580*	322 $_{18}$	279 $_{100}$	189 $_7$	121 $_{11}$	91 $_{15}$	723	1030
Warfarin TMS		2675*	380 $_{53}$	337 $_{100}$	261 $_{26}$	193 $_{28}$	73 $_{62}$	952	4970
Warfarin artifact (phenylbutenone)	P-I G	1440*	146 $_{56}$	145 $_{58}$	131 $_{90}$	103 $_{100}$	77 $_{58}$	124	1517
Warfarin enol 2TMS		2790*	452 $_{12}$	437 $_{10}$	409 $_{51}$	247 $_{13}$	73 $_{100}$	1116	4971
Warfarin-M (di-HO-) 3ET	UET	3225*	424 $_{49}$	381 $_{100}$	353 $_{26}$	231 $_{28}$	165 $_{15}$	1067	4835
Warfarin-M (di-HO-) 3ME	UME	3150*	382 $_{33}$	339 $_{100}$	325 $_{47}$	231 $_{44}$	151 $_{14}$	959	4830
Warfarin-M (dihydro-) ET	UET	2655*	338 $_{29}$	291 $_{100}$	215 $_{43}$	191 $_{47}$	121 $_{60}$	798	4836
Warfarin-M (dihydro-) ME	UME	2660*	324 $_{18}$	291 $_{62}$	215 $_{47}$	177 $_{62}$	91 $_{100}$	731	1032
Warfarin-M (dihydro-) 2TMS		2785*	454 $_{12}$	439 $_{40}$	364 $_{100}$	335 $_{74}$	73 $_{79}$	1119	4972
Warfarin-M (dihydro-) -H2O	UHY U+UHYAC UHYME	2550*	292 $_{100}$	263 $_{72}$	249 $_{43}$	198 $_{17}$	121 $_{32}$	575	1031
Warfarin-M (HO-) 2TMS	UTMS	3015*	468 $_{48}$	425 $_{100}$	337 $_{47}$	115 $_{14}$	73 $_{34}$	1136	4967
Warfarin-M (HO-) enol 3TMS	UTMS	3105*	540 $_{26}$	497 $_{100}$	395 $_{27}$	335 $_{20}$	73 $_{92}$	1181	4969
Warfarin-M (HO-) iso-1 2ET	UET	2810*	380 $_{44}$	337 $_{100}$	309 $_{61}$	233 $_{12}$	165 $_{13}$	952	4832
Warfarin-M (HO-) iso-1 2ME	UME	2810*	352 $_{21}$	309 $_{100}$	277 $_7$	151 $_5$	91 $_{14}$	853	1033
Warfarin-M (HO-) iso-1 2TMS	UTMS	2795*	468 $_{33}$	425 $_{100}$	268 $_{23}$	193 $_{49}$	73 $_{28}$	1136	4968
Warfarin-M (HO-) iso-2 2ET	UET	2870*	380 $_{73}$	337 $_{100}$	309 $_{66}$	187 $_{41}$	121 $_{59}$	952	4833
Warfarin-M (HO-) iso-2 2ME	UME	2830*	352 $_{32}$	309 $_{100}$	295 $_{31}$	201 $_{43}$	121 $_{56}$	853	4825
Warfarin-M (HO-) iso-3 2ET	UET	2870*	380 $_{35}$	337 $_{100}$	309 $_{54}$	165 $_{18}$	137 $_{28}$	953	4834
Warfarin-M (HO-) iso-3 2ME	UME	2870*	352 $_{21}$	309 $_{100}$	295 $_{12}$	206 $_8$	91 $_{21}$	853	4826
Warfarin-M (HO-dihydro-) iso-1 -H2O 2ME	UME	2780*	336 $_7$	293 $_{100}$	165 $_6$	150 $_{41}$	115 $_{13}$	788	4827
Warfarin-M (HO-dihydro-) iso-2 -H2O 2ME	UME	2805*	336 $_2$	293 $_{100}$	173 $_{10}$	145 $_{12}$	121 $_{13}$	788	4828
Warfarin-M (HO-dihydro-) iso-3 -H2O 2ME	UME	2830*	336 $_3$	293 $_{100}$	217 $_3$	151 $_{17}$	115 $_{10}$	789	4829
Xanthinol 2AC		2870	335 $_2$	322 $_{11}$	156 $_{28}$	130 $_{100}$	87 $_{80}$	1003	2724
Xipamide 2ME	UME	3350	382 $_{57}$	262 $_{100}$	168 $_{12}$	120 $_9$	91 $_8$	959	3082
Xipamide 4ME	UME	2780	410 $_{34}$	289 $_{100}$	276 $_{67}$	168 $_{19}$	134 $_{23}$	1041	3085
Xipamide iso-1 3ME	UME	2800	396 $_{100}$	365 $_{49}$	276 $_{58}$	121 $_{27}$	77 $_{24}$	1004	3083
Xipamide iso-2 3ME	UME	3320	396 $_{67}$	276 $_{100}$	233 $_{12}$	168 $_{24}$	77 $_{10}$	1004	3084
Xipamide -SO2NH	P U+UHYAC UME	2385	275 $_{16}$	155 $_{37}$	121 $_{100}$			488	3088
Xipamide -SO2NH ME	PME UME	2480	289 $_{20}$	169 $_{100}$	126 $_{12}$	111 $_7$	77 $_8$	556	3086
Xipamide -SO2NH 2ME	UME	2115	303 $_{89}$	272 $_{60}$	183 $_{100}$	118 $_{30}$	77 $_{40}$	627	3089
Xipamide-M (HO-) 4ME	UME	3000	426 $_{100}$	395 $_{36}$	275 $_{26}$	151 $_{22}$	134 $_{18}$	1071	3087
Xipamide-M (HO-) -SO2NH 2ME	UME	2550	319 $_{11}$	256 $_1$	169 $_{100}$	150 $_7$	126 $_9$	707	3419
Xylazine		1970	220 $_{56}$	205 $_{100}$	177 $_{20}$	145 $_{25}$	130 $_{25}$	274	5423
Xylazine AC		2150	262 $_{10}$	247 $_6$	220 $_{75}$	205 $_{100}$	77 $_{51}$	430	5424
m-Xylene		<1000*	106 $_{42}$	91 $_{100}$	77 $_{12}$	65 $_6$	51 $_{12}$	101	2966
o-Xylene		<1000*	106 $_{40}$	91 $_{100}$	77 $_{12}$	65 $_7$	51 $_{12}$	101	2967
p-Xylene		<1000*	106 $_{25}$	91 $_{100}$	77 $_8$	65 $_9$	51 $_{11}$	101	2965
Xylitol 5AC		1950*	289 $_4$	217 $_{27}$	145 $_{84}$	115 $_{100}$	103 $_{71}$	890	5606
Xylometazoline		2020	244 $_{100}$	229 $_{80}$	214 $_{23}$	119 $_{59}$	91 $_{12}$	357	1525
Xylometazoline AC		2260	286 $_{100}$	271 $_{66}$	229 $_{40}$	214 $_{39}$	128 $_{50}$	545	1521
Xylose 4AC	U+UHYAC	1745*	259 $_1$	170 $_{61}$	157 $_{31}$	128 $_{100}$	115 $_{60}$	702	1967
Xylose 4HFB		1235*	478 $_{12}$	465 $_7$	293 $_{62}$	265 $_{13}$	169 $_{100}$	1206	5811
Xylose 4PFP		1230*	378 $_{17}$	365 $_9$	147 $_{15}$	119 $_{100}$	69 $_{92}$	1203	5810
Xylose 4TFA		1315*	311 $_3$	278 $_{14}$	265 $_{10}$	197 $_{13}$	69 $_{100}$	1180	5809
Yohimbine		3140	354 $_{72}$	353 $_{100}$	184 $_{10}$	169 $_{20}$	156 $_{11}$	862	3995
Yohimbine AC		3190	396 $_{81}$	395 $_{100}$	353 $_{57}$	277 $_7$	169 $_{38}$	1007	4018
Zaleplone		2960	305 $_{41}$	277 $_5$	263 $_{38}$	248 $_{100}$	119 $_8$	639	5859
Zaleplone-M/artifact (deacetyl-)		2850	263 $_{48}$	248 $_{100}$	231 $_4$	221 $_5$	130 $_4$	434	5860
Zidovudine TMS		2280	339 $_{42}$	255 $_{18}$	129 $_{37}$	117 $_{70}$	73 $_{100}$	802	6211
Zidovudine 2TMS		2390	411 $_8$	216 $_{23}$	203 $_{90}$	147 $_{87}$	73 $_{100}$	1046	6212
Zimelidine		2270	316 $_{54}$	238 $_{25}$	193 $_{62}$	70 $_{74}$	58 $_{100}$	692	1475
Zinophos		1600	248 $_{42}$	192 $_{37}$	143 $_{59}$	107 $_{87}$	97 $_{100}$	370	3877
Zolazepam		2400	286 $_{62}$	257 $_{97}$	145 $_{13}$	285 $_{100}$	267 $_{76}$	543	7448

Zolmitriptan Table 1-8-1: Compounds in order of names

Name	Detected	RI	Typical ions and intensities					Page	Entry
Zolmitriptan		2850	287 $_2$	156 $_2$	143 $_4$	115 $_2$	58 $_{100}$	548	7508
Zolmitriptan AC		2755	329 $_3$	156 $_3$	143 $_8$	115 $_2$	58 $_{100}$	755	7509
Zolpidem	P G U+UHYAC	2715	307 $_{14}$	235 $_{100}$	219 $_9$	92 $_9$	65 $_{10}$	650	5280
Zolpidem-M (4'-HO-) AC	U+UHYAC	3095	365 $_{13}$	293 $_{100}$	234 $_8$	219 $_5$	72 $_8$	901	5107
Zolpidem-M (4'-HO-) -C2H6N MEAC	U+UHYAC	2670	352 $_{28}$	293 $_{100}$	235 $_{52}$	201 $_{35}$	187 $_{27}$	853	5281
Zolpidem-M (6-HO-) AC	U+UHYAC	3150	365 $_{18}$	293 $_{100}$	233 $_{56}$	219 $_{13}$	72 $_{10}$	901	5108
Zolpidem-M (6-HO-) -C2H6N MEAC	U+UHYAC	2720	352 $_{36}$	293 $_{100}$	233 $_{52}$	207 $_{25}$	92 $_{14}$	854	5282
Zolpidem-M (4'-HOOC-) ME	U+UHYAC	2905	351 $_7$	279 $_{100}$	251 $_{10}$	219 $_8$	72 $_{11}$	849	5733
Zolpidem-M (6-HOOC-) ME	U+UHYAC	2950	351 $_{11}$	279 $_{100}$	269 $_{28}$	219 $_{32}$	72 $_{11}$	849	5734
Zomepirac ME	PME-I UME	1835	305 $_{32}$	246 $_{100}$	139 $_{40}$	111 $_{38}$		638	1035
Zomepirac -CO2	P-I G U UHY	2040	247 $_{75}$	246 $_{99}$	211 $_{21}$	136 $_{15}$		366	1034
Zonisamide		1950	212 $_{36}$	132 $_{79}$	119 $_{20}$	104 $_{29}$	77 $_{100}$	248	7720
Zonisamide AC		2100	254 $_6$	212 $_{100}$	195 $_{22}$	132 $_{48}$	77 $_{80}$	397	7723
Zonisamide ME		1930	226 $_{27}$	133 $_{200}$	119 $_6$	77 $_{56}$		293	7721
Zonisamide MEAC		1980	268 $_{31}$	162 $_{13}$	132 $_{80}$	77 $_{150}$	56 $_{64}$	459	7722
Zonisamide 2TMS		1965	356 $_1$	341 $_{90}$	269 $_{42}$	206 $_{29}$	132 $_{100}$	869	7724
Zopiclone	G U+UHYAC	2950	245 $_{69}$	217 $_{28}$	143 $_{100}$	112 $_{37}$	99 $_{44}$	979	5314
Zopiclone-M (amino-chloro-pyridine)	U+UHYAC	1200	128 $_{100}$	101 $_{70}$	93 $_{15}$	73 $_{34}$		111	5315
Zopiclone-M (amino-chloro-pyridine) AC	U+UHYAC	1505	170 $_{22}$	128 $_{100}$	101 $_{34}$	93 $_3$	73 $_7$	157	5316
Zopiclone-M (HO-amino-chloro-pyridine) AC	U+UHYAC	1680	186 $_{20}$	144 $_{100}$	116 $_{16}$	109 $_{13}$	81 $_{13}$	187	6557
Zopiclone-M (HO-amino-chloro-pyridine) 2AC	U+UHYAC	1720	228 $_{12}$	186 $_{56}$	144 $_{100}$	111 $_9$	81 $_4$	298	6556
Zopiclone-M (piperazine) 2AC		1750	170 $_{15}$	85 $_{33}$	69 $_{25}$	56 $_{100}$		157	879
Zopiclone-M (piperazine) 2HFB		1290	478 $_3$	459 $_9$	309 $_{100}$	281 $_{22}$	252 $_{41}$	1146	6634
Zopiclone-M (piperazine) 2TFA		1005	278 $_{10}$	209 $_{59}$	152 $_{25}$	69 $_{56}$	56 $_{100}$	505	4129
Zopiclone-M/artifact	U+UHYAC	2060	246 $_{83}$	217 $_{78}$	191 $_{100}$	139 $_{34}$	113 $_{70}$	361	7801
Zopiclone-M/artifact (alcohol) AC	U+UHYAC	2390	304 $_{16}$	261 $_{100}$	217 $_{24}$	155 $_{43}$	112 $_{31}$	633	5317
Zopiclone-M/artifact (alcohol) ME	U+UHYAC	2080	276 $_{12}$	261 $_{100}$	246 $_{100}$	217 $_{54}$	191 $_{23}$	495	5318
Zotepine	P G U	2660	331 $_2$	299 $_3$	199 $_4$	72 $_{30}$	58 $_{100}$	762	4291
Zotepine artifact (desulfo-) HYAC	U+UHYAC	2395*	270 $_{13}$	228 $_{100}$	199 $_{20}$	165 $_{49}$	115 $_6$	467	6416
Zotepine HY	UHY	2310*	260 $_{100}$	231 $_{29}$	227 $_{56}$	199 $_{30}$	152 $_{22}$	418	4292
Zotepine HYAC	U+UHYAC	2440*	302 $_{19}$	260 $_{100}$	231 $_{12}$	199 $_{44}$	152 $_{18}$	622	4293
Zotepine-M (bis-nor-) HY	UHY	2310*	260 $_{100}$	231 $_{29}$	227 $_{56}$	199 $_{30}$	152 $_{22}$	418	4292
Zotepine-M (bis-nor-) HYAC	U+UHYAC	2440*	302 $_{19}$	260 $_{100}$	231 $_{12}$	199 $_{44}$	152 $_{18}$	622	4293
Zotepine-M (bis-nor-HO-) HY2AC	U+UHYAC	2735*	360 $_{24}$	318 $_{39}$	276 $_{100}$	243 $_{34}$	215 $_{15}$	883	4294
Zotepine-M (bis-nor-HO-) HYAC	U+UHYAC	2555*	318 $_{34}$	276 $_{100}$	259 $_{22}$	247 $_{42}$	184 $_{23}$	701	6278
Zotepine-M (bis-nor-HO-) iso-1 HY	UHY	2460*	276 $_{21}$	231 $_{41}$	228 $_{100}$	199 $_{14}$	165 $_{93}$	494	4296
Zotepine-M (bis-nor-HO-) iso-2 HY	UHY	2650*	276 $_{100}$	247 $_{37}$	243 $_{28}$	213 $_{10}$	184 $_{26}$	494	4297
Zotepine-M (bis-nor-HO-methoxy-) HY	UHY	2700*	306 $_{100}$	276 $_{39}$	264 $_{52}$	247 $_{23}$	171 $_{24}$	643	4298
Zotepine-M (bis-nor-HO-methoxy-) HY2AC	U+UHYAC	2915*	390 $_{30}$	348 $_{24}$	306 $_{100}$	273 $_{79}$	245 $_9$	984	4295
Zotepine-M (HO-) AC	U+UHYAC	2960	341 $_2$	72 $_{56}$	58 $_{100}$			982	4299
Zotepine-M (HO-) HY2AC	U+UHYAC	2735*	360 $_{24}$	318 $_{39}$	276 $_{100}$	243 $_{34}$	215 $_{15}$	883	4294
Zotepine-M (HO-) HYAC	U+UHYAC	2555*	318 $_{34}$	276 $_{100}$	259 $_{22}$	247 $_{42}$	184 $_{23}$	701	6278
Zotepine-M (HO-) iso-1 HY	UHY	2460*	276 $_{21}$	231 $_{41}$	228 $_{100}$	199 $_{14}$	165 $_{93}$	494	4296
Zotepine-M (HO-) iso-2 HY	UHY	2650*	276 $_{100}$	247 $_{37}$	243 $_{28}$	213 $_{10}$	184 $_{26}$	494	4297
Zotepine-M (HO-methoxy-) HY	UHY	2700*	306 $_{100}$	276 $_{39}$	264 $_{52}$	247 $_{23}$	171 $_{24}$	643	4298
Zotepine-M (HO-methoxy-) HY2AC	U+UHYAC	2915*	390 $_{30}$	348 $_{24}$	306 $_{100}$	273 $_{79}$	245 $_9$	984	4295
Zotepine-M (nor-) HY	UHY	2310*	260 $_{100}$	231 $_{29}$	227 $_{56}$	199 $_{30}$	152 $_{22}$	418	4292
Zotepine-M (nor-) HYAC	U+UHYAC	2440*	302 $_{19}$	260 $_{100}$	231 $_{12}$	199 $_{44}$	152 $_{18}$	622	4293
Zotepine-M (nor-HO-) HY2AC	U+UHYAC	2735*	360 $_{24}$	318 $_{39}$	276 $_{100}$	243 $_{34}$	215 $_{15}$	883	4294
Zotepine-M (nor-HO-) HYAC	U+UHYAC	2555*	318 $_{34}$	276 $_{100}$	259 $_{22}$	247 $_{42}$	184 $_{23}$	701	6278
Zotepine-M (nor-HO-) iso-1 HY	UHY	2460*	276 $_{21}$	231 $_{41}$	228 $_{100}$	199 $_{14}$	165 $_{93}$	494	4296
Zotepine-M (nor-HO-) iso-2 HY	UHY	2650*	276 $_{100}$	247 $_{37}$	243 $_{28}$	213 $_{10}$	184 $_{26}$	494	4297
Zotepine-M (nor-HO-methoxy-) HY	UHY	2700*	306 $_{100}$	276 $_{39}$	264 $_{52}$	247 $_{23}$	171 $_{24}$	643	4298
Zotepine-M (nor-HO-methoxy-) HY2AC	U+UHYAC	2915*	390 $_{30}$	348 $_{24}$	306 $_{100}$	273 $_{79}$	245 $_9$	984	4295
Zuclopenthixol	G U	3360	400 $_1$	221 $_{12}$	143 $_{100}$	100 $_{18}$	70 $_{24}$	1017	462
Zuclopenthixol AC	U+UHYAC	3460	442 $_1$	221 $_9$	185 $_{100}$	98 $_{24}$	70 $_{11}$	1102	319
Zuclopenthixol TMS		3490	472 $_1$	457 $_6$	221 $_{19}$	215 $_{100}$	98 $_{23}$	1140	4534
Zuclopenthixol-M (dealkyl-) AC	U+UHYAC	3490	398 $_2$	268 $_7$	141 $_{100}$	99 $_{30}$		1011	1261
Zuclopenthixol-M (dealkyl-dihydro-) AC	U+UHYAC	3450	400 $_{46}$	231 $_{44}$	141 $_{100}$	128 $_{16}$	99 $_{25}$	1018	1260

9 Table of Compounds in Order of Categories

9.1 Explanatory notes

This Table is arranged in order of category, as in some cases the anamnesis and/or the clinical symptoms suggest the intake of a drug or poison of a particular category (e.g., alkylphosphate insecticide). However, only the names of the parent compounds are listed. In order to determine which metabolite or derivative can be detected in which sample after which sample preparation, Table 1-8-1 should be used.

9.2 Table of compounds in order of categories

Table 1-9-1: Compounds in order of categories

Acaricides
Aminocarb
Aramite
Bromopropylate
Chlorbenside
Chlordimeform
Chlorfenson
Chlorobenzilate
Chloropropylate
Dicofol
Dinobuton
Flubenzimine
Phenkapton
Tetradifon
Tetrasul

Adenosine receptor agonists
Ethylcarboxamido-adenosine
Phenylisopropyl-adenosine
NECA

Alcohol deterrents
Disulfiram

Alkaloids
Californine
Chelerythrine
Cinnamolaurine
Cinnamoylcocaine
Coniine
Ecgonidine
Ecgonine
Ergometrine
Laudanosine
Lauroscholtzine
Methylpseudoephedrine
Nicotine
Norcinnamolaurine
Oxyberberine
Protopine
Reframidine
Sanguinarine
Strychnine
Trimethoxycocaine
Tropacocaine
Yohimbine

Anabolics
Clostebol
Drostanolone
Fluoxymesterone
Metandienone
Metenolone
Methylandrostanolone
Methyltestosterone
Nandrolone
Oxabolone
Oxymetholone
Stanozolol
Tetrahydrogestrinone
THG

Analgesics/Antiphlogistics
Acetaminophen
Acetanilide
Acetylsalicylic acid
Aminophenazone
Benzydamine
Bucetin
Bumadizone
Bumatizone
Carprofen
Cicloprofen
Diflunisal
Dipyrone
Ethenzamide
Famprofazone
Felbinac
Flupirtine
Flurbiprofen
Ibuprofen
Isopyrin
Lactylphenetidine
Lonazolac
Lonazolac
Mesalazine
Metamizol
Mofebutazone
Morazone
Nalbuphine
Naproxen
Nifenazone
Nimesulide
Oxapadol
Oxyphenbutazone
Paracetamol
Phenacetin
Phenazone
Phenopyrazone
Phenylbutazone
Pirprofen
Propyphenazone
Ramifenazone
Salacetamide
Salicylamide
Salsalate
Sulindac
Suxibuzone
Tenoxicam
Tiaprofenic acid
Zomepirac

Analytical standards
Amfetamine-D5
Amfetamine-D11
Buprenorphine-D4
Buprenorphine-M (nor-)-D3
Clomipramine-D3
Cocaine-D3
Codeine-D3
Diazepam-D5
Dronabinol-D3
Ecgonine-D3
Fentanyl-D5
Fluoxetine-D6
Haloperidol-D4
Heroin-D3
Ketamine-D4
Lysergic acid N,N-methylpropylamine
MDA-D5
MDE-D5
MDMA-D5
Mescaline-D9

Metamfetamine-D5
Methadone-D9
Methohexital-D5
Mianserin-D3
Morphine-D3
Nordazepam-D5
Oxycodone-D6
Paracetamol-D4
Pentobarbital-D5
Phenobarbital-D5
Tenamfetamine-D5
Tetrahydrocannabinol-D3
p-Tolylpiperazine
Trimipramine-D3

Androgens
Mesterolone
Testosterone
Testosterone acetate
Testosterone dipropionate
Testosterone propionate
Tibolone

Anesthetics
Chloroform
gamma-Butyrolactone
Diethylether
Etomidate
Halothane
Hexamid
Hexobarbital
gamma-Hydroxybutyric acid (GHB)
Ketamine
Methitural
Methohexital
Narcobarbital
Propofol
Thiobutabarbital
Thiopental
Tiletamine
Trichloroethylene

Anorectics
Amfepramone
Aminorex
Benfluorex
Cathine
Chlorphentermine
Clobenzorex
Fenbutrazate
Fenfluramine

Fenproporex
Mazindol
Mefenorex
d-Norpseudoephedrine
Orlistat
Pentorex
Phendimetrazine
Phenmetrazine
Phentermine
Propylhexedrine

Anthelmintics
Albendazole
Arecoline
Dichlofenthion
Fenbendazole
Fensulfothion
Mebendazole
Piperazine
Thionazine
Tiabendazole
Tinidazole
Zinophos

Antiamebics
Metronidazole
Ornidazole

Antiandrogens
Cyproterone

Antiarrhythmics
Ajmaline
Amiodarone
Aprindine
Detajmium bitartrate
Disopyramide
Flecainide
Lidocaine
Lorcainide
Lupanine
Mexiletine
Prajmaline
Procainamide
Propafenone
Quinidine
Sparteine
Tocainide

Antibiotics
Amoxicilline
Asulam

Azidocilline
Cefadroxil
Cefalexine
Cefazoline
Chloramphenicol
Clindamycin
Dapsone
Dichloroquinolinol
Dicloxacillin
Grepafloxacin
Lincomycin
Linezolide
Mafenide
Mezlocilline
Nitrofurantoin
Ofloxacin
Pentamidine
Piperacilline
Sparfloxacin
Sulfabenzamide
Sulfadiazine
Sulfadimethoxine
Sulfaethidole
Sulfaguanole
Sulfamerazine
Sulfamethizole
Sulfamethoxazole
Sulfametoxydiazine
Sulfanilamide
Sulfaperin
Sulfapyridine
Sulfathiourea
Tetroxoprim
Trimethoprim
Trovafloxacine

Anticholesteremics
Beclobrate
Bezafibrate
Clofibrate
Clofibric acid
Etiroxate
Etofibrate
Etofibrate
Etofylline clofibrate
Fenofibrate
Gemfibrozil
Probucol

Anticoagulants
Acenocoumarol
Phenprocoumon

Tioclomarole
Warfarin

Anticonvulsants
Beclamide
Carbamazepine
Clonazepam
Diisopropylidene-fructopyranose
Ethadione
Ethosuximide
Felbamate
Gabapentin
Lamotrigine
Levetiracetam
Loprazolam
Mephenytoin
Mesuximide
Oxcarbazepine
Paramethadione
Phenobarbital
Phenytoin
Pregabaline
Primidone
Sultiame
Topiramate
Trimethadion
Valproic acid
Valpromide
Vigabatrine
Zonisamide

Antidepressants
Amfebutamone
Amineptine
Amitriptyline
Amitriptylinoxide
Atomoxetine
Brofaromine
Bupropion
Citalopram
Clomipramine
Desipramine
Dibenzepin
Dimetacrine
Dosulepin
Doxepin
Duloxetine
Etryptamine
Fluoxetine
Fluvoxamine
Imipramine

Indeloxazine
Lofepramine
Maprotiline
Melitracene
Mianserin
Minaprine
Mirtazapine
Moclobemide
Nefazodone
Nomifensine
Nortriptyline
Noxiptyline
Opipramol
Paroxetine
Pirlindole
Pirlindole
Protriptyline
Reboxetine
Sertraline
Sibutramine
Sulpiride
Tranylcypromine
Trazodone
Trimipramine
Venlafaxine
Viloxazine
Zimelidine

Antidiabetics
Chlorpropamide
Glibenclamide
Glibornuride
Gliclazide
Glimepiride
Glipizide
Gliquidone
Glisoxepide
Metformine
Repaglinide
Rosiglitazone
Tolazamide
Tolbutamide

Antidiarrheals
Diphenoxylate
Loperamide

Antidotes
Dimethylaminophenol
(DMAP)
Flumazenil
Levallorphan

Methylene blue
Methylthionium chloride
Nalorphine
Naloxone
Naltrexol (beta-)
Naltrexone
Physostigmine
RO 15-4513

Antiemetics
Alizapride
Betahistine
Bromopride
Metoclopramide
Tetrahydrocannabinol
(Dronabinol)
Tropisetrone

Antiestrogens
Cyclofenil
Tamoxifen

Antifreezes (Glycols)
1,2-Butane diol
1,3-Butane diol
1,4-Butane diol
Diethylene glycol
Ethylene glycol
1,2-Propane diol
1,3-Propane diol
Tetraethylene glycol
Triethylene glycol

Antiglaucoma agent
Dorzolamide

Antigonadotropin
Danazole

Antihistamines
Adeptolon
Antazoline
Astemizole
Azatadine
Azelastine
Bamipine
Benzquinamide
Brompheniramine
Buclizine
Carbinoxamine
Cetirizine
Chlorcyclizine

Chloropyramine
Chlorphenamine
Chlorphenoxamine
Clemastine
Clemizole
Cyclizine
Desloratadine
Dimetindene
Dimetotiazine
Diphenhydramine
Diphenylpyraline
Doxylamine
Epinastine
Etoloxamine
Fexofenadine
Histapyrrodine
Isothipendyl
Ketotifen
Lodoxamide
Loratadine
Mebhydroline
Meclozine
Medrylamine
Mepyramine
Mequitazine
Mizolastine
Mizolastine
Orphenadrine
Oxatomide
Oxomemazine
Phenindamine
Pheniramine
Phenyltoloxamine
Pyrilamine
Pyrrobutamine
Terfenadine
Thiethylperazine
Tolpropamine
Tripelenamine
Triprolidine
Tritoqualine

Antihypertensives
Benazepril
Benazeprilate
Bunazosin
Captopril
Cilazapril
Cilazaprilate
Clonidine
Enalapril
Enalaprilate

Eprosartan
Guanfacine
Imidapril
Imidaprilate
Irbesartan
Ketanserin
Labetalol
Lisinopril
Lofexidine
Losartan
Methyldopa
Moexipril
Moexiprilate
Moxonidine
Perindopril
Perindoprilate
Phenoxybenzamine
Phentolamine
Quinapril
Quinaprilate
Ramipril
Ramiprilate
Reserpine
Spirapril
Trandolapril
Trandolaprilate
Trimethoxyhippuric acid
Urapidil
Valsartan

Antimalarials
Amodiaquine
Chloroquine
Cinchonidine
Cinchonine
Mefloquine
Pyrimethamine
Quinine

Antimigraines
Eletriptan
Frovatriptan
Naratriptan
Rizatriptan
Sumatriptan
Zolmitriptan

Antimycotics
Bifonazole
Chlorphenesin
Climbazole
Clotrimazole

Cloxiquine
Croconazole
Dichlorophen
Econazole
Enilconazole
Fenbuconazole
Fenticonazole
Fluconazole
Imazalil
Isoconazole
Miconazole
Omoconazole
Oxiconazole
Terbinafine
Tioconazole

Antineoplastics
Aminoglutethimide
Chlorambucil
Colchicine
Cyclophosphamide
o,p'-DDD
o,p'-DDE
o,p'-Dichlorophenylmethane
Fluorouracil
Mitotane

Antioxidants
Bis-tert-butyl-
methoxymethylphenol
Dilaurylthiodipropionate
Ethoxyquin
Ionol-4
Irganox
Laurylmethylthiodi-
propionate

Antiparkinsonians
Amantadine
Benzatropine
Benzhydrol
Biperiden
Bornaprine
Budipine
Carbidopa
(carboxylase inhibitor)
Levodopa
Memantine
Metixene
Pergolide
Phenglutarimide
Pramipexole

Pridinol
Procyclidine
Profenamine
Ropinirole
Selegiline
Tiapride
Trihexyphenidyl

Antirheumatics
Aceclofenac
Acemetacin
Azapropazone
Benoxaprofen
Bufexamac
Celecoxib
Diclofenac
Etodolac
Etofenamate
Etoricoxib
Fenbufen
Fenoprofen
Flufenamic acid
Hydroxyproline
Indometacin
Kebuzone
Ketoprofen
Ketorolac
Leflunomide
Meclofenamic acid
Mefenamic acid
Meloxicam-1
Nabumetone
Niflumic acid
Oxaceprol
Piroxycam
Proglumetacin
Proquazone
Rofecoxib
Tolfenamic acid
Tolmetin

Antiseptics
2-Benzylphenol
4-Benzylphenol
Benzalkonium chloride
5-Bromosalicylic acid
Chlorcarvacrol
Chlorocresol
4-Chlorophenol
Chloroxylenol
Clorofene
Dimethylbromophenol

Ethacridine
Hexylresorcinol
Hydroquinone
Menthol
Pentachlorophenol
1-Phenoxy-2-propanol
Sublimate
Triclosan

Antispasmotics
Adiphenine
Ambucetamide
Butinoline
Camylofine
Dicycloverine
Ethaverine
Ethoxyphenyldiethyl-
phenyl butyramine
Fencarbamide
Fenpipramide
Fluvoxate
Mebeverine
Moxaverine
Octamylamine
Oxybutynine
Papaverine
Phloroglucinol
Pitofenone
Pramiverine
Propivan
Propiverine
Tiropramide
Trimebutine
Veratric acid

Antitussives
Benproperine
Clobutinol
Clofedanol
Dropropizine
Eprazinone
Isoaminile
Narceine
Noscapine
Oxeladin
Pentoxyverine
Pholcodine

Aromatase inhibitors
Exemestane
Letrozole

Beta-Blockers
Acebutolol
Alprenolol
Atenolol
Befunolol
Betaxolol
Bisoprolol
Bunitrolol
Bupranolol
Carazolol
Carteolol
Carvedilol
Celiprolol
Esmolol
Levobunolol
Mepindolol
Metipranolol
Metoprolol
Nadolol
Nebivolol
Nifenalol
Oxprenolol
Penbutolol
Pindolol
Propranolol
Sotalol
Talinolol
Tertatolol
Timolol
Toliprolol

Biomolecules
Androst-4-ene-3,17-dione
Androstane-3,17-dione
Androsterone
Caffeic acid
Cholesta-3,5-dien-7-one
Cholestenone
Cholesterol
Cyanuric acid
Cytosine
Dehydroepiandrosterone
Dehydrotestosterone
Dihydrotestosterone
Dihydroxybenzoic acid
Dihydroxycinnamic acid
Dihydroxynorcholanoic acid
Dihydroxyphenylacetic acid
Dopamine
Endogenous biomolecule
Epiandrosterone
Epitestosterone enol

Hippuric acid
Homovanillic acid
Hydrocaffeic acid
Hydroxyandrostanedione
Hydroxyandrostene
Hydroxyandrosterone
Hydroxyeicosatetraenoic acid
Hydroxyetiocholanolone
Hydroxyindoleacetic acid
Hydroxyindolepropanoic acid
Hydroxy-methoxy-acetophenone
Hydroxytyramine
Indole acetic acid
Indole propionic acid
Methoxyhydroxyphenyl-glycol (MHPG)
Methycatechol
Methylcatechol
Phenylalanine
Phytanic acid
Proline
Pyridoxic acid lactone
Skatole
Tryptamine
Tryptophan
Urea
Vanillin mandelic acid

Bronchodilators
Bambuterol
Clenbuterol
Dioxethedrine
Diprophylline
Etamiphylline
Pirbuterol
Procarterol
Proxyphylline
Pseudoephedrine
Salbutamol
Terbutaline
Theophylline

Buffer agent
Trometamol (TRIS)

Ca Antagonists
Amlodipine
Barnidipine
Diltiazem
Felodipine
Gallopamil

Isradipine
Lacidipine
Nicardipine
Nifedipine
Nilvadipine
Nimodipine
Nisoldipine
Nitrendipine
Perhexiline
Verapamil

Capillary protectants
Benzarone
Quercetin

Cardiotonics
Digitoxigenin
Digitoxin
Enoximone
Eplerenone

Choline esterase inhibitors
Donepezil
Galantamine

Chemicals
Acenaphthene
Acenaphthylene
Acetaldehyde
Acetic acid
Acetyltriethylcitrate
Aminoethanol
Aminoethylphenol
Aminophenol
Aminothiophenol
Aniline
Anisic acid
Anisidine
Anthracene
Anthranilic acid
Barbituric acid
Benzamide
Benzil
Benzo[a]anthracene
Benzo[a]pyrene
Benzo[b]fluoranthene
Benzo[g,h,i]perylene
Benzo[k]fluoranthene
Benzoflavone
Benzofluorene
Benzophenone
Benzylacetamide

Benzylbutanoate
Benzylether
Bipyridine
Bis-(4-chlorophenyl-)sulfone
Bis-(hydroxy-dimethyl-phenyl-) methylpropane
Bisoctylphenylamine
Bis-tert.-butylmethylene-cyclohexanone
Bis-tert.-butylquinone
Bromobenzene
Bromofenoxim
Bromoquinoline
Bromothiophene
Butylamine
Catechol
Chlorobenzaldehyde
Chlorobenzoic acid
Chlorobenzyl alcohol
Chlorobenzylchloride
Chlorobiphenyl
Chlorophenol
Chlorotoluene
Chlorotrimethoxyhippuric acid
Chrysene
Citric Acid
Cyclohexanone
Cyclopentaphenanthrene
Deiquate
Dibenzo[a,h]anthracene
Dibenzofuran
Dibutylpentylpyridine
Dichlorobenzene
Diethylamine
Diethyldithiocarbamic acid
Dihexylamine
Dihydroxybenzylamine
Dimethoxybenzaldehyde
Dimethyl-3-phenyl-aziridine
Dimethylamine
Dimethylaniline
Dimethylnaphthalene
Dimethylphenol
Dinitrophenol
Diphenylethylamine
Diphenyloctylamine
Ethylamine
Ethylene oxide
Ethylenediaminetetraacetic acid (EDTA)
Ethylhexyldiphenylphosphate

Ethylpiperidine
Ethyltolylbarbital
Fluoranthene
Fluorene
Fluorophenylacetic acid
Formaldehyde
Formyl-phenazone
Heptafluorobutanoic acid
Hydroxy-3-methoxy-phenethylamine
Hydroxybenzoic acid
Hydroxybenzylalcohol
Hydroxyethylurea
Hydroxyindole
Hydroxymethoxy-benzylamine
Hydroxyphenylacetic acid
Indene
Indeno[1,2,3-c,d]pyrene
Indole
Ionol
Ionol-acetamide
IPCC
Isocitric acid
Isonicotinic acid
Isovanillic acid
Maleic acid
Methacrylic acid methylester
Methoxyaniline
Methoxybenzoic acid
Methoxyphenylpiperazine
Methylamine
Methylbenzoic acid
Methyldibenzofuran
Methylene-bis-(4-methyl-6-tert.-butylphenol)
Methylenedioxybenzoic acid
Methylenedioxybenzyl alcohol
Methylenedioxymethyl-nitrostyrene
Methylnaphthalene
Methylnitrostyrene
Methylphenanthrene
Methylphenoxyacetic acid
Methylpiperazine
Methylpyrene
Methylthiobenzoic acid
Morpholine
Naphthoflavone (alpha-)
Naphthol
N-Benzylethylenediamine

N-Benzylidenebenzylamine
N-Methyl-1-phenylethyl-amine
NMPEA
Octacosane
Octadecane
Octamethyldiphenylbicyclohexasiloxane
Perfluorotributylamine
Phenanthrene
Phenol
Phenyl-2-oxazoline
Phenylacetaldehyde
Phenylacetamide
Phenylacetic acid
Phenylacetone
p-Phenylenediamine
Phosphoric acid
Piperidine
Piperonol
Piperonylacetate
Piperonylic acid
Polychlorinated biphenyl (3Cl)
Polychlorinated biphenyl (4Cl)
Polychlorinated biphenyl (5Cl)
Polychlorinated biphenyl (6Cl)
Polychlorinated biphenyl (6Cl)
Polychlorinated biphenyl (7Cl)
Propiophenone
Propylamine
Propylene glycol dipivalate
Pyrene
Pyridine
Pyrocatechol
Pyrrolidine
Sulfuric acid
Tetrahydrofuran
Tetramethylcitrate
Thiophenylmethanol
Toluenesulfonic acid
Toluidine
Tributoxyethylphosphate
Tributylamine
Trichloroaniline
Triethylamine
Triethylcitrate
Trifluoroacetaldehyde
Trifluoroacetic acid
Trihexylamine
Trimethoxybenzaldehyd
Trimethoxybenzoic acid
Trimethoxybenzyl alcohol
Trimethoxyphenylnitroethene
Trimethoxyphenylnitropropene
Trimethylamine
Trimethylcitrate
Triphenylphosphine oxide
Vanillic acid
Vinyltoluene

Choleretics
Deoxycholic acid
Hymecromone

Coronary dilators
Amylnitrite
Etafenone
Fendiline
Prenylamine

Corticoids
Betamethasone
Deoxycortone
Fluocortolone
Hydrocortisone
Methylprednisolone
Prednisolone
Prednisone
Prednylidene
Triamcinolone

Counterirritant
3-Bromo-d-camphor
Chloropicrin
CN gas (chloroacetophenone)
CS gas (o-chlorobenzylidenemalonitrile)

Derivatizing agents
Bis-(trimethylsilyl-)trifluoroacetamide (BSTFA)
N-Methyl-trimethylsilyl-trifluoroacetamide (MSTFA)
Pentafluoropropionic acid
Pivalic acid anhydride
Propionic acid anhydride

Dermatics
Methylsalicylate
Panthenol
Salicylic acid
Sulfur
Trisalicyclide

Designer drugs
2C-B
2C-D
2C-E
2C-I
2C-P
2C-T-2
2C-T-7
BDB
BDMPEA
Benzylpiperazine
Brolamfetamine
BZP
Dimethoxyphenethylamine
DMA
DMCC
DOB
DOET
DOI
DOM
EBDB
ECC
Eticyclidine
gamma-Butyrolactone
gamma-Hydroxybutyric acid (GHB)
MBDB
MCC
m-Chlorophenylpiperazine (mCPP)
MDA
MDBP
MDE
MDMA
MDPPP
MECC
MeOPP
Methoxyetilamfetamine
Methoxymetamfetamine
Methoxyphenylpiperazine
Methylenedioxypyrrolidinopropiophenone
MMBDB
MMDA
MOPPP
MPBP
MPCP
4-MTA
N-Isopropyl-BDB
N-Methyl-DOB
PCC
PCDI
PCE
PCEEA
PCEPA
PCM
PCME
PCMEA
PCPIP
PCPR
PICC
Piperonylpiperazine
PMEA
PMMA
PPP
PRCC
PVP
PYCC
Pyrrolidinopropiophenone
Pyrrolidinovalerophenone
Rolicyclidine
TCDI
TCM
TCPY
Tenamfetamine
Tenocyclidine
TFMPP
TMA
Trimethoxyamfetamine

Diagnostic aids
Metyrapone

Disinfectants
2-Bromo-4-cyclohexylphenol
2-Chloro-4-cyclohexylphenol
m-Cresol
p-Cresol
2-Cyclohexylphenol
4-Cyclohexylphenol
Nitroxoline
Phenylethanol

Diuretics
Acetazolamide
Amiloride
Azosemide

Bemetizide
Bendroflumethiazide
Bumetanide
Butizide
Canrenoic acid
Canrenone
Carzenide
Chlorazanil
Chlorothiazide
Chlortalidone
Clopamide
Cyclopenthiazide
Cyclothiazide
Diclofenamide
Etacrinic acid
Etozoline
Furosemide
Hydrochlorothiazide
Indapamide
Mefruside
Metolazone
Monalazone
Muzolimine
Piretanide
Polythiazide
Quinethazone
Spironolactone
Thiophenecarboxylic acid
Tienilic acid
Torasemide
Triamterene
Trichlormethiazide
Xipamide

Emetics
Apomorphine
Emetine

Estrogens
Diethylstilbestrol
Estradiol
Estradiol undecylate
Estriol
Estrone
Ethinylestradiol
Mestranol
Quinestrol

Expectorants
Ambroxol
Bromhexine
Guaifenesin

Fats
Glyceryl tridecanoate
Glyceryl trioctanoate

Fatty acids
Behenic acid
Brassidic acid
Capric acid
Caprylic acid cetylester
Eicosanoic acid
Erucic acid
Glyceryl dimyristate
Glyceryl monomyristate
Glyceryl monooleate
Glyceryl monopalmitate
Glyceryl monostearate
Heptadecanoic acid
Lauric acid
Lignoceric acid
Linoleic acid
Linolenic acid
Methylstearate
Myristic acid
Nonadecanoic acid
Octanoic acid hexadecylester
Oleamide
Oleic acid
Palmitamide
Palmitic acid
Palmitoleic acid
Pentadecanoic acid
Ricinoleic acid
Stearamide
Stearic acid

Flavors
Benzaldehyde
Coumarin
Vanillin

Fungicides
Anilazine
Benomyl
Binapacryl
Biphenyl
Bisphenol A
Bitertanol
Bupirimate
Butylparaben
Captafol
Captan
Carbendazim
Carboxin
Chlorothalonil
Dazomet
Dichlofluanid
Dichloran
Ditalimfos
Dodemorph
Enilconazole
Ethirimol
Etridiazole
Fenarimol
Fenfuram
Fenpropemorph
Flusilazole
Folpet
Fuberidazole
Furalaxyl
Furmecyclox
Glycophen
Hexachlorobenzene
Hymexazol
Imazalil
Metalaxyl
Nitrothal-isopropyl
Nuarimol
Oxadixyl
Phenoxyacetic acid methylester
Prochloraz
Procymidone
Propamocarb
Propamocarb
Propiconazole
Pyrazophos
Quinomethionate
Quintozene
Tecnazene
Tetrabromo-o-cresol
Thiophanate
Thiram
Tolylfluanid
Triadimefon
Triadimenol
Triamiphos
Tridemorph
Vinclozolin

GC Background
GC septum bleed
GC stationary phase (methylsilicone)
GC stationary phase (OV-101)
GC stationary phase (OV-17)
GC stationary phase (UCC-W-982)

Gestagens
Allylestrenol
Chlormadinone
Gestonorone
Hydroxyprogesterone
Lynestrenol
Medroxyprogesterone
Norethisterone
Norgestrel
Progesterone

H2-Blockers
Famotidine
Ranitidine
Roxatidine

Hair dyes
p-Phenylenediamine

Heat transfer agents
2,2',3,4,4',5,5'-Heptachlorobiphenyl
2,2',3,4,4',5'-Hexachlorobiphenyl
2,2',4,4',5,5'-Hexachlorobiphenyl
2,2',4,5,5'-Pentachlorobiphenyl
Polychlorinated biphenyl (3Cl)
Polychlorinated biphenyl (4Cl)
Polychlorinated biphenyl (5Cl)
Polychlorinated biphenyl (6Cl)
Polychlorinated biphenyl (6Cl)
Polychlorinated biphenyl (7Cl)
2,2',5,5'-Tetrachlorobiphenyl
2,4,4'-Trichlorobiphenyl

Hemostatics
Tranexamic acid

Herbicides
Acetochlor
Alachlor
Allidochlor
Ametryne
Amitrole
ANTU
Atrazine
Aziprotryne
Barban
Benazolin
Bentazone
Benzthiazuron
Bromacil
Bromofenoxim
Bromoxynil
Buturon
Carbetamide
Chloramben
Chlorbromuron
Chlorbufam
Chloridazone
Chloroaniline
Chlorophenoxyacetic acid
Chloropropham
Chloroxuron
Chlorphenphos-methyl
Chlorthal-methyl
Chlorthiamid
Chlortoluron
Clopyralide
Cyanazine
Cycloate
Cycloxydim
Cycluron
Demedipham
Demetryn
Diallate
Dicamba
Dichlobenil
Dichloroaniline
Dichloromethoxybenzene
Dichlorophenol
Dichlorophenoxyacetic acid (2,4-D)
Dichlorophenoxybutyric acid
Dichlorprop
Diclofop-methyl
Difenzoquate
Diflubenzuron
Diflufenicam
Dimefuron
Dimethachlor
Dinitrophenol
Dinoseb
Dinoterb
Disugram
Diuron
Endothal
EPTC
Ethofumesate
Fenoprop
Fenoxaprop-ethyl
Fenson
Fenuron
Flamprop-isopropyl
Flamprop-methyl
Fluazifop-butyl
Fluchloralin
Flurenol
Flurodifen
Fluroxypyr
Glyphosate
Hexazinone
Ioxynil
Isocarbamide
Isoproturon
Isoxaben
Karbutilate
Lenacil
Linuron
MCPA
Mecoprop
Metamitron
Metazachlor
Methabenzthiazuron
Methoprotryne
Metobromuron
Metoxuron
Metribuzin
Monalide
Monolinuron
Monuron
Naphthoxyacetic acid methylester
N-1-Naphthylphthalimide
Napropamide
Naptalam
Neburon
Nitrofen
Oryzalin
Oxadiazon
Pencycuron
Penoxalin
Pentanochlor
Phenalenone
Phendipham
Phenmedipham
Picloram
Profluralin
Prometryn
Propachlor
Propazine
Propham
Propyzamide
Pyridate
Sebuthylazine
Sethoxydim
Simazine
Swep
Tebuthiuron
Terbacil
Terbumeton
Terbutryn
Terbutylazine
Thiazafluron
Triallate
Trichlorophenoxyacetic acid (2,4,5-T)
Trichlorophenoxyacetic acid isobutylester
Trichlorophenoxyacetic acid octylester
Triclopyr
Trietazine
Trifluralin

Hydrocarbons
Butane
Cyclohexadecane
Cyclotetradecane
DecaneDocosane
Dodecane
Eicosane
Heptadecane
Heptadecane
Hexacosane
Hexadecane
Nonadecane
Pentadecane
Tetradecane
Triacontane
Tricosane
Tridecane
Undecane

Hypnotics
Acecarbromal
Allobarbital
Allobarbital
Amobarbital
Aprobarbital
Barbital
Brallobarbital
Bromisoval
Butabarbital
Butalbital
Butallylonal
Butobarbital
Carbromal
Chloral hydrate
Chloralose
Clomethiazole
Crotylbarbital
Cyclobarbital
Cyclopentobarbital
Diethylallylacetamide
Dipropylbarbital
Ethinamate
Ethylloflazepate
Flunitrazepam
Flurazepam
Glutethimide
Heptabarbital
Hexethal
Idobutal
Loprazolam
Meprobamate
Methaqualone
Metharbital
Methylphenobarbital
Methylthalidomide
Methylthalidomide
Methyprylone
Midazolam
Narconumal
Nealbarbital
Nimetazepam
Nitrazepam
Paraldehyde
Pentobarbital
Phenallymal
Phenobarbital
Phenylmethylbarbital
Probarbital
Propallylonal
Pyrithyldione
Secobarbital

Sigmodal
Talbutal
Thalidomide
Triazolam
Trichloroethanol
Vinbarbital
Vinylbital
Zaleplone
Zolpidem
Zopiclone

Immunosuppressants
Mycophenolic acid

Ingredients of tar
Acenaphthylene
Anthracene
Benzofluorene
Chrysene
Cyclopentaphenanthrene
Dibenzofuran
Dimethylnaphthalene
Fluoranthene
Fluorene
Indene
Methyldibenzofuran
Methylnaphthalene
Methylphenanthrene
Methylpyrene
Naphthalene
Pyrene

Insect repellents
Diethyl toluamide (DEET)

Insecticides
Acephate
Aldicarb
Aldrin
Allethrin
Alphamethrin
Amitraz
Azamethiphos
Azinphos-ethyl
Bendiocarb
Bioallethrin
Bioresmethrin
Bromophos
Bromophos-ethyl
Butocarboxim
Butoxycarboxim
Carbaryl

Carbofuran
Carbophenothion
Chlordecone
Chlorfenvinphos
Chlormephos
Chlorpyrifos
Chlorthiophos
Coumaphos
Cyanophenphos
Cyanophos
Cyfluthrin
Cypermethrin
Cyphenothrin
DDD
DDE
DDT
Decamethrin
Deltamethrin
Demeton-S-methyl
Demeton-S-methylsulfone
Demeton-S-methylsulfoxide
Dialifos
Diazinon
Dichlorophenylacetate
Dichlorophenylethanol
Dichlorophenylmethane
Dichlorophenylmethanol
Dichlorvos
Dicrotophos
Dimethoate
Dimpylate
Dinocap
Dioxacarb
Dioxathion
Disulfoton
DNOC
Endosulfan
Endosulfan sulfate
Endrin
Ethiofencarb
Ethion
Ethoprofos
Etrimfos
Fenamiphos
Fenchlorphos
Fenitrothion
Fenpropathrin
Fenthion
Fenvalerate
Fonofos
Formetanate
Formothion

Heptachlor
Heptachlorepoxide
Heptenophos
Hexachlorocyclohexane (HCH)
Hydroxyquinoxaline
Iodofenphos
Isofenphos
Kadethrin
Kelevan
Lindane
Malaoxon
Malathion
Mercaptodimethur
Methamidophos
Methidathion
Methomyl
Methoxychlor
Mevinphos
Mirex
Mitotane
Monocrotophos
Naled
Naphthalene
Nitrophenol
Omethoate
Oxamyl
Oxydemeton-S-Methyl
Paraoxon
Parathion-ethyl
Parathion-methyl
Permethrin
Perthane
Phenothrin
Phorate
Phosalone
Phosdrin
Phosmet
Phosphamidon
Phoxim
Pirimicarb
Pirimiphos-methyl
Polychlorocamphene
Potasan (E838)
Profenofos
Promecarb
Propetamphos
Propoxur
Prothiofos
Quinalphos
Resmethrin
Rotenone

Sulfotep
Sulprofos
Temephos
TEPP
Terbufos
Tetrachlorophenol
Tetrachlorvinphos
Tetramethrin
Thiocyclam
Thiofanox
Thiometon
Tinox
Tolclophos-methyl
Toxaphene (TM)
Triazophos
Trichlorfon
Trichloronat
Trichlorophenol
Vamidothion

Laxatives
Aloe-emodin
Bisacodyl
Chrysophanol
Danthron
Frangula-emodin
Glycerol
Glyceryl triacetate
Mannitol
Phenolphthalein
Physcion
Picosulfate
Polyethylene glycol
Rhein

Local anesthetics
Articaine
Benzocaine
Bupivacaine
Butanilicaine
Cinchocaine
Etidocaine
Lidocaine
Mepivacaine
Oxetacaine
Oxybuprocaine
Prilocaine
Procaine
Pyrrocaine
Ropivacaine
Tetracaine

Molluscicides
Metaldehyde
Niclosamide

Muscle relaxants
Atracurium
Baclofen
Carisoprodol
Chlormezanone
Chlorzoxazone
Cyclobenzaprine
Dantrolene
Guaifenesin
Mephenesin
Methocarbamol
Tetrazepam
Tizanidine
Tolperisone
Xylazine

Neuroleptics
Alimemazine
Amisulpride
Amperozide
Aripiprazole
Azaperone
Benperidol
Bromperidol
Butaperazine
Chlorphenethazine
Chlorpromazine
Chlorprothixene
Clopenthixol
Clotiapine
Clozapine
Cyamemazine
Dixyrazine
Droperidol
Fluanisone
Flupentixol
Fluphenazine
Fluspirilene
Haloperidol
Homofenazine
Levomepromazine
Melperone
Mesoridazine
Methiomeprazine
Metofenazate
Moperone
Olanzapine
Oxypertine
Pecazine
Penfluridol
Perazine
Periciazine
Perphenazine
Phenothiazine
Pimozide
Pipamperone
Prochlorperazine
Promazine
Promethazine
Prothipendyl
Quetiapine
Remoxipride
Sulforidazine
Tetrabenazine
Thiopropazate
Thioproperazine
Thioridazine
Tiotixene
Trifluoperazine
Trifluperidol
Triflupromazine
Zotepine
Zuclopenthixol

Parasympatholytics
Atropine
Butethamate
Butylscopolaminium bromide
Chlorbenzoxamine
Cyclopentolate
Drofenine
Homatropine
Hyoscyamine
Mecloxamine
Oxyphencyclimine
Pinaverium bromide
Pirenzepin
Pseudotropine
Scopolamine
Tropicamide
Tropine

Parasympathomimetics
Aceclidine
Cisapride
Physostigmine
Pilocarpine
Pyridostigmine bromide

Pesticides
Amidithion
Ancymidol
Anthraquinone
Bifenox
Chlorflurenol
Dichloroaniline
Dichlorobenzophenone (DCBP)
Diphenylamine
Ethylene thiourea
Flurochloridone
Hexachlorophene
Maleic hydrazide
Metaldehyde
Naphthaleneacetic acid
Pentachloroaniline
Pentachlorobenzene
Phosphine
Piperonyl butoxide
Tetrachlorobenzene
Trichloromethoxypropion-amide

Plant ingredients
Arecaidine
Bulbocapnine
Buphanamine
Cannabidiol
Cannabidivarol
Cannabielsoic acid
Cannabigerol
Cannabinol
Cannabispirol
Cannabispirone
Caulophyllin
Chavicine
Clionasterol
Colchicine
Crinosterol
Cytisine
Dehydroabietic acid
Dihydrobrassicasterol
Elemicin
Ergost-3,5-ene
Ergost-5-en-3-ol
Ergosta-3,5,22-triene
Ergosta-5,22-dien-3-ol
Hydrocotarnine
Hydroxymethoxyflavone
Meconin
Methylcytisine
Methylpsychotrine
Myristicin
Pratol
Safrole
Stigma-3,5-dien-7-one
Stigmast-3,5-ene
Stigmast-5-en-3-ol
Stigmasterol
Thebaine
Thebaol
Tremulone

Plasticizers
Dioctylsebacate
Sebaic acid bisoctyl ester
Tribenzylamine
Tributylphosphate

Pollutants
Acenaphthene
Benzo[a]anthracene
Benzo[a]pyrene
Benzo[b]fluoranthene
Benzo[g,h,i]perylene
Benzo[k]fluoranthene
Dibenzo[a,h]anthracene
Indeno[1,2,3-c,d]pyrene

Potent analgesics
Acetylmethadol
Alfentanil
Buprenorphine
Cetobemidone
Dextromoramide
Dextropropoxyphene
Dihydromorphine
EDDP
Etonitazene
Fentanyl
Heroin
Hydromorphone
Hydroxypethidine
LAAM
Levacetylmethadol
Levorphanol
Meptazinol
Methadol
Methadone
Metonitazene
Morphine
Nefopam
Nicomorphine

Oxycodone
Oxymorphone
Parafluorofentanyl
Pentazocine
Pethidine
Phencyclidine
Piritramide
Propoxyphene
Remifentanil
Sufentanil
Tilidine
Tramadol
Viminol

Potent antitussives
Codeine
Dextromethorphan
Dextrorphan
Dihydrocodeine
Ethylmorphine
Hydrocodone
Methorphan
Norcodeine
Normethadone
Pholcodine
Thebacone

Preservatives
Benzoic acid
Chlorobenzoic acid
Ethylparaben
Ferulic acid
Methylparaben
N-Phenylalphanaphthylamine
Phenylmercuric acetate
Propylparaben

Psychedelics
BDB
BDMPEA
Brolamfetamine
2C-B
DMA
DMCC
DOB
DOET
DOM
Dronabinol
EBDB
ECC
Eticyclidine
Iso-LSD

N-Isopropyl-BDB
LAMPA
LSD
Lysergide
MBDB
MCC
MDA
MDE
MDMA
MDPPP
MECC
Mescaline
Methoxyetilamfetamine
Methoxymetamfetamine
N-Methyl-DOB
Methylenedioxypyrrolidino-
propiophenone
MMBDB
MMDA
MPCP
PCC
PCDI
PCE
PCM
PCME
PCPIP
PCPR
PICC
PMA
PMEA
PMMA
PRCC
Psilocine
Psilocybin
PYCC
Rolicyclidine
TCDI
TCM
TCPY
Tenamfetamine
Tenocyclidine
Tetrahydrocannabinol
TMA
Trimethoxyamfetamine
Trimethoxymetamfetamine

Refrigerants
Dichlorodifluoromethane
Frigen 11
Frigen 12
Trichlorofluoromethane

Rodenticides
Bromadiolone
Chloralose
Chlorophacinone
Coumachlor
Coumatetralyl
Crimidine
Cyclocumarol
Pindone
Pyranocoumarin
Sulfaquinoxaline

Rubber additives
Bis(2-hydroxy-3-tert-butyl-5-
ethylphenyl)methane
N-Phenylbetanaphthylamine
Squalene

Rubefacients
Benzylnicotinate
Bornyl salicylate
Capsaicine
Dihydrocapsaicine
Nonivamide

Scabicides
Crotamiton (cis)
Mesulphen

Sedatives
Acepromazine
Aceprometazine
Benactyzine
Benzilic acid
8-Chlorotheophylline
Etchlorvynol
Melatonin
Trimetozine

Serotonin antagonists
Cyproheptadine
Granisetron
Pizotifen

Softeners
Benzylbutylphthalate
Butyl stearate
Butyl-2-ethylhexylphthalate
Butyl-2-methylpropyl-
phthalate
Butylhexadecanoate
Butyloctadecanoate

Butyloctylphthalate
Decyldodecylphthalate
Decylhexylphthalate
Decyloctylphthalate
Decyltetradecylphthalate
Dibutyladipate
Diethylphthalate
Diisodecylphthalate
Diisohexylphthalate
Diisononylphthalate
Diisooctylphthalate
Dimethylphthalate
Dioctylphthalate
Ethylhexylmethylphthalate
Ethylmethylphthalate
Hexyloctylphthalate
Monoisooctyladipate
Triphenylphosphate

Solvents
Acetone
Acetonitrile
Benzene
Benzylalcohol
Benzylamine
Benzylbenzoate
Butanol
Butene
Carbon disulfide
Chloroform
Cyclohexane
Cyclohexanol
Cyclohexene
Dichloromethane
Diethylether
Dimethoxyethane
Dimethylbutane
Dimethylcyclopentane
Dimethylformamide
Dimethylsulfoxide
Dioxane
Ethanol
Ethylacetate
Ethyldimethylbenzene
Ethylmethylbenzene
Ethylmethylbutene
Heptane
Hexane
Isobutanol
Isobutylbenzene
Isooctane
Isopropanol

Isopropylbenzene
Methanol
Methylacetate
Methylbutene
Methylethenylcyclopropane
Methylhexane
Methylhexene
Methylpentane
Methylpropane
Methylpropanol
Nonane
Octane
Pentane
Propanol
Propylbenzene
Tetrachloroethylene
Tetrachloromethane
Tetramethylbenzene
Toluene
Trichloroethane
Trimethylbenzene
Xylene

Stimulants
Amfetamine
Amfetaminil
Amiphenazole
Bemegride
Bromantane
Brucine
Cafedrine
Caffeine
Cathinone
Cocaine
Cotinine
Cropropamide
Crotethamide
Dimetamfetamine
N,N-Dimethyl-5-methoxy-tryptamine
Diphenylprolinol
Etamivan
Etilamfetamine
Etofylline
Fencamfamine
Fenetylline
Harmaline
Harmine
Kavain
Lobeline
Meclofenoxate
Mefexamide

Metamfetamine
Methcathinone
Methylephedrine
Methylphenidate
Nicethamide
Pemoline
Pentetrazole
Pipradrol
Piracetam
Prolintane
Pyritinol
Serotonin

Sugars
Arabinose
Fructose
Galactose
Glucose
Lactose
Mannose
Rutinose
Saccharose
Xylose

Sugar alcohols
Erythritol
Inositol
Xylitol

Sweeteners
Cyclamate
Isosteviol
Saccharin
Sorbitol
Stevioside

Sympathomimetics
Dipivefrin
Dobutamine
Ephedrine
Etifelmin
Etilefrine
Fenoterol
Formoterol
Gepefrine
Heptaminol
Isoprenaline
Mephentermine
Metamfepramone
Metaraminol
Midodrine
Norephedrine

Norfenefrine
Octopamine
Orciprenaline
Oxedrine
Oxilofrine
Phenylephrine
Phenylpropanolamine
Pholedrine
Prenalterol
Protocatechuic acid
Tyramine

Thrombocyte aggregation inhibitors
Clopidogrel
Ditazol
Sulfinpyrazone
Ticlopidine

Thyreostatics
Carbimazole
Thiamazole

Toccolytics
Fenoterol
Ritodrine

Tranquilizers
Adinazolam
Alprazolam
Benzoctamine
Bromazepam
Brotizolam
Buspirone
Camazepam
Chlorazepate
Chlordiazepoxide
Clobazam
Clorazepate
Clotiazepam
Cloxazolam
Cyprazepam
Delorazepam
Diazepam
Estazolam
Etizolam
Etodroxizine
Fenazepam
Fludiazepam
Flutazolam
Fosazepam
Halazepam

Hydroxyzine
Ketazolam
Lorazepam
Lormetazepam
Loxapine
Medazepam
Metaclazepam
Methylpentynol
Mexazolam
Nordazepam
Oxazepam
Oxazolam
Pinazepam
Prazepam
Quazepam
Sulazepam
Temazepam
Tofisopam
Triflubazam
Zolazepam

Tuberculostatics
4-Aminosalicylic acid
Etambutol
Isoniazid

Uricosurics
Allopurinol
Benzbromarone
Probenecide

Urinary antiseptics
Methenamine
Phenazopyridine

UV Absorbers
Benzoresorcinol
Oxybenzone

Vasoconstrictors
Cyclopentamine
Dihydroergotamine
Ergotamine
Indanazoline
Naphazoline
Octodrine
Oxymetazoline
Synephrine
Tetryzoline
Tolazoline
Tramazoline
Xylometazoline

Vasodilators
Bamethan
Bencyclane
Buflomedil
Buphenine
Butalamine
Carbochromene
Cinnarizine
Cyclandelate
Flunarizine
Lidoflazine
Lisofylline

Naftidrofuryl
Nicergoline
Pentifylline
Pentoxifylline
Sildenafil
Theobromine
Trapidil
Viquidil
Xanthinol

Virustatics
Abacavir

Emtricitabine
Famciclovir
Indanavir
Nevirapine
Oseltamivir
Ribavarine
Valganciclovir
Zidovudine

Vitamins
Ascorbic acid
Colecalciferol

Nicotinamide
Nicotinic acid
Tocopherol
Thiamine
Vitamin B1
Gluconic acid
Pangamic acid
Pyridoxine
Vitamin B6